D1751998

**NMR Imaging
in Chemical Engineering**

*Edited by
Siegfried Stapf and Song-I Han*

Related Titles

Schorn, C., Taylor, B.

NMR-Spectroscopy: Data Acquisition
2nd Edition

2004
ISBN 3-527-31070-3

Kaupp, M., Bühl, M., Malkin, V. G. (eds.)

Calculation of NMR and EPR Parameters
Theory and Applications

2004
ISBN 3-527-30779-6

Sundmacher, K., Kienle, A., Seidel-Morgenstern, A. (eds.)

Integrated Chemical Processes
Synthesis, Operation, Analysis, and Control

2005
ISBN 3-527-30831-8

Wiley-VCH (ed.)

Ullmann's Processes and Process Engineering

2004
ISBN 3-527-31096-7

NMR Imaging in Chemical Engineering

Edited by
Siegfried Stapf and Song-I Han

WILEY-VCH

WILEY-VCH Verlag GmbH & Co. KGaA

The Editors

Dr. Siegfried Stapf
RWTH Aachen
ITMC
Worringerweg 1
52074 Aachen

Dr. Song-I Han
University of California
Dept. of Chem. and Biochem.
Santa Barbara, CA 93106-9510
USA

All books published by Wiley-VCH are carefully produced. Nevertheless, authors, editors, and publisher do not warrant the information contained in these books, including this book, to be free of errors. Readers are advised to keep in mind that statements, data, illustrations, procedural details or other items may inadvertently be inaccurate.

Library of Congress Card No.:
applied for

British Library Cataloguing-in-Publication Data
A catalogue record for this book is available from the British Library.

Bibliographic information published by Die Deutsche Bibliothek
Die Deutsche Bibliothek lists this publication in the Deutsche Nationalbibliografie; detailed biliographic data is available in the Internet at <http://dnb.ddb.de>.

© 2006 WILEY-VCH Verlag GmbH & Co. KGaA, Weinheim

All rights reserved (including those of translation into other languages). No part of this book may be reproduced in any form – by photoprinting, microfilm, or any other means – nor transmitted or translated into a machine language without written permission from the publishers.
Registered names, trademarks, etc. used in this book, even when not specifically marked as such, are not to be considered unprotected by law.

Typesetting Dörr + Schiller GmbH, Stuttgart
Printing betz-druck GmbH, Darmstadt
Binding Littges & Dopf
Cover Design Grafik-Design Schulz, Fußgönheim

Printed in the Federal Republic of Germany
Printed on acid-free paper

ISBN-13 978-3-527-31234-4
ISBN 10 3-527-31234-X

Nuclear Magnetic Resonance Imaging in Chemical Engineering

Forward

The field of nuclear magnetic resonance imaging (NMRI) has seen extraordinary technical advances since the seminal demonstrations of the technique by Paul Lauterbur and Peter Mansfield in the early 1970s. Driven by industrial and academic scientists and engineers, the advances in radiofrequency, magnet and gradient capabilities have been nothing short of remarkable. Most of these efforts have focused on biomedical applications, small animal and human imaging. The commercial (i.e., for profit) and research (i.e., grant funding) opportunities are unusually rich in the biomedical arena. Importantly for the life sciences, the primary imaging substrate is liquid-state water, which affords long NMR coherence times (tens to hundreds of milliseconds) and high spin densities (≈ 100 molar equivalent protons). The advantages conferred upon the field of NMRI by the $\approx 70\%$ water content of living systems cannot be overstated. Were water molecules NMR silent, it is unlikely that NMRI would have undergone such explosive technological developments and today it might be little more than a curiosity, pursued in a few academic and industrial laboratories.

Of course, water molecules are not NMR silent and NMRI engineering has, indeed, advanced at a remarkable pace to provide extraordinary technical capabilities. These capabilities now enable studies of systems beyond those in the biomedical arena, systems that are, in many respects, far more technically challenging. This has led to the development of innovative and fascinating strategies and tactics to deal with "NMRI-unfriendly" samples and conditions.

Coherence times in the solid-state can be distressingly short, tens to hundreds of microseconds, stimulating the development of novel spatial encoding methods. Samples to be examined can be fairly large, perhaps the wing of an aircraft or a truck tire or a gasket for a rocket engine, requiring the development of single-sided or *inside-out* NMRI scanners. Conversely, samples can be particularly small, for example, the output of a capillary separation column or a micro-fluidics reaction mixture, motivating the development of ultra high sensitivity micro-coils that can operate at very high magnetic field strengths. For samples composed of porous materials – filters, ceramics, concrete, etc. – the focus of interest is often the void structure within, which has lead to the development of diffusion and susceptibility sensitive methods that employ NMR active fluids and gasses. Reaction engineering is commonly presented with heterogeneous samples undergoing complex flow patterns, requiring the development of velocity and displacement-sensitive imaging strategies. Combustion and catalytic processes taking place at high temperatures have motivated the development of special NMRI probes for dynamic monitoring of samples under extreme conditions.

This monograph provides a snapshot of current state-of-the-art technology and applications by the leading practitioners of NMRI in the broadly defined field of chemical engineering. The Editors have chosen internationally respected laboratories to contribute to sections on Hardware and Methods, Porous Materials, Fluids and Flow, and Reactors and Reactions. The result is an excellent compilation for the NMRI student requiring an introduction to the field, the junior scientist looking for an NMRI solution to a chemical engineering problem, or the NMRI expert anxious to understand more fully what the competition is doing. Hopefully this volume will be viewed as a timely contribution to the field and will find a place on the bookshelves of NMR scientists and engineers interested in exploring the power of NMRI beyond its traditional applications.

Joseph J. H. Ackerman   Summer 2005
Associate Editor of Journal of Magnetic Resonance and
Professor and Chair of the Chemistry Department,
Washington University in St. Louis

The Discovery of Spin Echoes

If one enters a territory of thought or a field of scientific research which is relatively unexplored there is a good chance of making new discoveries by the happy event of accident in the laboratory or by a startling original combination of ideas in the mind. These occurences are favoured for those who apply themselves consistently and consequently do not dabble trivially in a wide variety of new fields as each emerges into the ephemeral limelight. Caution here, however, because persistent application within a restricted field is not in itself sufficient; there is even less chance for discovery if an attitude of hidebound rigour suppresses the imagination. I went through the experience of overcoming these conditions and was lucky enough to discover a new effect called "Spin Echo". This effect was found by accident in my laboratory because a particular combination of operations with my equipment happened to be just adequate to satisfy the conditions necessary for generation of the spin echo. At the time, I was in fact investigating another kind of signal, the occurrence of which was well established and the form of its response well known. An essential feature, however, contributing to the discovery of the new effect was that, on this occasion, the well known signal was to be obtained by a rather unorthodox experimental technique; the experiment used radiofrequency pulses of power instead of the continuous wave power normally employed.

As a postdoctoral research student at the University of Illinois I happened to be the first to use radiofrequency pulses of the right sort to look at nuclear magnetic resonance (NMR) signal transients. I learned about radar and sonar in the Navy during World War II, so I was inclined to work with pulse techniques when I carried out my physics thesis later in NMR at the Physics Department of the University of Illinois. My thesis did not reveal spin echoes however, but instead involved the measurement of NMR transient signals during the action of a driving pulse of radiowaves. The signal was seen only on the top of a pulse pedestal. My thesis problem was scooped by someone else who published the same experiment on mutations of the nuclear moment. I stayed at Illinois as a one-year postdoctoral research associate to make better measurements with apparatus improved to give shorter, sharper, and more intense pulses.

One day in July, 1949 a strange signal appeared on my oscilloscope display without any corresponding pulse excitation. So I kicked the apparatus and breathed a sigh of relief when the unexpected signal went away. After my first observation of the spin echo signal (which at that time was simply an unexplained spurious nuisance), I had to overcome the tendency to ignore it. In real life most of the disturbances that distract from our known goals turn out to be undesired and irrelevant. In the laboratory we call them artifacts or "false glitches". Consequently, the danger exists of overlooking the "significant glitch" and simply coercing the results to conform with the expected answer; in so doing perhaps missing out on a new discovery. A week later the unwanted signal returned and this time I was sufficiently intrigued to check it further and found that it could be explained as a real effect – a spontaneous spin echo from the protons in the test sample of glycerine which was being used for the experiment that I expected to be carrying through. In about three weeks I was able to predict mathematically what I suspected to be a constructive interference of precessing nuclear magnetism components by solving the Bloch nuclear induction equations. Here for the first time, a free precession signal in the absence of driving radiation was observed and accounted for afterwards. The spin echo began to yield information about the local atomic environment in terms of various amplitude and frequency memory beat effects, relaxation effects, all certainly not understood in the beginning.

In brief the spin echo is displayed by atomic nuclei which behave like spinning bar magnets. Species of a sample of nuclei (protons in water for example) are first lined up in a constant magnetic field. An applied radio frequency pulse to the sample causes the nuclei to tip in unison and after the pulse is removed the nuclei emit a coherent radio signal. The signal gradually disappears as the nuclei get out of phase while they precess individually about the constant magnetic field. The nuclei completely misalign, but yet there remains a hidden order of precession in the spin ensemble. The nuclei can be caused to realign and produce a spontaneous echo radio signal following the action of a second radiofrequency pulse, or subsidiary echoes after more than two pulses.

As I look back at this experience, it was an awesome adventure to be alone, during and for an interval of time after this discovery, with the apparatus showing one new effect after another, when there was no one in the Illinois Physics Department experienced in NMR with whom I could talk. Little did the early NMR resonance community realize that the analogue of spin echo hidden memory contained in excited phases of all kinds of states of matter, including plasmas, would be obtained in the future by use of optical laser, electric, and acoustic pulses as well. And now today the use of spin echoes is a standard procedure for magnetic resonance imaging of the human body for medical diagnosis.

Brasenose College Magazine (Oxford), 1997
E. L. Hahn (*Honorary Fellow*)
Professor of Physics, University of California,
Berkeley, USA

List of Contents

1	**Introduction**	*1*

Siegfried Stapf and Song-I. Han
- 1.1 A Brief Comment 1
- 1.2 The Very Basics of NMR 2
- 1.3 Fundamentals of NMR Imaging 8
- 1.4 Fundamentals of Detecting Motion 18
- 1.5 Bringing Them Together: Velocity Imaging 26
- 1.6 More Advanced Techniques I: Multiple Encoding and Multiple Dimensions 32
- 1.7 More Advanced Techniques II: Fast Imaging Techniques 34
- 1.8 Introducing Color into the Image: Contrast Parameters 39

2 Hardware and Methods

- 2.1 Hardware, Software and Areas of Application of Non-medical MRI 47
  D. Gross, K. Zick, T. Oerther, V. Lehmann, A. Pampel, and J. Goetz
- 2.1.1 Introduction 47
- 2.1.2 Hardware 48
- 2.1.2.1 Magnets 48
- 2.1.3 Software 56
- 2.1.4 Areas of Application 63
- 2.1.5 Outlook 69

- 2.2 Compact MRI for Chemical Engineering 77
  Katsumi Kose
- 2.2.1 Concept of Compact MRI 77
- 2.2.2 System Overview 78
- 2.2.3 Permanent Magnet 79
- 2.2.4 Gradient Coil 82
- 2.2.5 Rf coil 82
- 2.2.6 MRI Console 83
- 2.2.7 Typical Examples of Compact MRI Systems 86

- 2.3 Drying of Coatings and Other Applications with GARField 89
  P. J. Doughty and P. J. McDonald
- 2.3.1 Introduction 89

2.3.2	GARField Magnets	90
2.3.3	Applications	94
2.3.4	Human Skin Hydration	101
2.3.5	Further Developments	103
2.3.6	Conclusion	105

2.4 Depth Profiling by Single-sided NMR 107
F. Casanova, J. Perlo, and B. Blümich
- 2.4.1 Introduction 107
- 2.4.2 Microscopic Depth Resolution 108
- 2.4.3 Applications 113
- 2.4.4 Conclusions 122

2.5 Microcoil NMR for Reaction Monitoring 123
Luisa Ciobanu, Jonathan V. Sweedler, and Andrew G. Webb
- 2.5.1 Introduction 123
- 2.5.2 NMR Acquisition in Reaction Monitoring: Stopped- and Continuous-flow 124
- 2.5.3 Reaction Studies Using NMR 126
- 2.5.4 Small-scale NMR Reaction Monitoring 129
- 2.5.5 Multiple Microcoil NMR. Sensitivity and Throughput Issues 133
- 2.5.6 Conclusions 137

2.6 Broadening the Application Range of NMR and MRI by Remote Detection 139
Song-I Han, Josef Granwehr, and Christian Hilty
- 2.6.1 Introduction 139
- 2.6.2 Motivation 140
- 2.6.3 Principle of NMR Remote Detection 140
- 2.6.4 Sensitivity Enhancement by Remote Detection 145
- 2.6.5 Application of NMR Remote Detection 149
- 2.6.6 Concluding Remark 160

2.7 Novel Two Dimensional NMR of Diffusion and Relaxation for Material Characterization 163
Yi-Qiao Song
- 2.7.1 Introduction 163
- 2.7.2 Pulse Sequences and Experiments 164
- 2.7.3 Laplace Inversion 169
- 2.7.4 Applications 172
- 2.7.5 Instrumentation 179
- 2.7.6 Summary 181

2.8 Hardware and Method Development for NMR Rheology 183
Paul T. Callaghan
- 2.8.1 Introduction 183
- 2.8.2 Rheo-NMR Fundamentals 185
- 2.8.3 Apparatus Implementation 190
- 2.8.4 Applications of Rheo-NMR 193
- 2.8.5 Conclusions 203

2.9   Hydrodynamic, Electrodynamic and Thermodynamic Transport in Porous Model Objects: Magnetic Resonance Mapping Experiments and Simulations   205
      Elke Kossel, Bogdan Buhai, and Rainer Kimmich
2.9.1   Introduction   205
2.9.2   Spin Density Diffusometry   207
2.9.3   Flow Velocity and Acceleration Mapping   211
2.9.4   Hydrodynamic Dispersion   217
2.9.5   Thermal Convection and Conduction Mapping   221
2.9.6   Ionic Current Density Mapping   223
2.9.7   Concluding Remarks   228

3   Porous Materials
3.1   Diffusion in Nanoporous Materials   231
      Jörg Kärger, Frank Stallmach, Rustem Valiullin, and Sergey Vasenkov
3.1.1   Introduction   231
3.1.2   Measuring Principle   232
3.1.3   Intracrystalline Diffusion   236
3.1.4   Long-range Diffusion   237
3.1.5   Boundary Effects   243
3.1.6   Conclusion   247

3.2   Application of Magnetic Resonance Imaging to the Study of the Filtration Process   250
      R. Reimert, E. H. Hardy, and A. von Garnier
3.2.1   Filtration Principles   250
3.2.2   In-bed Filtration   251
3.2.3   Filtration Dynamics   254
3.2.4   Summary   262

3.3   Multiscale Approach to Catalyst Design   263
      Xiaohong Ren, Siegfried Stapf, and Bernhard Blümich
3.3.1   Introduction   263
3.3.2 ^{129}Xe Spectroscopy   265
3.3.3   NMR Relaxometry   267
3.3.4   Cryoporometry   269
3.3.5   Diffusometry   270
3.3.6   Flow Propagators   272
3.3.7   NMR Imaging   275
3.3.8   Conclusions and Summary   280

3.4   Pure Phase Encode Magnetic Resonance Imaging of Concrete Building Materials   285
      J. J. Young, T. W. Bremner, M. D. A. Thomas, and B. J. Balcom
3.4.1   Introduction   285
3.4.2   Single-Point Imaging – The SPRITE Techniques   286
3.4.3   Hydrogen (Water) Measurements   291
3.4.4   Chlorine and Sodium Measurements   298
3.4.5   Lithium Measurements   300
3.4.7   Conclusion   302

3.5	NMR Imaging of Functionalized Ceramics	304

S. D. Beyea, D. O. Kuethe, A. McDowell, A. Caprihan, and S. J. Glass

3.5.1	Introduction	304
3.5.2	Experimental Background	305
3.5.3	NMR Relaxation Behavior of Perfluorinated Gases	306
3.5.4	Results and Discussion	310
3.5.5	Conclusions and Future Research	319
3.5.6	Description of NMR Equipment	319
3.6	NMR Applications in Petroleum Reservoir Studies	321

George J. Hirasaki

3.6.1	Introduction	321
3.6.2	NMR Well Logging and Fluid Analysis	322
3.6.3	NMR Measurements	323
3.6.4	NMR Fluid Properties	324
3.6.5	Porosity	326
3.6.6	Surface Relaxation and Pore Size Distribution	328
3.6.7	Irreducible Water Saturation	330
3.6.8	Permeability	332
3.6.9	Fluid Identification	335
3.6.10	Exceptions to Default Assumptions	336
3.6.11	Conclusions	337
3.7	NMR Pore Size Measurements Using an Internal Magnetic Field in Porous Media	340

Yi-Qiao Song, Eric E. Sigmund, and Natalia V. Lisitza

3.7.1	Introduction	340
3.7.2	DDIF Concept	341
3.7.3	Applications	348
3.7.4	Summary	356

4 Fluids and Flows

4.1	Modeling Fluid Flow in Permeable Media	359

Jinsoo Uh and A. Ted Watson

4.1.1	Introduction	359
4.1.2	Modeling Multiphase Flow in Porous Media	360
4.1.3	System and Parameter Identification	362
4.1.4	Determination of Properties	364
4.1.5	Summary	381
4.2	MRI Viscometer	383

Robert L. Powell

4.2.1	Introduction	383
4.2.2	Theory	387
4.2.3	Experimental Techniques	389
4.2.4	Results	392
4.2.5	Conclusion	402

4.3	Imaging Complex Fluids in Complex Geometries	404
	Y. Xia and P. T. Callaghan	
4.3.1	Introduction	404
4.3.2	Rheological Properties of Polymeric Flow	404
4.3.3	NMR Microscopy of Velocity	408
4.3.4	NMR Velocity Imaging of Fano Flow	410
4.3.5	Other Examples of Viscoelastic Flows	414
4.4	Quantitative Visualization of Taylor–Couette–Poiseuille Flows with MRI+	416
	John G. Georgiadis, L.Guy Raguin, and Kevin W. Moser	
4.4.1	Introduction	416
4.4.2	Taylor–Couette–Poiseuille Flow	419
4.4.3	Future Directions	429
4.4.4	Summary	430
4.5	Two Phase Flow of Emulsions	433
	Nina C. Shapley and Marcos A. d'Ávila	
4.5.1	Introduction	433
4.5.2	NMRI Set-up and Methods	436
4.5.3	Complex Flows of Homogeneous Emulsions	444
4.5.4	Mixing of Concentrated Emulsions	447
4.5.5	Future Directions	451
4.6	Fluid Flow and Trans-membrane Exchange in a Hemodialyzer Module	457
	Song-I Han and Siegfried Stapf	
4.6.1	Objective	457
4.6.2	Methods	457
4.6.3	Materials	458
4.6.4	Results and Discussion	459
4.6.5	Conclusion	469
4.7	NMR for Food Quality Control	471
	Michael J. McCarthy, Prem N. Gambhir, and Artem G. Goloshevsky	
4.7.1	Introduction	471
4.7.2	Relationship of NMR Properties to Food Quality	473
4.7.3	Applications of NMR in Food Science and Technology	473
4.7.4	Summary	488
4.8	Granular Flow	490
	Eiichi Fukushima	
4.8.1	Introduction	490
4.8.2	NMR Strategies	493
4.8.3	Systems Studied	501
4.8.4	Future Outlook	505

5 Reactors and Reactions

5.1 Magnetic Resonance Microscopy of Biofilm and Bioreactor Transport 509
Sarah L. Codd, Joseph D. Seymour, Erica L. Gjersing, Justin P. Gage, and Jennifer R. Brown
5.1.1 Introduction 509
5.1.2 Theory 510
5.1.3 Reactors 516
5.1.4 Conclusions 531

5.2 Two-phase Flow in Trickle-Bed Reactors 534
Lynn F. Gladden, Laura D. Anadon, Matthew H. M. Lim, and Andrew J. Sederman
5.2.1 Introduction to Magnetic Resonance Imaging of Trickle-bed Reactors 534
5.2.3 Unsteady-state Hydrodynamics in Trickle-bed Reactors 542
5.2.4 Summary 549

5.3 Hyperpolarized ^{129}Xe NMR Spectroscopy, MRI and Dynamic NMR Microscopy for the *In Situ* Monitoring of Gas Dynamics in Opaque Media Including Combustion Processes 551
Galina E. Pavlovskaya and Thomas Meersmann
5.3.1 Introduction 551
5.3.2 Chemical Shift Selective Hp-^{129}Xe MRI and NMR Microscopy 552
5.3.3 Dynamic NMR Microscopy of Gas Phase 557
5.3.4 *In Situ* NMR of Combustion 561
5.3.5 High Xenon Density Optical Pumping 566

5.4 *In Situ* Monitoring of Multiphase Catalytic Reactions at Elevated Temperatures by MRI and NMR 570
Igor V. Koptyug and Anna A. Lysova
5.4.1 Introduction 570
5.4.2 Experimental 571
5.4.3 Results and Discussion 574
5.4.4 Outlook 587

5.5 *In Situ* Reaction Imaging in Fixed-bed Reactors Using MRI 590
Lynn F. Gladden, Belinda S. Akpa, Michael D. Mantle, and Andrew J. Sederman
5.5.1 Introduction 590
5.5.2 Spatial Mapping of Conversion: Esterification Case Study 592
5.5.3 ^{13}C DEPT Imaging of Conversion and Selectivity 603
5.5.4 Future Directions 606

Index 609

Preface from the Editors

Nuclear Magnetic Resonance, NMR, is now about 60 years of age. Over the years it has been declared dead surprisingly often, using the arguments that everything has been described theoretically, that all basic experiments had been done and that technological development would increase the power of NMR only marginally, at best. Most of these predictions have turned out to be so entirely wrong that the skeptics have now given up.

However, a certain discrepancy remains in the way NMR is understood, which can be explained by the various points of view that are assumed by different people. The year 1945 did not merely represent the birth of a new *method* for understanding the properties of matter; nor did it see the birth of an entirely new branch of *science*, although some may prefer to view it that way. Although NMR is based on fundamental aspects of physics and chemistry, the principles of which were mostly understood and described in seminal works during the first two decades of the lifetime of NMR, during this time, NMR has developed a whole toolbox of methods that can deal with almost any question arising in the context of structure and the dynamics of matter.

Of the many areas where NMR is applied these days, two can be considered as being established. The most important is certainly its use for structure elucidation, from small molecules up to medium-sized proteins in solution; no university with an analytical lab can afford to be without a liquid-state, high-resolution NMR system. Most chemistry students will come into contact with NMR at least once during their courses. Second, is diagnostic medical imaging, which many of us may have experienced personally. From the first crude and blurred NMR images that were acquired over 30 years ago, incredible developments have been achieved by the efforts of researchers and industry alike.

With all this progress taking place, it is somewhat surprising that the commercially important sector of industrial production, synthesis and quality control has only taken notice of NMR relatively recently. When we were collecting information and literature during the early stages of the preparation of this book, we realized that NMR spectroscopy has indeed gained widespread acceptance for analysis purposes, although this is still not usually on-line or in-line. However, the potential, from a scientific and more significantly also from an economic point of view, of MR imaging in the various fields of Chemical Engineering is vastly underexploited. After over a decade of systematic research in non-medical fields of transport phenomena by NMR, both hardware and measurement techniques and data interpretations have become sufficiently robust to enable routine applications to be made in the near future, so a compilation of current trends and case studies appeared absolutely necessary. Collating and combining this information from different sources seemed particularly appropriate to us, as there are no books, periodicals nor conferences which treat this rapidly expanding field in a dedicated manner.

As a result of all these considerations, we decided to include only those works that employ NMR imaging methods (in the wider sense, i.e., pulsed field gradient NMR) in a fashion that is not yet established, i.e., utilizing novel NMR methods and techniques, or demonstrating its great potential for applications that have not been routinely explored by NMR imaging. This is why only a few contributions are presented from the well established branch of pulsed field gradient NMR for porous media studies, utilizing diffusion for materials characterization. Even with this limitation, there remains a wide range of applications by so many great scientists that can not be included in one book. We have tried to present a variety of work with the purpose of giving a flavor of what we regard as state-of-the-art of non-medical NMR imaging. This book does not aim to highlight who is leading the field in non-medical NMR imaging, but rather intends to convince many scientists and chemical engineers how "cool" NMR imaging is and that it is worth looking into.

Hopefully the reader will benefit from this manifold approach, which casts light on the topic from a range of perspectives. Many of the contributors were originally physicists or chemists who became curious about particular applications, while an increasing minority of workers are contributing expertise from a Chemical Engineering aspect and introducing questions that are completely new to the fundamental researcher.

We considered the possibility that the audience of this book will probably share the same varied backgrounds as its authors. Even more generally, we want to reach researchers in academia and industry alike. For the academic sector, it is mainly postgraduate students but also faculty members who are addressed, all of whom are willing to gain an overview of existing techniques, limitations and strategies to solve individual problems from an engineering perspective. From many discussions with engineers we concluded that such an overview and a demonstration of the feasibility of the methods were desired. For the academic NMR researcher, who is often restricted to model systems and might lack insight into "real" problems, the various examples in the book provide a link to applications. Industrial researchers, or decision makers, will gain a sufficiently detailed view of the NMR toolbox, which can enable them to estimate the applicability of NMR to their particular problems, with respect to, for example, cost efficiency and output interpretability, and provides them with contact points to obtain further information. The Introduction and the various chapters are written in such a way that together they can help the reader understand the essential results without any prior knowledge of NMR.

Finally, we must confess that compiling this book was great fun. For once, we could collect together excellent pieces of work in this field without being jealous that we haven't done the particular work ourselves! We are most grateful to all contributing authors who shared our point of view that a demonstration of the power and, in addition, the beauty, of NMR imaging is the best way to spread the news that it is an exceptionally versatile tool. This book is about applications; it tells the reader what is possible, and how to solve a particular problem that he or she has encountered in the lab or the factory. It does not give a final recipe to the reader, but provides him with a lot of the necessary ingredients to allow him to find the best solution. If the book succeeds in doing this and makes the reader familiar with a technique or an application he or she hasn't thought of before, then the goal has been achieved.

Santa Barbara Song-I Han
Aachen Siegfried Stapf
Autumn 2005

List of Contributors

Belinda S. Akpa
University of Cambridge
Department of Chemical Engineering
Pembroke Street
Cambridge CB2 3RA
United Kingdom

Laura D. Anadon
University of Cambridge
Department of Chemical Engineering
Pembroke Street
Cambridge CB2 3RA
United Kingdom

Bruce J. Balcom
MRI Center, Department of Physics
University of New Brunswick
P.O. Box 4400
Fredericton
New Brunswick E3B 5A3
Canada

Bernhard Blümich
Institute for Technical Chemistry and
Macromolecular Chemistry
RWTH
52074 Aachen
Germany

Steven D. Beyea
Institute for Biodiagnostics (Atlantic)
National Research Council of Canada
Halifax
Nova Scotia
Canada and
New Mexico Resonance
Albuquerque
New Mexico
USA

Theodore W. Bremner
Department of Civil Engineering
University of New Brunswick
P.O. Box 4400
Fredericton
New Brunswick E3B 5A3
Canada

Jennifer R. Brown
Department of Chemical and Biological
Engineering and Center for Biofilm
Engineering
Montana State University
Bozeman
Montana 59717
USA

Bogdan Buhai
University of Ulm
Department of Nuclear Resonance
Spectroscopy
89069 Ulm
Germany

Paul T Callaghan
MacDiarmid Institute for Advanced
Materials and Nanotechnology
Victoria University of Wellington
Wellington
New Zealand

Arvind Caprihan
New Mexico Resonance
Albuquerque
New Mexico
USA

Federico Casanova
Institute for Technical Chemistry and
Macromolecular Chemistry
RWTH
52074 Aachen
Germany

Luisa Ciobanu
Beckman Institute
University of Illinois at Urbana-Champaign
405 N. Mathews
Urbana
Illinois 61801
USA
and Department of Electrical and
Computer Engineering
University of Illinois at Urbana-Champaign
1406 West Green Street
Urbana
Illinois 61801
USA

Sarah L. Codd
Department of Chemical and Biological
Engineering and Center for Biofilm
Engineering
Montana State University
Bozeman
Montana 59717
USA

Marcus d'Ávila
School of Chemical Engineering
State University of Campinas
Sao Paulo 13083–970
Brazil

Peter J. Doughty
Department of Physics
School of Electronics and Physical
Sciences
University of Surrey
Guildford
Surrey GU2 7XH
UK

Eiichi Fukushima
New Mexico Resonance
2301 Yale Blvd., SE; Suite C-1
Albuquerque
New Mexico 87106
USA

Justin P. Gage
Department of Chemical and Biological
Engineering and Center for Biofilm
Engineering
Montana State University
Bozeman
Montana 59717
USA

Prem Gambhir
Department of Food Science and
Technology
University of California
Davis
California 95616–8598
USA

John G. Georgiadis
Laboratory for Quantitative Visualization in Energetics
NSF Science & Technology Center of Advanced Materials for Purification of Water with Systems (CAMPWS)
Department of Mechanical & Industrial Engineering
University of Illinois at Urbana-Champaign
Urbana
Illinois 61801
USA

Erica L. Gjersing
Department of Chemical and Biological Engineering and Center for Biofilm Engineering
Montana State University
Bozeman
Montana 59717
USA

Lynn F. Gladden
University of Cambridge
Department of Chemical Engineering
Pembroke Street
Cambridge CB2 3RA
United Kingdom

S. Jill Glass
Sandia National Laboratories
Albuquerque
New Mexico
USA

Josef Granwehr
Department of Chemistry
University of California Berkeley and
Lawrence Berkeley National Laboratory
Material Sciences Divisions
Berkeley
CA 94720
USA

Joachim Goetz
Fraunhofer-Institute of Chemical Technology (ICT)
76327 Pfinztal
Germany

Artem G. Goloshevsky
Department of Food Science and
Technology
University of California
Davis
California 95616–8598
USA

Dieter Gross
Bruker-Biospin GmbH
76287 Rheinstetten
Germany

Song-I Han
Department of Chemistry and Biochemistry
University of California Santa Barbara
California 93106–9510
USA

Edme H. Hardy
University of Karlsruhe
Institute of Chemical Process Engineering
Kaiserstrasse 12
76128 Karlsruhe
Germany

Christian Hilty
Department of Chemistry
University of California Berkeley and
Lawrence Berkeley National Laboratory
Material Sciences Divisions
Berkeley
California 94720
USA

George J. Hirasaki
Rice University
P.O. Box 1892
Houston
Texas 77251–1892
USA

Jörg Kärger
Faculty of Physics and Earth Sciences
University of Leipzig
04103 Leipzig
Germany

Rainer Kimmich
University of Ulm
Department of Nuclear Resonance
Spectroscopy
89069 Ulm
Germany

Igor V. Koptyug
International Tomography Center
3A Institutskaya St.
Novosibirsk 630090
Russia

Elke Kossel
University of Ulm
Department of Nuclear Resonance
Spectroscopy
89069 Ulm
Germany

Dean O. Kuethe
New Mexico Resonance
Albuquerque
New Mexico
USA

Volker Lehmann
Bruker-Biospin GmbH
76287 Rheinstetten
Germany

Matthem H. M. Lim
University of Cambridge
Department of Chemical Engineering
Pembroke Street
Cambridge CB2 3RA
United Kingdom

Natlia V. Lisitza
Schlumberg-Doll Research
36 Quarry Road
Ridgefield
Connecticut 06877
USA

Anna A. Lysova
International Tomography Center
3A Institutskaya St.
Novosibirsk 630090
Russia

Mick D. Mantle
University of Cambridge
Department of Chemical Engineering
Pembroke Street
Cambridge CB2 3RA
United Kingdom

Michael J. McCarthy
Department of Food Science and
Technology
University of California
Davis
California 95616–8598
USA

Peter J. McDonald
Department of Physics
School of Electronics and Physical Sciences
University of Surrey
Guildford
Surrey GU2 7XH
UK

Andrew McDowell
New Mexico Resonance
Albuquerque
New Mexico
USA

Thomas Meersmann
Department of Chemistry
Colorado State University
Fort Collins
Colorado 80526
USA

Kevin W. Moser
Cardiovascular Imaging Technologies, L. L. C.
4320 Wornall Road
Suite 55
Kansas City
Missouri 64111
USA

Thomas Oerther
Bruker-Biospin GmbH
76287 Rheinstetten
Germany

André Pampel
University of Leipzig
Faculty of Physics and Earth Sciences
04103 Leipzig
Germany

Galina E. Pavlovskaya
Department of Chemistry
Colorado State University
Fort Collins
Colorado 80526
USA

Juan Perlo
Institute for Technical Chemistry and Macromolecular Chemistry
RWTH
52074 Aachen
Germany

Robert L. Powell
Department of Chemical Engineering & Materials Science
Department of Food Science & Technology
University of California
Davis
California 95616
USA

L. Guy Raguin
Laboratory for Quantitative Visualization in Energetics
NSF Science & Technology Center of Advanced Materials for Purification of Water with Systems (CAMPWS)
Department of Mechanical & Industrial Engineering
University of Illinois at Urbana-Champaign
Urbana
Illinois 61801
USA

Rainer Reimert
University of Karlsruhe
Institute of Chemical Process Engineering
Kaiserstrasse 12
76128 Karlsruhe
Germany

Xiaohong Ren
Institute for Technical Chemistry and
Macromolecular Chemistry
RWTH
52074 Aachen
Germany

Andrew J. Sederman
University of Cambridge
Department of Chemical Engineering
Pembroke Street
Cambridge CB2 3RA
United Kingdom

Joseph D. Seymour
Department of Chemical and Biological
Engineering and Center for Biofilm
Engineering
Montana State University
Bozeman
Montana 59717
USA

Nina C. Shapley
Department of Chemial Engineering
Columbia University
500 West 120th Street
New York
New York 10027
USA

Eric E. Sigmund
Schlumberg-Doll Research
36 Quarry Road
Ridgefield
Connecticut 06877
USA

Yi-Qiao Song
Schlumberg-Doll Research
36 Quarry Road
Ridgefield
Connecticut 06877
USA

Frank Stallmach
Faculty of Physics and Earth Sciences
University of Leipzig
04103 Leipzig
Germany

Siegfried Stapf
Institute for Technical Chemistry and
Macromolecular Chemistry ITMC
RWTH
52074 Aachen
Germany

Jonathan V. Sweedler
Beckman Institute
University of Illinois at Urbana-Champaign
405 N. Mathews
Urbana
Illinois 61801
USA
and Department of Chemistry
University of Illinois at Urbana-Champaign
600 S. Mathews
Urbana
Illinois 61801
USA

Michael D. A. Thomas
Department of Civil Engineering
University of New Brunswick
P.O. Box 4400
Fredericton
New Brunswick E3B 5A3
Canada

Jinsoo Uh
Department of Chemical Engineering
Texas A&M University
College Station
Texas 77843–3122
USA

Rustem Valiullin
Faculty of Physics and Earth Sciences
University of Leipzig
04103 Leipzig
Germany

Sergey Vasenkov
Faculty of Physics and Earth Sciences
University of Leipzig
04103 Leipzig
Germany

Agnes von Garnier
University of Karlsruhe
Institute of Chemical Process
Engineering
Kaiserstrasse 12
76128 Karlsruhe
Germany

A. Ted Watson
Department of Chemical Engineering
Colorado State University
Fort Collins
Colorado 80523–1370
USA

Andrew G. Webb
Beckman Institute
University of Illinois at Urbana-Champaign
405 N. Mathews
Urbana
Illinois 61801
USA
and Department of Electrical and
Computer Engineering
University of Illinois at Urbana-Champaign
1406 West Green Street
Urbana
Illinois 61801
USA

Yang Xia
Department of Physics
Oakland University
Rochester
Michigan 48309
USA

Joshua J. Young
MRI Center, Department of Physics and
Department of Civil Engineering
University of New Brunswick
P.O. Box 4400
Fredericton
New Brunswick E3B 5A3
Canada

Klaus Zick
Bruker-Biospin GmbH
76287 Rheinstetten
Germany

1
Introduction

Siegfried Stapf and Song-I Han

1.1

A Brief Comment

Whenever one wants to present an experimental technique for a certain application, one is faced with the question of how much depth the introduction should go into. This is particularly true for NMR, which has developed into an area of research of literally arbitrary complexity, while the basics of NMR imaging can be grasped sufficiently well with little background knowledge. Because this book focuses on applications of NMRI to chemical engineering problems, we will provide the reader with only the basic tools of NMR imaging that are necessary to appreciate the full potential and flexibility of the method as such, in fact its beauty. Much as the admiration for a masterpiece of music does not necessarily require the possession of the skills to reproduce it, but it benefits from a certain understanding of the structure and the context of the composition. Owing to the limited space that is available, however, it must remain beyond the scope of this Introduction to give a complete account of even the simplest relationships. Those readers who are already familiar with the basics of NMR may glance over this Introduction. Laypersons will obtain a crude overview but are directed to a limited number of standard textbooks given at the end of this chapter, a list which is not intended to be complete but should give a starting point for learning more about NMR and is kept short intentionally. Of the large number of NMR textbooks available, only relatively few concentrate on imaging techniques, and the majority of these are aimed at the medical researcher. In fact, medical imaging is in a very advanced state as far as applications are concerned, and it is worthwhile looking across the boundaries to get an overview of advances in, e.g., fast imaging, contrast factors, motion suppression or data processing techniques. For a more thorough understanding, the reader is referred to the lists of references at the end of each of the 29 chapters in this book, which together give an extensive account of the state-of-the-art of non-medical NMR imaging.

1.2
The Very Basics of NMR

If one dissects the abbreviation NMRI into its individual words, the essential features of the method are all covered: we are exploiting the interaction of the *Nuclear* magnetic properties with external static *Magnetic* fields that leads to *Resonance* phenomena with oscillating magnetic fields in the radiofrequency regime in order to obtain an *Image* of an object. The use of the term "nuclear" has become so unpopular in medical sciences that the abbreviation "MRI" is more common nowadays; although the "N" just states that the experiment is performed on the atomic nuclei, which make up all matter, including our body – to distinguish it from techniques that use the response of the electron. We should merely keep in mind that we are dealing with a quantum mechanical phenomenon. However, most of the systems that are relevant to NMR imaging are weakly coupled homonuclear systems of solution or soft materials where motional averaging and secular approximations are valid for simplified classical description to be sufficient. If one wants to exploit, e.g., the dipolar coupled spin network in ordered soft materials or large molecular assemblies for contrast imaging, a more detailed knowledge of the quantum mechanical relationships becomes inevitable.

We attempt to describe NMR Imaging in a simplified manner using only *three* essential equations that explain why we see a signal and what it looks like. The first equation describes the nuclear spin magnetization, thus the strength of the NMR signal (and indeed much more):

$$\boldsymbol{M_0} = N \frac{\gamma^2 \hbar^2 I(I+1)}{3 k_B T} \boldsymbol{B_0} \tag{1.1}$$

This equation is called the Curie law and relates the equilibrium magnetization $\boldsymbol{M_0}$ to the strength of the magnetic field $\boldsymbol{B_0}$. The constants have the following meaning: I is the nuclear spin quantum number (see below), γ is the gyromagnetic ratio specific for a given isotope, $\hbar$ is Planck's constant, k_B is Boltzmann's constant, N is the number of nuclei and T is the temperature.

What it essentially tells us is that the magnetization increases linearly with the number of nuclei and the magnetic field, and with the square of the gyromagnetic ratio γ. $\boldsymbol{M_0}$ is the quantity which after all translates into the NMR signal that we measure, so it should be as large as possible. In order to obtain maximum magnetization, one therefore wants to use a very strong magnetic field (although the ease with which weak fields can be generated possesses a significant attraction, see Chapters 2.2 and 2.4), and take advantage of a nucleus with a large γ. Of all stable isotopes, the nucleus of the hydrogen atom, ^1H, has the largest γ. Furthermore, it is contained in all organic matter and the ^1H isotope has almost 100% natural abundance, which is the reason why the vast majority of imaging experiments are done with ^1H nuclei. However, several advanced applications are presented in this book that exploit the additional information provided by other nuclei such as ^2H (Chapter 2.8), ^7Li (Chapter 3.4), ^{23}Na (Chapters 3.4 and 5.4), ^{35}Cl

(Chapter 3.4), the metals ^{27}Al, ^{51}V and ^{55}Mn (Chapter 5.4) or the monatomic gas ^{129}Xe (Chapters 2.6 and 5.3).

The origin of the Curie law is found in the nuclear equivalent of the magnetization, the *spin*. It resembles a compass needle that is located at the core of the nucleus; brought into a magnetic field, its energy state will depend on its orientation relative to the field direction. The spin is a quantum property, i.e., it can only assume quantized (half-integer or integer) values, given by the total spin quantum number: $I = 0, +1/2, +1, \ldots$. The main difference from a real compass needle is that when a spin is brought into the magnetic field, all the different orientations it can assume correspond to only a limited number of discrete energy levels, quantized between $-I$ and $+I$ in half-integer steps, so that $2I + 1$ possibilities result. One often symbolizes this effect by an arrow of a constant length that is oriented at well defined angles relative to B_0, but this is merely a crutch for visualizing the quantum mechanical property through a classical one. Independent of how one imagines the spin, one needs to keep in mind that only by bringing the spins into an external magnetic field can the different orientations of the spins differ in energy, which – as stated above – takes place in a non-continuous but quantized fashion, with energy differences of $\Delta E = \hbar\omega_0$, where ω_0 is the Larmor frequency. Its meaning will be discussed shortly. Keep in mind that not one but an ensemble of spins are present, which are distributed between the discrete energy levels. According to thermodynamics, the lower energy states are more likely to be populated – the average number of spins found in the different energy level states is given by the Boltzmann distribution:

$$\frac{n_{-\frac{1}{2}}}{n_{+\frac{1}{2}}} = \exp\left(-\frac{\hbar\omega_0}{k_B T}\right)$$

Here we have restricted ourselves to the case of two energy levels as are found for $I = 1/2$: $-1/2, +1/2$ (Figure 1.1). It describes the ^{1}H nucleus and many other isotopes that are important for imaging. The Curie law originates from the Boltzmann distribution. If we insert typical values for the magnetic field strength (say, 10 tesla) and room temperature ($T = 298$ K), we end up with a tiny fraction of 0.007 % in population difference for the ^{1}H nuclei. This difference is what provides the NMR signal. If we add up all (quantum mechanical) spins to a (classical) bulk magnetization, most of them cancel out, but only this very small population difference determines the actual value of M_0. For this reason NMR is traditionally regarded as an insensitive method. Although advanced techniques have made sample volumes down to nanoliters and concentrations down to micromolar detectable under certain circumstances, NMR spectroscopy and imaging can still not compete in terms of sensitivity with techniques such as fluorescence spectroscopy or atomic force microscopy, which are basically capable of detecting single atoms or molecules. However, the power of NMR lies in its unique ability to encode a cornucopia of parameters, such as chemical structure, molecular structure, alignment and other physical properties, interaction between atoms and molecules, incoherent dynamics (fluctuation, rotation, diffusion) and coherent flow

Energy E

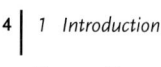

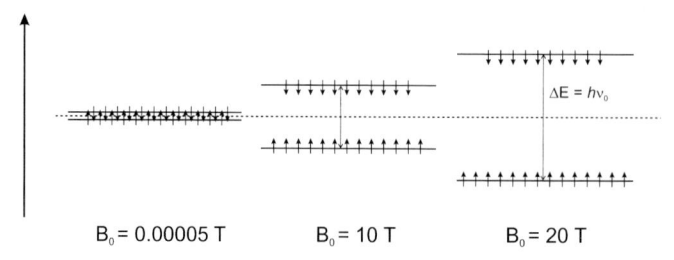

$B_0 = 0.00005$ T $B_0 = 10$ T $B_0 = 20$ T

Fig. 1.1 Schematic representation of the population difference of spins at different magnetic field strengths. The two different spin quantum number values of the ^{1}H spin, $+½$ and $-½$, are indicated by arrows. Spins assume the lower energy state preferentially, the ratio between "upper" and "lower" energy level being given by the Boltzmann distribution. The field strengths resemble typical values commonly used for high-resolution spectroscopy. The energy difference between the two states corresponds to the Larmor frequency, which is about 425 and 850 MHz for ^{1}H nuclei, respec-tively, at the two given fields of 10 and 20 T, respectively. For comparison, if the experiment had to be performed in the Earth's magnetic field (left part), the frequency would be as low as 2 kHz, which is in the audible range.

(translation) of the sample into the complex NMR signal, instead of simply measuring signal amplitudes of carriers that can be referred to distance or position information.

The longitudinal magnetization M_0, which we have just defined, is an equilibrium state for which a direct measurement would be of limited use for our purposes. The principle of NMR techniques, however, is not to measure this equilibrium quantity, but the response, thus the change of non-equilibrium transverse magnetization with time, which induces a voltage in a receiver coil enclosing the sample. When subjecting the spin system to an oscillating radiofrequency (rf) field, resonance phenomena can be utilized in such a way that, at the end of the irradiation, the magnetization M is manipulated to be oriented perpendicular to the magnetic field B_0, i.e., out of its equilibrium. In this state, the spin ensemble's net magnetization is precessing about B_0, and this precession takes place in a coherent manner ("in phase") among the spins in the ensemble as long as this coherence is not destroyed by natural or artificial influences. It turns out that controlling the time evolution of the coherence by means of a series of rf field and static magnetic field gradient pulses (pulse sequence) leaves a wealth of possibilities to encode information into the signal and extract it again at the time of acquisition. The time-dependence of M in the magnetic field, describing this precession motion, follows the second essential equation:

$$\frac{dM}{dt} = \gamma M \times B \tag{1.2}$$

The time dependence of the magnetization vector, $M(t)$, is thus related to the cross-product of M and B. Keep in mind also that the magnetic field can be time-dependent. We have replaced B_0 by B to indicate that the magnetic field can consist of different contributions. In particular, the rf field that interacts with the spins in the sample is a time-dependent magnetic field, B_{rf}, it is precisely this field that, when taken into account during the application of the rf pulse, results in the magnetization being rotated out of its equilibrium orientation. To a first approximation, however, we consider the magnetic field to be static. We can then solve Eq. (1.2) and obtain:

$$M_x = \boldsymbol{M_0} \sin \omega_0 t$$
$$M_y = \boldsymbol{M_0} \cos \omega_0 t$$
$$\boldsymbol{M_+} = \boldsymbol{M_0} \exp(i\omega_0 t)$$

The quantity of interest is the precession of the components perpendicular to B_0 that are measured in the experiment by induced voltage in the coil, which is subsequently amplified and demodulated. We can write them either as individual components M_x, M_y, or by a vector $\boldsymbol{M_+}$, which combines both of them. In the static field, the precession about $\boldsymbol{B_0}$ occurs with the Larmor frequency $\omega_0 = \gamma B_0$. If we neglect those processes which dampen the amplitude of the rotating transverse magnetization as precession proceeds, this already describes the frequency that we pick up with our receiver coil, and it is the third and perhaps the most important of our three fundamental equations of NMR:

$$\omega = \gamma |\boldsymbol{B}| \tag{1.3}$$

Now what do we learn from this? Given that the gyromagnetic ratio γ is known for all nuclei with a very high precision, the measurement of the signal frequency ω_0 allows us to determine the actual value of the magnetic field precisely! Indeed this is what NMR is basically doing, with one remarkable exception: $\boldsymbol{B}$ is not just the externally applied field, but it is *the magnetic field at the local position of the nucleus itself*, which may vary from one nucleus to the other. A large part of the toolbox of NMR is built around this simple dependence. There are two regimes that can be distinguished, providing totally different information about our system.

The *microscopic regime* is given by the immediate vicinity of the nucleus. It is surrounded by electrons the motion of which – just like the motion of any electric charge – induces a magnetic field that shields the nucleus from the external field, resulting in the nucleus specific local magnetic field. The single electron of the hydrogen atom shields a fraction of some 10^{-5} of the external field. Because the shielding depends on the actual charge distribution which, in turn, is a consequence of the molecular environment of the hydrogen atom, a particular moiety can be identified by the shielding effect it has on the nucleus. From $\omega = \gamma B$ we see that the resonance frequency of all nuclei varies proportionally with the field strength. The difference relative to a standard sample is called the *chemical shift* and is measured in unitless numbers, given as ppm (parts per million, 10^{-6}). Comparing all proton-containing chemical substances, the total range of ^1H nuclei resonance frequencies covers about 12 ppm. For instance, in an external magnetic field of 9.4 tesla, i.e., at a resonance frequency of about 400 MHz, the maximum

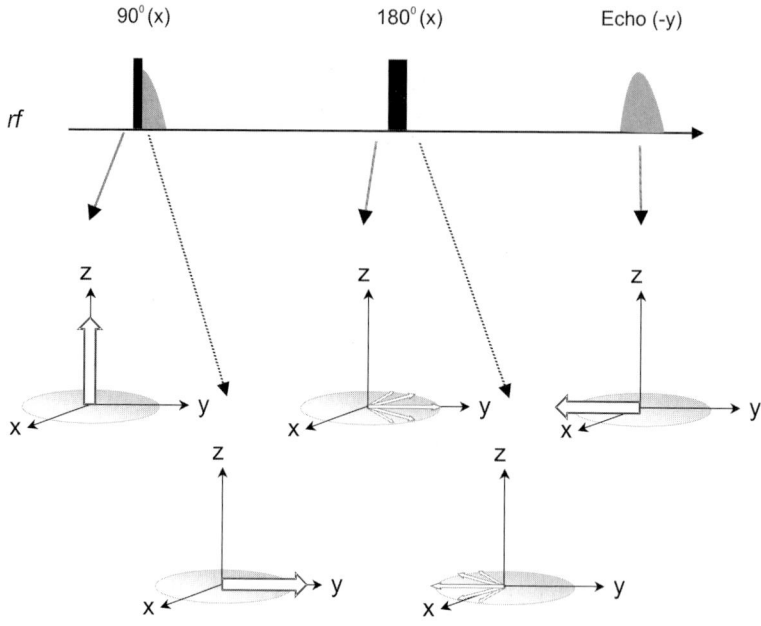

Fig. 1.2 Behavior of the magnetization in a simple echo experiment. Top: a free induction decay (FID) follows the first 90° pulse; "x" denotes the phase of the pulse, i.e., the axis about which the magnetization is effectively rotated. The 180° pulse is applied with the same phase; the echo appears at twice the separation between the two pulses and its phase is inverted to that of the initial FID. Bottom: the magnetization vector at five stages of the sequence drawn in a coordinate frame rotating at ω_0 about the z axis. *Before* the 90° pulse, the magnetization is in equilibrium, i.e., parallel to the magnetic field (z); immediately *after* the 90° pulse, it has been rotated (by 90°!) into the transverse (x,y) plane; as it is composed of contributions from different parts of the sample, the slightly different local fields lead to a loss of coherence, i.e., to a free induction decay in the pick-up coil; *before* the 180° pulse, the different contributions (narrower arrows) have lost part or all of their coherence; *after* the 180° pulse, each partial magnetization has been flipped (by 180°!) due to the effect of the electromagnetic field of the pulse, but still sees the same, slightly different local fields; *at a time corresponding to twice the pulse spacing*, the different phase shifts relative to the average value have been reverted, all partial magnetizations are *in phase* again, and their signal contributions add up *coherently* – so a *spin echo* is generated.

difference in frequencies observed is only about 4800 Hz. To distinguish subtle differences between the molecules, the resolution of a good spectrometer must be much better, often values of 10^{-10} (i.e., 0.4 Hz in our example) can be reached.

The *macroscopic* regime is the one that is directly accessible by the magnet designer. By introducing, on purpose, an inhomogeneity to the magnetic field by means of additional coils, B and therefore ω are made functions of the position. Of course, this only makes sense if each value of the field B occurs only once in the entire sample volume, so that an unambiguous assignment of the position is possible. The most obvious solution is the generation of a linear field dependence $B(\mathbf{r}) = B_0 + \mathbf{g}\mathbf{r}$ for which inversion of the frequency into position is directly applicable. The following chapter will address this relationship.

1.2 The Very Basics of NMR

What we have said up to now can be summarized in a simplified graphic representation of a basic NMR *pulse sequence*. The effects of either the microscopic or the macroscopic regime on the magnetization are visualized by the schematic picture given in Figure 1.2, which relates the behavior of the magnetization (the broad arrows in the sketches at bottom) with the events that take place in the transmitter/receiver unit of the spectrometer (top row). At the beginning, the *longitudinal* magnetization is in equilibrium and no signal is detected. The first rf pulse (it is called a 90° pulse as the pulse duration is just long enough to rotate the longitudinal magnetization by that angle) delivers M to the transverse plane and creates *transverse* magnetization. Here, the x in the subscript of 90° indicates a rotation about the x axis to place the transverse magnetization along the y direction. Immediately after such a 90° pulse it still possesses its original magnitude but is forced to precess in the transverse plane about B_0, inducing a signal in the receiver coil, the so-called *free induction decay* (FID). The transverse magnetization (also called coherence) is a sum of many contributions, originating from different spin positions inside a molecule as well as from different regions in space. Both effects – which we earlier termed *microscopic* and *macroscopic* – and also some others, generate a certain spread of local fields B and of Larmor frequencies ω, so that the individual contributions to the initial (maximum) magnetization start to "fan out" – note that the plot in Figure 1.2 must be understood as the *rotating frame*, i.e., we assume a point of view that is rotating with the carrier frequency of the transmitting radio field about the z axis. Some partial magnetizations (they are also called *isochromats*) precess faster, others precess slower than the average. Eventually, their vectors add up to zero and no signal is detected in the receiver. The 180° pulse "flips" all isochromats around, so that the faster ones suddenly find themselves at the trailing edge, but because the physical reason for their precession frequency has not changed, they begin to catch up with the slower ones. This process goes completely unnoticed by the operator of the NMR spectrometer, because the isochromats remain in a decoherent state relative to each other, and their vectorial sum is still zero. Only after a time that is exactly twice the separation between the two rf pulses do they get in phase again, and in this moment, the broad arrow, the complete magnetization, is regained, but now in the direction opposite to where it had pointed to immediately after the first rf pulse. This is because the 180° rotation was performed about the x axis, as depicted in Figure 1.2. By changing the *phase* of the rf pulses, i.e., by changing the axis about which the rotation of the magnetization vector takes place as discussed above, the final direction of this arrow can be chosen at will. The recovered magnetization induces a voltage in the receiver coil, seemingly out of nowhere – this is the so-called spin-echo that Erwin Hahn saw for the first time in 1950, much to his surprise (see his *memo about the discovery of the echo* at the beginning of this book). The second half of this echo is identical to the FID at the beginning of the sequence, apart from their initial signal amplitudes due to *microscopic* or *macroscopic* processes that have happened to the system during the time between the FID and echo acquisition. Comparing echo and FID amplitudes can give a first starting point for analyzing the dynamics that are taking place in the sample; using the echo and continuing the procedure with the application of more rf pulses allows one to recycle the very same magnetization many times and extract even more information from its behavior.

The *microscopic* and *macroscopic* field dependences are the two basic building blocks of NMR. The second one is essentially what this book is about, NMR Imaging. However, the first one – the basic principle of NMR spectroscopy – is always present, and can be exploited as additional information in an NMR image to distinguish different species in a heterogeneous sample, something that no other imaging technique is able to achieve with such large versatility and specificity.

One can probably guess that NMR is more complicated than this short description suggests. Just a few of the complications arising in a "real" experiment are: how to distinguish between the different influences (*microscopic* and *macroscopic*) on the resonance frequency; how to generate three-dimensional images from the simple linear dependence of the field strength; how to measure motion; and how to deal with the myriad of various microscopic effects arising from interactions of the nuclei with each other and with the external fields. We will address the most relevant ones in the following chapters, focusing on the *macroscopic* aspects and treating the *microscopic* influences mostly as markers that help us to introduce contrast into our images.

1.3
Fundamentals of NMR Imaging

In this section, we will describe three building blocks of NMR imaging: *phase encoding*, *frequency encoding* and *slice selection*. All three are related to the signal by the fourth equation:

$$\omega(r) = \gamma|B(r)| = \gamma(|B_0| + g \cdot r) \tag{1.4}$$

Here we have introduced the position dependence of the magnetic field through its first derivative, the gradient vector g, which renders the resonance frequency proportional to the position of the spin, r. More precisely, the gradient has the properties of a 3×3 tensor because the derivatives of all three orthogonal components of B need to be computed. However, B is usually taken as pointing in the z direction, and its x/y components are negligible just as their spatial derivatives are. Note that the product $g \cdot r$ indicates that only those components of the position vector which are parallel to the gradient direction possess a different resonance frequency.

We now have to face the question: how can we express the imaging experiment by a mathematical formalism? To begin with, a physical object can mathematically be described by a density function of position: $\varrho(r)$. If we apply an rf pulse to this object, and record the signal from the sample following the influence of the gradient, it will consist of contributions from all volume elements of the sample. In order to generate the image of the object, we will have to modify the weighting

1.3 Fundamentals of NMR Imaging

function (the effective gradient g turned on during the time t, hence gt) to reconstruct the actual spin-density distribution. This approach is equivalent to a scattering experiment which is known from many fields of materials research: a wave of defined properties (wavelength, polarization) is aimed at an object and the scattered waves are investigated in terms of the above mentioned weighting function, which corresponds to the reciprocal space. The principle is the same whether light, X-rays, phonons or particles such as electrons and neutrons are considered; for the last of these, it is their wave properties that define the scattering effect. For instance, the structure of a crystal is obtained from an X-ray or neutron scattering pattern, which is related to the Fourier transform of the density function of the scattering centers, averaged over the whole sample. The Fourier transformation translates information from the reciprocal space, or k-space, to the direct space, or r-space; k and r are therefore Fourier conjugate variables.

In a similar fashion, the information in NMR imaging is sampled in the reciprocal k-space and is then Fourier transformed into r-space, thereby reproducing the spin-density distribution $\varrho(r)$, or *shape*, of the object. There is, however, one major difference that makes NMR a particularly strong tool for investigating matter: unlike the classical scattering techniques, which only provide an *intensity* measure as their result, the NMR information is a complex number and also contains the *phase* of the signal. By applying field gradient pulses, the position can be encoded into the phase of the signal as a function of k, while its magnitude remains available for other information such as additional k-space dimensions, or chemical information via the modulation of the signal magnitude according to the gradient weighting function or the precession frequencies of the sample. This is the principle of multi-dimensional imaging, including imaging containing chemical shift resolution.

Let us now derive the equations that relate the spatial information to the signal behavior. As we have seen previously, a spin at position r possesses a Larmor frequency $\omega(r) = \gamma|B(r)| = \gamma(|B_0| + g \cdot r)$. It is convenient to subtract the reference value, given by the "average" field, $\omega_0 = \gamma|B_0|$, so that we obtain the frequency difference relative to an (arbitrarily chosen) position $r = 0$:

$$\omega(r) - \omega_0 = \gamma g \cdot r$$

In NMR, data acquisition is demodulated with the carrier frequency corresponding to the magnetic field strength in use, ω_{rf}, which is chosen usually close to ω_0. Now assume that the gradient g is applied only during a certain interval δ. We call this a gradient *pulse* – hence the frequently used term "pulsed field gradient (PFG) NMR" – and it can be of arbitrary shape, but a rectangle with sharp edges is often used and facilitates the following discussion (see Figure 1.3). (In fact, elaborate hardware development has been carried out over many years to produce short and stable gradient *pulses* with sharp edges, which are required for many applications – see, for instance, Chapter 2.1.) After the gradient pulse is switched off, the difference in frequencies leads to an accumulated phase shift, $\phi = [\omega(r) - \omega_0]\delta$, with respect to

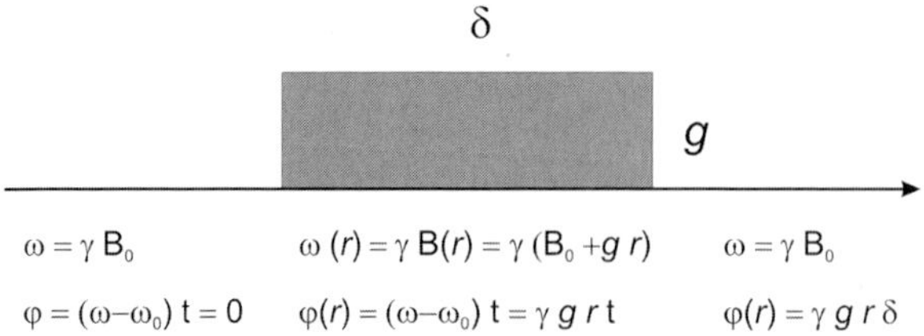

$\omega = \gamma B_0$ $\omega(r) = \gamma B(r) = \gamma (B_0 + g\, r)$ $\omega = \gamma B_0$
$\varphi = (\omega - \omega_0)\, t = 0$ $\varphi(r) = (\omega - \omega_0)\, t = \gamma\, g\, r\, t$ $\varphi(r) = \gamma\, g\, r\, \delta$

Fig. 1.3 Effect of a pulsed magnetic field gradient of strength g on the phase of a signal contribution originating from spins at position *r*. Prior to the gradient pulse, all spins experience the constant magnetic field B_0 and thus possess the same Larmor frequency; the phase shift relative to this average value is zero. During the gradient pulse, the field becomes position-dependent and a phase shift ϕ is accumulated that is proportional to position *r* and time *t*. After the gradient pulse (i.e., after completion of the *phase encoding* step), the spins memorize their individual phase shifts $\phi(r) = \gamma\, g\, r\, \delta$.

the reference. Thus, the resulting phase shift of spins at position *r* following a gradient pulse of duration δ and strength *g* can be written as

$$\phi(r) = \gamma\, \boldsymbol{g} \cdot \boldsymbol{r}\, \delta \tag{1.5}$$

If we consider that the signal generated by the precessing magnetization is a complex number, the phase shift created by this gradient pulse leads to the NMR signal, which is the multiplication of the unperturbed NMR signal with the phase factor $\exp[i\phi(r)]$. In order to compute the total signal arising from all spins within the sample with a spin-density distribution $\varrho(r)$, we simply have to integrate all phase factors over the entire volume:

$$\begin{aligned} S(\boldsymbol{k}) &= \int \varrho(r)\, \exp[i\phi(r)]\, dr = \int \varrho(r)\, \exp[i\gamma\, \boldsymbol{g} \cdot \boldsymbol{r}\, \delta]\, dr \\ &= \int \varrho(r)\, \exp[i 2\pi\, \boldsymbol{k} \cdot \boldsymbol{r}]\, dr \end{aligned} \tag{1.6}$$

In this equation, we have made the replacement $\boldsymbol{k} = (1/2\, \pi)\gamma g \delta$ in order to introduce the Fourier conjugate variable to *r*. This is because formally Eq. (1.6) is a Fourier transformation. What we really want to know is the shape of the sample, $\varrho(r)$, which we can derive by applying the inverse Fourier transformation to the signal function:

$$\varrho(r) = \int S(\boldsymbol{k})\, \exp[-i 2\pi\, \boldsymbol{k} \cdot \boldsymbol{r}]\, d\boldsymbol{k} \tag{1.7}$$

However, in order to be able to apply the inverse Fourier transformation, we need to know the dependence of the signal not only for a particular value of *k* (one gradient pulse), but as a continuous function. In practice, it is the Fast Fourier Transform (FFT) that is performed rather than the full, analytical Fourier Transform, so that the sampling of *k*-space at discrete, equidistant steps (typically 32, 64, 128) is being performed.

The recipe for the first building block of NMR imaging, the *phase encoding*, thus goes like this: apply a phase gradient of effective area *k*; acquire the signal S(*k*); repeat for a number of different equidistant values of *k*; perform the inverse

1.3 Fundamentals of NMR Imaging

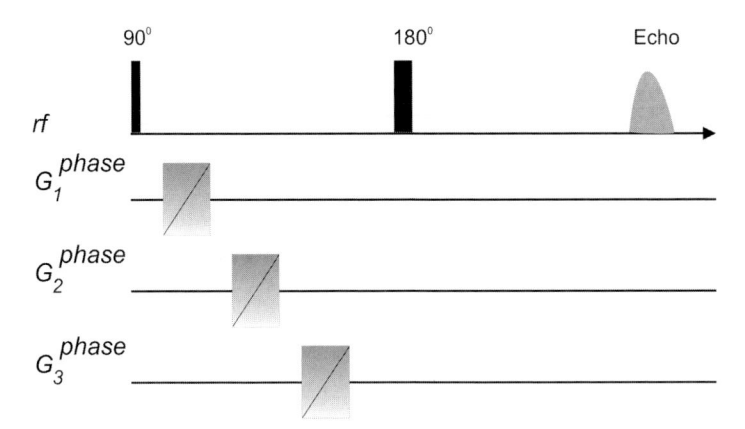

Fig. 1.4 Phase encoding scheme in three dimensions. Three pulsed gradients in orthogonal directions are applied and are varied independently of each other (symbolized by the diagonal line). The actual timing of the gradients is arbitrary provided they are placed between the rf pulses and before the acquisition of the echo signal. In practice, the gradients are often applied simultaneously. The indices 1, 2 and 3 represent orthogonal directions with no priority being given to a particular choice of combinations.

Fourier transformation to reconstruct the spin-density function of the sample, $\varrho(r)$. The variation of gradients is symbolized by diagonal lines in Figure 1.4.

Note that variation of the *phase encoding gradient* along one direction, say k_x, allows the reconstruction of the profile only in this direction, $\varrho(x)$. An example of this is shown in Figure 1.5, where the projection of a cylindrical object (such as a test tube filled with water) is depicted with the aid of simulated data. A series of FIDs is drawn in succession, a typical way of saving the data in a single, long vector file where each FID is being acquired in the presence of a particular gradient value, so that the range $-k_{x,\max} \ldots k_{x,\max}$ is covered and $k_x = 0$ is in the center, giving maximum signal intensity. Figure 1.5 also demonstrates the effect of an insufficient coverage of k-space on the image quality: if at the largest value of k_x used, the signal has not yet vanished, cut-off artifacts in the image do arise.

However, extension of the sequence towards two- or three-dimensional encoding is straightforward, the only requirement being that all gradient pulses are stepped *independently* of each other. They might be applied either simultaneously or sequentially, they only have to be placed somewhere between the first excitation of the sample by an rf pulse and the final signal acquisition (see Figure 1.4). Equation (1.7) already contains this possibility; however, in Eq. (1.8) we have split the three orthogonal components and express the same relationship by a three-fold integral to highlight the three-dimensional nature of the experiment:

$$\varrho(x,y,z) = \int S(k_x, k_y, k_z) \exp[-i2\pi k_x x] \exp[-i2\pi k_y y] \exp[-i2\pi k_z z] \, dk_x \, dk_y \, dk_z \tag{1.8}$$

The choice of gradient strength and duration can be optimized if the size of the object is known. From Eqs. (1.5 and 1.6) it becomes clear that a large gradient will lead to an ambiguous assignment of positions if the maximum phase angle exceeds 2π. One must therefore make sure that the difference between phase shifts generated for spins at the extreme ends of the sample is smaller than 2π for each different value of the gradient. In other words, $2\pi > 2(\gamma g_{max}\delta r_{max})/(n-1)$ if the gradient is varied in the range $-g_{max} \ldots +g_{max}$ in n steps. The coverage in r-space achieved with a particular set of gradient values is called the field of view (FOV) and is obtained from

$$FOV = \frac{n-1}{2} \frac{2\pi}{\gamma g_{max} \delta} \tag{1.9}$$

For n gradient steps (n different k values), the image function $\varrho(r)$ also possesses a resolution of n points as a consequence of the FFT procedure. Naturally, a larger g_{max} results in a better resolution if the number of steps n is chosen in such a way that the FOV slightly exceeds the object size itself. Note that a choice of a too small field of view will lead to so-called aliasing, i.e., those picture elements that correspond to a phase shift that exceeds the range π, will appear at the other side of the image so that superposition of two or more signal contributions into the

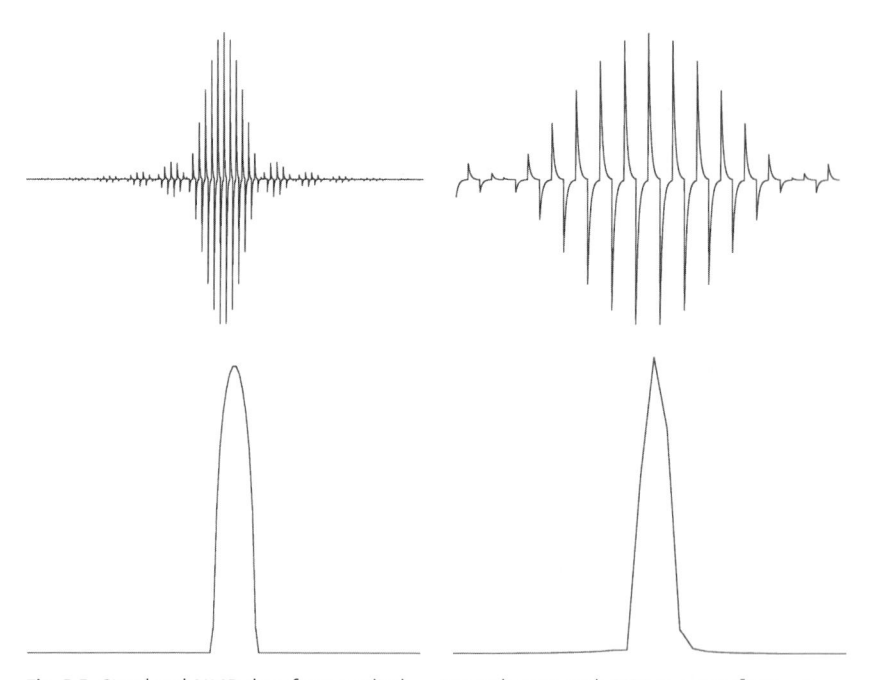

Fig. 1.5 Simulated NMR data from a cylindrical object projected onto a direction perpendicular to the cylinder axis. Left: 128 steps of k_x were applied and the succession of FIDs are plotted; k-space is covered sufficiently well, i.e., the signal has decayed to zero at the largest values of k_x. The Fourier transform of the series of FIDs (taken from the first points or from the integrals over each FID) is a satisfactory representation of the projection of the sample shape. Right: only the central 32 k_x values were acquired; the Fourier transform suffers from bad resolution, but also from a non-vanishing baseline due to the fact that the signal at the largest k_x values had not vanished.

1.3 Fundamentals of NMR Imaging | 13

correct field-of-view　　　　　　　　too small field-of-view

Fig. 1.6 Left: the field of view has been chosen sufficiently large to cover the whole object; a correct image is obtained. Right: the field of view is only half as wide; the top and bottom portions of the object are folded back into the opposite side of the image. In a real NMR image, the assignment of position becomes ambiguous.

image (backfolding) can occur. Figure 1.6 demonstrates the effect of backfolding for an FOV smaller than the object.

In order to demonstrate how the Fourier relationship works, Figure 1.7 shows what a two-dimensional object – a letter "i" in a circle – would look like in an NMR image, that is, its spin-density function $\varrho(x,y)$ is shown, and the original signal function which is obtained by varying $\bm{k}$, $S(k_x, k_y)$. While $S(k_x, k_y)$ is a complex function, only its real part is plotted. Because of the reciprocal relationship between $\bm{r}$ and $\bm{k}$, one can roughly say that the signal of high intensity in the vicinity of $\bm{k} = 0$ contains information about the general shape of the sample, and the small contributions at large $\bm{k}$ values is responsible for the fine structure, i.e., the resolution of the image. Although this is an oversimplified view, it is often helpful to keep this

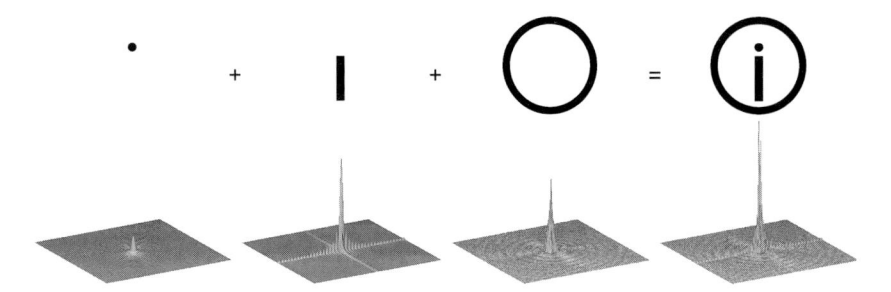

Fig. 1.7 Two-dimensional objects in position space (top row) and their Fourier transform, corresponding to the shape of the signal function $S(k_x, k_y)$ (bottom row). The actual object, a letter "i" inside a circle, is shown as the sum of three individual objects. Likewise, the Fourier transform is additive and the signal functions corresponding to each of the objects are shown for comparison.

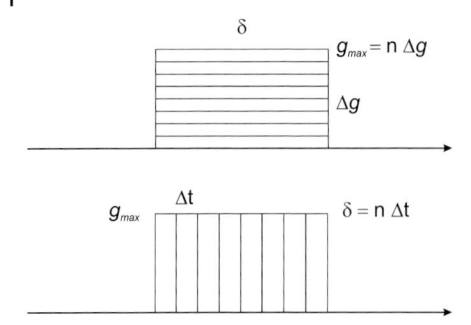

Fig. 1.8 Demonstration of the formal equivalence of *phase encoding* and *frequency encoding*. In the top figure (*phase encoding*), the gradient of duration δ and strength g_{max} is split into eight equal steps of Δg. The signal is acquired n times (here $n = 8$), following a gradient of strength $i\Delta g$ where i is varied from 1 to n. In the bottom figure (*frequency encoding*), the gradient strength g_{max} is maintained but the total duration δ is split into eight equal steps, Δt. The signal is acquired n times (here $n = 8$), following a gradient of duration $i\Delta t$ where i is varied from 1 to n. In both cases, the signal corresponding to the same $n = 8$ k values between zero and k_{max} is sampled, where $k_{max} = (2\pi)^{-1}\gamma g_{max}\delta$.

inverse relationship in mind and to estimate which effect the distortion of a particular part of the k-space data will have on the r-space image. In principle, only an analytical Fourier transformation of a continuous signal function $S(k)$ renders a perfect result; the compromise of a finite sampling of k-space gives rise to imperfections in the NMR image, which arise on top of potential physical problems, such as hardware imperfections. We will not discuss the manifold possibilities of how image artifacts can arise and how they can be identified and compensated, but refer the reader to the textbooks at the end of this chapter, where this is discussed in great detail.

The foremost advantage of *phase encoding* is the fact that the signal is acquired in the presence of a homogeneous magnetic field; the signal thus contains the full spectroscopic information and this can be exploited to generate an image with individual spectra contained in each voxel representing the local chemical environment. This is one of the most powerful implications of NMR imaging. Just to name one important example, an inhomogeneous distribution of a mixture of liquids can be imaged while obtaining the full spectroscopic dimension, distinguishing the different phases and/or chemical species in the mixture. By this means, chemically selective, spatially resolved monitoring can be employed to study complex mixtures. The disadvantage of *phase encoding* is its inherent slowness. To generate a three-dimensional image with a satisfactory resolution of, say, $64 \times 64 \times 64$ points, an equivalent number of 64^3 individual signals have to be acquired. Allowing for a repetition time of 1 s between acquisitions, this corresponds to an experimental time of three days! The repetition time depends on the relaxation parameters of the sample; for liquids, it is in the order of several seconds so that three-dimensional phase-encoded images become prohibitively slow to measure.

One way out of this dilemma is the so-called *frequency encoding* technique. It is often referred to as an approach that is very different from phase encoding, but actually, it is only the timing of k-space sampling which is modified: in frequency encoding, the signal is acquired while the field gradient is present (see Figure 1.8); it is then referred to as a *read gradient*. Equation (1.5) is still valid, so that each subsequent point of acquisition at a given time t is under the effect of a longer read gradient field and thus contains a different phase angle. The full acquisition of one line in k-space takes only a few milliseconds; Fourier transformation of this signal, $S(\mathbf{k})$, again yields the profile of the object in the corresponding direction, $\varrho(\mathbf{r})$, so that Eq. (1.7) also maintains its validity. In frequency encoding the time t is varied at constant gradient strength, thus it is also possible to write the signal of the frequency encoding technique as $S(t)$, and its Fourier transformation directly renders the frequency distribution of spins in the sample. Because $\omega = \gamma \mathbf{g r}$, this frequency distribution is proportional to the spin-density distribution of the object. The difference in the *phase encoding* method is that the signal is acquired *after* it was subjected to a gradient field of duration δ and strength $\mathbf{g}$. Normally the gradient strength $\mathbf{g}$ is varied, and for each such step a *new* signal is utilized, formed either by excitation or echo formation.

While frequency encoding is time-saving because it samples one dimension in k-space in a single acquisition, its disadvantage is that the spectral information of the spin system is masked by the spatial encoding; the result will be a superposition of chemical shift and positional information. In other words, an object filled with a substance that has a spectrum consisting of two lines will render a double image because the physical reason for the frequency shift cannot be separated in a straightforward manner. However, substances with simple spectra such as water are often considered, or the separation between different spectral lines is negligibly small. This has turned a combination of *frequency encoding* in one dimension and *phase encoding* in one or two further dimensions into the most widespread imaging technique. Figure 1.9 gives a schematic example of a suitable pulse sequence,

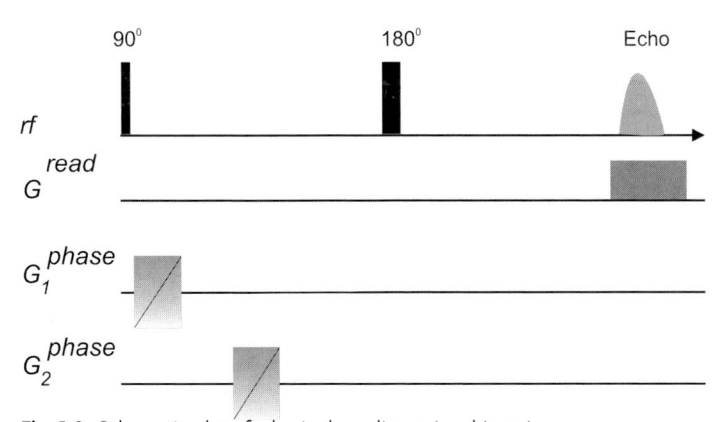

Fig. 1.9 Schematic plot of a basic three-dimensional imaging pulse sequence with frequency encoding along one axis (*read*), and phase encoding along the two remaining orthogonal directions. The choice of directions is arbitrary, as is the position of the phase gradients within the sequence.

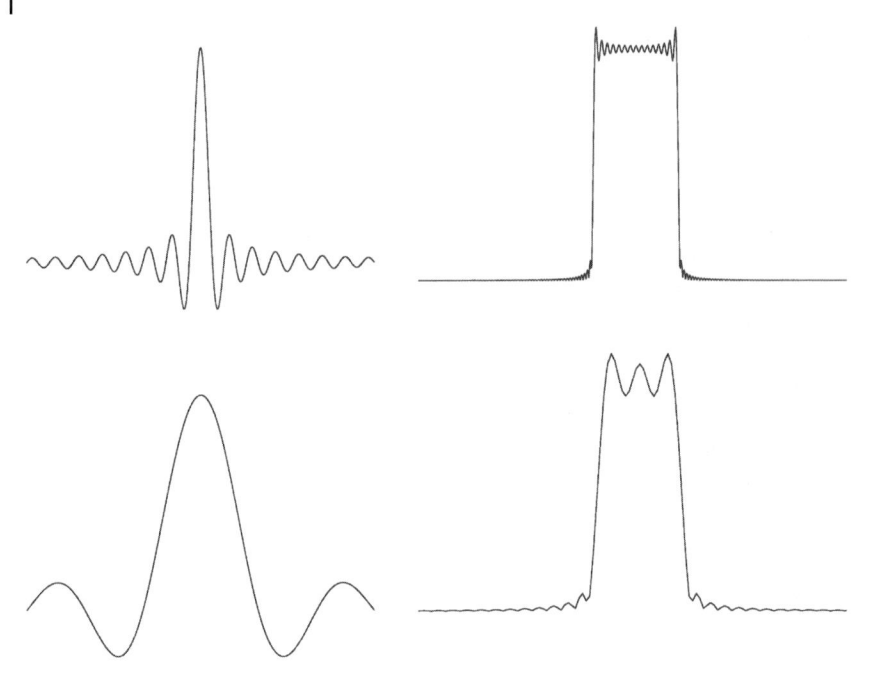

Fig. 1.10 Soft rf pulses (left) in the shape of a sinc (sin x/x) function, and their Fourier transforms (right), being equivalent to the excited slice in the presence of a constant magnetic field gradient. The well defined sinc function (top) produces an excitation that is a slice that deviates only slightly from a perfect rectangle. A cut-off sinc function, such as the one shown in the bottom part of the figure, produces a distorted slice profile which is, however, often sufficient for slice-selection in multi-dimensional imaging.

where the succession of gradients is arbitrary provided all phase encoding steps are completed before the beginning of the signal acquisition.

It must be mentioned that the reconstruction of a two- or even three-dimensional spin density distribution employing only frequency encoding is possible by varying the *direction* of the gradient, either by actual rotation of the sample or the gradient coil itself, or the rotation of the gradient fields by appropriate combinations of g_x and g_y gradients. The technique is known as *backprojection/reconstruction* and is similar to image processing routines used for generating 3D X-ray images in Computer Tomography (CT). While the advantage is that a crude picture of the sample can be obtained immediately, long before the image acquisition is completed, some disadvantages are the lower signal-to-noise ratio, resolution and image quality, and that it does not allow the assignment of spectral features in the object. Because of its limited use for materials and process studies, we will not discuss the backprojection technique further. One other important realization of imaging in inhomogeneous fields, namely the exploitation of the existing stray field found in every magnet, or the construction of a magnet with a well defined static constant field gradient, is described in some detail in Chapters 2.3 and 2.4.

Often, a cross-sectional image of an object contains all the necessary information that the researcher is interested in. However, such a two-dimensional cross section represents the integration over the third, non-encoded dimension. It is desirable to "cut out" a plane of the object, i.e., to produce an image of only a well defined slice without taking the remaining volume into account. *Slice selection* is indeed possible; again, it exploits the fundamental relationship $\omega = \gamma B$, but in a slightly different sense. So far we have implicitly assumed that all rf pulses in the imaging sequences affect the whole number of spins in the same way. This is strictly true only for a *hard pulse*, i.e., a pulse with a bandwidth greatly exceeding the range of Larmor frequencies in the sample. The bandwidth of an rf pulse is given by the reciprocal of its duration: a short pulse possesses a broad frequency spectrum. More precisely speaking, the frequency spectrum of an rf excitation is the Fourier transform of its shape function. However, because of $\omega(\mathbf{r}) - \omega_0 = \gamma\, \mathbf{g r}$, the frequency is proportional to the position in the presence of a magnetic field gradient $\mathbf{g}$! We now only have to define the shape of the slice we want to excite, and perform the inverse Fourier transformation to find out what the equivalent pulse should look like. For a rectangular slice, this is particularly simple – the Fourier transform of a rectangle is the sinc function, $\sin x/x$ (see Figure 1.10). Generating an rf pulse of the correct length that is modulated by a sinc function will thus excite a rectangular slice of the sample and leave the remaining spins unaffected. Today pulses of virtually arbitrary shape can be generated with high temporal resolution. In practice, the function is cut at a certain number of periods to limit its duration, resulting in a somewhat imperfect shape of the slice, which is usually tolerated (see Figure 1.10). In typical imaging, the *shaped pulse* or *soft pulse* has a duration in the order of milliseconds, compared with the *hard pulse* of some tens of microseconds. The terms "hard" and "soft" can thus also be understood in the sense of the power level of the pulses, which necessarily is much higher if a very short pulse must achieve the same 90° rotation of the magnetization.

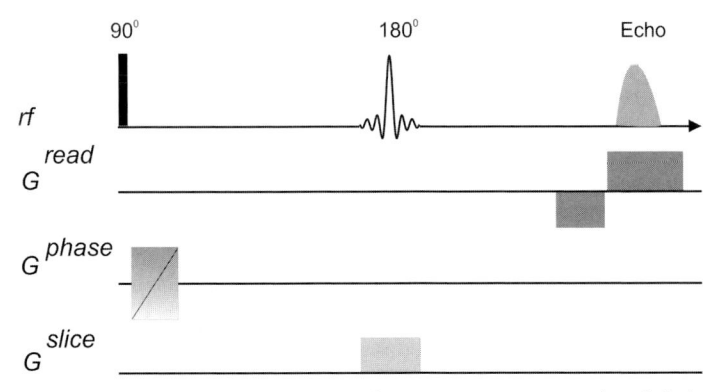

Fig. 1.11 Typical basic three-dimensional imaging sequence with slice selection, frequency encoding and phase encoding in three orthogonal directions. The compensating lobe for the read gradient is drawn as a rectangle with negative intensity directly before the actual read gradient. The shape of the 180° rf pulse is drawn schematically to indicate that a *soft* pulse is used.

The shape of any rf pulse can be chosen in such a way that the excitation profile is a rectangular slice. In the light of experimental restrictions, which often require pulses as short as possible, the slice shape will never be perfect. For instance, the commonly used 90° pulse is still acceptable, while a 180° pulse produces a good profile only if it is used as a refocusing pulse. Sometimes pulses of even smaller flip angles are used which provide a better slice selection (for a discussion of imaging with small flip angles, see Section 1.7).

From the three building blocks, we can construct a typical NMR imaging sequence that is routinely employed for acquiring slices of an extended object (see Figure 1.11). It consists of a slice selection gradient in one direction, a frequency gradient in a second and a phase encoding gradient in the third direction. The directions are orthogonal to each other and their order is arbitrary, hence the usual labeling by "read, phase, slice" instead of "*x, y, z*".

A final but important comment has to be added to the discussion of the sequence in Figure 1.11. The read gradient is preceded by a negative lobe of half its area. The reason for this is that one wants to scan the whole k-space dimension with the read gradient. If we could switch on the gradient and begin acquisition immediately, the first acquired point would represent $\boldsymbol{k} = 0$ and only positive values of $\boldsymbol{k}$ would be scanned. By applying a negative lobe before acquisition, the origin is moved towards a negative value before acquisition is begun; applying a lobe with half of the negative area of the read gradient means that the condition $\boldsymbol{k} = 0$ appears at the center of the acquisition window. The timing of the sequence is such that the echo generated by the rf pulses also appears in this instance. It is called *spin echo* and refocuses the effect of constant background gradients, and also the spreading in ω brought about by the properties of the sample spectrum (see also discussion of Figure 1.2). The combination of negative and positive lobes of g_{read} also produces an echo, the *gradient echo*. It appears at the point in time when $\boldsymbol{k} = 0$, or, in other words, when $\int g_{read} \, dt = 0$, i.e., the total effect of all gradients in the read direction vanishes. The integral is the mathematical definition of the area of the gradient, or the *zeroth* moment of the gradient, m_0. We will see in the following section why it makes sense to use this general nomenclature.

1.4
Fundamentals of Detecting Motion

With the three building blocks *phase encoding, frequency encoding* and *slice selection*, it is possible to map the spin density of arbitrary objects, with the main limitation given by the linewidth and the relaxation times (see Section 1.7). One of the main interests of NMR imaging in chemical engineering applications is the detection and quantification of motion. Simplified, motion can be expressed by an average velocity which, in turn, is computed from measuring the position difference, or *displacement*, divided by the time which has passed between the position-encoding events. One could, in principle, acquire two images and analyze the difference between their spin-distribution functions. With typical two-dimensional NMR images requiring acquisition times between 10^{-2} and 10^2 s, this would allow the determination of only relatively small velocities. Furthermore, there is no way to follow a particular spin under stationary flow conditions: the shape of a water-filled pipe will always look the same irrespective of the flow velocity. There are indeed ways to tag a certain subset of spins and follow their motion, which will be discussed in Section 1.5.

A much simpler and more flexible approach, however, extends the scheme of pulsed magnetic field gradients discussed above to include more complex time dependences. To see how this can be done, we write down the general dependence of the phase shift accumulated at time t subject to a space-dependent Larmor frequency:

$$\phi(\mathbf{r}, t) = \int_0^t \omega(t) \cdot \mathbf{r}(t) \, dt = \int_0^t \gamma \, \mathbf{g}(t) \cdot \mathbf{r}(t) \, dt$$

If we expand the position of a spin into a Taylor series:

$$\mathbf{r}(t) = \mathbf{r}_0 + \mathbf{v}_0 t + \tfrac{1}{2}\mathbf{a}_0 t^2 + \ldots$$

we obtain:

$$\phi(\mathbf{r}, t) = \gamma \int_0^t \mathbf{g}(t) \cdot [\mathbf{r}_0 + \mathbf{v}_0 t + \tfrac{1}{2}\mathbf{a}_0 t^2 + \ldots] \, dt$$

$$= \gamma \left[\mathbf{r}_0 \int_0^t \mathbf{g}(t) \, dt + \mathbf{v}_0 \int_0^t \mathbf{g}(t) t \, dt + \tfrac{1}{2}\mathbf{a}_0 \int_0^t \mathbf{g}(t) t^2 \, dt + \ldots \right] \quad (1.10)$$

$$= \gamma \left[\mathbf{r}_0 m_0 + \mathbf{v}_0 m_1 + \tfrac{1}{2}\mathbf{a}_0 m_2 + \ldots \right]$$

Terms of higher order than acceleration are usually not considered in NMR experiments. We can thus identify three integrals that are the respective Fourier conjugates to position $\mathbf{r}_0$, velocity $\mathbf{v}_0$ and acceleration $\mathbf{a}_0$.

The *zeroth* moment, m_0, was mentioned above. It is nothing more than the total area of the gradient [see Figure 1.12(a)]. If the area vanishes, the signal is independent of spin position. This is the case, for instance, for a bipolar gradient with positive and negative lobes of equal area [see Figure 1.12(b)].

The *first* moment, m_1, encodes velocity. One can easily show that the bipolar gradient in Figure 1.12(b) has a first moment of $m_1 = g\delta\Delta$, while $m_0 = 0$. Symmetrical gradient pairs are therefore the simplest way to measure velocities, while introducing no effect of position on the signal, i.e., the velocity encoded by such a gradient pair alone represents all spins in the whole sample. If phase encoding is achieved without an additional influence from velocity, the conditions $m_0 \neq 0$, $m_1 = 0$ need to be met simultaneously [see Figure 1.12(c), and the more detailed discussion of Figure 1.16].

The *second* moment, m_2, encodes acceleration. If *only* acceleration is to be measured, the conditions $m_0 = 0$, $m_1 = 0$ have to be fulfilled. Figure 1.12(d and

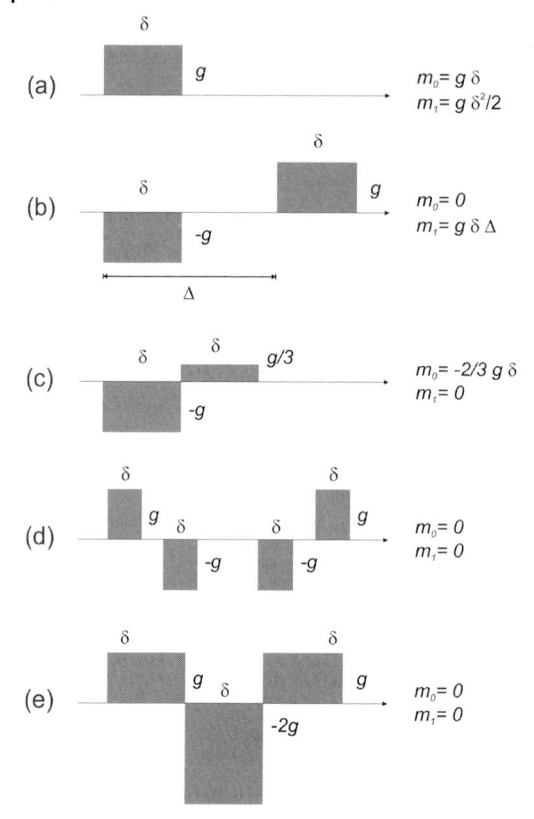

Fig. 1.12 Typical pulsed gradient shapes commonly used for imaging or measurements of motion. (a) A single gradient pulse, which is used for phase encoding of position, also has all other moments non-vanishing, so that the signal phase is affected by position, velocity and acceleration. (b) A bipolar gradient is the basic building block of most measurements of diffusion or flow; due to its vanishing zeroth moment, its application does not contain any position information. (c) A phase encoding gradient that is insensitive to motion needs to have a more complex shape, a minimum of two lobes are required. (d) Two bipolar gradients arranged back-to-back generate an encoding that is affected neither by position nor by velocity, but higher-order terms such as acceleration can be measured. The variant shown in (e) is equivalent to (d) but with the time intervals between pulses kept to a minimum. Note that in all cases, the *effective* gradient is shown; each 180° rf pulse applied anywhere within the sequence has the effect of reverting the sign of all gradient pulses that are preceding it so that the effective gradient becomes the negative equivalent to the physically applied gradient.

e) show two commonly used realizations. There is not as yet a wide range of applications where acceleration is actually the topic of interest; but see, for example, Chapters 2.9, 4.6 and 4.8. A detailed discussion of acceleration measurements and commonly used encoding schemes is given in Chapter 2.9.

In this section, we will focus on the different types of motion and their experimental determination, and will not consider imaging itself; Section 1.5 will then combine the two encodings of position and motion into velocity imaging sequences.

Let us first have a closer look at the dependence of the signal on the application of a pair of gradients with equal duration and intensity but opposite direction. This pair represents a two-fold encoding of position into the phase of the spins; the total phase can be written as

$$\begin{aligned}\phi(r_1, r_2) &= \gamma\delta\,(g_1\,r_1 + g_2\,r_2) \\ &= 2\pi\,(k_1\,r_1 + k_2\,r_2) \\ &= 2\pi\,k_1\,(r_1 - r_2) \\ &= -2\pi\,q\,R\end{aligned} \quad (1.11)$$

where r_1, r_2 are the positions at the times of the first and the second gradient pulses, respectively. Because of the interdependence of the gradients, only one variable wave vector remains, and it determines the phase shift proportional to the displacement, R. In order to symbolize that *displacements* are encoded rather than *positions*, the corresponding wave vector is renamed q. While it is obvious that $q = k_1 = -k_2$, a change of variable names often facilitates the discussion of more complex encoding schemes when distinguishing whether position or displacement ought to be encoded (Note that in the contributions of this book, the nomenclature may differ).

The signal, expressed as a function of q, can now be written as

$$S(q) = \int \overline{P}(R, \Delta)\,\exp[i2\pi\,q\,R]\,dR \quad (1.12)$$

Equation (1.12) is formally equivalent to Eq. (1.6), which defines the position encoding, but it contains as a kernel the distribution function of displacements, $\overline{P}(R, \Delta)$. This function is also called *propagator* and it should, if used correctly, include one further variable, the encoding time Δ between the gradient pulses over which the displacement R is determined. It is important to consider Δ because in many real situations, not only the average displacement but also the shape of the propagator will depend on the length of the interval during which it is being measured.

The bar above the symbol of the propagator denotes an ensemble average over the whole sample, or a selected volume of it – the position information is lost. The averaging is contained in the formal description of the propagator which can be written as

$$\overline{P}(R, \Delta) = \int \varrho(r_1)\,P(r_1|r_1 + R, \Delta)\,dr_1 \quad (1.13)$$

In this expression, the propagator is decomposed into the *spin density* at the initial position, $\varrho(r_1)$, and the *conditional probability* that a particle moves from r_1 to a position $r_1 + R$ during the interval Δ, $P(r_1|r_1 + R, \Delta)$. Note that this function cannot be directly retrieved because it is "hidden" in the integral. The propagator experiment is one-dimensional; however, by making use of more complex and higher-dimensional experiments, conditional probability functions such as the one mentioned here can indeed be determined (see Section 1.6 for examples).

Just as we obtain the spin density distribution $\varrho(r)$, and hence the projection of the investigated object, in phase encoded imaging, we can reconstruct $\overline{P}(R, \Delta)$ by

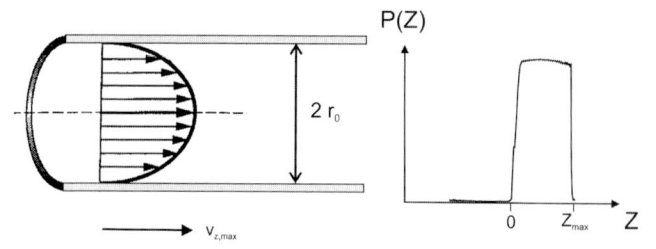

Fig. 1.13 Left: schematic plot of the distribution of flow velocities, v_z, for laminar flow of a Newtonian fluid in a circular pipe; the maximum value of the velocity, occurring in the center of the pipe, is shown for comparison. Right: experimentally obtained propagator of water flowing under laminar conditions in a circular pipe; the probability density of displacements is constant between 0 and $Z_{max} = v_{z,max}\Delta$, where Δ is the encoding time of the experiment.

applying the inverse Fourier transformation, where we have to scan q-space by varying the strength of the gradient *pair*. The resulting propagator might not be as intuitive as the function $\varrho(r)$, but it is a valuable tool to check distributions of displacements or velocities by comparing them with particular models. Some classical examples will be given here.

Perhaps the most simple flow problem is that of laminar flow along z through a cylindrical pipe of radius r_0. For this so-called Poiseuille flow, the axial velocity v_z depends on the radial coordinate r as $v_z(r) = v_{max}\left[1 - (\frac{r}{r_0})^2\right]$, which is a parabolic distribution with the maximum flow velocity in the center of the pipe and zero velocities at the wall. The distribution function of velocities is obtained from equating $\int P(r)dr = \int P(v_z)dv_z$ and the result is that $P(v_z)$ is a constant between

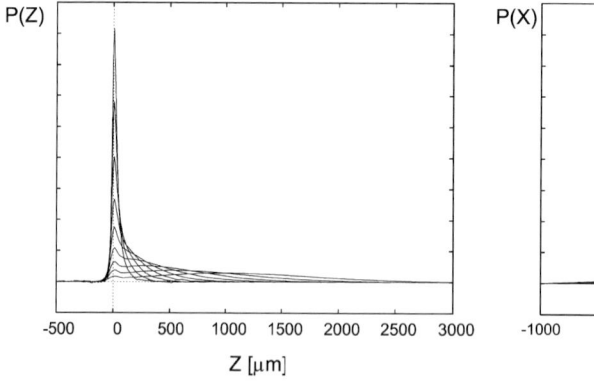

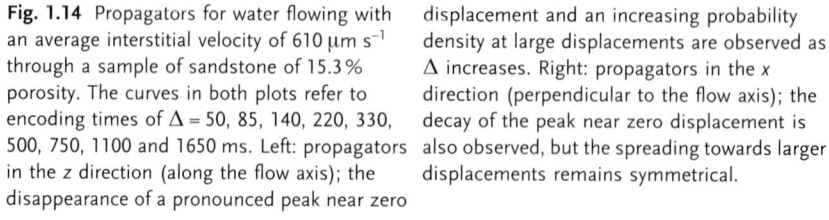

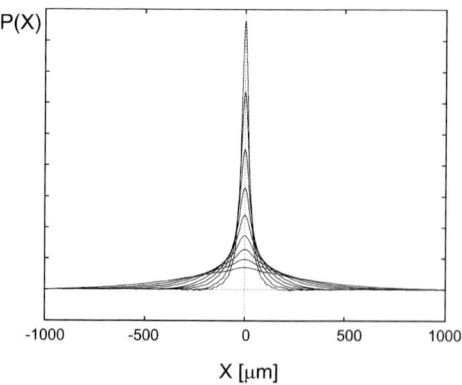

Fig. 1.14 Propagators for water flowing with an average interstitial velocity of 610 μm s^{-1} through a sample of sandstone of 15.3% porosity. The curves in both plots refer to encoding times of Δ = 50, 85, 140, 220, 330, 500, 750, 1100 and 1650 ms. Left: propagators in the z direction (along the flow axis); the disappearance of a pronounced peak near zero displacement and an increasing probability density at large displacements are observed as Δ increases. Right: propagators in the x direction (perpendicular to the flow axis); the decay of the peak near zero displacement is also observed, but the spreading towards larger displacements remains symmetrical.

0 and v_{max}. The propagator has the same shape, as it measures the distribution of downstream displacements, $Z = v_z \Delta$. The term *propagator* is often used synonymously for the distribution of velocities which is justified in this and similar cases, but not in general.

Figure 1.13 compares a diagram of the parabolic velocity profile with an example of an experimentally obtained propagator $P(Z)$ for flow in a pipe. It turns out that for this simple case, the shape of this distribution function is very sensitive to imperfections in the flow, which makes the determination of the propagator a suitable method to assess the quality of laminar flow. For instance, taking a closer look, the rectangular shape of the propagator is broadened by self-diffusion that takes place both parallel and perpendicular to the flow direction, and is thus dependent on the measurement time: for very short Δ, self-diffusion contributes a considerable broadening while for long Δ, radial Taylor dispersion leads to a change of flow velocities, also smoothing the edges of the propagator. It has been found that subtle changes in the propagator shape can be distinguished far better than direct determinations of the velocity at a given point.

A second example for a propagator is shown in Figure 1.14 where water was pumped through a core of sandstone. Flow through structured media is often characterized by a wide range of displacements, which might be decomposed into a stagnant and a mobile fraction. This also becomes observable for flow through rock, where the relatively low porosity of 15.3 % leads to a large fraction of the water remaining stagnant (the peak near zero displacement in Figure 1.14). More discussions of this phenomenon and the interpretation of propagators can be found, for instance, in Chapters 3.3, 4.1, 5.1 and 5.5.

In Figure 1.15, the propagator of free self-diffusion is shown. Actually, self-diffusion was the first case where the propagator terminology was discussed in the NMR context by Kärger (see also Chapter 3.1). The propagator of free diffusion is of

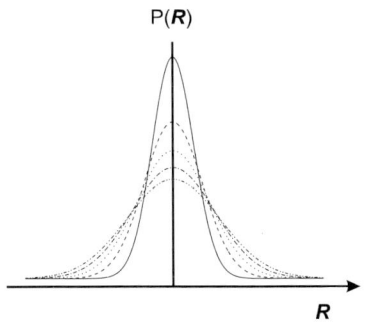

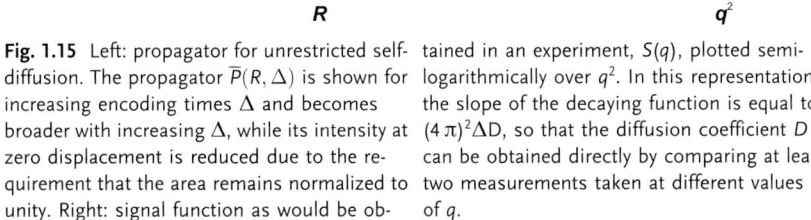

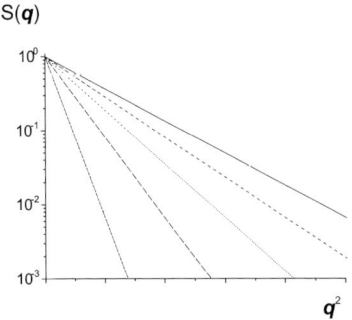

Fig. 1.15 Left: propagator for unrestricted self-diffusion. The propagator $\bar{P}(R, \Delta)$ is shown for increasing encoding times Δ and becomes broader with increasing Δ, while its intensity at zero displacement is reduced due to the requirement that the area remains normalized to unity. Right: signal function as would be obtained in an experiment, $S(q)$, plotted semi-logarithmically over q^2. In this representation, the slope of the decaying function is equal to $(4\pi)^2 \Delta D$, so that the diffusion coefficient D can be obtained directly by comparing at least two measurements taken at different values of q.

Gaussian shape; the mean-squared displacement of spins or molecules follows the same Gaussian evolution as the concentration of a pointlike source. It has the shape

$$\overline{P}(\boldsymbol{R},\Delta) = (4\pi D\Delta)^{-3/2} \exp[-\boldsymbol{R}^2/6D\Delta] \tag{1.14}$$

and is identical in all three directions. The result of an NMR experiment with an encoding wave vector q is as follows:

$$S(\boldsymbol{q}) = \exp[-(2\pi)^2 \boldsymbol{q}^2 \boldsymbol{R}^2/2] \tag{1.15}$$

This is, of course, again a Gaussian function.

The Gaussian shape of a propagator leads to one considerable advantage as far as NMR experiments are concerned. It is not necessary to resolve the full shape of the propagator, it is fully described by three parameters: amplitude, center position and width. While the amplitude is usually irrelevant and the average displacement of free diffusion is zero (which has the consequence that the signal function $S(\boldsymbol{q})$ contains no complex phase factor), only the width of the propagator is of interest. One can obtain this quantity by performing a simplified experiment. If one plots $S(\boldsymbol{q})$ as ln S over q^2, a straight line is obtained (right part of Figure 1.15). The slope of this line is proportional to the self-diffusion coefficient, D, and can – at least in principle – be obtained by measuring at just two different q values. Experimentally, a larger number of gradient steps are executed in order to verify the linear behavior of the function: any deviation from the linear shape is indicative for either restricted diffusion or multi-component diffusion. In medical imaging, for instance, where experimental time is critical and prior knowledge about the expected range of D is at hand, measurements at individual values of q with varying direction are combined to generate the full set of elements of the diffusion tensor, which is nowadays used routinely for fiber tracking in the human brain. This is already one example of the combination of motion encoding and NMR imaging which will be discussed further in the following chapters.

Restricted diffusion, correlated motion of spins, or any deviation from a "free" behavior of the molecules will result in a propagator shape different from a Gaussian one. A wide range of studies have dealt with such problems during the last two decades and NMR has turned out to be the method of choice for quantifying restricted diffusion phenomena such as for liquids in porous materials or dynamics of entangled polymer molecules.

It should be mentioned that the Gaussian propagator is also encountered in flow processes: the long-time limit of flow through a structured medium (see Figure 1.14), when the stagnant fraction has disappeared, is characterized by a displacement distribution of molecules which is centered about the average value, $<\boldsymbol{R}> = <v>\Delta$, and having a Gaussian shape which is given by the mean-squared displacement $<(\boldsymbol{R}-<\boldsymbol{R}>)^2> = 6D_{\text{eff}}\Delta$ [or, $<(Z-<Z>)^2> = 2D_{\text{eff}}\Delta$ if displacements in only one dimension are regarded, which is normally the case in NMR experiments]. D_{eff} is often called the dispersion coefficient. Contrary to the concept usually encountered in chemical engineering, as the result of a PFG-NMR measurement D_{eff} is frequently a time-dependent quantity which increases as Δ becomes larger, the asymptotic limit often not being available in the limited timescale of an NMR experiment. It is thus not unusual to find it to be considerably different in magnitude from the dispersion coefficient determined by classical, tracer methods often used in chemical engineering (see also Chapter 3.3).

The shape of the displacement, or velocity, distribution function alone just represents an average over the motion within the investigated fluid. It cannot be distinguished whether a broadened propagator is the result of a distribution of spins moving at different, but individually constant velocities, or the consequence of all spins undergoing the same dispersion process that makes them assume different velocities as a function of time. This distinction can be made with the help of additional encoding steps, thus making use of the next-higher moment of gradients, m_2. Consider the consecutive application of two gradient pulse pairs of equal magnitude, in other words, q_1 and $q_2 = q_1$: the total phase shift will be twice that being produced by a single gradient pair:

$$\phi_+(R_1, R_2) = 2\pi (q_1 R_1 + q_2 R_2) = 2\pi q_1 (R_1 + R_2)$$

Now the second encoding pair should be applied with opposite sign in another experiment, so that $q_2 = -q_1$, where the total phase shift depends on the difference between displacements during the two intervals only:

$$\phi_-(R_1, R_2) = 2\pi (q_1 R_1 + q_2 R_2) = 2\pi q_1 (R_1 - R_2)$$

(Note that this is almost the same relationship as we used to describe the result of a gradient pulse *pair*, but now it applies to a *pair of* gradient pulse *pairs* and we subtract displacements rather than positions.) By carrying out both experiments, which consist of a simultaneous variation of all four gradients, we obtain a measure of the average velocity ($q_2 = q_1$) or the change of velocities ($q_2 = -q_1$) between the two encoding events, which can also be interpreted in a sense as *acceleration*. It is measured with the second sequence because here, $m_1 = 0$, but $m_2 \neq 0$. Random motion will contribute to both decay functions, whereby coherent flow of spins will be refocused by the second sequence so that only the incoherent part of the flow affects the signal of the second sequence. So, comparing the results of both experiments allows us to single out the individual contributions to the displacement statistics. This procedure was suggested by Callaghan, Seymour and co-workers who discussed not only the distribution functions but also the different dispersion coefficients related to coherent and incoherent motion. It can be extended towards a general case where both gradient pairs are varied independently of each other. This method will be discussed along with other higher-dimensional techniques in Section 1.6.

1.5
Bringing Them Together: Velocity Imaging

With the tools to generate images and to measure different parameters of motion, we are now able to build more complicated pulse sequences that are capable of measuring a range of combinations of both parameters. Frequently, one wants to obtain a pictorial, and also quantitative, representation of a flow field. The same spins must then be encoded with respect to position *and* velocity. Other applications require spatially resolved maps of diffusion or dispersion coefficients. Motion can, on the other hand, also be determined indirectly by making use of the magnitude of magnetization, which is a consequence of polarization, excitation and relaxation effects. For instance, a tracer acting as a relaxation agent can be introduced into the fluid stream and will affect the signal inside the investigated object upon its arrival in the volume of interest, where a complete image of the instantaneous tracer distribution will be acquired. The technique is thus favorable for slow transport processes on time scales of seconds to hours and allows computation of diffusion and transport coefficients from a series of NMR images (see Section 1.8 for a discussion of relaxation contrast, see also Chapter 2.9). One other possibility is frequently used in angiography, the medical field of monitoring blood transport in vessels. If an image is performed in a particular slice of the

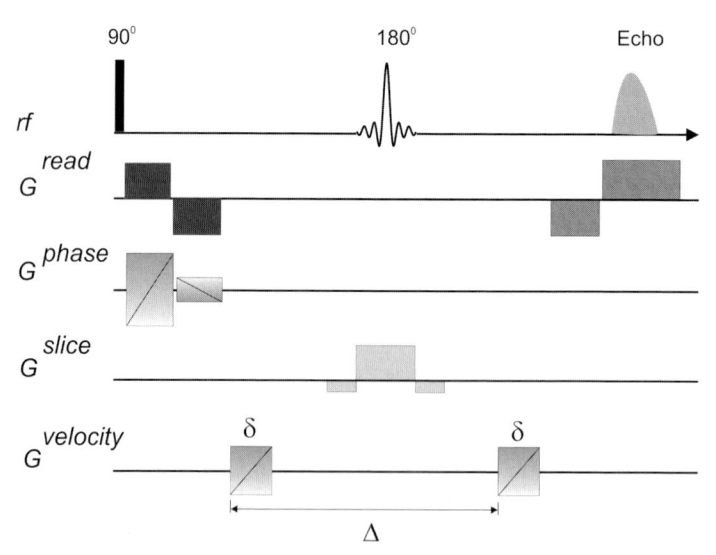

Fig. 1.16 Typical motion-compensated, velocity encoding imaging sequence. The basic scheme from Figure 1.11 is amended by additional compensating gradient pulses for the read, phase and slice directions. The velocity encoding pair of gradient pulses of duration δ and separation Δ is drawn on a separate line; in reality, it is applied along any of the three directions given for read, phase and slice. The diagonal lines indicate that the magnitude of this gradient pulse pair is varied in a number of steps; the experiment, however, can also be executed with a single magnitude value by comparison with an equivalent sequence without the velocity encoding gradients.

excitation volume of the resonator by using a slice selective rf pulse, blood leaving this slice during the pulse sequence before acquisition will lead to a loss of signal intensity, as new blood that enters into this slice has not been encoded by the pulse sequence. By applying various types of fast imaging sequences, the opposite contrast can be produced, i.e., fresh blood appears bright in the image because it carries full equilibrium polarization while the polarization of the spins in the blood fraction that has remained inside the sensitive volume has not recovered to its full amplitude due to fast repetition of excitation pulses. These strategies allow the estimation of slice-averaged or even pixel-averaged flow velocities but do not measure velocities directly. We refer the reader to standard textbooks of medical MRI for a more detailed description of these techniques. Although rarely used at present, some of the MRI techniques turn out to be worthwhile tools for NMRI in chemical engineering. The difference in medical applications compared with many chemical engineering applications is that it is mostly water and soft tissues that are the subject of the imaging so that one can rely on longer lifetimes of coherence and more homogeneous magnetization distribution inside the sample.

Before we elaborate on velocity imaging in the "proper" sense, we have to face the side-effects that flow has on our NMR image, i.e., the possibility of blurring or losing the signal of certain volumes because the spins move out of the detection coil before they can contribute to the signal. In fact, *blurring* – as we would observe in optical photography of a moving object – is an inadequate description of the image artifacts arising due to motion, which can be much more complicated than just reducing the spatial resolution of an image. A plethora of ghosts, double and multiple images, unwanted signal attenuation, etc. can arise, but once they become visible in the NMR image, they can be identified and taken care of. The best way to avoid artifacts is proper design of the pulse sequence that suppresses motion effects as effectively as possible. Figure 1.16 shows a sequence that exploits the moment formalism to generate a correct velocity encoded image. It is one of the simplest of such sequences and is known by the term "flow-compensated, velocity encoding, imaging sequence". The meaning of this seemingly paradoxical name is to introduce velocity encoding only in the direction where it is desired (for example, along the main axis of motion), and to suppress unwanted artifacts due to motion along other axes that might affect the image quality.

Following the discussion on moments, this leads us to the following requirements:

- $m_0 \neq 0$, $m_1 = 0$ in the phase direction – the position encoding needs to be preserved, but it will be insensitive to motion – realized by a pair of rectangular gradients with equal duration and intensity ratio 3:1.
- $m_0 = 0$, $m_1 = 0$ in the slice direction – the gradient will excite a slice but leave no residual phase shift depending on either position or velocity – realized by a pair of negative lobes on both sides of the actual slice gradient with the same total duration but 1/3 of the slice gradient strength (see Figure 1.12 c).
- $m_0 = 0$, $m_1 = 0$ in the read direction – these conditions can only be met for the center of the acquisition window! While $m_0 \neq 0$ before and after this echo so that position encoding is actually possible at all, it is unavoidable to have $m_1 \neq 0$ during

the same period because it is the evolution of the phase under the gradient that is determined. By rendering $m_1 = 0$ at the time of the echo, the influence of velocity on the signal is minimized, but in general the read direction will always be the one which is most severely affected by motion. Note that higher orders of motion can also be compensated if the gradient pulse trains fulfill conditions such as $m_2 = 0$ and so on, at the cost of more complicated and longer encoding schemes. For some typical combination of gradient pulses and the resulting moments, the review article by Pope and Yao provides a very useful overview.

In order to actually generate a velocity image, in addition to the compensated phase, slice and read gradients, a gradient pair is introduced that has the purpose of encoding velocity. It can be applied in any of the three directions, and at any time within the sequence. However, practical experience – or a thorough computation of the combined effect of all applied gradients as well as the hard-to-control background gradients and susceptibility effects – suggests short and symmetrical sequences, the flow encoding direction to coincide with the phase-encoding direction if image *and* flow information is to be encoded along the same direction and read gradients to be used along the direction of slowest motion. The best solution for a particular application cannot be given in general and will depend on many factors such as field strength, available hardware, sample shape and material, magnitude of velocities and much more. More advanced pulse sequences which allow the acquisition of images or the measurement of velocities in short times are discussed in the following chapters.

Figure 1.16 contains all that is necessary to generate a three-dimensional image with a three-dimensional velocity field overlaid to it. But how do we proceed to obtain the desired result? First of all, we need to remember that velocity information is brought into the signal by the bipolar encoding gradient via a phase shift according to Eq. (1.11). Often we are satisfied with the measure of an average velocity in a pixel, so we might want to use this phase shift directly and translate it into velocities with the proportionality $\phi(\mathbf{v}) = 2\pi \mathbf{q} \cdot \mathbf{v} \cdot \Delta$. (Note that instead of displacements, we are expressing our result in terms of velocities v, which is fully equivalent provided that during the encoding time Δ, velocities within the pixel do not change significantly.) However, as the image reconstruction process itself also relies on phase information, the bipolar velocity encoding gradient just adds an extra contribution that cannot easily be separated from the phase shifts resulting from the imaging gradients. The solution is to acquire two images, one with and one without ($\mathbf{q} = 0$) the velocity encoding gradient. Both images are processed, and the phases in each pixel are computed – because of the fact that the signal is complex, the phase angle is given by $\phi = \tan^{-1}[Im(S)/Re(S)]$, where Im and Re denote the imaginary and the real part of the signals, respectively. One can then subtract the phases of each pixel, and the difference renders the velocity in that pixel directly. In a real experiment, imperfect field and gradient distributions, as well as local susceptibility variations can, and usually will, give rise to uncontrollable shifts of the reference phase, which are also eliminated by this subtraction procedure. However, in practice one might deduce the difference phase from two experiments with finite but different q values, which takes just the same effort but can occasionally be a more robust approach. The result of such a procedure contains the information of the spin density of the object plus the average velocity per pixel in one direction. We can simply repeat the experiment with the vector q aligned in the other two orthogonal directions and obtain a set of data that can be used to reconstruct a vector plot, where each vector contains all velocity components in each pixel. Because of the reference measurement, we need four individual experiments to achieve this. However, only certain components of the velocity, or only two-dimensional images corresponding to selected slices of the object, are sufficient to describe the flow field of interest. The reader is referred to the many examples throughout this book where the experimental procedure to acquire velocity images is explained in more detail (see, e.g., Chapters 4.3, 4.6, 5.1, 5.2, 5.5).

Whenever a velocity *distribution* within a pixel is suspected, the propagator can be obtained by using the same pulse sequence, but varying the magnitude of the velocity encoding wave vector in full analogy to the measurement of the pure, spatially unresolved propagator described in the previous section. This case is actually shown in Figure 1.16 where the diagonal lines in the flow encoding gradient symbolize repeated application with different values. It should be noted that here, just as in all propagator and velocity encoding experiments, the range of gradients has to be considered. In complete analogy to encoding of position, the range of velocities covered, also called field of flow (*FOF*), is given by:

$$FOF = \frac{n-1}{2} \frac{2\pi}{\gamma g_{max} \delta \Delta}$$

As in imaging, aliasing can occur if velocities outside the *FOF* exist. For measurements of average velocities, the same condition holds if the total phase shift due to velocity exceeds 2π. For velocity fields of simple structure, this problem can be accounted for by so-called unwrapping algorithms (see e.g. Chapters 2.9 and 4.2).

The tagging technique is a different approach to determining velocities in a moving object. It is often used for visualization, but can also be extended to provide quantitative information about flow fields (see Chapter 4.5). The principle consists of "tagging", or labeling, a certain region of the sample by manipulating the magnetization inside this volume and acquiring the image after some delay during which the labeled volume is deformed. For flow systems with time-invariant flow patterns, movies can be reconstructed from a series of such images with different delay periods. A simple realization is the tagging of a single slice, which is achieved by applying a soft pulse in the presence of a pulsed field gradient, exactly as has been described in Section 1.3 for slice selection. This slice can be "dark", i.e., the spins in the slice are saturated by the rf pulse and the subsequent imaging sequence generates an image of the total volume minus the saturated spins, which show up as a dark stripe. Figure 1.17 (top) contains one possibility of what such a pulse sequence can look like, and snapshots for a vertically falling film of silicone oil are presented as an example for this method (Figure 1.18). By applying the slice selective pulse at a different position within the sequence, or simply by subtracting

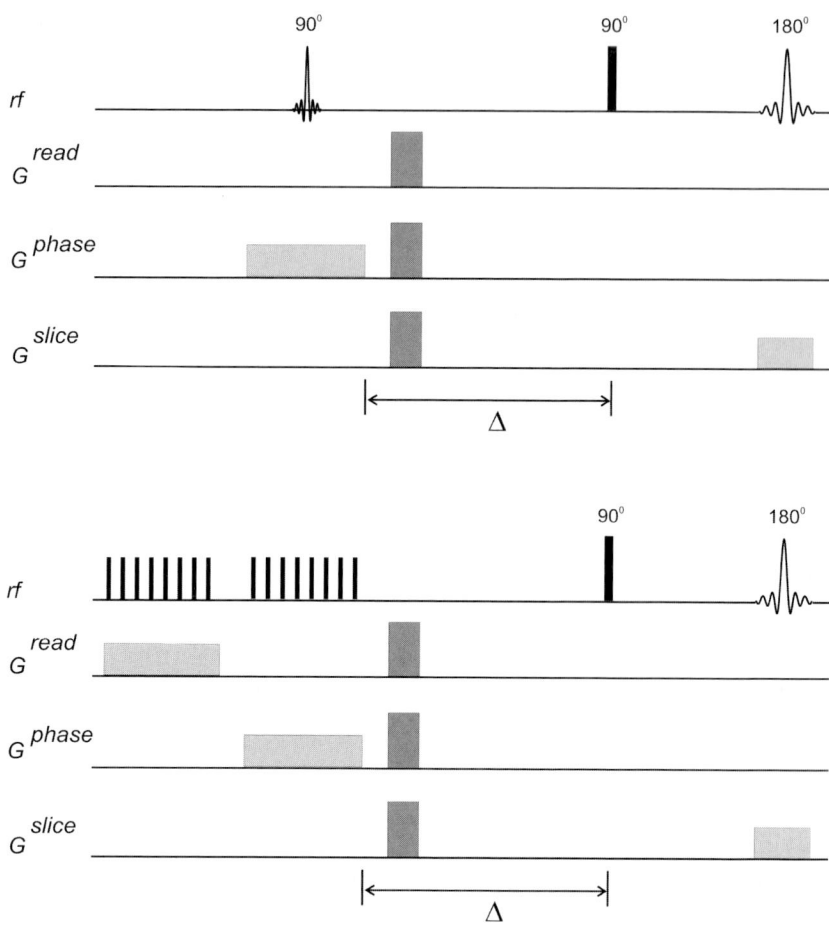

Fig. 1.17 Top: single-slice tagging sequence. The 90° soft pulse in the presence of a magnetic field gradient excites the magnetization inside a slice and transfers it to the transverse plane. The subsequent so-called spoiler gradients destroy any coherence of this magnetization so that it will not contribute to the signal later on in the sequence (the spoiler gradients are drawn in three orthogonal directions, but a single direction is often sufficient). The following imaging sequence acquires a signal of the whole object with the exception of the spins inside the slice that has been excited in the beginning and that has been deformed due to motion during the waiting time Δ; a single black stripe inside an otherwise undisturbed image is observed. Bottom: the DANTE (Delays Alternating with Nutations for Tailored Excitation) sequence excites a two-dimensional grid, which is the result of two trains of short pulses in the presence of two gradient pulses which must be in the same direction as the read and phase gradients of the imaging module. The final image contains a black grid that is deformed during the waiting time Δ by comparison with an equivalent sequence without the velocity encoding gradients.

the image from a reference spin-density image acquired without a tagging element, images can be obtained which highlight the bright moving volume in front of a dark background (see, for example, Chapter 3.3).

1.5 Bringing Them Together: Velocity Imaging

Another common approach is the generation of a grid of dark lines in the volume of interest of the object. It follows the same relationship that determines the shape of a single excited slice discussed in Figure 1.10. If a train of n short rf pulses of duration Δt, separation τ and a total flip angle of, say, 90° is applied in the presence of a gradient of magnitude g_0, the excitation profile is given by the Fourier transformation of this function. What one obtains is again a grid in frequency space which, because of the field gradient, translates into a grid in position space. In the image, n dark stripes at a distance $2\pi/\gamma g_0 \tau$ and with a width of $2\pi/\gamma n g_0 \tau$ will appear.

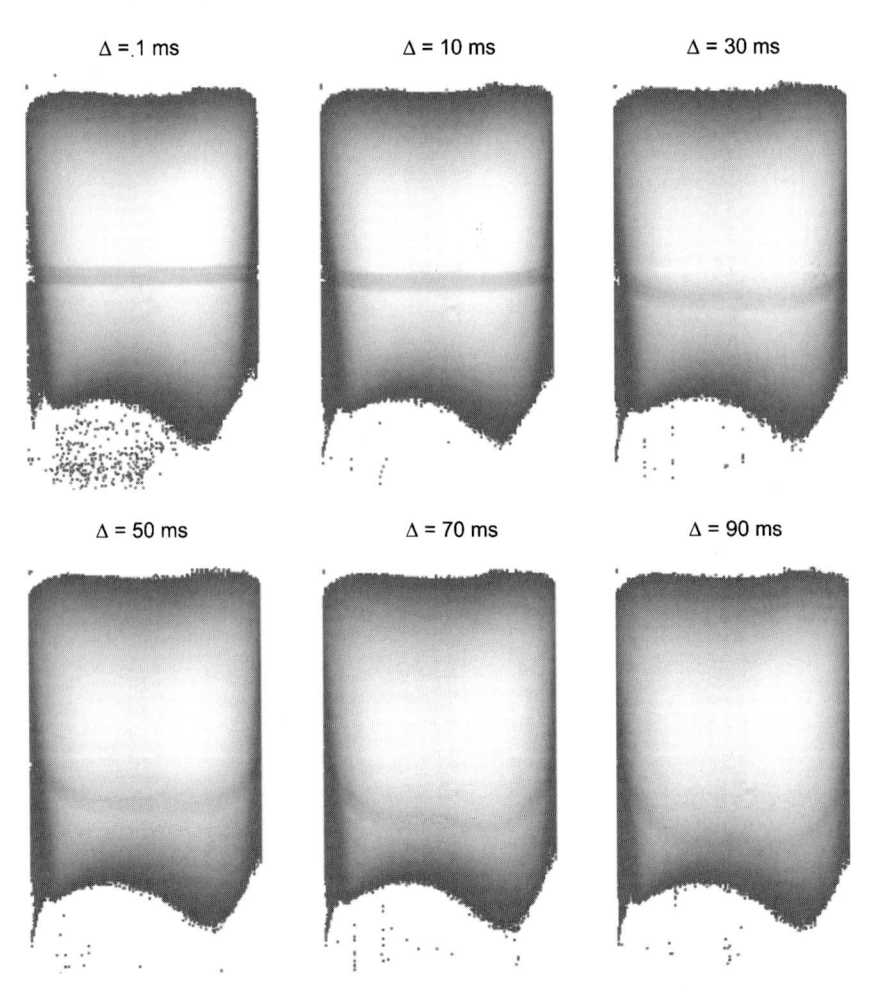

Fig. 1.18 A film of silicone oil of 1 mm thickness is flowing along a vertically oriented planer sheet of PMMA. In a tagging experiment, a horizontal slice of 2 mm thickness is marked and its deformation is recorded as a function of the separation time Δ between the tagging and imaging modules of the pulse sequence. Owing to the smaller film thickness at the edges of the plate, the falling velocity is smaller in the outer region as compared with the center of the film. Courtesy of M. Küppers, PhD thesis, RWTH Aachen, 2005.

The total effect of such a train is that of a 90° pulse on all spins outside of the dark stripes (although the total flip angle is not restricted to being 90°). The same element can be repeated with the gradient in an orthogonal direction, and the signal arising following these two composite pulses can further be used as the preparatory magnetization for an imaging sequence. The scheme of this so-called DANTE (Delays Alternating with Nutations for Tailored Excitation) pulse sequence is shown in Figure 1.17 (bottom). By shaping the envelope of the individual short rf pulses in a sinc-like fashion, as opposed to all pulses having the same intensity, rectangular stripes can be generated, while the sequence shown in Figure 1.17 results in sinc-shaped dark stripes with less defined boundaries. Examples of a DANTE tagging sequence applied to flow is nicely demonstrated in Chapter 4.4. A tagging experiment often makes complicated motion more readily available for the experimenter than a proper q-encoded image. While it does require a similar amount of experimental time, it is limited to displacements exceeding the image resolution. The ultimate time limit – for phase encoding as well as for tagging – is the longitudinal relaxation time of typically a few seconds in liquids; if it is greatly exceeded, the information encoded into the spin system is lost, or – for tagging – the contrast between tagged spins and the environment has disappeared.

1.6
More Advanced Techniques I: Multiple Encoding and Multiple Dimensions

The principle of how to design a hierarchy of pulse sequences to encode and correlate more *complex motional information* is discussed here using an example of a 2D correlation experiment called VEXSY (Velocity EXchange SpectroscopY) (see Section 4.6 for an example study). Based on the capability of NMR to correlate information in a multi-dimensional fashion, building blocks of NMR can be combined, correlated and extended to access higher order motions, such as acceleration. In a 2D VEXSY experiment, two independent pairs of bipolar gradient pulses (separation between the bipolar gradients is Δ) are used to encode for displacements. These displacements (or velocities: displacement divided by Δ) are correlated at two different times separated by t_m, termed the mixing time. This is in the form of a classical NMR exchange experiment, which is used to find correlations between features in the frequency spectrum by the identification of cross-signals in a two-dimensional density representation $S(f_1, f_2)$. Following the same principle that these off-diagonal cross-signals represent spins that are correlated and have changed or transferred their magnetization following various coupling mechanisms during a *mixing period* or pulse, also the off-diagonal cross-signal in a 2D VEXSY map represents spins that are correlated and have changed their initial velocity after a *mixing time* of t_m. A 2D VEXSY experiment can be based on a spin echo where transverse magnetization carries the signal of the moving spins during t_m, or on a stimulated echo where the longer living longitudinal magnetization carries the signal. When velocity change over a longer observation period is to be determined, or fast decaying spins due to short T_2 (inherent property of the spin) or T_2^* (caused

1.6 More Advanced Techniques I: Multiple Encoding and Multiple Dimensions

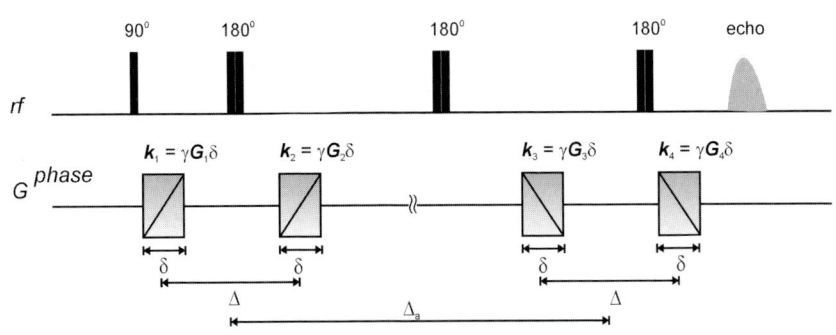

Fig. 1.19 Spin-echo based pulse sequence to encode velocity change. The gradients are stepped pair-wise independently (2D VEXSY) or simultaneously (1D VEXSY). For a VEXSY experiment, k_1 to k_4 are usually applied along the same spatial direction. δ is the duration of each gradient pulse, Δ the separation between each pair of bipolar gradient pulses and t_m the mixing time between the bipolar gradient pairs. The opposite polarity of the bipolar gradient pair is realized by an inversion 180° pulse.

by inhomogeneous magnetic fields) are carrying the signal, a longer t_m compared with the coherence lifetime needs to be employed. Then, a VEXSY sequence based on a stimulated echo is the method of choice.

Let us go into more detail about the VEXSY pulse sequence illustrated in Figure 1.19. Here, the pulsed gradient pairs based on a spin echo are incremented in parallel so as to phase-encode the spins for molecular translational motion. As discussed, a pulsed gradient pair based on a stimulated echo may be employed for longer t_m. Both pairs of field gradient pulses, the amplitudes of which are independently stepped, define a q vector. They are usually applied in the same spatial direction so that the set of spins traveling at constant velocity will have identical R_1 and R_2 displacements, thus contributing to the signal on the diagonal in the (R_1, R_2) space. By contrast, a migration of spins from one region of the displacement spectrum to another over the time t_m will lead to off-diagonal contributions. The off-diagonal signal represents all spins that have changed their initial velocity, thus revealing complex motion, including accelerated motion. Application of VEXSY, mainly to fluid flow and particle motion problems, has increasingly been reported in the literature in recent years and the interpretation of such correlations has been discussed by means of coordinate transformations and suitable 1D-experiments.

Following the concept of a VEXSY experiment that correlates displacements (usually) (see Section 4.6) the same direction but at different times and which allows the computation of velocity autocorrelation functions, the correlation between other quantities can be determined and visualized in terms of two- or higher-dimensional PFG experiments. For instance, applying two individual gradient pulses rather than two gradient pairs allows one to correlate positions at different times rather than displacements as given by the VEXSY scheme. This provides access to the conditional probability between positions at two times as was described in Eq. (1.13) but cannot be extracted from a one-dimensional experiment. In a similar fashion, displacements at the same time but along different *directions* have been measured. What is obtained is a two-dimensional propagator, $\overline{P}(X, Z, \Delta)$, which can, for instance, be used to investigate how displacements along a flow axis (Z) are related to those perpendicular to the flow axis (X). Some examples are discussed in more detail in the review by Stapf et al.

1.7
More Advanced Techniques II: Fast Imaging Techniques

In order to observe transient phenomena with NMR imaging, the use of rapid flow imaging techniques is necessary. Under these circumstances, it is essential to have an observation time that is short relative to the time scale of the physical property to be measured. Transient dynamics include, for example, turbulent flow at high Reynolds number, flow and dispersion through porous media and extensional deformation. There exist an overwhelming number of different pulse sequences and acronyms for fast imaging sequences, especially in the medical MRI literature. Here, we will present prominent and representative sequences out of a wide variety of fast imaging methods. However, once the principle is grasped, it is straightforward to understand other sequences or even create new sequences that are adjusted to your needs.

One means to overcome the dilemma of long experimental time is to employ small flip angles for the imaging sequence so that – because there is no longer any need to wait a multiple of the relaxation time T_1 for recovery of magnetization – much shorter repetition times can be used between the image encoding cycles. This method is termed Fast Low-Angle SHot imaging (FLASH). The fast repetitive excitation of the residual z-magnetization causes signal depletion but this is partly compensated to a certain degree by partial magnetization recovery due to spin–lattice relaxation. A typical flip angle (θ in Figure 1.20) is about 5°. In reducing the flip angle, the total image acquisition time of FLASH is reduced at the expense of

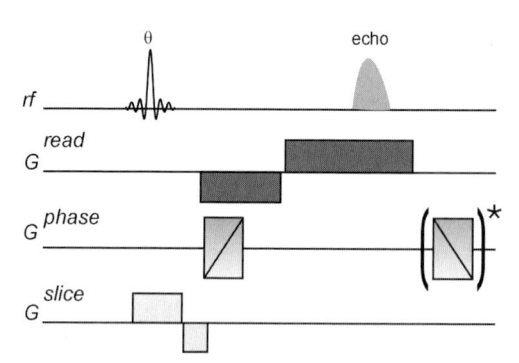

Fig. 1.20 Gradient-echo based pulse sequences based on low flip angles. When low flip angles and short image repetition times are employed at the expense of transverse magnetization during the course of the complete image acquisition, this represents a FLASH sequence (without *). The combination of flip angle and repetition time can be adjusted in such a way in a FAST sequence so that the transverse magnetization does not vanish during the completion of the sequence. In order to employ phase-encoding for 2D or 3D imaging, a "re-winding" phase-encoding gradient (*) has to be included at the end of each acquisition before the next excitation cycle.

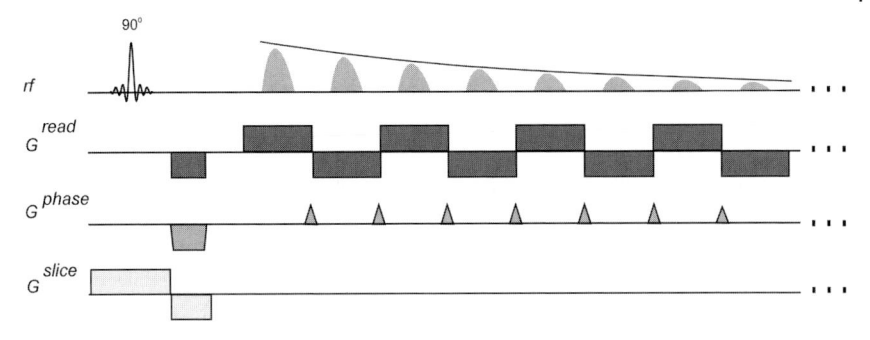

Fig. 1.21 Echo Planar Imaging (EPI) pulse sequence. Gradient-echo based multiple echoes are used for fast single-shot 2D imaging. Slice selection along G_s and frequency encoding along G_r are utilized. Phase encoding is realized using short "blipped" gradient pulses along G_p.

the image signal magnitude. Thus, it is possible to trade off imaging speed with signal amplitude ratio in a controllable manner. Another way of employing low flip angles for fast image acquisition is, rather than suppressing the transverse coherence as in FLASH, to preserve it under steady-state free precession conditions. Here, the initial magnetization for each phase encoding step along the indirect dimension remains the same throughout the course of the experiment. This, in combination with rewinding the phase gradient in each cycle, is utilized in a sequence known as the Fourier Acquired Steady-state Technique (FAST).

Another fast imaging technique tailored for examining solid samples with extremely short T_2^* relaxation times is to employ pure phase-encoding, acquire a single point instead of a full FID and to leave the phase encoding gradients turned on during the entire sequence in order to save the gradient switching and stabilization time. The modification of the weight function to access the k-vector is realized via a step-like change of the gradient function. The acquisition of only one point per phase encoding cycle, i.e., rf pulse and gradient step (SPI, SPRITE, see Chapter 3.4 for more details and specific illustrations of pulse sequences) with minimized dead and waiting time between the rf pulse and acquisition makes this sequence fast and robust.

Finally, instead of exciting the rf source prior to each phase encoding step, the rf source for the entire sequence can be obtained in a single step or a few steps utilizing gradient-echo or a spin-echo based multi echo trains. The latter is created by repeatedly refocusing the coherence created by a 90° pulse using a train of 180° pulses with a tailored phase relationship [known as a Carr–Purcell–Meiboom–Gill (CPMG) sequence]. Here, instead of exciting a fresh signal for each phase encoding step, multiple echoes created by one or a few echo trains are utilized so that one echo provides one k-line along the phase encoding step. One prominent example of gradient-echo based sequences is the Echo Planar Imaging (EPI) method. EPI is the fastest pulse sequence available, offering acquisition times of 100 ms and below to obtain T_2^* weighted two-dimensional images (see Section 1.8), but experiencing certain compromises regarding the image quality. Figure 1.21 shows one variant of a gradient-echo based EPI sequence. Here the entire echo train is acquired following a single rf excitation pulse, and the phase encoding of each line is achieved by adding a constant-amplitude gradient lobe, or "blip", after each frequency-encode gradient reversal. To run EPI sequences successfully, high quality gradients and gradient amplifiers are beneficial. EPI suffers from a number of limitations, among them are the fact that it cannot be applied to samples with short T_2^* relaxation times (e.g. due to variation in susceptibility in heterogeneous samples) and inhomogeneous magnetic field environment (non-perfect shimming). However, its speed makes it useful if, for example, dynamic events of rather homogeneous soft samples are to be examined.

One prominent multiple spin-echo based sequence, where portions of the k-space dataset are acquired within a single echo train, is termed Rapid Acquisition with Relaxation Enhancement (RARE). The so called Turbo Spin Echo (TSE) sequence is a commercial version of RARE. RARE is based on a CPMG echo train and an example sequence is displayed in Figure 1.22. The acquisition of the first echo signal is conventional and includes a phase encoding and read gradient. However, before the second echo can be acquired, the phase encoding has to be "rewound" to undo the de-phasing of the spins. The "rewinding" is realized by a phase encoding step of equal strength but opposite sign applied after the completion of the data acquisition. In addition, phase correction of the data is required. The whole sequence is repeated after a recovery time with different phase encoding steps. Within one echo train, a portion of k-space is covered. So, if l echoes are acquired per scan and n scans are acquired with varying phase encoding gradients, a total of $l \times n$ number of k-lines will be collected. In other words, each echo is responsible for a different portion of the k-space. One drawback is that different

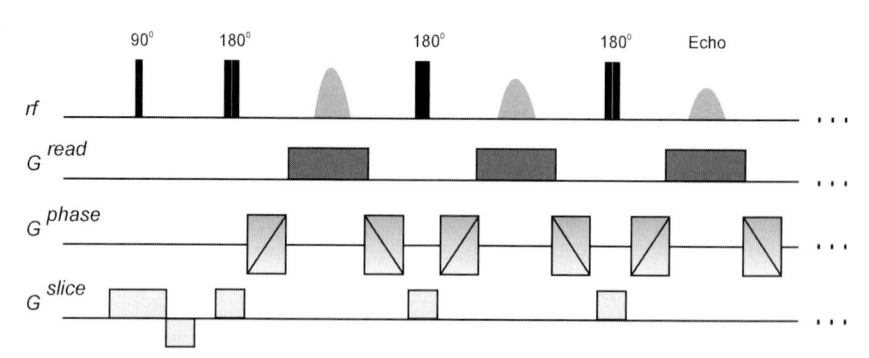

Fig. 1.22 RARE sequence. Here the formation of the first spin echo is conventional. The CPMG form of spin echo is used to avoid the accumulation of flip angle errors over the echo train. However, before the second echo can be acquired, the phase-encoding has to be "rewound" to undo the dephasing of the spins. Therefore, a phase encoding step of equal strength but opposite sign is applied after the completion of the data acquisition. In addition to rewinding, a phase correction of the data is required. If the entire k-space data set is to be acquired with multiple RARE echo trains, the whole sequence is repeated with different phase-encoding steps until the entire k-space is covered.

lines of k-space will have different T_2 weighting because they are acquired at different echo times. Depending on the exact design of the sequence, one can assign any of the echoes to acquire the center or outer part of the k-space which affects the contrast of the image. Examples of the use of RARE-type sequences can be found in Chapters 5.2 and 5.5.

Another spin-echo based fast imaging sequence is known as π-EPI, where the *entire* k-space dataset is acquired within *one* CPMG echo train. Each echo in the presence of a read gradient acquires one k-line. All lines in the 2D k-space are covered by walking along the second dimension by means of blip pulses along the phase encoding direction. The blip pulses have to be adjusted accordingly so that the spacing of the k-lines along the second dimension correctly represents the field of view. Because of the incorporated Carr–Purcell train of π pulses which generate

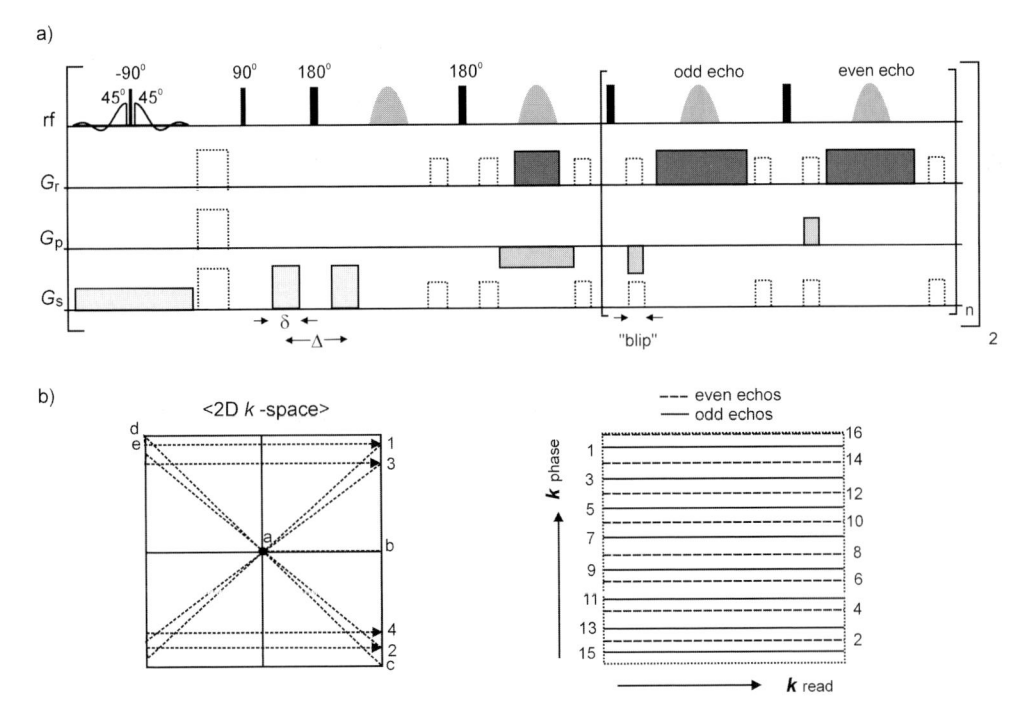

Fig. 1.23 (a) π-EPI pulse sequence with velocity encoding and pre-slice selection. The dotted lines represent crusher gradients. During acquisitions crusher gradients are used in the read and the slice directions to avoid unwanted phase encoding effects. (b) Schematic diagram of the trajectory followed by the sequence. "a" represents the time point of the sequence after slice selection, crusher gradients, velocity filter and 2π pulses prior to the beginning of the first read gradient. The k-line from "a" to "b" is acquired in the presence of the first read gradient, the negative phase gradient leads us to "c", the third π pulse to "d", the first negative phase blip to "e" and the second read gradient pulse during acquisition reads another k-line from "e" to "1". The next π pulse leads us to the lower part of the k-matrix. On the left, the trajectory is shown while on the right, the order (according to the numbers) of how the interleaving of odd and even echo data are acquired is depicted. (Taken from S. Han, P. T. Callaghan, **2001**, *J. Magn. Reson.* 148, 349–354.)

an echo train, each of which is the source of one k-line, the k-space trajectory jumps between the upper and the lower half of the 2D k-space, which necessitates a subsequent data sorting prior to Fourier transformation. One version of a π-EPI sequence including slice selection and a velocity encoding filter and the corresponding k-space raster is shown in Figure 1.23. The data are initially acquired at the borders of k-space, the low k-data being acquired at the middle of the echo-train. An inherent difficulty of π-EPI is the differing amplitude and phase modulation and the relative k-space shift of even- and odd-numbered echoes. One reason for this is the effect of imperfect rf pulses. One way to minimize cumulative effects due to rf pulse imperfections along the echo train is to replace the refocusing π pulse by a composite sequence of three rf pulses of $(\pi/2)_x - (\pi)_y - (\pi/2)_x$, which gives a better spin-state inversion if non-perfect pulses are employed. Another is the discrepancy between the precise area under the positive and negative k-space blips, resulting in a nonlinear k-space rastering after sorting the k-space data. These effects result in a ghost image which appears displaced along the phase encoding direction, and shifted by one-half of the field of view. The problem can be minimized by using different k-space trajectories, by manual adjustment of gradient values to reduce this artifact, or by employing appropriate post-processing techniques, for example phase and amplitude correction of either even or odd numbered echoes.

These fast imaging techniques, such as gradient-echo based EPI or spin-echo based RARE, or π-EPI sequences, can be applied to observe transient dynamic processes by taking a series of "snap shots". Each snap shot has to be fast enough that the motion appears "frozen" during one such image acquisition. A fundamentally different approach to obtaining motional information is to utilize the above discussed bipolar gradient pair as a velocity filter prior to a fast imaging sequence, e.g., a π-EPI sequence, as depicted in Fig. 1.23. Such a CPMG echo train based sequence is preferred if materials with short T_2^* are to be examined because static field and susceptibility variations due to the sample's inhomogeneity can be refocused. The following is an example of how velocity encoded images can be obtained in much less than one second using a PGSE filtered π-EPI sequence.

One experiment consists of two successive π-EPI imaging sequences, one with and the other without a velocity filter preceding the imaging sequence [see Figure 1.23(a)]. The phase contained in the ratio of the two signal matrix ideally provides a phase shift in each pixel due to translational motion alone as discussed in section 1.5. The slice selection can be realized by a self-refocusing selective storage sequence. A combination of a soft $(\pi/2)_x$ sinc-pulse and a hard $(\pi/2)_{-x}$ pulse applied prior to the actual q-encoding π-EPI sequence stores the signal from the desired slice as longitudinal magnetization, while leaving undesired magnetization in the transverse plane, where it is subsequently de-phased by crusher gradients applied in all three orthogonal directions. The advantage of this pre-slice selective technique is that it permits an arbitrarily short echo time in the subsequent EPI sequence. Using this sequence, a 2D velocity map with 64 × 64 pixels can be obtained in a few hundred milliseconds or even less, including two image acquisitions of about 50–100 ms duration and a waiting time in between to allow the spins to return to near equilibrium magnetization.

1.8
Introducing Color into the Image: Contrast Parameters

The image brightness is a direct measure of NMR signal intensity obtained under the particular pulse sequence used in the image acquisition. This brightness does not only reflect spin density, but is weighted by the chemical constitution, the hardness or softness of the material, the diffusion coefficient, the translational motion or motion on a much faster time scale, depending on the design of a sequence. The contrast of an image reflects the heterogeneity of the sample with respect to the chemical and physical properties mentioned. This is because these properties reflect on NMR parameters such as NMR signal amplitude, frequency, T_1 and T_2 relaxation times and line shape. So, the NMR sequence can be designed accordingly to obtain the desired contrast. The wide range of contrasts that can be imposed on a spin-density image is a unique power of NMR imaging.

A very different view of the understanding of the contrast is to regard any extra dimension in addition to the mere image in a multi-dimensional imaging experiment as "contrast". For example, instead of a velocity filtered imaging sequence, an extra dimension representing the velocity distribution function utilizing pulsed-field gradient methods can be acquired. Here, a series of velocity contrast images can be created from such data. We should also clarify the term weighted image and parameter image: a weighted image still contains the spin-density in the signal amplitude and represents a classical contrast image, whereas a parameter image only contains the pure parameter (e.g., velocity, diffusion coefficient, T_1, T_2) values in each pixel of the image. For example, Chapter 3.3 presents T_1 and diffusion coefficient parameter images.

Motional contrast is particularly prominent in imaging methods for chemical engineering applications. All methods to access diffusion, velocity or other translational motion that have been discussed in detail earlier can be employed as "filters" prior to an imaging sequence to obtain, for example, velocity or diffusion weighted images. It is important to distinguish between coherent and incoherent translational motion that can be accessed directly, and molecular rotational and tumbling motion on a much faster time scale, which reflect themselves on T_1 or T_2 relaxation times. In the following, we will discuss the characteristics of relaxation contrast in detail.

The relaxation times are actually the most basic contrast mechanism of NMR imaging. Medical NMR imaging makes great use of such relaxation contrast. Up to this point, we have mentioned the word "relaxation" occasionally but have generally either ignored the existence of relaxation or have described it as a limiting factor for the execution of an NMR experiment. However, relaxation times reveal a wealth of information about the system under study, and their computation and interpretation represent a considerable fraction of the wide range of NMR research fields. Without going into much detail (the reader is referred to the textbooks listed at the end of this Introduction, particularly to the book by Kimmich), it is worthwhile summarizing the essential features of relaxation.

First of all, one can introduce relaxation phenomenologically by amending the equation describing the time-dependence of the magnetization vector [Eq. (1.2)] by a decaying term:

$$\frac{dM}{dt} = \gamma M \times B - R(M - M_0) \tag{1.16}$$

Here, R denotes the relaxation matrix that has the diagonal terms $\mathbf{R}_{xx} = \mathbf{R}_{yy} = 1/T_2$, $\mathbf{R}_{zz} = 1/T_1$. Putting Eq. (1.16) into words, the transverse magnetization M_+ will decay from its initial value to zero with a time-constant T_2, while – at the same time – the longitudinal component M_z returns to its equilibrium value, M_0, with a time-constant T_1. The definition of these time-constants is based on the assumption that the relaxation process follows an exponential behavior. This is often the case, but not always; if a different time-dependence is found, the relaxation time loses its meaning in the stricter sense. It is then formally replaced by a function that can be split into a sum of many exponentially decaying functions, or a different measure is used; for practical reasons, the time during which the magnetization decays to e^{-1} of its initial value is often simply called the relaxation time T_1 or T_2.

What is the origin of the relaxation process? In principle, it is determined by the interaction between spins. This is usually a dipolar interaction in the case of protons, but other types of interactions can play a role, too. Once the spin system is brought out of equilibrium by an rf pulse, it begins to return to the initial state, i.e., the populations of the two energy levels shown in Figure 1.1, which have become jumbled, have to be re-established. This is only possible if energy is exchanged between neighboring spins, and an energy exchange generally requires a change of orientations, and therefore motion. A completely rigid, immobile sample – which is forbidden by Heisenberg's uncertainty relationship and the laws of thermodynamics, as it would be equivalent to zero temperature – will have an infinite relaxation time! Transverse relaxation happens simultaneously, but is somewhat different as it is described by a loss of coherence of magnetization which does not necessarily require a change in the total energy. Because of the additional mechanisms leading to this coherence loss, T_2 is often found to be shorter than T_1. The T_2^* relaxation time that is often described in the literature is *not* an intrinsic property of the material itself and is called the effective transverse relaxation time. Inhomogeneity in static magnetic fields and the sample's magnetization due to susceptibility variations are the cause of T_2^* and the resulting decoherence is reversible, while the true T_2 describes an irreversible decay process. T_2^* is always shorter than T_2 and often becomes important in imaging experiments where a "perfect" magnetic field cannot be achieved. It is then the time constant that describes the free induction decay, but the lost coherence can be recovered at the center of a spin echo, thus along the phase encoding direction in a spin echo sequence, as is demonstrated in Figure 1.2.

What needs to be known to discuss the feasibility of a particular imaging experiment is a good estimate for the relaxation times of the system under study, which then need to be compared with the times required for executing the sequence, frequently determined by hardware limitations. In a simple liquid, one usually has $T_1 = T_2$ and both are in the range of several seconds. This is long enough for any pulse sequence, but might have the disadvantage that repetition times, and therefore total measurement times, become prohibitively long. The addition of soluble salts that form a hydrate shell can help, because the additional relaxation mechanism introduced by the ions can reduce the relaxation times by orders of magnitude.

However, for liquids with a high concentration of ions, or very viscous liquids, the opposite problem can arise if the relaxation times become too short to complete a full pulse sequence and to acquire the remaining signal. The extreme case is represented by solids, which possess a very long longitudinal relaxation time T_1 (anything from seconds to hours depending on the rigidity of the molecular structure) while the transverse relaxation time T_2 becomes very short, often only being in the order of 10 μs. To a lesser degree, this also happens if liquids are in contact with large specific interfaces that can act as additional relaxation sources. For liquids in porous media, this is exploited in obtaining information about the pore structure (see Chapters 3.3, 3.5, 3.6 and 3.7). If such systems are to be imaged, dedicated strategies need to be developed – see Chapters 2.3, 3.4 and 5.4 for a discussion of this specific topic.

One further point needs to be mentioned when probing the feasibility of a particular experiment. Apart from its dependence on temperature and concentration (for instance of ions, solutes, impurities, isotopes), relaxation times – in particular the longitudinal relaxation time T_1 – depend on the field strength. This can be understood from the concept that energy exchange is most efficient if the timescale of molecular motion is equal to the Larmor frequency. Often, molecular motion takes place over a wide range of frequencies, so that the func-

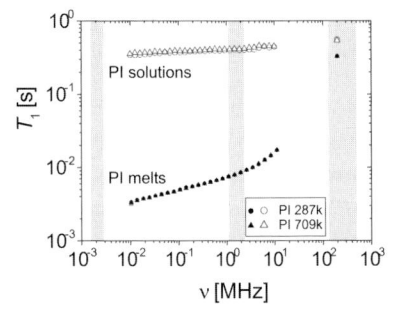

Fig. 1.24 Two examples of frequency-dependent relaxation times – T_1 is plotted as a function of the proton resonance frequency $v = \omega/2\pi$, which was obtained from measurements at different magnetic fields strengths. Left: polyisoprene (PI) melts and solutions of the same samples at 19 wt-% concentration in cyclohexane. Numbers indicate the average molecular weight. The difference between the melt and solution increases towards lower magnetic fields strengths, the frequency dependence is more pronounced for melts. (Courtesy of S. Kariyo, PhD thesis, RWTH Aachen, 2005). Right: relaxation times of water protons in a clean sand compared with a soil sample at different saturation levels. The relaxation time of the water protons is strongly dependent on the composition of the solid fraction of the sample, but also on the magnetic field strength. Vertical bars indicate, from left to right: Earth's magnetic field (about 2 kHz), typical field strength for well-logging apparatus (about 1–3 MHz), typical laboratory fields (several hundred MHz).

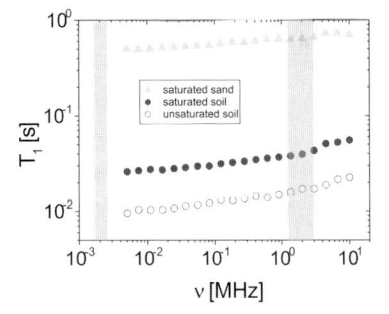

tional relationship can be rather complicated. However, T_1 is almost always decreasing with lower magnetic field strength. While for a simple liquid, this decrease is not measurable with field strengths accessible by commercial spectrometers, in solids, T_1 can decrease by as much as proportional to the square of the field strength. As a consequence, an experiment that is feasible at high fields might be more difficult at low fields, or *vice versa*, the relaxation contrast can become much more pronounced at low fields, which is one of the features that make experiments at low magnetic fields attractive (see also Chapters 2.2 and 2.4). Figure 1.24 demonstrates that the difference in T_1 can indeed be dramatic – for polymer melts, it is reduced from hundreds of milliseconds at high fields to a few milliseconds at low fields, where an imaging experiment would therefore require a different strategy. Fluids in soil, which are often investigated at rather low fields, also possess a frequency-dependent relaxation time, but the influence of the solid material's properties can be considerably larger.

One way of imposing T_1 contrast to obtain T_1-weighted images is by setting the repetition time of the sequence to an intermediate value between those of the different components of the heterogeneous sample. Pixels representing short T_1 material will appear with near full brightness, while those representing long T_1 material will be largely saturated and will be darker. Alternatively, we can obtain a

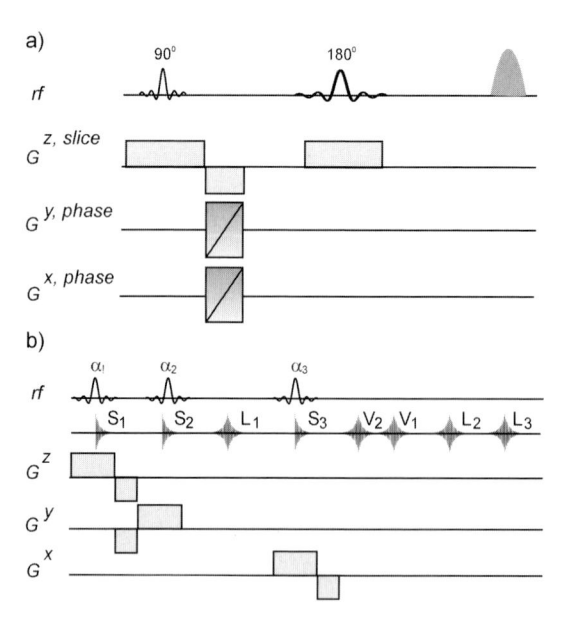

Fig. 1.25 (a) Chemical shift imaging (CSI) sequence with slice selection along z, and phase encoding along x and y spatial dimensions. The NMR signal is acquired in the ab-sence of gradients. The sequence prior to sig-nal acqui-sition can be regarded as a filter deliv-ering chemical shift spectra for each of the image pixels. (b) Volume selective spectroscopy (VOSY) sequence based on a stimulated echo sequence using three selective pulses of arbi-trary flip angle (α_1, α_2, α_3). These soft pulses are used for volume selection along x, y and z. The echoes L1, V1 and V2 can be exploi-ted for localized spectroscopy of the selected volume.

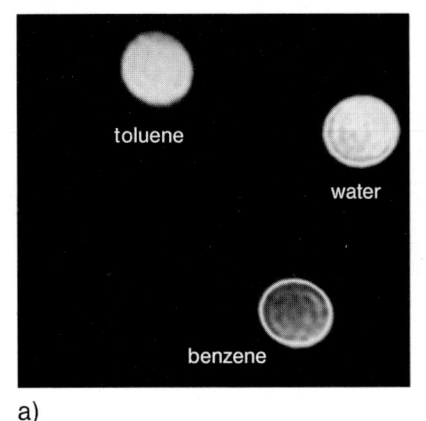

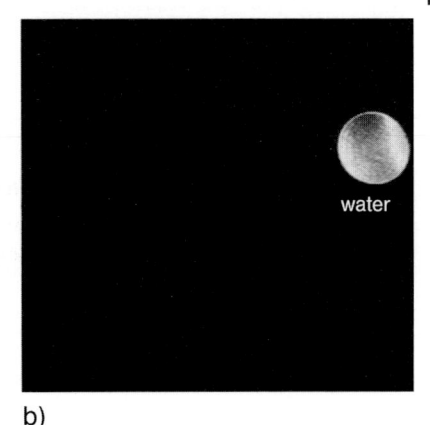

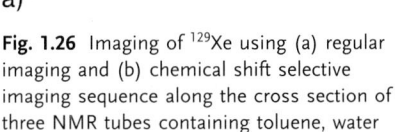

Fig. 1.26 Imaging of ^{129}Xe using (a) regular imaging and (b) chemical shift selective imaging sequence along the cross section of three NMR tubes containing toluene, water and benzene. All tubes contained dissolved hyperpolarized ^{129}Xe in the solvent. Selective excitation chose only the desired image, as is demonstrated for ^{129}Xe dissolved in water.

T_2-weighted image by setting the echo-time of the pulse sequence to a value in the middle of the range we are trying to discriminate (see for example Figure 5.1.1). The repetition time would be set long to prevent saturation in any signals with long T_1. An example of such T_2 contrast imaging is shown in Figures 5.1.4 and 5.1.8. Alternative possibilities of how to exploit the relaxation and viscosity or diffusivity properties in a sample at the same time are presented in Chapters 2.7 and 4.7.

The most powerful and also unique "contrast" of NMR imaging, is its capability to differentiate the chemical constitution of the material via the chemical shift dispersion of the NMR nuclei to be examined. The most prominent pulse sequence to obtain an image where each pixel contains a full chemical shift spectrum is called chemical shift imaging (CSI). Figure 1.25(a) illustrates a CSI imaging sequence in which the three dimensions of spatial encoding are imposed before acquiring the signal in the absence of a magnetic field gradient. Each NMR spectrum obtained corresponds to a single point in k-space and delivers full chemical shift information for each pixel in the image after Fourier transformation. CSI is the most informative, but also the most time-consuming approach. Instead of acquiring the entire chemical shift information as an extra dimension, one can reduce one dimension by utilizing frequency selective rf pulses, which only excite a narrow frequency bandwidth of interest. Here the frequency selective pulse acts as a filter prior to the imaging sequence and imposes a contrast to the image reflecting the chemical composition of the sample. Figure 1.26 illustrates nicely the effect of chemical shift selective imaging on three NMR sample tubes containing toluene, water and benzene placed inside an NMR imaging probe. Hyperpolarized ^{129}Xe was inserted into all three tubes and subsequently the ^{129}Xe NMR imaging signal was obtained. Note that ^{129}Xe has a large chemical shift range so that different solvents such as toluene versus benzene lead to a 10 ppm shift, which makes chemical shift selective imaging very easy. In (a), a regular spin warp imaging sequence was applied using full rf excitation and in (b), only the ^{129}Xe dissolved in water was selectively irradiated in order to obtain an image only from the water-containing part of the sample. See Chapter 5.5 for further investigations using CSI.

Another approach to obtain spatially selective chemical shift information is, instead of obtaining the entire image, to select only the voxel of interest of the sample and record a spectrum. This method called VOlume Selective spectroscopY (VOSY) is a 1D NMR method and is accordingly fast compared with a 3D sequence such as the CSI method displayed in Figure 1.25(a). In Figure 1.25(b), a VOSY sequence based on a stimulated echo sequence is displayed, where three slice selective pulses excite coherences only inside the voxel of interest. The offset frequency of the slice selective pulse defines the location of the voxel. Along the receiver axis (*rx*) all echoes created by a stimulated echo sequence are displayed. The echoes V2, V1, L2 and L3 can be utilized, where such multiple echoes can be employed for signal accumulation.

Discussion about contrast in NMR imaging can be continued extensively. We have merely mentioned the most basic contrast parameters for NMR imaging. There are endless varieties of "filter" sequences that can be set prior to an imaging sequence to specifically weight or select the NMR image according to the physicochemical characteristic of the sample. Some examples such as contrast representing dipolar coupling of various strength, order and alignment, viscosity and temperature can be created. However, here we will end our introduction in the hope that we have provided some basic ideas and tools of NMR imaging. This introduction reflects the way in which we started to understand the principle of NMR imaging when we were students and newcomers to NMR imaging. In a way, we are speaking out loud as if we are talking to a class. So, if a book could be written as interactively as teaching can be performed in a class, we are sure that many addenda, extensions, alterations and corrections would have been made to this introduction because, as always, we are still learning.

Bibliography

Textbooks

P. T. Callaghan **1994**, *Principles of Nuclear Magnetic Resonance Microscopy*, Clarendon Press, Oxford, 490 pp. Standard reference textbook for imaging, some examples but with a focus on theory.

E. Fukushima, S. B. W. Roeder **1986**, *Experimental Pulse NMR – a Nuts and Bolts Approach*, 10th edn, Perseus Publishing, Cambridge, MA, 556 pp. Standard textbook on basics of NMR, focus not on applications.

R. Kimmich **1997**, *NMR. Tomography, Diffusometry, Relaxometry*, Springer, Berlin, 510 pp. Standard textbook, theory and methods – applications mostly as examples.

B. Blümich **2000**, *NMR Imaging of Materials*, Clarendon Press, Oxford, 568 pp. Focus on theory and basics, collection of specific applications in the sense of examples; some flow topics covered.

D. W. McRobbie, E. A. Moore, M. J. Graves, M. R. Prince **2003**, *MRI – From Picture to Proton*, Cambridge University Press, Cambridge, 359 pp. Excellent textbook aimed at the medical user, but equally useful for beginners and as a reference.

M. T. Vlaardingerbroek, J. A. Den Boer **1999**, *Magnetic Resonance Imaging: Theory and Practice*, Springer, 481 pp. Very good textbook, mostly techniques, medical applications.

V. Kuperman **2000**, *Magnetic Resonance Imaging: Physical Principles and Applications*, Academic Press, New York, 182 pp. Textbook – hardware and technique, some flow.

R. S. Macomber **1998**, *A Complete Introduction to Modern NMR Spectroscopy*, Wiley Interscience, New York, 382 pp. Standard textbook mainly aimed at chemistry students.

H. Friebolin **2005**, *Basic One- and Two-Dimensional NMR Spectroscopy*, 4th edn, Wiley-VCH, 400 pp. Standard textbook for the chemistry student, a good introduction and overview.

P. Blümler, B. Blümich, R. E. Botto, E. Fukushima (eds.) **1998**, *Spatially Resolved Magnetic Resonance: Methods, Materials, Medicine, Biology, Rheology, Geology, Ecology, Hardware*, VCH, Weinheim, 774 pp. Collected lectures from an MR imaging conference, various fields, also contains chemical engineering and transport.

E. M. Haacke, R. W. Brown, M. R. Thompson, R. Venkatesan **1999**, *Magnetic Resonance Imaging: Physical Principles and Sequence Design*, Wiley & Sons, Chichester, 914 pp. Extensive textbook with focus on fundamentals, not on applications.

J. Kärger, D. M. Ruthven **1992**, *Diffusion in Zeolites*, Wiley & Sons, Chichester, 500 pp. Contains some NMR experiments, but mostly other techniques.

K. Albert **2002**, *On-line LC-NMR and Related Techniques*, Wiley & Sons, Chichester, 306 pp. Specialized, contains NMR spectroscopy but not imaging.

HalliburtonEnergyServices **2001**, *NMR Logging Principles and Applications*, Gulf Professional Publ., Butterworth-Heinemann, London, 234 pp. Specialized, aimed at petroleum industry, compares with other techniques.

Selected Review Articles

J. M. Pope, S. Yao **1993**, (Quantitative NMR imaging of flow), *Concepts Magn. Reson.* 5, 281–302.

L. F. Gladden, P. Alexander **1996**, (Applications of nuclear magnetic resonance imaging in process engineering), *Meas. Sci. Technol.* 7, 423–435.

P. J. McDonald **1997**, (Stray field magnetic resonance imaging), *Prog. Nucl. Magn. Reson. Spectrosc.* 30, 69–99.

W. S. Price **1997**, (Pulsed-field gradient nuclear magnetic resonance as a tool for studying translational diffusion: Part 1. Basic theory), *Concepts Magn. Reson.* 9, 299–336.

W. S. Price **1998**, (Pulsed-field gradient nuclear magnetic resonance as a tool for studying translational diffusion: Part 2. Experimental aspects), *Concepts Magn. Reson.* 10, 197–237.

E. Fukushima **1999**, (Nuclear magnetic resonance as a tool to study flow), *Annu. Rev. Fluid Mech.* 31, 95–123.

P. T. Callaghan **1999**, (Rheo-NMR: nuclear magnetic resonance and the rheology of complex fluids), *Rep. Prog. Phys.* 62, 599–670.

S. Stapf, S. Han, C. Heine, B. Blümich **2002**, (Spatio-temporal correlations in transport processes determined by multi-PFG experiments), *Concepts Magn. Reson.* 14, 172–211.

L. F. Gladden **2003**, (Recent advances in MRI studies of chemical reactors: ultrafast imaging of multiphase flows), *Top. Catal.* 24, 19–28.

2
Hardware and Methods

2.1
Hardware, Software and Areas of Application of Non-medical MRI
D. Gross, K. Zick, T. Oerther, V. Lehmann, A. Pampel, and J. Goetz

2.1.1
Introduction

NMR microscopy has become a well-established method in many different areas of research. The scope of the disciplines involved is extremely broad and is still expanding, encompassing chemical, petrochemical, biological and medical research, plant physiology, aerospace engineering, process engineering, industrial food processing, materials and polymer sciences.

One reason for the wide applicability of this method is the commercial availability of NMR microscopy accessories, which can be connected to commercial NMR spectrometers. In the early days of NMR microscopy, investigations had to be carried out to establish what types of objects were suited to the new method. What image quality could be achievable with different objects in terms of signal-to-noise ratio (SNR), contrast and resolution was not always predictable. The situation was easier for medical and pharmaceutical applications, because the NMR properties (T_1 and T_2 relaxation, diffusion) are always the same for the same types of tissue (fat, muscle, organs, cartilage, etc.) from humans and other mammals. Developing standard imaging methods for predictable contrasts by applying standard acquisition parameters was straightforward, e.g., for the repetition (TR) and echo times (TE). Even today such parameters and typical methods for the materials under study are not always clear from the outset, because the T_1 and T_2 relaxation times, the diffusion constants and the susceptibility properties of static and mobile composite materials are not usually known in advance.

The most difficult materials to study by NMR microscopy are those with short T_2 or T_2^* relaxation times and/or with low concentrations of the nuclear spins, which normally result in poor NMR signal intensities. One possibility for improving the image quality is to adapt the shape and size of the rf coils to the size of the objects in order to achieve the best possible filling factor and therefore the best sensitivity [1]. In addition, methods with short echo or detection times have been developed, such

NMR Imaging in Chemical Engineering
Edited by Siegfried Stapf and Song-I Han
Copyright © 2006 WILEY-VCH Verlag GmbH & Co. KGaA, Weinheim
ISBN: 3-527-31234-X

as backprojection reconstruction, strayfield imaging, constant time or single-point imaging. The study of fast dynamic processes requires gradient systems for switching the pulses with fast rise and fall times (20 to 100 µs). Such gradient systems are usually actively shielded to prevent eddy currents from the metal constructions in the shim systems and in the magnets, which would cause additional delays. Last but not least, the gradient amplifiers have to deliver extremely precise and stable pulses at a minimum noise and hum level. All of these requirements and properties, found by many research groups worldwide, have been taken into account by the manufacturers so that most objects can now be investigated with commercially available hardware and software, as described in the following chapters. Nevertheless, dedicated hardware and software still play an important role in new research and it is very useful to keep the hardware as modular and as easy to modify as possible. A recent example is the implementation of RheoNMR [2] to the existing NMR microscopy accessory. Other examples of applications and of the outlook for the future will be shown in this chapter.

2.1.2
Hardware

NMR microscopy systems are built as add-ons to NMR spectrometers. The general technology of an NMR spectrometer is not the subject of this chapter, only the parts that are of particular importance in NMR microscopy applications, such as the magnets, the shim systems, the gradient systems, the NMR microscopy probes and the gradient amplifiers, will be described in some detail.

2.1.2.1
Magnets

The first MRI experiments used resistive magnets with coils usually built from copper requiring a permanent power supply. With such magnets, only fairly low fields can be achieved, mainly because the heat generated by the current in the resistive coil limits the possible current strength. Nevertheless, such systems have been built commercially for clinical MRI. For NMR microscopy so-called iron magnets, resistive magnets with an iron yoke amplifying the magnetic field through the ferromagnetic properties of the iron, were used initially. These magnets have similar problems to the resistive type described above with respect to heat, and they usually run with permanent water cooling. In addition, the stability of the magnetic field is a problem in such systems. Usually the field is regulated by means of a hall generator, but for high resolution NMR a deuterium lock is needed in order to achieve sufficient field stability. Today these magnets are mainly used in EPR, because EPR requires a wide range of magnetic field strengths, which can easily be provided by the iron magnet.

Interest in permanent magnets is increasing in non-medical NMR imaging. One reason is obviously the lower costs, permanent magnets do not need a power supply or cryogenic liquids. Only the temperature has to be kept very stable, because the field strength is usually strongly temperature dependent. Another interesting feature of such magnets is that they can be fairly compact and can therefore be used in mobile systems. The well known example of a mobile MRI tool based on permanent magnets is the NMR Mouse [3]. Permanent magnets are used routinely in bench top NMR instruments.

Almost all high resolution (HR) spectroscopy and most MRI experiments use super conducting magnets. In a super conducting magnet the resistive wires are replaced by super conducting wires. These wires consist of thin filaments of a super conducting material (for example, Nb, NbSn) embedded in a copper matrix. With current technology the wires have to remain at a temperature of 4.2 K, in some cases even lower, in order to keep them super conducting. On the other hand, the samples should usually stay at ambient temperature inside the magnet. Therefore, super conducting magnets are usually built in the shape of a torus or a double cylinder.

NMR magnets are usually in a so-called persistent state, i.e., the resistance is almost zero. Such a magnet can stay on field for many years without being recharged, even the field drift is often negligible, and is much more stable than a magnet connected to a noisy power supply. Moreover, the law of conservation of the magnetic flux in a super conducting coil, which is not fully applicable in a type-2 super conductor, helps to keep the magnetic field constant. Only environmental influences, such as vibrations, temperature instabilities in the dewar or external magnetic fields, can disturb the stability of such a magnet. All these disturbing effects can usually be kept orders of magnitude less than the intrinsic instabilities of the other types of magnet.

While in iron magnets the magnetic field is mainly confined in the iron yoke and therefore the stray field is very small, super conducting magnets have a large stray field. This stray field leads to several safety problems in the surrounding environment of the magnet, i.e., bigger areas have to be restricted in access, computer screens are affected, and so forth. This problem was first addressed for clinical magnets using a so-called passive shield, an iron case made of tens of tons of iron. Modern so-called actively shielded magnets (ultra shield) containing an additional coil to reduce the stray field are now used. This technology has become more efficient and even cheaper than the passive shielding. Therefore, in general, it is now applied for all super conducting NMR magnets. Besides the reduction of the stray field, the active shielding also reduces the sensitivity of the magnet to magnetic fields in the environment.

Today all "standard" imaging applications use super conducting magnets, therefore from now on they will just be referred to as magnets. There are two major types, the so-called vertical bore magnets mainly used in spectroscopy (Figure 2.1.1), and the so-called horizontal bore magnets used in medical and biological imaging (Figure 2.1.2).

In general, vertical bore magnets are usually cheaper compared with horizontal systems of the same field strength. Moreover, vertical bore magnets can provide much higher magnetic fields, currently 21 T, 900 MHz with an open bore of 52 mm, compared with horizontal bore magnets, currently 11.7 T, 500 MHz with an open bore of up to 300 mm. From the scientific point of view, the main

Fig. 2.1.1 Vertical wide bore magnet, 16.44 T, 700 MHz, 89-mm inner diameter (Bruker-BioSpin AG, Fällanden, Switzerland).

difference between horizontal and vertical bore magnets is the direction of the magnetic field versus the direction of gravity. In a vertical bore system the direction of the magnetic field is parallel to the direction of gravity, in a horizontal bore it is perpendicular. This can have direct effects on the experiment, for example when looking at transport phenomena driven by the gravity field. Chemical reactions in a reactor may depend on the transport effect, which is driven by gravity.

In general, the optimal choice of the type of magnet depends on the application, for many different reasons. For example, in medicine, even knowing nothing about physiology, it seems clear that a patient should be lying down. Looking at the brain activity of monkeys, it appears favorable to have the monkey sitting in the magnet, hence a vertical magnet is required (Max-Planck-Institute for Biological Cybernetics, Tübingen, Germany). Another example in biology is the growing of plants, such as entire trees, which will only work properly in a vertical system (Landbow University, Department of Molecular Physics, Wageningen, The Netherlands).

Outside of biology and medicine similar arguments hold. A simulation reactor investigated by MRI must reproduce the physical properties of the original reactor, in particular the direction of the Earth's gravity relative to the direction of flow in the reactor can be very important. If the original reactor is a vertical cylinder, the simulation must be similarly orientated, then a vertical bore system is required.

Fig. 2.1.2 A 7-T and 300-MHz horizontal bore magnet with inner bore diameter of 160 mm (Bruker-BioSpin GmbH, Rheinstetten, Germany).

2.1.2.2 Shim Systems

One of the most important features of NMR magnets, besides the stability of the magnetic field, is it's homogeneity. State-of-the-art spectroscopy systems reach a field homogeneity of 10^{-10} in the volume of interest, imaging systems require, at least for localized spectroscopy applications, a homogeneity of 10^{-8}. This homogeneity is not only a feature of the magnet, but it includes the field disturbances introduced by the magnetic susceptibility of the various components inside the magnet bore. In order to fulfill these requirements a couple of different measures have to be taken.

Firstly, the magnet itself must be sufficiently homogeneous, which is basically provided by the coil design, the precision of the manufacturing and the quality of the materials used. After a magnet is built it is tested and optimized. This can be done by means of the so-called cryogenic shims, i.e., additional gradient coils built in the magnet dewar used to correct for field inhomogeneity. These coils are usually charged after the magnet has been brought to field and will stay at the same current for the lifetime of the magnet. Another technique is the so-called iron shim, where pieces of iron are fixed around or inside the magnet bore to correct for inhomogeneity by introducing additional field gradients. Secondly, a device is needed in order to correct for variable inhomogeneity introduced by exchangeable rf coils, gradient systems and mostly by the sample itself. This is usually achieved by a so-called room temperature shim system, a set of resistive gradient coils built into the room temperature bore of the magnet. Being resistive coils, these shim coils are permanently driven by high stability power supplies. As the current requirements of such shim coils are much smaller than that of the main coil (a couple of hundred milliamps, compared with several hundred amps), a much higher stability can be achieved here.

A set of shim coils usually represents the field expansion according to spherical harmonics. Depending on the required accuracy, more or fewer orders of the expansion are used. Clinical imaging systems usually use only the first order shims x, y and z, and the second order in the z direction, known as z^2. High resolution systems use up to six orders in the z direction. Traditionally each shim function was built as a separate coil, nowadays so-called matrix shim systems are used. In a matrix system, a shim gradient is produced by a combination of currents in different coils. This technique simplifies the manufacturing of the shim set, but it requires a higher precision of the currents. Therefore, combined systems are often used, where the most critical shims are still built conventionally.

2.1.2.3 Gradient Systems

In contrast to the shim gradient system used to homogenize the magnetic field, in MRI the term gradient system typically implies a set of gradients used to create an inhomogeneous magnetic field in a well defined manner. In some cases, particularly in horizontal bore systems, both functions (shim and imaging gradients) use the same physical gradient coil set. In NMR microscopy systems, a universal shim system, also used for spectroscopy, is usually used together with a separate, dedicated imaging gradient system. The major difference between a shim and an imaging gradient coil is the fact that the imaging coil must be fast switchable, while the shim coil is typically driven by a constant current.

The majority of all imaging systems use the so-called linear gradients, where only the first order of the expansion of the field is used. The gradients ($G_x = \delta B_z/\delta x$, $G_y = \delta B_z/\delta y$, $G_z = \delta B_z/\delta z$) should be constant over the field of view; this is known as gradient linearity.

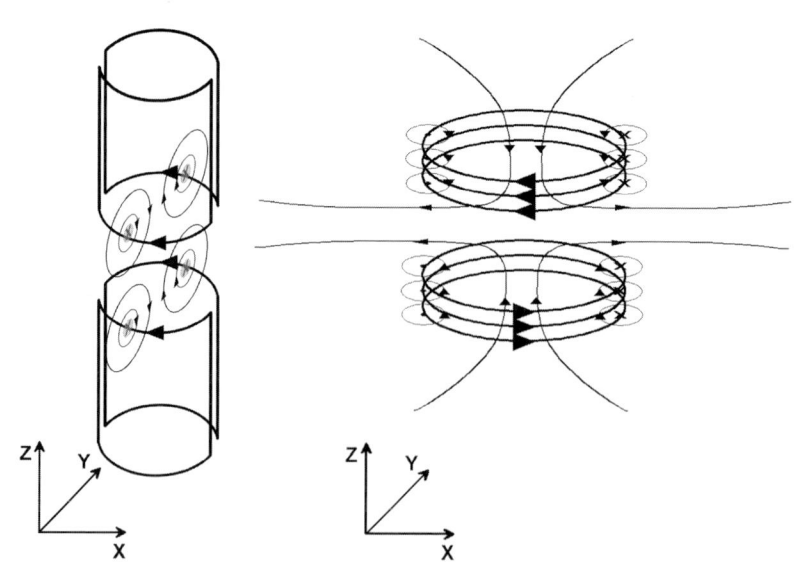

Fig. 2.1.3 A pair of saddle coils creates an x or y gradient (left), a pair of Maxwell coils creates a z gradient (right). The arrows on the coils indicate the current direction, the circles with arrows indicate the magnetic field created by one current element. The magnetic field of interest consists of the superposition of all z components of these field elements.

In super conducting magnets having a cylindrical bore in the z direction, the same basic coil designs are always used. In the z direction a so-called Maxwell Pair, a pair of turns with opposite current, saddle coils in the x and y directions, are used (Figure 2.1.3). The saddle coil design makes use of the fact that in the very high field in the z direction only derivatives of the z component need to be considered, all other components are negligible.

The basic designs for the gradient coils have been optimized over the years, and today so-called streamline designs are used. Current distributions are calculated numerically in order to optimize various parameters: the gradient strength, the gradient linearity, the active volume, the inductivity and the electrical resistance. These parameters often contradict each other, for example a high gradient strength (which is required) leads to a large inductance and a high resistance (not required). Hence, a compromise for the particular application of the gradient system has to be found. In addition, the calculated current distribution has to be in a shape that can be produced without too much effort. This leads to geometrical restrictions, and the current distribution is normally confined to the surface of a cylinder. The calculated current distribution is usually not achieved by winding wires round to give the coil, moreover the construction starts with a solid copper tube, where the structure is realized by cutting the copper out and filling the gaps with epoxy. Such a technology has two major advantages, the "thickness of the wire" can be varied along the coil, and the electrical resistance is lower because the amount of conductive material is bigger.

When the current is switched rapidly in a conventional gradient system, the time variable stray field of the gradient coil induces eddy currents in the metal components of the magnet dewar. In particular, metal at very low temperatures allows for very long decays of these currents on a time scale of milliseconds to seconds, because of the low resistance. These eddy currents create slowly decaying gradients opposite to the inducing gradient, which mainly increases the effective gradient switching times. The eddy current effects are usually compensated for by means of the so-called pre-emphasis, a method where the gradient pulse shapes are modified in order to compensate for eddy current effects. The pre-emphasis has to take account of the several different time constants of the various components, allowing for eddy currents. These time constants can be fairly long, even longer than the repetition time of the experiments, which makes the compensation very difficult.

A major breakthrough in gradient technology was therefore the development of actively shielded gradients. In an actively shielded gradient system the stray field outside is compensated for, in a manner similar to the magnet stray field in actively shielded magnets, by a second outer coil. The pre-emphasis only has to take account of the eddy currents generated inside the gradient coil, for example on the rf coil. The outer coil is electrically in series with the gradient coil and can therefore produce the same quality of compensation for each gradient strength. In the so-called streamline design (Figure 2.1.4), the gradient coil and the compensation coil can usually be calculated as a package [4–6]. Depending on the geometry the shield coil will reduce the effective gradient inside, so another compromise has to be found. In practice the

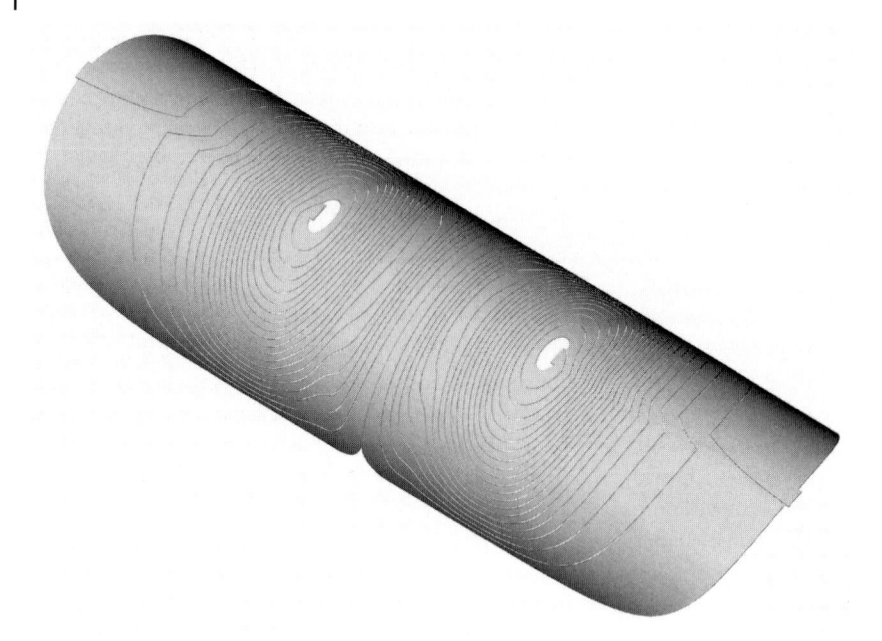

Fig. 2.1.4 Streamlined x or y gradients.

shield coil must have a significantly larger diameter than the gradient coil, the smaller this ratio the less effective the gradient system will be.

Another major achievement in gradient technology was the introduction of water cooling. Owing to the much better heat transport capabilities of water compared with air, the cooling could be improved significantly and therefore the possible duty cycles could be increased. Modern gradient systems require water cooling by design.

2.1.2.4 Rf Coils/Probes

In all NMR fields the sensitivity of the rf coils is a major issue, in NMR microscopy this is usually even more so than in clinical MRI, because the signal-creating volume scales with the third power of the geometrical dimension. In clinical systems the image resolution is in the range of millimeters, while in NMR microscopy resolutions down to less than 10 µm per pixel are sometimes required. Being a non-destructive method, the sample size is usually given and in order to achieve the optimal sensitivity the size of the rf coil must be adapted to the sample. Therefore a large range of different coil dimension for various applications, from micro coils (below 1-mm diameter) up to whole body coils for horses are in use. Depending on the dimensions, different coil designs are used.

Solenoid coils are the most efficient, the magnetic field strength in the coil for a given current and, as a result, the generated voltage of a given induction is about double compared with other designs. The disadvantage is that a solenoid must be oriented perpendicular to the main magnetic field, which is perpendicular to the bore of the magnet. This makes the access difficult, or even impossible, for bigger objects. Therefore, solenoid coils are typically used for small objects, up to 5-mm diameter. Another disadvantage of solenoid coils is the fact that the magnetic susceptibility of the wire cannot be neglected. Thus using susceptibility compensated wire for the coils, as in high resolution probes, improves the situation drastically. For medium sizes, 5 to 10 mm, the traditional saddle coils used in high resolution NMR have proved to be very good. Using modern susceptibility compensated wire they provide both high magnetic field homogeneity and good rf homogeneity. For larger diameters, saddle coils show characteristic artifacts in the images, therefore other designs have to be used. For coil diameters larger than 10 mm, initially the so-called Alderman Grant resonators, were used. They show a very good homogeneity provided electrically inactive samples are loaded. Electrically active samples, such as water, destroy the homogeneity of such resonators particularly at high fields. Finally, this coil design does not improve the image quality compared with saddle coils. Today so-called birdcage resonators are used virtually everywhere for diameters larger than 10 mm, at least at high frequencies. Originally birdcage resonators had 8 or 12 rungs. At higher fields it turns out that more rungs are required to provide good rf homogeneity. In principle the same approach as that used for the gradients is used, an optimal rf current distribution, usually on the surface of a cylinder, for a given volume is calculated numerically.

2.1.2.5 Hardware Set-up

The whole NMR imaging sensor system usually consists of a magnet, a shim system mounted inside the room-temperature bore of the magnet, a gradient system mounted inside the shim system and the rf coil mounted inside the gradient system. In the case of a saddle coil or a birdcage resonator, open access can be realized from the bottom to the top of the entire system with the coil diameter.

As already mentioned, the variation of even one geometrical parameter can change the performance or the possible applications of the whole system drastically. Two short examples are given below, when considering a wide bore (wb) magnet with an inner diameter of 89 mm equipped with a room-temperature shim system with an inner diameter of 72 mm. (a) For a maximum sample diameter of 30 mm the outer coil diameter can be limited to 40 mm. This outer coil diameter allows for an inner gradient system diameter of 40 mm. Taking into account the maximum possible outer diameter for the gradient system of 72 mm, one can reach a gradient strength of typically 1.5 T m^{-1} (Gradient System Micro2.5, Bruker-BioSpin GmbH, Rheinstetten, Germany). (b) For a maximum sample diameter of 38 mm the outer diameter of the rf coil can be limited to 57 mm. Building a gradient system of an inner diameter of 57 mm with an outer diameter of 72 mm, a maximum gradient strength of 300 mT m^{-1} can be reached (Gradient System Mini0.5, Bruker-BioSpin GmbH, Rheinstetten, Germany). The gradient strength in this case is smaller because of the larger inner diameter, and the less than optimal ratio between the inner and outer diameter of the gradient system.

2.1.2.6 Gradient Pulse Generation

Imaging experiments require the generation of a fully flexible gradient amplitude. Amplitude variations of any shape within the physical limits of the gradient hardware must be possible. Imaging methods often do not have long relaxation delays and may consist of many changing gradient shapes and switching events. Therefore, the gradient amplitudes cannot be preloaded to a memory, but must be calculated on the fly.

Gradient generation and amplification systems are generally particularly sensitive to low frequency noise and power supply hum. Other than in hifi systems they have to support frequencies down to zero hertz. As each analog amplification stage introduces noise one has to switch from digital to analog signals as late as possible. Analog voltage connections between the various hardware units with different power supplies introduce hum, because they reference to different ground levels. Currents flow between these ground levels and modify the transferred signal accordingly. In order to optimize the situation, the digital to analog (D/A) converter has been built into modern power amplifier housings to keep the ground levels as close as possible.

The final stage of the gradient generation must of course be the analog signals. For many reasons, mainly to avoid changes of the current due to temperature dependent changes of the resistance of the gradient system, the final stage must be a current controlled amplifier. Current regulated amplifiers are by design subject to oscillation problems, therefore a great deal of effort must be made to match the amplifier to the impedance of the gradient system. There are two principal technologies, so-called linear power supplies and switching power supplies. Linear power supplies usually provide better stability and less noise, but owing to the transistor technology they are limited to about 120 V. For higher voltages, switching amplifiers are the usual choice, but they can not be used at present for high gradient diffusion systems.

The linearity, the pulse reproducibility and the stability of the complete system must be in the range of a few parts per million, otherwise image distortions or the wrong results in the diffusion experiments would be created.

2.1.3
Software

Modern NMR software covers all facets of MR applications and assists the laboratory staff and the research groups not only in the standard procedures of scan preparation, data acquisition, reconstruction and analysis, but also offers an appropriate development environment for user defined measurement methods and data analysis algorithms and provides easy-to-use tools for data management, documentation, export and archiving. The software allows the user to run complex NMR machines in a routine manner and to integrate the spectrometer into the laboratory infrastructure [7].

Modern NMR software packages were developed more than 10 years ago. Since then, the programs have experienced numerous extensions, modifications and migrations into different computer platforms. Initially, the software had to run on dedicated systems or high performance workstations, whereas today the software mainly runs on personal computers, as a result of their steep rise in performance at reasonable prices. Such software is capable of handling a wide variety of applications, ranging from spectroscopy, NMR microscopy applications, small animal biological applications up to whole body applications in clinical systems.

2.1.3.1 User Community

The NMR user community is divided into two different groups, the routine and the research users. The routine users typically run predefined protocols with standardized parameter sets, requiring maximum sample throughput and the most efficient usage of the system's resources. To support these needs, modern NMR software offers parameter sets, which can be adapted for dedicated applications and then run in a well proven measurement workflow with just a few user interventions. Icon-based workflow support simplifies the user's view of the instrument, hiding its complexity, and highly versatile interactive graphic toolboxes offer all of the functionalities needed for convenient and precise geometric definitions, a decisive prerequisite for efficient set-up and scanning of the results. Scripts allow command sequences to be executed easily, typically empowered with flow control logic. Thus, the user can include data acquisition, reconstruction, analysis and data management tasks in fully or partially automated procedures, which can contain interactive stops to allow user decisions.

Research users need full access to the functional elements of the spectrometer system and require the most efficient and flexible tools for MR sequence and application development. If the measurement methods delivered with the software do not adequately address the specific investigational requirements of a research team, modern NMR software is an open architecture for implementing new and more sophisticated functionality, with full direct access to all hardware controlling parameters. After evaluation, the new functionality can be developed with the help of toolbox functions that allow rapid prototyping and final builds, to enable the new sequence to be executed by non experienced personnel and then used in routine applications. These toolboxes provide application oriented definitions and connect to standard mechanisms and routine interfaces, such as the geometry editor, configuration parameters or spectrometer adjustments.

2.1.3.2 Acquisition Software

The acquisition software sets up the right hardware parameters to begin an acquisition and controls the data flow during the acquisition. It can be controlled in two different ways, the routine way and the research based way with full access to all hardware based parameters. The routine way of starting an experiment requires the existence of a high level method, which is mainly a software module that translates high level, easily understandable parameters into low level, machine readable parameters.

2.1.3.3 Methods, Scripts and Automation Processes

Methods and scripts are software resources that help the operator to set up complex investigations without concentrating on the basic requirements. Furthermore, these software modules can handle the interaction between various parts of the NMR software package, to support the geometry editor, auto adjustments, routine parameter handling or pulse handling program.

Scripts are often fairly simple software components that can be generated very quickly to execute recurring operations or collect multiple operations into a large block for efficient usage. Scripts can be simple shell scripts with no graphical output capability, but the use of extended graphical toolbox scripts provides a simple way of implementing software features with easy user interaction.

Methods, on the other hand, are more complex pieces of software. They are written in a highly sophisticated programming language and are the connective link between the operator and the complex base level parameter system. One main advantage of the method programming is the possibility of using predefined toolbox functions for the common pulse program features. Additionally, the programmer can include all the experimental knowledge in the method and allow all users to profit from this experience without specific know how of all method and machine details. The possibility of a method of storing the high level parameters in protocols and loading them again in a very defined style enables the NMR software to run experiments in an automated way.

The automations are often controlled by scripts that define the protocols, the methods, the timing of an automation and the interaction of the methods and automation or geometry settings. The easy way to expand scripts with user functionality enables the research user to implement special functionality to support non-standard hardware and create complex laboratory-built automation scenarios.

2.1.3.4 Relaxation and Density Analysis

The measurement of relaxation times (spin–lattice relaxation time T_1 and spin–spin relaxation time T_2) and proton density is important for characterizing material properties. Quantitative knowledge of T_1 and T_2 relaxation times is a prerequisite for optimizing contrast in imaging and spectroscopic methods. Both parameters are dependent on the magnetic field strength, where the proton density is a sample specific parameter. T_1 measurements are often carried out using inversion recovery methods, T_2 measurements are mainly acquired by Carr–Purcell–Meiboom–Gill (CPMG) whereas spin density images can be obtained in many different ways.

To generate relaxation dependent parameter maps, a series in time of NMR images at the same position in the sample needs to be acquired. The pixel intensity in images acquired at different points in time varies, depending on the change of the spin states in the pixel considered, usually following exponential functions. The task of the post-processing software is to calculate parameter maps by fitting exponential functions for each individual pixel and representing the resulting values in parameter images as pixel intensities, whereas the exponential decay represents the relaxation time and the intensity at $t = 0$, the spin density. Modern software can be configured to run such post-processing calculations fully automatically and instantly display the results. Thus, the acquired images do not have to be displayed to the operator, but the desired parameter maps are produced without having to perform all of the time consuming routine post-processing steps.

2.1.3.5 Diffusion Analysis

Diffusion is the process by which molecules are transported as result of Brownian motion. The mean square of the distance covered by a diffusing molecule is proportional to the time and the diffusion coefficient of the investigated medium (Fick's first law). Magnetic resonance imaging makes use of magnetic field gradients to phase encode the spin system in order to detect phase shifts caused by motion. For coherent motion in the presence of a gradient pulse there is a phase shift of the received signal. For incoherent motion, the diffusion process, there is an amplitude reduction due to phase dispersion [8–10]. Consequently, image areas of reduced or restricted diffusion show relative signal hyper-intensity.

Diffusion weighted imaging (DWI) is the only non-invasive and non-contaminating method capable of self-diffusion imaging and quantification. DWI provides images with contrast enhancement between areas with low and high diffusion rates. The diffusion constant can be measured by acquiring a series of diffusion weighted images, each with a different level of diffusion labeling. Furthermore, from the same image series a diffusion parameter map can be calculated by fitting an exponential function to every single pixel along the diffusion dimension.

A specialty of diffusion weighted imaging is the diffusion tensor imaging (DTI). The prerequisite for diffusion tensor analysis is a set of diffusion images in numerous independent directions. In the analysis, diffusion tensor maps can be calculated and visualized using pseudo color or vector diagrams. Maps of special classification numbers can be calculated by post-processing software to enable diffusivity, anisotropy, apparent diffusion coefficient and eigenvector analysis.

DTI has already been demonstrated to be effective in analyzing the internal microstructure of different tissues. For instance, orientation of nerve fiber bundles in the white matter of the brain or hollow fiber orientations in material science can be visualized using DTI [11].

2.1.3.6 Velocity Maps, Rheology Analysis

Velocity maps of simple or complex liquids, emulsions, suspensions and other mixtures in various geometries provide valuable information about macroscopic and molecular properties of materials in motion. Two- and three-dimensional spin echo velocity imaging methods are used, where one or two dimensions contain spatial information and the remaining dimension or the image intensity contains the information of the displacement of the spins during an observation time. This information is used to calculate the velocity vectors and the dispersion at each position in the spatially resolved dimensions with the help of post-processing software. The range of observable velocities depends mainly on the time the spins stay in the rf coil, the T_2 relaxation time and the available gradient strength. From these velocity maps additional information of the investigated systems, such as shear rate or time dependent strain maps, can be obtained, which is used widely in NMR rheology applications to determine fluid properties. Typically, the results are displayed in pseudo color images, where the parameter intensities are color coded, or in vector diagrams, with the orientation of the flow represented by the direction of the vector and the velocity by its length.

2.1.3.7 Data Formats

Modern NMR software is capable of exporting a large variety of data formats to support external post-processing and third party software. The most basic format is the raw format, where the data are exported in integer or float format without any additional information about the image properties. This format allows the research developer to use the NMR data very easily for further investigations and calculations in post-processing software without hesitating over the right format or reordering of the data values. The disadvantage of the raw format is that information in addition to the data file is necessary to inform about size, orientation, field of view and so on. For long-term studies it is often necessary to combine data that have been acquired over a long period of time, with the help of external programs. Thus, the data export into well-defined, world-wide accepted software formats is a very important feature of modern NMR software. In a very general way, these exports support formats such as the well-known image formats bitmap and jpeg, or spreadsheet formats to support, for example, Microsoft Excel or Microsoft Word. The world standard for NMR imaging data and transfer is currently designated as Digital Imaging and Communications in Medicine (DICOM). It embodies a number of major enhancements to previous versions of the American College of Radiology and National Electrical Manufacturers Association (ACR-NEMA) standard, which was defined in the 1970 s [12]. DICOM specifies a hardware interface, a minimum set of software commands and a consistent set of data formats. The DICOM standard is an evolving, manufacturer independent, standard and it is maintained in accordance with the procedures of the DICOM Standards Committee. The current definition of the DICOM standard is described in "Digital Imaging and Communications in Medicine, National Electrical Manufacturers Association, 2001, Part 1,3,4". In addition to DICOM, the manufacturer independent standard for imaging, JCAMP-DX [13] is the standard for NMR spectroscopy. It is an evolving, open-ended, machine-independent, self-documenting file format for exchanging and archiving spectroscopic data. The name JCAMP-DX is used to describe data files that are compliant with the form and style described in the relevant protocols.

2.1.3.8 Regions of Interest, Statistics

A basic method of image analysis is the use of regions of interest (ROI). These ROI can have very simple shapes such as circles or rectangles, but can also have very complex outlines, which are defined by irregular shapes. The outlines for ROIs with irregular shapes can be defined manually by selecting the desired outline, or by more complex automatic and semi-automatic algorithms, such as region growing. In the ROIs, the mean value of the intensities, standard deviations, histograms, areas and volumes can be calculated for use in further analysis steps.

2.1.3.9 Image Algebra

Image algebra is an easy way to apply simple mathematical operations to images. These operations can be calculations with any constants or other images from a measurement series. Typical applications for this easy but important feature are difference images, increase of signal-to-noise by summing up echo images, user defined image scaling or detection of local contrast changes by quotient calculation.

2.1.3.10 3D Visualization

A three-dimensional (3D) dataset can be obtained by acquisition of concatenated two-dimensional (2D) data sets or real 3D imaging methods. Post processing offers the possibility of investigating such data sets in different ways, depending on the desired information. One can select between different types of 3D visualizations. These types can be multi-planar formatting, surface rendering, volume rendering, maximum intensity projection, vector displays, movies, or pseudo color visualizations. Multi-planar formatting allows the data to be rotated and shifted in any direction and stores an aligned data set in new data sets. Consequently, any data set can be aligned to a sample specific coordinate system without having to consider the right positioning of the sample during the experiment, which is a big time benefit in the sample preparation period. All calculations are carried out by fairly simple matrix manipulations, followed by data interpolation to the new pixel grid.

In comparison with the multi-planar formatting, surface rendering is a 2D projection feature, which reconstructs a surface of any selected object for surface structure investigations or volume determinations. It is an indirect method of obtaining an image from a volume dataset. The surfaces are produced by mapping data values onto a set of geometric primitives in a process known as "iso surfacing". Typical NMR images are not binary images but have a wide range of intensities. Owing to partial volume effects and limited resolution the exact detection of a surface is often very difficult. Dedicated algorithms such as region growing or marching volumes are applied for an optimum surface detection to support the surface detection process. After this process, the surface can be reconstructed by rendering the data sampled onto a regular, 3D grid to create a smooth, even contoured surface. This final surface is a 3D object, which can be rotated along any axis and by using shading techniques, a 3D impression in a 2D projection is generated.

Volume rendering is a technique for displaying a sampled 3D scalar field directly, without first fitting geometric primitives to the samples. It is a reconstruction of a continuous function from this discrete data set, and projection of it onto the 2D viewing plane from the desired point of view. Iso-surfaces can be shown by mapping the corresponding data values almost to opaque values and the rest to transparent values. The appearance of surfaces can be improved by using shading techniques. These interiors appear as clouds with varying density and color. A big advantage of volume rendering is that the interior information is not discarded, so that it enables the 3D data set to be looked at as a whole. Disadvantages are the difficult interpretation of the cloudy interiors and the long time, compared with surface rendering, required to perform the volume rendering.

Maximum intensity projection is a 3D rendering technique that is extremely effective for the visualization of 3D angiographic image data. It is a technique used to produce 2D images of 3D data from different viewpoints, using advanced methods such as illumination, shading and color, by integrating the 3D data along the line of sight. Calculated data sets can be presented as 2D projections from any viewpoint, so moving images from different rotations angles can be recorded.

2.1.3.11 External Programs

In addition to the system controlling and data processing NMR software, a huge variety of third party software is available for any type of NMR imaging analysis and visualization. The following list is a selection of some of the commercially available software and software that is free, which can import NMR data in different formats.

- IDL – software for data analysis, visualization and cross-platform application development (http://www.rsinc.com/)
- PV Wave – development solutions that allow users to import, manipulate, analyze and visualize data rapidly (http://www.vni.com/)
- ImageJ – this is a public domain Java image processing program inspired by NIH Image that can be extended by plugins (http://rsb.info.nih.gov/ij/)
- 3D Doctor – advanced 3D modeling, image processing and measurement software (http://www.ablesw.com)
- Matlab – high-level technical computing language and interactive environment for algorithm development, data visualization, data analysis and numerical computation (http://www.mathworks.com/)
- Scilab – scientific software package for numerical computations (Freeware) http://www.scilab.org
- MAWI – software tool for 3D image processing developed especially for the analysis of images of microstructures (http://www.itwm.fraunhofer.de/mab/projects/MAVI)
- AVS – Advanced Visual Systems interactive data visualization software (http://www.avs.com)
- Amira – Advanced 3D Visualization and Volume Modeling (http://amira.zib.de/).

2.1.4
Areas of Application

The types of objects and the corresponding areas of research where NMR microscopy is involved are manifold [14]:
- hard materials such as rocks, oil cores, timber, polymers, composite materials [15]
- plant materials, including living plants, seeds, leaves, roots, berries and flowers
- initial, intermediate and final food products [15]
- animal tissue samples such as excised perfused organs (kidney, heart), eyes, embryos
- living animals such as mice, rats, fish and insects.

Results are published in various scientific journals, not always in the typical NMR journals. An efficient overview of the existing areas of application is given at the ICMRM (International Conference for Magnetic Resonance Microscopy), where contributions from all of these areas are usually presented. Figure 2.1.5 represents all oral and poster presentations at the ICMRM 2003 at Snowbird, UT, USA [16], sorted according to the major topics or keywords of the individual contributions. The assignments into the various categories cannot always be made definitively, because sometimes there are overlapping criteria, e.g., in cases of flow investigations in porous media.

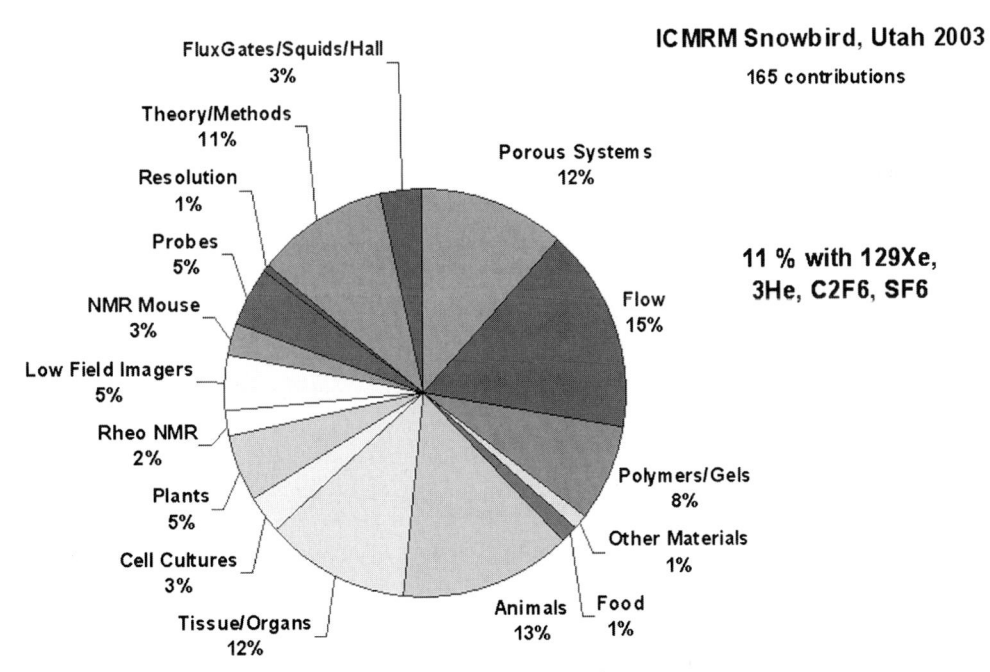

Fig. 2.1.5 Oral presentations and posters contributions of the ICMRM 2003, sorted according to the major topics.

Some topics and the number of contributions per selected category vary from conference to conference as a result of changes in the major areas of research and by the number of conference participants per research area. Other conferences take place covering specialized topics where NMR microscopy plays an important role, e.g., the "International Conference on Magnetic Resonance in Porous Media", the "International Society for Magnetic Resonance in Medicine" or the "International Conference on Application of Magnetic Resonance in Food Science".

NMR microscopy is ultimately an innovative method of research and it is not surprising that most of the commercially installed systems, approximately 80%, are installed in public scientific research centers, where new applications are continuously being developed. The method is not particularly widely distributed in industry, where standardized methods are more often used. However, NMR microscopy is mainly used in the pharmaceutical industry for the development of new drugs, in the food industry for the development of new types of food, in the chemical industry for creating and characterizing new materials and in the polymer industry, e.g., for creating new mixtures for tires.

Many investigations are made using the commercially available "standard" NMR microscopy hardware and software, although in some cases this hardware is modified in order to fulfill specific requirements and to expand the number of possible applications. Such modifications and expansions then become part of new commercially available hardware and software if they are useful for a larger number of users, as was the case recently with the development of the Rheo-NMR [17].

Some applications are shown in the following sections, where the standard hardware, software and methods have been partially modified or connected to special experimental constructions. In the past, other applications have been performed by specific groups, who built their own dedicated rf probes and/or gradient systems. These originally specialized products have now found their way into the commercially available NMR microscopy products of today [18, 19].

2.1.4.1 Polymers at High Temperatures

The properties of polymers such as hardness, stability, elasticity and ageing are caused by the composition, the cross-link density and by the mobility of the molecular groups or molecular chains. The mobility is linked directly to the applied temperature. Higher mobility at higher temperature results in longer T_2 relaxation times and, as a consequence, in an increase in the signal-to-noise ratio in the NMR images [20]. Therefore, the probes used in NMR microscopy studies should cover a wide temperature range. However, the typical imaging probes used usually cover a temperature range between −20 and +80 °C. This range can be extended from −100 to +200 °C with dedicated rf inserts or dedicated probes (Figure 2.1.6). At the same time it is also important to keep the water-cooled gradient systems at room temperature.

The performance of such a probe can be demonstrated for polymer studies by heating a phantom sample up to 460 °C. The sample was made from a PTFE plug, two pieces of a PVC hose and a silicon rubber hose. The deformation and the

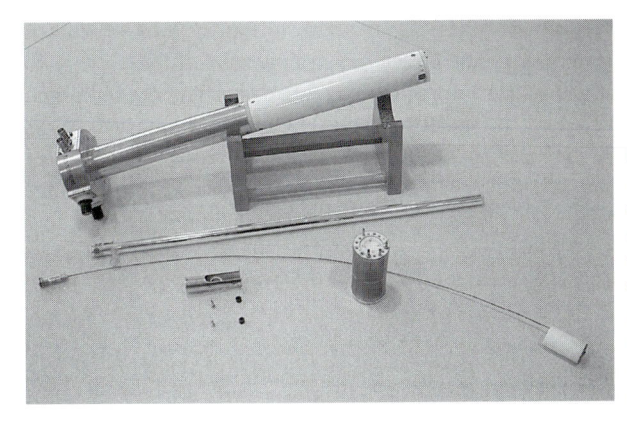

Fig. 2.1.6 Unassembled NMR microscopy probe with dedicated 15 mm resonator for expanded temperature ranges between −100 and +200 °C (probe base, glass dewar, rf resonator, temperature sensor and fixation parts).

decomposition were shown using a 3D spin-echo imaging method. The results are shown in Figure 2.1.7. No image intensity was visible at room temperature. The PVC and the silicon rubber material became visible at temperatures higher than 350 K, and the polyethylene material at 415 K, caused by the higher mobility of the molecules with increasing temperature. The decomposition of the individual parts of the sample started at approximately 440 K. Gas bubbles and voids became visible first in the silicon rubber, then in the PVC and the PTFE material.

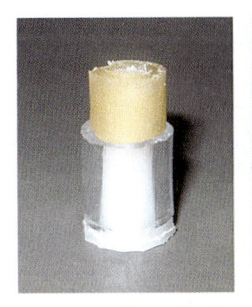

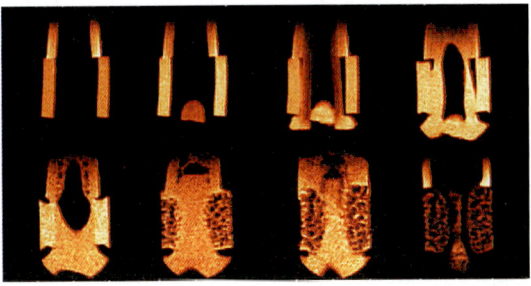

Fig. 2.1.7 Effect of high temperatures on a composite polymer phantom made from a polyethylene plug (white), a PVC hose (transparent) and a silicon rubber hose (yellow) shown at room temperature (left) and after heating (right). The NMR images (middle) are taken as 2D cross sections from 3D data sets through the phantom at 353, 390, 415, 440, 440, 460, 460 and 350 K over a time period of approximately 80 min for the complete process.

2.1.4.2 Processing of Bread in Food Research

NMR microscopy is used in the food industry to study the properties and quality of the starting materials, changes to the properties during the production, mixing/stirring, fermentation, cooking processes and the properties of the final food products, e.g., the moisture distributions can be monitored [21]. Dynamic processes such as re-hydration, drying, freezing, melting, crystallization and gelation have been investigated [14].

Using the following example of bread from the fermentation of the dough to the bread during and after baking, some of the aspects mentioned above can be demonstrated. An important parameter for the characterization of the fermentation process is the porosity and the mass density of the fresh bread yeast dough [22]. PTFE tubes with an inner diameter of 15 mm were filled with various types of dough. The fermentation was observed for 20 h. Figure 2.1.8 (left) shows four images of the interior of the yeast dough after different fermentation times. The evolution of the decreasing average mass density during the fermentation depends on the composition of the dough and the fermentation temperature (Fig. 2.1.8, right). There are phases of decreasing and again increasing mass density, which is combined with a different structure of the final baked bread.

The baking process of dough can be observed directly in an NMR microscopy probe at temperatures up to 473 K [23]. Figure 2.1.9 shows two dough samples baked at 423 and 438 K. The thickness of the crust and the frequency distribution of the pore diameter differ significantly. Histograms of the signal intensity were derived as a measure of the homogeneity of the product quality after baking. The frequency distribution of the signal intensity is clearly correlated with the baking temperature.

2.1.4.3 Flow in Complex Systems

NMR microscopy is appropriate to study the flow behavior of complex materials, the flow in complex geometric structures and processes such as extrusion, injection moulding, flow in nozzles, pipes, etc., because the velocity vectors can be directly

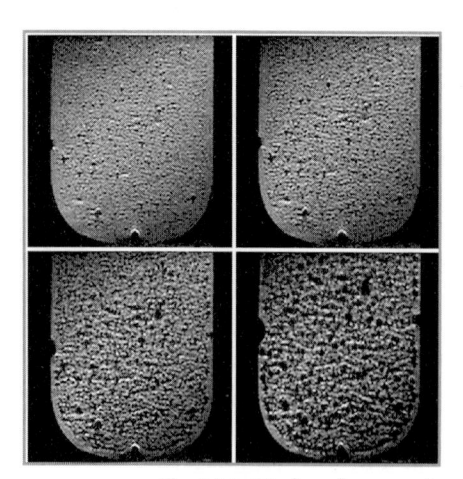

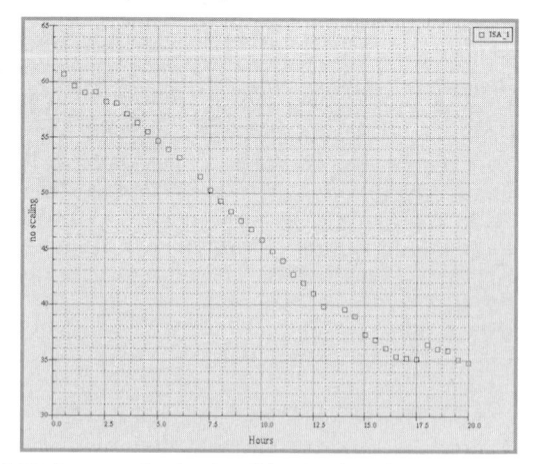

Fig. 2.1.8 2D slices from a series of 3D data sets, acquired every 30 min (left). The average signal intensities from a region of interest from the center area of all 3D data sets were determined and are shown for the various times (right). The intensities are proportional to the density of the dough and contain information about the change in porosity during the process. [400 MHz, 9.4 T, 3D spin echo 128 × 128 × 128 pixels, FOV (field of view) 20 × 20 × 20 mm, resolution 156 mm, total time 27 min per 3D data set, 40 data sets].

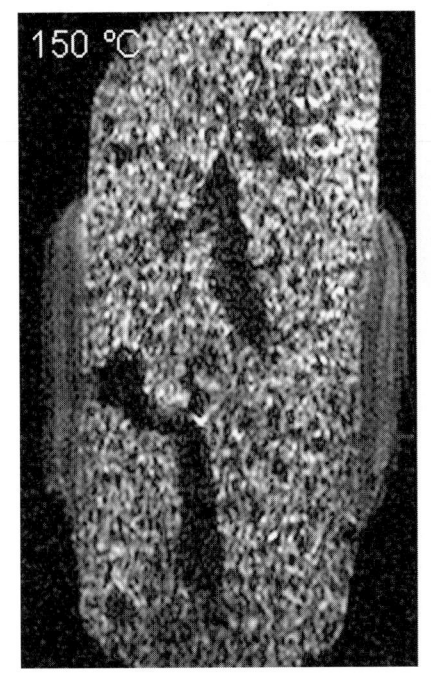

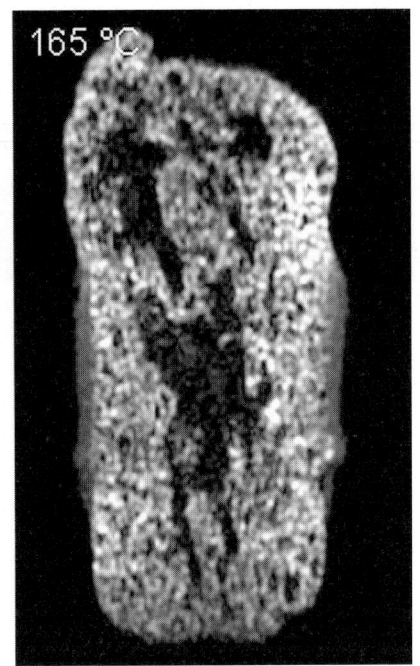

Fig. 2.1.9 MRI (400 MHz, 9.4 T) of the cake dough (12.5-mm outer diameter) after baking, measured at room temperature. [3D spin echo (matrix 256 × 128 × 128, FOV 36 × 18 × 18 mm, TE 5.3 ms, TR 200 ms, 12 averages, experiment time 11 h, spatial resolution 140.6 µm]. Baking temperature and time: 423 K, 46 min (left), 438 K, 41 min (right). PTFE tape was used to stabilize the position of the samples during the MRI.

determined at each location [24–26]. The flow through porous media or complex structures plays an important role in process engineering and in nature. Examples are filters, catalysts, separation column–trickle-bed reactors, flow and storage of ground water, and flow of water and oil in rocks, where the homogeneity of the flow, bypass flows and chemical reactions can be studied. The influence of wall effects that are dependent on the fluid and the properties of particles has been quantified for the flow in porous media [27].

In another type of flow experiment the flow of fluids driven by rotating elements is studied by Rheo-NMR [2, 17, 28]. Rheo-NMR encompasses all NMR applications dealing with the combination of flow behavior, flow-induced structural changes and NMR data. Thus, it is possible to realize flow experiments (viscometric flows: tube-, Couette-, plane shear-flow, plate/plate, cone/plate) in NMR devices in order to determine the corresponding velocity profiles. Through additional information (tube-flow: pressure drop, Couette: torque), assuming constitutive-laws, it is possible to derive the corresponding viscosity- and wall-slip functions and to study and quantify flow-induced structural changes by means of appropriate NMR experiments.

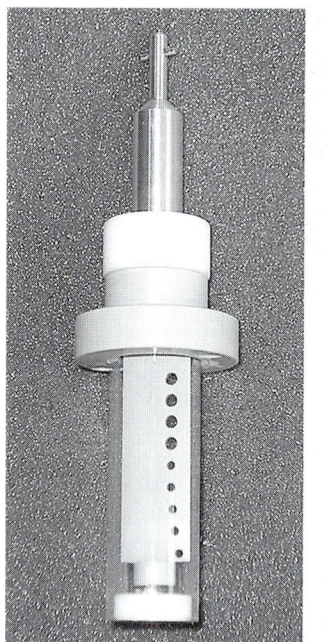

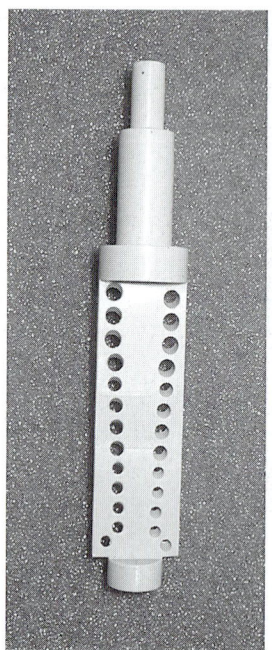

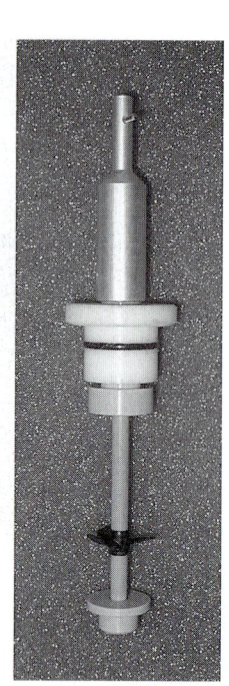

Fig. 2.1.10 Complete cell with a paddle with one row of holes (blade stirrer), mounted in a glass tube (left), a second paddle with two rows of holes (middle) and a propeller mounted on a rotation axis (right) used to study the velocity distribution in different types of mixers.

The rheo cells can easily be replaced by various types of mixers, propellers or paddles (Figure 2.1.10). It is then possible to analyze the temporal evolution of chemical/physical reactions of mixing, demixing and sedimentation of materials in process engineering, e.g., during the mash process or fermentation [28, 29]. The stirring mechanics and speed can be optimized for various materials of different particle sizes and viscosity.

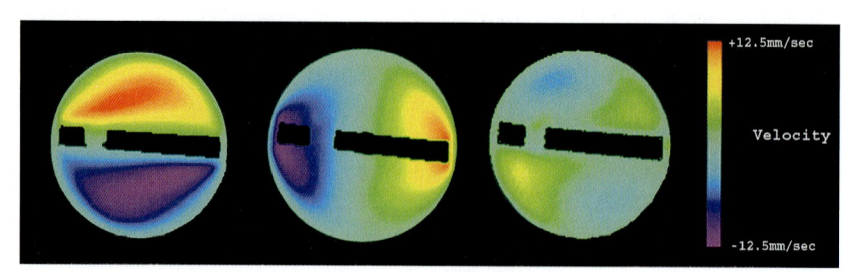

Fig. 2.1.11 The horizontal in-plane (left and middle) and the vertical (right) velocity components of the water are shown in a color encoded display style. The black structure in the images is the material of the paddle. The plane shown here crosses one of the holes of the paddle.

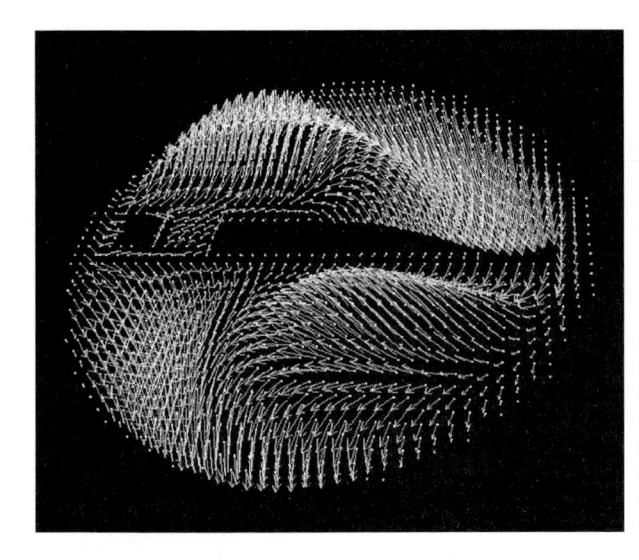

Fig. 2.1.12 Vector diagram of the velocity distribution of flowing water driven by a paddle with holes. The diagram was calculated by the AMIRA software (http://amira.zib.de).

The typical observable velocity range is, theoretically, between 0.1 and 1000 mm s^{-1}. Experimental constraints may decrease this range.

Early experiments were carried out with a water or a beer-mash filled "paddle cell" [29]. The holes are drilled into the paddle at an angle of approximately 15° in order to generate a vertical mass transportation during the rotation of the paddle. The velocity of the water is determined by a 3D spin-echo method. The data acquisition has to be triggered with the help of the position of the rotating paddle. The velocity components are shown in Figure 2.1.11. For an easier interpretation of the data, velocity vectors can be calculated (Figure 2.1.12). Further experiments with mixtures of liquids and with a different mixer geometry are in progress.

2.1.5
Outlook

For a manufacturer the difficulty is to estimate future developments or trends in NMR microscopy. Based on dedicated laboratory-made hardware developed by the NMR microscopy users and on their requests for new commercial hardware and software, the following topics could become more important: micro-coil applications, multiple receiver systems and multi-coil arrangements, NMR microscopy at very high magnetic fields, MAS imaging and localized ^{1}H MAS spectroscopy and localized single-shot 2D spectroscopy. There are no clear-cut distinctions between most of the individual topics, as will be discussed in the following sections.

2.1.5.1 Micro-coil Applications

Micro-coils can be used to reach higher spatial resolutions, as a result of the increase in sensitivity of very small rf coils. The possible resolution in NMR

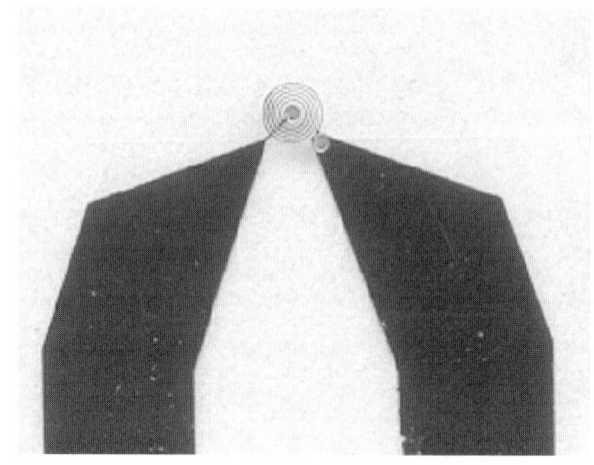

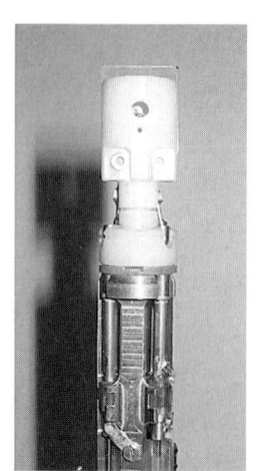

Fig. 2.1.13 Planar spiral micro-coil of 1.2 mm outer diameter (left), mounted as an exchangeable rf insert on a standard NMR microscopy probe (right) (Bruker-Biospin GmbH, Rheinstetten, Germany). The micro-coil is embedded, together with a glass sample holder, in a slotted PTFE holder.

microscopy is limited physically by the diffusion of the nuclei between the excitation and the detection of the NMR signals. Therefore, resolutions of 5 µm can in principle be attained for objects with high water contents in experiments with echo times of a few milliseconds. In small objects with a slower diffusion or with restricted diffusion, resolutions down to 1 µm may be reached. The commercially available gradient systems, providing about 300 G cm^{-1} gradient strength, are just strong enough. However, in practice the limiting parameter is the sensitivity of the rf coils. The relatively small number of spins per voxel prevents the acquisition of images with reasonable signal-to-noise and contrast-to-noise ratios. The sensitivity may be improved by further developing the micro-coils. Nano-technology is be-

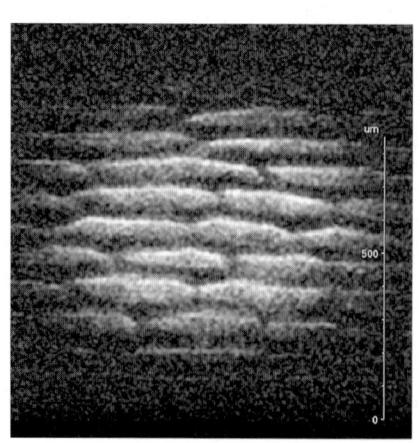

Fig. 2.1.14 Single-cell image from an onion, acquired with a 1.2 mm spiral planar micro-coil at 500 MHz, 11.7 T. A 2D spin-echo technique without slice selection was used, 8 averages were accumulated in 7 min. The in-plane resolution is 7 µm per pixel.

coming more and more available for the development and production of small structures such as micro-coils of high precision and high sensitivity [30]. The materials of the micro-coils, the sample containers and the samples themselves have to be selected very carefully in order to minimize distortions caused by the different susceptibility properties of the individual components. Figure 2.1.13 shows a spiral shaped flat micro-coil of 1.2 mm diameter, which was mounted as an exchangeable rf insert on a commercial NMR microscopy probe.

A "single cell layer" of an excised piece of onion epidermis was selected as a test object for this micro-coil, (Figure 2.1.14), and an imaging experiment was carried out, similar to the one by Mansfield et al., who used a laboratory-made micro-coil probe and gradient system [31]. Other micro-coil types, e.g., volume coils or coils that are immersed into the objects, can be adapted to specific applications and mounted on commercial imaging probes.

2.1.5.2 Multiple Receive Systems and Multi-coil Arrangements

Multiple receive systems with multi-coil arrays have become widely distributed in medical MRI. The benefit is the enhancement of the signal-to-noise ratio per time or a reduction of the acquisition time. This technique is not used in NMR microscopy for objects of intermediate size in standard bore (52-mm id) and wide bore (89-mm id) magnets, which are the most widely distributed magnet types for NMR microscopy. The main reason is the restricted space in such magnets for the shim

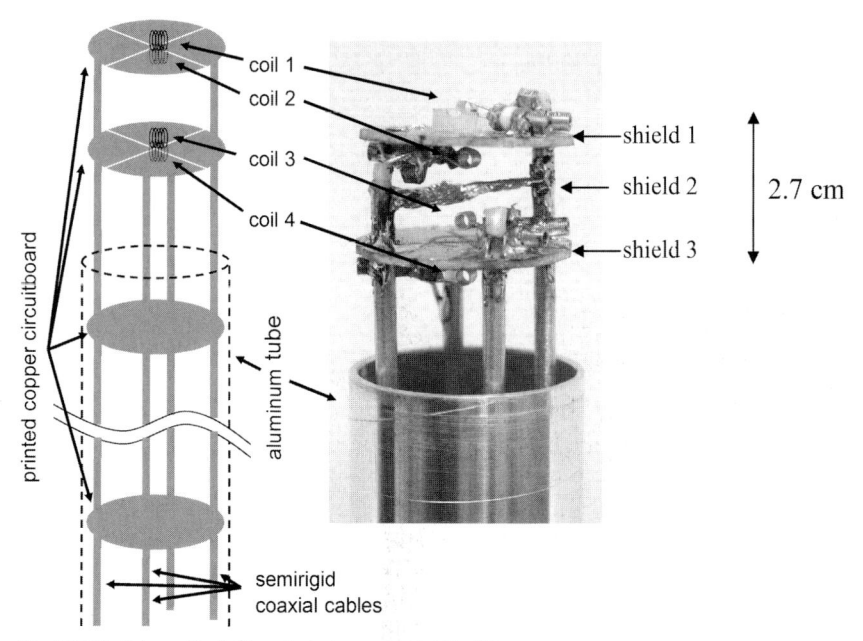

Fig. 2.1.15 Schematic (left) and photograph (right) of the 750 MHz four-coil probe (courtesy of A. Purea, T. Neuberger, A. G. Webb, Institute of Physics, University of Wuerzburg, Germany).

system, the gradient system and finally the rf coil array with integrated preamplifiers. However, going to really small objects such as single cells, tumor spheroids or tissue samples, allows the use of numerous small rf coils or micro-coils, where many objects, one per coil are investigated in a parallel or interleaved acquisition mode. This can be extremely valuable, because many such small coils can be mounted in the homogeneous volume of the magnets [approximately 30 to 45 mm along the B_0 direction (the static magnetic field)] and in the linear range of the gradient systems (approximately 20 to 40 mm along the B_o direction). The simultaneous investigations can therefore increase the sample throughput enormously, which is an important issue, because the experiment time for such small samples is usually in the range of several hours and the system access time especially at high field systems is always limited. Multi-channel NMR microscopy systems are commercially available, but the multi-coil probes are still custom built. Purea et al. [32] have described the development of a four-channel probe that consists of four individual solenoid coils (length 2.8 mm, outer diameter 2 mm) stacked along the z axis (Figure 2.1.15). The probe was designed for a 17.6 T (750 MHz) wide-bore magnet, and operates within a 40-mm inner diameter gradient set with a maximum strength of $1 \, \text{T m}^{-1}$. Practical use of the probe was demonstrated by obtaining four three-dimensional T_1 maps of chemically fixed *Xenopus laevis* oocytes simultaneously at a spatial resolution of $30 \times 60 \times 60 \, \mu\text{m}$.

2.1.5.3 Very High Magnetic Fields

Magnets with extremely high magnetic fields of from 17.6 T (750 MHz) to 21.14 T (900 MHz) have been installed in facilities throughout the world within recent years. These magnets are mainly used for high resolution (HR) and solid state NMR, and only seldom for NMR microscopy investigations, although a gain in

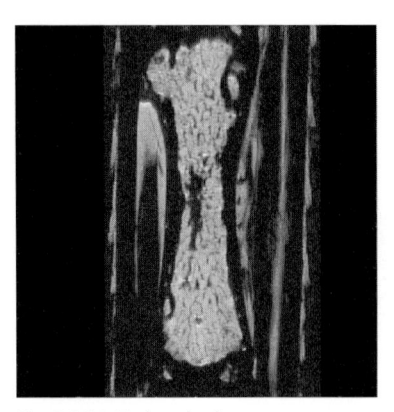

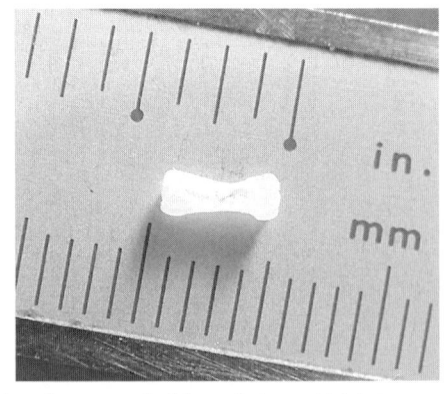

Fig. 2.1.16 Trabecular bone structure, muscle and tendon of a mouse tail *in vitro* at 21.14 T, 900 MHz (left). The trabecular structure and the muscle tissue are clearly resolved, the tendon tissue is not visible because of the short T_2 relaxation rime. The bone is shown after removal of the soft tissue (right). Image parameters: multi-slice spin-echo method, 512 × 256 pixels, FOV 6 × 3 mm, resolution 11.7 μm per pixel, slice thickness 64 μm, TR (repetition time) 2 s, TE (echo time) 8.7 ms.

sensitivity is observed in imaging applications at the higher fields. Experiments on fixed chicken embryos at 17.6 T resulted in a gain in SNR of a factor of 3.5 compared with at 7 T, if all experimental parameters were kept the same [33]. The short relaxation time of the 3D spin-echo experiment caused some saturation, otherwise the gain in SNR at the higher field would have been even larger, but at the expense of a longer experiment time.

Inhomogeneities in the objects are critical at the very high magnetic fields. They can cause strong local magnetic field gradients due to the different magnetic susceptibility properties of the various materials. Such strong local field gradients result in geometrical image distortions. The susceptibility changes are proportional to the magnetic field strength. No air bubbles or air filled voids must be included in the objects. Special care must be taken over the choice of the objects and the preparation of the samples to be investigated at the very high magnetic fields (Figure 2.1.16). It is anticipated that spectroscopic imaging experiments (chemical shift imaging, localized spectroscopy) of small objects (cell clusters, tumor spheroids, histology samples) will deliver valuable results in the near future.

2.1.5.4 MAS Imaging and MAS Localized Spectroscopy

The investigation of "semi-solid" objects such as lipid membranes [34], liquid crystalline dispersions [35, 36], drug delivery systems [37], food [38] and also

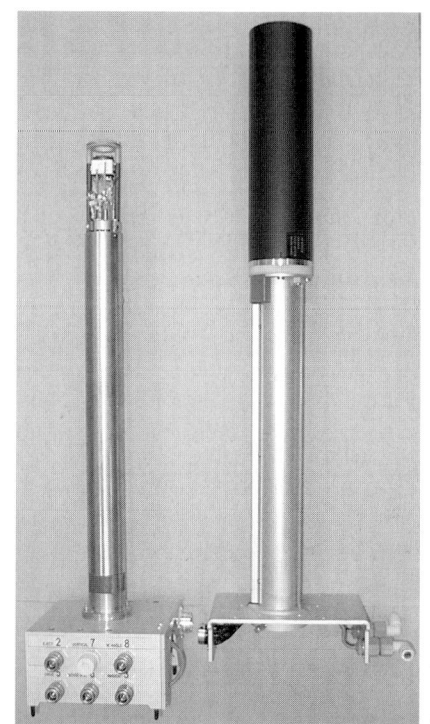

Fig. 2.1.17 Conventional NMR microscopy gradient system (Micro2.5, 2.5 G cm^{-1} A^{-1}, 40-mm inner diameter) (right) and a conventional MAS probe (left). The probe is mounted into the separate gradient system instead of the original imaging probe.

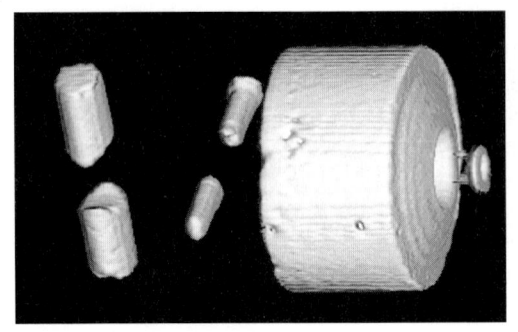

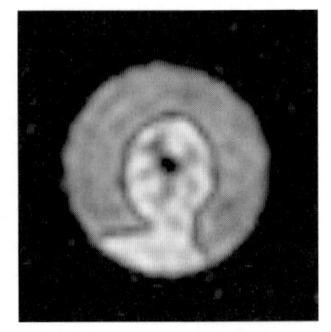

Fig. 2.1.18 Image of the water in cylinders of a phantom (left). The image was rendered from a 3D data set using the software ImageJ [42] in combination with volumeJ [43]. Some water was centrifuged into the small cylinders perpendicular to the rotation axes. The water that does not fit into the cylinders is centrifuged into the space left between the phantom and the sealing Teflon plug and from there it is centrifuged to the rotor walls. Also shown is an image of a piece of cable (right), cut at one side, where the inner copper wire has been removed from the insulation and replaced by adamantine.

biological tissues [39, 40] can be very efficient using NMR methods. However, these objects are difficult to study by HR NMR and MRI because of the very broad resonance lines caused by dominating anisotropic NMR interactions, such as magnetic susceptibility, chemical shift anisotropy or even dipolar coupling.

The combination of magic angle sample spinning (MAS) with spatially localized methods used in MRI provides an enormous reduction in the linewidths simultaneously with spatially resolved images or spectra from localized volumes of "semi-solid" samples. Such experiments were initially developed by pioneering workers [18, 19] who built dedicated hardware and created dedicated acquisition methods. Today it is possible to combine conventional triple axes gradient systems used in NMR microscopy applications with conventional magic angle sample spinning (MAS) probes. No further modifications are required (Figure 2.1.17).

In previous studies, probes with special designs of gradient coils wrapped around the stator were used. However, the combination of a separate gradient system and an MAS probe has some important advantages: (a) The gradient coils are separated from the radiofrequency components of the probe thus minimizing any impact of the gradient coils on the radiofrequency properties of the circuits of the probes. (b) An external cooling of the gradient coils can be used, thus enabling high currents, high gradient strengths and high linearity. (c) There is no torque acting on the stator caused by high currents in the gradient coils, which could otherwise compromise the mechanical stability and disorient the sample, thus leading to distorted NMR results.

Some experiments were carried out on an Avance 400-MHz wide-bore spectrometer (Bruker-Biospin GmbH, Rheinstetten, Germany) to demonstrate the quality of this new combination. The probe was inserted into the gradient system in such a way that the sample is located in the center of the gradient system and the rf coil axis was aligned along the direction of the room diagonal, defined by the gradient axes system. Samples were spun between 3500 and 7000 Hz.

The accuracy of the triggered MAS images was shown by a KelF cylinder (diameter 3 mm) in which three interconnecting and rectangularly arranged cylinders of diameters 1, 0.7 and 0.3 mm, respectively, were drilled. This phantom was inserted into a 4-mm MAS rotor and spun at 3500 Hz. The cylinder along the spinning axes was filled with a drop of water. The rotor was tightly closed with a Teflon plug. The image obtained using a 3D gradient-echo experiment is shown in Figure 2.1.18 (left). It is clear that no blurring effects are visible, which could be to due to instability of sample spinning or inaccurate matching of the sample spinning with the spinning of the gradient. In fact, it is impossible to tell whether this image is made by a rotating sample or not. A second phantom was made from a 5-mm long and 2.5-mm thick piece of cable, where the inner copper wire was removed from the insulation and replaced by adamantine. The cable insulation was slotted on one side. The acquisition parameters were the same as for the KelF experiment. Figure 2.1.18 (right) shows the precise image obtained from the semi-solid materials of the phantom.

Spatially localized NMR spectroscopy was performed using a sample composed of a monoolein–D_2O dispersion forming a cubic liquid-crystalline phase [41]. The spectra were selected from a slice of 300-µm thickness. The direction of the gradient was along the spinning axis. A spectral resolution was obtained that had never been observed before when selecting the whole volume of the sample.

References

1 P. T. Callaghan **1991**, *Principles of Nuclear Magnetic Resonance Microscopy*, Clarendon Press, Oxford.
2 P. T. Callaghan **2002**, (Rheo-NMR: A new window on the rheology of complex fluids, encyclopedia of nuclear magnetic resonance), in *Advances in NMR*, (vol. 9), eds. D. M. Grant, R. K. Harris, John Wiley & Sons, Ltd, Chichester.
3 G. Eidmann, R. Savelsberg, P. Blümler, B. Blümich **1996**, (The NMR MOUSE®: A mobile universal surface explorer), *J. Magn. Reson. A* 122, 104–109.
4 R. Turner **1988**, (Minimum inductance coils), *J. Phys. E: Sci. Instrum.* 21, 948–952.
5 R. Turner, R. M. Bowley **1986**, (Passive screening of switched magnetic field gradients), *J. Phys. E: Sci. Instrum.* 19, 876–879.
6 H. Schmidt, M. Westphal et al. **1993**, Bruker Analytische Messtechnik, (Method for the construction of an optimized magnet coil), Patent Application EP0563647.
7 W. Ruhm **2002**, Bruker Report 150/151.
8 J. E. Stejskal, J. E. Tanner **1965**, (Spin echoes in the presence of a time-dependent field gradient), *J. Chem. Phys.* 42, 1.
9 E. L. Hahn **1950**, *Phys. Rev.* 80, 580.
10 H. Y. Carr, E. M. Purcell **1954**, *Phys. Rev.* 94, 630.
11 D. LeBihan, J. F. Mangin, C. Poupon, C. A. Clark, S. Pappata, N. Molko, H. Chabriat **2001**, (Diffusion tensor imaging: concepts and applications), *J. Magn. Reson. Imag.* 13, 534–546.
12 http://medical.nema.org, DICOM.
13 http://www.jcamp.org, JCAMP.
14 J. Götz **2004**, *Applications of NMR to Food and Model Systems in Process Engineering*, Habilitation Thesis, Wissenschaftszentrum Weihenstephan der Technischen Universität München, (http://tumb1.biblio.tu-muenchen.de/publ/diss/ww/2004/goetz.html)
15 J. Götz, N. Eisenreich, A. Geißler, E. Geißler **2002**, (Characterization of the structure in highly filled composite materials by means of MRI), *J. Propell., Explos., Pyrotechn.* 27 (6), 1–6.

16 7th *International Conference on Magnetic Resonance Microscopy*, **2003**, Snowbird, UT, USA.
17 P. T. Callaghan, E. Fischer **2001**, Rheo-NMR: A New Application for NMR Microscopy and NMR Spectroscopy, Bruker Report, (p.) 149.
18 R. A. Wind, C. S. Yanoni **1981**, US Patent 4,301,410.
19 D. G Cory, J. W. M. van Os, W. S. Veeman **1988**, *J. Magn. Reson.* 76, 543.
20 N. Eisenreich, A. Geißler, E. Geißler, J. Götz **2004**, (Structure characterisation of foams and filled polymers by means of MRI), *Proceedings of the 8th International Conference on Magnetic Resonance in Porous Media*, Palaiseau, France, July 7– 9, *J. Magn. Reson.*
21 S. Ablett, Unilever Research **1992**, *Trends Food Science & Technology* (vol. 3), August/September.
22 J. Götz, D. Groß, P. Köhler **2003**, (On-line observation of dough fermentation by magnetic resonance imaging and volumetric measurements), *Zeitschrift Lebensm.-Unters. F A (Eur. Food Res. Technol.)* 217, 504–511.
23 J. Götz, E. Geißler, P. Köhler, D. Groß, V. Lehmann **2005**, (On-line observation of dough baking by magnetic resonance imaging), *Lebensmittel Wissenschaft und Technologie (Food Science and Technology)*, submitted for publication.
24 S. Laukemper-Ostendorf, K. Rombach, P. Blümler, B. Blümich **1997**, Bruker Application Note, NMR/1200/2/97/8-HA.
25 J. Götz, K. Zick **2003**, (Local velocity and concentration of the single components in water/oil mixtures monitored by means of MRI flow experiments in steady tube flow), *Chem. Eng. Technol.* 26 (1), 59–68.
26 J. Götz, K. Zick, W. Kreibich **2003**, (Possible optimisation of pastes and the according apparatus in process engineering by MRI flow experiments), *Chem. Eng. Process.* 42 (7), 515–534.
27 J. Götz, K. Zick, C. Heinen, T. König **2002**, (Visualisation of flow processes in packed beds with NMR imaging: Determination of the local porosity, velocity vector and local dispersion coefficients), *Chem. Eng. Process.* 41 (7), 611–630.
28 J. Götz, K. Zick **2005**, (Rheo NMR with applications on food), in *Handbook of Modern Magnetic Resonance*, ed. G. Webb, Kluwer Academic Publishers, London.
29 J. Götz, K. Zick, V. Lehmann, D. Groß, M. Peciar, (Analysis of the mixing and flow behaviour of beer mashes in a stirring vessel by means of NMR flow experiments), in preparation.
30 C. Massin **2004**, *Microfabricated Planar Coils in Nuclear Magnetic Resonance*, Series in Microsystems, (vol. 15), Hartung Gorre Verlag, Constance.
31 R. Bowtell, P. Mansfield, J. C. Sharp, G. D.Brown, M. McJury, P. M.Glover **1992**, *Magnetic Resonance Microscopy*, VCH, Weinheim, (pp.) 427–439.
32 A. Purea, T. Neuberger, A. G. Webb **2004**, *Concepts Magn. Reson. Part B (Magn. Reson. Eng.)*, 22B (1) 7–14.
33 B. Hogers, D. Gross, V. Lehmann, H. de Groot, A. de Roos, A. C. Gittenberger-de Grot, R. E. Poelmann **2001**, *J. Magn. Reson. Imag.*14, 83.
34 F. Volke, A. Pampel **1995**, (Membrane hydration and structure on a subnanometer scale as seen by high resolution solid state nuclear magnetic resonance: POPC and POPC/$C_{12}EO_4$ model membranes), *Biophys. J.* 68, 1960–1965.
35 A. Pampel, F. Volke **2003**, (Studying lyotropic crystalline phases using high-resolution MAS NMR spectroscopy), in *Lecture Notes in Physics*, eds. R. Haberlandt, D. Michel, A. Pöppl, R. Stannarius, Springer, Berlin.
36 A. Pampel, E. Strandberg, G. Lindblom, F. Volke **1998**, (High-resolution NMR on cubic lyotropic liquid crystalline phases), *Chem. Phys. Lett.* 287, 468–474.
37 A. Pampel, R. Reszka, D. Michel **2002**, (Pulsed field gradient MAS NMR studies of the mobility of carboplatin in cubic liquid crystalline phases), *Chem. Phys. Lett.* 357, 131–136.
38 M. E. Amato, G. Ansanelli, S. Fisichella, R. Lamanna, G. Scarlata, A. P. Sobolev, A. Segre **2004**, (Wheat flour enzymatic amylolysis monitored by in situ H-1 NMR spectroscopy), *J. Agric. Food. Chem.* 52, 823–831.
39 B. Sitter, T. Bathen, B. Hagen, C. Arentz, F. E. Skjeldestad, I. S. Gribbestad **2004**,

(Cervical cancer tissue characterized by high-resolution magic angle spinning MR spectroscopy, MAGMA), *Magn. Reson. Mater. Phys., Biol. Med.* 16, 174–181.
40 A. R. Tate, P. J. D. Foxall, E. Holmes, D. Moka, M. Spraul, J. K. Nicholson, J. C. Lindon **2000**, (Distinction between normal and renal cell carcinoma kidney cortical biopsy samples using pattern recognition of H-1 magic angle spinning (MAS) NMR spectra), *NMR Biomed.* 13, 64–71.
41 A. Pampel, F. Engelke, D. Gross, T. Oerther, K. Zick **2005** (Pulsed Field Gradient NMR in Combination with Magic Angle Spinning), *Spin Report* 157 ISSN 1612-4898.
42 W. S. Rasband **1997–2004**, *ImageJ*, 1.33 u ed. Bethesda. W. Burger, M. J. Burger **2005** (Digitale Bildverarbeitung, Eine Einführung mit Java und ImageJ), *Springer Verlag.*
43 M. D. Abramoff, M. A. Viergever **2002**, (Computation and visualization of three-dimensional soft tissue motion in the orbit), *IEEE Trans. Med. Imag.* 21, 296–304.

2.2
Compact MRI for Chemical Engineering
Katsumi Kose

2.2.1
Concept of Compact MRI

More than 10 000 clinical MRI systems are now routinely used in hospitals throughout the world. MRI systems for engineering purposes, however, are less common, for several reasons. The first is that the samples and objectives in engineering cover a wide range and a single MRI system cannot meet the variety of demands. The second is that existing commercially available MRI systems which usually use superconducting magnets are very expensive, require large installation spaces and cannot be moved. MRI systems for engineering, therefore, should have design flexibility and be inexpensive, compact and portable. The concept of "compact MRI" was proposed based on these background considerations (1–3).

Figure 2.2.1 shows a functional block diagram of an MRI system. As shown in the figure, the MRI system can be divided into two functional sub-systems: electrical and magnetic sub-systems. The magnetic sub-system consists of a magnet, gradient coil set and rf coil(s). The dimensions and specifications of the magnetic sub-system

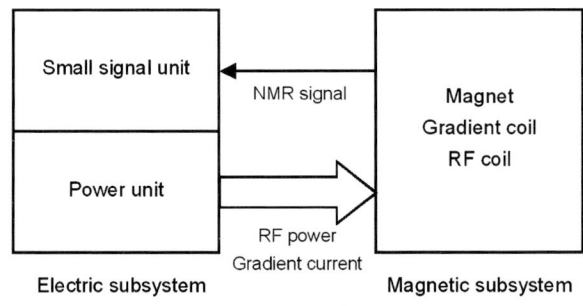

Fig. 2.2.1 Functional block diagram of an MRI system.

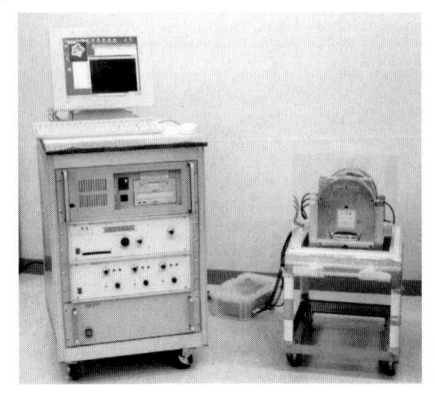

Fig. 2.2.2 Compact and portable MRI system with a 0.3-T, 8-cm gap permanent magnet.

units depend on the sample size and shape, and its physical and/or chemical properties. Those of the electrical sub-system units, however, do not depend on the sample properties for most MRI applications. After consideration of the above situation, the concept of "compact MRI" was proposed: the electrical sub-system was installed in a single portable rack and the magnet was a dedicated permanent magnet. Figure 2.2.2 shows a typical example of a compact and portable MRI system.

2.2.2
System Overview

Figure 2.2.3 shows a typical block diagram for a compact MRI system. The square box surrounded by the dotted lines shows the electrical sub-system, which is stored in a single portable rack as shown in Figure 2.2.2. The electrical sub-system is thus often called the (portable) MRI console. In accord with Figure 2.2.3, an MR image acquisition process is described as follows.

Imaging parameters such as the repetition time, spin-echo time, image matrix size, number of signal accumulations, and so on, are input from the keyboard or mouse to the PC (personal computer), and the pulser (pulse programmer) outputs the rf pulse shape (and phase) and gradient waveforms for the image acquisition pulse sequence. The modulator generates the rf pulse by mixing the rf pulse shape and the Larmor frequency reference signal. The transmitter amplifies the rf pulse to excite the nuclear spins of the sample placed in the rf coil. The gradient drivers amplify the gradient waveforms to drive the gradient coils to generate magnetic field gradients over the sample.

The NMR signal is detected by the rf coil (single coil configuration) and amplified by the preamplifier. The amplified signal is then demodulated using the Larmor frequency reference signal to generate the NMR signal in the rotating frame. The detected NMR signal is digitized using the AD converter (ADC) and stored in the PC. After a complete matrix of NMR data has been collected, the image reconstruction (essentially 2D or 3D FFT) is performed and the MR image is shown on the display.

In the following sections, each hardware unit will be described in detail.

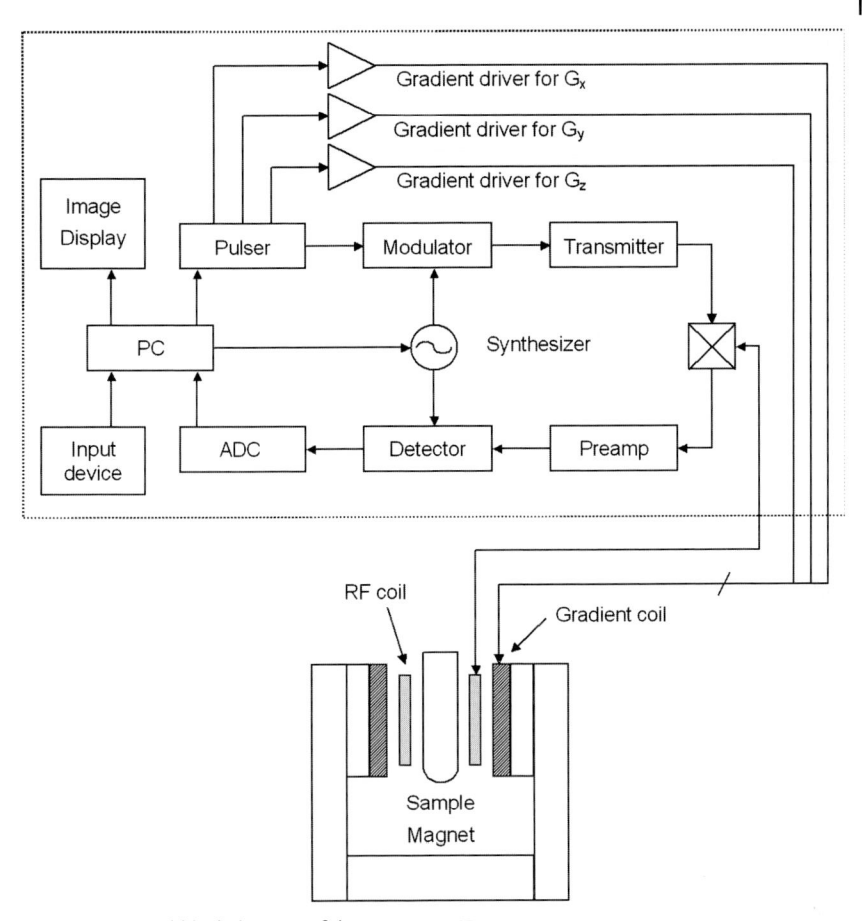

Fig. 2.2.3 Typical block diagram of the compact MRI.

2.2.3
Permanent Magnet

The discovery of high performance permanent magnetic materials, such as Nd-Fe-B compounds (4), has made mid-field (ca. 0.3–0.4 T) whole-body MRI permanent magnets very compact (5). This magnet technology has been applied to the permanent magnets of the compact MRI system (6). In addition to the clinical whole-body MRI magnet design, a novel permanent magnet design was introduced for the compact MRI system (7–9).

Figure 2.2.4 shows two typical magnet configuration designs. The left figure shows a conventional magnet configuration design with an iron yoke. This design is usually applied to relatively low field (ca. 0.2–0.5 T) compact MRI magnets. The right figure shows a novel design without a yoke (7–9). This design originates from the "magic ring" or "magic sphere" proposed for particle accelerators (7). However, in the actual manufacturing process, the ring or sphere was replaced by a square or

Pole piece

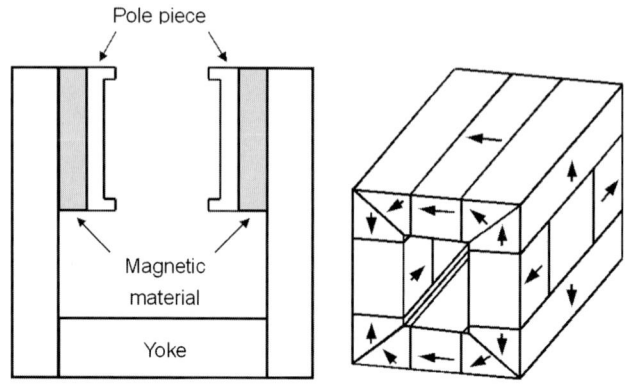

Fig. 2.2.4 Yoked permanent magnetic circuit (left). Yokeless permanent magnetic circuit (right).

cube and the magnet configuration was constructed using blocks of permanent magnets with flat faces as shown in Figure 2.2.4. When Nd-Fe-B compound materials are used, the magnetic field between the magnet gap can be ca. 1–4 T because the residual magnetic flux density of such materials is around 1 T.

Figure 2.2.5 shows examples of the permanent magnets: the left figure shows a 0.2-T, 25-cm gap yoked permanent magnet and the right figure shows a 1.0-T, 6-cm gap yokeless permanent magnet. Although a 1.0-T, 6-cm gap permanent magnet can be made using the yoked magnet design (6), the magnet's weight would be about 1400 kg. However, when the yokeless magnet design is used, the weight of the 1.0-T, 6-cm gap permanent magnet reduces to about 200 kg, as we can imagine by comparing the photographs in Figure 2.2.5.

The magnetic field homogeneity is the next important requirement for MRI magnets. In many MRI applications, the magnetic field homogeneity of around

Fig. 2.2.5 Yoked magnet, 0.2-T, 25-cm gap (left). Yokeless magnet, 1.0-T, 6- cm gap (right).

100 Hz (in resonant frequency) is sufficient, because the T_2 of protons of most liquids is longer than several ms. When the magnetic field strength is 0.2 T or the Larmor frequency of protons is about 8 MHz, the magnetic field homogeneity should be better than 10 ppm. When the magnetic field strength is 1.0 T or the Larmor frequency of protons is about 43 MHz, the magnetic field homogeneity should be better than 2 ppm.

To obtain a homogeneous magnetic field for MRI, magnetic field shimming is necessary. The first step in this process is "passive shimming", performed by attaching iron or permanent magnetic material chips onto the surface of the pole pieces. This shimming is usually performed in the manufacturing process of the permanent magnets. If the homogeneity achieved after this first step is not sufficient for a specific MRI application, a second step is required. The second step in the shimming process is usually performed by adjusting the currents in gradient coils and higher-order shim coils.

The stability of the magnetic field is another critical requirement for MRI permanent magnets. Because the temperature coefficient of the residual magnetic flux of the Nd-Fe-B material should be about -1100 ppm $°C^{-1}$, the magnetic fields of permanent magnets made with this material exhibit a similar temperature dependence. The temperature drift of the magnetic field of permanent magnets is serious, especially for high magnetic field magnets such as 1.0 T, because the temperature induced drift of the Larmor frequency increases proportionally with the magnetic field strength.

Figure 2.2.6 shows the Larmor frequency and temperature of a 1.0 T permanent magnet measured over approximately half a day while a 3D spin-echo imaging sequence was applied. The temperature was measured at three locations; on each of the two sides of the magnet, using Pt resistance thermo-sensors, and in the middle of the magnet gap space, using a thermocouple. In Figure 2.2.6, the

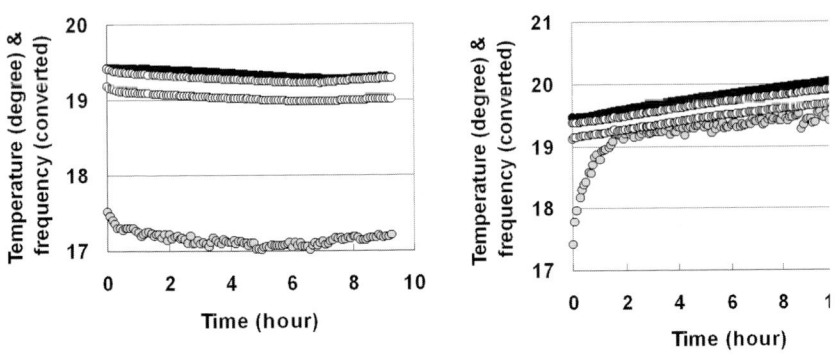

- Frequency
- magnet gap
- Upper part
- Lower part

Fig. 2.2.6 Temperature drift of a 1.0-T yokeless permanent magnet.

left graph shows the measurements when the electric currents in the field gradient coils were off and the right graph shows the measurements when the currents in the gradient coils were on.

The Larmor frequency $f(T)$ MHz was plotted as a temperature (T) by assuming the following relationship: $f(T) = 44.064[1 - 0.0011(T - 25.0)]$. This equation is based on the fact that the Larmor frequency is 44.064 MHz at 25.0 °C and the assumption that the temperature coefficient of the magnetic flux density of the permanent magnet is -1100 ppm °C^{-1}. The right-hand graph in Figure 2.2.6 clearly shows that the electric current of the magnetic field gradients considerably affect the magnet temperature and Larmor frequency (magnetic field strength). The Larmor frequency change of ≈ 3.4 kHz h^{-1} due to this heating can be corrected using a time-sharing internal NMR lock technique with a computer controlled direct digital synthesizer board (6). By using the NMR lock technique, MR microscopy images were successfully acquired at a high magnetic field using permanent magnets.

2.2.4
Gradient Coil

The compact MRI uses permanent magnets with planar pole pieces. The gradient coils, therefore, have planar structure. For construction of planar gradient coils, Anderson's classical paper is very instructive (10). However, to obtain a large homogeneous region for linear magnetic field gradients between the magnet pole pieces, some optimization or inversion technique is required. Among several design techniques for gradient coils, the target field approach is the most straightforward and widely used (11–13).

Figure 2.2.7 shows planar gradient coils designed and manufactured using the target field method. These gradient coils were designed for a 0.2 T permanent magnet with a 16-cm gap, 40-cm diameter pole pieces and 12-cm diameter spherical homogeneous region. Although this gradient coil set produced better gradient fields than those produced by a classical design gradient coil set, the effects of magnetic materials used for the magnetic circuit were not considered. A design method that includes static and dynamic properties of the magnetic materials of the permanent magnet is challenging and will be developed in the future.

2.2.5
Rf coil

The compact MRI system has an advantage for rf coil design because solenoid coils can be used in most applications. The solenoid coil has about three times better SNR than that of the saddle-shaped coil (14). Even if the saddle-shaped or birdcage coil is used in the quadrature mode, the solenoid coil will still have better SNR because an SNR gain of only about 1.4 times is obtained in that mode.

Figure 2.2.8 shows a typical tuning and matching circuit for an rf coil. If the self-inductance of the rf coil is too large for the tuning and matching, the rf coil wire

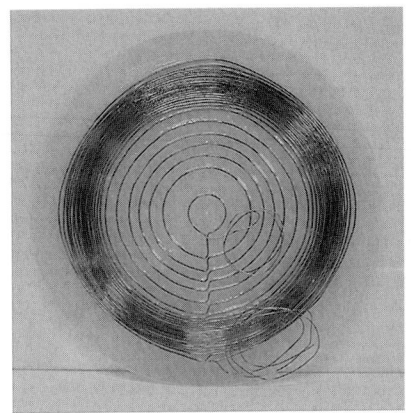

Fig. 2.2.7 Planar gradient coils designed and manufactured using the target field method and genetic algorithm [18]. Left: G_x or G_y coil. Middle: G_z coil. Right: three axis gradient coil set made by accumulating three flat gradient coils.

should be separated into several segments using capacitors that are serially connected. With this technique, the solenoid coil and tank circuit can be used for a wide frequency range and various sample dimensions.

2.2.6
MRI Console

Figure 2.2.9 shows a typical portable MRI console. The MRI console consists of a digital system (PC), MRI transceiver, 3 channel gradient driver and rf transmitter.

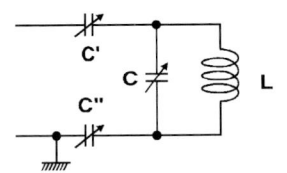

Fig. 2.2.8 Tuning and matching circuit for an rf coil.

All of the units are installed in a single 19-in portable rack and the total weight is about 80 kg.

2.2.6.1 Digital System (PC)

All of the digital modules are installed inside the industrial PC as shown in Figure 2.2.10. The major functions of the MRI digital system are implemented by three commercially available ISA boards.

The pulse programmer is implemented using a DSP (digital signal processor) board (DSP6031, MTT Corp., Kobe, Japan) with a TMS320C31 DSP chip running at the 40 MHz clock frequency (15). The pulse programmer has 100-ns time resolution using an internal-timer interrupt sequence. The DSP board has four channels with 16-bit DA converters: three of them are used for the gradient waveforms and one for the rf pulse shape (optional). It has 8-bit digital outputs, which are used for trigger signals for the rf pulses and data acquisition.

The Larmor frequency rf reference signal is generated using a DDS (direct digital synthesizer) board (FSW01, DST Inc., Asaka, Japan). This board can generate coherent and spectrally pure rf signals from 5 to 200 MHz, of which the frequency is controlled via the ISA bus. This board has an essential role in the NMR lock process.

The data acquisition is performed using an ADC board (PC-414G3, DATEL, USA). Because the ADC board has 14-bit resolution, simultaneous two-channel acquisition capability up to 1 MHz sampling rate and 32 Kword FIFO buffer memory, it has sufficient performance for most MRI applications.

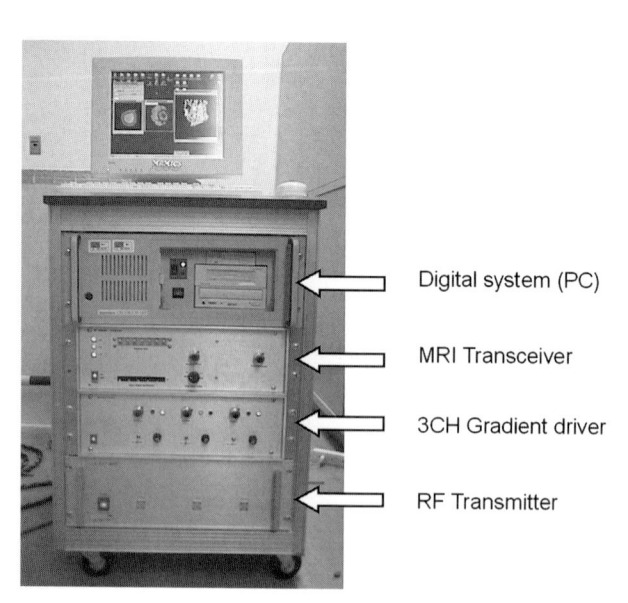

Fig. 2.2.9 Portable MRI console.

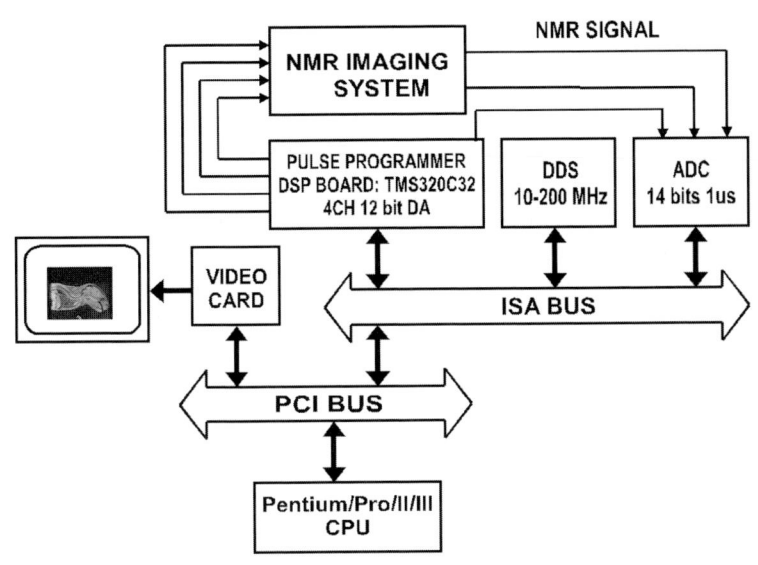

Fig. 2.2.10 Block diagram of the digital system of the compact MRI constructed within a PC.

The PC system runs under a Windows95/98/2000 operating system and is capable of real-time image acquisition, reconstruction and display (16, 17).

2.2.6.2 MRI Transceiver

Figure 2.2.11 shows a typical block diagram of the MRI transceiver used for the compact MRI system. The waveform generator can be replaced by the DA converter on the DSP board described in the previous section. Because the typical input/output power level for the transmitter and from the preamplifier is about 1 mW, a commercially available transmitter and preamplifier are directly connected.

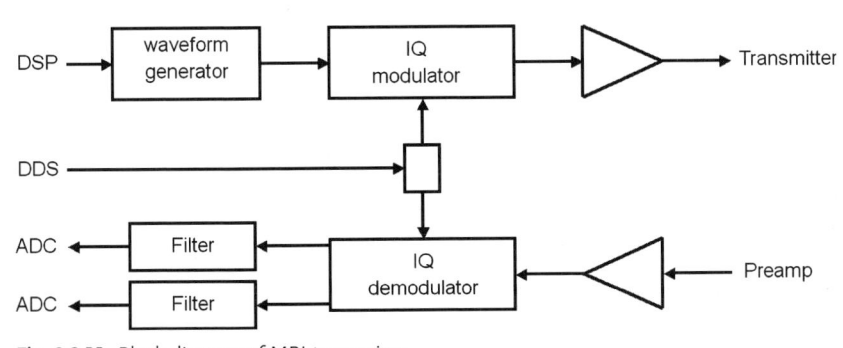

Fig. 2.2.11 Block diagram of MRI transceiver.

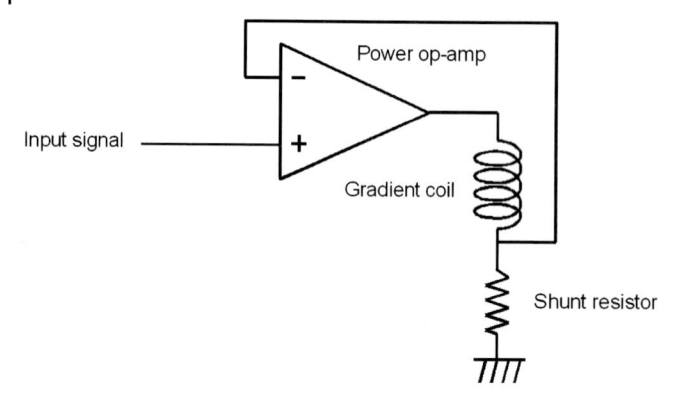

Fig. 2.2.12 Gradient driver using a power operational amplifier.

2.2.6.3 Gradient Driver

Figure 2.2.12 shows the output circuit of the gradient driver. The power operational amplifier (op-amp) drives the gradient coil in the constant current mode, because it operates so that the input signal voltage and the voltage at the shunt resistor become equal. However, if the operation voltage of the op-amp is not sufficiently high during the gradient current switching, the op-amp drives the gradient coil in the constant voltage mode. The gradient driver, therefore, should be designed or chosen by considering the electrical property of the gradient coils and requirements for the gradient field switching.

2.2.6.4 Rf Transmitter

The rf transmitter amplifies an rf pulse signal of about 1 mW up to several W or up to several kW. The amplifier should work in a linear mode (class AB) because excitation pulse shape for slice selection must be reproduced. Class AB rf transmitters such as these with blanking gates are widely available commercially.

2.2.7
Typical Examples of Compact MRI Systems

Up until now, several tens of compact MRI systems have been constructed and used for various purposes. Among them, two typical systems that use a yoked magnet and a yokeless magnet are described below.

2.2.7.1 MRI for On-line Salmon Selection

Salmon roe is a favorite food for Japanese people, often served with sushi. Salmon are captured in the ocean (about 1000 salmon in one fishing expedition) and separated into female and male in the fishing ports, because female salmon sell at a much higher price than male salmon. However, as selection by external

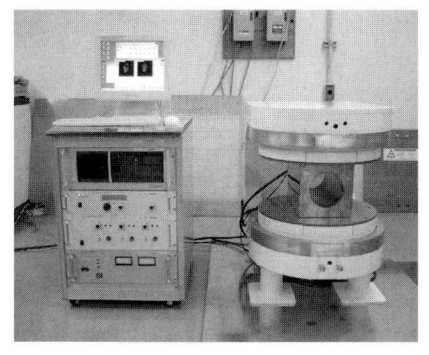

Fig. 2.2.13 Compact MRI for on-line salmon selection.

appearance is a very time consuming task (at least 30 s for one salmon even by an expert selector), development of a high-speed and automatic salmon selection system was desirable. MRI is a promising tool for salmon selection because the T_2 of salmon roe is expected to be short compared with the surrounding tissues.

Figure 2.2.13 shows an overview of the MRI system developed for salmon selection. A 0.2-T C-shaped yoked permanent magnet with a 25-cm gap [50-ppm homogeneity for 15-cm DSV (diameter spherical volume), weight 1.4 tons] is used for the magnet. For the rf coils, two solenoid coils with a 14-cm circular aperture and 14 cm × 18 cm oval aperture were developed.

Figure 2.2.14 shows cross-sections of salmon acquired with a gradient echo sequence ($TR = 30$ ms, $TE = 15$ ms, slice thickness = 4 cm, image matrix = 128 × 128, total acquisition time = 3.84 s). For the female salmon, roe were clearly visualized as dark regions because the T_2 is shorter than that of the surrounding tissues. For the male salmon, milt was observed instead of roe. However, the image intensity of the milt was similar to that of the surrounding tissues.

The required speed for salmon selection is one salmon per second. To achieve this speed, projection data of a specific slice of a salmon should be used and the imaging capability is thus indispensable in developing a selection strategy for the projection data.

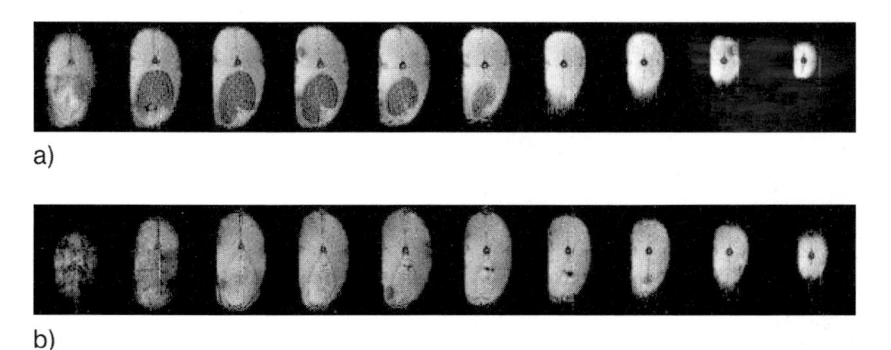

Fig. 2.2.14 Cross-sectional images of (a) female and (b) male salmon displayed from head (left) to tail (right).

2.2.7.2 Desktop MR Microscope

If a small magnet for MR microscopy of small samples (<1 cm^3) could be developed, a handy or desktop MR microscope could be constructed. The yokeless permanent magnet is best suited for this purpose.

Figure 2.2.15 shows a desktop MR microscope using a small yokeless permanent magnet. The magnet has the following specifications: magnetic field strength 0.98 T; dimensions 28 cm (width) × 24 cm (height) × 18 cm (depth); gap width 41 mm; homogeneity 14 ppm over 13 mm DSV; weight ≈ 85 kg.

The bottom image in Figure 2.2.15 is a maximum intensity projection of a 100 μm cubic-resolution 3D-image dataset of a blue berry acquired with this system. The network structure of the fruit is clearly visualized. If the MRI console and permanent magnet can be made smaller, a true desktop or handy MR microscope will then be constructed.

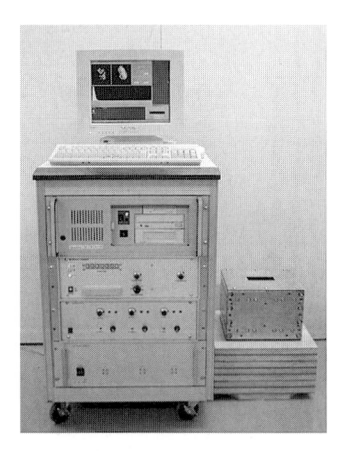

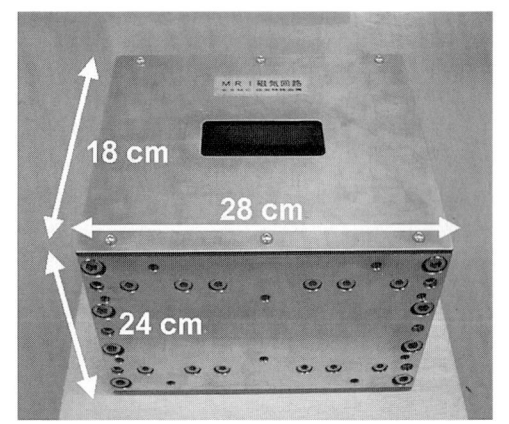

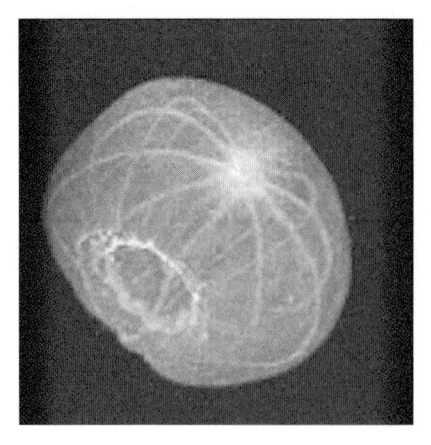

Fig. 2.2.15 Desktop MR microscope: system overview (top left); yokeless permanent magnet (top right); and maximum intensity projection image of a blue berry (bottom).

References

1 K. Kose **1999**, *Portable MRI Systems*, presented at the 5th International Conference on Magnetic Resonance Microscopy, Heidelberg, Germany.

2 K. Kose, Y. Matsuda, T. Kurimoto, S. Hashimoto, Y. Yamazaki, T. Haishi, S. Utsuzawa, H. Yoshioka, S. Okada, M. Aoki, T. Tsuzaki **2004**, (Development of a compact MRI system for trabecular bone volume fraction measurements), *Magn. Reson. Med.* 52, 440–444.

3 K. Kose (ed.) **1984**, *Compact MRI*, Kyoritsu Publishing Company, Tokyo.

4 M. Sagawa, S. Fujimura, N. Togawa, H. Yamamoto, Y. Matsuura **1984**, (New material for permanent magnets on a base of Nd and Fe), *J. Appl. Phys.* 55, 2083–2087.

5 T. Miyamoto, H. Sakurai, H. Takabayashi, M. Aoki **1989**, (A development of a permanent magnet assembly for MRI devices using Nd-Fe-B material), *IEEE Trans. Magn.* 25, 3907–3909.

6 T. Haishi, T. Uematsu, Y. Matsuda, K. Kose **2001**, (Development of a 1.0 T MR microscope using a Nd-Fe-B permanent magnet), *Magn. Reson. Imag.* 19, 875–880.

7 K. Halbach **1980**, (Design of permanent multipole magnets with oriented rare earth cobalt material), *Nucl. Instrum. Meth.* 169, 1–10.

8 H. Zijlstra **1985**, (Permanent magnet for NMR tomography), *Philips J. Res.* 40, 259–288.

9 H.A. Leupold, E. Potenziani II, M. G. Abele **1988**, (Applications of yokeless flux confinement), *J. Appl. Phys.* 64, 994–5990.

10 W. A. Anderson **1961**, (Electrical current shims for correcting magnetic fields), *Rev. Sci. Instrum.* 32, 241–250.

11 R. A. Turner **1986**, (A target field approach to optimal coil design), *J. Phys. D: Appl. Phys.* 19, 147–151.

12 R. Turner **1993**, (Gradient coil design: a review of methods), *Magn. Reson. Imag.* 11, 903–920.

13 J.-M. Jin **1998**, *Electromagnetic Analysis and Design in Magnetic Resonance Imaging*, CRC Press, Boca Raton.

14 D. I. Hoult, R. E. Richards **1976**, (The signal to noise ratio of the nuclear magnetic resonance), *J. Magn. Reson.* 24, 71–85.

15 K. Kose, T. Haishi **1998**, (Development of a flexible pulse programmer for MRI using a commercial digital signal processor board), in *Spatially Resolved Magnetic Resonance*, eds. P. Blumler, B. Blumich, R. Botto, E. Fukushima, Wiley-VCH.

16 K. Kose, T. Haishi, A. Caprihan, E. Fukushima **1996**, (Real-time NMR image reconstruction systems using high-speed personal computers), *J. Magn. Reson.* 124, 35–41.

17 T. Haishi, K. Kose **1998**, (Real-time image reconstruction and display system for MRI using a high-speed personal computer), *J. Magn. Reson.* 134, 138–141.

18 S. Handa, F. Okada, K. Kose **2005**, (Effects of Magnetic Circuits on Magnetic Field Gradients Produced by Planar Gradient Coils, Proc 13th ISMRM, 851).

2.3
Drying of Coatings and Other Applications with GARField
P. J. Doughty and P. J. McDonald

2.3.1
Introduction

Conventional magnetic resonance imaging, using spectrometers with switched magnetic field gradients, is of great utility in elucidating many processes in chemical engineering and related materials sciences. Inevitably, however, there are situations that exceed the limitations of conventional apparatus, whether for reasons of linewidth, sample geometry or spatial resolution. The drying and curing of layers of liquid dispersions and solutions used as coating materials, such as paints and varnishes, or alternatively as bonding materials, such as glues and adhesives, is one such example. This forms the primary focus of the applications discussed in this chapter. In typical use, paint or glue layers are at most a few hundred microns thick. They begin life as liquids, but rapidly become solid. Issues of interest include the through-depth uniformity of drying and the process of film formation from coalesced particles or concentrated solution. In principle, MRI can follow these processes. However, high spatial and sometimes temporal resolution is required over a wide range of T_2 times. This requirement for high spatial resolution (few microns) across a thin planar sample, encompassing a range of T_2 values, points towards the use of a strong permanent gradient profiling methodology. GARField, which stands for Gradient At Right-angles to Field, is a bench-top permanent-magnet system designed for such a niche application. GARField magnets form the instrumentation focus of the chapter. Although GARField was developed for the through depth characterization of paint layers, it is also finding other niche applications, including the characterization of skin-care products, such as cosmetics. This too will be explored in this chapter.

The relatively low entry level instrumentation cost and the relatively simple experimental methods associated with GARField – both comparable to a standard bench-top relaxation analysis spectrometer as commonly used by the food industry, for example, for water/fat ratio determinations – offer potential advantages to the industrial based user. Indeed, the overwhelming majority of the applications development work described here has been carried out in collaboration with major multi-national industrial corporations such as ICI Paints, Unilever and Uniqema, with industry sponsored research laboratories and associations such as Traetek, and with a range of small–medium sized enterprises.

2.3.2
GARField Magnets

2.3.2.1 Historical Origin

GARField magnets owe their development to Stray Field Imaging (STRAFI) [1–3]. Both replace the conventional pulsed field gradients of an MR imaging system with a static gradient. STRAFI exploits the large static field gradient necessarily surrounding a high-field super conducting NMR spectroscopy magnet. The primary *raison d'etre* of STRAFI is profiling (imaging) centimeter sized samples with very broad resonance linewidths, i.e., short T_2 relaxation times, usually solids. It is also a solution to profiling through paint layers, and as such facilitated early feasibility studies and development work. However, for such applications there are significant problems and downsides. Firstly, the field lines around a high field magnet are necessarily curved. Consequently, the excitation plane represented by a surface of constant field magnitude is shaped like a saucer. This presents an inherent limit to the spatial resolution through a planar sample. Secondly, resolution is also limited by sample curvature resulting from, e.g., meniscus and edge drying effects. A small, radiofrequency, surface sensor coil sensitive only to a central region of the layer is preferable, but this cannot be made efficiently for the inherent geometry of STRAFI with B_0 parallel to G. Lastly, using a magnet designed and manufactured to generate a highly homogeneous magnetic field for use in NMR spectroscopy applications to then produce a large field gradient for profiling through paint layers instead, is a waste of a good resource. It is not cost effective and, in any event, it is far from an optimum technological solution.

2.3.2.2 GARField Magnet Geometry

The attraction of a magnet with an in-built static gradient to create a simple MR profiling system coupled with consideration of the limitations of STRAFI, just discussed, led to the design of the GARField magnet [4]. The magnet is designed specifically to produce the optimum magnetic field to profile through thin planar sample layers with T_2 within the 50 μs to 1 ms range, such as paints, and as such is an example of a growing class of application specific magnet designs. Others include magnets for down-borehole logging, or quality control on or in a production line, as well as the NMR MOUSE; these are variously discussed elsewhere in this volume. Such magnets need not be unduly expensive. GARField magnets, for instance, can be of comparable cost to those supplied with conventional bench-top instrumentation.

The GARField magnet consists of two shaped pole pieces providing a static magnetic field, B_0, as illustrated in Figure 2.3.1. The field is oriented approximately in a horizontal direction, z. By design, the magnitude of the magnetic field, $|B_0|$, is constant in the horizontal plane, xz. This is more important for good resolution than ensuring high field homogeneity – the field curves slightly. There is a strong magnetic field gradient, G_y, in the vertical direction, y. The shape of the pole pieces is determined analytically, using a scalar potential method to solve the Laplace equation, $\nabla^2 \phi = 0$. The exact shape is given by a contour of ϕ, as described else-

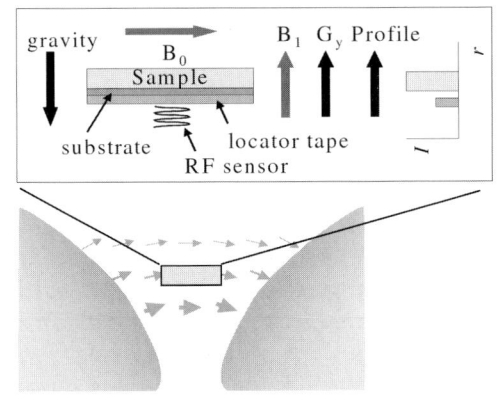

Fig. 2.3.1 A schematic diagram of GARField magnet pole pieces and the field pattern they produce together with a magnified sketch of the sample and sensor mounting showing the relative field, gradient and profile [$I(r)$] orientations.

where [4]. The resultant magnet is characterized by the parameter $G_y/|B_0|$, which is a constant across the inter-pole piece volume for any given implementation.

The field orientation allows the excitation field, B_1, of the sensor also to be oriented in the vertical direction parallel to the gradient. This differs from the situation in STRAFI. The advantage of the GARField layout is that a B_1 excitation/sensor coil can be made from a small surface winding below the sample, able to excite/sense a well defined central region of the sample away from edge effects.

It is the authors' experience that the improved geometry of the magnet and sensor coil fully compensates for the much lower resonance frequency of a GARField system compared with a standard STRAFI set-up, and as such means that the sensitivity of the two is comparable. Equally, while STRAFI can offer larger gradient strengths, the improved geometry of the GARField magnet and the fact that only a small region of sample is examined (lever effect, meniscus, field profile) all conspire to indicate that the best resolution achievable in practice with GARField is 2 or 3 times as good as with STRAFI.

2.3.2.3 GARField Implementation

The GARField magnet was first implemented using a combination of NdFeB magnet blocks and shaped steel pole pieces to produce a magnet characterized by $G_y/|B_0| = 25$ m^{-1}, Figure 2.3.2. The minimum pole piece separation is 10 mm [4]. In use, the NMR probe sensor coil is mounted on a cradle between the poles at a position where the gradient is 17.5 T m^{-1} and the pole piece separation is ≈20 mm. The corresponding field strength is 0.7 T, giving a ^{1}H resonance frequency of around 30 MHz. The sample is inserted from above onto the cradle, above the NMR coil. The height and leveling of the cradle is adjustable to allow the sample to be positioned accurately within the magnetic field. Optimization is achieved by adjusting the cradle so as to sharpen the edge of a profile recorded from the base of a flat sample, such as a drop of oil on a glass cover slip. With care, the probe and sample can be leveled to within a few microns over the excited region (approximately 3 mm in diameter).

Spin-echo Fourier transform imaging methods are used with a quadrature echo sequence: $\alpha_x-\tau-(\alpha_y-\tau-\text{echo}-\tau)_n$. Here $\alpha_{x/y}$ is a radiofrequency pulse of nominal flip angle 90° and relative phase x or y at the center of the profile, τ is the pulse gap and n is the number of echoes recorded. Each NMR echo train signal is averaged over a number of scans, each separated by a repetition delay time, τ_{RD} (normally $\geq 3T_1$ to allow the initial magnetization to be restored). Each averaged echo signal is Fourier transformed to produce a profile. Owing to the increasing distance from the coil, the pulse flip angle, α, associated with the pulses varies across the sample. To compensate for this, the experimental profile intensity is generally normalized to that recorded from an elastomer standard of known uniform composition. As in conventional T_2 weighted MRI, image contrast is provided by a combination of ^{1}H concentration and nuclear spin relaxation signal attenuation. The latter is sensitive to molecular mobility, a fact that must be taken into account when analyzing GARField profiles. If desired, profiles resulting from different echoes of the

Fig. 2.3.2 A wire frame drawing and photograph of a Mark I GARField magnet as manufactured by Resonance Instruments Ltd. The third picture is a close up of the sample space.

sequence can be co-added to improve the signal-to-noise ratio. If this is not done, then they can be analyzed separately on a point-by-point basis to yield spin–spin, T_2, relaxation maps from across the sample. Unusually for MRI, very mobile liquid samples characterized by a high self-diffusion coefficient can exhibit shorter apparent T_2 values and hence less signal than less mobile samples: this is because the data are recorded in a strong field gradient [5]. The precise parameters used during a given experiment are varied to suit the circumstances, with consequent implications for performance. Pixel resolution of around 6–12 µm is routinely possible, with temporal resolution of from one to a few minutes. The spatial resolution is usually determined by the sample condition, whether it is truly planar or not.

2.3.3
Applications

2.3.3.1 Coatings and Glues

Chemical coatings cast from dispersions or solutions are used very widely in the modern world for both functional and decorative purposes. The need for a better understanding of the processes by which they are created is driven by the desire for continuously improved material performance and properties and increasingly by legislation restricting the release of volatile organic compounds (VOCs) into the atmosphere. There are many approaches to the restriction or elimination of the release of VOCs during film formation worthy of investigation, some of which have been studied using GARField. The products used in this area consist of three basic types. The first are typically dispersions of either oil drops (emulsions) or polymer particles (latexes) suspended in a solvent. During film formation, the solvent gradually evaporates, promoting particle compaction and coalescence, leading to a homogeneous film. Subsequent cross-linking can harden the layer. To produce a uniform film, these processes, illustrated in Figure 2.3.3, must occur in an orderly fashion. The second are polymer solutions that increase in concentration by solvent evaporation, eventually drying and film forming. The third are multi-component systems that flow and then cross-link on mixing. For the first two cases in particular, GARField MR profiling experiments have allowed the development and testing of models of the film formation process at a level of detail that has not previously been possible.

2.3.3.2 Latex Dispersions

Multi-component cross-linking latex dispersions are a potential replacement for many film-forming compounds. These overcome many of the obstacles to obtaining waterborne coatings at room temperature without the release of VOCs, and form hard, mechanically strong and chemically resistant films of great utility. For convenience and efficiency, photoinitiated free-radical polymerization is appealing. A study of such a photoinitiated cross-linking latex system provides a particularly powerful demonstration of the utility of GARField [6]. The particular materials

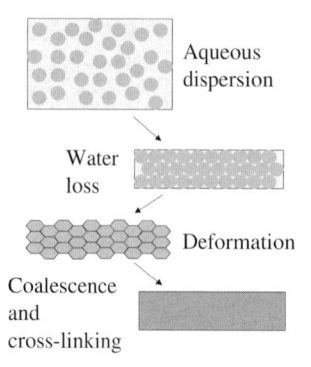

Fig. 2.3.3 The stages of film formation from an aqueous dispersion.

studied consist of a latex dispersion with a dissolved cross-linker and photoinitiator. The main contributory processes to film formation were confirmed to be water evaporation and photoinitiated free-radical polymerization. However, the relative rates of interdiffusion of each of the components across the layer, oxygen inhibition of free-radical polymerization reactions and the ingress of atmospheric oxygen into the surface as well as formulation turbidity were all shown to play an important role in determining the detailed behavior of the system. It proved possible to investigate each of these processes individually using GARField by systematically varying the formulation and/or experimental conditions. Each was studied in an essentially similar fashion.

A layer of the sample was cast onto a glass cover-slip, which, in turn, was placed on the GARField sample cradle and exposed to light and air [Figure 2.3.4(a)]. Profiles were recorded at regular intervals throughout the drying and film formation, allowing the changes in the sample to be monitored continuously. The

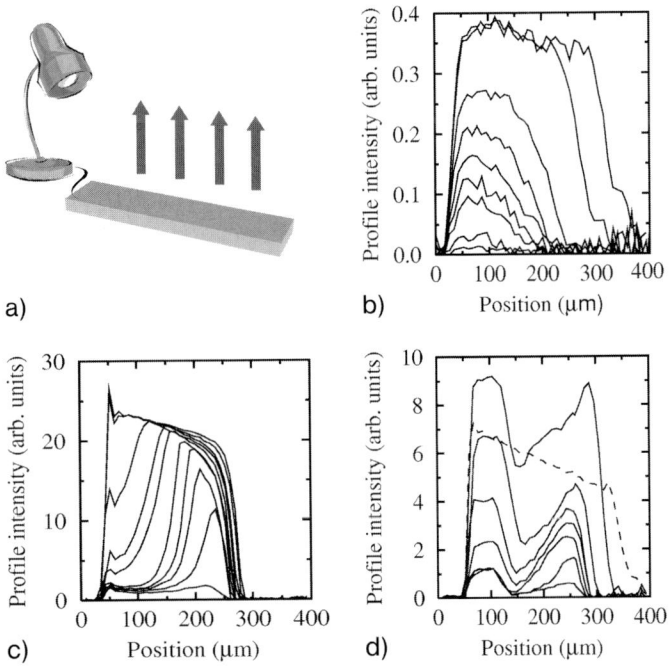

Fig. 2.3.4 Film formation of a photoinitiated cross-linking latex coating as measured by GARField. (a) The coating is exposed to air (evaporation) and light from above. (b) A sample comprising a combination of only polymer and water dries from the upper surface (right) as shown by a time series of profiles, recorded at 10, 20, 30, 40, 50, 60, 70, 100 and 120 min after casting the layer. (c) A combination of polymer and photoinitiator only cures from the lower surface (left) after a 90 min induction period due to oxygen absorption. The profiles shown were recorded 10, 90, 100 and 110 min and 2, 3, 4, 5, 6 and 17 h after casting the layer. (d) The full formulation film forms in the central layers first. In this final time series, the profiles shown were recorded after 10 min (dotted trace, T_1 attenuated) and then, from the top down, 30, 60 and 90 min and 2, 3, 6 and 17 h after casting the layer.

various mixtures could be profiled with the same MR parameters, allowing direct comparison of profiles for different samples.

Figure 2.3.4(b) shows ^{1}H profiles of the drying of the pure latex. The pulse gap, τ, is 95 μs. The profile resolution is 9 μm. The signal intensity primarily reflects the mobile water content. The initial profile of the layer is uniform and shows the thickness of the layer to be approximately 400 microns. The glass substrate is to the left, and the air is to the right. As water evaporates away from the top surface, the upper surface recedes, providing a check on the evaporation rate. Also, the profiles develop a negative gradient across the layer, reflecting water loss and particle compaction at the surface.

Figure 2.3.4(c), shows a similar experiment involving the photoinitiator and cross linker only, exposed to light from above (right-hand side). The initial profile is essentially uniform, and does not change appreciably for the first 90 min. This is due to polymerization inhibition by oxygen dissolved in the mixture. After this time, the intensity begins to drop rapidly at the base of the layer, indicating that polymerization occurs here first. The lower signal intensity is due to the reduced mobility (shorter T_2) of the cross-linked polymer. The reason for the slower progression near the surface is the inhibition of polymerization by additional atmospheric oxygen, which continues to ingress into the layer throughout the experiment.

A time series of profiles for the complete formulation of the three components (latex, cross linker and photoinitiator) exposed to light and allowed to evaporate is shown in Figure 2.3.4(d). The series clearly illustrates both the complexity of the processes occurring and the ability of GARField profiling to visualize them. The curious behavior of losing the signal first in the center of the layer, i.e., polymerization initially occurs in the centre of the layer, is explained by a combination of the two major factors just described: evaporation and (oxygen inhibited) photoinitiated cross linking, and the fact that, unlike the photoinitiator and cross linker only mixture, the complete formulation is turbid. Light cannot penetrate so easily to the base of the layer.

Detailed quantitative analyses of the data allowed the production of a mathematical model, which was able to reproduce all of the characteristics seen in the experiments carried out. Comparing model profiles with the data enabled the diffusion coefficients of the various components and reaction rates to be estimated. It was concluded that oxygen inhibition and latex turbidity present real obstacles to the formation of uniformly cross-linked waterborne coatings in this type of system. This study showed that GARField profiles are sufficiently quantitative to allow comparison with simple models of physical processes. This type of comparison between model and experiment occurs frequently in the analysis of GARField data.

2.3.3.3 Alkyd Emulsions

A separate approach to the problem of VOC release during film formation is the use of polymer films cast from aqueous emulsions. Alkyd emulsions in particular have been proposed as new environmentally friendly paints and have therefore attracted study by GARField [7]; this work has aspects that are quite distinct from that already discussed. The GARField study complements work already done on lateral or in plane drying of alkyds [8] and other colloidal layers [9] using conventional MR microscopy, not described here. The combination of these two techniques is particularly powerful.

Routh and Russel [10] proposed a dimensionless Peclet number to gauge the balance between the two dominant processes controlling the uniformity of drying of a colloidal dispersion layer: evaporation of solvent from the air interface, which serves to concentrate particles at the surface, and particle diffusion which serves to equilibrate the concentration across the depth of the layer. The Peclet number, Pe is defined for a film of initial thickness H with an evaporation rate E (units of velocity) as HE/D_0, where $D_0 = k_B T/6\pi\mu r$ – the Stokes–Einstein diffusion coefficient for the particles in the colloid. Here, r is the particle radius, μ is the viscosity of the continuous phase, T is the absolute temperature and k_B is the Boltzmann constant. When $Pe \gg 1$, evaporation dominates and particles concentrate near the surface and a skin forms, Figure 2.3.5, lower left. Conversely, when $Pe \ll 1$, diffusion dominates and a more uniform distribution of particles is expected, Figure 2.3.5, upper left.

Routh and Russel's ideas were demonstrated experimentally using GARField profiling of drying alkyd aqueous emulsion (mean droplet size 133 nm) layers of varying thickness, under different conditions of evaporation. The profiles, Figure 2.3.5, right, show that during the early stages the layers dry as anticipated. In these profiles, the signal originates predominantly, but not exclusively, from the water. It is possible to calibrate for the water content across the layer experimentally given certain assumptions, but this has not been done here [7]. With low Peclet number

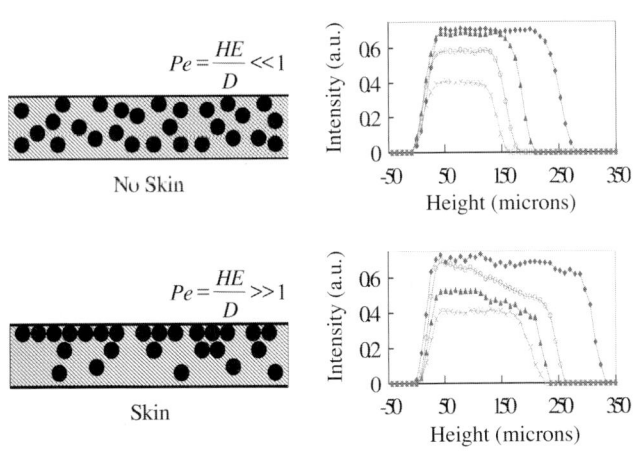

Fig. 2.3.5 Profiles recorded from a drying alkyd emulsion layer are shown on the right. At low Peclet number (upper set of profiles), drying is uniform whilst at high Peclet number (lower set), particles concentrate at the surface. Schematics of the process are shown on the left. The Peclet number is defined as HE/D where H is the film height, E the evaporation rate and D the particle diffusivity. The upper set of profiles are recorded 62, 602, 821, 956 and 1061 min after the layer was cast; the lower set 2, 7, 13 and 31 min after the layer was cast.

layers (Figure 2.3.5, upper right, where $Pe = 0.2$), signal intensity remains uniform across the layer as it dries, confirming a uniform concentration of water across the dispersion. With high Peclet number, (Figure 2.3.5, lower right, where $Pe = 16$), a negative gradient in signal intensity develops across the layer as water evaporates away from the top surface faster than it is replenished by diffusion across the layer. During this period, irrespective of the Peclet number, the layer thins uniformly with time indicating that the drying rate is evaporation limited. The surprising aspect of these profiles, not predicted by the theory, is that the layers attain uniform concentration once again before the layers are completely dry. Uniform concentration typically recurs at a water fraction of about 15%. At this point, a second, much slower stage of drying commences during which, irrespective of the initial Peclet number, a concentration gradient once again develops. This too has been observed by GARField [7], although the data are not presented here. During the second stage, the drying rate, as revealed by the sample thinning, follows the square root of time and is therefore diffusion rather than evaporation rate limited. The "final" water content is about 2%. An inordinate amount of time is required to remove this final water fraction and the films remain tacky at the end of the experiment. In further profiling experiments (again the data are not reproduced here), it was found that the "dry" layer could be easily rewetted by casting a fresh coat on top of the first. In such experiments the water fraction recovered to, but never exceeded, 15%, the value that characterizes the boundary between the first and second stages of drying.

The fact that rewetting is possible at all suggests that the alkyd particles do not fully coalesce at the end of the second stage of drying. It has been proposed that these experiments provide evidence that the alkyd emulsion adopts a bi-liquid foam structure during the second stage of drying in which the deformed alkyd droplets are separated by thin layers of water trapped in surfactant bilayers. Moreover, it is suggested that the start of the second stage characterized by a water fraction of 15% and the final state, 2%, represent two distinct metastable thermodynamic phases of the bilayer: a carbon black film and a Newton black film [11]. Numerical models of water transport across a layer of this form, during the second stage of drying, give good qualitative agreement with experiment.

2.3.3.4 Polymer Solutions

The Peclet number concept has been extended to the investigation of the process of glassy film formation from polymer solutions. Skin formation in drying polymer layers can slow solvent evaporation, trap solvent in the film and lead to surface wrinkling, thereby affecting the film formation process. The situation is exacerbated in the case of semicrystalline polymers, as crystalline regions may form during film formation as well as amorphous (glassy) regions, producing a film with regions of varying physical properties. GARField has provided one of the first viable means of assessing the through depth composition of a polymer solution layer during drying, thereby enabling some advances in the understanding of such processes. The specific system studied is poly(vinyl alcohol), PVOH [12]. Akin to the dispersions, layers of different thickness were observed to dry as a function of evaporation rate. Once again the Peclet number, $Pe = HE/D$, was introduced to gauge the relative strength of water evaporation against polymer diffusivity in determining the uniformity of polymer concentration in the drying layer. Additionally, a second dimensionless parameter, a reduced drying time τ_D, defined as the ratio of the typical time required for the polymer to dry to a glass and the time for the layer to crystallize, was introduced. This led to a two-dimensional parameter space, $Pe - \tau_D$ within the four quadrants of which different characteristics of the dried layers are expected. GARField profiling experiments similar to those carried out with the alkyd emulsion described above have confirmed this description of the behavior of the drying of PVOH layers. Figure 2.3.6 shows profiles recorded from drying layers for the three of the four quadrants that are accessible experimentally. When $\tau_D \ll 1$, the reduced drying time constant is low and a glassy film is expected. Experimentally, the profiles – which predominantly reflect the water content – show that the layers dry to little or no residual water. When $\tau_D \gg 1$, the crystallization process dominates, and a more crystalline film is expected. This traps water in the lower layers in the later stages of drying, as revealed by the intense peak which remains at long times in the lower right profile set. When $Pe \ll 1$, drying is uniform and when $Pe \gg 1$ it is not. Experimentally, the profiles in the lower part of the figure show much more uniform water concentration than those in the upper part. In the upper part, they show a strong water concentration gradient.

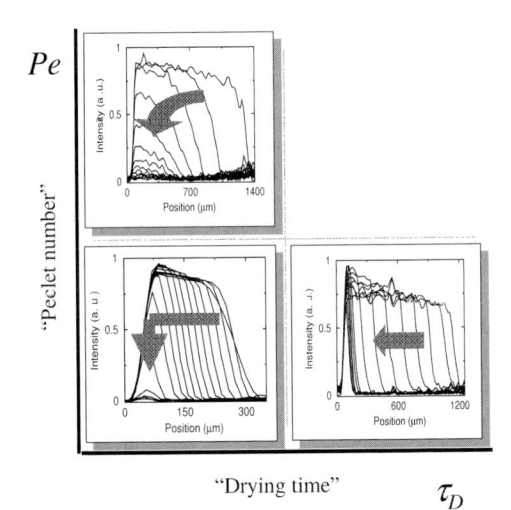

Fig. 2.3.6 The two dimensional $Pe-\tau_D$ space, which characterizes the drying of semi-crystalline polymer solutions, in this case PVOH. In the lower right quadrant ($Pe = 0.18$, $\tau_D = 5.8$), the drying is uniform but, because it is slow (the profiles shown were recorded at approximately 12 h intervals), a crystalline layer forms that traps residual water. In the lower left quadrant ($Pe = 0.24$, $\tau_D = 0.32$), the drying is uniform but no water is trapped due to faster drying (the profiles shown were recorded approximately every 21 min). In the upper left quadrant ($Pe = 1.46$, $\tau_D = 0.17$), the increased Peclet number causes the drying to be non-uniform, (the profiles shown were recorded every 77 min).

2.3.3.5 Glue Layers

GARField profiling has been applied to the curing of wood glue layers and also to their characterization in terms of moisture penetration and transport. Systems studied consist of commercial and model wood glues and resin and hardener components. These typically cure on application or mixing within tens of minutes to hours, depending on formulation, composition and temperature. As the glue and hardener molecules become less mobile as they cure, it is possible to use GARField profiling to monitor the curing process. Curing is revealed by a loss of signal. The time taken for the signal intensity from a given glue layer to halve provides a measurable characteristic parameter related to the cure time. Comparison of measurements made in this way with published working times for a range of glues has been used to validate the method [13]. This type of non-destructive testing offers a new means of measuring the cure time for wood glues under different conditions.

An extension of this work is the investigation of water transport across fully cured glue lines bonding pieces of wood. Figure 2.3.7 shows an example. In these experiments, two thin layers of wood, in this case pine, around 500 µm thick [13], were glued together. The top layer was then constantly exposed to water while GARField profiles were recorded across the bonded sample. The lower part of Figure 2.3.7 shows exemplar profiles. The water reservoir is to the extreme right beyond 1300 µm on the scale. The signal-to-noise ratio is low in this region as it is at the limit of the profile field of view. The upper wood layer is from 800 to 1300 µm and the lower layer from 300 to 800 µm. The glue layer is at 800 µm and a marker tape is positioned at 300 µm. During the experiment, the signal intensity was observed to grow in the upper layer of pine as water diffused into the wood. However, there was no comparable increase in the lower level indicating that the glue, in this case urea formaldehyde, acts as a water barrier. The upper part of the figure shows the time dependence of the average signal intensity in the upper and lower layers of glued samples for three different glues: urea formaldehyde, phe-

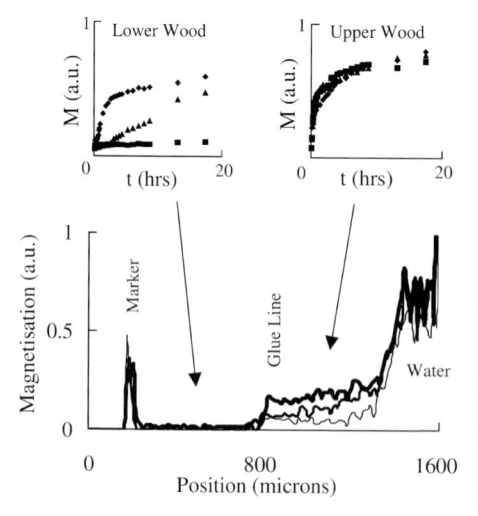

Fig. 2.3.7 Lower: GARField profiles showing a UF (urea formaldehyde) glue line acting as a barrier to water transport for up to 24 h. The glue line is at 800 µm on the scale. Wood is above and below this. The water reservoir is beyond 1300 µm. The profiles shown were recorded after 20 (thin line), 100 and 1400 (thick line) min of exposure to water. Upper: plots of the magnetization signal intensity in the lower and upper wood layers as a function of time for three glues: urea formaldehyde (squares), phenolic resorcinol formaldehyde (triangles), and poly(vinyl acetate) (diamonds).

nolic resorcinol formaldehyde and poly(vinyl acetate). The different glues studied showed various behaviors, from dissolving on contact with water [poly(vinyl acetate)], to acting as a water resistant barrier for up to 24 h (urea formaldehyde).

2.3.4
Human Skin Hydration

Human skin is the largest organ in the human body. It is fundamentally important to health as the semi-permeable barrier – the first line of defence – between the body and the external world. However, it remains relatively inaccessible to conventional magnetic resonance imaging, firstly because it is thin and therefore requires high spatial resolution, and secondly because it is characterized by relatively short T_2 relaxation times, particularly in the outermost *stratum corneum*. Conventional studies have not usually achieved a resolution better than 70–150 μm, with an echo time of the order of a millisecond or so. As a planar sample, skin has proved amenable to GARField study where it has been possible to use both a shorter echo time and achieve a better spatial resolution, albeit in one direction only. Such studies have attracted the interest of the pharmaceutical and cosmetic industries that are interested in skin hydration and the transport of creams and lotions across the skin.

Following preliminary experiments to test the feasibility of profiling skin with GARField [14], studies have been conducted using skin both *in vitro* [15] (abdominal tissues, obtained following cosmetic surgery, cleaned and stored frozen until required for use) and *in vivo* [16] (of the finger, hand and lower arm). It has proved possible not only to detect a signal from these samples, but also to distinguish regions in the skin with different characteristics. Profiling experiments on *in vitro* skin layers using two different pulse gaps, τ = 500 and 150 μs, illustrate this. Results

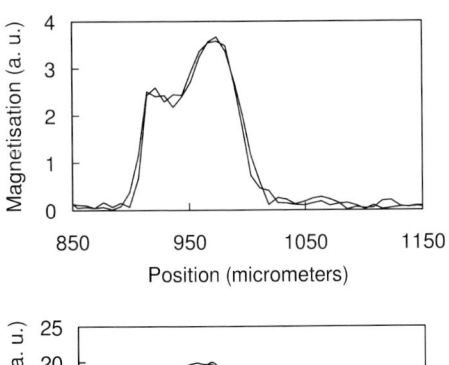

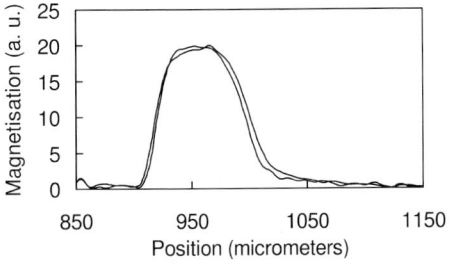

Fig. 2.3.8 Lower: GARField profiles of a human skin sample sandwiched between two glass slides, recorded immediately after the sample was floated onto the first slide and again approximately 90 min later. Upper: increasing the pulse gap τ from 150 to 500 μs increases mobility contrast and allows discrimination between the *stratum corneum* (right) and viable epidermis (left). Again two profiles are shown, recorded approximately 90 min apart.

are shown in Figure 2.3.8. The *in vitro* samples were prepared for use by slowly defrosting in a fridge overnight before use, followed by partial hydration by flotation on a bath of water for around 30 min. They were then placed on glass coverslips on the GARField probe cradle. The shorter echo time permits detection of the signal from both slow and fast moving hydrogen nuclei. The two profiles shown in the lower part of Figure 2.3.8 were recorded some 90 min apart and show good reproducibility. They reflect primarily the hydrogen density, which is strongly correlated to the water density, in the skin layer. The density is more or less uniform. The two profiles in the upper part of Figure 2.3.8 were recorded similarly, but with a longer echo time. This leads to attenuation of the more mobile signal components (due to faster diffusion in the field gradient) and the resultant profiles therefore preferentially visualize less mobile regions, and show that the skin is divided into two layers. These have been identified with the viable epidermis and the *stratum corneum*. The ability to distinguish these regions has been tested by repeated experiments, and a high degree of reproducibility has been achieved in the results.

In vitro skin work has been extended to look at the ingress of liquids into skin in order to explore the possibility of using GARField profiling to characterize the efficacy of skin-care product ingredients in crossing the skin barrier. Liquids studied in these experiments include decanol, glycerine and squalane. These liquids produce a large magnetization signal that can be observed as the liquid diffuses into the skin layer. Typically, the skin layer was prepared as before, and profiles were taken at appropriate time intervals after a layer of the liquid was applied to the top of the skin. The profiles for glycerine [see Figure 2.3.9(a)] show extremely clearly the progress of a glycerine front into a skin layer. The first profile (with the lowest magnetization) is that of the mobile moisture in the initial skin layer with no glycerine present. Once again, the viable epidermis (to the left) and the *stratum corneum* (to the right) can just be distinguished as regions of differing intensity: the whole layer is around 110-μm thick. The remaining profiles were taken at intervals after application of the glycerine layer. They show the ingress of a glycerine front into the skin from right (*stratum corneum*) to left (viable epidermis). These data demonstrate the diffusion of glycerine into the skin, and allow the diffusion coefficient to be estimated by determining the gradient of a plot of the displacement of the glycerine front against the square root of elapsed time for each profile [see Figure 2.3.9(b)]. The fact that this plot is linear indicates that a Fickian diffusion process is occurring with a diffusion coefficient of the order of $1.3 \pm 0.5 \times 10^{-9}$ cm^2 s^{-1} (based on a series of such experiments). The ability to determine this type of information is almost unique to GARField.

The fingertip skin can be profiled *in vivo* using the Mark I GARField magnet. Using a larger Mark II GARField magnet, see Section 2.3.5, Further Developments, it is possible to profile lower-arm skin of volunteers *in vivo*. Not only is this area of greater biological/biomedical interest, but the ability to gain sufficient access to mark and place the volunteer's arm with far greater precision in the magnet has greatly enhanced the reproducibility of the experiments. The lower curvature of the arm compared with the fingertip also improves the profiles obtained, reducing the occurrence and the scale of sample leveling artifacts.

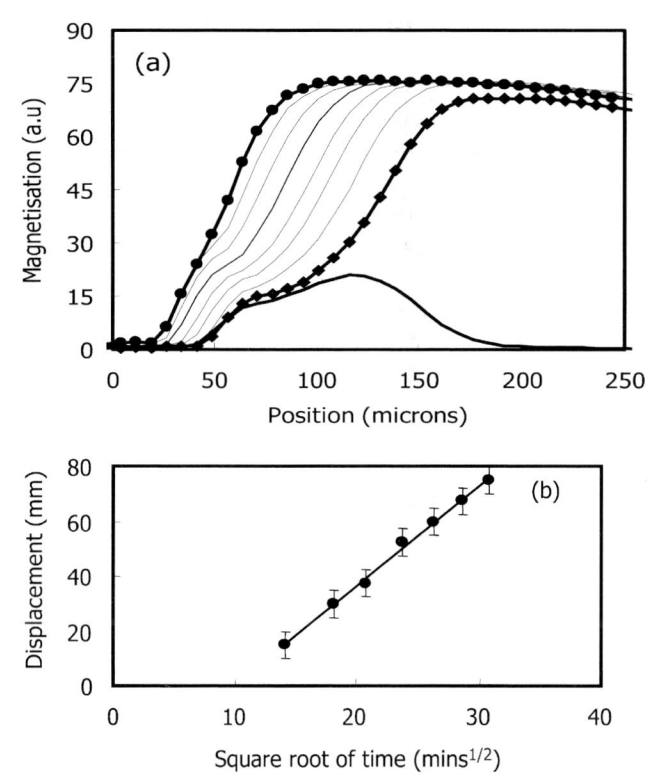

Fig. 2.3.9 A time series of profiles showing the ingress from right (*stratum corneum*) to left (viable epidermis) of glycerine into human skin *in vitro*. The skin before application of glycerine is shown by the lowest trace. The inset shows the advance of the glycerine front against the square root of time from which Fickian diffusion is inferred.

Preliminary studies, which have repeated many of the *in vitro* experiments *in vivo*, reported that the differentiation between *stratum corneum* and viable epidermis is at least as good, if not better *in vivo* and that many of the other experiments are similarly reproduced [16].

2.3.5
Further Developments

The original GARField magnet was built with a "belt-and-braces" approach to ensure that, on completion, it met its design specification as closely as possible. In particular, access to the sample volume space is from above only, due to the symmetrical design of the yoke, which surrounds the poles from the sides. In the design of a second GARField magnet, a less cautious approach has allowed many improvements. A mark-II GARField has been built with a C-frame yoke that allows access from the side as well as from above, and the whole magnet has been scaled-up in size by a factor of 3/2, so that the pole piece separation, and therefore the maximum sample (and probe mounting) size, have been increased correspondingly. This has been achieved without a reduction in field strength. The design of the pole pieces has also been changed so that they are now curved on both their upper and lower sides, so as to provide two different G_y/B_0 ratios: 16.67 and 33.33 m^{-1}. This allows measurement of samples at two different gradient strengths, but at the same field strength, allowing the separation of signal attenuation due to diffusion and normal spin–spin relaxation.

The increased size of the NMR probe mounting space has also been used to advantage. The increased space permits the inclusion of a second, switched magnetic field gradient based on a conventional current winding. The switched gradient allows in-plane spatial resolution. In a typical measurement using this approach, first demonstrated with STRAFI [17], the vertical dimension is imaged with standard GARField profiling techniques, while the orthogonal in-plane direction is imaged at much lower resolution using the switched gradient. In this manner, compensation can be made for minor sample leveling errors and planar inhomogeneities due, for example, to meniscus effects.

The GARField philosophy of designing the magnetic field to meet the requirements of experiment has also been applied to the development of another permanent magnet based MRI profiling system, designed to study the cements and concretes widely used in the construction and civil engineering industry. In this case, the intention is to be able to profile non-destructively the moisture within the surface layers of a sample too large to place between the poles of a magnet *in situ*, with a portable instrument similar to the well known NMR Mobile Universal Surface Explorer (MOUSE) [18]. The magnet design objective is a planar field of constant magnitude (but not necessarily constant direction) displaced from the magnet surface with an orthogonal gradient in field strength. The design criteria can be met by a linear array of permanent magnets with their north–south axis rotated 180° periodically one from the next. The system is further improved by the use of a sensor coil constructed of windings with the same periodic pitch, but displaced from the magnets by one quarter of the spatial period. Such a pattern ensures that the B_1 excitation magnetic field is everywhere normal to the B_0 measurement magnetic field [19]. A system has been constructed with three permanent magnets positioned horizontally next to each other and two excitation windings as illustrated in Figure 2.3.10. As constructed, the magnetic field corresponds to a ^{1}H resonance frequency of about 3 MHz at a distance of 50 mm from the surface. The magnets can be moved relative to the sample surface using stepper motors visible in the lower part of the figure, which shows the magnet as built. By raising the sensitive plane, a profile of signal against depth can be built up. In addition to the MOUSE, other approaches to achieving similar ends for comparable applications have been proposed by Callaghan and coworkers [20], Fukushima and coworkers [21] and Marble and coworkers [22].

A potential application of this system is the assessment of the curing and efficacy of hydrophobic coatings applied to cementitious and other building materials. *In situ* work has yet to be carried out, but laboratory studies suggest that this will be possible [23, 24]. The left part of Figure 2.3.11 shows the ingress of a commercial

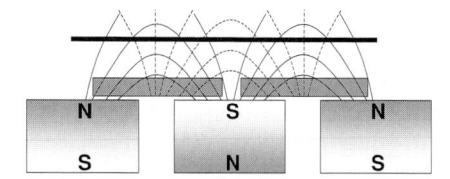

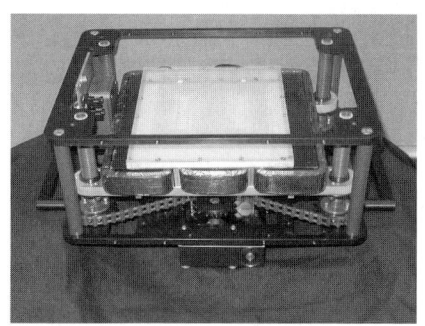

Fig. 2.3.10 A schematic (top) and photograph (bottom) of the surface GARField magnet. In the schematic the detector coil (hatched boxes) and its associated field lines are shown (dotted) and are seen to be always orthogonal to the permanent magnet field lines (solid). The magnitude of the static field is constant in the horizontal plane (thick black line). A stepper motor and gear-drives for raising the whole assembly relative to the external surface are visible in the photograph.

hydrophobic silane coating into building sandstone. The subsequent loss of signal is due partly to evaporation of the carrier solvent and partly to curing of the coating on the internal pore surfaces. Subsequent experiments show that the coating initially impedes the ingress of water, but ultimately allows water pumping through the treated layer (Figure 2.3.11, right part). Although water pumping through treated surfaces is often suspected, and can lead to serious degradation, few techniques can reveal the process in so dramatic a fashion.

2.3.6
Conclusion

GARField has found a niche application area in the characterization of drying and film forming from aqueous dispersions and in skin-care. As a bench-top perma-

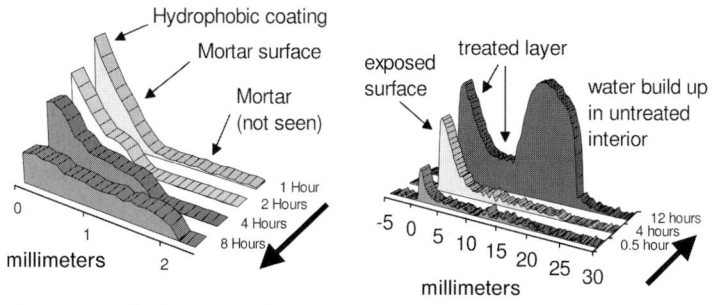

Fig. 2.3.11 Left: the ingress of a hydrophobic silane coating into building sandstone. Right: subsequent water ingress and pumping through a treated surface.

nent magnet which can be readily coupled to any one of a range of low-cost, low-frequency and low-resolution commercial spectrometers it affords an entry level route to routine coatings characterization. The range of possible applications continues to grow and advances continue to be made in the basic instrumentation. The future for GARField looks bright.

Acknowledgements

P. J. D. and P. J. M. thank The UK Engineering and Physical Sciences Research Council, The European Commission, The Royal Society, ICI Paints, Unilever Research, Uniqema and Disperse Technologies for financial support. They acknowledge the considerable input and help of their colleagues, especially Dr. Paul Glover (University of Nottingham), Dr. Joe Keddie (University of Surrey) and Dr. Peter Aptaker (Laplacian Ltd.) in carrying out much of the work reviewed here.

References

1 A. A. Samoilenko, D. Y. Artemov, L. A. Sibeldina **1988**, *JETP Lett.* 47, 417–419.
2 P. J. McDonald, B. Newling **1998**, *Rep. Progr. Phys.* 61, 1441–1493.
3 P. J. McDonald **1997**, *Prog. Nucl. Magn. Reson. Spectrosc.* 30, 69–99.
4 P. M. Glover, P. S. Aptaker, J. R. Bowler, E. Ciampi, P. J. McDonald **1999**, *J. Magn. Reson.* 139, 90–97.
5 P. T. Callaghan **1991**, *Principles of Nuclear Magnetic Resonance Microscopy*, Clarendon Press, Oxford.
6 M. Wallin, P. M. Glover, A. C. Hellgren, J. L. Keddie, P. J. McDonald **2000**, *Macromolecules* 33, 8443–8452.
7 J. P. Gorce, D. Bovey, P. J. McDonald, P. Palasz, D. Taylor, J. L. Keddie **2002**, *Eur. Phys. J. E* 8, 421–429.
8 E. Ciampi, U. Goerke, J. L. Keddie, P. J. McDonald **2000**, *Langmuir* 16, 1057–1065.
9 J. M. Salamanca, E. Ciampi, D. A. Faux, P. M. Glover, P. J. McDonald, A. F. Routh, A. Peters, R. Satguru, J. L. Keddie **2001**, *Langmuir* 17, 3202–3207.
10 A. F. Routh, W. B. Russel **1999**, *Langmuir* 15, 7762.
11 O. Sonneville-Aubrun, V. Bergeron, T. Gulik-Krzywicki, B. Jönsson, H. Wennerström, P. Lindner, B. Cabane **2000**, Surfactant films in biliquid foams, *Langmuir* 16, 1566–1579.
12 E. Ciampi, P. J. McDonald **2003**, *Macromolecules* 36, 8398–8405.
13 G. Bennett, J. P. Gorce, J. L. Keddie, P. J. McDonald, H. Berglind **2003**, *Magn. Reson. Imaging* 21, 235–241.
14 M. Dias, J. Hadgraft, P. M. Glover, P. J. McDonald **2003**, *J. Phys. D, Appl. Phys.* 36, 364–368.
15 L. Backhouse, M. Dias, J. P. Gorce, K. Hadgraft, P. J. McDonald, J. W. Wiechers **2004**, *J. Pharm. Sci.* 93, 2274–2283.
16 P. J. McDonald, A. Akhmerov, L. J. Backhouse, S. Pitts **2005**, *J. Pharm. Sci.* 94, 1850–1860.
17 J. Godward, E. Ciampi, M. Cifelli, P. J. McDonald **2002**, *J. Magn. Reson.* 155, 92–99.
18 G. Eidmann, R. Savelsberg, P. Blümler, B. Blümich **1996**, *J. Magn. Reson. A* 122, 104–109.
19 UK Patent Application 0426957.7, **2004**.
20 US Patent Application 6489872, **2002**.
21 New Zealand Patent Application 520114, **2003**.
22 A. E. Marble, C. Mastikhin, B. G. Colpitts, B. J. Balcom **2004**, *An Analytical Methodology for Magnetic Field Optimi-*

zation in Unilateral NMR, abstract presented at 4th Colloquium on Mobile NMR, RWTH, Aachen, Germany, 2004.

23 S. Black, D. M. Lane, P. J. McDonald, D. J. Hannant, M. Mulheron, G. Hunter, M. R. Jones **1995**, *J. Mater. Sci. Lett.* 14, 1175–1177.

24 A. Chowdhury, A. Gillies, P. J. McDonald, M. Mulheron **2001**, *Mag. Concrete Res.* 53, 347–352.

2.4
Depth Profiling by Single-sided NMR
F. Casanova, J. Perlo, and B. Blümich

2.4.1
Introduction

Nuclear magnetic resonance (NMR) is an established analytical tool, widely used for determining structures and conformations in chemistry, biology, medicine and material science [1, 2]. Material characterization is carried out by measuring NMR parameters such as chemical shift, nuclear spin relaxation times, dipolar couplings and self-diffusion coefficients. NMR methods are usually developed to work in the strong and homogeneous fields of superconducting magnets, but the limited working volume of these devices, where the sample must be accommodated, dramatically limits the areas of application of NMR techniques for *in situ* studies. In contrast to conventional NMR, where the sample is adapted to fit into the probe, inside-out NMR uses open magnet geometries specially adapted to the object under study. Such magnets are inexpensive and can be used as portable sensors, which offer access to a large number of applications that are inaccessible if using closed magnet geometries. Historically the inside-out concept was conceived as a procedure for examining geological formations [3], but in recent years a number of tool geometries have been designed and optimized, depending on the application, for material analysis and quality control [4, 5], moisture detection in composites [6, 7], medical diagnostics [8, 9] and the analysis of objects of cultural heritage [10, 11].

The price paid in gaining this open access is that it is impossible to attain high and homogeneous magnetic fields, a fact that reduces the number of available experimental techniques which can be implemented with these sensors. Although the presence of a static gradient is in general a disadvantage, it can be exploited to obtain depth resolution into the material. The procedure is wholly equivalent to that used by the STRAFI technique [12], where the strong stray field gradient of superconducting magnets is used to obtain profiles with high spatial resolution. However, the lateral gradients of the magnetic field generated by single-sided magnets determine the rather poor depth resolution. Several attempts have been made to increase the gradient uniformity by tailoring the magnet geometry, but spatial resolution of better than half a millimeter is hard to achieve. Previous designs demands required the magnet to generate planes of constant field over a large depth range. This field profile is convenient because it allows the selection of slices at different depths into the object, simply by electronically switching the tuning frequency. However, this strong requirement forces the optimization procedure to vary the position and orientation of a large number of permanent block magnets, and results in complicated magnet arrays. Although the retuning procedure is simple and fast, important distortions are introduced into the profile when contrasts by relaxation or self-diffusion are used to improve the discrimination of heterogeneities in the material. The static field changes by several MHz in a few millimeters, restricting the use of the T_1 contrast in samples with frequency dependent T_1. Moreover, the transverse relaxation time $T_{2\text{eff}}$ measured by a Carr–Purcell–Meiboom–Gill (CPMG) sequence in inhomogeneous fields, which is a mixture of T_1 and T_2 [13], changes with the depth as a consequence of the variation of the B_0 and B_1 field profiles. Even the contrast by self-diffusion is distorted as a consequence of the variation of the static gradient magnitude as a function of depth. These problems can be circumvented if the sample profiling is performed by changing the relative position of the sample with respect to the sensitive slice, keeping the excitation frequency constant. Besides being a distortion free procedure, it only requires a plane of constant magnetic field at a fixed distance from the sensor surface to be generated, a fact that considerably reduces constraints imposed on the design of the magnet. Taking advantage of this simplification we have recently presented a new single-sided NMR sensor based on a novel geometry that offers the possibility of obtaining microscopic depth resolution [14]. By repositioning the sample, profiles with a spatial resolution of better than 5 μm have been achieved. An important fact to point out is that the magnet is of extremely simple construction and is inexpensive to manufacture, which is an important factor when such tools are intended for quality control. In the following sections the sensor is described briefly, and a number of important applications are presented where the high spatial resolution plays a determining role in obtaining valuable information about the sample structure.

2.4.2
Microscopic Depth Resolution

2.4.2.1 Sensor Design

The magnet geometry optimized to generate a plane of a highly constant magnetic field is shown in Figure 2.4.1. It is based on the classical U-shaped geometry [5], where two block magnets with opposite polarization are placed on an iron yoke separated by a gap d_B. The main difference between the U-shape and the new design is the splitting of the two block magnets leaving a small gap d_S, which is a crucial factor in the improvement of the uniformity of the gradient. In the first approach the value of d_B defines the distance from the magnet surface where the field magnitude is constant along z. Once this has been chosen, the value of d_S is adjusted to define a constant field along x at the same depth.

The prototype described in this work uses NdFeB permanent magnet blocks of $40 \times 45 \times 50$ mm^3 along x, y and z, respectively, placed on an iron yoke 25 mm thick. The gap d_B was set to 14 mm, and d_S was determined to be 2 mm by

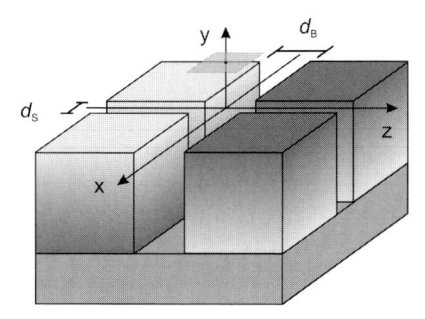

Fig. 2.4.1 Magnet geometry used to generate a highly flat sensitive volume. It consists of four permanent block magnets positioned on an iron yoke. The direction of polarization of the magnets is indicated by the gradient color, two of them are polarized along y and two along −y. Magnets with opposite polarization are positioned to leave a gap d_B of 14 mm and magnets with the same polarization are separated by a small gap d_S of 2 mm. The drawing also shows the position of the plane where the field is constant, which for these dimensions is at 10 mm from the magnet surface.

accurately scanning B_0 along x via the NMR resonance frequency of a thin oil film. For these dimensions the plane of constant field is generated at 10 mm from the magnet surface (Figure 2.4.1). At this position the magnetic field is 0.411 T (along z) and the gradient is 20 T m^{-1} along the depth direction. It must be pointed out that the spot where the field can be considered constant has a limit of 1×1 cm^2, after which the lateral gradients increase dramatically. The objects to be scanned are in general larger than this spot, so that the lateral selection of the sensitive volume poses a major problem in the design of a sensor intended for high depth resolution. This selection is achieved by appropriate choice of the dimensions of the rf coil, which, for the present sensor, is a two-turn rectangular coil 14 mm along x and 16 mm along z wound from 1 mm diameter copper wire. Besides the lateral

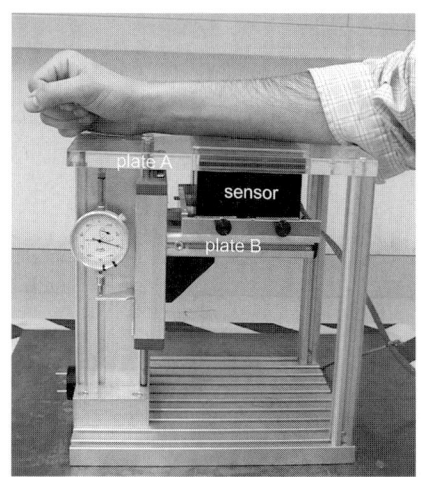

Fig. 2.4.2 Photograph of the lift used to reposition the sensitive slice across the sample with a precision of 10 μm. The object (in this case the lower surface of the arm) is placed on top of the plate A, which is parallel to the movable plate B where the sensor is mounted. In this instance the surface of the object is precisely aligned with the flat sensitive slice.

selection, an important characteristic of the rf coil is a small inductance to prevent detuning due to loading changes during the scanning procedure.

2.4.2.2 Scanning Procedure

The profiling method requires the sensitive slice to be shifted through the object. Figure 2.4.2 shows the mechanical lift used to move the sensor with respect to the sample. The object under study, for instance the lower surface of the arm in the picture, is positioned on top of a flat holder (A) and the NMR sensor is placed under it on a movable plate (B). The mechanical construction allows one to move the sensor up and down with a precision of 10 µm. The distance between the rf coil and the sensitive slice defines the maximum penetration depth into the sample (maximum field of view of the 1D image). Depending on the application, the position of the rf coil with respect to the sensitive slice can be changed to maximize the sensitivity.

2.4.2.3 Spatial Resolution Tests

The maximum spatial resolution achievable with the new magnet can be determined by measuring the Point Spread Function (PSF) [15]. It can be obtained as the image of a sample much thinner than the expected width of the PSF, but for the present resolution limit it is impractical to build such a sample slice. As an alternative, the PSF can be obtained as the derivative of the step image of a flat oil–glass interface. Figure 2.4.3(a) shows the 1D image of the interface obtained by centering the oil surface along the sensitive slice, and using 5-µs long rf pulses to excite a slice much thicker than the expected width of the PSF. Figure 2.4.3(b) shows the PSF, which is symmetrical and has a width of 2.3 µm; this is the maximum spatial resolution achievable with the present sensor.

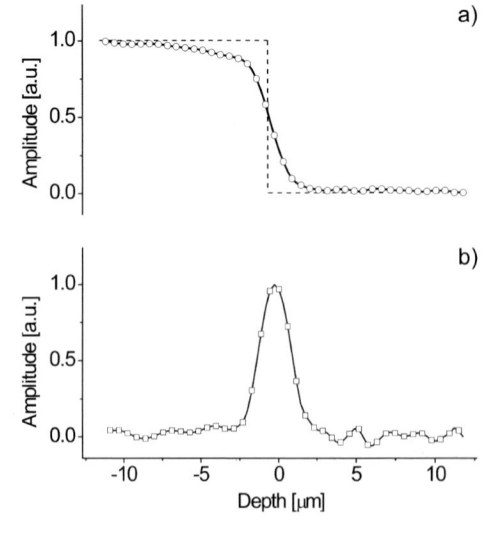

Fig. 2.4.3 (a) Image of a flat oil–glass interface centered along the sensitive volume. The length of the pulses was set to 5 µs to excite a slice thicker than the PSF width. The dashed line displays the step sample interface. (b) PSF obtained as the derivative of the step image shown in (a). It is symmetrical and has a linewidth of about 2.3 µm.

Depending on the transmitter power available and the bandwidth of the rf circuit, a thick slice can be excited and imaged by direct Fourier transformation of the acquired signal. However, when a CPMG-like sequence is applied for sensitivity improvement or to produce relaxation time contrast, the thickness that can be effectively excited is reduced by both the off-resonance excitation and the B_1 magnitude variation across the slice. If not having amplitude variations in the profile of a uniform sample is the criterion taken, then the maximum slice thickness that can be imaged is estimated to be 50 μm. When a large region needs to be scanned, the slice is repositioned across the sample by moving the sensor in steps equal to the maximum slice thickness. A full profile is then obtained by combining the set of slices necessary to cover the desired field of view.

To illustrate the performance of the method, a phantom consisting of a sandwich of two latex sheets 70 μm thick separated by a 150 μm glass slide was scanned by moving the position of the sensor in steps of 50 μm (Figure 2.4.4). It is important to point out that the image reconstruction is straightforward: portions of 50 μm are obtained in each step and then plotted one next to the other. No correction, interpolation or smoothing is required at the places where two consecutive portions join. The high resolution can be appreciated from the sharp edges of each latex sheet.

2.4.2.4 Relaxation Contrast

The slice selection procedure can be combined with a number of pulse sequences to spatially resolve NMR parameters or to contrast the profiles with a variety of filters. The most commonly used acquisition schemes implemented to sample echo train decays are the CPMG $[(\pi/2)_0-(\pi)_{90}]$ or a multi-solid echo sequence $[(\pi/2)_0-(\pi/2)_{90}]$. In these instances, the complete echo train can be fitted to determine

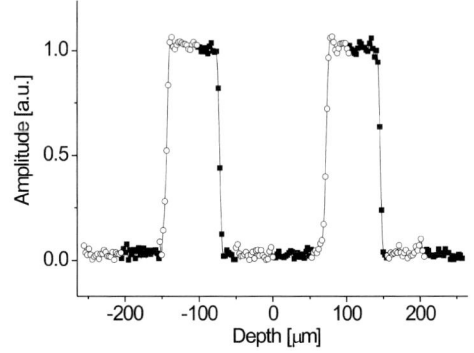

Fig. 2.4.4 Profile of a sample made from two latex layers 70 μm thick separated by a 150 μm thick glass spacer. The full profile is the combination of 10 images with an FOV of 50 μm, covering a total depth of 500 μm. Each of these images is the FT of the echo signal obtained as the addition of the first 16 echoes acquired during a CPMG sequence and 512 scans with an echo time of t_E = 345 μs. Using a recycle delay of 50 ms each image was acquired in 30 s and the total profile in 5 min. In this experiment the acquisition window was set to 300 μs, which in the presence of a gradient of 20 T m^{-1} defines a nominal resolution of 4 μm. For this particular instance, the rf coil was positioned 2 mm from the sensitive plane.

the $T_{2\text{eff}}$, or different parts of the train can be co-added to obtain relaxation time weighted profiles. We have found this last option more convenient, especially when the echo envelope has a complicated time dependence, or when $T_{2\text{eff}}$ is comparable to the echo time. Therefore, a weighting function is defined as follow:

$$w(i_i, i_f, j_f, t_E) = \frac{i_i - i_f}{j_i - j_f} \sum_{j=j_i}^{j_f} S(j\,t_E) \bigg/ \sum_{i=i_i}^{i_f} S(i\,t_E)$$

where $S(t)$ is the intensity of the signal at time t, and the integration limits i_i, i_f, j_i and j_f are adjusted to obtain the optimum contrast. In the following sections, and depending on the application, the acquired echo train decay is co-added for improvement in the sensitivity or the function w is evaluated to produce profiles weighted by relaxation or by diffusion.

2.4.2.5 Scanning Large Depths

In many cases not only high spatial resolution but also a large penetration depth is desired. This can be achieved by increasing the dimensions of the magnet blocks and gaps keeping their proportion fixed. When the sensor is scaled up the depth where the plane of constant field is generated as well as the maximum resolution are scaled up by the same factor. Although magnet scaling is the natural procedure to reach larger depths, it leads to a non-desired increase in the sensor size. A more elegant solution is to keep the sizes of the magnet blocks constant and to increase the gaps keeping their proportions fixed. Setting $d_B = 20$ mm and $d_S = 3$ mm, the sensitive volume is defined at 18 mm from the magnet surface, where $B_0 = 0.25$ T and $G_0 = 11.1$ T m^{-1}. Combining this magnet with an rf coil 22 × 25 mm² a working depth or maximum field of view of about 10 mm is defined. An object made of three rubber layers separated by glass spacers was scanned, and the results are shown in Figure 2.4.5. It is important to stress the excellent sensitivity of the device even at this large penetration depth. Each point in the profile corresponds to a 100-µm thick slice at 10 mm from the sensor and requires a measuring time of only 3 s.

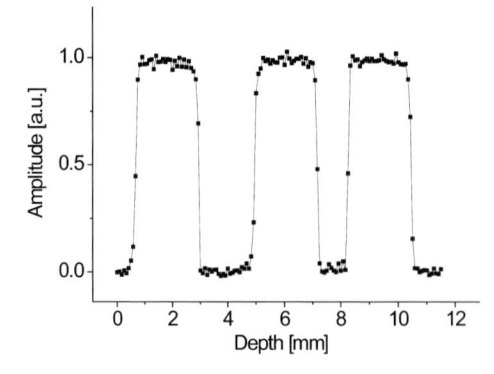

Fig. 2.4.5 Profile of a phantom made of three 2-mm thick rubber layers separated by glass slides of 2- and 1-mm thick. The CPMG sequence was executed with the following parameters: repetition time = 50 ms, $t_E = 0.12$ ms, number of echoes = 48 and 64 accumulations. The profile was scanned with a spatial resolution of 100 µm in 5 min.

2.4.3
Applications

The possibility of obtaining high-resolution depth profiles with single-sided sensors opens up the possibility of resolving the near-surface structure of arbitrarily sized samples *in situ*. In this section a number of applications in material science, medicine and artwork analysis are presented to illustrate the power of the described profiling technique. Depending on the application, different experimental parameters were critical in obtaining the desired information. While high spatial resolution was required, for example, to resolve thin multi-layer structures in PE materials, a short experimental time was needed to follow the absorption of cream into the skin, the appropriate contrast was required to resolve different paints and a large penetration depth was needed to resolve the full structure of polymer coatings on cement samples. For each particular situation the sensor was adapted to obtain the best performance.

2.4.3.1 Solvent Ingress into Polymers

An important example of sample change is the ingress of liquids in contact with the object surface. Even small amounts of solvent may invoke major changes in material properties. In this work the water uptake in polyethylene (PE) was followed as a function of the absorption time. A 3 mm thick PE sample was initially dried in an oven at 100 °C over 48 h and then immersed in water at room temperature. A series of profiles was measured for 0, 3, 6, 27 and 65 days of exposure to water, where 0 refers to immediately after drying [Figure 2.4.6(a)]. The profiles were obtained by scanning the sample with a resolution of 200 µm and assigning to each position the amplitude resulting from the addition of the first 8 echoes acquired during a solid-echo train. The vertical scale in Figure 2.4.6 was converted into water content (% by weight) via calibration. It is important to stress that the variation in the signal intensity in the profiles is due to a lengthening of the $T_{2\text{eff}}$ of the PE, which is influenced by the presence of water. This allows the indirect quantification of tiny amounts of water having NMR signals smaller than the detectable limit. The results show that the technique can be used to follow the penetration front of solvents into hard materials. The sensitivity of the technique is high enough even to detect the moisture of the sample due to the ambient humidity. Figure 2.4.6(b) compares the profile of the dried sample with that of a sample exposed for a long period of time to the humidity of the air at room temperature. The profile reveals that the sample contains 1 % moisture distributed almost uniformly over its thickness.

2.4.3.2 Multi-layer Plastic Sheets

To reduce the permeation of specific compounds through plastic sheets, barriers are introduced in multi-layer structures. Each layer provides its own performance in terms of heat sealing, barrier properties, chemical resistance, stiffness, etc., and

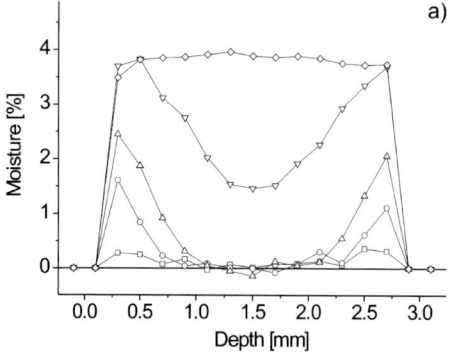

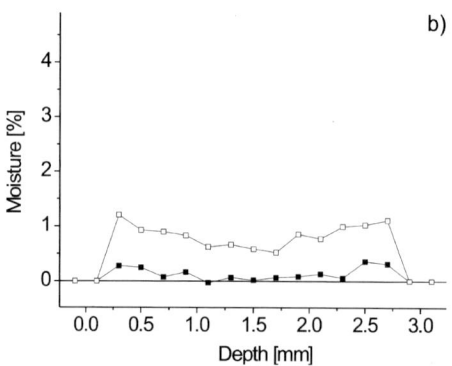

Fig. 2.4.6 (a) Profiles showing the ingress of water in a 3 mm thick PE sample as function of time for 0 (□), 3 (○), 6 (△), 27 (▽) and 65 (◇) days. The profile amplitude corresponds to the addition of the first 8 echoes acquired with a solid-echo train using $t_E = 30$ μs and 4-μs long rf pulses. A nominal spatial resolution of 200 μm was set by acquiring an echo window 6-μs long. The position of the sensor was moved in steps of 200 μm requiring 15 points to cover the complete sample thickness. Using 1024 scans and a recycle delay of 40 ms a total time per point of 40 s was required. (b) Moisture profiles obtained after drying the sample for 2 days at 100 °C and after subsequently exposing the sample to the humidity of air for 4 weeks.

their combination yields a material with properties superior to the sum of the individual ones. The barriers may be made of ethylene–vinyl alcohol copolymer (EVOH), poly(vinylidene chloride) (PVDC), amorphous nylons, poly(ethylene terephthalate) (PET) and high-density polyethylene (HDPE), which are glued by different types of resins. To tailor the properties of such composite materials, test procedures are required. The single-sided sensor described in this work is a new analytical tool in this regard.

The profiling technique was used to scan the multi-layer wall of a PE gasoline tank. For this particular application, multiple layers are required to keep the vapor emission below the limits imposed by environmental regulations. The tank structure consists of five layers, and starting from the outside they are regrind (recycled HDPE), resin, EVOH, resin and white HDPE. This material has a short T_{2eff} of the order of 300 μs that remains constant across the different layers, with the exception of EVOH layer, where T_{2eff} drops to 30 μs. Figure 2.4.7 shows a profile through the tank wall. The amplitude of the profile has been computed as the sum of the first 8 echoes acquired applying a multi-solid-echo train using $t_E = 40$ μs and 4-μs long rf pulses. The profile clearly reveals the structure of the material, and the position and width of each layer can be determined precisely. Besides providing structural information, the technique can be applied to study the efficiency of the barriers to stop the permeation of specific substances.

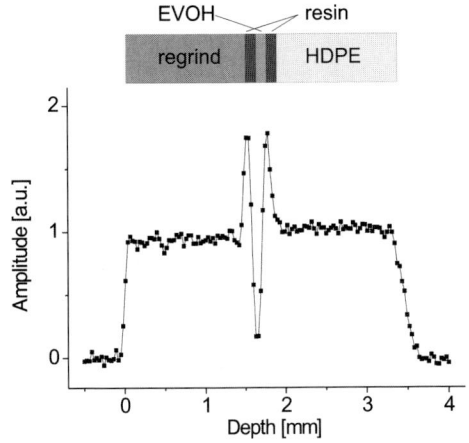

Fig. 2.4.7 Profile of a PE gasoline tank wall. T_1 and T_2 were measured across the sample, and uniform values of about 90 ms and 300 μs, respectively, were obtained, except for EVOH where T_2 drops to 30 μs. The profile amplitude is the coefficient at zero frequency of the FT of the signal obtained as the direct addition of the first 8 echoes generated with a solid-echo train. A nominal spatial resolution of 50 μm was set, acquiring an echo window 20-μs long. The position of the sensor was moved in steps of 25 μm requiring 160 points to cover the complete sample thickness. Using 512 scans per point and a repetition time of 150 ms the acquisition time per point was 75 s.

2.4.3.3 Human Skin In Vivo

Clinical diagnostics represents one of the most important applications of NMR imaging. However, the spatial resolution achieved in medical systems is of the order of 1 mm, so that studies of skin are difficult. A particular topic of interest is the effect of cosmetics and anti-aging products on skin. In this section we take advantage of the high spatial resolution achieved with the described single-sided sensor to study the skin structure *in vivo*. By choosing the right contrast, the profiling technique can be used to follow the ingress of creams into the skin.

Skin consists of several layers, the two most important being a superficial epithelial layer, the epidermis, and a deep connective tissue layer, the dermis. The epidermis is itself stratified beginning from the *stratum basale*, where the cells are generated, to the outer *stratum corneum*, where the cells are keratinized in a squamous structure. The dermis is sub-divided in two layers, the thinner more superficial one that lies adjacent to the epidermis (*papillare*) and a deeper one known as the *stratum reticulare*. Finally, the hypodermis lies deeper than the dermis and consists of adipose tissue. This is not part of the skin, although some of the epidermal appendages including hairs and sweat glands often appear as if they penetrate this layer. The hypodermis does allow the skin to move relatively freely over the underlying tissues.

The skin layers from the palm of the hand were scanned *in vivo*. A CPMG sequence was applied to sample the echo train decays as a function of depth. The decay was determined by both the relaxation time and the diffusion coefficient. To improve the contrast between the layers, a set of profiles was measured as a function of the echo

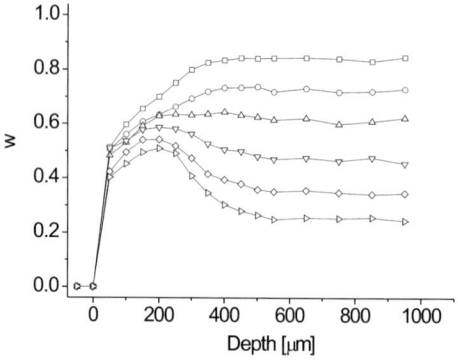

Fig. 2.4.8 Skin profiles measured in the palm region as a function of the echo time. The profile amplitude is the value of the weighting function obtained for different echo times: t_E = 35 (□), 70 (○), 100 (△), 140 (▽), 180 (◇) and 220 μs (▷). The indices i_i, i_f, j_i and j_f were recalculated for each subsequent experiment: keeping $t_E i_f$ = 2.1 ms, $t_E j_i$ = 2.1 ms, and $t_E j_f$ = 10 ms, in this way w is calculated at constant time. A nominal spatial resolution of 100 μm was set by acquiring a window 20-μs long. The position of the sensor was moved in steps of 50 μm in the first 500 μm and the in steps of 100 μm up to 1000 μm. Using 64 scans and a repetition time of 300 ms a total time per point of 20 s was required.

time to increase the signal attenuation due to diffusion (Figure 2.4.8). It can be observed that for t_E short enough to neglect attenuation by diffusion, the amplitude of the profile changes only in the region of the epidermis. This is in agreement with an expected change from young soft cells (*stratum basale*) to dead cells (*stratum corneoum*). By increasing t_E, the amplitude of the profile is weighted by diffusion in a controlled way. The results in Figure 2.4.8 show that the region corresponding to the epidermis is the one less affected by diffusion, while the dermis reticulare is the one most affected.

The technique has been used to resolve the skin structure in various regions of the body. The best contrast was obtained using i_i = 1, i_f = 50, j_i = 51 and j_f = 300 as integration limits for the weighting function, setting the echo time to 70 μs and spatial resolution to 20 μm. Figure 2.4.9(a) shows the profiles measured on the palm of the hand and on the lower-arm region. As expected, the profile measured in the palm shows a thicker epidermis than that measured on the lower arm. Apart from this difference, both profiles present the same trend for the subsequent layers. Both reach a maximum in the amplitude with comparable values in the region associated with the dermis papilare, and both decrease in the region corresponding to the dermis reticulare. The last parts of the profiles again shows increasing values as a consequence of the smaller diffusivity expected in the adipose tissue layer. The assignment of the different strata was made based on literature data of histological methods. To check the reproducibility of the profiles a number of volunteers were examined, and a systematic difference was observed between male and female subjects. Figure 2.4.9(b) shows the comparison of two representative examples where the difference in the thickness of the epidermis can clearly be appreciated.

Skin functions as both, an important physical barrier to the absorption of toxic substances and simultaneously as a portal of entry of such substances. The *stratum corneum* of the epidermis is most significant in providing some degree of physical

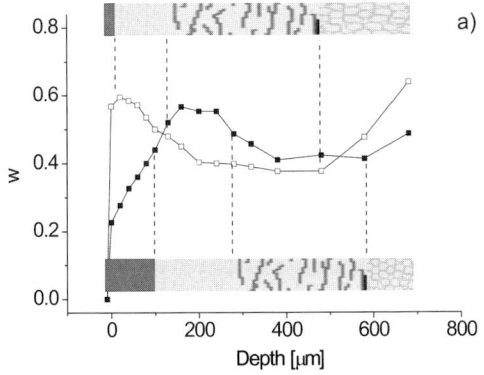

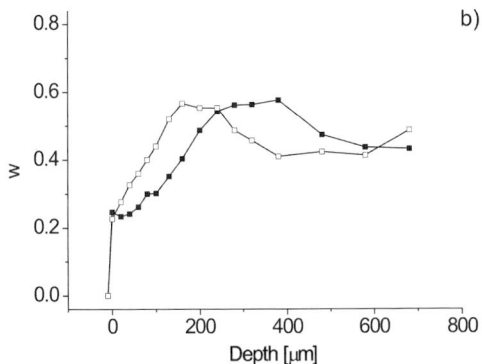

Fig. 2.4.9 (a) Comparison of skin profiles measured on the palm (■) and the lower arm (□). The profile amplitude is the value of the weighting function w (1, 60, 61, 300, 70 μs). A nominal spatial resolution of 20 μm was set acquiring an echo window 50-μs long. The position of the sensor was moved in steps to cover the thickness of the epidermis (E), the dermis papilare (DP), the dermis reticulare (DR) and the sub-cutis (SC). Using 64 scans per point and a repetition time of 300 ms the total acquisition time per point was 20 s. (b) Comparison of skin profiles measured in the palm of a male (■) and a female (□) volunteer.

protection from percutaneous absorption of chemicals. The thickness of the skin, especially the *stratum corneum*, also determines the degree to which substances are absorbed. Thicker skin is a greater barrier to the passage of foreign substances. Depending on the skin thickness, variable absorption of substances is expected in different regions of the body.

The technique used to measure the profiles of Figure 2.4.9 was used to study the absorption of cosmetic creams through the skin. For these experiments a higher resolution is required and a faster scanning is needed to follow the absorption as a function of the application time. For this application a penetration depth always lower than 1 mm is sufficient. This allowed us to use a smaller rf coil, generating a stronger B_1 to increase the maximum thickness that can be properly excited to 100 μm by shortening the excitation pulses – this is in contrast to the 50 μm shown in the experiment in Figure 2.4.4. A series of profiles showing the cream absorption in the palm for different application times is plotted in Figure 2.4.10. The scanning procedure was the same as the one described in the context of Figure 2.4.4, but now the amplitude corresponds to w (1, 50, 51, 300, 70 μs). The change in the profile shape as a function of time clearly shows the ingress of the cream into the skin.

This type of experiment allows quantification of the cream absorption time, as well as the evolution after saturation and returning to its original state. These

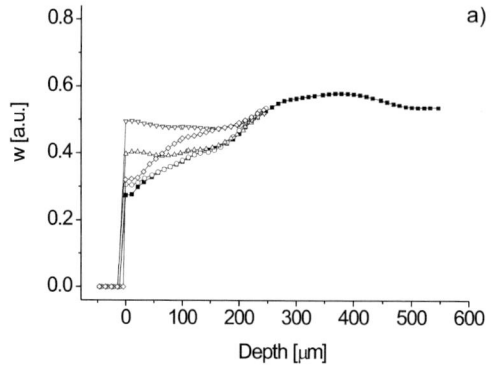

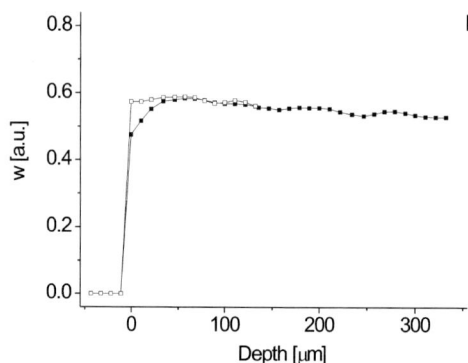

Fig. 2.4.10 Profiles showing the cream absorption in the skin. (a) Set of profiles acquired in the palm region as a function of the cream application time. The profiles at 0 (■), 1 (○), 3 (△) and 25 (▽) min show how fast the cream penetrates the epidermis. The profiles are the combination of three images with an FOV of 100 μm each, which meant it was possible to cover the region with appreciable changes in a short time. Using 64 scans per position and a repetition time of 300 ms the total acquisition time per profile was 1 min. (b) Profiles of the lower arm comparing the effect of cream applied for 5 min (□) with the normal profile (■).

experiments have been performed on the palm where the epidermis is thicker, but Figure 2.4.10(b) shows that the method also allows detection of the effect of skin creams in regions where the epidermis is thinner, such as in the lower arm.

The results presented in this section demonstrate that this powerful profiling technique is suitable for studying the effect of external agents on the skin as well as how disease or environmental damage may alter the barrier properties of skin and modify the absorption of substances. Although *in vitro* experiments were initially carried out on skin samples obtained from surgery, we have found important discrepancies with measurements carried out *in vivo* on corresponding regions of the body. Given the free access offered by the open geometry of the sensor such investigations can be conveniently performed *in vivo*.

2.4.3.4 Characterization of Paintings

The conservation of cultural heritage is a particular field where the use of non-destructive characterization methods is mandatory, considering the uniqueness of the objects under study. Even in cases where sampling of tiny volumes is allowed, non-destructive testing offers the possibility of applying complementary techniques to obtain more information on one specific sample. In this section we focus on the application of the described profiling technique to study paintings. The analysis of paintings presents a number of challenges because of the high complexity of these systems. Depending on the style, they are made up of several superimposed layers on a support, normally a wooden panel or canvas. Several paint layers are often applied. A layer of varnish is usually applied after drying to protect and saturate the colors. In general, the study of such multi-layered paintings presents difficulties for most analytical techniques. The chemical properties of adjacent areas can be completely different and should ideally be analyzed individually. It is actually impossible to physically dissect these structures by means of a scalpel, as different layers are merged together in thicknesses ranging between 10 and 100 µm. Paints are a mixture of pigments and a binder, which cures into a solid paint layer once applied to a surface [16].

For the studies in this work we used a set of reference samples that includes two types of binders, tempera and oil, combined with a number of well-known pigments. Following the techniques of the old masters, these paints were applied on a standard wood support covered by a layer of gypsum to assure a homogeneously colored and smooth surface. The materials used during the various periods and by the different masters are well known and can be reproduced to calibrate different experimental techniques. Conservators are interested in obtaining information that helps to define the type of paint used in the artwork as well as the structure and preservation of the support of the painting. As a first step the profiling technique was applied to resolve the structure of reference samples with a single paint layer prepared using the same pigment and two different binders, tempera and oil (Figure 2.4.11). Besides distinguishing the type of binder, the method resolves the wood, the gypsum and even the canvas layer between the wood and gypsum. Such detailed information may be helpful in assessing the state of conservation of the paintings, including the support structure.

In general paintings are composed of more than one layer of paint, complicating the characterization by the possible mixing of the pigments. By focusing only on the paint layer structure and using higher spatial resolution the method was used

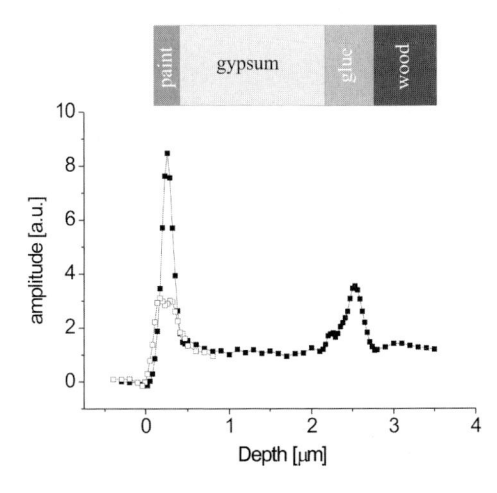

Figure 2.4.11 Profiles of paintings where different layers can clearly be resolved. A solid-echo train was used with $t_E = 40$ µs, and the first 4 echoes were used to calculate the amplitude. The profiles were reconstructed by moving the sensor in steps of 50 µm in the paint and canvas regions, and 100 µm in the gypsum and wood layers. Using 128 scans per point and a repetition time of 100 ms the total acquisition time per point was 16 s. Profiles of paint based on tempera (■) and oil (□) binders show appreciable difference.

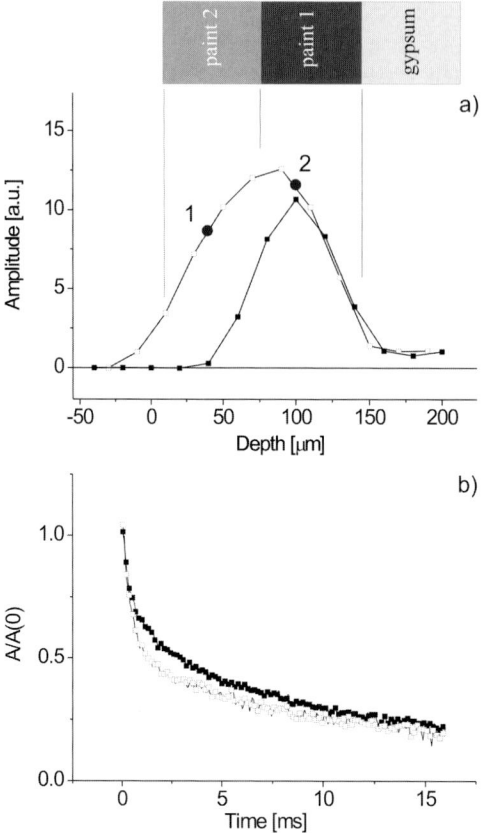

Fig. 2.4.12 Discrimination between paint layers with different pigments. (a) Profiles of a mono- (■) and a double-layer (□) painting where the difference in thickness can clearly be observed. (b) CPMG decays measured at the points 1 (■) and 2 (□) indicated in (a). The position of these points was selected to measure independently each of the two paint layers with different pigments (*terra rossa* and *terra verde*, respectively). The change in the decay time is apparent, making it possible to distinguish between the layers via the relaxation time.

to resolve two paint layers 50 μm thick that were of the same binder but different pigments. Figure 2.4.12(a) compares two profiles obtained for a single and a double paint layer. For these experiments the resolution was set to 20 μm and the samples were scanned in the first 200 μm only. The binder used in these samples was tempera, and the pigments were *terra verde* and *terra rossa*. The profiles clearly show the different thicknesses of the two samples. Figure 2.4.12(b) shows the CPMG decays measured at the depths indicated as 1 and 2 in the figure. The position of each of these points was chosen to be inside each of the two paint layers.

The echo decays show various characteristic time dependences, which can be due to the different pigments. To study the dependence of the relaxation time on the type of pigment, a series of samples with single paint layers was measured. Figure 2.4.13 shows the results obtained for five different pigments. The surprisingly large variation in decays observed allows the color of the paint to be determined. Besides being of relevance for conservators, the possibility of resolving pictures hidden behind the superficial paint layer non-destructively offers an exciting potential for the investigation of paintings.

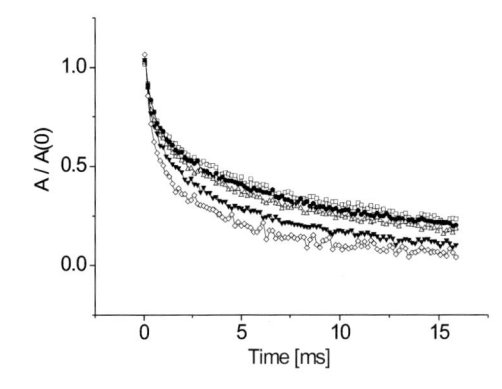

Fig. 2.4.13 CPMG decays of the pigments *terra rossa* (□), *azurro oltremare* (●), *terra verde* (△), *ocra giana* (▼) and *terra siena bruciata* (◇). Based on the different decay times the color of the paint can be determined from the NMR signal.

The results presented in this section are part of a preliminary study where the performance of single-sided NMR as a non-destructive method of characterizing paintings was tested. In contrast to other techniques, such as X-rays, which is only sensitive to heavy nuclei, NMR detects the signals from hydrogen nuclei present in every organic compound. A correct assessment of the current chemical and physical state of the painting is crucial for making decisions with respect to conservation and restoration efforts. The possibility of resolving multi-layer structures in paintings, distinguishing different pigments, provides conservators with valuable information about the choice of conservation strategies or restoration materials.

2.4.3.5 Multi-layer Cement Materials

As a final application of the profiling technique, the sensor for large depth measurements described in Section 2.4.2.5 was used to resolve multi-layer polymer coatings on concrete samples. Such coatings are used to protect concrete from degradation and corrosion. They are applied to the concrete surface to reduce the porosity in the upper first millimeters to prevent the penetration of water and

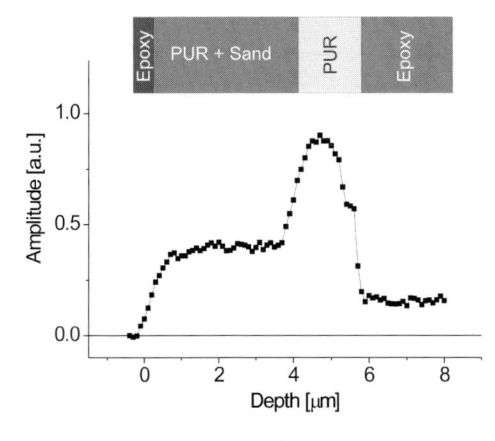

Fig. 2.4.14 Profile of a multi-layer polymer coating used to protect concrete surfaces from environmental corrosion. The profile is the signal amplitude resulting from the addition of the first 32 echoes acquired with a CPMG sequence with $t_E = 50$ μs. It has an FOV of 8 mm and was measured with a spatial resolution of 100 μm. Using 256 scans per point and a repetition time of 100 ms, the total acquisition time per point was 25 s.

decrease the rise of sub-soil water by capillarity forces. Figure 2.4.14 shows a profile of a sample that consists of four layers of epoxy plus sand, polyurethane (PUR) plus sand, elastic polyurethane and epoxy. The application of this technique to the systematic study of the effect of the layers to the attack of various substances will be helpful in improving the quality of the material and to detect *in situ* the conservation state of buildings.

2.4.4
Conclusions

In this work we have taken advantage of a recently proposed magnet geometry that provides microscopic depth resolution by generating a magnetic field with an extremely uniform gradient. Thanks to the open geometry of the sensor, the near surface structure of arbitrary large samples was scanned with a resolution better than 5 µm. The scanning procedure is based on repositioning a flat sensitive slice across the sample keeping the measurement conditions constant, a fact that eliminates systematic errors which can otherwise be introduced into the relaxation times and diffusion weighted profiles. The potential use of the profiling technique was demonstrated in a number of applications, such as the measurement of moisture profiles, the effects of cosmetics in human skin, solvent ingression in polymers and the assessment of the state of paintings in the conservation of objects of interest in cultural heritage.

Acknowledgments

We are grateful to K. Kupferschläger for the design and construction of the mechanical lift used to reposition the sensor with a high precision, and to Prof. A. Sgamellotti from the university of Perugia for providing us with the samples of the paintings. Support of this project by the DFG Forschergruppe FOR333 Surface NMR of Elastomers and Biological Tissue is gratefully acknowledged.

References

1 P. T. Callaghan **1991**, Principles of Nuclear Magnetic Resonance Microscopy, Clarendon Press, Oxford.
2 B. Blümich **2000**, NMR Imaging of Materials, Clarendon Press, Oxford.
3 R. L. Kleinberg **1996**, (Well logging), in Encyclopedia of NMR, eds. D. M. Grant, R. K. Harris, Wiley, New York, pp. 4960–4969.
4 G. Eidmann, R. Savelsberg, P. Blümler, B. Blümich **1996**, (The NMR-MOUSE, A mobile universal surface explorer), J. Magn. Reson. A122, 104–109.
5 G. A. Matzkanin **1998**, (A review of nondestructive characterization of composites using NMR) in Nondestructive Characterization of Materials, Springer, Berlin, pp. 655.
6 P. J. Prado **2001**, (NMR hand-held moisture sensor), Magn. Reson. Imag. 19, 505–508.

7 A. E. Marble, I. V. Mastikhin, B. G. Colpitts, B. J. Balcom **2005**, (An analytical methodology for magnetic field control in unilateral NMR), J. Magn. Reson. 174, 78–87.

8 US Patent 5,959,454, **1999**, Bruker Analytic, Magnet arrangement for an NMR tomography system, in particular for skin and surface examinations.

9 R. Haken, B. Blümich **2000**, (Anisotropy in tendon investigated in vivo by a portable NMR scanner, the NMR-MOUSE), J. Magn. Reson. 144, 195–199.

10 B. Blümich, S. Anferova, S. Sharma, A. L. Segre, C. Federici **2003**, (Degradation of historical paper: nondestructive analysis by the NMR-MOUSE), J. Magn. Reson. 161, 204–209.

11 N. Proietti, D. Capitani, E. Pedemonte, B. Blümich, A. L. Segre **2004**, (Monitoring degradation in paper: non-invasive analysis by unilateral NMR. Part II), J. Magn. Reson. 170, 113–120.

12 P. J. McDonald **1997**, (Stray field magnetic resonance imaging), Prog. Nucl. Magn. Reson. Spectrosc. 30, 69–99.

13 M. D. Hürlimann, D. D. Griffin **2000**, (Spin dynamics of Carr–Purcel–Meibohm–Gill-like sequences in grossly inhomogeneous B_0 and B_1 fields and applications to NMR well logging), J. Magn. Reson. 143, 120–135.

14 J. Perlo, F. Casanova, B. Blümich **2005**, (Profiles with microscopic resolution by single-sided NMR), J. Magn. Reson. 176, 64–70.

15 A. G. Webb **2004**, (Optimizing the point spread function in phase-encoded magnetic resonance microscopy), Concept Magn. Reson. 22A, 25–36.

16 L. Carlyle **2001**, The Artist's Assistant, Oil Painting Instruction Manuals and Handbooks in Britain 1800–1900 with Reference to Selected Eighteenth-Century Sources, Archetype Publications, London.

2.5
Microcoil NMR for Reaction Monitoring
Luisa Ciobanu, Jonathan V. Sweedler, and Andrew G. Webb

2.5.1
Introduction

Knowledge of kinetics is an important component of investigating reaction mechanisms. Such information greatly benefits drug design [1], drug metabolism [2] and a variety of industrial processes [3]. A number of different analytical techniques can be employed to investigate kinetic data. Spectroscopic tools are particularly useful as, besides reaction rates, they can provide chemical information at the molecular level. Nuclear magnetic resonance (NMR) is one of the most information-rich techniques, able to provide a high degree of structural information. In this chapter we describe past work, recent advances and future prospects for NMR as a tool for studying chemical reaction kinetics, with an emphasis on small-scale reactions and correspondingly small NMR detectors. In Section 2.5.2 both stopped- and continuous-flow NMR techniques, the effects of flow on the NMR signal and the engineering challenges faced when designing continuous-flow NMR probes are discussed. Section 2.5.3 surveys applications of NMR in studies of equilibria and kinetics of multicomponent mixtures, concentrating on the use of standard commercial and continuous-flow NMR probes. Section 2.5.4 is dedicated to studies performed using "microcoil" NMR probes and micromixers. It also summarizes the properties of micro-receiver coils and the sensitivity enhancements that can be attained from their use, in addition to the basic characteristics of the most commonly used micromixers. The focus of Section 2.5.5 is on the more recently developed multiple microcoil NMR probes and their applications. Finally, in Section 2.5.6, we summarize the present state and potential future applications of NMR spectroscopy in the study of reaction kinetics.

2.5.2
NMR Acquisition in Reaction Monitoring: Stopped- and Continuous-flow

The process of acquiring time-dependent NMR spectra can be carried out using two basic approaches. In the stopped-flow mode the system being studied remains within the NMR detector for the entire data acquisition period. The chemical reactants are either mixed in the NMR tube itself or are mixed outside and introduced into the tube rapidly. This approach has the advantage of simplicity, but is limited by the need for very rapid mixing when the time constant for the reaction process is short. As diffusional mixing is relatively slow, some active form of mixing is often needed, which itself can interfere with data acquisition. For example, a settling period can be required so that liquid vortices subside to allow a stable shim.

The alternative method is "continuous-flow", in which the reactants flow through the detection coil during data acquisition. Continuous-flow NMR techniques have been used for the direct observation of short-lived species in chemical reactions [4–6]. The main difference between stopped- and continuous-flow NMR is that in the latter the sample remains inside the detection coil only for a short time period, termed the residence time, τ [7], which is determined by the volume of the detection cell and the flow rate. The residence time alters the effective relaxation times according to the relationship in Eq. (2.5.1):

$$\frac{1}{T_{n,\text{effective}}} = \frac{1}{T_n} + \frac{1}{\tau} \tag{2.5.1}$$

where $n = 1$ or 2, which represents the longitudinal and transverse relaxation times, respectively. It is clear from the above equation that as the flow rate increases (τ decreases), the longitudinal relaxation time of the system, $T_{1,\text{effective}}$, will decrease. Therefore, a shorter recycle delay can be used to allow complete relaxation to occur between pulses, and more transients in a given period of time can be accumulated. The theoretical maximum sensitivity is obtained when the repetition rate is equal to the residence time in the NMR flow cell, providing that the sample plug is several times longer than the detector length. This condition assumes that the "fresh" incoming nuclei have been fully polarized by the time that they enter the flow cell. In order to achieve the full equilibrium Boltzmann magnetization it is necessary for the nuclei to have been in the main magnetic field, before entering the NMR cell, for a long period of time relative to the spin lattice relaxation time (T_1). This becomes a concern for nuclei with long spin lattice relaxation times (e.g., ^{13}C with $T_1 = 10$–100 s). One way to solve this problem is to install a large equilibration volume inside the magnet, in front of the detection cell.

It is also clear from Eq. (2.5.1) that the linewidth of the observed NMR resonance, limited by $1/T_2$, is significantly broadened at high flow rates. The NMR line not only broadens as the flow rate increases, but its intrinsic shape also changes. Whereas for stopped-flow the line shape is ideally a pure Lorentzian, as the flow rate increases the line shape is best described by a Voigt function, defined as the convolution of Gaussian and Lorentzian functions. Quantitative NMR measurements under flow conditions must take into account these line shape modifications.

The experimental picture of continuous-flow NMR becomes more complicated when one considers the presence of laminar, turbulent or more complex flow patterns. The flow pattern is determined by the geometry of the flow cell and the flow rate. Several techniques such as chromatography, laser Doppler velocimetry and hot-wire anemometry can be used to visualize flow. These techniques have the disadvantage that they require doping of the fluid, dye (chromatography) or scattering centers (laser Doppler velocimetry) or insertion of a probe (anemometry). Given the use of NMR spectroscopy in reaction monitoring, the use of dynamic NMR microimaging is a promising approach to measuring flow [8]. Compared with conventional methods, NMR microimaging has the advantage that is totally non-invasive, can be used for measurements of both flow and diffusion, can measure flow in three dimensions and is able to provide velocity resolution of a few tens of microns per second.

There are several MR flow imaging methods including spin-tagging, time-of-flight, phase-contrast and q-space imaging [9]. Using q-space imaging in the form of pulsed-field gradient spin-echo (PGSE) experiments, Zhang and Webb [10] have investigated the flow characteristics of Newtonian fluids in different flow-cell geometries with flow-cell volumes of ≈ 1 µL. In the PGSE scheme the spins are dephased by imposing a magnetic field gradient. After a time delay Δ they are rephased by applying an identical pulse gradient after a phase inverting 180° pulse. The signal from stationary spins is refocused whereas any motion occurring during Δ which transfers spins to a new location results in a net phase shift. This leads to the expression given in Eq. (2.5.2) for the signal in the PGSE experiment:

$$E_\Delta(q) \propto (e^{j2\pi q v \Delta})(e^{-4\pi^2 q^2 D \Delta}) \tag{2.5.2}$$

where v is the flow velocity, Δ is the time delay between the two gradient pulses, D is the diffusion coefficient and q is defined as:

$$q = \frac{\gamma g \delta}{2\pi} \tag{2.5.3}$$

where γ is the gyromagnetic ratio, and g and δ are the amplitude and the duration of the gradient pulses, respectively. In order to measure spatially resolved velocity maps, a series of MR images are acquired with gradient pulses successively incremented by a value g_{inc} and then inverse Fourier transformation is performed for each pixel in the image. The result is a Gaussian-shaped function, usually denoted $\bar{P}_s$, centered at k_v and with full width at half maximum k_{FWHM} where k_v and k_{FWHM} are given by:

$$k_v = \frac{N}{2\pi n_D} \gamma \delta g_{max} v \Delta \qquad (2.5.4)$$

$$k_{FWHM} = \frac{2N}{\pi n_D} \sqrt{\ln 2 \gamma^2 \delta^2 g_{max}^2 D \Delta} \qquad (2.5.5)$$

where n_D is the number of increments in the PGSE gradient, N is the digital array size for the q-space transformation and g_{max} is the maximum PGSE gradient applied. The experimental results obtained by Zhang and Webb, supported by computational simulations, showed that in the case of NMR cells with diameters larger than the connecting tubing (e.g., the bubble-type flow cell), a gradual tapering of the NMR cell should be employed to avoid the formation of eddy flow effects.

2.5.3
Reaction Studies Using NMR

This section contains a brief survey of NMR spectroscopic investigations of chemical reaction kinetics and mechanisms. One of the goals of reaction kinetics studies is to measure the rate of the reaction (or rate constant) – the rate at which the reactants are transformed into the products. Another goal is to determine the elementary steps that constitute a multi-step reaction. Finally, and perhaps the most important goal is to identify transitory intermediate species. NMR, in common with other spectroscopic techniques, is especially valuable in achieving this

Fig. 2.5.1 Schematic of an NMR flow cell used to introduce and mix reactants and follow product formation used in a conventional 5-mm diameter NMR probe [11].

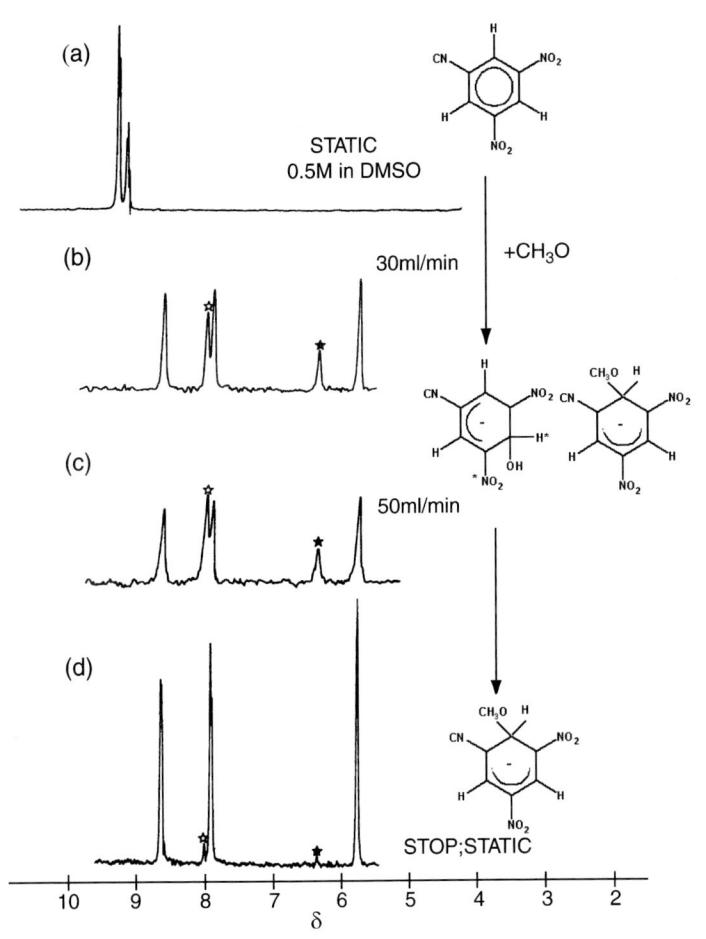

Fig. 2.5.2 ¹H NMR spectra recorded at 100 MHz showing the ring proton resonances of 3,5-dinitrocyanobenzene (a), and spectral changes during its reaction with methoxide ion (in 87.5% DMSO–12.5% MeOH) under the flow rate conditions indicated (b, c and d). Reprinted with permission from Ref. [5]. Copyright (1975) American Chemical Society.

final goal, allowing the monitoring of intermediates without the necessity of isolation from reaction mixtures. Both continuous- and stopped-flow NMR experiments have been used to monitor reaction intermediates and to perform kinetic measurements down to the millisecond timescale.

The use of NMR for studying chemical reactions began about 30 years ago. In 1972, Asahi and Mizuta [11] reported a study of the performance of five different types of flow cells used in NMR studies of chemical reactions: (a) a straight glass tube, (b) a pipette-type tube, (c) a spiral capillary in a conventional sample tube, (d) a jet in the base of a conventional tube, and (e) a conventional spinning tube with an inlet at the base and an outlet at a height of about 50 mm. The best results were obtained using a flow cell of type (e) (Figure 2.5.1). The authors presented flow ^{1}H NMR results for the reaction of thiamine with the hydroxide ion.

In 1975, Fyfe et al. [5] reported studies on the reaction of 3,5-dinitrocyanobenzene with a methoxide ion. As illustrated in Figure 2.5.2(a), the NMR spectrum of 3,5-dinitrocyanobenzene exhibits two NMR peaks. The NMR spectra of the mixture for two different flow rates, 30 and 50 mL min^{-1}, are shown in Figure 2.5.2(b) and (c), respectively. They were able to detect the presence of a reaction intermediate [NMR peaks marked with an * in Figure 2.5.2(b) and (c)]. At higher flow rates the relative intensities of the NMR peaks resulting from the unstable isomer increased. When the flow was stopped [Figure 2.5.2(d)] only signals due to the stable isomer (the final reaction product) were detected. The rate constant for the reaction intermediate was found to be 0.94 s^{-1}.

Several other organic and inorganic reaction intermediates have been studied using NMR methods. Trahanovsky et al. [12] reported a series of experiments in which they studied unstable molecules, such as benzocyclobutadiene, using flow NMR. Tan and Cocivera [13, 14] studied the reaction of 4-formylpyridine with amino acids, imidazole and D,L-alanylglycine using stopped-flow proton NMR.

Although limited by sensitivity, chemical reaction monitoring via less sensitive nuclei (such as ^{13}C) has also been reported. In 1987 Albert et al. monitored the electrochemical reaction of 2,4,6-tri-*t*-butylphenol by continuous flow ^{13}C NMR [4]. More recently, Hunger and Horvath studied the conversion of vapor propan-2-ol (^{13}C labeled) on zeolites using ^{1}H and ^{13}C *in situ* magic angle spinning (MAS) NMR spectroscopy under continuous-flow conditions [15].

The application of NMR to the study of chemical reactions has been expanded to a wide range of experimental conditions, including high pressure and temperatures. In 1993, Funahashi et al. [16] reported the construction of a high pressure ^{1}H NMR probe for stopped-flow measurements at pressures <200 MPa. In the last decade, commercial flow NMR instrumentation and probes have been developed. Currently there are commercially available NMR probes for pressures of 0.1–35 MPa and temperatures of 270–350 K (Bruker) and 0.1–3.0 MPa and 270–400 K (Varian). As reported recently, such probes can be used to perform quantitative studies of complicated reacting multicomponent mixtures [17].

In terms of biochemical applications, NMR can provide unique information regarding protein folding, ligand binding and conformational changes. Grimaldi and Sykes performed a study of conformational changes of concanavalin A in the presence of Mn^{2+}, Ca^{2+} and α-methyl-D-mannoside using stopped-flow NMR [18]. More recently, van Nuland et al. used NMR to study non-equilibrium events of protein folding by recording NMR spectra after initiation of the reaction in the NMR probe [19]. Various protein folding studies have been performed using high pressure, high resolution NMR spectroscopy. In an excellent review article, Jonas described a series of experiments dedicated to the analysis of pressure-assisted cold denaturation and detection of protein folding intermediates as well as to the investigations of local perturbations in proteins and the effects of point mutations on pressure stability [20].

2.5.4
Small-scale NMR Reaction Monitoring

Poor sensitivity and the consequent long measurement times, characteristic of NMR, often preclude the acquisition of real-time data for reactions involving small amounts of material. Other spectroscopic techniques, such as fluorescence and mass spectrometry, have a significant sensitivity advantage. In fact most spectroscopic techniques have a 10^2–10^6 higher intrinsic sensitivity compared with NMR. The signal-to-noise ratio (SNR) of the NMR experiment, assuming a coil having uniform B_1 field over a well defined volume, is directly proportional to the number of nuclear spins in the system, as given by Hoult and Richards [21]:

$$\mathrm{SNR} \propto \frac{k_0 \frac{B_1}{i} V_s C \gamma \frac{h^2}{4\pi^2} I(I+1) \frac{\omega_0^2}{kT3\sqrt{2}}}{\sqrt{4kT\Delta f (R_{\mathrm{coil}} + R_{\mathrm{sample}})}} \qquad (2.5.6)$$

where V_s is the sample volume, k_0 is a constant which accounts for spatial inhomogeneities in the B_1 field produced by the probe, C is the sample concentration ω_0 is the Larmor frequency, Δf is the measured bandwidth, R_{coil} and R_{sample} are the coil and sample resistances, respectively, and the factor of $\sqrt{2}$ is introduced because the noise measurement is the root mean square (rms).

Equation (2.5.6) is valid for any coil geometry: however, we will focus on coils of solenoidal geometry. The reason for this is that most microcoils are solenoidal, given the intrinsically high sensitivity of this geometry and the relative ease of fabrication at small sizes compared with alternative geometries, such as saddle coils. Sample losses, R_{sample}, have two contributions, namely inductive and dielectric losses. Inductive losses arise from radiofrequency eddy currents induced in the sample in response to the rf excitation from the coil: these dissipate power in conductive samples. Inductive losses can be modeled as an effective resistance, R_m, in series with the coil. For a cylindrical sample of length $2g$ and radius b, and a solenoidal coil with n turns, length $2g$ and radius a, Gadian and Robinson [22] have shown that the value of R_m is given by:

$$R_\mathrm{m} = \frac{\pi \omega_0^2 \mu_0^2 n^2 b^4 g \sigma}{16(a^2 + g^2)} \qquad (2.5.7)$$

Dielectric losses arise from the direct capacitive coupling of the coil and the sample. Areas of high dielectric loss are associated with the presence of axial electric fields, which exist half way along the length of the solenoid, for example. Dielectric losses can be modeled by the circuit given in Figure 2.5.3. The other major noise source arises from the coil itself, in the form of an equivalent series resistance, R_{coil}. Exact calculations of noise in solenoidal coils at high frequencies and small diameters are complex, and involve considerations of the proximity and skin depth effects [23].

The factor B_1/i in Eq. (2.5.6), the magnetic field per unit current, is defined to be the coil sensitivity. For almost all coil geometries, the value of B_1/i is inversely

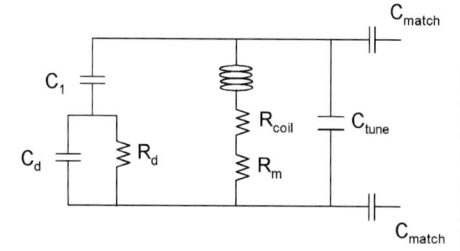

Fig. 2.5.3 Typical NMR resonant tank circuit, showing coil loss mechanisms. This LC circuit is then placed in series with two matching capacitors (C_{match}). The resistance of the circuit is represented by R_{coil}, the inductive losses by R_m and the dielectric losses by C_1, C_d and R_d.

proportional to the coil diameter. The net result is that for solenoidal coils of diameter greater than about 100 μm, the SNR per unit volume of sample is inversely proportional to the coil diameter, and therefore proportional to $V_{coil}^{-1/3}$ for a coil with a fixed length to diameter ratio.

The development of microcoil techniques has been reviewed by Minard and Wind [24, 25] and by Webb [26]. In a more recent publication Seeber et al. reported the design and testing of solenoidal microcoils with dimensions of tens to hundreds of microns [27]. For the smallest receiver coils these workers achieved a sensitivity that was sufficient to observe proton NMR with an SNR of unity in a single scan of ≈10 μm³ (10 fL) of water, containing 7×10^{11} proton spins. Reducing the diameter of the coil from millimeters to hundreds of microns thus increases its sensitivity greatly, allowing analysis of pL to μL sample volumes.

Reducing the coil diameter improves the SNR, but also decreases the NMR spectral resolution [26]. This is because the static magnetic field distortions (arising from magnetic susceptibility mismatches between the coil windings, NMR tube and surrounding air) within the sample become more pronounced the closer the coil windings are to the sample. Static field homogeneity can be improved by increasing the coil length: theoretically an infinitely long cylinder produces a perfectly uniform magnetic field, irrespective of the magnetic susceptibilities of the windings and surroundings. In 1995, Olson et al. [28] obtained high resolution, proton nuclear magnetic resonance spectra on 5-nL samples by immersing the NMR coil in a fluid with a magnetic susceptibility very close to that of copper, thus approaching the ideal "infinite cylinder" geometry. In order to avoid background signal from the surrounding fluid, a perfluorocarbon FC-43 (fluorinert) was chosen. The coil used in their studies was a tightly wound 17-turn solenoid, with an od of 357 μm, id of 75 μm and length of 1 mm. The coil was wound directly onto a polyimide-coated fused silica capillary in which the sample was introduced. The development of such probes has proved very useful in expanding the use of NMR to the study of mass-limited samples. Such NMR microcoils have also been successfully used as on-line detectors in capillary separation [29–31].

Although one can potentially attempt to cool the rf microcoil, which according to Eq. (2.5.2) would further improve the SNR by reducing the R_{coil} term in the denominator, in the case of microcoils the sample is extremely close to the coil and if the sample has to be kept at room temperature, cooling the coil alone is extremely difficult.

The coupling of small coils with small-scale mixing devices is particularly attractive, as very rapid and efficient mixing can be achieved at small size scales. A number of devices have been designed to improve mixing at the micron scale. These micromixers fall into one of two classes: (a) active mixers in which the flow is controlled by using moving parts or varying pressure gradients [32–34] and (b) passive mixers which utilize only a pump or pressure head to ensure a constant flow rate [35, 36]. Passive mixers are more common given that they are easier to manufacture and to interface with microfluidic systems. The fabrication and design of such mixers have been described by Bessoth et al. [35]. A schematic of such a micromixer is shown in Figure 2.5.4: the design is based on the principle of distributive mixing. The diffusion time, t, of a molecule is proportional to the square of the diffusion distance, L, and inversely proportional to the diffusion coefficient, D: $t \propto L^2/D$. Therefore, splitting the flows of two liquids (A and B) to be mixed into n partial flow regions and rearranging them in thin laminae reduces the mixing time by a factor of n^2. The main component of the micromixer is a microchip, which is made from a glass/silicon/glass sandwich. On the silicon wafer, the inlet channel of the liquid A is split into 16 partial flows. On the reverse side (layer 2) liquid B is also split into 16 partial flows. The two liquids come together at the confluence point and the neighboring channels are repeatedly combined and eventually united into a single outlet channel. For small molecules with a diffusion coefficient $D \approx 10^{-9}$ m^2 s^{-1} the mixing time achieved with such a mixer is of the order of 25 ms, much shorter than it can be obtained using a single Y-connector ($\approx$1–2 s).

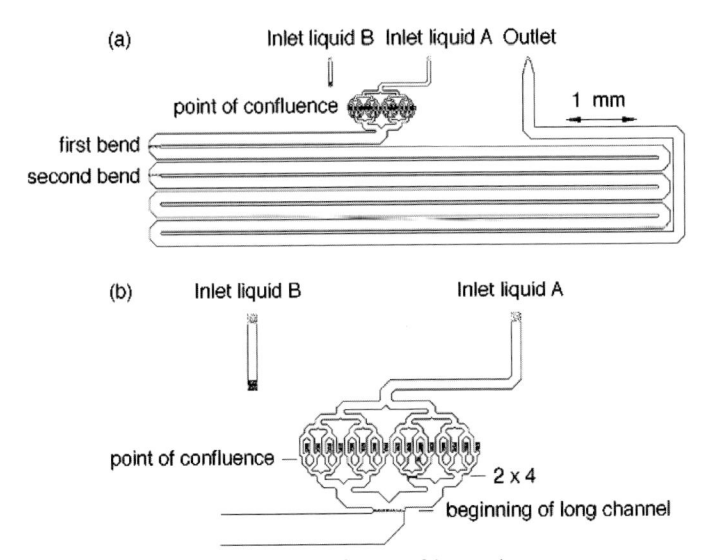

Fig. 2.5.4 (a) Schematic of layer 1 of a microfabricated micromixer that can be used to initiate reactions. (b) Details of layer 1. Reprinted from Ref. [35]. Copyright (1999) with permission from The Royal Society of Chemistry.

Using flow microcoil NMR in combination with a micromixer, Kakuta et al. expanded the technique previously used by Nuland et al. and performed studies of protein unfolding kinetics [37]. The use of continuous-flow and a micromixer allows a longer observation time when compared with stopped-flow NMR. Kakuta et al. studied real-time methanol-induced conformational changes in ubiquitin. Under acidic conditions, methanol induces a transition from a native to a partially unfolded state, the A-state. Certain amino acids in ubiquitin, His-68 and Tyr-59, are well suited as proton NMR probes to follow such conformational changes. The experiments were performed by mixing two solutions: 20% CD_3OD–80% D_2O and 7 mM ubiquitin (pD = 2.4) and 80% CD_3OD–20% D_2O (pD = 2.4). After mixing, the composition of the resultant solution should be 50% CD_3OD–50% D_2O at pD = 2.4. The conformational change takes place from the point of solution mixing and continues until the solution reaches the NMR microcoil cell. The time-dependent behavior of the two states, characterized by the intensities of the peaks of native- and A-state His-68 and Tyr-59, is shown in Figure 2.5.5(a). Up to t = 40 s the population ratio, A/N, increases exponentially and reaches a plateau up to ≈80 s after which it increases again. This behavior suggests that the conformational

Fig. 2.5.5 A study examining the conformational changes of the protein ubiquitin, showing the population ratio of the A-state to the native-state as a function of time. (a) The reaction from 0 to 120 s. (b) The reaction for the first 40 s, including curves fit to a single exponential. Reprinted with permission from Ref. [37]. Copyright (2003) American Chemical Society.

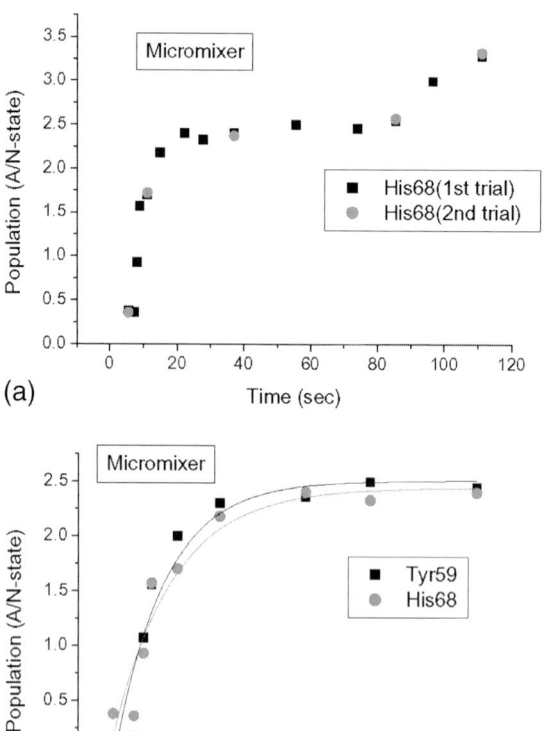

change in ubiquitin is at least second order, and that the protein may have more than two states during this transition. A detailed analysis of the behavior within the first 40 s after mixing is shown in Figure 2.5.5(b). The population ratio, A/N, can be fitted to a single exponential with a rate constant of 0.21 s^{-1}.

2.5.5
Multiple Microcoil NMR. Sensitivity and Throughput Issues

Another exciting area of instrumental integration is the development of multiple microcoil NMR probes, which allows multiple detection points (time points) to be monitored simultaneously, therefore increasing the amount of data that can be acquired per unit time. The applicability of this technique to the study of chemical reactions has been demonstrated by Ciobanu et al. [38]. These workers measured the rate constant of the xylose–borate reaction using an NMR probe containing three individual microcoils. Two fluid flows, containing the reactants, were mixed and they then flowed through a capillary around which are wound multiple, physically distinct NMR detector coils. The distance between the mixer and each individual NMR coil, together with the flow rate used, determined the post-reaction time being monitored. As in the experiments performed by Kakuta et al., signal averaging can be performed for as long as necessary to obtain an adequate SNR, as data measurement time and reaction time are decoupled via this experimental method. However, the longer the required total data acquisition time, the greater the amounts of reactants needed. The use of three small-volume microcoils (Figure 2.5.6) minimized the sample amount. In this particular application, as the reaction progressed, a decrease in the height of the α and anomeric peaks characteristic to

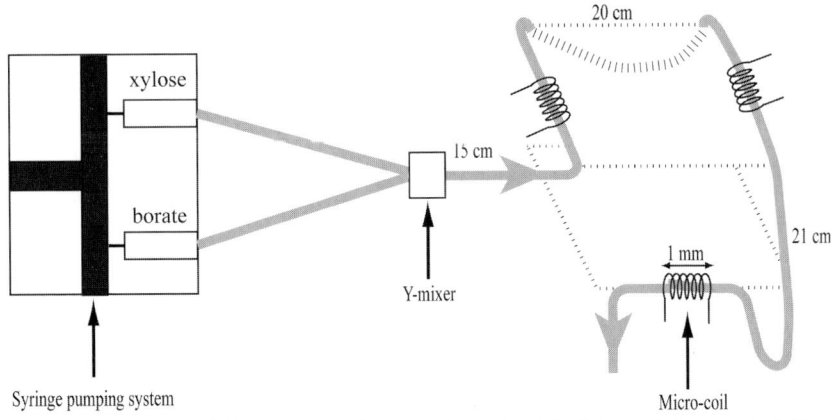

Fig. 2.5.6 Schematic of the experimental set-up used to monitor reaction kinetics with a multiple microcoil system. Two syringes on the pump inject the reactants into two capillaries. The reactants are mixed rapidly with a Y-mixer. After mixing, the solution flows through the microcoils while data are being acquired. The total reaction time observed at each coil depends on the flow rate and distance from the mixer. Reprinted with permission from Ref. [38]. Copyright (2003) John Wiley & Sons.

Fig. 2.5.7 The ratio of the peak amplitudes, product/reactant, as a function of time with the data obtained from the system illustrated in Figure 2.5.5. Symbols: squares, coil 1; triangles, coil 2; and circles, coil 3. The error bars represent the standard error obtained from ten measurements. Reprinted with permission from Ref. [38]. Copyright (2003) John Wiley & Sons.

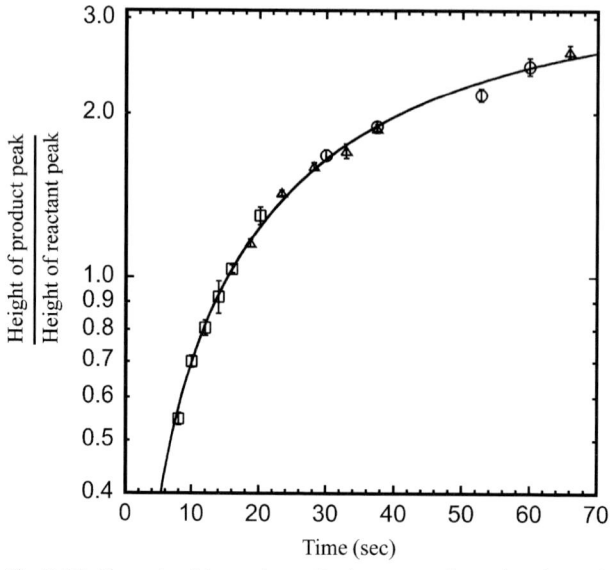

xylose (reactant) and the appearance of a new peak (product) were observed. NMR spectra, corresponding to successive reaction times, were recorded independently and simultaneously with all three rf coils. The results from a quantitative analysis of the spectra are presented in Figure 2.5.7. The data from all three probes were plotted on the same axis in this figure and fell on the same curve, validating the use of distinct NMR coils at multiple locations along the capillary/reaction time coordinate. The ratio of the peak heights, product peak/reactant peak, was plotted as a function of time. By fitting the experimental data to the theoretical product/reactant ratio, a reaction rate constant, $k_1 = 0.077$ 0.004 s^{-1} mol^{-1} L, for the second-order reaction was obtained. The same method can also be used to perform continuous-flow two-dimensional NMR experiments at a particular reaction time, which demonstrates the capability of examining intermediate species. Figure 2.5.8 shows COSY spectra of the D-xylose–borate mixture for continuous- and stopped-flow, respectively. The spectrum in Figure 2.5.8(a) was recorded for a flow rate of 2 µL min^{-1} corresponding to a reaction time of $t \approx 165$ s. The on-flow COSY is highlighted by the presence of intense reactant peaks in the region of 3.0–3.5 ppm. The spectrum in Figure 2.5.8(b) shows very weak reactant cross-peaks (3.0–3.5 ppm) and strong cross-peaks for the xylose–borate product. Interestingly, the cross-peak observed at 3.45–5.10 ppm in the continuous-flow COSY disappeared in the stopped-flow COSY. This suggests the presence of a reaction intermediate, which, however, was not identified.

Further expansion of the probehead to more coils is also possible, with a concomitant reduction in the total amount of material needed for a kinetic study.

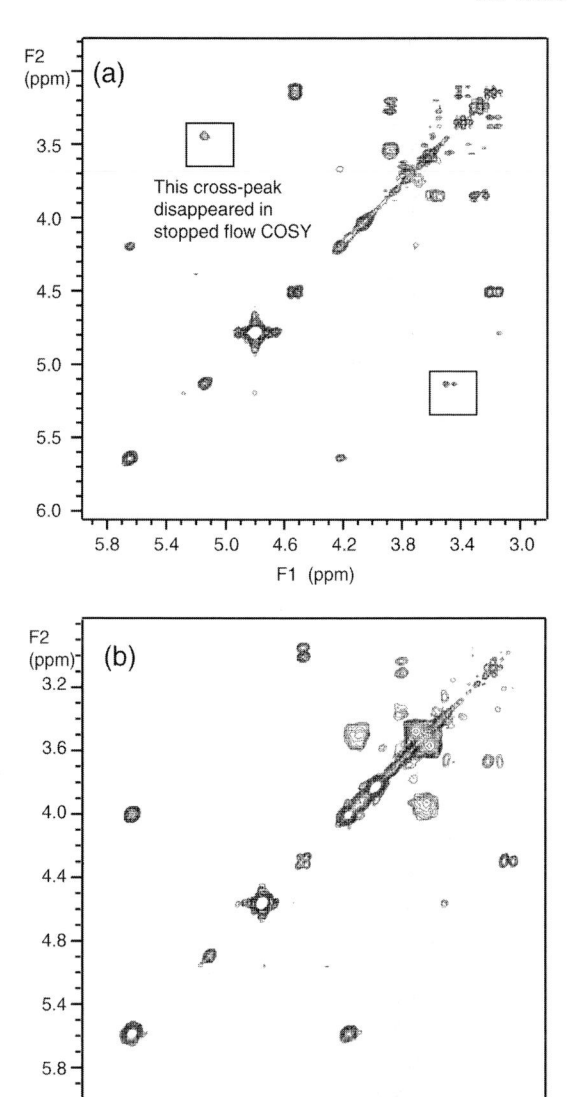

Fig. 2.5.8 COSY spectra of 300 mM D-xylose plus 400 mM borate at pD = 10. The spectra were recorded at 300 MHz with a single NMR microcoil using the instrumentation shown in Figure 2.5.5. (a) Continuous-flow. Flow rate = 2 μL min^{-1}, corresponding to a reaction time $t \approx 165$ s. The on-flow COSY is highlighted by the presence of intense reactant peaks in the region of 3.0–3.5 ppm. (b) Stopped-flow. The spectrum shows very weak reactant cross peaks (3.0–3.5 ppm) and strong cross peaks for the xylose–borate product. Interestingly, the cross peak observed at 3.45–5.10 ppm in the continuous-flow COSY disappeared in the stopped-flow COSY. Acquisition: 1000(t_2) × 512(t_1) data points; 16 scans per t_1 increment, 0.256 s acquisition time; relaxation delay 3 s (stopped-flow) and 1 s (continuous-flow). Reprinted with permission from Ref. [38]. Copyright (2003) John Wiley & Sons.

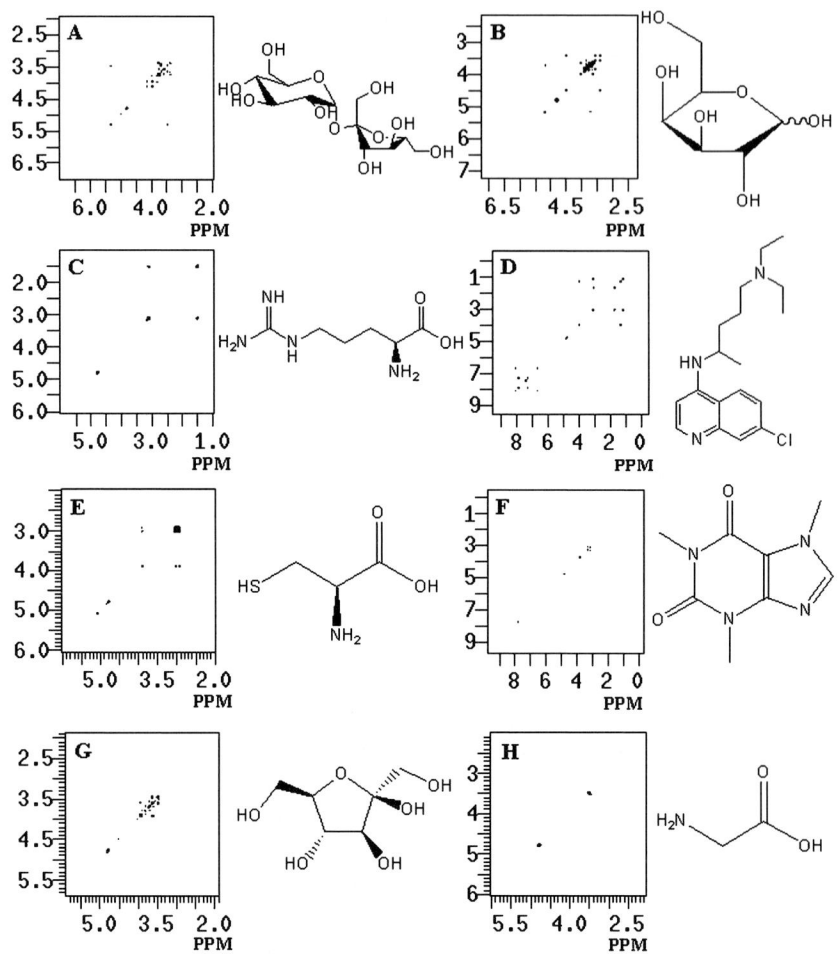

Fig. 2.5.9 COSY spectra acquired at 600 MHz with an eight-coil probe along with the chemical structures of the compounds used. Each sample was a 10 mM solution in D_2O loaded into the coil via the attached Teflon tubes, with the samples being: (A) sucrose, (B) galactose, (C) arginine, (D) chloroquine, (E) cysteine, (F) caffeine, (G) fructose and (H) glycine. Data acquisition parameters: data matrix 2048 × 256, 8 scans, sw = 6000 Hz, sw_1 = 6000 Hz. Data were zero filled in t_1 to 2048 points, processed with shifted sine-bell window functions applied in both dimensions, symmetrized and displayed in magnitude mode. Reprinted from Ref. [39]. Copyright (2004) with permission from Elsevier.

Recently, Wang et al. [39] reported the construction of an eight coil probehead operating at 600 MHz. By using four receivers and radiofrequency switches, combined with careful timing of data acquisition, spectra from eight different chemical solutions were acquired in the time normally required for one. Figure 2.5.9 shows two-dimensional COSY spectra of eight different compounds (sucrose, galactose, arginine, chroloquine, cysteine, caffeine, fructose and glycine) obtained using the above mentioned NMR probe. Chemical reaction studies using such a probe are planned.

2.5.6
Conclusions

Advances in NMR spectroscopy over the last ten years have revolutionized its applications to the study of different chemical processes. The earliest experiments involved the sequential accumulation of spectra following initiation of the reaction. However, this approach was limited to relatively slow reactions. In order to monitor rapid reactions one has to initiate the reaction within the active volume of the NMR probe. This can be accomplished using various injection and continuous-flow methods. Continuous-flow is particularly valuable as it allows the acquisition of data at a particular reaction time for as long as is necessary. However, the longer the acquisition time the more material is needed. The amount of material is minimized using microcoil NMR probeheads. In order to increase the throughput, probes that incorporate multiple microcoils have been designed. The recent development of such probes is transforming NMR into a powerful technique for the study of a wide range of chemical and biochemical reactions.

Besides basic research into reaction kinetics, the monitoring of product formation in a continuous fashion using NMR is particularly intriguing for industrial applications in the chemical and pharmaceutical industries. As the components in such large scale synthetic processes are often present at high millimolar to molar concentrations, the relatively poor sensitivity of NMR is less of an issue. We expect such applications to be demonstrated in the near future. Advantages of the capillary sampling and microcoil approach to such applications include the ease of sampling with a capillary, and the elevated temperature and pressure capabilities of typical fused-silica capillaries, which will allow industrial processes to be monitored continuously over a broad range of conditions.

Acknowledgments

The authors wish to acknowledge the support of the National Institutes of Health through grant R01 GM53030 for supporting this research.

References

1 R. Almog, C. A. Waddling, F. Maley, G. F. Maley, P. Van Roey **2001**, (Crystal structure of a deletion mutant of human thymidylate synthase Delta(7–29) and its ternary complex with Tomudex and dUMP), *Protein Sci.* 10(5), 988.

2 J. M. Hutzler, T. S. Tracy **2002**, (Atypical kinetic profiles in drug metabolism reactions), *Drug Metab. Dispos.* 30(4), 355.

3 X. S. Chai, Q. Luo, J. Y. Zhu **2002**, (Multiple headspace extraction-gas

chromatographic method for the study of process kinetics), *J. Chromatogr. A* 946(1–2), 177.

4 K. Albert, E.-L. Dreher, H. Straub, A. Rieker **1987**, (Monitoring electrochemical reactions by ^{13}C NMR spectroscopy), *Magn. Reson. Chem.* 25, 919.

5 C. A. Fyfe, M. Cocivera, S. W. H. Damji **1975**, (High resolution nuclear magnetic resonance studies of chemical reactions using flowing liquids. Investigation of the kinetic and thermodynamic intermediates formed by the attack of methoxide ion on 1-X-3,5-dinitrobenzenes), *J. Am. Chem. Soc.* 97, 5707.

6 C. A. Fyfe, S. W. H. Damji, A. Koll **1979**, (Low-temperature flow NMR investigation of the transient intermediate in the nucleophilic aromatic substitution of 2,4,6-trinitroanisole by n-butylamine), *J. Am. Chem. Soc.* 101, 951.

7 K. Albert **2002**, in *On-line LC-NMR and Related Techniques*, ed. K. Albert, John Wiley & Sons, New York.

8 P. T. Callaghan, C. D. Eccles, Y. Xia **1988**, (NMR microscopy of dynamic displacements: k-space and q-space imaging), *Phys. E* 21, 820.

9 E. M. Haake, R. W. Brown, M. R. Thompson, R. Venkatesan **1999**, *Magnetic Resonance Imaging*, John Wiley, New York.

10 X. Zhang, A. G. Webb **2005**, (Magnetic resonance microimaging and numerical simulations of velocity fields inside flowcells for coupled NMR microseparations), *Anal. Chem.*, 77, 1338.

11 Y. Asashi, E. Mizuta **1972**, (Flow method in high-resolution nuclear magnetic resonance), *Talanta* 19, 567.

12 W. S. Trahanovsky, D. R. Fisher **1990**, (Observation of benzocyclobutadiene by flow nuclear magnetic resonance), *J. Am. Chem. Soc.* 112, 4971.

13 L. K. Tan, M. Cocivera **1982**, (Amino acid addition to 4-formylpyridine and its quaternized salt: transamination), *Can. J. Chem.* 60, 778.

14 L. K. Tan, M. Cocivera **1982**, (Flow ^{1}H NMR study of the rapid nucleophilic addition of amino acids to 4-formylpyridine), *Can. J. Chem.* 60, 772.

15 M. Hunger, T. Horvath **1996**, (Conversion of propan-2-ol on zeolites LaNaY and HY investigated by gas chromatography and in situ MAS NMR spectroscopy under continuous-flow conditions), *J. Catal.* 167, 187.

16 S. Funahashi, K. Ishihara, S. Aizawa, T. Sugata, M. Ishii, Y. Inada, M. Tanaka **1993**, (High-pressure stopped-flow nuclear magnetic resonance apparatus for the study of fast reactions in solution), *Rev. Sci. Instrum.* 64, 130.

17 M. Maiwald, H. H. Fischer, Y.-K. Kim, K. Albert, H. Hasse **2004**, (Quantitative high-resolution on-line NMR spectroscopy in reaction process monitoring), *J. Magn. Reson.* 166, 135.

18 J. J. Grimaldi, B. D. Sykes **1975**, (Concanavalin A: a stopped flow nuclear magnetic resonance study of conformational changes induced by Mn^{++}, Ca^{++}, and alpha-methyl-D-mannoside), *J. Biol. Chem.* 250, 1618.

19 N. A. J. van Nuland, V. Forge, J. Balbach, C. M. Dobson **1998**, (Real-time NMR studies of protein folding), *Acc. Chem. Res.* 31, 773.

20 J. Jonas **2002**, (High-resolution nuclear magnetic resonance studies of proteins), *Biochim. Biophys. Acta* 1595, 145.

21 D. I. Hoult, R. E. Richards **1976**, (The signal-to-noise ratio of the nuclear magnetic resonance experiment), *J. Magn. Reson.* 24, 71.

22 D. G. Gadian, F. N. H. Robinson **1979**, (Radiofrequency losses in NMR experiments on electrically conducting samples), *J. Magn. Reson.* 34, 449.

23 T. L. Peck, R. L. Magin, P. C. Lauterbur **1995**, (Design and analysis of microcoils for NMR microscopy), *J. Magn. Reson. B* 108, 114.

24 K. R. Minard, R. A. Wind **2001**, (Solenoidal microcoil design – Part II: Optimizing winding parameters for maximum signal-to-noise performance), *Concepts Magn. Reson.* 13, 190.

25 K. R. Minard, R. A. Wind **2001**, (Solenoidal microcoil design. Part I: Optimizing RF homogeneity and coil dimensions), *Concepts Magn. Reson.* 13(2), 128.

26 A. G. Webb **1997**, (Radiofrequency microcoils in magnetic resonance), *Progr. Nucl. Magn. Reson. Spectrosc.* 31, 1.

27 D. A. Seeber, R. L. Cooper, L. Ciobanu, C. H. Pennington **2001**, (Design and testing of high sensitivity micro-receiver coil apparatus for nuclear magnetic resonance and imaging), *Rev. Sci. Instrum.* 72, 2171.

28 D. L. Olson, T. L. Peck, A. G. Webb, R. L. Magin, J. V. Sweedler **1995**, (High resolution microcoil ^{1}H-NMR for mass limited nanoliter-volume samples), *Science* 270, 1967

29 M. E. Lacey, Z. J. Tan, A. G. Webb, J. V. Sweedler **2001**, (Union of capillary high-performance liquid chromatography and microcoil nuclear magnetic resonance spectroscopy applied to the separation and identification of terpenoids), *J. Chromatogr. A* 922(1–2), 139.

30 M. E. Lacey, A. G. Webb, J. V. Sweedler **2000**, (Monitoring temperature changes in capillary electrophoresis with nanoliter-volume NMR thermometry), *Anal. Chem.* 72(20), 4991.

31 A. M. Wolters, D. A. Jayawickrama, C. K. Larive, J. V. Sweedler **2002**, (Insights into the cITP process using on-line NMR spectroscopy), *Anal. Chem.* 74(16), 4191.

32 J. Evans, D. Liepmann, A. P. Pisano **1997**, (Planar laminar mixer), in *Proc. IEEE MEMS Workshop*, Nagoya, Japan, (p.) 96.

33 H. T. Evensen, D. R. Meldrum, D. L. Cunningham **1998**, (Automated fluid mixing in glass capillaries), *Rev. Sci. Instrum.* 69, 519.

34 R. M. Moroney, R. M. White, R. T. Howe **1991**, (Ultrasonically induced microtransport), in *Proc. IEEE MEMS Workshop*, Amsterdam, The Netherlands.

35 F. G. Bessoth, A. deMello, A. Manz **1999**, (Microstructure for efficient continuous flow mixing), *Anal. Commun.* 36, 213.

36 J. Branebjerg, P. Gravesen, J. P. Krog, C. R. Nielsen **1996**, (Fast mixing by lamination), in *Proc. IEEE MEMS Workshop*, San Diego, CA.

37 M. Kakuta, D. A. Jayawickrama, A. M. Wolters, A. Manz, J. V. Sweedler **2003**, (Micromixer-based time-resolved NMR: Applications to ubiquitin protein conformation), *Anal. Chem.* 75(4), 956.

38 L. Ciobanu, D. A. Jayawickrama, X. Zhang, A. G. Webb, J. V. Sweedler **2003**, (Measuring reaction kinetics by using microcoil NMR spectroscopy), *Angew. Chem., Int. Ed. Engl.* 42(38), 4669.

39 H. Wang, L. Ciobanu, A. S. Edison, A. G. Webb **2004**, (An eight-coil high-frequency probehead design for high-throughput nuclear magnetic resonance spectroscopy), *J. Magn. Reson.* 170, 206.

2.6
Broadening the Application Range of NMR and MRI by Remote Detection
Song-I Han, Josef Granwehr, and Christian Hilty

2.6.1
Introduction

NMR remote detection is a new technique in which the signal detection location is physically separated from the sample location where NMR or MRI information is encoded. A mobile NMR-active sensor medium (or "nucleus") interacts with the host sample without altering it, and information about that sample environment is encoded with radiofrequency (rf) and field gradient pulses as longitudinal magnetization of the sensor. After the sensor has left the sample, its magnetization is read out with an optimized detector, which can often be made much more sensitive than the circuit used to encode the information. Remote detection capitalizes fully on the unique strength of NMR – the wealth of encoding possibilities – while providing a means to overcoming its inherent weak point, the low sensitivity.

2.6.2
Motivation

NMR spectroscopy and magnetic resonance imaging (MRI) techniques are particularly powerful and versatile in providing information about atomic and macroscopic structures over a large length and time scale. Chemical constitution and molecular conformations can be reconstructed, rotational and translational dynamics measured and thermal diffusion, coherent flow and macroscopic densities mapped. However, NMR techniques are intrinsically insensitive compared with other analytical and visualization tools. One reason is that for a thermally polarized sample, even at high field, only a few ppm of the total number of NMR active nuclei contribute to the overall signal. Furthermore, conventional rf detection scales up unfavorably for use with large, bulky sample volumes and low magnetic fields. Much effort has been invested in seeking techniques that provide for higher nuclear spin polarization and higher signal detection efficiency to overcome this sensitivity limitation. High magnetic fields would serve both aspects (the signal-to-noise ratio scales with $B_0^{7/2}$, where B_0 is the static magnetic field), however the upper limit field strength for commercially available NMR magnets lies currently at 21 T. Various hyper-polarization techniques [1–6] can provide percent-range nuclear polarization, with signal enhancements of several orders of magnitude. Cryo-cooled rf circuits give about four times higher detection sensitivity [7], and when working with nanoliter to microliter range sample volumes, advantage can be taken of the better mass sensitivity offered by rf microcoils [8–10]. However, most of these techniques are only applicable to specific nuclei and samples or enhance the detection sensitivity by just a small factor, despite a considerable cost implication.

An important consideration for making NMR and MRI suitable for applications in chemical engineering is to provide for flexibility in handling samples of various sizes, materials, shapes or temperatures, while ensuring sufficiently high NMR detection sensitivity. In this chapter, we present a novel technique known as NMR and MRI remote detection [11, 12], which addresses these problems in a fundamental fashion, as well as a few examples where remote detection has been applied successfully.

2.6.3
Principle of NMR Remote Detection

2.6.3.1 Spatial Separation of NMR Encoding and Signal Detection

The idea of NMR remote detection is to spatially separate the encoding of information about a sample from the detection of the signal. The information is stored as longitudinal magnetization M_z of a mobile sensor medium, which can be transferred between the sample and the detector (Figure 2.6.1). The encoding is technically optimized to provide a high quality of the encoded information, while the detector is optimized to measure M_z as sensitively as possible. The sample is no longer confined to conditions that meet both high encoding quality and detection sensitivity. Optimizing for high detection sensitivity in conventional NMR or MRI typically means using high magnetic fields and small volume rf coils. On the other hand, conditions for optimized encoding include a good rf and magnetic field homogeneity, a sufficiently large rf coil volume to encompass bulky samples and low magnetic fields if samples with large susceptibility gradients, due to paramagnetic impurities and metallic components, are to be imaged.

As remote detection spatially separates the detection from the encoding, optimized detection can be realized by a variety of means to sensitively read out M_z. Detection can be achieved not only by an rf coil of optimized size – the detection sensitivity is inversely proportional to the coil diameter [8–10] – but also by alternative methods or devices. Examples of the latter include Superconducting Quantum Interference Devices (SQUIDs) [13, 14], spin-exchange optical detection [2] or atomic magnetometers [15–17] (Figure 2.6.1). This approach to optimizing encoding quality and detection sensitivity individually by transporting only the NMR information, but not the sample itself, to the detector can enhance the sensitivity of NMR and MRI by orders of magnitude.

2.6.3.2 NMR Information Carrier

The question to be answered next is how the NMR information on the sample is being stored and carried to the detection location. The information can be transported by any fluid that has NMR active nuclei with a sufficiently long spin-lattice relaxation time, T_1, to enable reading out of the encoded information after the time of travel to the detector. Furthermore, the carrier should be chemically inert so that it does not alter the sample when passing through. Depending on the application, the distance that the carrier travels during this time can span a wide range from tens of millimeters to several meters. A variety of nuclei are suitable to act as NMR

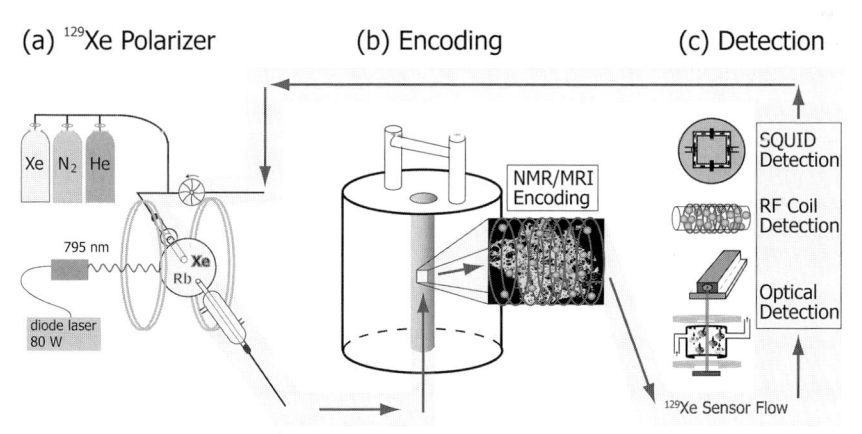

Fig. 2.6.1 Schematic of an experiment with remote detection. The basic steps are: (a) the polarization of the sensor medium, (b) NMR or MRI encoding using rf pulses and magnetic field gradients and (c) signal detection. The NMR or MRI information travels between the locations (b) and (c).

signal carriers, among them laser-polarized ^{3}He or ^{129}Xe gas [1, 2], ^{13}C nuclei of solvents or ^{1}H nuclei of oil or water. The T_1 of the carriers listed decreases in the order in which they are given. If not in contact with any relaxation causing substance or surface, this is up to 40 h for ^{3}He, tens of minutes for ^{129}Xe gas and a few seconds for ^{1}H of water. Accordingly, the traveling distances covered can range from many meters down to only a few millimeters, depending on the choice of the carrier. The long-lived inert noble gases ^{3}He or ^{129}Xe have to be laser-polarized in order to deliver high initial polarization, whereas for the shorter lived ^{1}H of water, sufficiently high thermal polarization is achieved in the presence of magnetic fields with medium to high strength. This discussion also shows that the carrier can be in the gas or in the liquid phase, so that complex flow properties of the carrier itself through porous samples can be studied.

Depending on the choice of the carrier medium, NMR spectra as well as NMR images can be detected remotely using the same basic principle that will be discussed shortly. However, for remote detection of NMR spectra, the information carrier has to fulfill one more criteria. It has to show a chemical shift upon non-covalent contact with the sample environment. Thus, differences in surface chemistry, pore size or other physicochemical properties of the material can be detected. The best candidate nucleus from this point of view is ^{129}Xe, and other nuclei such as ^{15}N and ^{13}C also have potential to serve as sensor atoms, given that mechanisms for hyperpolarization have been developed to overcome the signal limitation due to their low density, low natural abundance and low gyro-magnetic ratio. The spin polarization of ^{129}Xe can be enhanced by 3–5 orders of magnitude relative to thermal polarization by Rb-Xe spin exchange optical pumping using circularly polarized laser light with the frequency of the Rb D_1 line [1, 2]. Figure 2.6.1 shows schematically the basic steps of remote detection when using continuous-flow laser-polarized ^{129}Xe gas as the information carrier. All example data presented in this chapter use this fundamental scheme.

The most basic experimental manifestation of travel information can be obtained by tracing the flow of tagged spins from the encoding to the detection location. The spins in the encoding coil are tagged by applying a π-pulse to invert the ^{129}Xe magnetization. Next, the arrival of the traveling spins is monitored continuously at the detection location by repeated pulsing and the subsequent recording of the free induction decay (FID) at the detection coil. An example of a travel time curve obtained in this way is shown in Figure 2.6.2. At short times after the tagging, the spin-encoded fluid packets have not yet reached the detector, therefore the detection signal shows maximum intensity. When encoded fluid begins to reach the detector, the signal intensity becomes reduced, finally showing a negative intensity. Later the signal slowly recovers to its maximum. The signal never reaches an inverse value of the unencoded signal in this example, mainly due to dispersion and mixing with unencoded spins (positive amplitude), but also because of an rf field that is not perfectly homogeneous, B_1, and the corresponding inhomogeneous distribution of flip angles of the tagging pulse, which is a common problem with large encoding volumes.

2.6 Broadening the Application Range of NMR and MRI by Remote Detection | 143

(a)

(b) π Inversion pulse   π/2 Detection pulses

SS

Travel time

(c)

Traveled NMR signal amplitude [a.u.]

Travel time [ms]

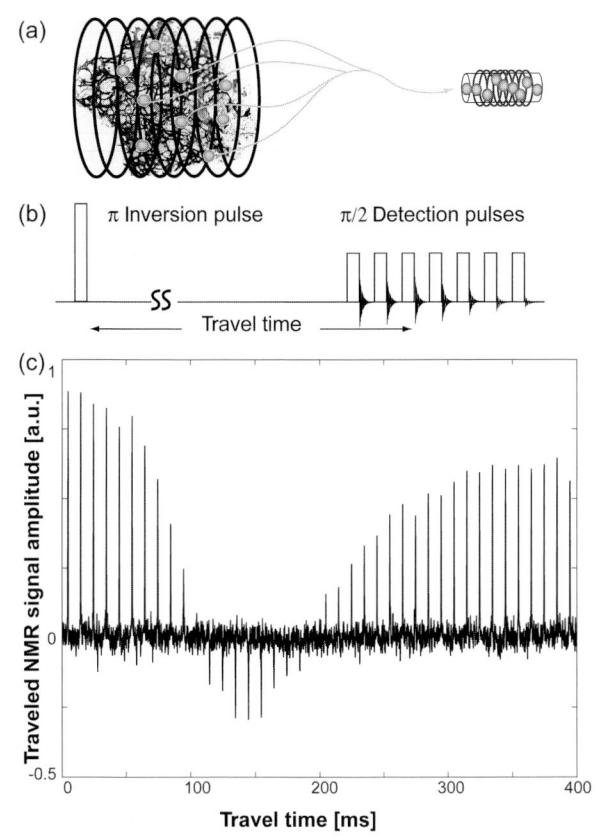

Fig. 2.6.2 (a) Transportation of tagged spins from the encoding rf coil to the detection rf coil using a remote detection set-up. (b) Basic pulse sequence used to invert the magnetization of spins in the encoding coil, and subsequently to detect the magnetization at the detector location as a function of the travel time. (c) A typical travel time curve where encoding and detection were performed at high field inside the same magnet

2.6.3.3 Encoding and Reconstruction of NMR and MRI Information

In general, a detectable NMR signal is induced by transverse magnetization that precesses coherently with characteristic amplitude and frequency distributions in response to the applied pulse sequence, a tailored series of rf and gradient pulses. Any NMR signal that can be observed directly as a transient signal during an evolution time t_2 can also be read out indirectly in a point-by-point fashion. When using remote detection, at each time increment, t_1, during precession, a projection of the magnetization along the x or y axis is stored as longitudinal magnetization using a π/2 storage pulse. As such, it is not affected by dephasing due to the presence of inhomogeneous magnetic fields. It is then physically transported to the detector during the travel time, where its amplitude is read out. Both the x and the y components of the magnetization can be obtained in two subsequent experiments and constitute one complex point of the NMR signal. This procedure is repeated for every time evolution increment t_1, and the series of the signal points collected corresponds to the indirectly detected NMR signal in the time domain. Its Fourier transformation (FT) is the reconstructed NMR spectrum or image, depending on the applied pulse sequence. One drawback is that the indirect acquisition inherently adds one dimension to the remote experiment. However, as discussed above, in many cases this is outweighed by a large gain in sensitivity due to the optimized detection, whether it is achieved conventionally with an inductive coil or by other means. When using inductive detection, the bandwidth of detection can be narrowed, for example, by the application of pulse sequences such as spin lock detection [18] or the use of multiple echoes with pulse train refocusing [19], which can considerably enhance the detected signal. This is possible because in the detected dimension, the signal amplitude, and not the frequency, carries the indirectly detected frequency information.

To demonstrate these principles, the remote detection of the spectrum of ^{129}Xe gas adsorbed onto the surface of aerogel fragments is presented [11]. The encoding rf coil encompasses the sample, laser-polarized ^{129}Xe gas passes through and is detected as it travels to the rf detection coil. Both coils are contained in the same magnetic field and are built into the same NMR probe [Figure 2.6.3(a)]. Figure

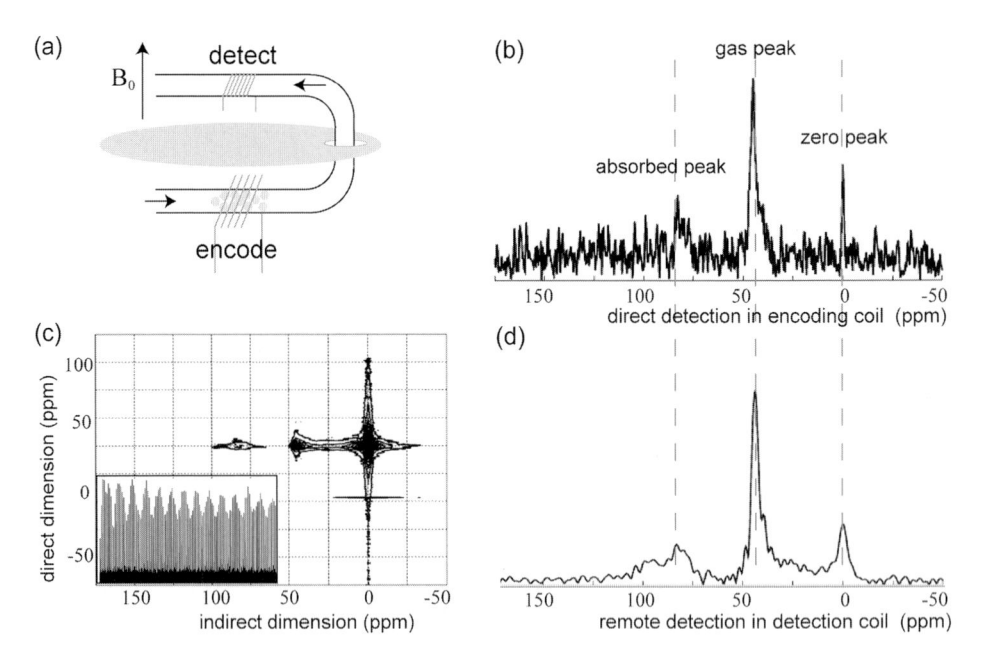

Fig. 2.6.3 (a) Set-up for remote detection with two rf coils, one coil containing an aerogel sample where ^{129}Xe spectroscopic information is encoded, and a detection coil through which the signal carrier passes. (b) ^{129}Xe spectrum, detected directly in the encoding coil obtained with one $\pi/2$ pulse and one signal acquisition. (c) and (d) Remotely detected spectrum. The amplitude modulation of the signal arriving at the detector along the indirect dimension [inset in (c)] after FT gives the spectrum from the sample location (d). The transiently detected signal arriving at the remote detector consists of only one peak for pure ^{129}Xe gas [figure taken from 11].

2.6.3(b) shows the ^{129}Xe spectrum of the sample detected directly in the encoding coil. It contains peaks for the free ^{129}Xe gas and the ^{129}Xe adsorbed inside the 100 Å small aerogel pores. Figure 2.6.3(c) correlates the transiently detected signal in the detection coil with the remotely reconstructed ^{129}Xe spectrum of the sample. The inset of Figure 2.6.3(c) shows the amplitude modulation of the detected signal in the indirect dimension of the remote data, which contains the spectral information from the location of the sample. These experiments show the feasibility of remote detection as a conceptually new approach to address the sensitivity limitations in NMR spectroscopy.

2.6.4
Sensitivity Enhancement by Remote Detection

The separation of encoding and detection allows the use of different types of detectors. While at high field it might be most convenient to use pulsed or continuous wave inductive detection, where sensitivity improvements could originate from the possibility of concentrating the sensor spins into a smaller volume than the encoding volume, at low fields alternative detection methods such as spin-exchange optical detection [2] or magnetometers [13, 17] might be preferred. The discussion of the sensitivity of remote detection can be split into a detector independent part, which only includes the relative sensitivity between the remote detector and the circuit that is used for encoding, and a detector dependent part, which discusses the ability of the different detectors to measure the polarization or the magnetic moment of the sensor spins.

2.6.4.1   Detector-independent Sensitivity

The most fundamental aspect of a sensitivity discussion of remote detection is the fact that it is inherently a point-by-point technique. Each spectrum recorded by the detector does not contain any information other than its amplitude. Conceptually, a remote NMR experiment is very similar to a 2D NMR experiment with a z filter between encoding and detection, which causes all transverse magnetization to dephase. For 2D NMR experiments, it has been shown that the signal-to-noise ratio (SNR) per square root time, which will be denominated as sensitivity in the following, is the same as in the 1D case when neglecting T_2 relaxation [20, 21]. To compare the sensitivity of a remotely detected spectrum [Figure 2.6.4(b)] with an equivalent experiment with direct detection [Figure 2.6.4(a)], we can use an expression similar to the discussion in Ref. [20]:

$$\left(\frac{\text{remote sensitivity}}{\text{direct sensitivity}}\right)_{\text{NMR}} = \frac{\Lambda}{\sqrt{2}} \sqrt{\left[\frac{t_r^{max}/\tau_r}{t_d^{max}/\tau_d}\right] \left[\frac{1}{t_r^{max}} \int_0^{t_r^{max}} [s_r^e(t_r)]^2 dt_r\right]} \quad (2.6.1)$$

$$= \frac{\Lambda}{\sqrt{2}} \sqrt{\frac{t_r^{max}/\tau_r}{t_d^{max}/\tau_d}} \sqrt{\frac{T_2^r}{2 t_r^{max}} \left[1 - \exp\left(-\frac{2 t_r^{max}}{T_2^r}\right)\right]} \approx \frac{\Lambda}{\sqrt{2}} \sqrt{\frac{T_2^r}{2 t_d^{max}}}$$

where τ_d and τ_r are the recycle delays and t_d^{max} and t_r^{max} are the detection times in the direct and the remote experiment, respectively [22]; Λ is the relative sensitivity between the remote detector and the encoding circuit; $s_r^e(t_r) = \exp(-t_r/T_2^r)$ is the envelope of the signal in the remote detection dimension, which decays with T_2^r. Because the encoding dimension of the remotely detected experiment samples the same data points as the direct experiment, the signal decay in the encoding volume, $s_d^e(t_d) = \exp(-t_d/T_2^d)$, cancels from the sensitivity ratio. A matched filter for optimum sensitivity in the remote dimension was assumed, which causes the bandwidth of the detection to be $\Delta f = 4/T_2^r$. The factor $1/\sqrt{2}$ in Eq. (2.6.1) is because only one component of the transverse magnetization can be stored and read out at a time with remote detection, requiring two cycles to obtain the complex signal. t_d^{max}/τ_d and t_r^{max}/τ_r represent the duty cycle in the remote and the direct experiment, respectively. For the approximation it was assumed that $\tau_d = \tau_r$ and $t_r^{max} > T_2^r$. As is visualized in Figure 2.6.4, if $\Lambda = 1$ and the number of signal averaging steps of an experiment with direct detection is equal to the number of points detected indirectly with remote detection, the relative sensitivity is given by the integral of all the remotely detected signals relative to the integral of all the directly detected signals. An equal sensitivity is obtained only if the remote signal does not decay during detection, because, in addition, the plain signal averaging in

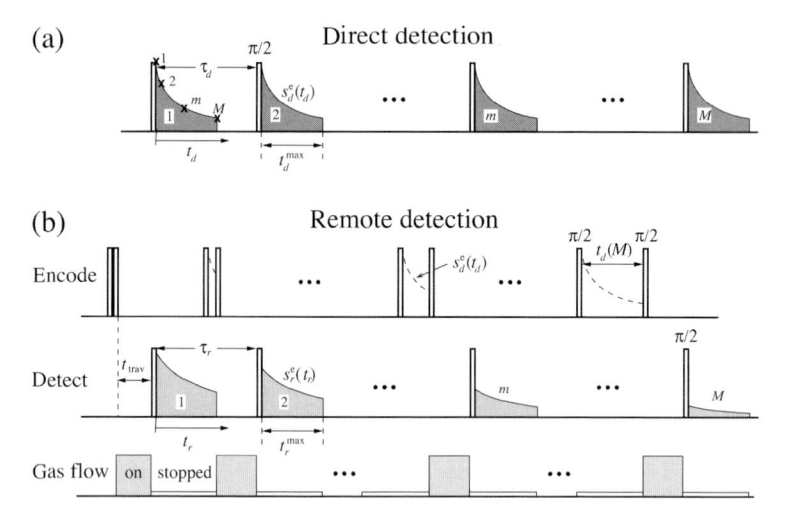

Fig. 2.6.4 Sensitivity comparison between direct (a) and remote detection (b). With direct detection, an FID is recorded transiently with M data points, which are marked with the symbols "x" in the first FID in (a). Remotely, M encoding steps are necessary to obtain the same data set, which allows one to perform M signal averaging steps in the direct dimension in the same time. The encoding and detection steps in the remote experiment are intermingled, therefore only a time overhead corresponding to one travel time occurs, which will be neglected. The stored magnetization of the mth encoding step corresponds to the mth data point with direct detection and marks the magnetization at the beginning of the remote detection. The sensitivity is proportional to the total area under all the FIDs in both cases. With identical detectors, we would obtain an equal sensitivity with remote detection only if the remote signal did not decay.

the direct dimension does not show any signal decay and has therefore an infinitely narrow bandwidth.

The situation is different in an experiment where the acquisition dimension of the conventional, direct detection experiment itself measures only the magnitude of a signal, as opposed to its time evolution. This is the case for example in an MRI experiment with phase encoding in all three dimensions. Here, it is the 1D sensitivity that has to be compared between remote and direct detection, because the direct FID is no longer sampled point-by-point. Also, by following the treatment of 1D sensitivity in Ref. [20], this yields:

$$\left(\frac{\text{remote sensitivity}}{\text{direct sensitivity}}\right)_{\text{MRI}} = \frac{\Lambda}{\sqrt{2}} \sqrt{\left[\frac{t_r^{max}/\tau_r}{t_d^{max}/\tau_d}\right] \frac{\frac{1}{t_r^{max}}\int_0^{t_r^{max}} [s_r^e(t_r)]^2 dt_r}{\frac{1}{t_d^{max}}\int_0^{t_d^{max}} [s_d^e(t_d)]^2 dt_d}}$$

$$= \frac{\Lambda}{\sqrt{2}} \sqrt{\frac{\frac{T_2^r}{\tau_r}\left[1 - exp\left(-\frac{2t_r^{max}}{T_2^r}\right)\right]}{\frac{T_2^d}{\tau_d}\left[1 - exp\left(-\frac{2t_d^{max}}{T_2^d}\right)\right]}} \approx \frac{\Lambda}{\sqrt{2}} \sqrt{\frac{T_2^r}{T_2^d}}$$

(2.6.2)

For the approximation it was again assumed that $\tau_d = \tau_r$ and $t_r^{max} > T_2^r$, $t_d^{max} > T_2^d$. A comparison between Eqs. (2.6.1) and (2.6.2) shows that the potential signal advantage of remote detection in the case of imaging experiments is considerably larger than for spectroscopy experiments. This is because for imaging, the bandwidth of the direct experiment is determined by the decay time and not by the signal averaging. This is particularly important for samples with large susceptibility gradients, which cause fast dephasing of the transverse magnetization.

In the above calculations it was assumed that all of the sensor medium is fully regenerated between different repetitions of the experiment, which is reasonable because it is not a T_1 decay that determines this "relaxation", but a flow that forces the encoded sensor to move ahead. Furthermore, multiplicative noise, or t_1 noise, was not considered. Depending on the type of experiment, this noise source can have a considerable influence on the sensitivity [22–25]. Another factor that has been left out is longitudinal relaxation of the sensor medium between encoding and detection. This simply causes the remote sensitivity to be multiplied by a factor $exp(-t_t/T_1)$, with t_t the travel time, which is close to unity for ^{129}Xe, but can be considerably smaller for a different sensor medium. Another aspect that has not been covered is that the sensor medium could disperse between encoding and detection, thereby being diluted with unencoded fluid. This would either require a detector with a larger active volume so that all the encoded fluid can still be read out in one experiment, or it requires multiple experiments to catch all the encoded gas. In the case of a spectroscopy experiment without spatial dependence of the encoded information, it is not necessary to gather all the encoded fluid, and the signal would just be scaled proportionally to the amount of encoded fluid in the detector. However, if spatial information is encoded, the entire encoded sensor medium is required to be detected. If it is diluted, the sensitivity is reduced, but the image can still be reconstructed accurately. However, if some of the encoded fluid remains undetected, the image will be weighted unevenly.

In experiments where the flow of the sensor medium is studied transiently, an additional decay of the signal is present due to the continuous flow of the fluid out of the detector. In a first approximation, this effect, which depends on the flow rate and the detector volume, may also be included in the apparent transverse relaxation time T_2^*, and the above formulae can be applied to estimate the sensitivity. If detection is carried out inductively with a train of n pulses, typically the time between subsequent detection pulses should be chosen to be no less than $2T_2^*$ to maintain good sensitivity and avoid artifacts. To increase the temporal resolution of the flow detection further, the detector volume has to be reduced. If the number of acquisitions n is increased, the noise of the integrated signal will scale with $\sqrt{n}$. On the other hand, the smaller detection volume allows for a more sensitive coil, again compensating for the lost sensitivity. These parameters all have to be considered when designing a remote NMR acquisition system. However, this calculation does not take into account the potential gain of information by such an experiment, which will be demonstrated in Section 2.6.5.3. Furthermore, if phase encoding is performed in the flow direction of the sensor medium, it is possible that the sinusoidal pattern of the longitudinal magnetization after the storage pulse is preserved during the flow. If the temporal resolution of the detection is fast enough to resolve this pattern, the sensitivity could be somewhat recovered. However, this strongly depends on the sample object to be studied and is outside the scope of this discussion.

2.6.4.2 Optimized Rf Coil Detection

In order to use Eqs. (2.6.1) and (2.6.2) to estimate the sensitivity gain obtained by remote detection, knowledge of the relative sensitivity of the detector and the encoding circuit, Λ, is required. Here we discuss the sensitivity of an rf coil detector as an example, because all the experiments presented in this text use inductive detection at high field. The signal-to-noise ratio of inductive NMR detection can be approximated by the following simplified equation [12]:

$$(SNR)_{LCR} = K\eta M_0 V_c \sqrt{\frac{Q\omega_0}{T\Delta f V_c}} = Km_0 \sqrt{\frac{Q\omega_0}{T\Delta f V_c}} \qquad (2.6.3)$$

where K is a numerical factor that depends on the coil geometry, the noise figure of the preamplifier, and also takes into account various physical constants; η is the filling factor of the sample in the detection coil; M_0 is the nuclear magnetization, which takes into account the concentration of the target spins in the sample volume; V_c is the volume of the coil; Q is the quality factor and ω_0 is the resonance frequency of the rf circuit; T is the temperature of the probe; and Δf is the detection bandwidth [26–28]. Δf in a pulsed NMR experiment is inversely proportional to T_2 and has already been included in Eqs. (2.6.1) and (2.6.2). Therefore it is not discussed any further in this section. m_0 is the net magnetic moment of the spins inside the coil volume, which is transported without loss from the encoding to the detection location in an ideal remote experiment. The SNR in a remote experiment can be enhanced by optimizing the factors η, M_0, V_c, Q, ω_0 and T independently for the encoding and detection steps, which is not possible in a conventional NMR experiment. For example, an rf coil with better filling factor η and quality factor Q can be used for detection than may be possible for the encoding coil due to constraints imposed by the presence of the sample. A higher magnetic field can be employed for more sensitive detection (ω_0), while a lower field may be used for encoding, for example to decrease susceptibility artifacts in imaging. Alternatively the detection coil can be cooled to enhance detection sensitivity while still preserving the optimal temperature conditions for the sample.

Finally, to estimate Λ at a given B_0 and T, it is sufficient to consider the coil's Q and V_c. The B_1 field, and thus the $\pi/2$ pulse length t_{90}, are directly related to these parameters, as expressed in Ref. [12]

$$(SNR)_{LCR} \propto m_0 \sqrt{\frac{Q}{V_c}} \propto m_0 B_1 \propto \frac{m_0}{t_{90}} \qquad (2.6.4)$$

Consequently, knowledge of the t_{90} of both coils – for the same applied rf power – allows a rough calculation of the expected signal-to-noise ratio, in agreement with the principle of reciprocity [27]. The sensitivity ratio between the two coils is thus given by

$$\Lambda = \frac{(SNR)_r}{(SNR)_d} = \sqrt{\frac{Q_r/V_c^r}{Q_d/V_c^d}} = \frac{t_{90}^d}{t_{90}^r}, \qquad (2.6.5)$$

where d and r denote the corresponding parameters of the encoding and the remote detection circuit, respectively.

2.6.5
Application of NMR Remote Detection

2.6.5.1 Broadening the Application Range

The possibility of detecting the NMR signal remotely from the sample location in a more sensitive manner opens up a wealth of new applications. One prominent example is NMR imaging at low magnetic field [29–33], including the earth field [34]. Advantages are the narrower linewidth in the presence of a given relative inhomogeneity, less distortion due to inherent susceptibility variations, which occur for example at interfaces, better performance in the presence of paramagnetic and metallic objects, greatly enhanced T_1 contrast, the possibility of accommodating larger samples and lower cost. Sensitive detection can be achieved, e.g., by inductive detection at high magnetic field, or by using alternative detection techniques as named above, which scale more favorably at low magnetic fields. Low field encoding together with high field detection implies a long travel distance of the carrier nuclei between the different magnetic fields. It has already been demonstrated that NMR images encoded at magnetic fields of 4–7 mT can be accurately reconstructed at 4.2 T utilizing laser-polarized ^{129}Xe gas for transporting the encoded information over a distance of 5 m [11]. Although high field detection using a superconducting magnet in combination with low field encoding is not cost effective, new applications for NMR imaging can be opened up. A much simpler set-up could be possible with a permanent magnet for detection, which can have fields up to about 1 T. The possibility of remote low field NMR imaging together with sensitive detection would free the remote detection technique from the expensive and immobile high field NMR magnet and enable measurements to be made outside a laboratory. Possible applications in chemical engineering include the NMR imaging and flow study on a full-size, running chemical reactor, separation column or extraction apparatus.

Although it may appear controversial, remote detection of NMR is not uniquely applicable to large samples and reactors, but it can also be applied to miniaturized microfluidic lab-on-a-chip devices [35–37]. This is because the chip device itself is large and bulky compared with the micron scale fluid channels that are embedded in it. The best sensitivity when working with such small volumes can be achieved by using microsolenoid or microsurface coils [8, 38], but a much larger rf coil is needed to encompass the entire chip device. The transportation of fluid through the channel structure is an integral part of the operation of a microfluidic chip device, which naturally fulfills the prerequisite for the application of remote detection. A possible scenario is to use a large volume coil that encompasses the entire chip for encoding, and subsequently leading the sensor molecules out of the chip into a solenoidal microcoil for detection. Common to the applications of remote detection in large and small samples is the fact that flow is an integral part of the technique. It is not only essential for signal transportation, but can also be studied in itself, whereby unprecedented information about the flow inside of channels or porous samples can be obtained on a much broader time and length scale than is possible by conventional NMR flow imaging.

2.6.5.2 Hardware for High Field NMR Remote Detection

In order to perform NMR encoding and remote detection in a high field, two rf coils need to be accommodated within the same field. The first proof of principle experiments for remote detection NMR spectroscopy and imaging wee performed with probes containing both coils at once, as can be seen in Figure 2.6.5(a and b) [11, 12]. Such a design requires the construction of a new probe for each type of sample and experiment, and is therefore inconvenient. This section contains some general suggestions for probe hardware design, showing how remote detection can be made easily accessible for use with any high field NMR set-up [39].

Most of the commercially available NMR imaging probes have an accessible clear bore above and below the coil, which can be used for remote signal encoding. A detection-only probe can be inserted from the top into the bore of the magnet in such a way that the detection coil sits immediately above the imaging coil [Figure 2.6.5(c)]. Similarly, an rf probe with a narrow body can be built that can be inserted into the

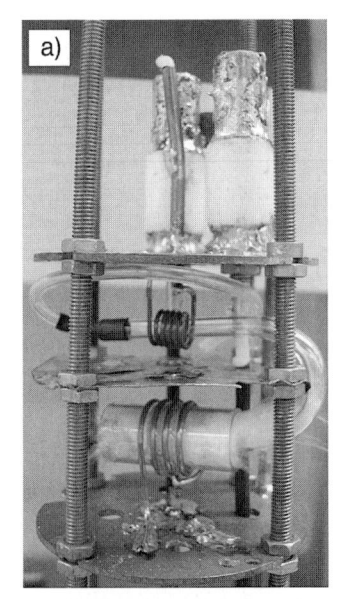

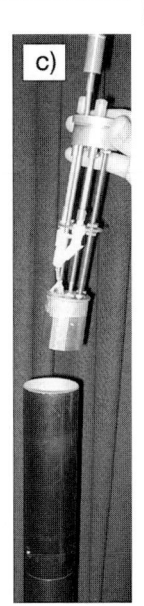

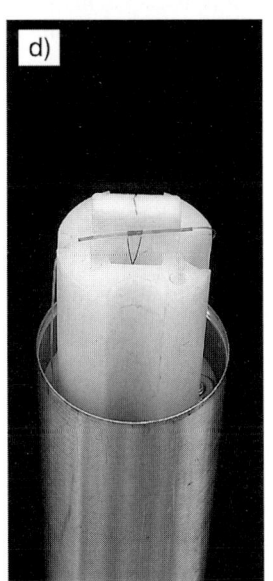

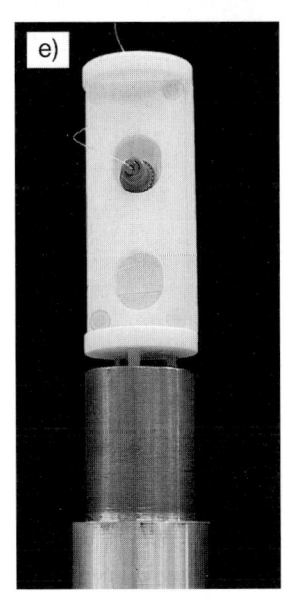

Fig. 2.6.5 Hardware for high field NMR remote detection. Photographs (a) and (b) show laboratory-built remote detection probes with both rf coils built into the same body; (c), (d) and (e) are detector-only remote probes that can be inserted from the top or bottom into the NMR imaging assembly, so that the well shielded detection rf coil is placed immediately above or below the encoding coil. The detector probe in (c) contains a relatively large saddle-coil and is used for (flow) imaging. The detector probe in (d) contains a microsolenoid coil for optimized mass sensitivity, which is particularly useful for microfluidic NMR applications. The same probe is shown in (e) with a mounted holder for a microfluidic chip that is inserted into an imaging probe.

NMR imaging probe and gradient stack assembly from below [Figure 2.6.5(d and e)] so that the detection coil sits just below the imaging coil. This detection-only probe can also be used together with other NMR probes with different characteristics for encoding. In all cases, care has to be taken to adequately shield and decouple the encoding and detection rf coil. In the probes presented in Figure 2.6.5, a well grounded copper or aluminum "hat" on top of the detector serves this purpose. Both saddle-coil [Figure 2.6.5(c)] and microsolenoid coil [Figure 2.6.5(d)] detectors have been built for ^{1}H, ^{129}Xe and ^{13}C nuclei and have been successfully and reproducibly applied for various studies including imaging and flow studies in porous media, plants and microfluidic chip devices. The detector-only probes are optimized for highest sensitivity, but the B_0 homogeneity at the detector location may be compromised, because the two probes could be too large to fit inside of the homogeneous spot of a high field magnet, requiring a trade-off when shimming the magnet. Coils with the axial direction along the B_0 field would in principle be preferable to facilitate transport of the carrier nuclei in a straight tube from the encoding coil to the detector coil, but are not always easiest to manufacture and suffer from a somewhat lower sensitivity compared with solenoid coils [26].

Low field encoding coupled with high field or non-inductive detection requires a separate design. At present, considerable effort is being devoted to developing and implementing low field and ultra low field NMR and MRI equipment [29–34]. A few examples are reviewed in this book (Chapters 2.2–2.4). In principle, there is no obstacle to coupling these possibilities to design commercially available remote NMR and MRI equipment for low field studies.

2.6.5.3 High Field NMR Imaging with Remote Detection

In this example, high-field remote detection NMR imaging of a phantom with the engraved letters "CAL" is performed with a continuous-flow of a gas mixture containing 1% of xenon with natural isotope abundance. It is demonstrated how a non-uniform flow pattern can influence the resulting image features, but also how, in spite of this, an image without distortions can be reconstructed. In particular, the non-uniform flow pattern and a mismatch in the sample volumes of the encoding and detection cell lead to a partitioning of the encoded gas volume. If individual partitions are detected separately and Fourier transformed, parts of the image are obtained corresponding to the respective times of travel, as shown in the right part of Figure 2.6.6. However, if the detection steps are repeated over a time period sufficiently long to collect all the information in several batches, the complete image can be recovered, as can be seen in the left image of Figure 2.6.6. The spreading of the transported signal along the flow path in remote detection is not necessarily reflected in a degradation of the image quality, as long as the entire signal is collected to complete an individual signal point in the k-space. However, a distorted flow path lengthens the experimental time. The experimental time may be vastly reduced, while improving the sensitivity and resolution, by utilizing fast pumping or injection devices and employing stop-and-go flow control systems. A similar approach can be used not only to recover the image of an object, but in

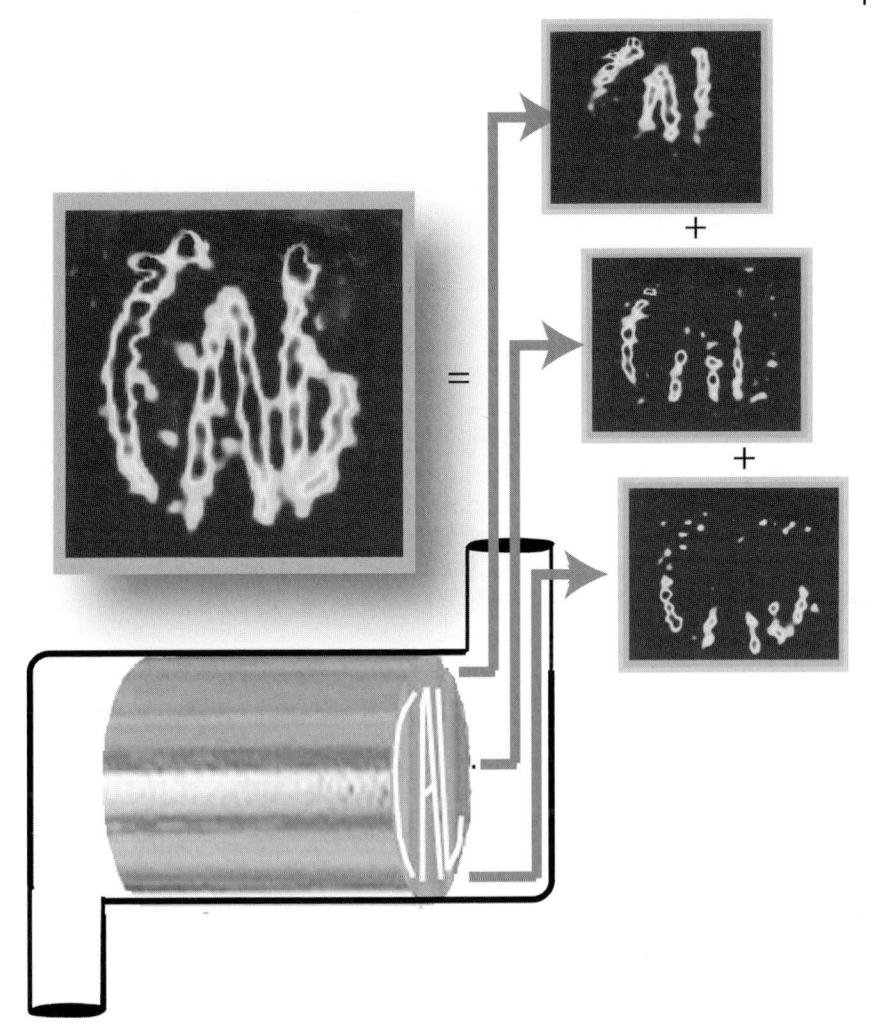

Fig. 2.6.6 Remotely reconstructed high field NMR images of laser-polarized ^{129}Xe gas in the hollow "CAL" pores. Owing to the flow pattern where the spins have to flow around two corners [see the probe design in Figure 2.6.5(b)], earlier batches of spins reconstruct the upper part and later batches the lower part of the porous sample. After all batches have been added up, a complete image is reconstructed [figure taken from 12].

addition to study the properties of the flow itself in a unique fashion, as discussed in the next section.

2.6.5.4 Flow Detection and Visualization

The flowing sensor medium as an integral part of remote detection naturally leads to the study of flow through porous media [40, 41]. In addition to carrying the information about an image of the sample, as in the previous section, the arrival of the encoded sensor in the detector depends on the flow properties and the flow path of the gas through the entire object [42]. Therefore it permits characterization of flow with a time-of-flight (TOF) experiment [43], the principle of which is shown in Figure 2.6.7(a). The difference with an experiment where remote detection is used solely for sensitivity optimization is that in the present case, the detection volume is typically chosen smaller than the effective void space volume of the sample, to improve the temporal resolution. Also, the encoding and the detection volumes should be placed as close to each other as possible to avoid additional dispersion of the flowing sensor medium outside of the sample. If detection is done inductively, for example using a probe of the same type as the one depicted in Figure 2.6.5(c), a pulse train can be used as shown in Figure 2.6.2(b) to sample the magnetic moment of the spins inside the detection volume in a strobe like manner. Because the chemical shift of the sensor signal in the detector is known, continuous wave detection is also possible. In combination with MRI encoding [44], the dispersion curve for a fluid in each volume element of the sample can be measured. A general pulse sequence is shown in Figure 2.6.7(b). As the flow between encoding and detection removes any transverse magnetization, only one data point can be encoded spatially at a time, but provided that the temporal resolution of the

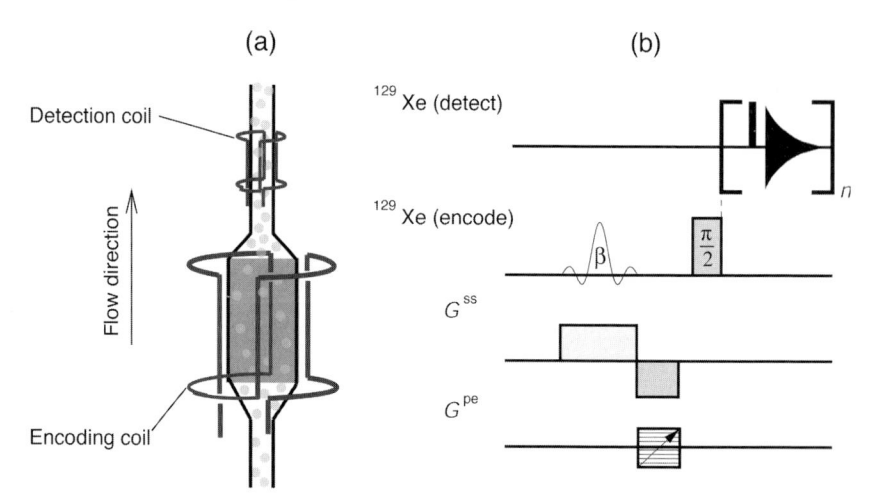

Fig. 2.6.7 General principle of time-of-flight flow detection. (a) Schematic of a set-up for TOF experiments. An object of interest is placed inside an environment optimized for encoding (field gradients not shown). As the sensor medium flows out of the analyte object, its magnetization is recorded with a second coil with a smaller volume, which is placed as close to the encoding volume as possible. (b) Generic pulse sequence used for TOF experiments. Encoding along one dimension can be done by inverting the magnetization of a slice through the sample. In this case, β is 180°, and no storage pulse is needed. For 2D encoding, a selective pulse with $\beta = 90°$ and a bipolar field gradient G^{ss} is used for slice selection in one dimension, and phase encoding gradients G^{pe} are used to resolve the other spatial dimensions. 3D encoding can be done with a hard 90° preparation pulse and phase encoding in 3D. The spacing between the n detection pulses defines the temporal resolution of the TOF experiment.

detection technique is fast enough, the complete TOF curve for the sensor can be recorded transiently for each encoding step. This experimental approach to measuring hydrodynamic dispersion is similar to the technique of an initial narrow-pulse tracer injection, with the subsequent observation of the effluent concentration of the tracer [42, 45]. The difference is that when using MRI techniques the point of injection can be defined non-invasively anywhere inside the porous medium. Equally important is that the spin magnetization behaves as an ideal tracer, as it does not affect the properties of the flowing medium.

Two fundamentally different approaches are possible for encoding. One employs slice selective pulses to modify only the polarization of the sensor medium as a function of position. The other uses phase encoding, where transverse magnetization, M^+, coherently precesses for a given time under the influence of magnetic field gradients, followed by a 90° storage pulse to transfer one component of M^+ to longitudinal magnetization, M_z, which is resistant to all types of decays except T_1 relaxation as it flows or diffuses through inhomogeneous magnetic fields. This second scheme, which employs a storage pulse, in principle allows the use of any desired encoding sequence. However, the approach with a slice-selective pulse only, can be used for encoding even if M^+ dephases too fast to allow for phase encoding, for example if a sample with large susceptibility gradients is to be characterized in a high magnetic field.

An example for gas flow using hyperpolarized ^{129}Xe gas in a mixture of 1% Xe (natural isotope abundance), 10% N_2 and 89% He through a cylindrical glass phantom is shown in Figure 2.6.8. Encoding was carried out with slice-selective pulses to invert the magnetization of slices either perpendicular or parallel to the flow direction. The sample was scanned by changing the rf offset frequency of the inversion pulse. The magnetization was detected with a train of 90° pulses, spaced by 100 ms. Each slice was recorded four times, with the position of the detection pulses shifted by 25 ms between subsequent experiments. This interleaved data acquisition allowed for a smoother representation of the TOF data and corresponds to signal averaging, but while the time between detection pulses is not longer than the time it takes the sensor medium to flow through the detection volume, the temporal resolution is not improved. In the experiment with slice selection parallel to the flow direction, the data are not perfectly symmetrical, as would be expected from the symmetry of the sample. This is because the z gradient coils were designed for smaller samples, and the phantom reaches outside the linear region of the gradients. In the experiment with the inverted slices perpendicular to the flow direction, the slope of the data in Figure 2.6.8(b) corresponds to the flow velocity of the gas. As can be seen in Figure 2.6.8(a), the flow in this object changes its direction twice, which is reproduced in the flow curve. At the turning point at a TOF of about 3 s, a splitting of the flow into two branches can be observed, indicating that the gas is not homogeneously mixed at this point, but that one fraction of the gas flows along a relatively well defined path, while a second fraction is trapped in an inactive volume with only slow exchange.

The understanding of mass transfer in porous media is highly relevant not only for the oil and gas industry, but also for process control or in chromatographic

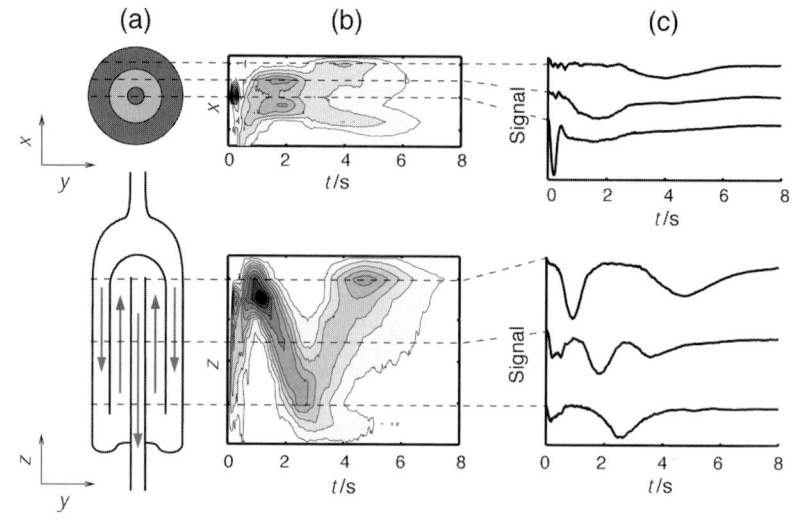

Fig. 2.6.8 Time-of-flight dispersion curves versus the encoding position of gas flowing through a cylindrically symmetrical glass phantom with large "pores" on the order of 1-cm diameter, obtained with slice selective inversion of magnetization. The flow direction changes twice as the gas is flowing from inlet to outlet. Slices parallel (upper) and perpendicular (lower) to the flow direction were inverted. (a) Cross section of the object perpendicular to the inverted slices. The arrows in the lower graph depict the direction of the gas flow. (b) Contour plots of the signal. (c) TOF signal of selected slices, as indicated by the dashed lines.

columns. While corresponding NMR techniques have been developed and successfully used for the study of liquids (e.g., Refs. [40, 44, 46]), measurements of diffusion and flow of gases in porous media are still challenging [47–49] (see also Chapters 2.5 and 5.3 for references). In the next example, gas flow and dispersion in a porous Bentheimer sandstone rock is visualized. The same gas mixture as above was used. The rock was cylindrically shaped with a height of 39 mm and a diameter of 24 mm. It had an effective porosity of about 15%, and its pore size was on the order of 100 μm. The full three-dimensional space was resolved using phase encoding. No signal averaging was carried out, except for a four-step phase cycle to subtract the baseline, because M_z of ^{129}Xe reaches a maximum in the absence of encoding, and to obtain frequency discrimination, because only one component of M^+ is retained with each storage pulse [25]. In Figure 2.6.9 the isochronal regions, *i.e.*, areas of the rock from where it takes the gas an equal amount of time to reach the detector, are displayed. Figure 2.6.9(a) shows a 3D representation, indicating that despite the homogeneity of the rock the flow is neither perfectly homogeneous nor symmetrical. However, there are also no regions that are blocking the flow path. Figure 2.6.9(b) shows a slice through the same data in a 2D diagram. One can see that mass flow is present through the entire cross section of the rock sample without interruption, except near the walls. No channeling could be observed. The observation that the regions surrounded by the contour lines do not get significantly broader for longer distances between the

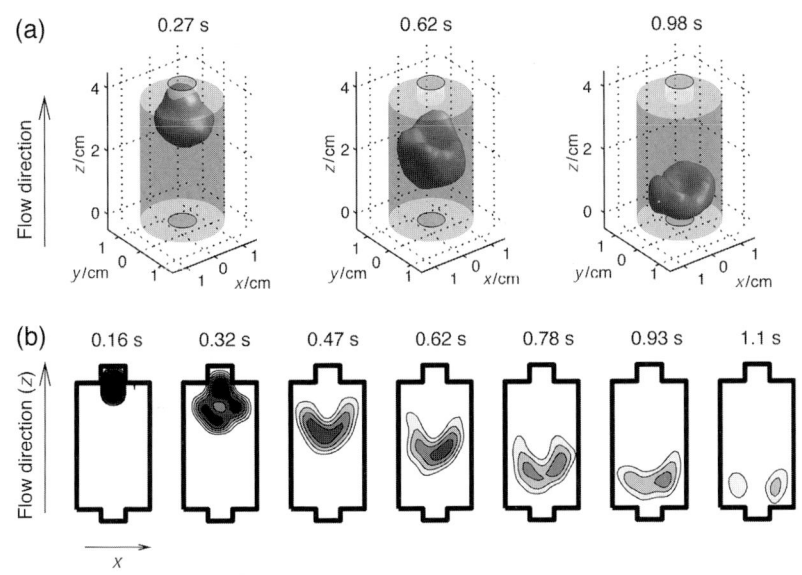

Fig. 2.6.9 Visualization of gas flow through a porous sandstone rock. A 3D phase encoding sequence with a hard encoding pulse was used. (a) 3D representation of an isochronal surface at different times after the encoding step. The cylindrical surface represents the rock. (b) Only a slice through the center of the rock is displayed, showing the origin of the gas that is flowing through the detector at different times after the tagging [figure taken from 43].

position where the sensor medium was encoded and the detector indicates that dispersion of the gas inside the rock is low.

These data show the great potential of remote detection especially for flow studies, where the loss of a transient spectral or spatial dimension is compensated by a transient flow dimension. If T_2^* of the mobile sensor medium inside a porous object is long enough so that phase encoding is possible, this technique offers unprecedented applications of MRI, because detection is carried out in a location with no significant susceptibility gradient, where the signal decay is basically limited by the residence time of the gas inside the active volume of the detector. Even if M^+ dephases too quickly to allow any manipulation, it is still possible to saturate the magnetization with slice-selective pulses and observe the dispersion as a function of position, and full 3D information can be obtained with projection reconstruction, applying multiple slice-selective pulses in different directions.

2.6.5.5 Microfluidic Flow and Miniaturized Devices

Miniaturized fluid handling devices have recently attracted considerable interest and gained importance in many areas of analytical chemistry and the biological sciences [50]. Such microfluidic chips perform a variety of functions, ranging from analysis of biological macromolecules [51, 52] to catalysis of reactions and sensing in the gas phase [53, 54]. They commonly consist of channels, valves and reaction chambers that are able to handle nanoliters to microliters of fluid. To enable precise fluid handling, accurate knowledge of the flow properties within these devices is important. Owing to the small channel size and the associated low Reynolds number, laminar flow is usually assumed. However, in small channels, fluid flow characteristics can be dominated by surface interactions, or viscous and diffusional effects. In many instances, devices are designed to take advantage of such properties, or exhibit geometries engineered to disrupt fluid flow, and are not necessarily amenable to modeling in a straight-forward way [55, 56]. Currently, methods used to gain experimental insights into microfluidic flow rely on optical detection of markers [57, 58]. Here, we show that NMR in the remote detection modality can be applied as an alternative method to profile flow in capillaries and microfluidic devices.

The NMR signal is acquired remotely, as described in the previous section. In particular, image information is encoded into the nuclear spins of the fluid contained inside a microfluidic device. Subsequently, the fluid flows out of the chip into a capillary, where it is detected using a microcoil with an inner diameter that matches the capillary dimensions. The experimental set-up contains an encoding assembly consisting of radiofrequency and pulsed field gradient coils designed for imaging of macroscopic objects [lower part in Figure 2.6.5(c)]. A detection probe [Figure 2.6.5(d and e)] containing the microcoil is placed immediately below (Figure 2.6.10) [59]. The benefit of using remote detection is, on the one hand, that it overcomes the low filling factor inside of the imaging coil, which is on the order of 10^{-5}. In the case presented here, direct NMR detection would not achieve sufficiently high SNR, while remote detection allows for the sensitive detection of the liquid or gas flowing out of the microfluidic chip device into the detector. On the other hand, through the separation of the detection and the encoding steps, remote

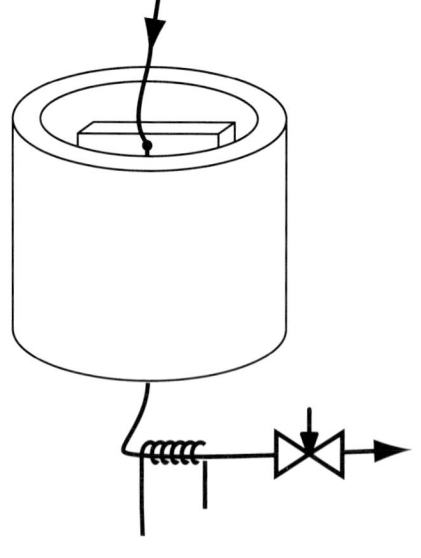

Fig. 2.6.10 Specialized experimental set-up for microfluidic flow dispersion measurements. Fluid is supplied from the top, flows via a capillary through the microfluidic device to be profiled and exits at the bottom. The whole apparatus is inserted into the bore of a superconducting magnet. Spatial information is encoded by MRI techniques, using rf and imaging gradient coils that surround the microfluidic device. They are symbolized by the hollow cylinder in the figure. After the fluid has exited the device, it is led through a capillary to a microcoil, which is used to read the encoded information in a time-resolved manner. The flow rate is controlled by a laboratory-built flow controller at the outlet [59, 60].

detection enables the measurement of the time-of-flight for an ensemble of molecules originating at any position on the microfluidic device without any additional effort. In this way, it yields time-resolved dispersion images [43].

As an example, profiles of gas flow in different model microfluidic devices are shown in Figure 2.6.11. In order to obtain adequate signal in the gas phase, hyperpolarized xenon was chosen as the signal carrier. In the profiles, the vertical axis corresponds to a spatial dimension that is resolved along the z axis parallel to B_0, and the horizontal axis shows the time-of-flight from the corresponding location inside of the chip. The NMR pulse sequence applied is shown in Figure 2.6.7(b), whereby for these profiles, the phase encoding option with non-selective pulses was used.

In Figure 2.6.11(a), a straight capillary with 150-μm inner diameter was profiled. The gas flow direction was from the top to bottom of the figure. No divergence is observed in the dispersion curve of the capillary, indicating that under the given conditions the dispersion of flow is small, and that this scheme is thus adequate to study the dispersion within a device of interest. This may appear unexpected, as microfluidic devices are usually assumed to exhibit laminar flow, however it can be explained by the fast lateral diffusion of individual gas molecules, which uniformly sample the whole cross section of the tube in a very short time compared with the travel time. Below each image, its projection is shown together with an independ-

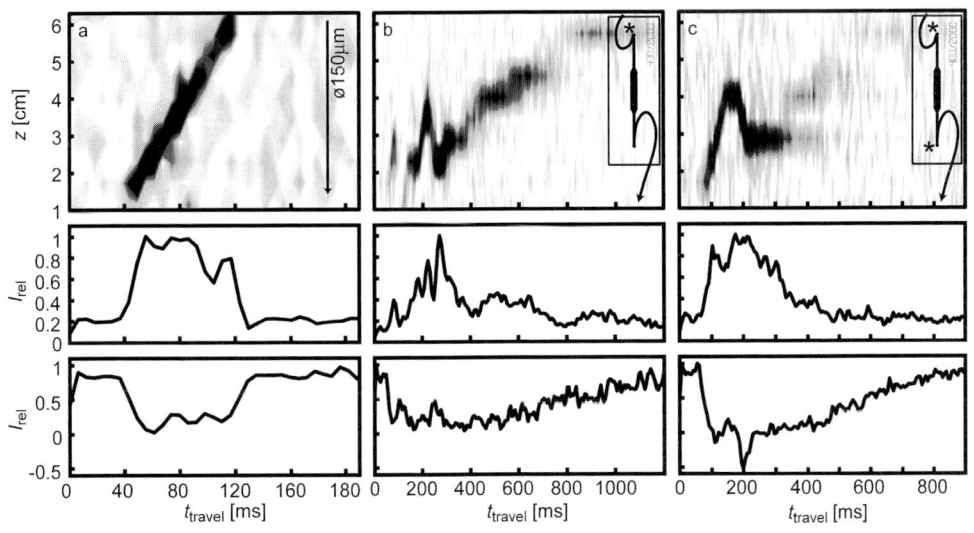

Fig. 2.6.11 Flow dispersion profiles obtained with: (a) a capillary, (b) with a model microfluidic chip device containing a channel enlargement, directly connected to a capillary and (c) with the same microfluidic chip connected to a capillary via a small mixing volume. A sketch of the model microfluidic device is placed at the right side of each image, drawn to scale along the z axis, and the presence of a cavity is marked with an asterisk (*). The flow is in the direction of the negative z axis. Below each image, traces are plotted showing the skyline projection of the image (top), and the corresponding overall time of travel curve (bottom) [figure taken from 60].

ently acquired overall travel curve (as in Figure 2.6.2), which was used to determine the maximum travel time for the imaging experiment.

The profiles in Figure 2.6.11(b and c) were obtained with a prototype microfluidic chip that had a channel depth of 55 µm, and contained a horizontal enlargement in the center, as shown by the schematic next to the experimental data. The difference between the two measurements is that in the case of (b), a direct connection was used between the chip and the outlet capillary, while in (c), a cylindrical mixing volume of 900 nl was present in between. In both images, the loop of the connecting capillary is visible at low travel times. Subsequently, in (b) the different flow velocities in the narrow (travel times 400–500 ms) and wide (500–700 ms) sections can directly be observed due to the different slopes. Above 800 ms, gas that remains for a long time in a mixing volume at the connection to the inlet capillary can be observed. It could be speculated as to whether the relative weakness of the signal at the site where the channel widens (400–500 ms and 700–800 ms) is merely due to fast flow in combination with limited spatial resolution, or due to some other effect. The difference between the images in (b) and (c) is, however, very clear. In (c), a residence time of up to 300 ms can be observed at the mixing volume of the chip outlet. Moreover, in (c) the signal shows a forking at a travel time of 350 ms, with one part coming from the enlarged section of the channel and the other still from the mixing volume. While one population of gas molecules passes this volume rather quickly, another population apparently remains at that point for at least twice as long, indicating a non-uniform mixing of the gas that is passing through this volume.

These examples illustrate that NMR imaging with remote detection is a viable alternative for the study of flow properties inside microfluidic chip devices. While an application to gas flow is presented here, the same experiments can be used with liquids. In the latter case, hyperpolarization is not needed due to the higher spin density, and thus the experimental set-up is simplified further. The method is only limited by the intrinsic nuclear spin-lattice relaxation time of the fluid, which needs to be in the order of seconds or more. Fortunately, this is the case for a large number of fluids, including water and organic solvents. The present approach has several advantages over other methods, such as the versatility in the choice of the flow medium, the ability to work with opaque devices, and, depending on the application, the absence of the need to introduce tracer substances into the fluid flow. Foremost, however, the present method intrinsically provides additional information contained in the time-of-flight dimension, which elucidates patterns that may not be easily accessible with other methods presently available.

2.6.6
Concluding Remark

The main implications of the new remote detection methodology are the sensitivity enhancement, which can be several orders of magnitude in certain applications, the option to study flow transiently with time-of-flight techniques and the broadening of the application range of NMR and MRI to samples that were previously impossible or extremely challenging to examine. Examples are large geological samples, materials with paramagnetic impurities, reactors with heterogeneous, paramagnetic catalysts or metallic components, porous media with a wide distribution of pore sizes leading to flow and image properties over a large temporal and spatial range and bulky microfluidic chip devices interlaced with small channels. We have presented the principle and merit of the remote detection methodology on several examples, and have provided ideas for future applications in chemical engineering that can be developed by remote detection of NMR and MRI.

Acknowledgements

We would like to thank Alex Pines for his ideas, support and continued belief in the seemingly "impossible"; a great deal of all this was needed, especially at the beginning of the remote detection project. We would also like to acknowledge Juliette Seeley, Adam Moulé, Sandra Garcia, Elad Harel, Erin McDonnell and Kimberly Pierce, whose efforts are reflected in many of the presented data, most of which are experimental results. Furthermore, we would like to thank Yi-Qiao Song for providing us with the rock sample. C. H. acknowledges support from the Swiss National Science Foundation (SNF) through a post-doctoral fellowship.

References

1 W. Happer, E. Miron, S. Schäfer, D. Schreiber, W. A. Van Wijngaarden, X. Zeng **1984**, *Phys. Rev. A* 29, 3092–3110.
2 D. Raftery, H. Long, D. Shykind, P. J. Grandinetti, A. Pines **1994**, *Phys. Rev. A.* 50, 567–574.
3 A. Abragam **1961**, *The Principles of Nuclear Magnetism*, Clarendon Press, Oxford.
4 K. H. Hausser, D. Stehlik **1968**, *Adv. Magn. Reson.* 3, 79–139.
5 A. R. Lepley, G. L. Closs **1973**, *Chemically Induced Magnetic Polarization*, Wiley, New York.
6 C. R. Bowers, D. P. Weitekamp **1987**, *J. Am. Chem. Soc.* 109, 5541–5542.
7 C. Hilty, C. Fernández, G. Wider, K. Wüthrich **2002**, *J. Biomol. NMR* 23, 289–301.
8 E. W. McFarland, A. Mortara **1992**, *Magn. Reson. Imag.* 10, 279–288.
9 D. L. Olson, T. L. Peck, A. G. Webb, R. L. Magin, J. V. Sweedler **1995**, *Science* 270, 1967–1970.
10 A. G. Webb **1997**, *Prog. Nucl. Mag. Res. Spectrosc.* 31, 1–42.
11 A. J. Moulé, M. Spence, S.-I Han, J. A. Seeley, K. Pierce, S. Saxena, A. Pines **2003**, *Proc. Natl. Acad. Sci. USA* 100, 9122–9127.
12 J. A. Seeley, S.-I Han, A. Pines **2004**, *J. Magn. Reson.* 167, 282–290.
13 K. Schlenga, R. F. McDermott, J. Clarke, R. E. de Souza, A. Wong-Foy, A. Pines **1999**, *IEEE T. Appl. Supercon.* 9, 4424–4427.
14 R. McDermott, A. H. Trabesinger, M. Mück, E. L. Hahn, A. Pines, J. Clarke **2002**, *Science* 295, 2247–2249.
15 I. K. Kominis, T. W. Kornack, J. C. Allred, M. V. Romalis **2003**, *Nature* 422, 596–599.
16 D. Budker, D. F. Kimball, V. V. Yashchuk, M. Zolotorev **2002**, *Phys. Rev. A* 65, 055403.
17 V. V. Yashchuk, J. Granwehr, D. F. Kimball, S. M. Rochester, A. H. Trabesinger, J. T. Urban, D. Budker, A. Pines **2004**, *Phys. Rev. Lett.* 93, 160801.

18 D. P. Weitekamp **1983**, *Adv. Magn. Reson.* 11, 111–274.
19 K. H. Lim, T. Nguyen, T. Mazur, D. E. Wemmer, A. Pines **2002**, *J. Magn. Reson.* 157, 160–162.
20 R. R. Ernst, G. Bodenhausen, A. Wokaun **1987**, *Principles of Nuclear Magnetic Resonance in One and Two Dimensions*, Clarendon Press, Oxford.
21 M. H. Levitt, G. Bodenhausen, R. R. Ernst **1984**, *J. Magn. Reson.* 58, 462–472.
22 J. Granwehr, J. A. Seeley **2005**, submitted.
23 A. F. Mehlkopf, D. Korbee, T. A. Tiggelman, R. Freeman **1984**, *J. Magn. Reson.* 58, 315–323.
24 J. Granwehr **2005**, *Concept. Magnetic Res.*, in press.
25 J. Granwehr, J. T. Urban, A. H. Trabesinger, A. Pines **2005**, *J. Magn. Reson.* 176, 125–139.
26 D. I. Hoult, R. E. Richards **1976**, *J. Magn. Reson.* 24, 71–85.
27 E. Fukushima, S. B. W. Roeder **1981**, *Experimental Pulse NMR: A Nuts and Bolts Approach*, Addison-Wesley Publishing Company, Boston.
28 R. Freeman **1997**, *A Handbook of Nuclear Magnetic Resonance*, 2nd edn, Longman, Singapore.
29 A. Macovski, S. Conolly **1993**, *Magn. Reson. Med.* 30, 221–230.
30 C. H. Tseng, G. P. Wong, V. R. Pomeroy, R. W. Mair, D. P. Hinton, D. Hoffmann, R. E. Stoner, F. W. Hersman, D. G. Cory, R. L. Walsworth **1998**, *Phys. Rev. Lett.* 81, 3785–3788.
31 M. P. Augustine, A. Wong-Foy, J. L. Yarger, M. Tomaselli, A. Pines, D. M. TonThat, J. Clarke **1998**, *Appl. Phys. Lett.* 72, 1908–1910.
32 R. McDermott, S-K. Lee, B. Ten Haken, A. H. Trabesinger, A. Pines, J. Clarke **2004**, *Proc. Natl. Acad. Sci. USA* 101, 7857–7861.
33 A. N. Matlachov, P. L. Volegov, M. A. Espy, J. S. George, R. H. Krauss **2004**, *J. Magn. Reson.* 170, 1–4.
34 J. Stepišnik, V. Eren, M. Kos **1990**, *Magn. Reson. Med.* 15, 386–391.
35 J. M. Ramsey **1999**, *Nat. Biotechnol.* 17, 1061–1062.
36 D. Figeys, D. Pinto **2000**, *Anal. Chem.* 72, 330A–335A.
37 A. Groisman, M. Enzelberger, S. R. Quake **2003**, *Science* 300, 955–958.
38 C. Massin, F. Vincent, A. Homsy, K. Ehrmann, G. Boero, P.-A. Besse, A. Daridon, E. Verpoorte, N. F. de Rooij, R. S. Popovic **2003**, *J. Magn. Reson.* 164, 242–255.
39 S. Han, J. Granwehr, S. Garcia, E. McDonnell, A. Pines **2005**, submitted.
40 E. Fukushima **1999**, *Annu. Rev. Fluid Mech.* 31, 95–123.
41 J. D. Seymour, P. T. Callaghan **1997**, *AIChE J.* 43, 2096–2111.
42 J. Bear **1972**, *Dynamics of Fluids in Porous Media*, American Elsevier, New York.
43 J. Granwehr, E. Harel, S. Han, S. Garcia, A. Pines, P. Sen, Y. Song **2005**, *Phys. Rev. Lett.* 95, 075503-/075503-4.
44 P. T. Callaghan **1991**, *Principles of Nuclear Magnetic Resonance Microscopy*, Oxford University Press, Oxford.
45 M. H. G. Amin, S. J. Gibbs, R. J. Chorley, K. S. Richards, T. A. Carpenter, L. D. Hall **1997**, *Proc. R. Soc. London A* 453, 489–513.
46 S. Stapf, S.-I. Han, C. Heine, B. Blümich **2002**, *Concept. Magnetic Res.* 14, 172–211.
47 S. L. Codd, S. A. Altobelli **2003**, *J. Magn. Reson.* 163, 16–22.
48 R. W. Mair, R. Wang, M. S. Rosen, D. Candela, D. G. Cory, R. L. Walsworth **2003**, *Magn. Reson. Imag.* 21, 287–292.
49 I. V. Koptyug, S. A. Altobelli, E. Fukushima, A. V. Matveev, and R. Z. Sagdeev **2000**, *J. Magn. Reson.* 147, 36–42.
50 T. Chovan, A. Guttman **2002**, *Trends Biotechnol.*, 20, 116–122.
51 M. A. Burns, B. N. Johnson, S. N. Brahmasandra, K. Handique, J. R. Webster, M. Krishnan, T. S. Sammarco, P. M. Man, D. Jones, D. Heldsinger, C. H. Mastrangelo, D. T. Burke **1998**, *Science* 282, 484–487.
52 D. L. Huber, R. P. Manginell, M. A. Samara, B. I. Kim, B. C. Bunker **2003**, *Science* 301, 352–354.
53 Y. Ueno, T. Horiuchi, T. Morimoto, O. Niwa **2001**, *Anal. Chem.* 73, 4688–4693.
54 J. Kobayashi, Y. Mori, K. Okamoto, R. Akiyama, M. Ueno, T. Kitamori, S. Kobayashi **2004**, *Science* 304, 1305–1308.

55 J. A. Pathak, D. Ross, K. B. Migler **2004**, *Phys. Fluid.* 16, 4028–4034.
56 T. J. Johnson, D. Ross, L. E. Locascio **2002**, *Anal. Chem.* 74, 45–51.
57 A. K. Singh, E. B. Cummings, D. J. Throckmorton **2001**, *Anal. Chem.* 73, 1057–1061.
58 S. Devasenathipathy, J. G. Santiago, K. Takehara **2001**, *Anal. Chem.* 74, 3704–3713.
59 E. McDonnell, S. Han, C. Hilty, K. Pierce, A. Pines **2005**, in press.
60 C. Hilty, E. McDonnell, J. Granwehr, K. L. Pierce, S. Han, A. Pines **2005**, *Proc. Natl. Acad. Sci. USA* 102 (42), 14960–14963.

2.7
Novel Two Dimensional NMR of Diffusion and Relaxation for Material Characterization
Yi-Qiao Song

Abstract

Many materials are complex mixtures of multiple molecular species and components and each component can be in multiple chemical or physical states. Realtime determination of the components and their properties is important for the understanding and control of the manufacturing processes. This paper reviews a recently developed technique of 2D NMR of diffusion and relaxation and its application to identify components of materials. This technique may have further applications for the study of biological systems and in industrial process control and quality assurance.

2.7.1
Introduction

NMR spectroscopy is one of the most widely used analytical tools for the study of molecular structure and dynamics. Spin relaxation and diffusion have been used to characterize protein dynamics [1, 2], polymer systems[3, 4], porous media [5–8], and heterogeneous fluids such as crude oils [9–12]. There has been a growing body of work to extend NMR to other areas of applications, such as material science [13] and the petroleum industry [11, 14–16]. NMR and MRI have been used extensively for research in food science and in production quality control [17–20]. For example, NMR is used to determine moisture content and solid fat fraction [20]. Multicomponent analysis techniques, such as chemometrics as used by Brown et al. [21], are often employed to distinguish the components, e.g., oil and water.

The usefulness of NMR in such analysis is because the proton spin-relaxation time constants are different for different components, such as water, liquid fat and solid fat. For example, the signal from solid fat is found to decay rapidly while the liquid signals decay much slower. This phenomenon is the basis for an NMR technique to determine the solid fat content [20]. However, as the relaxation time constant of water, for example, could depend on its local environment, such as protein concentration, it may overlap with that of oil and other components. As a result, it could be difficult to formulate a robust and universal relaxation analysis. It is often dependent on the specific ingredient feeds and may need to be calibrated frequently. In this chapter, a new type of two-dimensional (2D) NMR of relaxation and diffusion will be discussed that might improve the characterization of food products and materials in general, and other chemical engineering applications.

Conventional 2D NMR spectroscopy [22] is usually performed by measuring a signal matrix as a function of two independent time variables. Then, 2D Fourier transform is performed with respect to the two variables to obtain the 2D spectrum as a function of the two corresponding frequencies. The new 2D NMR of relaxation and diffusion is similar conceptually in that the signal matrix is measured as a function of two variables. However, as relaxation and diffusion often cause the spin magnetization to decay exponentially, the data matrix is analyzed by Laplace inversion instead of Fourier inversion. The result of such an experiment is a 2D joint probability distribution of T_1-T_2, or D-T_2, or D-T_1, T_2-T_2, etc. Such experiments were reported many years ago [23–25], however, its application was not widespread due to the difficulty in data analysis using conventional Laplace inversion techniques. A new algorithm (Fast Laplace Inversion, FLI) was developed in 2000 [26, 27] which enables rapid 2D Laplace inversion using contemporary desktop computers.

This chapter reviews all aspects of the 2D NMR of relaxation and diffusion. Firstly, numerous pulse sequences for the 2D NMR and the associated spin dynamics will be discussed. One of the key aspects is the FLI algorithm and its fundamental principle will be described. Applications of the technique will then be presented for several example systems.

2.7.2
Pulse Sequences and Experiments

A number of NMR pulse sequences for the 2D experiments can be generalized into one that consists of two segments, segment 1 and segment 2. Each segment may be composed of RF pulses, free evolution periods and magnetic field gradients. The sequence is parametrized by two variables, x_1 and x_2, for the two segments, respectively. The spin system will in general evolve differently in the two segments. Measurements will be performed as a function of x_1 and x_2 forming a 2D matrix,

$$M(x_1, x_2) = k(x_1, x_2, R_1, R_2) f(R_1, R_2) \mathrm{d}R_1 \mathrm{d}R_2 \qquad (2.7.1)$$

where R_1 and R_2 are the spectroscopic properties of the sample. The function k is the response of the molecules with specific properties, e.g., R_1 and R_2 and is known as a kernel function. The function $f(R_1, R_2)$ is the probability density (distribution) of molecules with the properties R_1 and R_2, which is often the quantity to be measured. For conventional multi-dimensional Fourier spectroscopy, R_1 and R_2 are frequency variables and the kernel k often consists of sine and cosine functions. As a result, an inverse Fourier transform can be applied to the data $M(x_1, x_2)$ to obtain F.

However, some aspects of the spin dynamics are better described using functions other than Fourier series. For example, the magnetization decay in a CPMG [28] experiment follows an exponential form,

$$k(\tau, T_2) = exp(-\tau/T_2) \tag{2.7.2}$$

where τ is the total time elapsed at the time of a given echo and T_2 is the spin-spin relaxation time constant. Notice that this is a one-dimensional kernel. Thus, a numerical Laplace inversion [29–31] is required to obtain F as a function of T_2.

In the following sections, we will discuss several examples of the 2D NMR pulse sequences to illustrate the essential aspects required to obtain the correlation functions of relaxation and diffusion.

2.7.2.1 Relaxation Correlation Experiment

The idea of exploration of relaxation correlation was first reported in 1981 by Peemoeller et al. [23] and later by English et al. [24] using an inversion-recovery experiment detected by a CPMG pulse train. This pulse sequence is shown in Figure 2.7.1.

The data were obtained by repeating the sequence with a series of τ_1 and acquiring a series of echoes for each experiment. Thus, $\tau_2 = 2nt_{cp}$ where n is the echo number and t_{cp} is the time period between the first 90 degree pulse and the next pulse. The first pulse inverts the spin magnetization from its equilibrium. Over the time period τ_1, the magnetization recovers along the z axis, a T_1 process. The recovery time, τ_1, and the echo time, τ_2, are two independent variables and the acquired data can be written as a two-dimensional array, $M(\tau_1, \tau_2)$. The range of τ_1 is determined by the T_1 of the sample and often spans from a fraction of the minimum T_1 of the sample to several times that of the maximum T_1. At the end of the τ_1 period, the CPMG sequence is initiated and a series of echoes is produced between every two adjacent 180° pulses. During τ_2, the decay is due to T_2. The echo spacing t_{cp} is often chosen to be short, for example, 100 μs, in order to acquire more echoes for better signal-to-noise ratio and to reduce the potential diffusion effect due to background gradients. This signal relates to the probability density $f(T_1, T_2)$ via Eq. 2.7.1 and the kernel is:

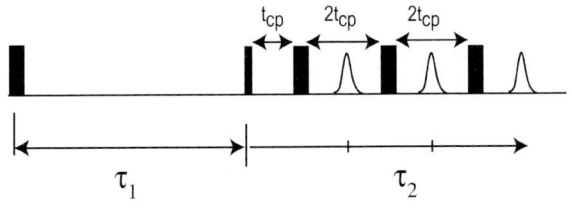

Fig. 2.7.1 Pulse sequence of the T_1-T_2 correlation experiment. The wide and narrow bars are π and $\pi/2$ pulses, respectively. During the τ_1 and τ_2 periods, the spin system experiences spin relaxation along the direction of and transverse to the magnetic field, respectively.

$$k(\tau_1, \tau_2, T_1, T_2) = (1 - 2e^{-\tau_1/T_1})e^{-\tau_2/T_2} \qquad (2.7.3)$$

The function $f(T_1, T_2)$ corresponds to the density of molecules with relaxation times of T_1 and T_2. Therefore, $f(T_1, T_2) \geq 0$ for all T_1 and T_2.

English et al. [24] recognized that a two-dimensional Laplace inversion can be applied directly to the data to obtain f without specific modeling of the relaxation mechanism in the samples. However, owing to limited computing resources and the numerical algorithm, they were able to analyze only data sets with a small number of τ_1 and τ_2. Recently, a more efficient algorithm of FLI has been developed [26, 27] to allow the analysis of 2D data sets that are much larger than the previous ones using desktop computers. The essence of the computation difficulty and the FLI algorithm will be addressed in later sections.

2.7.2.2 Diffusion-relaxation correlation experiments

Diffusion-relaxation correlation has been utilized to study biological tissues, e.g., compartmentalization in tissues [32–35]. In many reports, a sequence that combines a stimulated echo-type sequence with a pulsed field gradient and a CPMG as a detection has been described [35]. Other pulses sequences have also been used to study the diffusion-relaxation correlation, e.g., Ref. [36].

Figure 2.7.2 illustrates two implementations of the diffusion-relaxation experiment using the pulsed field gradient. In the first implementation, a spin-echo

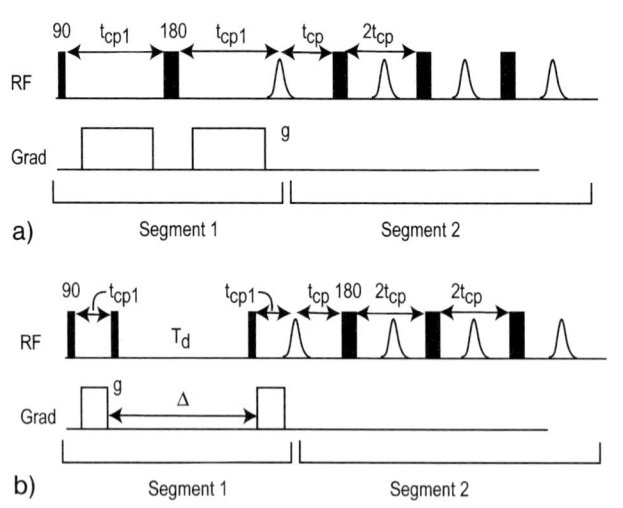

Fig. 2.7.2 Diffusion-relaxation correlation sequences using pulsed field gradients. (a) The first segment is a spin-echo with the echo appearing at a time $2t_{cp1}$ after the first pulse. (b) The first segment is a stimulated echo appearing at a time t_{cp1} after the third pulse. The detection (2nd) segment for both is a CPMG pulse train that is similar to that in Figure 2.7.1. The amplitude or the duration of the gradient pairs in both sequences is incremented to vary the diffusion effects.

2.7 Novel Two Dimensional NMR of Diffusion and Relaxation for Material Characterization

sequence is used for segment 1 with a gradient pair before and after the second pulse. The first echo spacing t_{cp1} is often fairly long, several milliseconds, to allow sufficient time for the gradient pulses to yield appropriate diffusion decay given the diffusion constant of the sample. The second echo spacing during segment 2 is often very short, for example, 100 µs, to increase the number of echoes and to reduce the diffusion effect due to background gradients. The phases of the RF pulses are 0 and 180° for the first pulse, and 90 for the remaining 180° pulses. This phase cycling ensures that the first echo is the result of a spin-echo. The diffusion decay caused by this segment is known [37] as

$$\exp\left[-D\gamma^2 g^2 \delta^2 (\Delta + 2\delta/3)\right] \tag{2.7.4}$$

where γ is the gyromagnetic ratio, D the diffusion constant, g the amplitude of the gradients, δ the during of the gradient pulse and Δ the time period between the two gradients. This equation is valid within the Gaussian phase approximation, such as diffusion in bulk fluids.

Stimulated echo with a pulsed gradient can also be used for the diffusion weighting segment, Figure 2.7.2(B). The first three $\pi/2$ pulses and the two associated gradient pulses produce the stimulated echo. The π pulses then refocus the magnetization repeatedly producing a train of echoes. Similar to the spin-echo case for the diffusion weighting Figure 2.7.2(A), the period between the first two pulses is often several milliseconds to allow sufficient gradient-induced modulation. In addition, the diffusion time between the 2nd and 3rd pulses ($\approx \Delta$) allows a further increase of diffusion weighting. A long Δ period can be particularly useful for samples with a short T_2 but a long T_1.

For the case of the spin-echo for diffusion weighting, the full kernel can be written as

$$\begin{aligned} k = {} & \exp(-2t_{cp1}/T_2) \\ & \cdot \exp\left[-D\gamma^2 g^2 \delta^2 (\Delta + 2\delta/3)\right] \\ & \cdot \exp(-2nt_{cp}/T_2) \end{aligned} \tag{2.7.5}$$

where the first factor is the T_2 decay during the first segment which can be a fixed constant provided t_{cp1} is kept constant for experiments with different gradient g.

For the case of the stimulated echo for diffusion weighting, the full kernel can be written as

$$\begin{aligned} k = {} & \exp(-2t_{cp1}/T_2 - T_d/T_1) \\ & \cdot \exp\left[-D\gamma^2 g^2 \delta^2 (\Delta + 2\delta/3)\right] \\ & \cdot \exp(-2nt_{cp}/T_2) \end{aligned} \tag{2.7.6}$$

where the first factor is the T_1 and T_2 decay during the first segment, which can be a fixed constant provided t_{cp1} and T_d are kept constant for experiments with different gradient g.

2.7.2.3 Correlation Experiments in Static Field Gradients

Similar to the pulsed field gradient, static field gradient can also be used to measure diffusion. Static field gradient is always found away from the geometric center of the NMR magnet, and in specifically designed magnets using permanent magnetic materials, such as an NMR well-logging tool [38], a Garfield magnet [39] and the NMR MOUSE [40]. Such a magnet can be much less expensive than magnets with high field homogeneity and can offer a very stable magnetic field gradient.

In a constant field gradient, one can use the RF pulses to control and select the desired coherence pathways. For example, the T_1-T_2 sequence of Figure 2.7.1 can be performed in the static field gradient. The presence of the static gradients has two effects. The first is that all pulses can be slice selective and may not excite the entire sample. Thus the echo signal amplitude and echo shape can be dependent on the RF power and the pulse durations. The second complication is the presence of multiple coherence pathways due to non-ideal pulses and off-resonance effects. For instance, the nutation angle of an RF pulse will depend on the resonance frequency offset from the pulse frequency. As a result, the CPMG echoes will have contributions from many coherence pathways [41], some that are similar to spin-echo, some similar to stimulated echoes and still other higher order echoes [42]. As the diffusion decay is different for different coherence pathways, the CPMG echoes may not follow a single exponential decay in static field gradient [41, 42]. Thus, it is preferable to use a very short echo time t_{cp} to minimize the diffusion effect during the CPMG segment. This scheme is particularly beneficial for the use with the FLI algorithm.

Hürlimann and Venkataramanan [43] have derived the kernel for the inversion-recovery CPMG experiment in an inhomogeneous field:

$$k = (1 - 2e^{-\tau_1/T_1})e^{-\tau_2/T_{2\mathit{eff}}}, \qquad (2.7.7)$$

where $T_{2\mathit{eff}}$ is a mixture of T_2 and T_1 because the contribution from the longitudinal magnetization during the CPMG sequence due to the off-resonance effects.

For the spin-echo diffusion-relaxation experiment, they showed

$$\begin{aligned} k = &\exp(-2t_{cp1}/T_2) \\ &\cdot \exp\left[-2D\gamma^2 g^2 t_{cp1}^3/3\right] \\ &\cdot \exp(-2nt_{cp}/T_{2\mathit{eff}}). \end{aligned} \qquad (2.7.8)$$

They also showed that a spin-echo segment with two π pulses improves the echo signal due to the inclusion of a stimulated-echo coherence pathway.

The stimulated echo diffusion-relaxation experiment exhibits a kernel that is similar to that of the one with the pulsed field gradients:

$$\begin{aligned} k = &\exp(-2t_{cp1}/T_2 - T_d/T_1) \\ &\cdot \exp\left[-D\gamma^2 g^2 t_{cp1}^2(T_d + 2t_{cp1}/3)\right] \\ &\cdot \exp(-2nt_{cp}/T_{2\mathit{eff}}). \end{aligned} \qquad (2.7.9)$$

2.7 Novel Two Dimensional NMR of Diffusion and Relaxation for Material Characterization

In an experiment, t_{cp1} is to be varied systematically to obtain the 2D data matrix. For the spin-echo and stimulated-echo based sequences, molecular diffusion causes signal decay in the first segment, thus both are called diffusion-editing sequences. As spin-relaxation also occurs during t_{cp1}, the signal decay during the first segment is not purely due to diffusion.

2.7.2.4 Mixed Diffusion and Relaxation Experiment

When T_{cp} is longer in a CPMG sequence under a constant gradient, the diffusion effect can be observed. For the Hahn echo coherence pathway where every π pulse refocuses the dephasing, the diffusion contribution is,

$$k = \exp\left[-2nD\gamma^2 g^2 t_{cp}^3/3\right]$$
$$\times \exp(-2nt_{cp}/T_{2eff}) \quad (2.7.10)$$

Hence, a series of measurements with several T_{cp} values will provide a data set with variable decays due to both diffusion and relaxation. Numerical inversion can be applied to such data set to obtain the diffusion-relaxation correlation spectrum [44–46]. However, this type of experiment is different from the 2D experiments, such as T_1-T_2. For example, the diffusion and relaxation effects are mixed and not separated as in the PFG-CPMG experiment Eq. (2.7.6). Furthermore, as the diffusion decay of CPMG is not a single exponential in a constant field gradient [41, 42], the above kernel is only an approximation. It is possible that the diffusion resolution may be compromised.

2.7.2.5 Summary

Exponential decay often occurs in measurements of diffusion and spin-relaxation and both properties are sensitive probes of the electronic and molecular structure and of the dynamics. Such experiments and analysis of the decay as a spectrum of T_1 or D, etc., are an analog of the one-dimensional Fourier spectroscopy in that the signal is measured as a function of one variable. The recent development of an efficient algorithm for two-dimensional Laplace inversion enables the two-dimensional spectroscopy using decaying functions to be made. These experiments are analogous to two-dimensional Fourier spectroscopy.

2.7.3
Laplace Inversion

The general strategies to solve this problem have been discussed extensively in the literature on mathematics [47]. Numerical Recipes [48] and other NMR literature [30, 31, 49] are a good introduction. Even though there are well-established algorithms for performing a numerical Laplace inversion [29–31], its use is not necessarily trivial and requires considerable experience. It is thus useful to understand the essential mathematics involved in the analysis as a better guide to its application and to be aware of its limitations. The basic theory for the 2D inversion shares many essential characteristics with the 1D algorithm and thus we will review the 1D algorithm first and then discuss the unique aspects of the 2D algorithm.

2.7.3.1 General Theory

The 1D Laplace inversion can be approximated by a discretized matrix form:

$$M = KF + N, \tag{2.7.11}$$

where the data vector M is from a series of measurements, N is the noise, matrices K and F are discretized versions of the kernel and f respectively. The true solution to F should satisfy Eq. (2.7.11) by $||M - KF|| < \sigma$, where $||.||$ is the norm of a vector and σ is the noise variance. However, given a finite signal-to-noise ratio, many solutions satisfy this criterion – this is the manifestation of the ill-conditioned nature of the Laplace inversion.

One technique uses regularization and obtains a fit to the data through minimization of the following expression:

$$||M - KF||^2 + \alpha ||F||^2 \tag{2.7.12}$$

The first term measures the difference between the data and the fit, KF. The second term is a Tikhonov regularization and its amplitude is controlled by the parameter α. The effect of this regularization term is to select a solution with a small 2-norm $||F||^2$ and as a result a solution that is smooth and without sharp spikes. However, it may cause a bias to the result. When α is chosen such that the two terms are comparable, the bias is minimized and the result is stable in the presence of noise. When α is much smaller, the resulting spectrum F can become unstable.

In order to accelerate the minimization of Eq. (2.7.12), the data and the kernel can be compressed to a smaller number of variables using singular value decompositions (SVD) of K,

$$K = U\Sigma V^T \tag{2.7.13}$$

where Σ is a diagonal matrix with singular values in a descending order, and U and V are unitary matrices. For the exponential kernel, the singular values typically decay quickly. We limit our algorithm to the sub-space spanned typically by the 10 largest singular values. Such a sub-space is adequate for the limited signal-to-noise ratio (SNR) of the usual experimental data. Using SVD of K, (Eq. 2.7.12) can be rewritten in an identical structure but with a compressed data $\tilde{M} = U'MU$ and kernels of much smaller dimensions. Then, the optimization of Eq. (2.7.12) is performed in this reduced space.

2.7.3.2 Fast Laplace Inversion – FLI

The 2D Laplace inversion, such as Eq. (2.7.1), can in fact be cast into the 1D form of Eq. (2.7.11). However, the size of the kernel matrix will be huge. For example, a T_1-T_2 experiment may acquire 30 τ_1 points and 8192 echoes for each τ_1 assuming that 100 points for T_1 and T_2 are used, respectively. Thus, the kernel will be a matrix of $(30 \cdot 8192) \cdot 10\,000$ with $2.5 \cdot 10^9$ elements. SVD of such a matrix is not practical on current desktop computers. Thus the 1D algorithm cannot be used directly.

For many of the 2D NMR experiments, such as most of those outlined earlier, the experimental design determines a unique structure of the kernels so that the full kernel can be written as a product of two independent kernels:

$$k(x_1, x_2, R_1, R_2) = k_1(x_1, R_1)k_2(x_2, R_2) \tag{2.7.14}$$

For example, the kernels for the T_1-T_2 experiment are

$$k_1 = [1 - 2exp(-\tau_1/T_1)] \tag{2.7.15}$$

$$k_2 = exp(-\tau_2/T_2) \tag{2.7.16}$$

Thus, the matrix form of Eq. (2.7.1) is then

$$M = K_1 F K_2^T + N. \tag{2.7.17}$$

Here M and N are the data and the noise in their 2D matrix form. The matrix elements are defined as $(K_1)_{ij} = k_1(x_{1i}, R_{1j})$ and $(K_2)_{ij} = k_2(x_{1i}, R_{1j})$.

With this tensor structure of the kernel, 2D Laplace inversion can be performed in two steps along each dimension separately [50]. Even though such procedure is applicable when the signal-to-noise ratio is good, the resulting spectrum, however, tends to be noisy [50]. Furthermore, it is not clear how the regularization parameters should be chosen.

The major benefit of the tensor product structure of the kernels is that SVD of K_1 and K_2 is fairly manageable on desktop computers and will take from a fraction of a second to tens of seconds, depending on the matrix sizes, using Matlab (MathWorks). Once the SVD of K_1 and K_2 have been obtained, the SVD of the product matrix can be evaluated. For example, the product of the singular values of K_1 and K_2 will be the singular value of the product matrix K. As a result, for a 2D experiment with similar SNR as in a 1D experiment, for example, $SNR \sim 1000$, 50–100 singular values are found to be useful for the 2D experiment while typically 10 are found for the 1D experiment. Once the SVD of the combined kernel is obtained, the 2D inversion problem [Eq. (2.7.17)] is converted into a 1D problem [Eq. (2.7.11)] and the existing algorithm for inversion, e.g., Ref. [31] can be used directly. The detailed mathematics are presented in Ref. [26].

2.7.4
Applications

Sedimentary rocks from oil reservoirs exhibit significant porosity where crude oils and water often coexist to share the pore space. The characterization of the pore structure and the fluids *in situ* is essential in the development of oilfields and specifically in the design of the production strategy and the facility. NMR has become an increasingly important well-logging and laboratory technique to quantify rock and fluid properties. 2D NMR has recently been introduced to the petroleum industry as a commercial well-logging service [58]. We will first review a few examples of the 2D NMR experiments on the sedimentary rocks in laboratory and well-logging applications.

In addition, the 2D NMR could find applications for other materials, in particular, as a tool for *in situ* characterization. The measurements of the relaxation and diffusion do not require a high magnetic field, and they can be performed in a grossly inhomogeneous field. Thus, a mobile sensor similar to the NMR MOUSE [40] and the NMR logging tools [38], can be built to perform measurement *in situ* and non-invasively. For example, many food products are primarily a mixture of water and fat, such as dairy products. Quantifying the amount of water and fat is important in the production facility as they are associated with the quality of the products, such as texture and taste. NMR T_2 measurement has already become an important quality control technique to quantify the moisture and fat content. We will show a few examples of 2D NMR results from dairy products to suggest that the 2D NMR technique could become a substantial improvement over the existing technique.

2.7.4.1 Rocks

Figure 2.7.3 illustrates an example of a T_1-T_2 experiment showing the raw data for τ_2 decays for several values of τ_1. The existence of fast and slow relaxations is

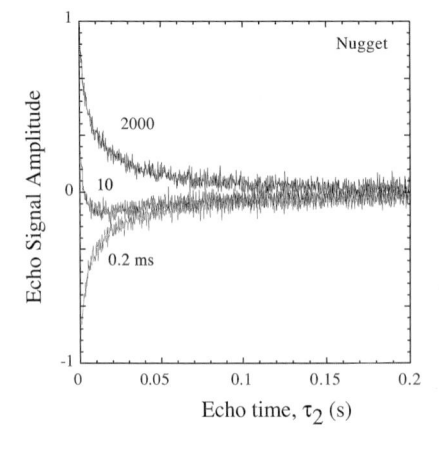

Fig. 2.7.3 Echo signals as a function of τ_1 and τ_2 for a Nugget sandstone sample obtained using the T_1-T_2 pulse sequence. The decays are shown for τ_1 of 0.2, 10 and 2000 ms. The experiment was performed at 2 MHz on a Maran Ultra spectrometer. Figure is taken from Ref. [27] with permission.

Fig. 2.7.4 T_1-T_2 spectrum for oolitic limestone. The dashed lines in B are for $T_1 = T_2$ and $T_1 = 4T_2$. The solid thick line is theoretical behavior of the sum of surface and bulk contributions to T_1 and T_2. The inset is a 30-micron thin-section micrograph of the oolitic rock. The dark elliptical structures are grains of 200 microns and the gray regions are open pores. The experiment was performed at 2 MHz on a Maran Ultra spectrometer. Figure is taken from Ref. [27] with permission. The A and B panels are two different presentations of the same spectrum.

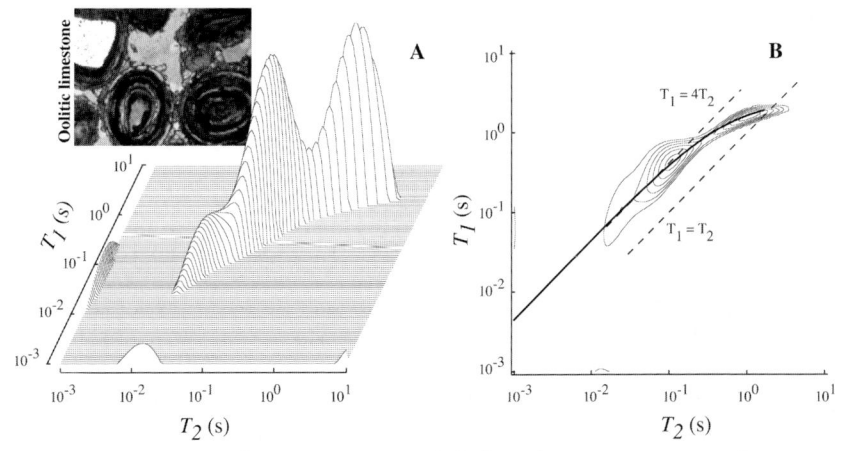

apparent from this graph. For instance, the decay with short $\tau_1 = 0.2$ ms is inverted compared with that for long $\tau_1 = 2$ s, however, the shapes of both are similar. For $\tau_1 = 10$ ms, the fast relaxing components have already recovered substantially to give rise to a positive signal at early τ_2 while the long T_1 components remain negative.

The T_1–T_2 correlation spectrum and a cross-section image for oolitic limestone are shown in Figure 2.7.4. From the image, one can identify several features. The dark circular structures are porous grains (ooids), and inter-ooid pores are about 100 μm (the gray areas). In addition, there are pores within each grain, apparently of much smaller sizes. As the relaxation times in rocks are known to be strongly affected by the mineral surfaces, one could expect to observe a range of relaxation times from this type of rock.

The T_1–T_2 spectrum in Figure 2.7.4 clearly demonstrates two spectral features, one at $T_2 \approx 0.1$ s and the other at $T_2 \approx 1$ s. These two peaks have been interpreted as the evidence of two distinct pore environments [42] with the peak of $T_2 = 1$ s to be the intergranular large pores and that of $T_2 = 0.1$ s to be the intragranular pores. Further analysis of the surface contributions to the spin relaxation showed that both peaks are consistent with a single T_1^s/T_2^s ratio for surface relaxation, suggesting that there is only one source of surface relaxation for both types of pores (intragranular and intergranular) because they are bound by the same solid material.

The D-T_2 experiments were performed [51] for a Berea sandstone sample at a proton Larmor frequency of 1.74 MHz and a static gradient of 13 G cm^{-1}. The sample was first vacuum saturated with brine, then centrifuged when immersed in oil resulting in a saturated mixture of oil and water. A D–T_2 map was obtained using FLI, shown in Figure 2.7.5. The T_2 and D distributions were obtained by

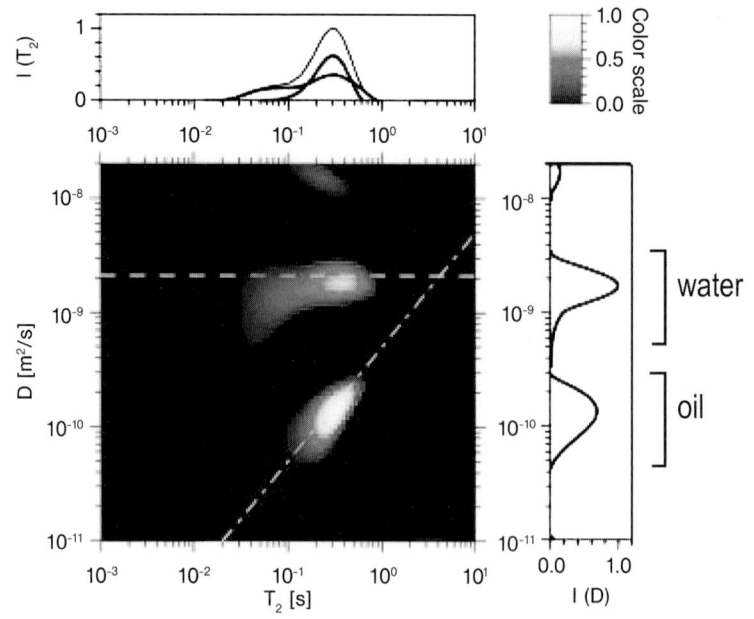

Fig. 2.7.5 Two-dimensional D–T_2 map for Berea sandstone saturated with a mixture of water and mineral oil. Figures on the top and the right-hand side show the projections of $f(D, T_2)$ along the diffusion and relaxation dimensions, respectively. In these projections, the contributions from oil and water are marked. The sum is shown as a black line. In the 2D map, the white dashed line indicates the molecular diffusion coefficient of water, whereas the white dot-dashed line shows the correlation between the average diffusion coefficient and average relaxation times of many oil samples. The contributions of the water and oil phase strongly overlap in the T_2 dimension, but are clearly separated in the 2D map. From the relative contributions of the two signals, the water saturation was found to be 53.5%. Figure taken from Ref. [51] with permission.

projecting $f(D, T_2)$ along the D or T_2 dimension and are shown on the top and the side, respectively. The D–T_2 distribution allows clear separation of the NMR signals from the different fluid phases and reveals interesting correlations between the relaxation and diffusion properties. In sedimentary rocks, T_2 has contributions from the surface and bulk relaxation. For a broad range of materials where the diffusion contrast might be smaller between the multiple components, the 2D method is in general advantageous compared with the 1D method of T_2 or D alone.

The effects of restricted diffusion are noticeable for water in Figure 2.7.5 in that the water signal is slightly below the diffusion constant for bulk water (dashed line). In other rock samples with smaller pores, the reduction of D can be more pronounced, e.g., for Indiana limestone [51]. In these experiments, molecules diffuse over length scales of up to about 10 μm. The pore space of Indiana limestone has a structure at or below this length scale, where diffusion can be strongly restricted. In contrast, the typical pore body size of Berea sandstone is significantly larger, about 80 μm [52]. For this reason, diffusion in this rock is not severely restricted for the parameters of the experiment, except in the pore throat area. Theoretical studies of the $D-T_2$ behavior due to restricted diffusion in simple pore geometry have been reported by Callaghan et al. [53] and Marinelli et al. [54].

2.7.4.2 NMR Well-logging

NMR has become one of the important services for oilfield development by providing a direct measurement of porosity independent of the rock lithology and an estimate of the pore size and permeability [14, 55]. Several designs of the NMR logging instruments, e.g. Refs. [38, 56], often called inside-out NMR, are currently used commercially primarily to perform measurements of T_2 and T_1 using CPMG and saturation-recovery types of sequences, respectively. For the history of NMR well-logging, please refer to Ref. [57]. Since 2002, 2D NMR sequences, such as $D-T_2$ and T_1-T_2 correlation, have been used on the Schlumberger NMR tools [58].

The current NMR logging tools contain a permanent magnet inside the tool housing and a magnetic field is projected outward, see Figure 2.7.6 for an illustration. A rf coil is built around the magnet to create an rf field outside the tool housing. The combination of the rf field and the static field selects a region outside the tool to satisfy the Larmor condition, which is indicated by the various grayish

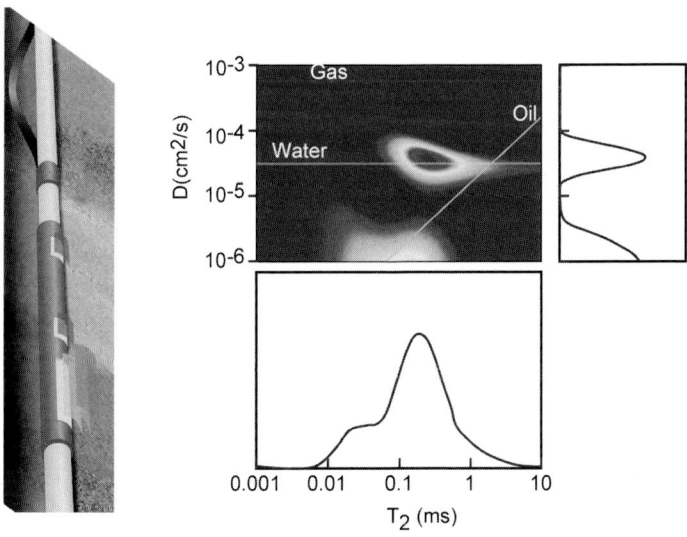

Fig. 2.7.6 Left: A representation of a Schlumberger NMR well-logging tool [56]. The long cylinder is the tool body and the shaded areas contain permanent magnets. The multiple sensitivity regions are shown as the colored sheets that are outside the tool body. Right: Two-dimensional $D-T_2$ map generated from the diffusion-editing measurements and analyzed using FLI. There are two distinct peaks that clearly separate and correspond to the oil and water signals, respectively. The lower peak corresponds to the oil phase, the higher peak corresponds to the water phase. Note that the T_2 distribution of the oil and water peaks overlap significantly. From the map, a water saturation of 53% was measured. Figure taken from Ref. [58] with permission.

sheets outside the tool body. The various sheets are resonance regions at different frequencies, subsequently at different distances from the tool surface. These resonance regions can be selected by simply switching the synthesizer frequency.

The diffusion-editing technique was tested in a oil well in the East Mount Vernon field, IN, USA [58]. The NMR tool was positioned at a depth of about 2900 feet, in a zone that was expected to show a water saturation of about 50%. A suite of data consisting of nine diffusion-editing sequences was acquired with spacings of the first two echoes varied between 2 and 12 ms in addition to the standard CPMG measurement. Figure 2.7.6 shows the diffusion-relaxation map extracted from these measurements by FLI. There are two clearly separated peaks. The diffusion coefficient of the upper peak is close to the molecular diffusion coefficient of water, and is therefore the water peak. The second peak, the oil peak, has a much smaller diffusion coefficient, indicating that the oil in this well has a moderate viscosity. The integral under each peak in Figure 2.7.6 corresponds to the saturation of the respective phase.

2.7.4.3 Milk, Cream and Cheeses

Food products can generally be considered as a mixture of many components. For example, milk, cream and cheeses are primarily a mixture of water, fat globules and macromolecules. The concentrations of the components are important parameters in the food industry for the control of production processes, quality assurance and the development of new products. NMR has been used extensively to quantify the amount of each component, and also their states [59, 60]. For example, lipid crystallization has been studied in model systems and in actual food systems [61, 62]. Callaghan et al. [63] have shown that the fat in Cheddar cheese was diffusion-restricted and was most probably associated with small droplets. Many pioneering applications of NMR and MRI in food science and processing have been reviewed in Refs. [19, 20, 59].

In many products, the spin-relaxation properties of the components can be different due to molecular sizes, local viscosity and interaction with other molecules. Macromolecules often exhibit rapid FID decay and short T_2 relaxation time due to its large molecular weight and reduced rotational dynamics [18]. Mobile water protons, on the other hand, are often found to have long relaxation times due to their small molecular weight and rapid diffusion. As a result, relaxation properties, such as T_2, have been used extensively to quantify water/moisture content, fat contents, etc. [20]. For example, oil content in seeds is determined via the spin-echo technique as described according to international standards [64].

However, T_2 is sensitive to the molecular interactions of spins and dependent on the molecular environment [60]. Thus, T_2 may overlap for different components in certain materials and this technique alone may not be sufficient to identify the components. The relaxation time distributions are often broad, e.g., in meat [21], thus making it more difficult to associate the relaxation time constants with the components.

2.7 Novel Two Dimensional NMR of Diffusion and Relaxation for Material Characterization

Our approach is to use the two-dimensional relaxation and diffusion correlation experiments to further enhance the resolution of different components. It is important to note that the correlation experiment, e.g., the T_1–T_2 experiment, is different from two experiments of T_1 and T_2 separately. For instance, the separate T_1 and T_2 experiment, in general, cannot determine the T_1/T_2 ratio for each component. On the contrary, a component with a particular T_1 and T_2 will appear as a peak in the 2D T_1–T_2 and the T_1/T_2 ratio can be obtained directly. For example, small molecules often exhibit rapid rotation and diffusion in a solution and T_1/T_2 ratio tends to be close to 1. On the other hand, the rotational dynamics of larger molecules such as proteins can be significantly slow compared with the Larmor frequency and resulting in a T_1/T_2 ratio significantly larger than 1.

In early 2004, Hürlimann studied several cheese samples using D–T_2 correlation experiments. The D–T_2 spectrum shows predominantly two signals, one with a diffusion coefficient close to that of bulk water, and the other with a D about a factor of 100 lower. The fast diffusing component is identified as water and the other as fat globules. Two components of cheese in the D–T_2 map has also been observed by Callaghan and Godefroy [65]. Recently, Hürlimann et al. have performed a systematic 2D NMR study of milk, cream, cheeses and yogurts [66]. Some of the preliminary results are discussed here.

T_1–T_2 and D–T_2 2D spectra were acquired using the inversion-recovery-CPMG (Figure 2.7.1) or the stimulated echo-CPMG [Figure 2.7.2(B)] sequence, respectively, in the fringefield outside a 2-T superconducting magnet. The RF frequency was 5 MHz and the static magnetic gradient 55 G cm^{-1}. The two-dimensional datasets, shown in Figure 2.7.3, were analyzed using the FLI algorithm [26, 27]. For milk, a single is peak is observed. For cream and cheese, two features are present.

The spectra of milk are dominated by the water signal due to the low fat content in regular milk, ≈ 3%. In cream and cheese, the water is removed from the milk during the production giving rise to a much stronger fat signal. Many interesting features are observed from the 2D spectra. For example, the two components are easily identified from the 2D maps and we found that the water and fat contents can be readily determined by integrating the corresponding regions in the maps. The water T_1 and T_2 decrease from the regular milk, to cream and cheese, probably due to the higher casein (protein) concentration. However, the T_1/T_2 ratio of the water component is found to be kept the same, ≈ 5, for all three samples. The reason for this unique value is not clear at present, possible due to the particular water-protein dynamics.

On the contrary, the fat signal and its relaxation time and diffusion constants appear to be unaffected in different samples, except for the relative signal amplitude increasing with the fat content. This is consistent with the knowledge of fat in dairy products that it forms small droplets (globules) of a few microns in size and the globules are well separated due to the coating of proteins at the surface. The size of such globules can restrict the diffusion of the lipid molecules within the globules [63].

The diffusion constant of water is consistent with the bulk water value for milk, however, it is slightly reduced in the cream sample and significantly reduced in the

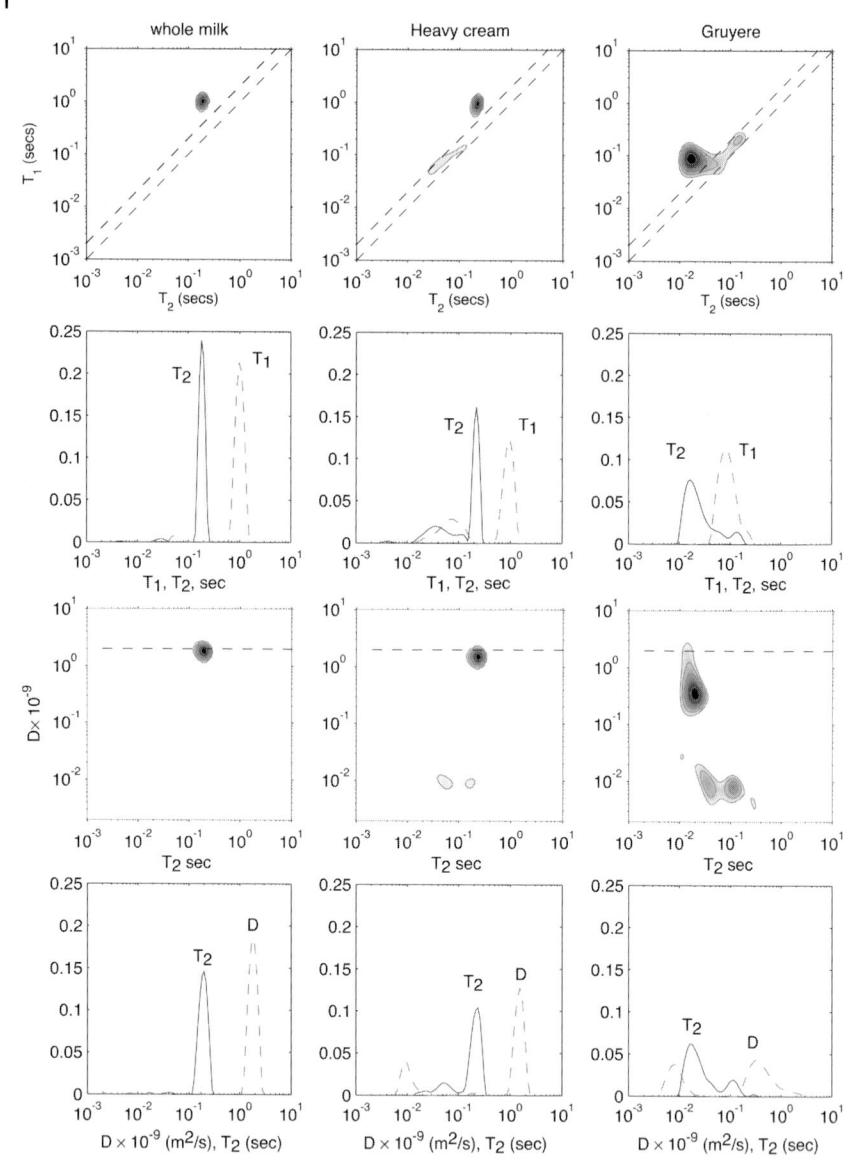

Fig. 2.7.7 T_1–T_2 and D–T_2 spectra for milk, cream and cheese. Data for each sample are shown in one column and the various experimental results are in the different rows. *First row*: T_1–T_2 spectra. The two lines in each spectrum indicate the constant T_1–T_2 ratios of 1 and 2. The peak to the left of the lines correspond to the water signal with a T_1–T_2 of 4–5. The features between the two lines are due to the fat and exhibit a T_1–T_2 ratio of ≈ 1.5. *Second row*: Projected T_1 and T_2 spectra. *Third row*: D–T_2 spectra. The peaks with a diffusion constant close to $10^{-9}\,m^2\,s^{-1}$ is due to water and the other features with $D \approx 10^{-11}\,m^2\,s^{-1}$ are due to fat. *Fourth row*: Projected D and T_2 spectra.

cheese. It is possible that the reduction in D is due to the presence of the fat globules as a barrier to water diffusion. Such a phenomenon has been observed in inorganic porous media [67, 68], such as granular materials and rocks, as well as the biological tissues [69]. In cream and cheese where the concentration of the fat globules is higher, the water space is correspondingly reduced and surrounded by the fat globules which are impermeable to water movement, thus restricting the flow and the diffusion of water.

In summary, the new 2D experiments of relaxation and diffusion appear to offer a new method to identify and quantify the components in dairy products. The two components are well separated in the 2D maps while they can be heavily overlapped in the 1D spectrum. We find that some microscopic properties of the products can be reflected in the relaxation and diffusion properties. These new techniques are likely to be useful to assist the characterization of the products for quality control and quality assurance.

2.7.4.4 Solid Samples: Candies, Chocolates and Pills

The new 2D NMR methods can be applied to many other materials. Even though this technique is mostly sensitive to the mobile species of protons, we have found a strong signal in many solid samples, such as hard candies, chocolates and pills. In Figure 2.7.8, 2D relaxation correlation spectra are shown for a few samples to demonstrate the feasibility of the new 2D technique. Although a specific interpretation of the data is not yet available, one may observe several interesting differences of the spectra. Firstly, the peaks of the hard candy (A) are mostly along the diagonal lines with a low T_1/T_2 ratio, while the signals for the chocolates are significantly away from the diagonal line with larger T_1/T_2 ratios. Secondly, the spectra of the two chocolate samples (one from the USA made by Hershey and the other Swiss dark chocolate) show a very similar pattern. The Swiss dark chocolate exhibits a relatively stronger peak at 5 ms than the peaks at longer T_2 s, whereas for the Hershey chocolate, the peak height at 5 ms, T_2 s is slightly lower than the peak at 20 ms. It is possible that this difference in the relative peak heights is related to the difference in the ingredients.

2.7.5
Instrumentation

The 2D techniques discussed in this chapter can certainly be performed on conventional high field NMR systems with the highly uniform magnetic fields. However, many examples reported, for example, the work on milk and cheeses [66], are performed at low magnetic fields with constant field gradients. This aspect is in fact a strength of the new method, in that it does not require a highly uniform high magnetic field. For instance, a magnetic field of 400 G can be easily achieved using a permanent magnet with a cost far below that of a superconducting magnet. Furthermore, because a constant field gradient is compatible with the method, simple magnet designs can be employed compared with the complex designs

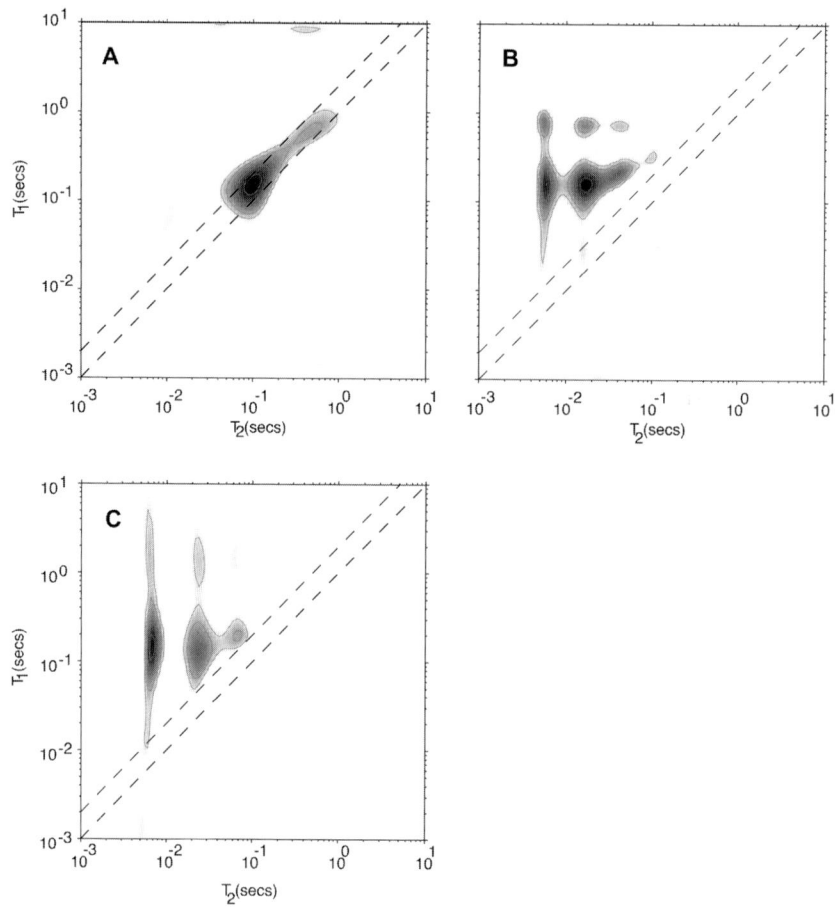

Fig. 2.7.8 T_1–T_2 spectra for a cough drop candy (A), Hershey chocolate (B) and Swiss Thins, dark chocolate (70% chocolate) (C). Experiments were performed at 85.1 Mhz in a uniform magnetic field and at room temperature. The lines are contours at 10, 30, 70 and 90% of the maximum peak height. The two dashed lines are for the T_1/T_2 ratio to be 1 and 2.

required for the uniform fields, e.g., a Halbach design. This simplicity coupled with the flexibility and the lower cost makes it possible to fit such a magnet into existing production lines, for example, a pipeline or a conveyer belt.

The time required for the 2D experiment is another important consideration for the feasibility of such techniques. As it is a 2D method, the total experimental time is often determined by the number (N_1) of values along the first dimension, e.g., τ_1 in a T_1–T_2 experiment. The required N_1 is a constraint to the range of the parameter, the signal-to-noise ratio and the desired resolution in the spectrum [70]. For example, one of the first 2D logging experiments used 9 values of t_{cp1} and was executed in about 10 min [58]. With a higher magnetic field and a better filling-factor coil, the need for signal averaging can be reduced and the experimental time minimized further. It is possible to perform such 2D experiments in 10–20 s.

2.7.6
Summary

This chapter reviews the recent development of a new 2D NMR technique that measures the correlation functions of relaxation and diffusion constants. The essential elements of the spin dynamics during such experiments are described and the basic concept of the 2D technique is explained. In particular, the data processing using the Fast Laplace Inversion (FLI) algorithm is discussed, which is one of the most important aspects of the 2D methodology. Many examples of the 2D experiments are reviewed, e.g., the study of the brine-saturated rocks and the evolution from milk to cheese. We believe that one of the key advantages of this technique lies in its simplicity and the significantly less stringent requirement on the magnetic field strength and uniformity making it potentially suitable for broad applications in industrial processes as a monitoring and quality control device. It might also be able to assist the biological research on compartmentalization of plant and animal tissues.

Acknowledgements

The author thanks L. Venkataramanan, M. D. Hürlimann, C. Flaum, M. Flaum, P. Frulla and C. Straley for their contributions in the development and applications of the 2D NMR and FLI.

References

1. A. G. Palmer 3rd, C. D. Kroenke, J. P. Loria **2001**, *Methods Enzymol.* 339, 204.
2. D. M. Korzhnev, K. Kloiber, L. E. Kay **2004**, *J. Am. Chem. Soc* 126 (23), 7320.
3. R. Bachus, R. Kimmich **1983**, *Polymer* 24, 964.
4. D. W. McCall, D. C. Douglass, E. W. Anderson **1959**, *J. Chem. Phys.* 30 (3), 771.
5. M. H. Cohen, K. S. Mendelson **1982**, *J. Appl. Phys.* 53, 1127.
6. P. G. de Gennes **1982**, *C. R. Acad. Sci. II* 295, 1061.
7. P. Z. Wong **1999**, *Methods in the Physics of Porous Media*, Academic Press, London.
8. W. P. Halperin, F. D'Orazio, S. Bhattacharja, J. C. Tarczon **1989**, in *Molecular Dynamics in Restricted Geometries*, eds. J. Klafter, J. Drake, John Wiley & Sons, New York.
9. S.-W. Lo, G. J. Hirasaki, R. Kobayashi, W. V. House **1998**, *The Log Analyst* Nov.–Dec. 43–47.
10. S.-W. Lo, G. J. Hirasaki, W. V. House, R. Kobayashi **2002**, *Soc. Petrol. Eng. J.* March 2002, 24.
11. E. von Meerwall, E. J. Feick **1999**, *J. Chem. Phys.* 111 (2), 750.
12. J. A. Zega, W. V. House, R. Kobayashi **1989**, *Physica A* 156, 277.
13. B. Blümich **2000**, *NMR Imaging of Materials*, Oxford University Press, Oxford.
14. R. Kleinberg **1995**, in *Encyclopedia of Nuclear Magnetic Resonance*, eds. D. M. Grant, R. K. Harris, John Wiley & Sons, New York.
15. D. Freed, L. Burcaw, Y.-Q. Song **2005**, *Phys. Rev. Lett.* 94, 067602.
16. A. L. Van Geet, A. W. Adamson **1964**, *J. Phys. Chem.* 68 (2), 238.
17. B. Hills **1998**, *Magnetic Resonance Imaging in Food Science*, John Wiley & Sons, Chichester.

18 M. J. McCarthy **1994**, *Magnetic Resonance Imaging in Foods*, Chapman and Hall, London.

19 M. J. McCarthy, S. Bobroff **2000**, in *Encyclopedia of Analytical Chemistry*, ed. R. A. Meyers, John Wiley & Sons, Chichester, pp. 8264–8281.

20 J. S. de Ropp, M. J. McCarthy **2000**, in *Encyclopedia of Analytical Chemistry*, ed. R. A. Meyers, John Wiley & Sons, pp. 4108–4130.

21 R. J. S. Brown, F. Capozzi, C. Cavani, M. A. Cremonini, M. Petracci, G. Placucci **2000**, *J. Magn. Reson.* 147 (1), 89.

22 R. R. Ernst, G. Bodenhausen, A. Wokaun **1994**, *Principles of Nuclear Magnetic Resonance in One and Two Dimensions*, Oxford University Press, New York.

23 H. Peemoeller, R. K. Shenoy, M. M. Pintar **1981**, *J. Magn. Reson.* 45, 193.

24 A. E. English, K. P. Whittall, M. L. G. Joy, R. M. Henkelman **1991**, *Magn. Reson. Med.* 22, 425.

25 J.-H. Lee, C. Labadie, C. S. Springer Jr, G. S. Harbison **1993**, *J. Am. Chem. Soc.* 115, 7761.

26 L. Venkataramanan, Y.-Q. Song, M. D. Hürlimann **2002**, *IEEE Trans. Signal Proc.* 50, 1017.

27 Y.-Q. Song, L. Venkataramanan, M. D. Hürlimann, M. Flaum, P. Frulla, C. Straley **2002**, *J. Magn. Reson.* 154, 261.

28 S. Meiboom, D. Gill **1958**, *Rev. Sci. Instrum.* 29, 688.

29 S. W. Provencher **1982**, *Comput. Phys. Commun.* 27, 229.

30 G. C. Borgia, R. J. S. Brown, P. Fantazzini **1998**, *J. Magn. Reson.* 132, 65.

31 E. J. Fordham, A. Sezginer, L. D. Hall **1995**, *J. Magn. Reson. Ser. A* 113, 139.

32 C. Beaulieu, F. R. Fenrich, P. S. Allen **1998**, *Magn. Reson. Imag.* 16 (10), 1201.

33 G. J. Stanisz, R. M. Henkelman **1998**, *Magn. Reson. Med.* 40 (3), 405.

34 G. Laicher, D. C. Ailion, A. G. Cutillo **1996**, *J. Magn. Reson. Ser. B* 111 (3), 243.

35 S. Peled, D. Cory, S. A. Raymond, D. A. Kirschner, F. A. Jolesz **1999**, *Magn. Reson. Med.* 42 (5), 911.

36 D. van Dusschoten, P. A. de Jager, H. Van As **1995**, *J. Magn. Reson. Ser. A* 116 (22–28).

37 E. L. Hahn **1950**, *Phys. Rev.* 80, 580.

38 R. L. Kleinberg, A. Sezginer, D. D. Griffin, M. Fukuhara **1992**, *J. Magn. Reson.* 97, 466.

39 P. M. Glover, P. S. Aptaker, J. R. Bowler, E. Ciampi, P. J. McDonald **1999**, *J. Magn. Reson.* 139, 90.

40 G. Eidmann, R. Savelsberg, P. Blümler, B. Blümich **1996**, *J. Magn. Reson. A* 122, 104.

41 M. D. Hürlimann **2001**, *J. Magn. Reson.* 148, 367.

42 Y.-Q. Song **2002**, *J. Magn. Reson.* 157, 82.

43 M. D. Hürlimann, L. Venkataramanan **2002**, *J. Magn. Reson.* 157, 31.

44 R. Freedman, S.-W. Lo, M. Flaum, G. Hirasaki, A. Matteson, A. Sezginer **2001**, *Soc. Petro. Eng. J.* December 2001, 452.

45 N. J. Heaton, R. Freedman **2005**, US Patent US 6, 859, 032 (B2).

46 N. J. Heaton **2002**, *Method for Analysing Multi-measurement NMR Data*, unpublished data.

47 C. L. Lawson, R. J. Hanson **1974**, *Solving Least Squares Problems*, Prentice-Hall, Englewood Cliffs, NJ.

48 W. H. Press, S. A. Teukolsky, W. T. Vetterling, B. P. Flannery **1997**, *Numerical Recipes in C*, Cambridge University Press, Cambridge.

49 R. M. Kroeker, R. M. Henkelman **1986**, *J. Magn. Reson.* 69, 218.

50 Y.-Q. Song et al. **2000**, unpublished data.

51 M. D. Hürlimann, L. Venkataramanan, C. Flaum **2002**, *J. Chem. Phys.* 117, 10223.

52 Y.-Q. Song **2003**, *Concept Magn. Reson.* 18A (2), 97.

53 P. T. Callaghan, S. Godefroy, B. N. Ryland **2003**, *J. Magn. Reson.* 162, 320.

54 L. Marinelli, M. D. Hürlimann, P. N. Sen **2003**, *J. Chem. Phys.* 118 (19), 8927.

55 W. E. Kenyon **1992**, *Nucl. Geophys.* 6, 153.

56 N. J. Heaton, R. Freedman, C. Karmonik, R. Taherian, K. Walter, L. DePavia **2002**, *Soc. Petro. Eng. J.* paper 77400.

57 J. A. Jackson, (ed.) **2001**, *Concepts in Magnetic Resonance, Special Issue: The History of NMR Well Logging* (vol. 13), Wiley, Chichester.

58 M. D. Hürlimann, L. Venkataramanan, C. Flaum, P. Speier, C. Karmonik, R. Freedman, N. Heaton **2002**, *43rd*

Annual SPWLA Meeting in Oiso, Japan, paper FFF.
59 S. J. Schmidt, X. Sun, J. B. Litchfield 1996, *Crit. Rev. Food. Sci. Nutr.* 36 (4), 357.
60 B. P. Hills, S. F. Takacs, P. S. Belton 1989, *Mol. Phys.* 67 (4), 903.
61 S. L. Duce, T. A. Carpenter, L. D. Hall 1990, *Lebensm.-Wiss. Technol.* 23, 565.
62 M. J. McCarthy, S. Charoenrein, J. B. German, K. L. McCarthy, D. S. Reid 1991, *Adv. Exp. Med. Biol.* 302, 615.
63 P. T. Callaghan, K. Jolley, R. Humphrey 1983, *J. Coll. Interf. Sci.* 93, 521.
64 AOCS Official Method Ak 4–95, ISO 10565, ISO 10632 for Residues, USDA, GIPSA Approval, FGIS00–101.
65 S. Godefroy, P. T. Callaghan 2003, *Magn. Reson. Imag.* 21, 381.
66 M. D. Hürlimann, L. Burcaw, Y.-Q. Song 2005, presented at ENC 2005 Providence, RI; manuscript submitted to J. Colloid Interf. Sci.
67 D. E. Woessner 1963, *J. Phys. Chem.* 67, 1365.
68 P. P. Mitra, P. N. Sen, L. M. Schwartz, P. Le Doussal, Phys 1992. *Rev. Lett.* 68, 3555.
69 L. L. Latour, K. Svoboda, P. P. Mitra, C. H. Sotak 1994, *Proc. Nat. Acad. Sci.* 91 (4), 1229.
70 Y.-Q. Song, L. Venkataramanan, L. Burcaw 2005, *J. Chem. Phys.* 122, 104104.

2.8
Hardware and Method Development for NMR Rheology
Paul T. Callaghan

2.8.1
Introduction

It is common to regard condensed matter in terms of two basic phases of solid and liquid. Under stress a solid will deform by a fixed amount and store energy elastically, whereas a liquid flows and dissipates energy continuously in viscous losses. However, many interesting materials in their condensed phase possess both solid and liquid-like properties. These include polymer melts and solutions, lyotropic and thermotropic liquid crystals, micellar surfactant phases, colloidal suspensions and emulsions. Most biological fluids, most food materials and many fluids important in industrial processing or engineering applications exhibit such complexity. Complex fluids [1] exhibit storage (elasticity) and energy loss (viscosity) properties and they generally exhibit "memory", which means that the stress which they exhibit at any moment will depend on their history of prior deformation. Most complex fluids have non-linear mechanical behavior, which means that their mechanical properties may change as the deformation increases, an effect which is generally attributed to molecular re-organization. A particularly interesting aspect of such materials is their wide range of characteristic time scales, from the rapid (ps to ns) local Brownian motion of small molecules or molecular segments, to the very slow (ms to s) motions associated with the reorganization or reorientation of large molecular assemblies or macromolecules. The study of the mechanical properties of complex fluids is known as "rheology" [2], a name which derives from the Greek word "rheo" which means "to flow".

Traditionally, rheology was a subject concerned principally with mechanical properties. The principal instrument used for mechanical analysis of complex fluids is the benchtop rheometer, a device in which the fluid is subject to shear or extension and either the strain is measured under controlled stress conditions or the stress is measured under controlled strain. The most common applications concern shearing strains in which the sensing devices are based on a cone-and-plate, parallel plates or a concentric cylindrical Couette cell. Under small oscillatory strains the linear viscoelastic response may be studied whilst in steady-state flow, the non-linear properties of the fluid are exhibited. One particular measurement of importance in non-linear viscoelasticity concerns the flow behavior under steady shearing conditions, for example in the measurement of stress versus rate-of-strain behavior, the so-called flow curve.

Recently attention has focused on trying to determine the molecular basis of complex mechanical properties. If we can better understand this basis then we can better design desirable fluidic properties, we can better process modern materials based on polymers and organized molecular states, we can better produce foods of the right texture and we can better understand how nature works, for example in the way synovial fluid protects our joints, or in the way a spider can extrude a protein silk whose strength surpasses that of steel. The interest in the molecular–mechanical link has led to the amalgamation of a number of spectroscopic and rheological techniques in which a flow or deformation cell is incorporated within the spectrometer detection system. Examples include the use of neutron scattering, light scattering, birefringence and dichroism techniques [3]. The most recent addition, NMR, has provided a number of new and valuable features. For example, it can be used to study materials which are optically opaque. The imaging capability of NMR means that it can be used to measure local velocity profiles and molecular densities directly. In addition, the wide ranging spectroscopic tools available to NMR makes it possible to measure molecular order and dynamics.

The term "rheo-NMR" seems to have been coined by Samulski and coworkers [4] who performed an ingenious experiment in which they sheared a polymer melt using a cone-and-plate device while simultaneously observing the NMR spectrum. However, the earliest suggestion of the use of NMR methods to measure rheological properties was by Martins et al., in 1986 [5]. These workers pointed out that physically reorienting a sample of a nematic liquid crystal inside the NMR magnet would lead to a slow evolution of the spectrum during subsequent director realignment, interpretation of which could yield information about rotational viscoelasticity. This first type of rheo-NMR method was later applied by Goncalves et al. [6], who used proton NMR spectra to monitor the magnetic field reorientation of an initially aligned sample of a nematic polymer liquid crystal, following the sudden physical rotation. By carefully fitting these spectra, rotational (Leslie) viscosities were calculated along with elastic constants associated with defect formation. In the same spirit, Schmidt and coworkers and Kornfield and coworkers have used deuterium NMR to observe the reorientation of liquid crystalline directors under shear flow [7–11].

A quite different approach to rheo-NMR was taken by Xia and Callaghan [12], in an NMR microscopy measurement of the velocity profile of a high molecular weight polymer solution flowing through a capillary. In this study anomalous polymer diffusion was found at a radius within the pipe at which the local shear rate exceeded the polymer terminal relaxation time. These alternative rheo-NMR approaches access different facets of the problem, one addressing the issue of molecular alignment in shear and the other gaining insight regarding the local rate-of-strain inside the shearing geometry. The experiment in which molecular self-diffusion is correlated with local shear rate, points to the potential that NMR has to combine precise velocimetry with detailed information at the molecular level. Ultimately, rheo-NMR offers its greatest advantage by being able to provide a simultaneous visualization of flow, while measuring spectroscopic properties which reveal molecular organization and dynamics. At heart the rheo-NMR method is ideally a dual focus on NMR microscopy and NMR spectroscopy, the former giving insight at the mechanical length scale and the latter at the molecular length scale. An earlier review of the science of rheo-NMR can be found in a chapter in Ref. [13] in *Reports on Progress in Physics*. In this chapter we will review some of the basic principles of rheo-NMR, providing a little more emphasis on apparatus considerations and recent applications.

2.8.2
Rheo-NMR Fundamentals

2.8.2.1 The Rheo-NMR Flow Cell

Mechanical rheometry requires a measurement of both stress and strain (or strain rate) and is thus usually performed in a simple "rotating" geometry configuration. Typical examples are the cone-and-plate and "cylindrical Couette" devices [1, 14]. In "stress-controlled" rheometric measurements one applies a known stress and measures the deformational response of the material. In "strain-controlled" rheometry one applies a deformation flow and measures the stress. "Stress-controlled" rheometry requires the use of specialized torque transducers in conjunction with low friction air-bearing drive in which the control of torque and the measurement of strain is integrated. By contrast, strain-controlled rheometry is generally performed with a motor drive to rotate one surface of the cell and a separate torque transducer to measure the resultant torque on the other surface.

With rheo-NMR, measurement of stress is difficult, because of the problem of transducer placement within the magnet. However, Grabowski and Schmidt have achieved this by means of a simple coil spring attached to the baseplate of a cone-and-plate device [7]. Of course, complementary stress–strain relationships can be determined by mechanical rheometry. Freed of the need to measure stress, rheo-NMR enables the use of a wide range of geometries in which only the flow is externally controlled. Of course, the simplest NMR rheometer consists of a pipe through which fluid can pass during an NMR experiment. However, the capillary or pipe geometry, while simple in construction and application, has a number of disadvantages. Firstly, there is a limit to the viscosity of the fluid that may be pumped through the NMR magnet. Secondly, there is a need for large volumes of fluid (i.e., in excess of tens of milliliters) to supply the flow loop and reservoir, thus curtailing the use of expensive monodisperse or isotopically labelled materials. Thirdly, there is the strong variation in strain rate across the capillary diameter, a disadvantage in experiments where a uniform deformation rate is required. For this reason it is desirable to employ rheological cells that incorporate a small volume of enclosed fluid and which are situated inside the NMR probe but driven by an external motor and drive shaft. The development of rheo-NMR techniques has been associated with the construction of such specialized cells [15].

The ability to obtain velocity images of the fluid under investigation, and to be able to selectively excite any region of the sample for subsequent NMR spectroscopy, is central to the total rheo-NMR approach. This "micro-imaging" facility distinguishes rheo-NMR from scattering methodologies (for example, as in neutron scattering). While some reference is made here to bulk measurements of NMR spectroscopic parameters obtained from sheared fluids, in which the entire sample contributes to the signal [4, 7] the focus of this chapter will be on the combined imaging–spectroscopy concept. The problem with the "bulk fluid approach" is that such measurements are based on the assumption of spatially uniform fluid behavior, which underpins the whole of conventional mechanical rheometry. However, there are many fluids for which this assumption breaks down, even under conditions of uniform stress. The problem is even more severe in the case of extensional flow for which the cell geometry inevitably produces only a select region in which the desired extensional behavior is manifest, and where some means of spatial localization is essential if meaningful NMR spectroscopy is to be performed. Finally, in NMR spectroscopic studies of molecular ordering under shearing strain, the orientation of the principal velocity, vorticity and gradient axes with respect to the polarizing field direction is crucial, and here selective excitation or microimaging methods are essential if such orientations may be independently accessed. Figure 2.8.1 shows a typical rf and gradient pulse sequence used to obtain a velocity image. Further details on NMR microscopy methodology can be found in Ref. [16].

While it is possible to perform rheo-NMR in large scale MRI systems, the use of microscopic geometries with sub-millilitre volumes is highly desirable, especially where expensive speciality materials are to be examined. Such sample volumes

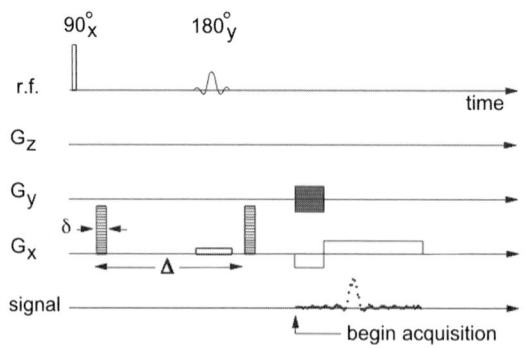

Fig. 2.8.1 Rf and gradient pulse sequence used to encode for velocity in a two-dimensional image in the (x, y) plane normal to the slice direction z. In the example shown the velocity encoding is along the x axis. By changing the direction of the velocity encoding gradient pulse pair of duration δ other directions for velocity components can be measured.

require the use of NMR microscopy in which one images the fluid within the shearing space of a small scale rheological cell at a resolution of a few tens of microns. Such measurements are usually performed in high field (> 5 T) superconducting magnets with vertical narrow (< 90 mm) room temperature bores. When combined with translational encoding methods, NMR microscopy [16] allows one to obtain velocity maps at a resolution of a few tens of microns. Typically such measurements take tens of minutes so that steady-state flow conditions are usually required but by limiting the numbers of dimensions imaged, or by using multi-echo rapid scale methodology, the image acquisition time can be speeded up by two orders of magnitude to sub-second intervals. Of course, there is an additional advantage in using small (i.e., millimeter scale) geometries in rheo-NMR, because at any given flow velocity (generally limited to below 1 m s^{-1} in NMR velocimetry), the smaller the cell used, the higher the velocity derivative and hence the strain rate that can be achieved. For example, for cells of mm dimensions strain rates in the order of several 100 s^{-1} can be studied, thus providing access to sub-10 ms molecular relaxation times.

2.8.2.2 Shear and Extensional Flow Fields

We will now describe the basic hydrodynamic relationships applicable in the case of steady-state flow in which the Eulerian velocity field is time-independent and written as $v(r)$. Here the rate of strain elements are given by [1]

$$\varkappa_{\alpha\beta} = \partial v_\alpha / \partial r_\beta \tag{2.8.1}$$

where α and β represent components of the Cartesian axis frame. For a Newtonian fluid the stress tensor may be written in terms of the rate of strain tensor via the simple constitutive equation

$$\sigma_{\alpha\beta} = \eta(\varkappa_{\alpha\beta} + \varkappa_{\beta\alpha}) + p\delta_{\alpha\beta} \tag{2.8.2}$$

where η is the constant viscosity, p is the hydrostatic pressure and $(\varkappa_{\alpha\beta} + \varkappa_{\beta\alpha})$ is the symmetric rate of strain tensor. In steady flow rheology the isotropic pressure term conveys no insight regarding the fluid properties and is neglected. Note that Newton's third law requires that the stress be uniform across the gap in planar Couette flow.

2.8.2.2.1 Planar Couette

Simple shear (also known as planar Couette flow) is achieved when fluid is contained between two plane parallel plates in relative in-plane motion. If the velocity direction is taken to be x, one has $\varkappa_{xy} = \dot{\gamma}$, all other $\varkappa_{\alpha\beta}$ zero and

$$v_x = \dot{\gamma} \cdot y, \quad v_y = 0, v_z = 0 \tag{2.8.3}$$

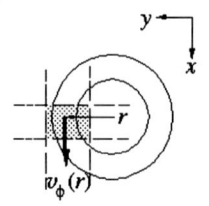

Fig. 2.8.2 Geometry pertaining to cylindrical Couette cell.

The coordinates (x, y, z) define the (velocity, gradient, vorticity) axes, respectively. For non-Newtonian viscoelastic liquids, such flow results not only in shear stress, but in anisotropic normal stresses, describable by the first and second normal stress differences $(\sigma_{xx} - \sigma_{yy})$ and $(\sigma_{yy} - \sigma_{zz})$. The shear-rate dependent viscosity and normal stress coefficients are then [1]

$$\eta(\dot{\gamma}) = \sigma_{xy}/\dot{\gamma}, \quad \Psi_1(\dot{\gamma}) = (\sigma_{xx} - \sigma_{yy})/\dot{\gamma}^2 \quad \Psi_2(\dot{\gamma}) = (\sigma_{yy} - \sigma_{zz})/\dot{\gamma}^2 \qquad (2.8.4)$$

2.8.2.2.2 Cylindrical Couette

Planar Couette flow is difficult to maintain in a steady state. Cylindrical Couette flow is much easier, in which the fluid is contained in the annulus between two cylinders in relative angular motion about their common axes, as shown in Figure 2.8.2.

This is most easily achieved by rotating the inner cylinder and keeping the outer fixed in the laboratory frame. Note, however, that this geometry leads to the formation of Taylor vortex motion if inertial effects become important (Reynolds number $Re \gg 1$). Most rheo-NMR experiments are actually performed at low Re. In the cylindrical Couette, the natural coordinates are cylindrical polar (ϱ, ϕ, z) so the shear stress is denoted $\sigma_{\phi\varrho}$ and is radially dependent as ϱ^{-2}. The strain rate across the gap is given by [2]

$$\dot{\gamma} = \varrho \frac{\partial \omega}{\partial \varrho} = \frac{\partial v_\phi}{\partial \varrho} - \frac{v_\phi}{\varrho} \qquad (2.8.5)$$

2.8.2.2.3 Cone-and-plate

Another approximation to planar Couette conditions can be found in the cone-and-plate cell, shown in Figure 2.8.3. The angular speed of rotation of the cone is taken to be Ω (in radians per second) while the angle of the cone is α (in radians) and is generally small, say 4–8°. A point in the fluid is defined by spherical polar (r, θ, ϕ), cylindrical polar (ϱ, z, ϕ) or Cartesian (x, y, z) coordinates, where $\varrho = y = r\sin\theta$ and $z = r\cos\theta$.

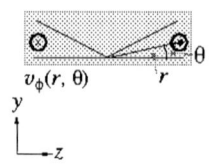

Fig. 2.8.3 Geometry pertaining to cone-and-plate cell.

The azimuthal component of the fluid velocity, v_ϕ, is identical to v_x and the local fluid angular velocity is $\omega = v_\phi/\varrho$. This azimuthal velocity is $\gamma \sin\theta \omega(\theta) = \varrho\omega(\theta)$ and the local shear rate, $\dot{\gamma}$, is $-\sin\theta \frac{d\omega}{d\theta}$ which, for no slip boundary conditions, is Ω/α for small α. Under uniform shear with no slip, it may be shown that $\partial v_x/\partial y \approx 0$ and $\partial v_x/\partial y \approx \dot{\gamma}$ [2, 17].

2.8.2.2.4 Uniaxial Elongation

Finally we will consider the two flow fields that yield extensional deformation. The first is axial elongational in which the fluid is contained between two planar surfaces in relative motion along their planar normals. With the axial extension direction taken as z, one has $\varkappa_{zz} = \dot{\varepsilon} = -2\varkappa_{xx} = -2\varkappa_{yy}$ and

$$v_x = -\frac{1}{2}\dot{\varepsilon}x \qquad v_y = -\frac{1}{2}\dot{\varepsilon}y, \qquad v_z = \dot{\varepsilon}z. \tag{2.8.6}$$

and the shear rate dependent extensional viscosity is [18]

$$\eta_E(\dot{\gamma}) = (\sigma_{zz} - \sigma_{xx})/\dot{\varepsilon} \tag{2.8.7}$$

This flow field is somewhat idealized, and cannot be exactly reproduced in practice. For example, near the planar surfaces, shear flow is inevitable, and, of course, the range of x and y is consequently finite, leading to boundary effects in which the extensional flow field is perturbed. Such uniaxial flow is inevitably transient because the surfaces either meet or separate to laboratory scale distances.

2.8.2.2.5 Planar Elongation

By contrast, in planar axial flow, with the extension direction taken as x, one has $\varkappa_{xx} = \dot{\varepsilon} = -\varkappa_{yy}$ and $\varkappa_{zz} = 0$

$$v_x = \dot{\varepsilon}x, \qquad v_y = -\dot{\varepsilon}y, \qquad v_z = 0 \tag{2.8.8}$$

This flow field can be maintained in a steady state, at least in the Eulerian sense, either by use of a four-roll mill [18] as in Figure 2.8.4(a) or by means of opposed jet flow as in Figure 2.8.4(b). However, it is important to note that the flow is still transient in the Lagrangian sense. That is, pure planar extension is confined to the central stagnation

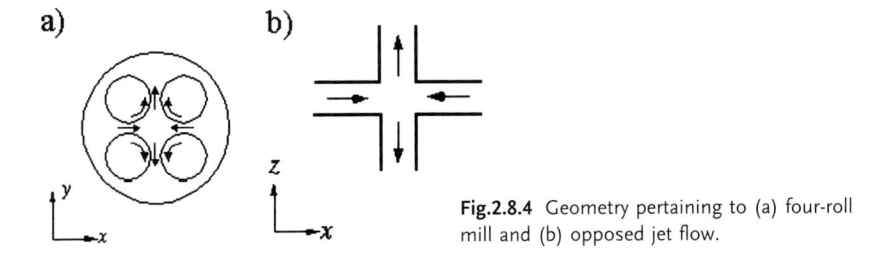

Fig.2.8.4 Geometry pertaining to (a) four-roll mill and (b) opposed jet flow.

flow point, except at that precise point molecules in its vicinity, passing along the flow lines, experience the extensional strain, only while they are near the central region.

2.8.3
Apparatus Implementation

In the case of the cylindrical Couette, cone-and-plate and four-roll mill, a drive shaft provides the required torque for rotation whilst in the case of the opposed jet, fluid is pumped into each of the two inlet ports and removed from the outlet ports via flexible Teflon pipes. In order to realize these geometries in an NMR microimaging probehead each cell can be adapted either by simply fixing it within an existing rf coil, or, where a specialized additional rf coil is required, plugging the combined cell-coil unit directly on top of the probe tuning stage. The cell drive shaft can then be easily coupled to a vertical drive shaft (or feed pipes) as required. The upper end of this shaft can then connect to the gearbox and motor, the rotation speed being thus placed under spectrometer control. The materials used need to be "NMR invisible" in the vicinity of the receiver coil (i.e., non-magnetic and transparent to rf). Suitable materials are Teflon, PEEK [poly(ethyl ether ketone)] or MACOR (Corning, New York). A typical rheo-NMR system with driver, motor, cells and drive shaft is shown in Figure 2.8.5.

Figures 2.8.6 to 2.8.9 are diagrams for each apparatus. One type of Couette cell shown in Figure 2.8.6(a) is made from two NMR tubes (10 and 5 mm), one inside the other, and positioned using a Teflon spacer at the bottom and a Teflon cap at the top. The inner tube is attached to its coupling shaft by a rubber sleeve. Fluid in the inner tube acts as a marker which rotates as a rigid body so that the velocity at the outer surface of the inner tube can be accurately extrapolated. This Couette cell is a sealed system and only a small amount of fluid is needed (approximately 5 cm^3). Another Couette cell is shown in Figure 2.8.6(b) in which PEEK is used as the containment and a small 1-mm gap separates cylinders of 17-mm od and 19-mm id, respectively. Note that the vorticity axis is parallel to the NMR polarizing field, B_0, in these systems. Figure 2.8.6(c) shows another Couette cell in which the 5.0-mm diameter shaft orientation is transverse to the polarizing field, thus placing the gradient direction parallel to B_0 at the top and bottom of the 0.5-mm gap region

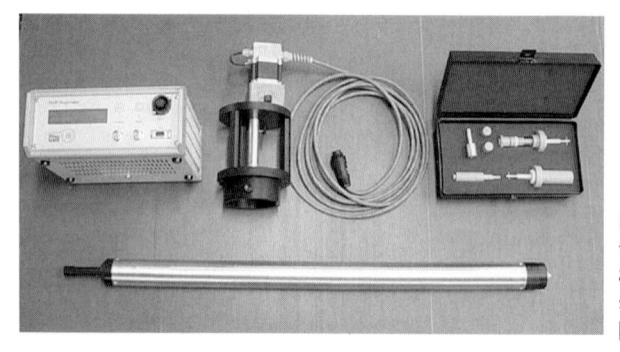

Fig. 2.8.5 Rheo-NMR kit for use with the Bruker 89 mm vertical bore superconducting magnet [20].

2.8 Hardware and Method Development for NMR Rheology

i)

Motor drive

B_0

Teflon bush

Magnet bore
Gradient coils
Rf coils
Sample
Marker fluid

10 mm

ii) 5.0 mms^{-1}

0 mms^{-1}

iii) Velocity (mm/s) vs Radius (mm)

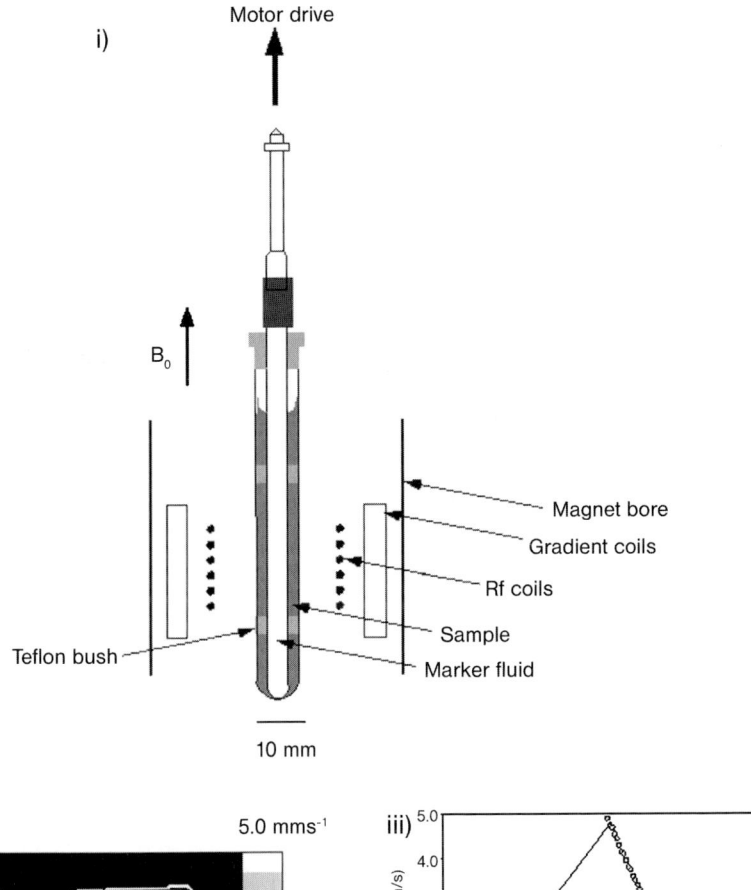

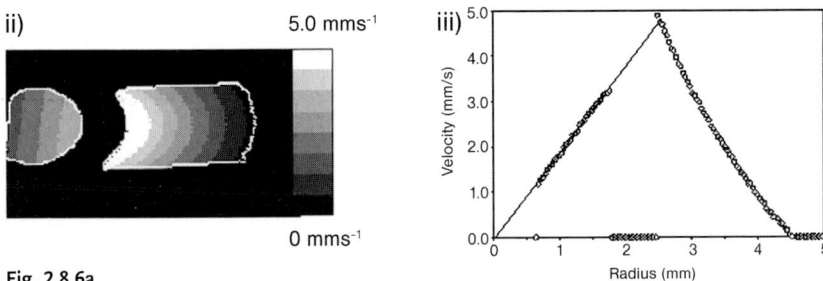

Fig. 2.8.6a

and the velocity direction parallel to B_0 at the sides. The ability to select different regions of fluid in which the molecular order is measured along different directions in the principal axis frame of the flow is crucial in such studies.

In the Couette cell the shear stress varies signficantly with radial position across the gap as r^{-2}. Should a more uniform stress environment be required then the cone-and-plate geometry may be used [17]. An example apparatus is shown in Figure 2.8.7.

Typically gap angles of between 7° and 4° are used. The cone is rotated in a similar manner to the Couette cell, however, unlike the Couette cell the cone-and-

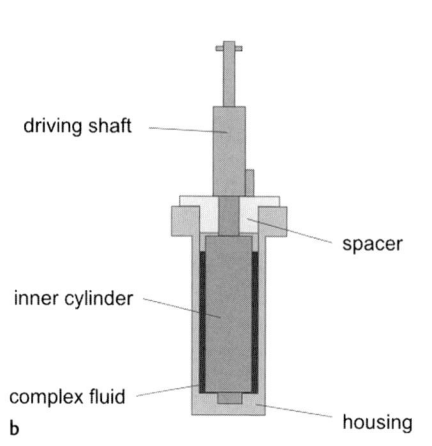

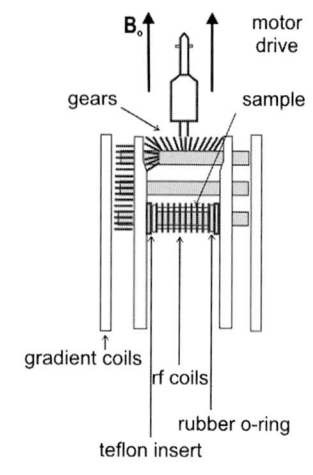

Fig. 2.8.6 (a) (i) Implementation of vertical cylindrical Couette cell using concentric glass tubes; (ii) velocity image taken across a horizontal slice; and (iii) velocity profile taken across the cell. Note that the marker fluid in the inner cylinder exhibits rigid body motion and allows extrapolation to find the velocity of the inner cylinder outer surface (adapted from Ref. [15]). (b) Implementation of vertical cylindrical Couette cell using PEEK (see Ref. [21]). (c) Implementation of horizontal cylindrical Couette cell (see Ref. [21]).

plate is not a sealed system and so when cone velocities are sufficiently high, fluid can spurt out of the gap. To compensate for this a Teflon wall can be positioned around the plate. However, the gap between the cone-and-plate must be filled so that the outer edge of the fluid does not touch the wall.

A four-roll mill is a more complicated construction, as shown in Figure 2.8.8. The four vespel rods, which may be kept in position by a block of Teflon at the bottom of the cylinder, are rotated simultaneously, drawing fluid in between one pair of rods and expelling fluid between the other pair. The drive shaft couples through the rod at the top and all rods are connected through a set of gears, the respective directions of rotation being shown schematically in Figure 2.8.4. This is a sealed system, with only about 5 mL of fluid required to fill the apparatus.

The opposed jet is a much simpler apparatus as shown in Figure 2.8.9. In the system shown [15], two perpendicular, intersecting cylindrical holes of 2.0-mm radius are drilled through a block of Teflon to produce a cross flow junction. At the end of each hole is a threaded socket into which a pipe fitting can be connected. Fluid is pumped into the cell from opposing sides, using two separate HPLC pumps. The fluid collides in the center of the junction and then exits via an orthogonal path. Instead of placing this system inside an exterior rf coil, it is possible to use an *in situ* solenoid coil wound directly around the vicinity of the junction (but not in contact with the fluid), thus optimizing sensitivity. The cross flow system involves steady-state pumping and hence the amount of fluid needed is significantly higher than for the enclosed cells, typical experiments requiring approximately 100 mL.

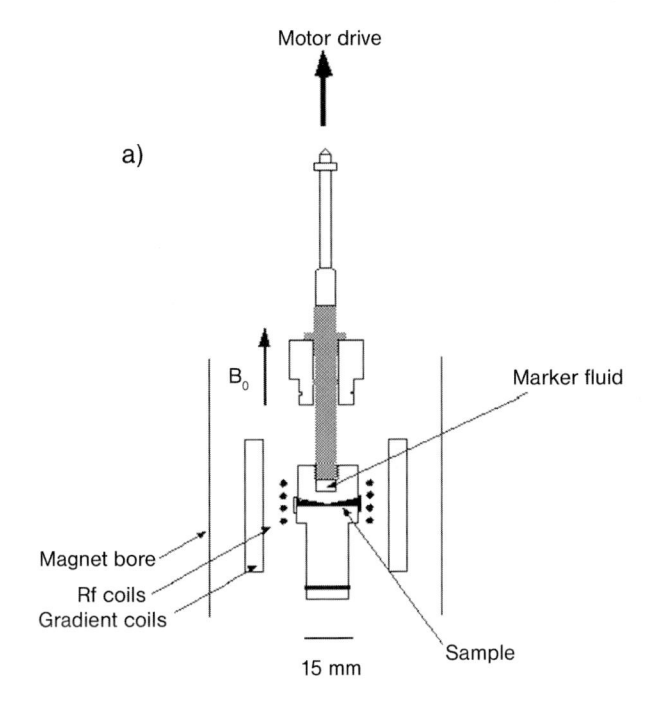

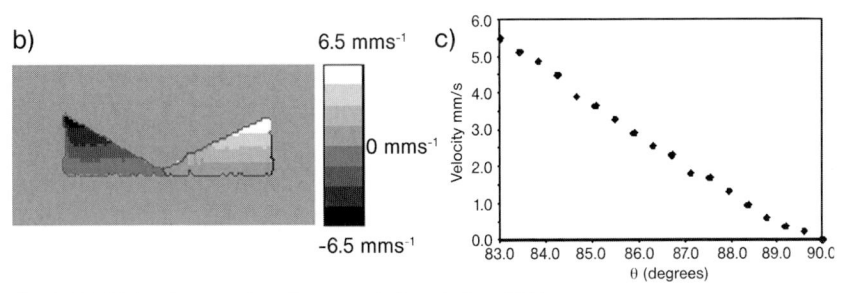

Fig. 2.8.7 (a) Implementation of cone-and-plate device. (b) Velocity image taken across a vertical slice. (c) Velocity profile taken across the cell with 7° cone angle. Note the highly linear variation of velocity (adapted from Ref. [15]).

2.8.4
Applications of Rheo-NMR

In order to illustrate the potential applications of rheo-NMR five examples have been chosen. The first example deals with wormlike micelles [22] in which NMR velocimetry is used to profile anomalous deformational flow and deuterium NMR spectroscopy is used to determine micellar ordering in the flow. The second example concerns flow in a soft glassy material comprising a solution of intermittently jammed star polymers [23], a system in which flow fluctuations are apparent. The third

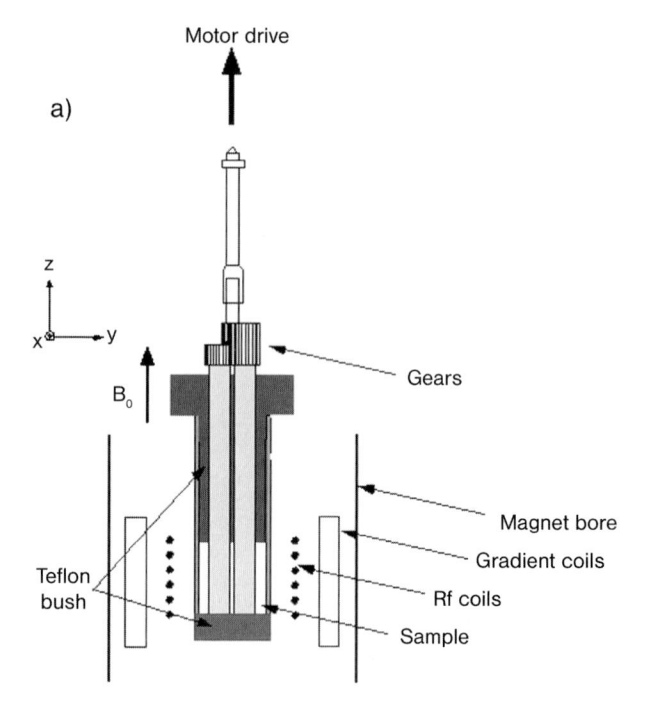

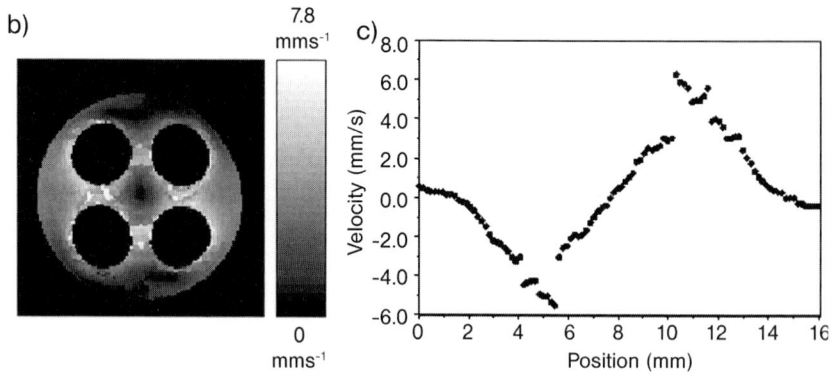

Fig. 2.8.8 (a) Implementation of four-roll mill device. (b) Speed image obtained across a horizontal slice by combining separate x- and y-velocity images. (c) Velocity profile taken across the x-axis of the cell showing the uniform extensional strain rate $\dot{\varepsilon}$, such that $v_x = \dot{\varepsilon} x$ (adapted from Ref. 15]).

example concerns the measurement of nuclear spin relaxation times during shearing flow, of a semi-dilute polymer solution [24], while the remaining two examples [10, 25] deal with the use of deuterium NMR spectroscopy to measure molecular orientation during flow, in polymer melts and in liquid crystalline side-chain polymers.

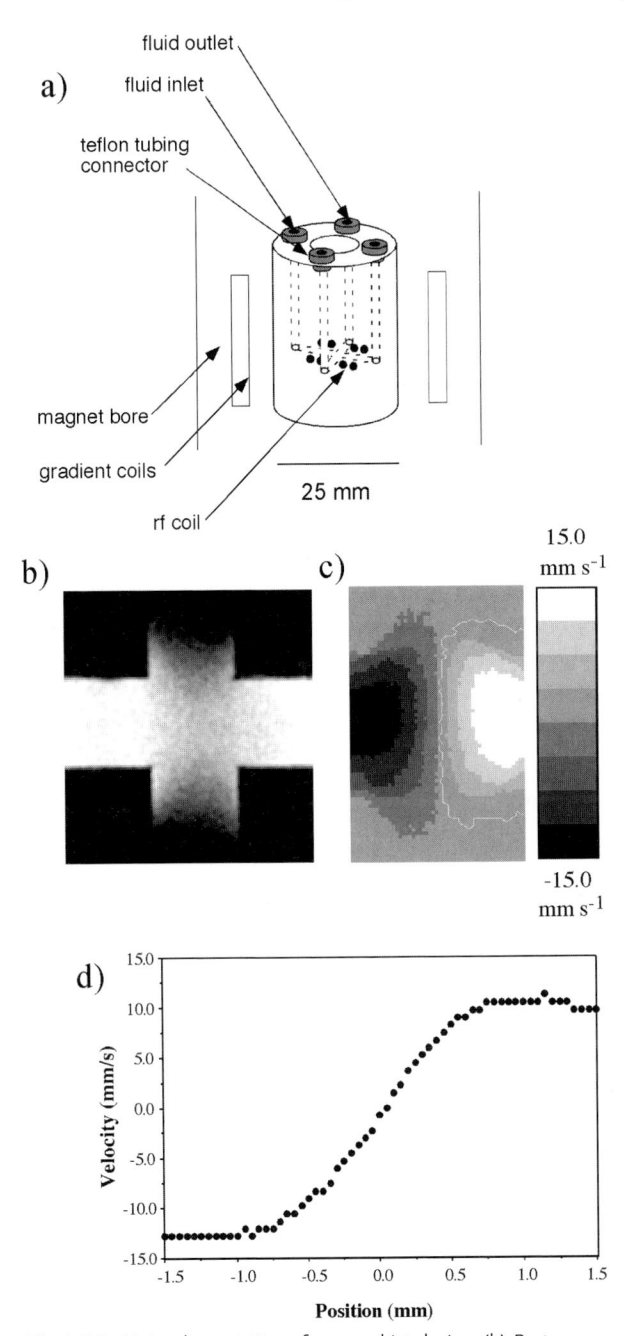

Fig. 2.8.9 (a) Implementation of opposed jet device. (b) Proton density image obtained across a horizontal slice. (c) Velocity image obtained across a horizontal slice. (d) Velocity profile taken along the x axis of the cell showing the uniform extensional strain rate $\dot{\varepsilon}$ such that $v_x = \dot{\varepsilon} x$ (adapted from Ref. [15]).

2.8.4.1 Shear-banded Flow in Wormlike Micelle Solutions

A significant contribution of rheo-NMR has been to show that the uniform shear-rate assumption may be violated in the case of certain classes of fluids in which pathological flow properties are exhibited. Figure 2.8.10 shows shear-rate maps [26] obtained for the wormlike surfactant system, cetylpyridinium chloride–sodium salicylate in water. While the velocity gradients show no deviation from uniformity at very low shear rates, above a certain critical value $\dot{\gamma}_c$ a dramatic variation in the rate-of-strain across the 7° cone gap is found. In particular a very high shear-rate band is found to exist at the mid-gap.

This micelle system exhibits a broad stress plateau in the flow curve (stress versus shear rate). The term "plateau" means a region of changing shear rate for which the stress is constant. Such a plateau can be explained by the shear banding phenomenon [27–28]. The key property of such systems is the constitutive behavior shown schematically in Figure 2.8.11 where the region of decreasing stress as the shear rate increases finds its origin in the reduction in entanglements as the worm chains align in the flow, a behavior which is predicted by the reptation-reaction model of Cates [29]. The declining stress section of the flow curve is associated with unstable flow. For average shear rates in excess of $\dot{\gamma}_c$ corresponding to the stress maximum, σ_c, in the schematic flow curve, separation of distinct shear bands may occur, in the manner of a first-order phase separation [28]. These bands correspond to fluid residing at the intersections of the selected stress [30] tie line with the upper and lower branches of the underlying flow curve, and the proportions of each band will be as required to satisfy the average shear rate. That the NMR results are consistent with this picture is clear in Figure 2.8.10(b), where a series of profiles show that as the gap apparent shear rate is increased the high shear rate band expands in width at approximately a constant maximum shear rate.

Banding effects have also been seen in these wormlike micelle materials via optical birefringence [31, 32], although it is not clear that birefringence banding necessarily corresponds to shear banding [33]. Of course, the anisotropy of bi-

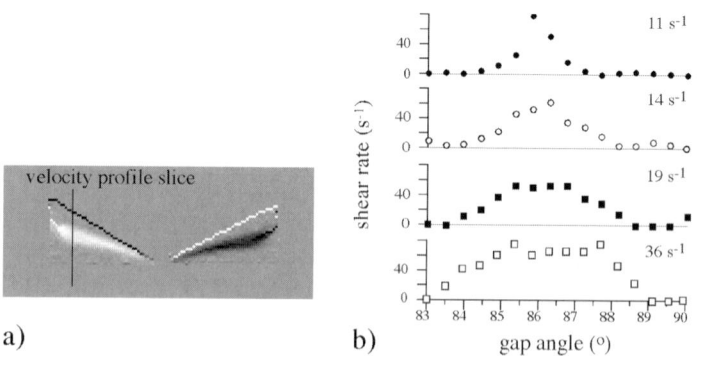

Fig. 2.8.10 (a) Grey scale map of shear taken across gap of 7° cone-and-plate device, for the semi-dilute wormlike micelle system 60 mM cetylpyridinium chloride– 100 mM sodium salicylate. (b) Shear profiles for differing applied shear rates showing the growth of the high shear bandwidth (adapted from Ref. [26]).

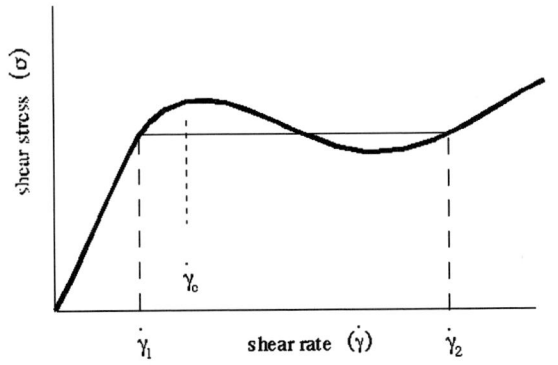

Fig. 2.8.11 Schematic constitutive relationship (shear stress versus shear rate) for a wormlike micelle system exhibiting constitutive instability. When the average shear rate exceeds $\dot{\gamma}_c$ the fluid subdivides into two coexisting shear bands residing on stable branches of the constitutive relationship.

refringence is associated with molecular alignment. NMR is also capable of investigating order and alignment through utilizing inter-nuclear dipole interactions or nuclear quadrupole interactions [34]. This interaction results in a two-line NMR spectrum in which the splitting is proportional to the local order parameter, and to the alignment with respect to the magnetic field via the second rank Legendre polynomial, $P_2(\cos\theta) = (3\cos^2\theta - 1)/2$. Note that θ is the angle between the principal axis of molecular order and the magnetic field. By combining spectroscopic methods with NMR imaging it also becomes possible to measure local order [35, 36].

Figure 2.8.12 shows D_2O ^2H NMR spectra for a wormlike micelle system in which birefringence banding had been observed (20% CTAB–D_2O at 41 °C), a system which is close to an isotropic-nematic transition at rest and in thermal equilibrium, but for which an imposed shear flow can induce a nematic phase [37, 38]. These spectra are plotted as a function of radial position across the gap of a cylindrical Couette cell, the magnetic field being aligned with the vorticity axis [38]. At the inner wall, where the stress is highest, a splitting is observed, indicative of a finite quadrupole interaction and the formation of a nematic phase, while at the outer wall a single peak is observed, consistent with an isotropic phase. These data also suggest transition between the phases through a mixed phase region, over the region of intermediate stress. Recent work on wormlike micelles has also shown the induction of a nematic phase even when the system at equilibrium resides far from an isotropic–nematic phase transition.

The use of deuterium NMR methods holds promise for the investigation of a wide class of materials where shear deformation is likely to result in molecular ordering.

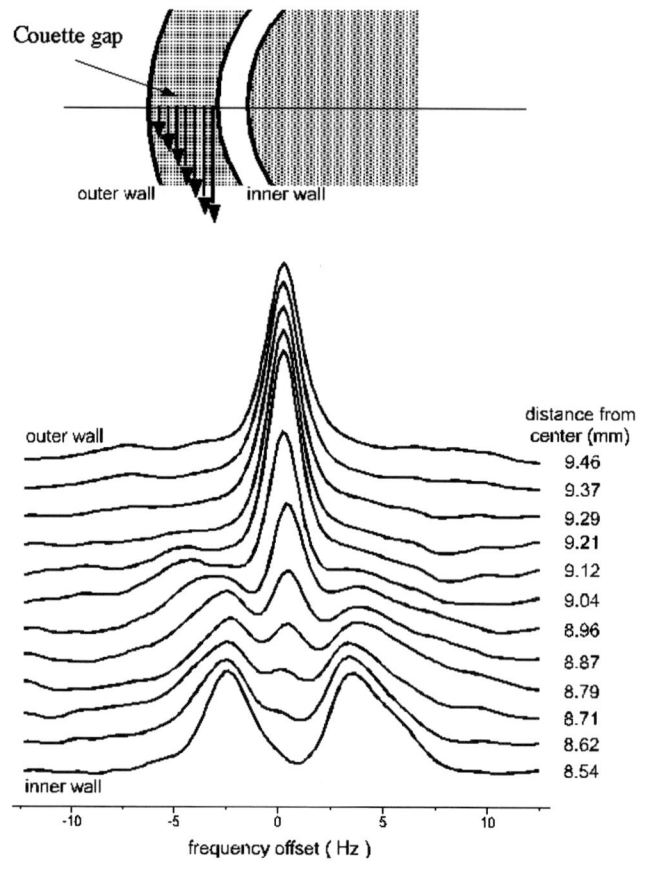

Fig. 2.8.12 ^{2}H NMR spectra obtained at 46 MHz, from 20% w/w CTAB–D$_2$O (41 °C) at different positions across the annular gap of a cylindrical Couette cell and at an apparent shear rate of 20 s^{-1}. Near the inner wall, where the stress is highest, a quadrupole splitting is observed, consistent with an ordered phase, while near the outer wall the single peak of an isotropic phase is seen. In between, a mixed phase region exists (adapted from Ref. [38]).

2.8.4.2 Velocity Field Fluctuations for a Soft Glassy Material

One characteristic of shear banded flow is the presence of fluctuations in the flow field. Such fluctuations also occur in some glassy colloidal materials at colloid volume fractions close to the glass transition. One such system is the soft gel formed by crowded monodisperse multiarm (122) star 1,4-polybutadienes in decane. Using NMR velocimetry Holmes et al. [23] found evidence for fluctuations in the flow behavior across the gap of a wide gap concentric cylindrical Couette device, in association with a degree of apparent slip at the inner wall. The timescale of these fluctuations appeared to be rapid (with respect to the measurement time per shear rate in the flow curve), in the order of tens to hundreds of milliseconds. As a result, the velocity distributions, measured at different points across the cell, exhibited bimodal behavior, as apparent in Figure 2.8.13. These workers interpreted their data

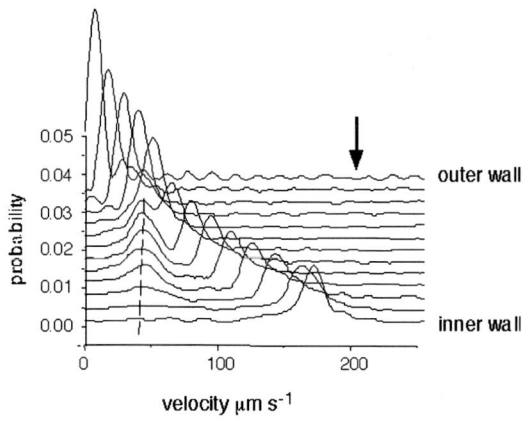

Fig. 2.8.13 Velocity probability distributions at different positions across the gap in a 5 mm–9 mm Couette cell at a shear of 0.101 s^{-1} and following long pre-shearing at high shear rate. The arrow indicates the inner wall velocity. The double peaks indicate a bimodal behavior associated with velocity fluctuations (adapted from reference [23]).

in terms of a qualitative model in which intermittent changes due to jamming/unjamming transitions occur, analogous to cage dynamics in colloidal glasses.

2.8.4.3 Shear-banded Flow in a Semi-dilute Polymer Solution – T_2 Effects

In another example [24], this time involving semi-dilute solutions of polyacrylamide (PAA) in water, rheo-NMR ^{1}H spectroscopy has been performed to show not only shear banding effects in velocimetry, but also associated heterogeneity in spin–spin relaxation times, T_2. The experiment was performed using a cylindrical Couette cell in which the solution in the annulus is imaged in order to investigate spatially dependent NMR parameters. A number of interesting properties emerge, as are apparent in Figure 2.8.14. Shear-banded flow is observed, in contrast to rheo-NMR studies of other semi-dilute polymers in which simple power-law shear thinning is found. Remarkably, the PAA chain protons exhibit a significant T_2 reduction under shear. In the same study, shear cessation experiments were carried out, revealing a T_2 recovery, which is multi-exponential. These results were interpreted in terms of shear-induced deformation and relaxation of a Doi–Edwards tube [39].

2.8.4.4 Deformation of Entangled Polymer Molecules Under Shear

A longstanding conjecture in polymer physics concerns the nature of the molecular deformation undergone by entangled random coil polymer molecules under shear flow [39, 40]. As a result of the deformation, the chain entropy is reduced leading to an increase in the free energy, the basis of the elastic response of polymer melts. One useful way to describe the deformation is by means of the averaged segmental alignment tensor [39]:

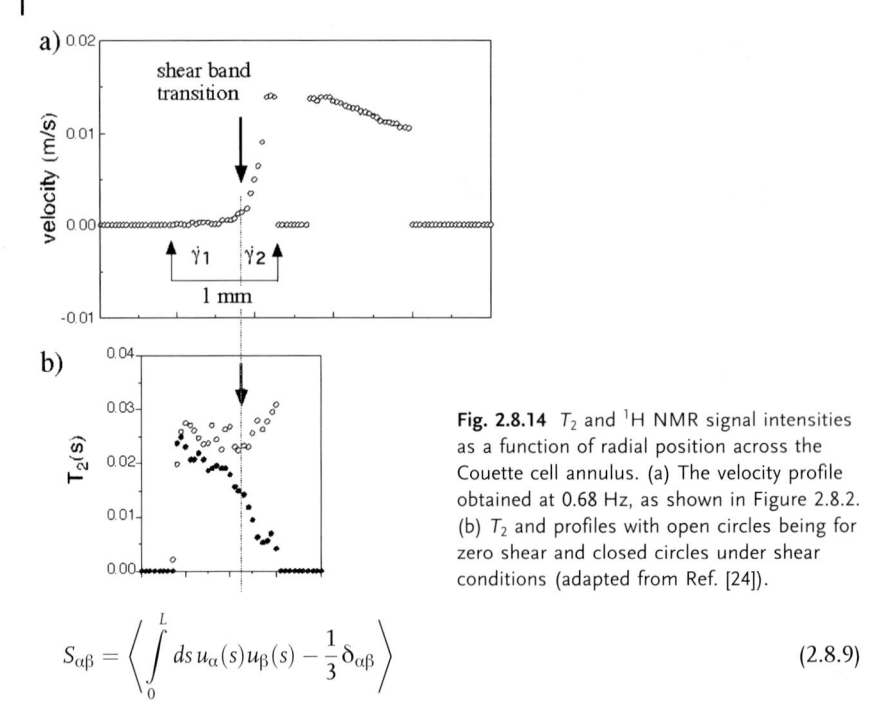

Fig. 2.8.14 T_2 and ^{1}H NMR signal intensities as a function of radial position across the Couette cell annulus. (a) The velocity profile obtained at 0.68 Hz, as shown in Figure 2.8.2. (b) T_2 and profiles with open circles being for zero shear and closed circles under shear conditions (adapted from Ref. [24]).

$$S_{\alpha\beta} = \left\langle \int_0^L ds\, u_\alpha(s) u_\beta(s) - \frac{1}{3}\delta_{\alpha\beta} \right\rangle \quad (2.8.9)$$

where <...> represents the ensemble average and the integral is taken over the curvilinear path of s chain segments along the chain length L. $S_{\alpha\beta}$ is predicted by the Doi–Edwards formulation of entangled polymer dynamics [39, 40]. This model is based on a statistical mechanical depiction of polymer chains topologically confined by surrounding chains.

Using deuterium nuclear magnetic resonance it is possible to measure the alignment tensor in a high molecular weight poly(dimethyl siloxane) melt, as shown in Figure 2.8.15. The work was carried out [21, 35] using a small horizontal Couette cell (gap 0.5 mm) in which both the velocity axis element S_{xx} and the gradient element, S_{yy} of the alignment tensor can be projected along the magnetic field. In order to introduce deuterons into the sample, a benzene probe molecule was introduced as a dilute species. The tumbling probe molecule undergoes steric interactions with the segment whose orientation u defines the local director, and experiences an anisotropic mean orientation. Thus the probe exhibits a scaled down quadrupole splitting associated with that local site via a "pseudo-nematic" interaction. The process of Brownian diffusion around the polymer results in a overall sensitivity to the ensemble average <...>, exactly as desired [21].

As shown in Figure 2.8.15(a), NMR microimaging is used to excite a desired region of the sample for spectroscopy experiments during steady-state shear. Figure 2.8.15(b) shows the measured alignment tensor elements along with fits using the Doi–Edwards model. Interesting features of the data are the biaxiality of the deformation, as seen in the difference between S_{yy} and S_{zz}, and the discrepancy between the Doi–Edwards prediction of a plateau in S_{xy} and the measured behavior.

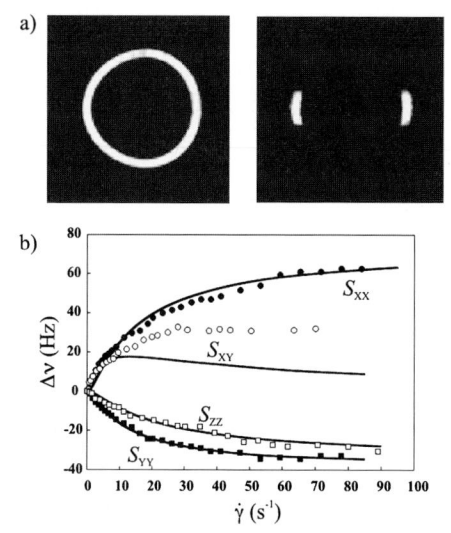

Fig. 2.8.15 (a) ^{1}H NMR images of poly(dimethylsiloxane) polymer in the annular gap of the horizontal Couette cell, along with that obtained using a selective storage pulse sequence used to localize a part of the fluid. (b) Deuterium quadrupole splittings, Δv as a function of shear rate obtained from the ^{2}H NMR spectra derived using the same selective storage pulse sequence. These localized spectroscopy experiments were derived from selected regions of the horizontal Couette cell in which the velocity direction (solid circles) and gradient direction (solid squares) are parallel to the magnetic field and at a 45° angle giving S_{xx}, S_{yy} and S_{xy} (open circles), respectively. Similar measurements were done in a vertical Couette cell in which the vorticity axis is parallel to the B_0 field (open squares) giving S_{zz}. The solid lines are fits using Doi–Edwards theory and a polymer tube disengagement time of $\tau_d = 210$ ms (adapted from Ref. [21]).

2.8.4.5 Director Reorientation of Liquid Crystalline Polymers Under Shear and Extension

Nematic liquid crystals under shear stress experience not only deformational flow but also director reorientation [41, 42], a process involving at least three independent rotational viscosities when describing the linear viscous response. Such nematic systems, which have intrinsic diamagnetic anisotropy, in addition usually experience a torque due to the magnetic field. In consequence, deformational flow results in a competitive director reorientation. Siebert et al. [10] studied such effects using thermotropic side-chain liquid crystal polymers, materials which present Leslie viscosities sufficiently large that significant alignment is possible at quite low shear rates, in the order of 1 s^{-1}. In a range of experiments they have continuously sheared deuterated liquid crystal polymers in a cone-and-plate cell, observing the quadrupole splitting, and hence the equilibrium director orientation [via $\Delta v = AP_2(\cos\theta)$]. Using Leslie–Ericksen theory [41, 42] these workers showed that under a uniform shear rate $\dot\gamma$ the director settles at an equilibrium angle to the magnetic field given by [10]:

$$\tan\theta = \mp \frac{\chi_a B_0^2}{2\mu_0 |\mu_3| \dot\gamma} \pm \sqrt{\left(\frac{\chi_a B_0^2}{2\mu_0 |\mu_3| \dot\gamma}\right)^2 \pm \frac{|\mu_2|}{|\mu_3|}} \qquad (2.8.10)$$

where μ_2 and μ_3 are Leslie coefficients, and χ_a is the anisotropic part of the diamagnetic susceptibility tensor. The positive sign in the equation corresponds to a flow aligning ($\mu_2/\mu_3 > 0$) nematic while the negative to a tumbling ($\mu_2/\mu_3 < 0$) nematic. For a flow aligning system a stable angle is found for all shear rates, while for a tumbling nematic, a stable angle exists provided that $\dot\gamma$ is sufficiently small that the term inside the square root sign is positive.

Siebert et al. demonstrated Eq. (2.8.10) using two different nematic systems [10]. The "flow-aligning" system was a nematic side-chain polysiloxane obtained by the addition of 2,3,5,6-tetradeuterio-4-methoxyphenyl-4-butenyloxybenzoate to poly(hydrogen methyl siloxane). As shown in Figure 2.8.16, the splittings form an asymptote with increasing shear rate at a θ value of around 70°. This angle, associated as it is with the dominance of mechanical torque over magnetic torque, corresponds to the alignment angle that would occur at all shear rates in the case of zero magnetic field.

The tumbling system was a mixture of 65% (w/w) poly{4-[4-(4-methoxyphenylazo)phenoxl]butylmethacrylate} in 4-tri-deuterio-methoxybenzoic acid-[4-n-hexoxyphenylester]. At shear rates beyond $\theta \approx 50°$, no splitting could be found, a result consistent with the onset of director tumbling. Figure 2.8.16 shows the respective alignment angles as a function of shear rates along with fits to the data using Eq. (2.8.10). In the case of the two different materials, opposite signs for μ_2 and μ_3 were necessary in order to reproduce the observed curvature of the data, confirming the respective flow aligning and tumbling character of the different polymer liquid crystals.

The same "flow-aligning" side-chain liquid crystalline polymer has been studied [43] in extensional flow using a rheo-NMR method in which selective excitation of

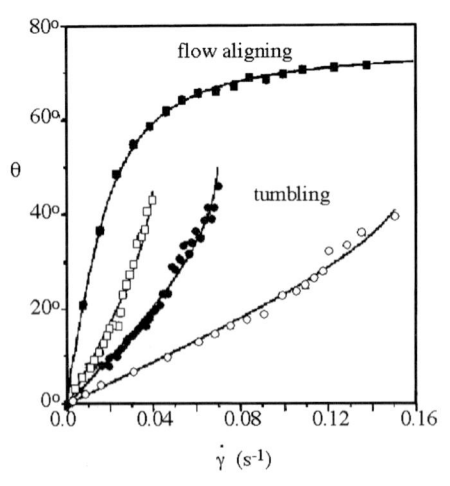

Fig. 2.8.16 Director orientation, θ, as a function of shear rate for both flow aligning (solid squares) and tumbling (open squares 325 K, solid circles 328 K and open circles 333 K) nematic polymers. (From Siebert et al. [10].)

the stagnation region at the center of a four-roll mill was used to provide ^{2}H NMR spectra from a deuterated DMSO probe molecule dissolved in the polymer, while undergoing pure planar extension. By contrast, with the case of shear flow, the director does not undergo a continuous reorientation but instead a sudden flip at a critical extensional strain rate given by:

$$\dot{\varepsilon}_c = \frac{\chi_a B_0^2}{\mu_0 |\mu_3 + \mu_2|} \tag{2.8.11}$$

This is precisely the observed behavior, as seen in Figure 2.8.17.

2.8.5
Conclusions

Rheo-NMR is an area of research in which only a handful of groups presently participate. However, the potential exists for a substantial increase in rheo-NMR research activity. The applications illustrated in this chapter concern polymer melts and semi-dilute solutions, colloidal glasses, lyotropic micellar systems and liquid crystalline polymers. The method has wide applicability across the entire class of materials known as soft solids or complex fluids. At this point in its development, the rheo-NMR methods are not straightforward and requires specialist NMR expertise. Nonetheless NMR rheology offers some unique advantages over other forms of rheo-spectroscopy, including the ability to work with opaque materials, the ability to combine velocimetry with localized spectroscopy and the ability to access a wide range of molecular properties relating to organization, orientation and dynamics. To date, rheo-NMR has been able to provide valuable information

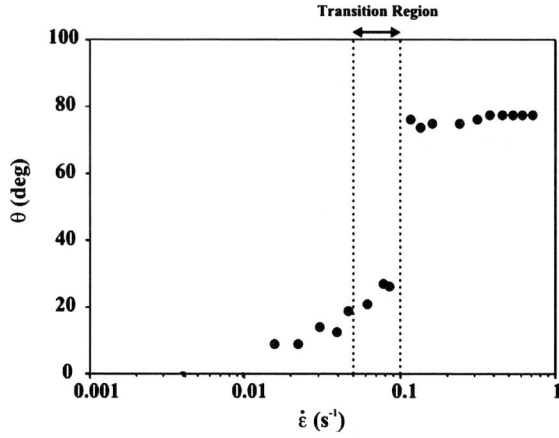

Fig. 2.8.17 Director orientation θ obtained from ^{2}H NMR spectra from the polysiloxane backbone polymer with liquid-crystal side-chain 4-methoxyphenyl-4'-butenyloxybenzoate, as a function of the apparent extension rate, $\dot{\varepsilon}$. The experiment is carried out using the signal selectively excited from the stagnation region of a four-roll mill. Extensional strain rates are experimentally determined by NMR velocimetry (adapted from Ref. [43]).

on a wide range of behaviors, including slip, shear-thinning, shear banding, yield stress behavior, jamming/unjamming, flow fluctuations, nematic director alignment and shear-induced phase transitions. As the methodology develops further as the technology becomes more routine, and as the measurement strategies become increasingly reliable, the community of users is likely to grow. Ultimately, the rheo-NMR approach offers remarkable promise as a technique yielding those macroscopic/molecular relationships essential to a deeper understanding of complex fluid and soft matter properties.

References

1 R. Larson **1999**, The Structure and Rheology of Complex Fluids, Oxford University Press, Oxford.
2 H. A. Barnes, J. J. Hutton, K. Walters **1989**, An Introduction to Rheology, Elsevier, Amsterdam.
3 G. G. Fuller **1995**, Optical Rheometry of Complex Fluids, Clarendon Press, Oxford.
4 A. I. Nakatani, M. D. Poliks, E. T. Samulski **1990**, Macromolecules 23, 2686.
5 A. F. Martins. P. Esnault, F. Volino **1986**, Phys. Rev. Lett. 57, 1745.
6 L. N. Goncalves, J. P. Casquilho, J. Figueirinhas, C. Cruz, A. F. Martins **1993**, Liq. Crys.14, 1485.
7 D. A. Grabowski, C. Schimdt **1994**, Macromolecules, 27, 2632.
8 M. Lukaschek, S. Müller, A. Hansenhindl, D. A. Grabowski, C. Schmidt **1996**, Colloid Polym. Sci. 274, 1.
9 M. Lukaschek, D. A. Grabowski, C. Schmidt **1995**, Langmuir 11, 3590.
10 H. Siebert, D. A. Grabowski, C. Schmidt **1997**, Rheolog. Acta 36, 618.
11 M. D. Kempe, J. A. Kornfield **2003**, Phys. Rev. Lett. 90, 115501-1-4.
12 Y. Xia, P. T. Callaghan **1991**, Macromolecules 24, 4777.
13 P. T. Callaghan **1999**, Rep. Progr. Phy. 62, 599.
14 K. Walters **1975**, Rheometry, Chapman and Hall, London.
15 M. M. Britton, P. T. Callaghan, M. L. Kilfoil, R. W. Mair, K. Owens **1998**, Appl. Magn. Reson. 15, 287.
16 P. T. Callaghan **1991**, Principles of Nuclear Magnetic Resonance Microscopy, Oxford University Press, Oxford.
17 M. M. Britton, P. T. Callaghan 1997, J. Rheol. 41, 1365–1386.
18 R. G., Larson **1987**, Constitutive Equations for Polymer Melts and Solutions, Butterworths, Boston.
19 G. G. Fuller, G. Leal **1980**, Rheol. Acta 19, 580.
20 http://www.rheo-nmr.com
21 R. J, Cormier, M. L. Kilfoil, P. T. Callaghan **2001**, Phys. Rev. E 6405, 1809.
22 R. Rehage, H. Hoffmann **1991**, Mol. Phys. 74, 933.
23 W. M. Holmes, P. T. Callaghan, D. Vlassopoulos, J. Roovers **2004**, J. Rheol. 48, 1085–1102.
24 P. T. Callaghan, A. M. Gil **2000**, Macromolecules 33, 4116–4124.
25 R. J. Cormier, C. Schmidt, P. T. Callaghan **2004**, J. Rheol. 48, 881–894.
26 M. M. Britton, P. T. Callaghan **1997**, Phys. Rev. Lett 78, 4930–4933.
27 T. C. B. McLeish, R. C. Ball **1986**, J. Polym. Sci. Polym. Phys. Ed. 24, 1735.
28 M. E. Cates, T. C. B. McLeish, G. Marrucci **1993**, Europhys. Lett. 21, 451.
29 M. E. Cates **1990**, J. Phys. Chem. 94, 371.
30 P. D. Olmsted, C.-Y. D. Lu **1997**, Phys. Rev. E 56, R55.
31 J. P. Decruppe, R. Cressely, R. Makhoufli, E. Cappelaere **1995**, Colloid Polym. Sci. 273, 346.
32 R. Makhoufli, J. P. Decruppe, A. Ait-Ali, R. Cressely **1995**, Europhys. Lett. 32, 253.
33 E. Fischer, P. T. Callaghan **2000**, Europhys. Lett. 50, 803.
34 A. Abragram **1961**, The Principles of Nuclear Magnetism, Clarendon Press, Oxford.

35 M. L. Kilfoil, P. T. Callaghan **2000**, Macromolecules 33, 6828.
36 M. L. Kilfoil, P. T. Callaghan **2001**, J. Magn. Reson. B150, 110–115.
37 E. Cappelaere, J.-F. Berret, J. P. Decruppe, R. Cressely, P. Lindner **1997**, Phys. Rev. E 56, 1869.
38 E. Fischer, P. T. Callaghan **2001**, Phys. Rev. E 6401, 1501.
39 M. Doi, S. F. Edwards, **1987** The Theory of Polymer Dynamics, Oxford University Press, Oxford.
40 M. Doi, S. F. Edwards **1978**, J. Chem. Soc., Faraday Trans. 2 74, 1802 and 1818.
41 F. M Leslie **1966**, Quart. J. Appl. Math. 19, 357.
42 J. L Ericksen 1961, Trans. Soc. Rheol. 5, 23.
43 R. J. Cormier, C. Schmidt, P. T. Callaghan **2004**, J. Rheol. 48, 881–894.

2.9
Hydrodynamic, Electrodynamic and Thermodynamic Transport in Porous Model Objects: Magnetic Resonance Mapping Experiments and Simulations
Elke Kossel, Bogdan Buhai, and Rainer Kimmich

2.9.1
Introduction

In this chapter, a number of transport phenomena with entirely different natures are compared for liquids filling porous systems. Here "transport" can refer to flow, diffusion, electric current or heat transport. Corresponding NMR measuring techniques will be described. Applications to porous model objects will be juxtaposed to computational fluid dynamics simulations.

"Magnetic resonance microscopy" and "imaging" are unspecific terms usually associated with rendering morphologies of objects. These types of images, however, are rarely what the chemical engineer has in mind when applying NMR techniques. "Spatial resolution" in this context rather refers to quantitative "mapping" of physical or physico-chemical parameters. We therefore speak of "velocity mapping", "acceleration mapping", "electric current density mapping" or "temperature mapping", and avoid terms of a more qualitative specificity such as "microscopy" or "imaging".

Mapping of transport parameters in complex pore spaces is of interest for many respects. Apart from "classical" porous materials such as rock, brick, paper and tissue, one can think of objects used in microsystem technology. Recent developments such as "lab-on-a-chip" devices require detailed knowledge of transport properties. More detailed information can be found in new journals such as *Lab on a Chip* [1] and *Microfluidics and Nanofluidics* [2], for example, devoted especially to this subject. Electrokinetic effects in microscopic pore spaces are discussed in Ref. [3].

In the light of the wide variety of complex pore spaces where microscopic transport features are of interest, we decided to concentrate on a special type of microporous systems suitable for testing NMR mapping methods and to examine microscopic transport laws [4]. Artificial porous model objects have been fabricated with a spatial resolution down to 50 μm [5]. The pore space structures were based on computer generated random site percolation clusters [6].

The advantages of this type of system are obvious: the pore space is of sufficient complexity to represent any natural or technical pore network. As the model objects are based on computer generated clusters, the pore spaces are well defined so that point-by-point data sets describing the pore space are available. Because these data sets are known, they can be fed directly into finite element or finite volume computational fluid dynamics (CFD) programs in order to simulate transport properties [7]. The percolation model objects are taken as a transport paradigm for any pore network of major complexity.

This strategy permits one to test the validity of macroscopic theories on microscopic length scales, the reliability of experimental techniques and, *vice versa*, the appropriateness of the CFD treatment. Furthermore, having put the simulations on a safe basis also enables one to predict transport features outside the experimentally accessible parameter range with some confidence of reliability [8].

For NMR microscopy using micro-coil devices, spatial resolutions even below 1 μm have been reported in the literature [9]. However, experiments probing transport such as flow or electric current need an experimental set-up much larger than the micro-coils in use in those experiments. In practice, it turned out that flow experiments, for instance, become very difficult in radiofrequency coils smaller than a few millimeters. For sensitivity reasons and due to limitations of the magnetic field gradients, resolutions below 20 μm are scarcely feasible under such circumstances [10]. This fact appears to form the basis of some restriction intrinsic to the NMR technology at the present state-of-the-art. However, in this context it should be mentioned that by sacrificing spatial resolution and accepting indirect measuring protocols, there are other NMR methods based on relaxation [11] and diffusion phenomena [12] that can provide information which is relevant even for nanometer length scales.

In the following, diverse NMR methods for probing transport phenomena will be described. Ordinary magnetic resonance imaging, or – in this context – spin density mapping, is already a simple tool that is used for long-time, long-distance diffusion studies, as will be demonstrated in Section 2.9.2. The description of flow velocity and acceleration mapping will follow in Section 2.9.3. Hydrodynamic dispersion, that is the superimposed effect of self-diffusion and coherent flow, can be mapped with the pulsed field gradient technique delineated in Section 2.9.4. Thermal effects such as convection and heat conduction are the subject of Section 2.9.5. Finally, in Section 2.9.6 we will specify an NMR technique for ionic current density mapping. The basic radiofrequency (rf) and field gradient pulse schemes to be discussed below are summarized in Figures 2.9.1 and 2.9.2 while more specific versions will be discussed in later sections.

2.9.2
Spin Density Diffusometry

This type of diffusion measurement is based on ordinary magnetic resonance imaging. The transverse and longitudinal relaxation times in a fluid filling a porous medium can be assumed to be effectively homogeneously distributed in the whole pore space. Fast exchange tends to average surface and bulk relaxation features (see Ref. [13], for instance). Under such circumstances, any ordinary magnetic resonance image recorded with a spin-echo technique [for typical three- and two-dimensional imaging sequences see Figures 2.9.1(a) and 2.9.2(a), respectively] can be considered to be equivalent to a spin density map, at least as long as the edge enhancement due to echo attenuation effects by diffusion [14, 15] can be

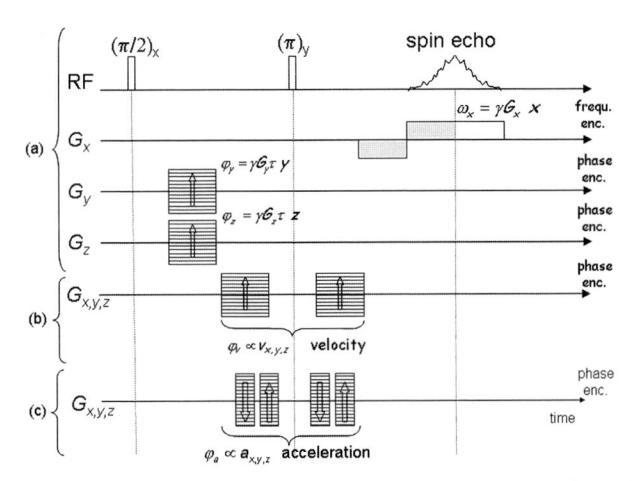

Fig. 2.9.1 (a) Radiofrequency (rf) and field gradient pulse sequences for three-dimensional imaging, (b) gradient pulse train that allows for velocity mapping if combined with an imaging sequence and (c) gradient pulse train that allows for acceleration mapping if combined with an imaging sequence. The techniques are based on Hahn spin-echo signals. (Idealized) expressions for the shifts due to frequency (ω_x) and phase encoding gradients (φ_y, φ_z, φ_v, φ_a) are given. τ is the length of a single gradient lobe, $\Delta\tau$ is the delay between two such lobes. The gradient pulses are assumed to have a rectangular shape in this schematic representation. Under practical conditions, switching "ramps" must be taken into account (see Ref. [36]), which, however, do not change the linear relationships between phase shifts and experimental parameters. All phase encoding pulses must be incremented independently in subsequent transients of the experiment. The arrows indicate the relative direction of increments (represented by horizontal ladder lines). The "area" of the negative lobe of the frequency encoding gradient is equal to that of the left section of the positive lobe extended until the echo reaches its maximum. Note that a full three-dimensional velocity or acceleration vector map of a three-dimensional system requires three three-dimensional NMR experiments, one for each of the three components of the velocity or acceleration vectors. In order to avoid excessively time consuming experiments [37], it is often more favorable to restrict oneself to the recording of two-dimensional maps. A typical two-dimensional imaging pulse sequence is shown in Figure 2.9.2(a). The pulse sequences provide a multidimensional "hologram", which is then analyzed with the aid of correspondingly multidimensional Fourier transforms [28].

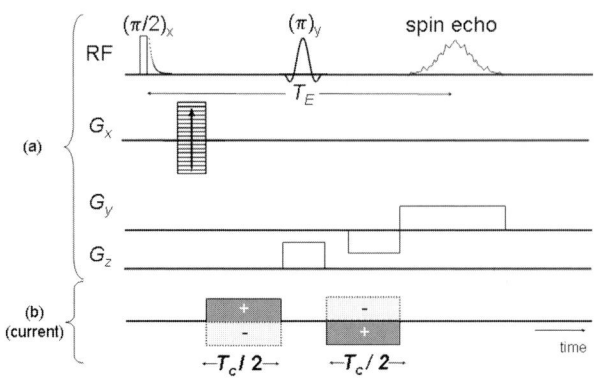

Fig. 2.9.2 Radiofrequency, field gradient and ionic current pulse sequences for two-dimensional current density mapping. T_E is the Hahn spin-echo time, T_c is the total application time of ionic currents through the sample. The 180°-pulse combined with the z gradient is slice selective. The pulse sequences shown are suitable for two-dimensional current distributions. The evaluation of three-dimensional current distributions requires a three-dimensional imaging sequence [see Figure 2.9.1(a)] and multiple experiments with the orientation of the sample relative to the magnetic field incremented until a full 360°-revolution is reached. The polarity of the current pulses is alternated in subsequent scans in order to eliminate offset artifacts according to $j_{x,y} = (j_{x,y\pm} - j_{x,y\mp})/2$.

excluded or neglected. Of course, the spin density in the fluid filled space will normally also be homogeneously distributed. However, the situation addressed here refers to interdiffusion of deuterated and undeuterated liquid species. "Spin density" thus indicates the local concentration of the resonant species.

Spin density diffusometry is to be distinguished from the more popular field gradient NMR diffusometry techniques [16], [17]. The time scale of such experiments is limited by the transverse and the longitudinal relaxation times, respectively, of the liquid under investigation. Such studies therefore refer to diffusion times of seconds, at most, under ordinary conditions. That is, the root mean squared displacement achieved by water molecules at room temperature, for instance, in a field gradient NMR diffusometry experiment, is only a few micrometers. In a homogeneous ("bulk") medium, the diffusion behavior on these relatively short length or time scales will be the same as for much longer experimental probing intervals.

The situation becomes quite different in heterogeneous systems, such as a fluid filling a porous medium. Restrictions by pore walls and the pore space microstructure become relevant if the root mean squared displacement approaches the pore dimension. The fact that spatial restrictions affect the echo attenuation curves permits one to derive structural information about the pore space [18]. This was demonstrated in the form of "diffraction-like patterns" in samples with micrometer pores [19]. Moreover, "subdiffusive" mean squared displacement laws [20], $\langle r^2 \rangle \propto t^\gamma$ with $\gamma < 1$, can be expected in random percolation clusters in the so-called scaling window,

2.9 Hydrodynamic, Electrodynamic and Thermodynamic Transport in Porous Model Objects

$$a < \sqrt{\langle r^2 \rangle} < \xi \qquad (2.9.1)$$

where a is an elementary length unit of the pore space structure and ξ is the correlation length [4]. The correlation length is the typical length scale up to which the pore structure has fractal (i.e., self similar) properties. It may be desirable to exploit this type of indirect information on the confining pore space structure [21–23]. However, mean pore sizes are often much larger than micrometers so that no restriction effect will show up in field gradient diffusometry experiments [16].

Under such circumstances, spin density diffusometry turned out to be more favorable. We are not discussing solvent intrusion experiments with solid-like materials, as they are typically carried out with the aid of stray field imaging [24]. In those experiments, the solvent is distinguished from the host material by the huge difference in the transverse relaxation times. The technique to be described here monitors interdiffusion between two sample compartments initially filled with deuterated and undeuterated liquids (or gels) of the *same* chemical species. Bringing the compartments into contact initiates interdiffusion. Mapping of the proton spin density thus permits the evolution of the corresponding concentration profiles to be followed.

The time resolution of the experiment is restricted by the time needed for the recording of one map. This is seconds for fast two-dimensional imaging [25–28] and minutes for conventional methods at signal intensities and spatial resolutions typical for NMR microscopy. If only (one-dimensional) profiles are recorded, the time resolution will be much less. The propagation of the diffusion front, on the other hand, can be recorded over periods as long as are acceptable in the frame of the investigation. Spatial resolution in the order of a few hundred μm can routinely be achieved with typical NMR microscopy systems. It depends on the experimental conditions, such as field strength, gradient strength, radiofrequency coil performance and electronic noise [19, 28]. A resolution of a few tens of μm can be obtained typically for samples with a diameter of a few cm or smaller. Under special conditions, even micrometer resolutions have been reported [9, 29, 30].

Figure 2.9.3 shows typical maps [31] recorded with proton spin density diffusometry in a model object fabricated based on a computer generated percolation cluster (for descriptions of the so-called percolation theory see Refs. [6, 32, 33]). The pore space model is a two-dimensional site percolation cluster: sites on a square lattice were occupied with a probability p (also called "porosity"). Neighboring occupied sites are thought to be connected by a pore. With increasing p, clusters of neighboring occupied sites, that is pore networks, begin to form. At a critical probability p_c, the so-called percolation threshold, an "infinite" cluster appears. On a finite system, the "infinite" cluster connects opposite sides of the lattice, so that transport across the pore network becomes possible. For two-dimensional site percolation clusters on a square lattice, p_c was numerically found to be 0.592746 [6].

Obviously, the diffusion coefficient of molecules in a porous medium depends on the density of obstacles that restrict the molecular motion. For self-similar structures, the fractal dimension d_f is a measure for the fraction of sites that belong

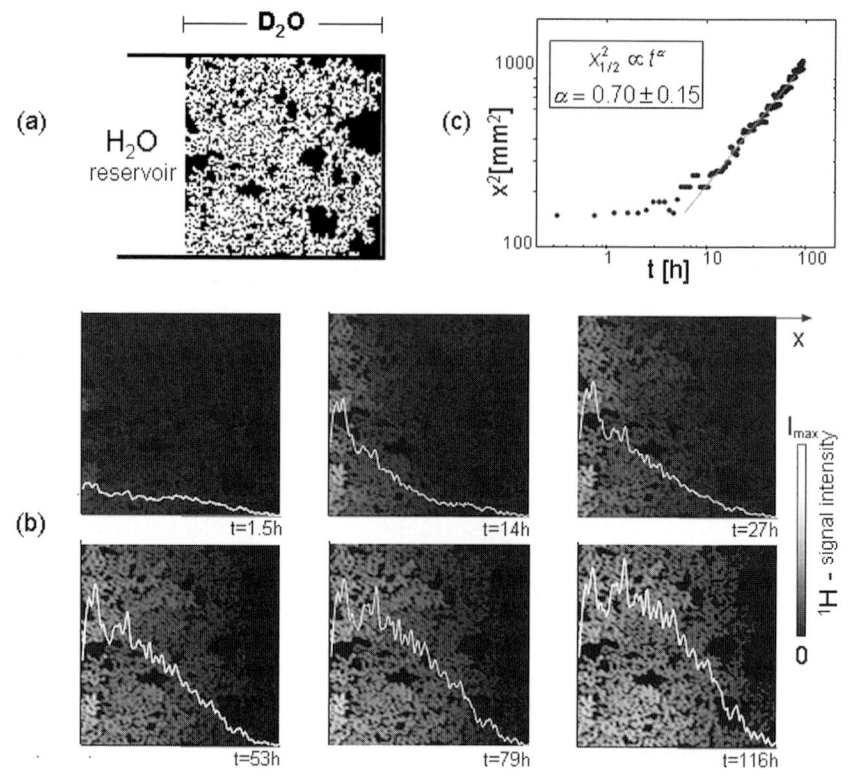

Fig. 2.9.3 Proton spin density diffusometry in a two-dimensional percolation model object [31]. The object was initially filled with heavy water and then brought into contact with an H$_2$O gel reservoir. (a) Schematic drawing of the experimental set-up. The pore space is represented in white. (b) Maps of the proton spin density that were recorded after diffusion times t varying from 1.5 to 116 h. Projections of the proton concentration distributions on the main diffusion direction (= x axis) are overlaid to the maps (white lines). (c) The squared displacement of the diffusion front (measured at half height) reproduces the power law according to Eq. (2.9.2). The experimental parameters are: sample size, 4 cm × 4 cm × 1 mm; magnetic flux density, 4.7 T; digital spatial resolution, 230 µm × 230 µm; time resolution, 26 min.

to the structure [32]. According to the Alexander–Orbach conjecture [34], the mean squared displacement of the molecules over the time t is proportional to

$$\langle r^2 \rangle \propto t^{4/(3d_f)} \tag{2.9.2}$$

where t is restricted to times during which the displacement is larger then a and smaller than ξ [see Eq. (2.9.1)]. The data shown in Figure 2.9.3(c) verify this prediction very well [31].

2.9.3
Flow Velocity and Acceleration Mapping

2.9.3.1 Pulse Sequences

Supplementing three- or two-dimensional imaging pulse sequences [Figures 2.9.1(a) or 2.9.2(a)] by additional phase-encoding gradient pulses according to Figure 2.9.1(b) and (c) permits one to record maps of velocity and acceleration components along the gradient direction, respectively. If the full vector or magnitude information is needed, experiments for each of the components must be combined.

The principles of such techniques are explained in the monographs in Refs. [19, 28, 35]. A nuclear magnetization with a component perpendicular to the external magnetic flux density B_0 precesses around the field direction with an angular frequency $\omega_0 = \gamma B_0$ where γ is the gyromagnetic ratio. The main magnetic field is assumed to point along the z axis of the laboratory reference frame. If a spatially constant field gradient $\hat{G}$ is applied, for example along the x axis, the precession angular frequency changes to $\omega = \gamma(B_0 + \hat{G}_x x)$. If the gradient is switched off after a time interval τ, the precession continues with the original frequency. By this time, the precession of the magnetization and therefore the induction signal has gained a position dependent phase shift $\varphi(x)$:

$$\varphi(x) = \int_0^\tau \gamma \hat{G}_x x(t) dt \qquad (2.9.3)$$

If the nuclei are displaced by a flow field, their positions on the gradient axis become time dependent:

$$x(t) = x_0 + v_{x0} t + \frac{1}{2} a_{x0} t^2 + \ldots \qquad (2.9.4)$$

That is, the phase shift depends on the initial position x_0, the initial velocity v_{x0} and the initial acceleration a_{x0}. Higher order terms vanish if the flow field is stationary on the time scale of the NMR experiment (i.e., time-dependent accelerations do not occur in this case). For a gradient pulse of duration τ and strength $\hat{G}_x$ the total phase shift is [see Figure 2.9.4(a)]

$$\varphi = \gamma \hat{G}_x \left(x_0 \tau + \frac{1}{2} v_{x0} \tau^2 + \frac{1}{6} a_{x0} \tau^3 \right) = \varphi(x_0) + \varphi(v_{x0}) + \varphi(a_{x0}) \qquad (2.9.5)$$

This phase shift depends on x_0, v_{x0} and a_{x0} at the same time, i.e., on the parameters needed for spin density, velocity and acceleration mapping, respectively. In order to distinguish the three quantities of interest, more sophisticated gradient pulse sequences have been designed, resulting in phase shifts that essentially depend only on a single parameter or a selection of parameters.

For many systems and flow fields, the position dependent term dominates the phase shift while the velocity dependent term is still significantly larger than the acceleration dependent term. As a direct consequence, terms with a higher order of τ than the term including the parameter to be evaluated may often be neglected,

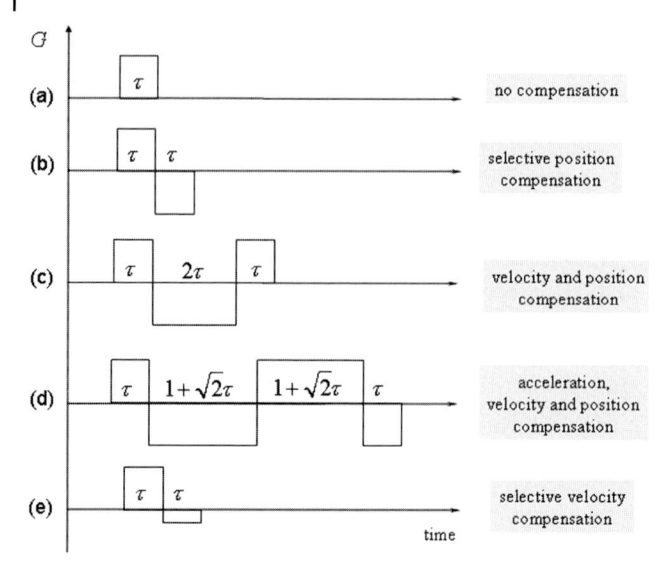

Fig. 2.9.4 Typical gradient pulse trains with and without compensation of phase shifts due to position, velocity and acceleration. (a) No compensation; (b) compensation of phase shifts due to the position coordinate in gradient direction; (c) compensation of phase shifts due to the velocity component along the gradient and the position coordinate in gradient direction; (d) compensation of phase shifts due to the acceleration and velocity components and the position coordinate; and (e) compensation of phase shifts due to velocity but not for position dependent shifts. The magnitude of the gradient strength is $\hat{G}$ in all cases except for the pulse train (e), where the gradient strength of the first lobe is $\hat{G}$, but $-\hat{G}/3$ for the second. For a detailed description and explanation see Refs. [28, 36]. If a 180°-rf pulse is applied in the middle of the gradient pulse train [compare Figures 2.9.1(b and c)], all signs of gradients in the second moiety must be inverted relative to the first one. Under practical conditions, the gradient pulse lobes have a trapezoidal shape, and there may be delays between the lobes. This must be taken into account in theoretical treatments [36].

while terms with a lower order of τ have to be compensated by the gradient pulse sequence. If the phase shifts due to higher order terms cannot be neglected, their contributions also have to be compensated for. In Figure 2.9.4, a number of gradient pulse sequences are shown which either compensate for low order terms or for high order terms. In the first case, the evaluation of the non-compensated term with the lowest order in τ becomes possible while in the second case, spin density maps without motion artifacts can be achieved.

Figure 2.9.4(b–e) shows typical gradient lobe trains suitable for the compensation of phase shifts due to any of the above quantities [28, 36]. If an intermittent 180°-rf pulse is used (such as in the Hahn echo sequences shown in Figures 2.9.1 and 2.9.2), the gradient lobes after the 180°-pulse must be inverted relative to the lobes before the 180°-pulse. The bipolar gradient pulse in Figure 2.9.4(b) compensates phase shifts due to the position, but produces phase shifts by velocity and acceleration components along the gradient direction. The variant in Figure 2.9.4(c) compensates phase shifts due to the position and to the velocity compo-

2.9 Hydrodynamic, Electrodynamic and Thermodynamic Transport in Porous Model Objects

nent, but generates phase shifts by the acceleration component. Finally, the gradient lobe train shown in Figure 2.9.4(d) compensates for phase shifts due to the position, the velocity and acceleration components, but provides phase shifts by higher order terms (if existent). These are the tools that can be exploited for flow parameter mapping.

For phase encoding of a velocity component, a gradient pulse sequence similar to the sequence in Figure 2.9.4(b) can be used. Two gradient pulses of duration τ and of the same magnitude of the amplitude $\hat{G}_x$ but with opposite sign are applied. The position dependent term of the phase shift vanishes whereas the velocity dependent term provides a finite phase shift proportional to the velocity component [see Figure 2.9.4(b)]:

$$\varphi(x_0) = 0; \quad \varphi(v_{x0}) = \frac{1}{2}\gamma\hat{G}_x v_{x0}(2\tau^2 - 4\tau^2) = -\gamma\hat{G}_x v_{x0}\tau^2 \qquad (2.9.6)$$

This gradient sequence is often modified by a delay $\Delta\tau$ between the two gradient lobes, as indicated in Figure 2.9.1(b). The position dependent term is not affected by the additional interval, while the velocity dependent term then results in

$$\varphi(v_{x0}) = -\hat{G}_x\gamma(\tau^2 + \tau\Delta\tau)v_{x0} \qquad (2.9.7)$$

Phase encoding of an acceleration component can be based on a gradient pulse train, as given in Figure 2.9.4(c). Both the position dependent term and the velocity dependent term are compensated. The first and the third lobe of this gradient pulse train have the same sign, duration and amplitude, whereas the second lobe has the same amplitude, an opposite sign and twice the duration of the first pulse. The induced phase shift then depends on the acceleration alone in a linear relationship [see Figure 2.9.4(c)]:

$$\varphi(x_0) = \varphi(v_{x0}) = 0; \quad \varphi(a_{x0}) = 2\gamma\hat{G}_x a_{x0}\tau^3 \qquad (2.9.8)$$

The phase shifts by gradient pulses with delays in between [such as shown in Figure 2.9.1(c)] are represented by more complex expressions as published in Ref. [36].

Phase encoding of higher order terms (locally time dependent accelerations) is feasible in principle with pulse trains such as the example shown in Figure 2.9.4(d). However, the relatively long acquisition times of NMR parameter maps conflict with "snapshot" records that would be needed for such maps. In any case, in Ref. [27] fast imaging techniques are described and discussed that could be employed for moderate versions of "snapshot" experiments.

Combining the two- or three-dimensional imaging sequences in Figures 2.9.1(a) and 2.9.2(a), respectively, with the phase encoding pulse train given in Figure 2.9.4(b) (modified for an intermittent 180°-rf pulse) provides complete velocity component mapping sequences as shown in Figure 2.9.1(b), for instance. Likewise, acceleration component mapping sequences are configured by combinations of Figure 2.9.4(c) (modified for an intermittent 180°-rf pulse again) with Figures 2.9.1(a) or 2.9.2(a) [see Figure 2.9.1(c)].

Referring to phase encoding gradient pulses according to Figure 2.9.4(b), the digital resolution and the velocity "field of view" are given by [28]

$$\Delta v_x = \frac{\pi}{\gamma \tau^2 \hat{G}_x} \qquad (2.9.9)$$

and

$$-\frac{\pi}{\gamma \tau^2 \Delta \hat{G}_x} < v_x < \frac{\pi}{\gamma \tau^2 \Delta \hat{G}_x} \qquad (2.9.10)$$

respectively, where $\Delta \hat{G}_x$ is the increment of the gradient pulses, $\hat{G}_x$ is the maximum field gradient strength [compare with Figure 2.9.1(b)]. Likewise, the digital resolution and the acceleration "field of view" are given for the phase-encoding pulse shown in Figure 2.9.4(c) by

$$\Delta a_x = \frac{\pi}{2 \gamma \tau^3 \hat{G}_x} \qquad (2.9.11)$$

and

$$-\frac{\pi}{2 \gamma \tau^3 \Delta \hat{G}_x} < a_x < \frac{\pi}{2 \gamma \tau^3 \Delta \hat{G}_x} \qquad (2.9.12)$$

respectively.

2.9.3.2 Flow Compensation and Flow Artifacts

If an ordinary image or a spin density map is to be recorded in the presence of a relatively fast stationary flow with constant velocities, the spatial phase encoding gradient pulses in Figures 2.9.1(a) and 2.9.2(a) should be replaced by the "velocity compensated" gradient pulse shown in Figure 2.9.4(e). Otherwise, flow artifacts will arise. This gradient pulse provides spatial phase encoding according to [28]

$$\varphi(x_0) = \frac{2}{3} \gamma \hat{G}_x \tau x_0; \quad \varphi(v_{x0}) = 0 \qquad (2.9.13)$$

whereas there is no phase shift by the flow velocity.

Another type of flow artifact is due to voxel inflow and outflow problems: large velocities make a spin leave its designated voxel in the time between signal encoding and signal readout. As the traveled distance d and the velocity v are related by the expression $d = vt$, it is obvious that the effect is more severe, the smaller the voxel size is. Slow velocities, on the other hand, may be masked by diffusive displacements.

2.9.3.3 Velocity Mapping of Strongly Heterogeneous Flow Velocity Fields

A flow experiment should be set up in a way that the largest detectable velocity present in the sample is covered by the corresponding velocity field of view. This avoids distortions by back-folding of velocities into the experimental velocity scale.

However, the larger the field of view, longer and stronger gradients are required to maintain the required resolution and the larger is the attenuation by diffusion. In the case of velocity encoding with a bipolar gradient pair of amplitude $\hat{G}$ as that shown in Figure 2.9.4(b), the signal amplitude is decreased due to diffusion by the factor

$$A_{\text{diff}} = e^{-\frac{2}{3}\gamma^2 \hat{G}^2 D \tau^3} \qquad (2.9.14)$$

where D is the diffusion coefficient in the liquid. Additionally, voxel inflow and outflow artifacts increase in severity for long gradient times.

Therefore, instead of performing one experiment with a large number of gradient steps, the entire range should be better covered by at least two experiments with a significantly smaller number of gradient steps. The first experiment has a velocity field of view that covers the entire velocity range in a small number of steps. The velocity map is not distorted but small velocities are identified as "zero" due to the coarse resolution.

The second experiment covers the velocity range between the minimum measurable velocity and the minimum velocity detected by the first experiment. It is distorted because of back-folding of high velocities into the spectrum, but it shows the small velocities properly. All voxels from the first map that show zero velocity can now be compared with the corresponding (undistorted) voxels of the second map. If those show a finite velocity, the value can be transferred to the first map. The combination of the two maps is then a map that covers the full velocity range and includes small velocities, but has a fine resolution at low velocities and a coarse resolution at high velocities. With this two step strategy, shorter gradient pulses can be used and a substantial decrease in the diffusive attenuation of the signal can be achieved [10].

2.9.3.4 Applications to Flow Through Percolation Model Objects

Figure 2.9.5(a) shows a two-dimensional site percolation cluster with $p = 0.6207$ [7, 37]. It was generated on a 300×300 sites square lattice. The pore space is represented in white. Quasi two-dimensional model objects of different size of this cluster have been fabricated for experimental investigations [5, 7, 10, 37, 38]. The sample shown in the photograph Figure 2.9.5(b) has been milled into polystyrene sheets with a circuit board plotter. One lattice site corresponds to 400 µm by 400 µm and the depth of the pores is 2 mm. Figure 2.9.5(c) shows the same structure almost a factor of ten smaller. These samples have been made from PMMA sheets with deep X-ray lithography [5, 39]. From left to right, a lattice site corresponds to 80×80, 60×60 and 50×50 µm^2, resulting in total sample sizes of 24×24, 18×18 and 15×15 mm^2, respectively. The pore depth is about 1 mm. The samples were flooded with water after first having evacuated them in vacuum. Flow through the samples was driven by a pericyclic pump or by gravity from a reservoir above the sample. The reservoir level was kept constant by pumping.

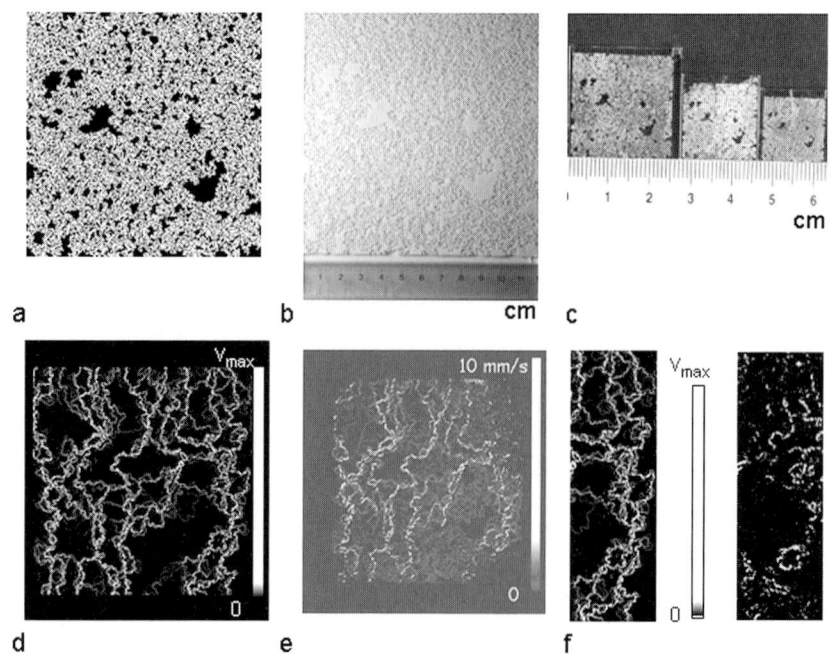

Fig. 2.9.5 (a) Typical computer generated percolation cluster that served as a template for the sample fabrication. (b) Photograph of a model object milled 1 mm deep into a polystyrene sheet. The total object size is 12 × 12 cm². (c) Photographs of model objects etched 1 mm deep into PMMA sheets by X-ray lithography. The total object sizes are 15 × 15 mm², 18 × 18 mm² and 24 × 24 mm² from right to left. (d) Numerically simulated map of the flow velocity magnitude. (e) Experimental map of the flow velocity magnitude for the same percolation cluster. (f) Details of the experimental flow velocity magnitude map derived for the 18 × 18 mm² cluster (right side). For comparison: the same cut-out of the simulated map (left side). The size of the displayed section is 6 × 18 mm².

Figure 2.9.5(d–f) shows maps of the flow velocity magnitude $v = \sqrt{v_x^2 + v_y^2}$. Flow occurs in a branched channel network while stagnant zones indicate the dead ends of the cluster. The map in Figure 2.9.5(e) was recorded with the 12 cm by 12 cm model object shown in Figure 2.9.5(b). This experiment was performed with a 4.7 T Bruker magnet in a 40 cm horizontal bore. Eight identically patterned polystyrene sheets were stacked in order to guarantee a good signal-to-noise ratio. The digital spatial resolution as defined by the pulse sequence parameters was 300 μm by 300 μm, while the digital velocity resolution was 60 μm s^{-1}.

Flow through porous media is a phenomenon of high complexity. Generally, the flow field can be calculated by solving the Navier–Stokes equations and the continuity equation for the boundary conditions given by the pore space. Numerical solutions of the flow field were obtained with the FLUENT 5.5 [40] finite element software package. The experimental map, Figure 2.9.5(e), compares well with the numerically obtained map, Figure 2.9.5(d).

Smaller pore sizes require a better spatial resolution, i.e., smaller voxels, which in turn conflict with a good signal-to-noise ratio. Moreover, field distortions at material boundaries then become more significant and even spins with moderate velocities will leave their designated voxel in the time between phase encoding and signal readout.

Figure 2.9.5(f) shows details of a flow map obtained from the 18 mm by 18 mm sample [central photograph in Figure 2.9.5(c)]. It was measured with a 9.4-T Bruker DSX spectrometer with a microimaging gradient unit. Gradients up to 0.8 T m^{-1} were used for space and velocity encoding, resulting in a resolution of 26 µm by 37 µm for space and 0.67 mm s^{-1} for velocity. The spatial resolution was close to the resolution limit of the spectrometer. The velocity map was measured in two steps to weaken inflow/outflow artifacts and diffusive attenuation, as explained in the previous section. The size of the part shown is 6 mm by 18 mm. Some details of the flow paths are nicely reproduced. The deviations of the measured map from the simulated map may be caused by the poor signal-to-noise ratio, susceptibility distortions or inflow/outflow artifacts. It is also possible that tiny air bubbles remained inside the pore space and blocked some of the flow paths or that the sample structure deviates from the template. After the lithographic process, the samples were inspected under a microscope but the inspection was limited to a small number of spots.

For acceleration mapping, a percolation cluster was generated on a 200 × 200 square grid with $p = 0.6$. Figure 2.9.6(c) shows a photograph of the corresponding object. The cluster was milled into a polystyrene sheet. Its size is 10 × 10 cm^2 with a depth of 2 mm. Acceleration mapping experiments were carried out with a Bruker 4.7-T magnet with a 40-cm horizontal bore. The maximum applied field gradient was 50 mT m^{-1}. A detailed discussion of the effect of gradient ramps on the phase encoding and of the four pulse gradient sequence [Figure 2.9.1(c)] that was used for acceleration encoding can be found in Ref. [36]. Experimental results for velocity and acceleration magnitudes are compared in Figure 2.9.6(a and b). The positions where the acceleration is finite (i.e., above the experimental resolution) are obviously localized spots in contrast to the more continuous flow patterns visible in the velocity map. That is, acceleration maps specifically reveal the distribution of flow bottlenecks and changes of the flow resistance. Note that experimentally obtainable velocity maps are too crude to deliver an acceleration map by derivation of the flow field. A distinct method for the direct measurement of acceleration is therefore essential.

2.9.4
Hydrodynamic Dispersion

Hydrodynamic dispersion represents the combined action of "flow" and molecular self-diffusion, resulting in superimposed displacements of the molecules. The situation is characterized by the Péclet number [41] defined by

$$Pe = \frac{\bar{u}l}{D} \tag{2.9.15}$$

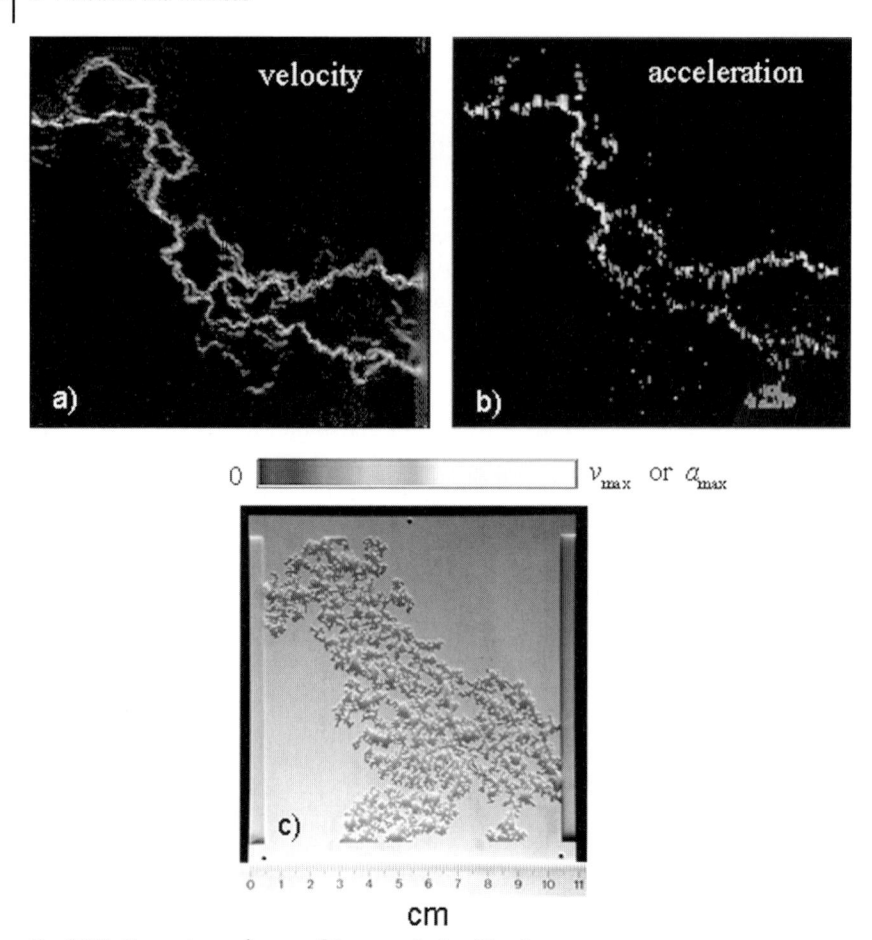

Fig. 2.9.6 Comparison of maps of the magnitude of the flow velocity (a) and of the flow acceleration (b). The maps have been determined with the NMR techniques outlined in the text. A photograph of the model object is shown in (c).

where l is the length scale on which the two transport processes are to be compared, $\bar{u}$ represents the mean velocity corresponding to that length scale and D is the Brownian self-diffusion coefficient. Note that the mean velocity $\bar{u}$ in percolation clusters depends on the averaging volume and therefore on the length scale l [7]. If $Pe \ll 1$, all transport is dominated by self-diffusion. On the other hand, if $Pe \gg 1$ the governing transport mechanism is coherent flow. In percolation models, the characteristic length scale is defined macroscopically by the correlation length ξ, so that Eq. (2.9.15) reads

$$Pe = \frac{\bar{u}\xi}{D} \qquad (2.9.16)$$

The tortuosity of flow streamlines causes an additional source of incoherence that must be accounted for at large Péclet numbers.

2.9 Hydrodynamic, Electrodynamic and Thermodynamic Transport in Porous Model Objects

Figure 2.9.7 shows a typical rf and field gradient pulse sequence for the acquisition of two-dimensional (slice selective) maps of the hydrodynamic dispersion coefficient D_{disp} [42]. That is, phase shifts due to coherent flow are compensated by a gradient pulse train corresponding to Figure 2.9.1(c). Assuming a field gradient along the z direction, the attenuation curve of the echo amplitude, A_{echo}, by incoherent displacements can be expanded according to [28]

$$
\begin{aligned}
A_{echo}(\delta, \Delta \gg \delta) &= \langle e^{iqz} \rangle \\
&= \left\langle 1 + \frac{iqz}{1!} - \frac{q^2 z^2}{2!} - \frac{iq^3 z^3}{3!} + \ldots \right\rangle \\
&\stackrel{qz \ll 1}{\approx} 1 - \frac{1}{2} q^2 \langle z^2(\Delta) \rangle \\
&\approx e^{-\frac{1}{2} q^2 \langle z^2(\Delta) \rangle}
\end{aligned}
\qquad (2.9.17)
$$

where the wave number is given by $q = \gamma G \delta$. The interval lengths are defined in Figure 2.9.7. In the limit $\Delta \gg \delta$, the effective displacement time is $t \approx \Delta$. The mean squared displacement by hydrodynamic dispersion along the gradient direction obeys

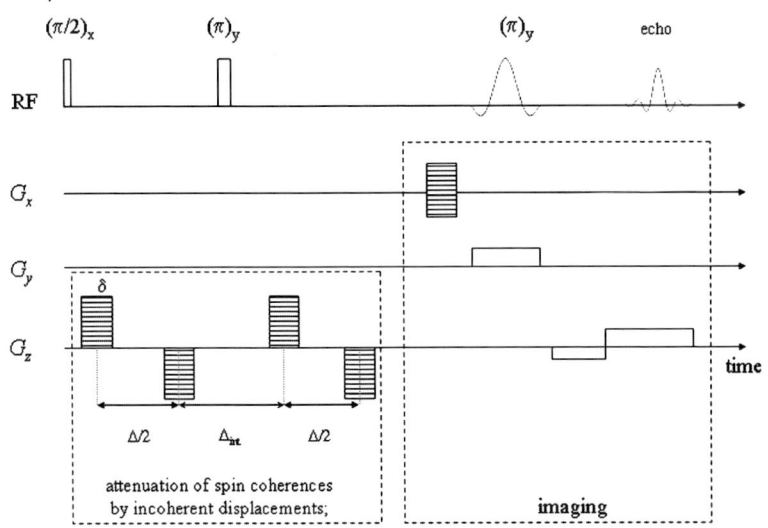

Fig. 2.9.7 Hahn spin-echo rf pulse sequence combined with bipolar magnetic field gradient pulses for hydrodynamic-dispersion mapping experiments. The lower left box indicates field-gradient pulses for the attenuation of spin coherences by incoherent displacements while phase shifts due to coherent displacements on the time scale of the experiment are compensated. The box on the right-hand side represents the usual gradient pulses for ordinary two-dimensional imaging. The latter is equivalent to the sequence shown in Figure 2.9.2(a). In order to avoid flow artifacts it may be advisable to replace the spatial encoding pulses (right-hand box) by "velocity compensated" pulses such as shown in Figure 2.9.4(e) for phase encoding. The amplitude of the Hahn spin-echo is attenuated by hydrodynamic dispersion. Evaluation of the echo attenuation curve for fixed intervals but varying preparation gradients (left box) permits the allocation of a hydrodynamic dispersion coefficient to each voxel, so that maps of this parameter can be rendered.

$$\langle z^2 \rangle = 2D_{\text{disp}} t \tag{2.9.18}$$

For displacements shorter than the mean pore dimension, $\sqrt{\langle z^2 \rangle} < a$, where flow velocities tend to be spatially constant and homogeneously distributed, Brownian diffusion is the only incoherent transport phenomenon that contributes to the hydrodynamic dispersion coefficient. As a direct consequence, the dispersion coefficient approaches the ordinary Brownian diffusion coefficient,

$$D_{\text{disp}} \approx D = \text{const} \tag{2.9.19}$$

in this limit.

Time intervals permitting displacement values in the scaling window $a < \sqrt{\langle z^2 \rangle} < \xi$ are related with the tortuous flow as a result of random positions of the obstacles in the percolation model [4]. Hydrodynamic dispersion then becomes effective. For random percolation clusters, an anomalous, i.e., time dependent dispersion coefficient is expected according to

$$D_{\text{disp}} = D_{\text{disp}}(t) \propto t^f \tag{2.2.20}$$

where $0 < f < 1$.

In the long-range limit, $\sqrt{\langle z^2 \rangle} \gg \xi$, a superposition of displacements by a constant drift velocity occurs including all retardations by the tortuosity of the percolation cluster. The dispersion coefficient then becomes stationary again and takes an effective value,

$$D_{\text{disp}} \approx D_{\text{eff}} = \text{const} \gg D \tag{2.9.21}$$

Hydrodynamic dispersion is illustrated in Figure 2.9.8 which represents numerical simulations of the trajectory of a tracer particle in a medium flowing through a random site percolation cluster. The results for different Péclet numbers are compared. Differences in the flow velocities stipulate large changes in the tortuous trajectories. For low Péclet numbers, i.e., relatively low flow velocities, Brownian diffusion dominates [Figure 2.9.8(a), right-hand trajectory]. On the other hand, if the flow is fast, coherent displacements along the main streamlines show up [Figure 2.9.8(a), left-hand trajectory].

In Figure 2.9.8(b), the mean traveling time of a neutral tracer across the percolation cluster is plotted versus the Péclet number [43]. Two regimes can be identified, namely "isotropic dispersion" ($Pe \ll 1$) and "mechanical dispersion" ($Pe \gg 1$) with a crossover close to $Pe \approx 1$. In the latter case, the data can be represented by a power law

$$\tau_t \propto (Pe)^{-\chi} \text{ for } 10^0 < Pe < 10^4 \tag{2.9.22}$$

The exponent turned out to be $\chi \approx 1$. This finding demonstrates that coherent flow determines transport in the mechanical dispersion regime and that diffusion is negligible under such conditions. For a discussion also see Ref. [43].

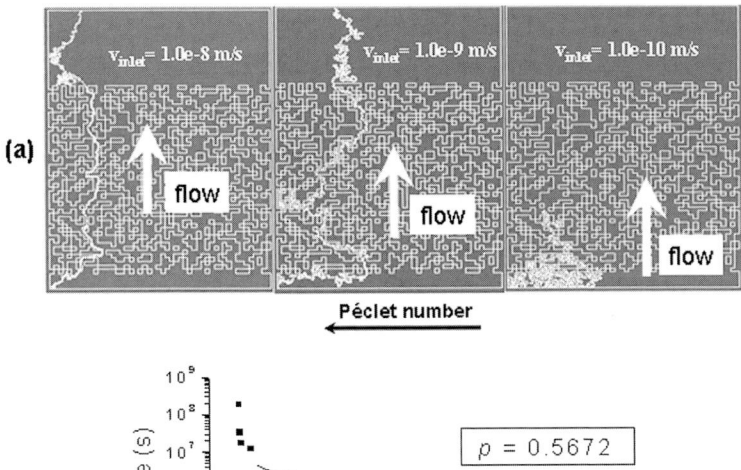

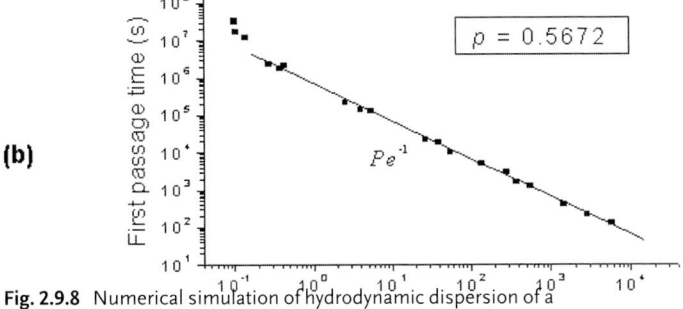

2.9 Hydrodynamic, Electrodynamic and Thermodynamic Transport in Porous Model Objects

Fig. 2.9.8 Numerical simulation of hydrodynamic dispersion of a neutral tracer particle in a random site percolation cluster for different Péclet numbers. The porosity of the percolation network was assumed to be $p = 0.5672$. (a) Representations of tracer trajectories for $Pe = 0.09$ (right), $Pe = 1.17$ (middle) and $Pe = 40.35$ (left). (b) First passage time across the cluster as a function of the Péclet number.

2.9.5
Thermal Convection and Conduction Mapping

An entirely different type of transport is formed by thermal convection and conduction. Flow induced by thermal convection can be examined by the phase-encoding techniques described above [8, 44, 45] or by time-of-flight methods [28, 45]. The latter provide less quantitative but more illustrative representations of thermal convection rolls. The origin of any heat transport, namely temperature gradients and spatial temperature distributions, can also be mapped with the aid of NMR techniques. Of course, there is no direct encoding method such as those for flow parameters. However, there are a number of other parameters, for example, relaxation times, which strongly depend on the temperature so that these parameters can be calibrated correspondingly. Examples are described in Refs. [8, 46, 47], for instance.

The temperature mapping method used in Ref. [8] is based on measurements of the spin-lattice relaxation time T_1 of a suitable liquid such as ethylene glycol filling the pore space of a percolation model object. At a 200-MHz Larmor frequency, T_1 of this particular liquid increases with increasing temperature as a result of the mobility enhancement. Varying the repetition time of the ordinary imaging sequence [see Figure 2.9.1(a) or 2.9.2(a)] permits one to evaluate the local spin-lattice relaxation times T_1 [28] and render them in the form of a temperature map. The spin-echo amplitude as a function of the repetition time T_R is given by

$$A_{\text{echo}}(T_R) \sim 1 - ce^{-T_R/T_1} \qquad (2.9.23)$$

where $c \approx 1$ is a constant. For ethylene glycol protons at a magnetic field of 4.7 T the following relationship for the absolute temperature T was found:

$$T(T_1) = 36.7 \ln\left(\frac{T_1}{0.18\text{ms}}\right) \text{K} \pm 2\text{K} \qquad (2.9.24)$$

This calibration relationship was verified in a range 190 K < T < 360 K.

Figure 2.9.9(a) shows a schematic representation of a thermal convection cell in Rayleigh–Bénard configuration [8]. With a downward temperature gradient one expects convection rolls that are more or less distorted by the tortuosity of the fluid filled pore space. In the absence of any flow obstacles one expects symmetrical convection rolls, such as illustrated by the numerical simulation in Figure 2.9.9(b).

The spatial temperature distribution established under steady-state conditions is the result both of thermal conduction in the fluid and in the matrix material and of convective flow. Figure 2.9.10, top row, shows temperature maps representing this combined effect in a random-site percolation cluster. The convection rolls distorted by the flow obstacles in the model object are represented by the velocity maps in Figure 2.9.10. All experimental data (left column) were recorded with the NMR methods described above, and compare well with the simulated data obtained with the aid of the FLUENT 5.5.1 [40] software package (right-hand column). Details both of the experimental set-up and the numerical simulations can be found in Ref. [8]. The spatial resolution is limited by the same restrictions associated with spin

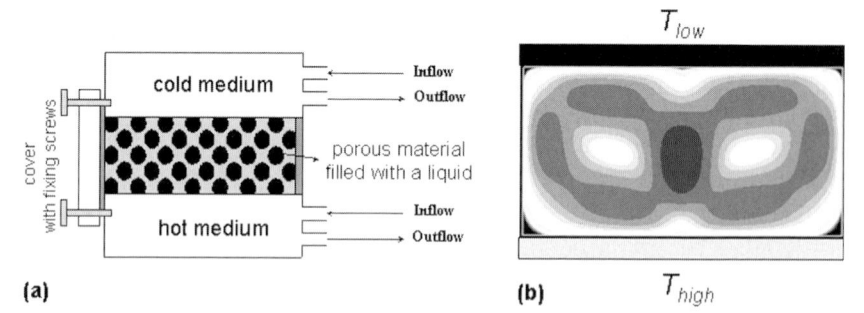

Fig. 2.9.9 (a) Schematic cross section of a convection cell in Rayleigh–Bénard configuration. In the version examined in Refs. [8, 44], a fluid filled porous model object of section 6 × 4 cm² was confined by cooling and heating compartments at the top and bottom, respectively. (b) Velocity contour plot of typical convection rolls expected in the absence of any flow obstacles (numerical simulation).

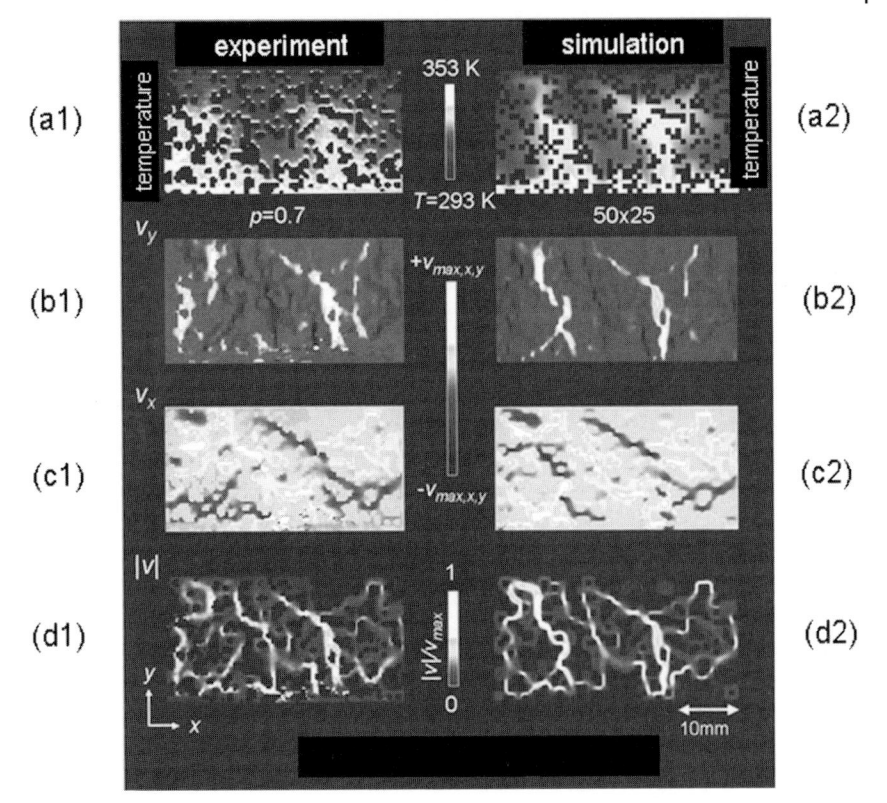

2.9 Hydrodynamic, Electrodynamic and Thermodynamic Transport in Porous Model Objects | 223

Fig. 2.9.10 Maps of the temperature and of the convection flow velocity in a convection cell in Rayleigh–Bénard configuration (compare with Figure 2.9.9). The medium consisted of a random-site percolation object of porosity $p = 0.7$ filled with ethylene glycol (temperature maps) or silicon oil (velocity maps). The left-hand column marked with an index 1 represents experimental data. The right-hand column refers to numerical simulations and is marked with an index 2. The plots in the first row, (a1) and (a2), are temperature maps. All other maps refer to flow velocities induced by thermal convection: velocity components v_x (b1) and (b2) and v_y (c1) and (c2), and the velocity magnitude (d1) and (d2).

density mapping in the presence of flow. In particular, voxel inflow and outflow may be a problem for the precise localization of T_1 in view of the relatively long relaxation intervals needed for this type of measurement.

2.9.6
Ionic Current Density Mapping

Ionic current density maps can be recorded with the aid of the pulse sequence shown in Figure 2.9.2. The principle of the technique [48–52] is based on Maxwell's fourth equation for stationary electromagnetic fields,

$$\vec{j}(\vec{r}) = \frac{1}{\mu_0} \vec{\nabla} \times \vec{B}(\vec{r}) \tag{2.9.25}$$

which, in Cartesian components, reads

$$j_x = \frac{1}{\mu_0}\left(\frac{\partial B_z}{\partial y} - \frac{\partial B_y}{\partial z}\right), \quad j_y = \frac{1}{\mu_0}\left(\frac{\partial B_x}{\partial z} - \frac{\partial B_z}{\partial x}\right), \quad j_z = \frac{1}{\mu_0}\left(\frac{\partial B_y}{\partial x} - \frac{\partial B_x}{\partial y}\right) \quad (2.9.26)$$

where $\vec{B}(\vec{r})$ is the magnetic flux density at the position $\vec{r}$ produced by the local current density $\vec{j}(\vec{r})$. Note that this flux density is superimposed on that of the external magnetic field, $\vec{B}_0$. The quantity μ_0 is the magnetic field constant.

In the following, we restrict ourselves to quasi two-dimensional current distributions invariant along the z direction, for simplicity [51]. Under such conditions we have $j_z = 0$ and $\partial B_y/\partial z = \partial B_x/\partial z = 0$. The only finite current density components are then

$$j_x(\vec{r}) = \frac{1}{\mu_0}\left(\frac{\partial B_z(\vec{r})}{\partial y}\right), \quad j_y(\vec{r}) = \frac{1}{\mu_0}\left(-\frac{\partial B_z(\vec{r})}{\partial x}\right) \quad (2.9.27)$$

The problem now is to determine the derivatives $\partial B_z/\partial x$ and $\partial B_z/\partial y$. This can be done by evaluating the local precession phase shifts caused by the currents, that is

$$\varphi_c = \gamma B_z T_c \quad (2.9.28)$$

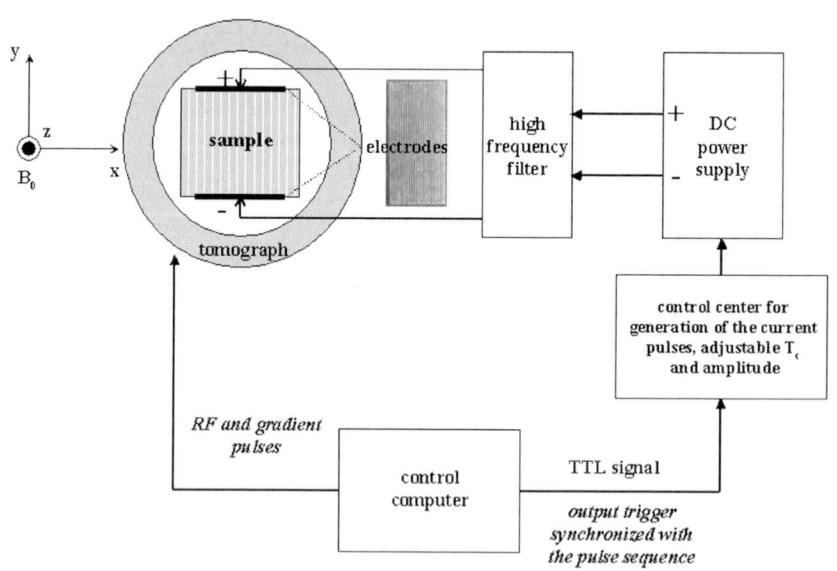

Fig. 2.9.11 Schematic representation of the experimental set-up feeding electric currents into the sample for ionic current density mapping. The (quasi two-dimensional) sample is placed perpendicular to the main magnetic field, B_0. The rf, field gradient and current pulse program is shown in Figure 2.9.2. The current density components in the x and y direction are evaluated.

where T_c is the total evolution time in the presence of the currents [see Figure 2.9.2(b)]. Numerically deriving such phase shift maps provides the desired local derivatives of the current induced magnetic flux density.

The optimal current pulse length, $T_{c,opt}$, in the sequence, Figure 2.9.2(b), depends on the transverse relaxation time, T_2, the diffusion coefficient of the ions, D, and the digital resolution of the NMR imaging experiment, e.g., in the x direction, $\Delta x = x_{FOV}/N$, where x_{FOV} is the field of view range and N is the number of pixels. The resulting expression is [52]

$$\frac{1}{T_{c,opt}} = \frac{1}{T_2} + 4D\left(\frac{4\pi N}{5 x_{FOV}}\right)^2 \tag{2.9.29}$$

Figure 2.9.11 shows schematically the experimental set-up of the ionic current device. The sample is packed between two electrodes that are shaped meander-like in order to avoid eddy currents due to pulsed field gradients. The high frequency filter is essential to prevent an "antenna" effect of the electrode cables inside the magnet.

The porous medium under investigation must be filled with an electrically conducting liquid, that is, an electrolyte solution. The electric conductivity is given by the empirical Kohlrausch law [53]

$$\sigma(c) = \Lambda_0 c - k_0 c^{3/2} \tag{2.9.30}$$

where σ represents the conductivity, Λ_0 the molecular conductance characteristic for the salt dissolved, c the salt concentration and k_0 is a constant depending on the salt valence. The salt concentration of a few grams per liter is chosen as a compromise between conservation of the electrode performance by low concentrations and the experimental advantage of a good electric conductivity. The conductivity can be assumed to be constant in the entire sample. The total electric current is typically varied in the range 10–1000 mA for percolation model objects of the size used in this context and corresponds to a maximum current density of about 1 mA mm^{-2}. To avoid distortions arising from the cables connecting the electrodes with the dc power supply, all current conducting leads should be kept parallel to the main magnetic field.

The phase shift map shown in Figure 2.9.12(a) was evaluated directly as the arcus tangent of the quotient of the imaginary and real signal parts. It still needs to be "unwrapped" with respect to periodicity ambiguities. A suitable method can be found in Ref. [54] as the two-dimensional Goldstein algorithm, which turned out to be more reliable than the one-dimensional Itoh variant [48]. Figure 2.9.12(b) shows the unwrapped result, which then can be derived numerically in two directions in order to obtain the current density components according to Eqs. (2.9.27) and (2.9.28). Figure 2.9.13(b2) shows a typical current density map of a quasi two-dimensional random site percolation model object evaluated in this way.

Another problem that should be considered carefully is the fact that the evaluated phase shift data set is discrete, so that conventional numerical derivation methods

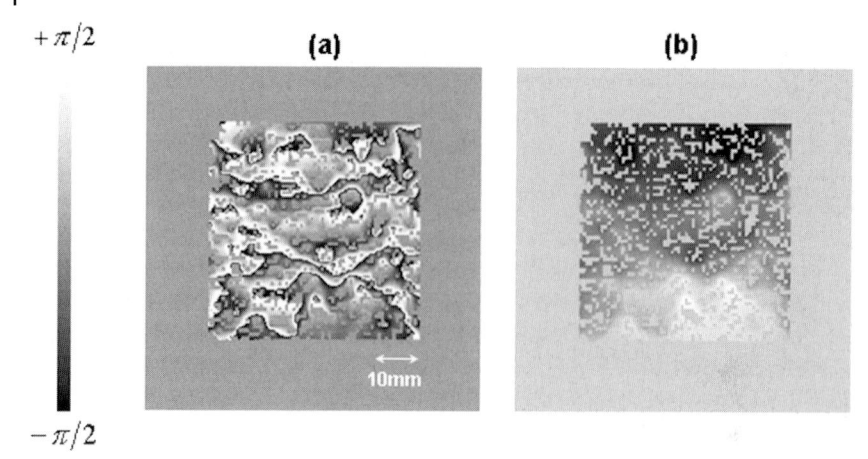

Fig. 2.9.12 Phase shift maps measured as the difference in the signal phases with and without current pulses. The rf, field gradient and current pulse program are shown in Figure 2.9.2. The phase shift maps must be "unwrapped" as the original evaluation yields phase angles only in the principal range between $-\pi/2$ and $+\pi/2$ as a consequence of the π periodicity of the tangent function formed as the quotient of the imaginary and real signal component [map (a)]. "Unwrapping" means that the phase shifts are spread over the real angular range with the aid of continuity considerations. Map (b) shows the phase shift map unwrapped on the basis of Goldstein's algorithm [54]. Numerical derivatives of the phase shift map provide the in-plane components of the current density as explained in the text (see Figure 2.9.13).

fail. The derivatives are best obtained using difference filters, such as the Sobel filter of second order [48]. Other known filters such as the Laplace filter or the Roberts filter may also be applicable.

The spatial current distribution can be simulated by solving Laplace's equation with Neumann boundary conditions at the pore walls,

$$\nabla^2 \Phi(\vec{r}) = 0 \qquad (2.9.31)$$

From this one obtains the distribution of the electric field strength according to

$$\vec{E}(\vec{r}) = -\vec{\nabla}\Phi(\vec{r}) \qquad (2.9.32)$$

Finally, the current density is provided by Ohm's law,

$$\vec{j}(\vec{r}) = \sigma \vec{E}(\vec{r}) \qquad (2.9.33)$$

Resulting maps of the current density in a random-site percolation cluster both of the experiments and simulations are represented by Figure 2.9.13(b2) and (b1), respectively. The transport patterns compare very well. It is also possible to study hydrodynamic flow patterns in the same model objects. Corresponding velocity maps are shown in Figure 2.9.13(c1) and (c2). In spite of the similarity of the

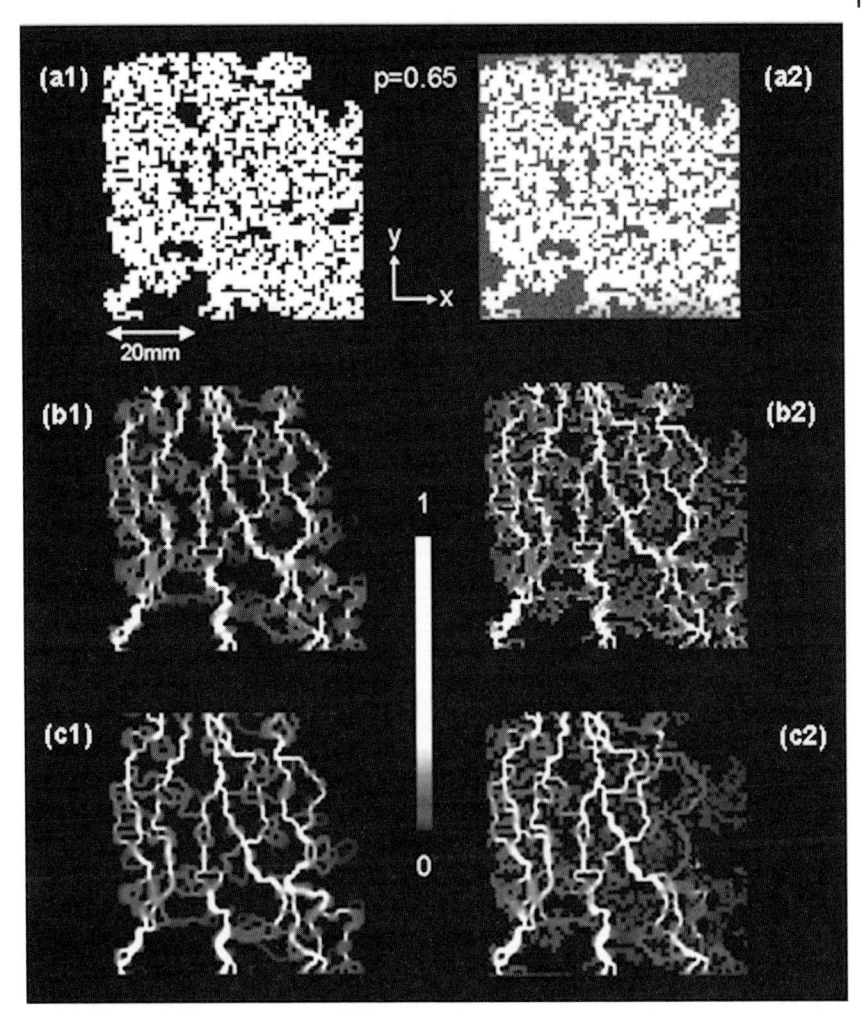

Fig. 2.9.13 Quasi two-dimensional random site percolation cluster with a nominal porosity $p = 0.65$. The left-hand column refers to simulated data and the right-hand column shows NMR experiments in this sample-spanning cluster (6×6 cm^2). (a1) Computer model (template) for the fabrication of the percolation object. (a2) Proton spin density map of an electrolyte (water + salt) filling the pore space of the percolation model object. (b1) Simulated map of the current density magnitude relative to the maximum value, j/j_{max}. (b2) Experimental current density map. (c1) Simulated map of the velocity magnitude relative to the maximum value, v/v_{max}. (c2) Experimental velocity map. The potential and pressure gradients are aligned along the y axis.

transport patterns there is a remarkable difference between the electric current and hydrodynamic flow. Examining the details of the maps shown in Figure 2.9.13 one realizes that thin pore bottlenecks hinder hydrodynamic flow more strongly than ionic currents. This is a manifestation of the different transport resistance laws applicable to electric currents and Poiscuille flow [51].

2.9.7
Concluding Remarks

The unmatched strength of the NMR technology is its versatility and specificity in probing and mapping parameters of practical interest. In this context, we have demonstrated that the transport phenomena, namely flow, interdiffusion, hydrodynamic dispersion, electric currents, thermal conduction and convection, and the respective experimental parameters, can be examined in complex porous systems by NMR methods. We have considered random-site percolation clusters that were generated on a computer as a well defined model system. Fabricating model objects on this basis provides samples where the full details of the pore space structure are known quantitatively. This in turn permits finite-element or finite-volume computational fluid dynamics studies, which, in combination with the experimental NMR results, provide an unsurpassed conclusiveness concerning the relationship of the pore space structure and the transport properties. There is a vast variety of NMR applications that have not yet been realized in the general microscale community [55]. Fundamental investigations with "known" porous media as outlined in this chapter will reveal the laws required to understand transport at microscopic length scales. Knowledge elaborated on this basis is expected to trigger studies of interest for chemical engineering and microsystem technology applications.

References

1 *Lab on a Chip*, a Royal Society of Chemistry journal, Cambridge.
2 *Microfluidics and Nanofluidics*, a Springer-Verlag journal, Heidelberg.
3 D. Li, *Electrokinetics in Microfluidics 2004*, Elsevier, Oxford.
4 R. Kimmich **2002**, *Chem. Phys.* 284, 253–285.
5 E. Kossel, M. Weber, R. Kimmich **2004**, *Solid State NMR* 25, 28–34.
6 D. Stauffer, A. Aharony **1992**, *Introduction to Percolation Theory*, Taylor & Francis, London.
7 A. Klemm, R. Kimmich, M. Weber **2001**, *Phys. Rev. E* 63, 041514–1–041514–8.
8 M. Weber, R. Kimmich **2002**, *Phys. Rev. E* 66, 056301–1–056301–13.
9 R. Subramanian, M. Lam, A. G. Webb **1998**, *J. Magn. Reson.* 133, 227–231.
10 E. Kossel, R. Kimmich **2005**, *Magn. Reson. Imag.* 23, 397–401.
11 C. Mattea, R. Kimmich **2005**, *Phys. Rev. Lett.* 94, 024502–1–024502–4.
12 N. Fatkullin, E. Fischer, C. Mattea, U. Beginn, R. Kimmich **2004**, *ChemPhysChem* 5, 884–894.
13 S. Stapf, R. Kimmich, R.-O. Seitter **1995**, *Phys. Rev. Lett.* 75, 2855–2858.
14 P. T. Callaghan, A. Coy, L. C. Forde, C. J. Rofe **1993**, *J. Magn. Reson. A* 101, 347–350.
15 N. Nestle, K. Rydyger, R. Kimmich **1997**, *J. Magn. Reson.* 125, 355–357.
16 I. Ardelean, R. Kimmich **2003**, *Annu. Rep. N. M. R. Spectrosc.* 49, 43–115.
17 J. Kärger, H. Pfeifer, W. Heink **1988**, *Advan. Magn. Reson.* 12, 1.
18 D. G. Cory, A. N. Garroway **1990**, *Magn. Reson. Med.* 14, 435–444.
19 P. T. Callaghan **1991**, *Principles of Nuclear Magnetic Resonance Microscopy*, Clarendon Press, Oxford.
20 R. Metzler, J. Klafter **2000**, *Phys. Rep.* 339, 1.
21 R. Orbach **1986**, *Science*, 231, 814.
22 P. Levitz **1997**, *Europhys. Lett.* 39, 593.

23 S. Havlin, D. Ben-Avraham **1987**, *Advan. Phys.* 36, 695.
24 P. J. McDonald, B. Newling **1998**, *Rep. Prog. Phys.* 61, 1441.
25 A. Haase, J. Frahm, D. Matthaei, W. Hänicke, K.-D. Merbold **1986**, *J. Magn. Reson.* 67, 258.
26 J. Hennig, A. Nauerth, H. Friedburg **1986**, *Magn. Reson. Med.* 3, 823.
27 M. D. Mantle, A. J. Sederman **2003**, *Prog. N. M. R. Spectrosc.* 43, 3–60.
28 R. Kimmich **1997**, *NMR Tomography, Diffusometry, Relaxometry*, Springer, Berlin.
29 S.-C. Lee, K. Kim, J. Kim, S. Lee, J. H. Yi, S. W. Kim, K.-S. Ha, C. Cheong **2001**, *J. Magn. Reson.* 150, 207.
30 L. Ciobanu, D. A. Seeber, C. H. Pennington **2002**, *J. Magn. Reson.* 158, 178.
31 A. Klemm, R. Metzler, R. Kimmich **2002**, *Phys. Rev. E* 65, 021112.
32 M. Sahimi **1993**, *Application of Percolation Theory*, Taylor & Francis, London.
33 A. Bunde, S. Havlin (eds.) **1996**, *Fractals and Disordered Systems*, Springer-Verlag, Berlin.
34 S. Alexander, R. Orbach **1982**, *J. Phys. (France) Lett.* 43, L-625.
35 B. Bluemich **2003**, *NMR Imaging of Materials*, Oxford University Press, Oxford.
36 B. Buhai, A. Hakimov, I. Ardelean, R. Kimmich **2004**, *J. Magn. Reson.* 168, 175–185.
37 H.-P. Müller, J. Weis, R. Kimmich **1995**, *Phys. Rev. E* 52, 5195–5204.
38 R. Kimmich, A. Klemm, M. Weber **2001**, *Magn. Reson. Imag.* 19, 353.
39 M. Madou **1997**, *Fundamentals of Microfabrication*, CRC Press, Boca Raton, FL.
40 Fluent Inc., Lebanon, NH, USA.
41 M. Sahimi **1995**, *Flow and Transport in Porous Media and Fractured Rock*, VCH Verlag, Weinheim.
42 S. L. Codd, B. Manz, J. D. Seymour, P. T. Callaghan **1999**, *Phys. Rev. E* 60, R3491–R3494.
43 H. A. Makse, J. S. Andrade Jr., H. E. Stanley **2000**, *Phys. Rev. E* 61, 583.
44 M. Weber, A. Klemm, R. Kimmich **2001**, *Phys. Rev. Lett.* 86, 4302–4305.
45 J. Weis, R. Kimmich, H.-P. Müller **1996**, *Magn. Reson. Imag.* 14, 319–327.
46 A. G. Webb **2001**, (Temperature measurement using nuclear magnetic resonance), *Annu. Rep. N. M. R. Spectrosc.* 45, 1–67.
47 R. Kimmich, K. Bühler, A. Knüttel **1991**, *J. Magn. Reson.* 93, 256–264.
48 G. C. Scott, M. L. G. Joy, R. L. Armstrong, R. M. Henkelman **1992**, *J. Magn. Reson.* 97, 235–254.
49 K. Beravs, A. Demsar, F. Demsar **1999**, *J. Magn. Reson.* 137, 253–257.
50 Y. Manassen, E. Shalev, G. Navon **1988**, *J. Magn. Reson.* 76, 371.
51 M. Weber, R. Kimmich **2002**, *Phys. Rev. E* 66, 026306.
52 I. Sersa, O. Jarh, F. Demsar **1994**, *J. Magn. Reson. Ser. A* 111, 93–99.
53 J. H. Morre, D. N. Spencer **2001**, *Encyclopedia of Chemical Physics and Physical Chemistry, Volume I: Fundamentals*, Institute of Physics, London.
54 D. C. Ghiglia, M. D. Pritt **1998**, *Two Dimensional Phase Unwrapping*, Wiley, New York.
55 D. Sinton **2004**, *Microfluid Nanofluid* 1, 2–21.

3
Porous Materials

3.1
Diffusion in Nanoporous Materials
Jörg Kärger, Frank Stallmach, Rustem Valiullin, and Sergey Vasenkov

3.1.1
Introduction

Most contributions to this book deal with the application of NMR to deduce space-dependent information from the samples under study. In most cases, this information refers to the distribution of different components. Similarly, however, the space dependence of diffusivity, velocity or even acceleration [1, 2] of the components involved may also be studied. In the present chapter, a quasi-homogeneous distribution of the host, i.e., the nanoporous material, and the guest molecules is implied. Upgrading of these molecules, that is the transformation of the starting material to value-added products, is one of the key issues of chemical engineering. In many technical applications, including mass separation and catalytic conversion, molecular diffusion in the host system could become rate limiting. NMR with Pulsed Field Gradients [PFG NMR, also referred to as "pulsed gradient spin-echo" (PGSE NMR)] [3–5] is able to provide quantitative information on this process. Hence, as a consequence of the quasi-homogeneity of the sample, the necessity of space-dependent measurement is avoided. Therefore, with respect to the observation of molecular displacements, a much higher spatial resolution can be attained than is so far accessible in conventional NMR imaging. In this chapter, after a short introduction to the method, examples of the benefit of this procedure will be presented. Although the method is not restricted to diffusion in equilibrium with the gas phase, for the sake of simplicity, we shall essentially confine ourselves to this case. As a consequence, it is justified to assume that the NMR signal observed is essentially a result of the guest molecules within the host material.

3.1.2
Measuring Principle

By monitoring the difference in the location of molecules in two subsequent instances of time, rather than the location itself, diffusion measurement by PFG NMR can be understood as an extension of NMR imaging. The method is essentially based on the application of an inhomogeneous field $\vec{B}_{add} = \vec{g}\vec{r}$ (the so-called field gradient pulses) over two short intervals of time of identical duration δ, in addition to the constant magnetic field $\vec{B}_0$. In what follows, for reasons of simplicity we will consider a situation when the constant field $\vec{B}_0$ is elongated along the z axis, and the field gradient $\vec{B}_{add}$ has only one component in the same direction, i.e., $B_{add} = gz$. The function of two gradient pulses applied at times t_1 and t_2 is to encode and decode the spins by means of the acquired phases φ in the applied inhomogeneous magnetic field according to their positions z. If a spin, located at a position $z(t_1) \equiv z_1$ at the time t_1 (encoding period) is shifted to position $z(t_2) \equiv z_2$ at t_2 (decoding period), the phase difference Δφ, gained by the spin due to the mismatch of the effective magnetic fields at these two positions, results in

$$\Delta\varphi = \gamma\delta g(z_2 - z_1) \tag{3.1.1}$$

where γ denotes the gyromagnetic ratio of the nucleus under study. Spins will only contribute with the cosine of this phase difference to the measured total signal, known as the spin-echo. Introducing the so-called mean propagator $P(z, t)$ [4, 6–9], i.e., the probability (density) that during time t an arbitrarily selected molecule (spin under study) will be shifted over a distance z, the total spin-echo attenuation may therefore be represented as

$$\Psi = \int P(z, t)\cos(\gamma\delta gz)dz \tag{3.1.2}$$

where t denotes the time interval between the two gradient pulses. The PFG NMR signal attenuation, which is the experimentally directly accessible quantity, is thus found to be nothing other than the Fourier transform of the propagator, which in turn results in the reversed Fourier transform of the spin-echo attenuation. In Ref. [7], this way of analyzing PFG NMR data was introduced for visualizing the internal molecular dynamics in beds of nanoporous particles. The application of the information thus obtained to probing the geometrical details of the sample (the method of "dynamic" imaging) and the detection of its analogy to diffraction [1, 3, 4, 6] are important developments of this principle. Typical space and time scales over which the displacements may be followed are micrometers and milliseconds up to seconds, respectively.

The wealth of information accessible by analyzing this type of PFG NMR data is reflected in Figure 3.1.1. This shows a representation of the (smoothed) propagators for ethane in zeolites NaCaA (a special type of nanoporous crystallite) at two different temperatures and for two different crystal sizes. Owing to their symmetry in space, it is sufficient to reproduce only one half of the propagators. In fact,

Figure 3.1.1 reflects all relevant situations of molecular transport that we will be referring to in the following subsections, as is summarized schematically in Figure 3.1.2. At sufficiently low temperatures and for the larger crystals (top left of Figure 3.1.1), the observed molecular trajectories are essentially unaffected by the crystal boundaries and the intercrystalline space. This is the measuring regime of intracrystalline diffusion [case (i) in Figure 3.1.2, considered in more detail in Section 3.1.3]. Under identical measuring conditions, the propagator in the small crystallites exhibits a most spectacular pattern (top right of Figure 3.1.1): the propagator does not change with time! This is the condition for ideal dynamic imaging, characterized by the fact that (a) the molecules remain restricted to a certain range and (b) the displacements without this restriction would be much larger than the restricted range. In the present case, restriction is brought about by the step in the potential energies between the intra- and intercrystalline space (comparable to the heat of adsorption), which at this low temperature is still too large to be overcome by the thermal energy of the diffusing molecules. At higher temperatures (right bottom of Figure 3.1.1) the molecules are able to overcome this step in the potential energy with sufficient frequency during the observation time. Under the given conditions, PFG NMR yields the so-called long-range diffusivity [case (ii) in Figure 3.1.2, Section 3.1.4). In some cases (left bottom of Figure 3.1.1) it is possible to

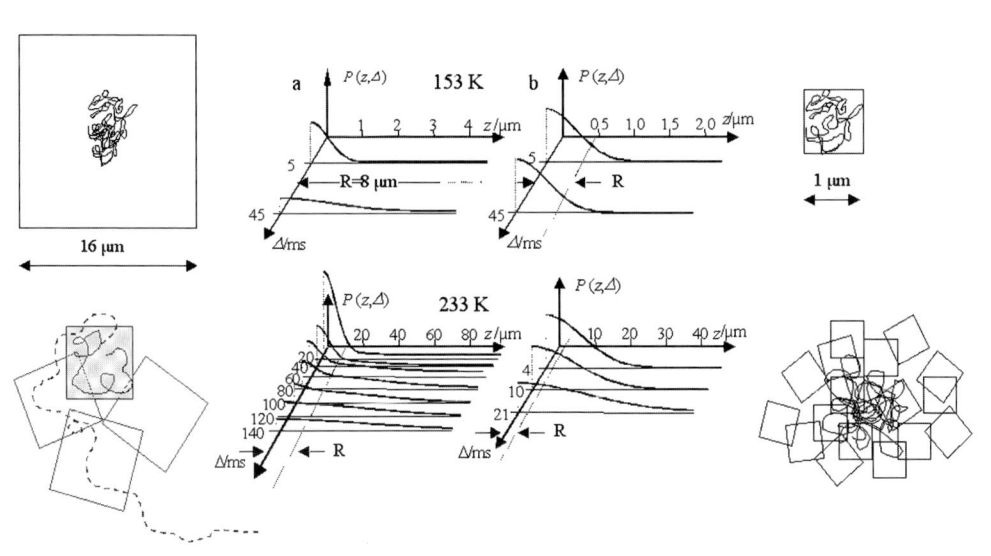

Fig. 3.1.1 Mean propagator (i.e., probability distribution of molecular displacements z during the observation time $t \equiv \Delta$) for ethane in NaCaA crystallites of different size. Top left: unrestricted intracrystalline diffusion with the Gaussian propagator observed for large crystallites and low temperatures. Top right: ideal case of restricted diffusion in small crystallites at low temperatures with a propagator, whose shape does not change with time. Bottom right: intercrystalline or long-range diffusion regime with a propagator, which is of Gaussian form. Bottom left: intermediate case between the ideally restricted intracrystalline (solid trajectory) and long-range intercrystalline diffusion (dashed trajectory). The propagator is of a complex form.

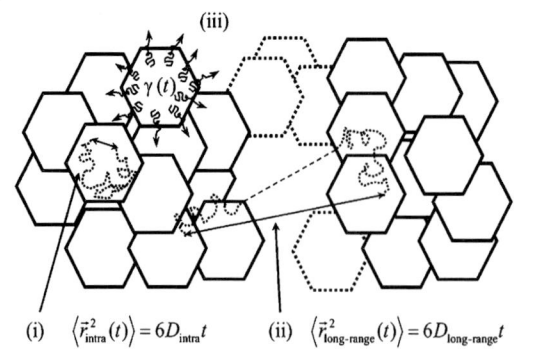

Fig. 3.1.2 Details of molecular transport in beds of nanoporous particles as accessed by PFG NMR. Three different situations inherent to different regimes of molecular motion are shown: (i) intracrystalline diffusion, (ii) long-range diffusion and (iii) intermediate exchange regime, giving access to the tracer exchange function $\gamma(t)$.

(i) $\langle \vec{r}_{intra}^2(t) \rangle = 6 D_{intra} t$ (ii) $\langle \vec{r}_{long\text{-}range}^2(t) \rangle = 6 D_{long\text{-}range} t$

differentiate between the molecules under study with respect to their trajectories, viz., between those which have left their crystallites during the given observation time and those which have not. The information thus provided [case (iii) in Figure 3.1.2, Section 3.1.5] coincides with that of conventional tracer exchange experiments, though with much smaller space and time scales.

The propagator is often Gaussian (including the present cases of intracrystalline and long-range diffusion)

$$P(z,t) = (4\pi Dt)^{-1/2} \exp(-z^2/4Dt) \tag{3.1.3}$$

where in the situations considered above the diffusivities D refer to either intracrystalline [case (i)] or long-range [case (ii)] diffusion. Diffusion phenomena following this relationship are referred to as normal diffusion and can be represented by Fick's law. Inserting Eq. (3.1.3) into (3.1.2) yields the attenuation formula commonly used in PFG NMR:

$$\Psi(\delta g, t) = \exp(-\gamma^2 \delta^2 g^2 Dt) = \exp(-\gamma^2 \delta^2 g^2 \langle z^2(t) \rangle / 2) \tag{3.1.4}$$

where in the second equation use has been made of the Einstein relationship

$$\langle z^2(t) \rangle = \int z^2 P(z,t) dz = 2Dt \tag{3.1.5}$$

In many cases, including transport phenomena which differ from normal diffusion, Eq. (3.1.4) turns out to be a good approximation, in particular if only small magnitudes of the "gradient intensity" $\gamma\delta g$ (sometimes referred to as the generalized scattering vector of PFG NMR) are considered. Under these circumstances, the genuine diffusivity D is replaced by an effective diffusivity, D_{eff}

$$D_{eff} = \langle z^2(t) \rangle / 2t \tag{3.1.6}$$

as a simple inversion of the Einstein equation [Eq. (3.1.5)]. According to this definition, in the case of normal diffusion D_{eff} clearly coincides with D.

As a typical example, in the long time limit, $t \to \infty$, of diffusion confined to spheres of radius R, the effective diffusivity of PFG NMR measurements is found to be [4, 8, 10]

$$D_{\text{eff}} \equiv D_{\text{restricted}} = R^2/5t \quad (3.1.7)$$

The internal dynamics in complex systems can sometimes be described by different states of varying mobility, where molecular propagation in each of these states is described by the propagator as given by Eq. (3.1.4). In the limiting case where molecular exchange between all of these states is fast in comparison with the observation time, PFG NMR spin-echo attenuation is again given by Eq. (3.1.4) with an effective diffusivity

$$D = \sum p_i D_i \quad (3.1.8)$$

where D_i and p_i denote the diffusivity and the relative number of molecules in state i. A more general treatment can be found in Ref. [8] for example. This approach, originally introduced for a quantitative description of diffusion in beds of zeolite crystallites, has turned out to be of fairly general use for diffusion in composed systems [11–13].

In order to put the present section into the general context of the book, it is worthwhile referring to the first NMR imaging studies of molecular uptake by beds of zeolite crystallites [10, 14]. In these studies, NMR imaging was applied to follow the evolution of the concentration profile of n-butane in beds of crystallites of activated zeolite NaCaA. Depending on the bed geometry, two distinctly different evolution patterns have been observed: in a bed of compacted, small crystallites, molecular uptake was accompanied by a gradually diminishing concentration gradient towards the bed center, while in a loose bed of big crystals concentration increased uniformly over the whole sample. These two situations are referred to as the limiting cases of uptake control by long-range diffusion and by intracrystalline diffusion. From the solution of Fick's second law, the diffusion equation, with the relevant initial and boundary conditions, and by simple dimension arguments, the mean time of diffusion-limited molecular uptake may be easily determined to be proportional to L^2/D, where L is a characteristic extension of the system [10, 15]. Thus, with d and h denoting the crystallite diameter and the bed height, respectively, and with the coefficients of intracrystalline and of long-range diffusion, as specified by Figures 3.1.1 and 3.1.2, the characteristic time constants can be introduced for molecular adsorption/desorption on an individual zeolite crystallite or on the whole bed:

$$\tau_{\text{intra}} \approx d^2/D_{\text{intra}}$$
$$\tau_{\text{long-range}} \approx h^2/D_{\text{long-range}} \quad (3.1.9)$$

Obviously, such limiting cases as monitored by NMR imaging refer to the conditions $\tau_{\text{intra}} \ll \tau_{\text{long-range}}$ (limited by bed, i.e., long-range diffusion) and $\tau_{\text{intra}} \gg \tau_{\text{long-range}}$ (limited by intracrystalline diffusion).

Most of the diffusion studies referred to in this section have been performed by means of subsequent generations of laboratory-built PFG NMR diffusion spectrometers in the Department of Interface Physics at Leipzig University. They consist of narrow-bore superconducting magnets and the relevant electronic parts inherent to standard commercially available NMR spectrometers. A unique feature of the spectrometers is ensured by the specially designed PFG unit enabling the production of ultra-high-intensity ($g_{max} = \pm 35$ T m^{-1} in a 7.5 mm od NMR tube) field-gradient pulses with short rise and fall times (≈ 120 µs to reach the maximum intensity and *vice versa*). These two factors, in fact, determined our success in examining the slow motion in polymer systems with diffusivities down to 10^{-15} m^2 s^{-1} and probing diffusion processes for molecular species with very short transverse NMR relaxation times, which is often the case in nanoporous materials. A more elaborate description of the apparatuses and details of the experimental procedure can be found, in Refs. [16–18] for example.

3.1.3
Intracrystalline Diffusion

Introduction of PFG NMR to zeolite science and technology has revolutionized our understanding of intracrystalline diffusion [19]. In many cases, molecular uptake by beds of zeolites turned out to be limited by external processes such as resistances, surface barriers or the finite rate of sorbate supply, rather than by intracrystalline diffusion, as previously assumed [10, 20–24]. Thus, the magnitude of intracrystalline diffusivities had to be corrected by up to five orders of magnitude to higher values [25, 26].

However, in more recent studies, most interestingly, where a correct interpretation of conventional uptake measurements may be implied, still notable discrepancies (of up to two orders of magnitude) between the results of different techniques of diffusion measurement can be observed [26, 27]. Recent PFG NMR diffusion studies of short hydrocarbons in MFI-type zeolites (silicalite-1) have opened up new vistas on their explanation [28, 29]. As an example, Figure 3.1.3 shows the intracrystalline diffusivities of n-butane at different temperatures, measured with different observation times, plotted as a dependence on the mean displacement according to Eq. (3.1.5). The most probable explanation of the dramatic dependence of the measured diffusivities on the displacement is the existence of internal transport barriers. Indeed, the full lines as revealed by Monte Carlo (MC) simulations using the simplest assumptions [barrier spacing uniform and equal to 3 µm steps over the internal transport barriers differ from the remaining ones by only an (by 21.5 kJ mol^{-1}) enhanced activation energy] fit the experimental data well. Thus, simulation data from the literature [30] and the measured data become compatible with the model [28].

Some micro- and mesoporous materials exhibit anisotropic pore structures, which may yield different values for the diffusivities in the three orthogonal spatial directions. In such systems, the self-diffusion should be described by a diffusion tensor rather than by a single scalar self-diffusion coefficient. By measuring over a

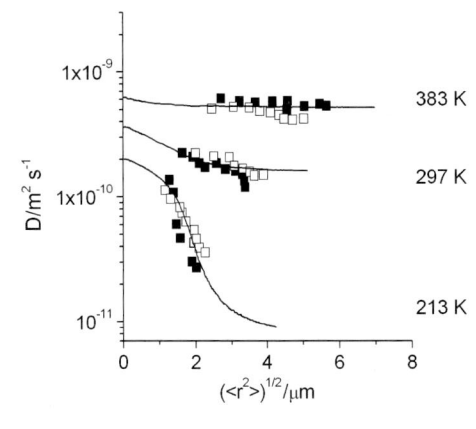

Fig. 3.1.3 Dependence of the diffusion coefficients on the root mean square displacements at different temperatures measured using PFG NMR (points) and determined in dynamic MC simulations (lines) for n-butane in silicalite-1.

sufficiently large range of gradient pulse intensities, anisotropic diffusion within such a sample yields characteristic deviations from the single exponential spin-echo attenuations given in Eq. (3.1.4). Implying a powder average over the sample and rotational symmetry of the diffusion tensor in each particle one obtains [4, 8]

$$\Psi(\delta g, t) = \frac{1}{2}\int_0^\pi exp[-\gamma^2\delta^2 g^2 t(D_{par}\cos^2\Theta + D_{perp}\sin^2\Theta)]\sin\Theta d\Theta \quad (3.1.10)$$

with D_{par} and D_{perp} denoting the two orthogonal principal elements of the diffusion tensor. Mesoporous materials of type MCM-41 are known to consist of hexagonally arranged channels with diameters of several nanometers. Figure 3.1.4 shows the results of PFG NMR diffusion measurements of water in such systems, where specifically synthesized MCM-41 particles with diameters of a few micrometers were used [31]. It was found [Figure 3.1.4(a)] that the diffusivity in channel direction (D_{par}) is higher by about one order of magnitude than the direction perpendicular to it (D_{perp}) and smaller by more than one order of magnitude than in the free water. Thus, for the first time, clear evidence was provided that in MCM-41 matter transfer is also possible in the direction perpendicular to the channels. Measuring these diffusivities (and with them the corresponding displacements) as a function of time [Figure 3.1.4(b)] the agreement between the maximum displacements and the particle dimensions as illustrated by the insert nicely confirms this analysis. The situation for the largest observation times reflected by Figure 3.1.4(b) are those illustrated by Figure 3.1.1 at the top right.

3.1.4
Long-range Diffusion

Under the measuring regime of long-range diffusion, molecular trajectories are much larger than the individual crystallites (bottom right of Figure 3.1.1). As a consequence, they consist of parts alternating between the intracrystalline and intercrystalline space. With Eq. (3.1.8), the effective diffusivity of the overall process may thus be expressed as [8, 32, 33]

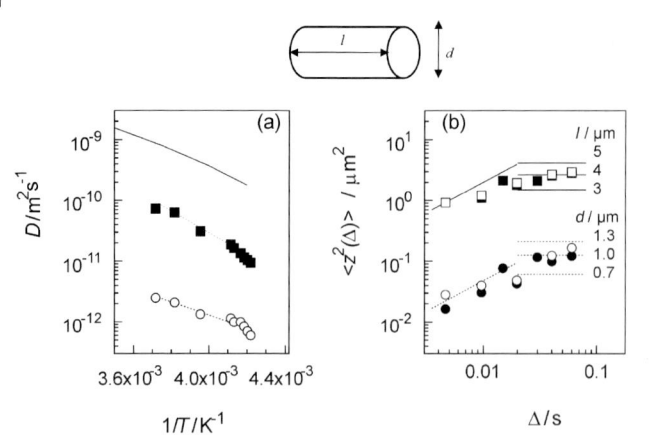

Fig. 3.1.4 Anisotropic self-diffusion of water in MCM-41 as studied by PFG NMR. (a) Dependence of the parallel (filled rectangles) and perpendicular (circles) components of the axisymmetrical self-diffusion tensor on the inverse temperature at an observation time of 10 ms. The dotted lines can be used as a visual guide. The full line represents the self-diffusion coefficients of super-cooled bulk liquid water. (b) Dependence of the parallel (rectangles) and perpendicular (circles) components of the mean square displacement on the observation time at 263 K in two different samples (open and filled symbols, respectively). The horizontal lines indicate the limiting values for the axial (full lines) and radial (dotted lines) components of the mean square displacements for restricted diffusion in cylindrical rods of length l and diameter d. The oblique lines, which are plotted for short observation times only, represent the calculated time dependences of the mean square displacements for unrestricted (free) diffusion with $D_{par} = 1.0 \times 10^{-10}$ m^2 s^{-1} (full line) and $D_{perp} = 2.0 \times 10^{-12}$ m^2 s^{-1} (dotted line), respectively.

$$D_{long-range} = p_{inter} D_{inter} + p_{intra} D_{intra,eff} \approx p_{inter} D_{inter} \quad (3.1.11)$$

where $p_{inter/intra}$ and $D_{inter/intra,eff}$ denote the relative amounts of molecules in the inter- and intracrystalline spaces and their (effective) diffusivities. Alternatively, $p_{inter/intra}$ can be related to the relative lifetimes of molecules in the corresponding phases. By noting $D_{intra,eff}$ rather than D_{intra} we have indicated that, in this context, the contribution of intracrystalline displacements to the overall transport is not represented correctly by the genuine intracrystalline diffusivity. Instead, the intracrystalline part of the trajectories is notably influenced by encounters with the surface. Recall that we are discussing the long-range limit, implying that the observation time notably exceeds τ_{intra}. As a consequence of the large intracrystalline concentration, mass balance at the crystal surface does not allow all molecules approaching the crystal boundary from inside to escape into the intercrystalline space. Thus, molecules become trapped for some period of time, leading to a reduction of the effective intracrystalline diffusivity, $D_{intra,eff}$, in comparison with D_{intra}. With Eq. (3.1.7), $D_{intra,eff}$ is thus found to be inversely proportional to the trapping time. Therefore, it turns out that $D_{intra,eff}$ is negligibly small in the long-range limit we are interested in. As a consequence, irrespective of the fact that p_{inter} is much smaller than p_{intra} (which, under the conditions of gas-phase adsorption being considered, is essentially close to 1), long-range diffusion in beds of zeolite crystallites may be assumed to be controlled exclusively by the rate of mass transfer outside of the individual crystallites [second part of Eq. (3.1.11)]. The corresponding diffusivity is generally given by the simple gas-kinetic approach [19, 34–37]

$$D_{inter} = \frac{1}{3}\frac{\lambda_{eff} u}{\alpha} \quad (3.1.12)$$

where λ_{eff}, u and α denote the mean free path, the thermal velocity and the tortuosity of the intercrystalline space, respectively. With Eq. (3.1.11) and the inherent assumption of fast exchange between the two regions, the inequality

$$\tau_{intra} \ll \tau_{long-range} \quad (3.1.13)$$

is implied [cf. Eq. (3.1.9)], where $\tau_{long-range}$ is now assumed to mean the observation time.

Figure 3.1.5 shows the Arrhenius plot of the long-range diffusivities of ethane in a bed of crystallites of zeolite NaX [34]. At sufficiently low temperatures the gas phase concentration is so small that within the intercrystalline space mutual encounters of the molecules essentially do not occur (Knudsen diffusion, insert on the right). Therefore, the effective mean free path λ_{eff} in Eq. (3.1.12) is given by the mean diameter of pores formed by the intercrystalline space. Thus, bearing in mind that the thermal velocity u increases only with the square root of temperature, the temperature dependence of $D_{long-range} = p_{inter}D_{inter}$ is essentially given by that of p_{inter}. As the relative concentrations in the gaseous and adsorbed phases are

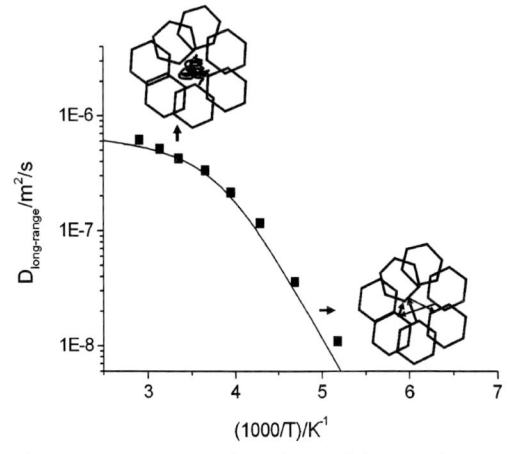

Fig. 3.1.5 Temperature dependence of the coefficient of long-range self-diffusion of ethane measured by PFG NMR in a bed of crystallites of zeolite NaX (points) and comparison with the theoretical estimate (line). The theoretical estimate is based on the sketched models of the prevailing Knudsen diffusion (low temperatures, molecular trajectories consist of straight lines connecting the points of surface encounters) and gas phase diffusion (high temperatures, mutual collisions of the molecules leading to the Brownian-type trajectories in the intercrystalline space).

interrelated by the Boltzmann factor, one has $p_{inter} \propto \exp(-E_{des}/kT)$, where E_{des} (the isosteric heat of adsorption) is the difference between the potential energies of a molecule in the gaseous and adsorbed states. In Figure 3.1.5, this situation is reflected by the straight part of the analytically calculated curve at sufficiently low temperatures. The activation energy of long-range diffusion, as obtained from the slope of the curve in this temperature range, is equal to 27 kJ mol^{-1} and coincides with literature data for the heat of adsorption [39].

With further increasing temperature, however, molecular concentration in the intercrystalline space is likewise increasing so that mutual encounters of the molecules become more and more probable (bulk diffusion, insert at the top). Eventually, the effective mean free path coincides with that in the gas phase, becoming inversely proportional to the gas phase pressure and hence to p_{inter}. As a consequence, the increase of p_{inter} with increasing temperature is compensated by the corresponding decrease of λ_{eff} and hence of D_{inter}, so that now $D_{long\text{-}range} = p_{inter} D_{inter}$ is increasing only slightly with the temperature, following the $T^{1/2}$ dependence of the thermal velocity.

A quantitative analysis [34], based on the adsorption isotherms and the intercrystalline porosity, yielded the remarkable result that a satisfactory fit between the experimental data and the estimates of $D_{long\text{-}range} = p_{inter} D_{inter}$ following Eqs. (3.1.11) and (3.1.12) only lead to coinciding results for tortuosity factors α differing under the conditions of Knudsen diffusion (low temperatures) and bulk-diffusion (high temperatures) by a factor of at least 3. Similar results have recently been obtained by dynamic Monte Carlo simulations [39–41].

Figure 3.1.6 provides an example demonstrating the practical relevance of comparative studies of long-range diffusion with intracrystalline diffusion, considered in the previous section. The diffusivity data refer to a technical catalyst applied for Fluid Catalytic Cracking (FCC) [42]. FCC catalysts consist of zeolite crystals of type Y of about 1 μm diameter, which, by means of a binder, are compacted into particles with a diameter of about 40 μm. As the sorption capacity of the binder is negligibly small, long-range diffusion within these particles can be described in essentially the same way as those accomplished just with beds of crystallites. n-Octane has been considered as an adsorbate, as a typical molecule involved in FCC reactions. The long-range diffusivities have been determined for mean displacements of the order of 5 μm so that they were sufficiently large in comparison with the individual crystallites, but still small enough that boundary effects caused by the particle surfaces are still of minor significance. In order to allow measurements over a larger range of temperatures, the measurements of intracrystalline diffusion have been performed with specially synthesized Y-type zeolite crystals with mean diameters of about 3 μm [43].

The practical relevance of the data presented in Figure 3.1.6 results from their use for an estimate of the respective mean lifetimes as predicted by Eq. (3.1.9). Bearing in mind that the extensions of the range of intraparticle (i.e., long-range) diffusion (the FCC particle diameter) is larger than that of intracrystalline diffusion by a factor of 40, with a ratio of $D_{long\text{-}range}/D_{intra}$ of about 10^2 at the typical temperature of FCC catalysis (800 K), the intracrystalline lifetime is smaller than the

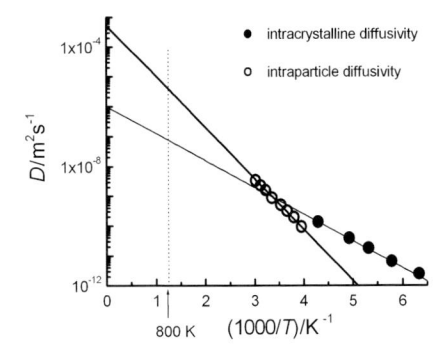

Fig. 3.1.6 Temperature dependence of the intraparticle diffusivity of n-octane in an FCC catalyst and the intracrystalline diffusivity of n-octane in large crystals of USY zeolite measured by PFG NMR. The concentration of n-octane in the samples was in all cases 0.62 mmol g^{-1}. Lines show the results of the extrapolation of the intracrystalline diffusivity and of the intraparticle diffusivity of n-octane to higher temperatures, including in particular a temperature of 800 K, typical of FCC catalysis.

molecular lifetime within the particles by at least one order of magnitude. One has to conclude, therefore, that the rate of molecular exchange of the reactant and product molecules in FCC catalysis is controlled by transport within the catalyst particles rather than within the zeolite crystals. This conclusion has been confirmed by comparative measurements of differently formulated FCC catalysts with identical zeolite contents: the catalytic performance was found to be enhanced in parallel with an enhancement of the long-range diffusivities. With these PFG NMR diffusion studies, for the first time long-lasting speculations about the dominating transport mechanisms in FCC catalysts [44–46] could be replaced by a set of directly measured key parameters describing the various diffusion processes involved.

In the context of Eq. (3.1.11) it has been discussed that long-range diffusion in beds of zeolite crystallites is described by the relationship of multi-state diffusion [Eq. (3.1.8)] only semi-quantitatively. While the contribution of intercrystalline diffusion to the overall diffusivity is well defined, the contribution of intracrystalline diffusion is far more difficult to quantify. The problem is caused by the possibility that molecular exchange from an intracrystalline to intercrystalline space may not be completely uncorrelated with the displacements in the intracrystalline space, as assumed by deriving Eq. (3.1.8) [47]. It is obvious that molecules are only able to propagate into the intercrystalline space if they have already reached the crystallite surface. As a consequence, it is generally accepted that, as a good approximation, long-range diffusion is exclusively determined by intercrystalline transport.

The situation is completely different for mass transfer within the pore network of monolithic compounds. Here mass transfer can occur both on the pore surface or in the pore volume and molecular exchange between these two states of mobility can occur anywhere within the pore system, being completely uncorrelated with the respective diffusion paths. As a consequence, Eq. (3.1.11) is applicable, without any restrictions, to describing "long-range" diffusion in the pore space. Equation (3.1.14) is thus obtained,

$$D_{\text{long-range}} = p_{\text{surface}} D_{\text{surface}} + p_{\text{gas}} D_{\text{gas}} \tag{3.1.14}$$

with $p_{\text{surface/gas}}$ and $D_{\text{surface/gas}}$ denoting the relative number of molecules on the pore surface (in the gas phase) and their diffusivities. A qualification, however, is that the foregoing is true if the adsorbed and the gaseous phases do not macroscopically phase separate within the pore network [48]. As demonstrated at the beginning of this section, the magnitudes of p_{gas} and D_{gas} can be calculated using the information on the adsorption isotherm and the pore geometry [49]. With these data, due to $p_{surface} = 1 - p_{gas}$, the surface diffusivity $D_{surface}$ can be calculated from the experimentally accessible long-range diffusivities. Figure 3.1.7 shows the results of this type of investigation with a sample of porous silicon [50], consisting of an array of parallel channel pores of 3.6 nm diameter, with cyclohexane and acetone as probe molecules [51, 52]. The results shown in the figure refer to room temperature. By connecting the sample volume to a gas reservoir containing the probe molecules under study, sample loading could be easily varied by simply changing the pressure in the reservoir. A relative pore filling (or concentration of molecules in pores) of $\theta \approx 0.5$ corresponds to about one monolayer of molecules adsorbed onto the pore wall.

A remarkable feature for both sorbates under study is shown in Figure 3.1.7, which exhibits a most pronounced concentration dependence of the diffusivity. The fact that the diffusivity increases rather than decreases with increasing loading indicates that this dependence is not caused by a mutual molecular hindrance on the surface, as in this case the reverse behavior should be observed. Instead, one has to argue that the observed behavior is a consequence of the guest–host rather than of the guest–guest interaction. In fact, it could imply that any surface heterogeneity tends to direct the first molecules towards the strongest adsorption sites. Consequently, with increasing loading, sites with decreasing adsorption energies will be occupied, which results in a steadily decreasing average effective activation energy of diffusion [53]. This supposition is confirmed by the observed temperature dependences shown in Figure 3.1.8. In complete agreement with the predicted behavior, the slope of the Arrhenius plots of the diffusivities decreases with increasing loading.

The experimental arrangement chosen in these studies allows the diffusion processes in the region of the adsorption hysteresis to be followed. Adsorption hysteresis is the phenomenon of history-dependent adsorption and describes the effect that, in addition to the pressure, the concentration also depends on whether the given pressure has been attained from lower values (i.e., on the "adsorption branch") or from higher values (the "desorption branch") [54]. Irrespective of its

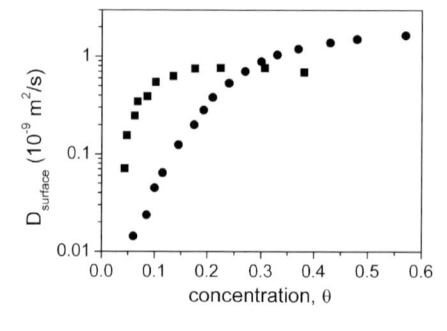

Fig. 3.1.7 The surface diffusion coefficient $D_{surface}$ of cyclohexane (squares) and acetone (circles) in porous silicon with 3.6-nm mean pore diameter at pore loadings up to about one monolayer. The measurements were performed at $T = 297$ K.

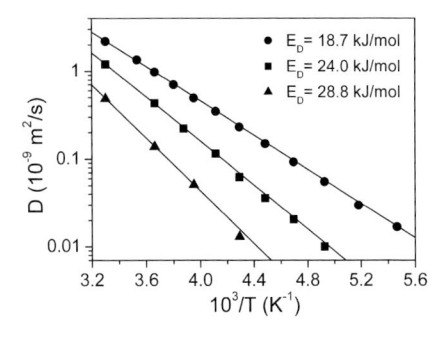

Fig. 3.1.8 The Arrhenius plots of the self-diffusion coefficients of acetone in porous silicon with 3.6-nm pore diameter at different pore concentrations $\theta = 0.6$ (circles), $\theta = 0.27$ (squares) and $\theta = 0.18$ (triangles). The solid lines show the fits to the experimental data using the Arrhenius relationship $D \propto \exp(-E_D/RT)$ with the activation energies for diffusion E_D indicated in the figure.

great technical relevance for porosimetry [55] and the fact that the phenomenon of adsorption hysteresis over several decades is known already, its microscopic origin is still discussed with some controversy [56–59]. Owing to the PFG NMR data shown in Figure 3.1.9, for the first time information about the inherent diffusivities can now also be included in this discussion. For cyclohexane as a probe molecule, Figure 3.1.9(a) shows both the total amount adsorbed and the respective diffusivity as a function of the pressure applied. In parallel with the amount adsorbed, the diffusivities are also found to differ on the adsorption and desorption branches for one and the same pressure. Even more interestingly, Figure 3.1.9(b) shows the diffusivities on the adsorption and desorption branches as a function of the respective loadings. In addition to cyclohexane [redrawn from Figure 3.1.9(a)], Figure 3.1.9(b) also displays the data for acetone. It is most interesting that in both cases the diffusivities on the adsorption and desorption branches differ notably from each other for one and the same loading. It is important to check that this behavior is observed over many adsorption–desorption cycles. This experimental finding strongly suggests that the differences in the adsorption and desorption branches are associated with differences in molecular arrangement and dynamics, which appear in the different diffusivities. This typically has to be linked with the development of metastable states often accompanying phase transitions under mesoscalic confinement.

3.1.5
Boundary Effects

Figure 3.1.1, bottom left, illustrates a situation where PFG NMR may provide immediate evidence about the existence and intensity of additional transport resistances on the surface of the individual crystallites, the so-called surface barriers [60, 61]. This option is based on the sensitivity of PFG NMR towards molecular displacements. Molecules traveling over distances exceeding the typical crystallite sizes have to leave the individual crystallites (and are captured by some other crystallite(s) on their further trajectory). This fraction of molecules contributes to the broad part of the propagator. Plotting the relative intensity of the broad part of the propagator as a function of t we thus obtain the relative number $\gamma(t)$ of molecules, which have left their (starting) crystallites at time t. The function $\gamma(t)$ is

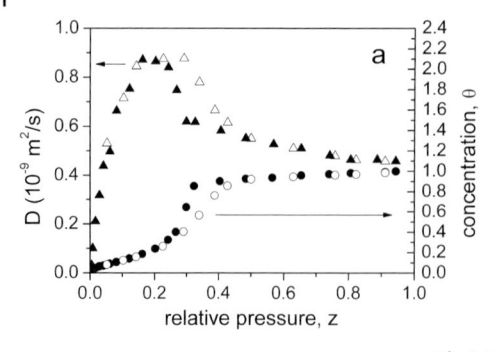

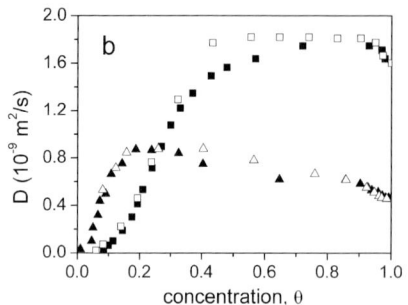

Fig.3.1.9 (a) The adsorption–desorption isotherm (circles, right axis) and the self-diffusion coefficients D (triangles, left axis) for cyclohexane in porous silicon with 3.6-nm pore diameter as a function of the relative vapor pressure $z = P/P_s$, where P_s is the saturated vapor pressure. (b) The self-diffusion coefficients D for acetone (squares) and cyclohexane (triangles) as a function of the concentration θ of molecules in pores measured on the adsorption (open symbols) and the desorption (filled symbols) branches.

simply the tracer exchange curve [10, 15], which is why this method of analyzing PFG NMR data has been called the NMR tracer desorption technique [60, 61]. As a measure of the rate of tracer exchange it is convenient to introduce the tracer exchange time τ_{intra} (or intracrystalline mean lifetime) as the first moment of the tracer exchange curve,

$$\tau_{\text{intra}} = \int_0^\infty [1 - \gamma(t)] dt \quad (3.1.14)$$

Approaching the crystallites by spheres of radius R, in the case of diffusion-limited exchange Eq. (3.1.14) can be shown to lead to [10, 15]

$$\tau_{\text{intra}}^{\text{diff}} = \frac{R^2}{15 D_{\text{intra}}} \quad (3.1.15)$$

Comparison between $\tau_{\text{intra}}^{\text{diff}}$ as determined on the basis of Eq. (3.1.15) from the microscopically determined crystallite radius and the intracrystalline diffusivity measured by PFG NMR for sufficiently short observation times t (top left of Figure 3.1.1), with the actual exchange time τ_{intra} resulting from the NMR tracer desorption technique, provides a simple means for quantifying possible surface barriers. In the case of coinciding values, any substantial influence of the surface barriers can be excluded. Any enhancement of τ_{intra} in comparison with $\tau_{\text{intra}}^{\text{diff}}$, on the other side, may be considered as a quantitative measure of the surface barriers.

As an example, Figure 3.1.10 illustrates the use of this procedure for elucidating the location of coke deposits on zeolite catalysts [62]. Samples of zeolites H-ZSM-5 have been subjected to an atmosphere of either n-hexane (filled symbols) or mesitylene (open systems) over different time intervals ("coking times"). Subsequently, the samples pre-treated thus have been loaded with methane as a probe molecule. Figure 3.1.10 displays the values of τ_{intra} (measured by NMR tracer desorption technique) and τ_{intra}^{diff} [calculated via Eq. (3.1.15)] for methane at room temperature in these samples as a function of their coking time. The inserts accompanying the representation characterize the different situations: with mesitylene as a coking chemical the intracrystalline diffusivity and hence τ_{intra}^{diff} remain essentially unaffected. Hence the increase of τ_{intra} with increasing coking time may be attributed to increasing coke depositions on the outer surface, while there is essentially no coke deposition in the intracrystalline space (insert bottom right). In contrast to this behavior, coking with n-hexane leads to a parallel increase of τ_{intra} and τ_{intra}^{diff}. One may conclude, therefore, that coke deposition in the zeolite crystallite (bottom left) simultaneously reduces the intracrystalline diffusivity and the rate of tracer exchange. Only during a final stage (top right), does preferential coke deposition on the crystallite surface lead to a further increase of the exchange time while the intracrystalline diffusivity (and hence τ_{intra}^{diff}) remains unchanged. It is worthwhile mentioning that the observed difference in the coking patterns of n-hexane and mesitylene is in good agreement with an important difference in their sizes. While n-hexane is small enough to be accommodated by the pore network, because of its larger size mesitylene is kept outside the intracrystalline space.

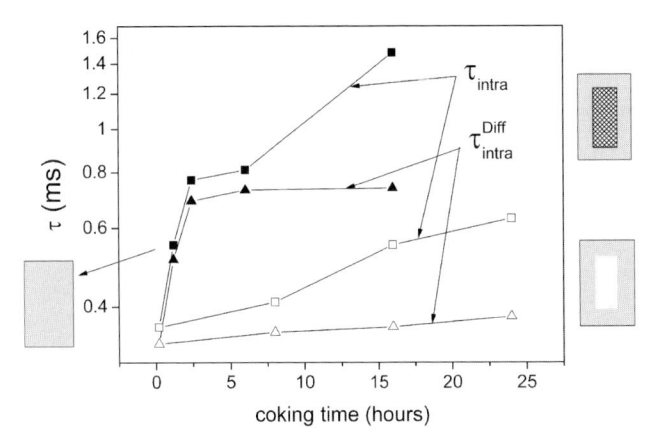

Fig. 3.1.10 Molecular lifetimes τ_{intra} and τ_{intra}^{diff} in H-ZSM-5 crystallites obtained using the NMR tracer desorption technique and calculated via Eq. (3.3.15), respectively. Tracing by probe molecules (methane, measurement at 296 K) after an H-ZSM-5 catalyst has been kept for different coking times in a stream of n-hexane (filled symbols) and mesitylene (open symbols) at elevated temperature. The inserts present the evidence provided by a comparison of τ_{intra} and τ_{intra}^{diff} with respect to the distribution of the coke deposits on the zeolite crystallites. With mesitylene as a coking molecule, the coke is found to be predominantly deposited close to the crystallite surface (bottom right). With n-hexane, coke deposition is found to proceed in essentially two stages: initially, the coke is deposited over the whole of the crystallites (bottom left), while in a second stage (top right) coke is deposited close to the crystallite surface.

Therefore, the difference in the location of coke deposition may be rationalized as a simple consequence of the different sizes of the starting substances for coking.

For intracrystalline diffusion paths sufficiently small in comparison with the crystallite radii, the effective diffusivity as defined by Eq. (3.1.6) may be expanded in a power series [9, 63, 64], leading to

$$D_{\text{eff}}(t)/D_0 = 1 - \frac{4}{3\sqrt{\pi}}\frac{1}{R}\sqrt{D_0 t} - \frac{1}{2R^2}(D_0 t) \tag{3.1.16}$$

and

$$D_{\text{eff}}(t)/D_0 = 1 - \frac{2}{3\sqrt{\pi}}\frac{1}{R}\sqrt{D_0 t} - \frac{1}{R^2}(D_0 t) \tag{3.1.17}$$

respectively, where D_0 stands for the genuine intracrystalline diffusivity. As to be expected (and implied throughout Section 3.1.3), the experimentally accessible quantity D_{eff} coincides with the true intracrystalline diffusivity D_0 in the limiting case of sufficiently short displacements, i.e., for $(D_0 t)^{1/2}$ much less than the typical crystallite size. Equation (3.1.16) describes the situation of ideal confinement to an intracrystalline space (Figure 3.1.1, top right). In this case the crystallite surface acts as an (ideally) reflecting boundary for the molecules in the intracrystalline space. Equation (3.1.17) has been derived for absorbing boundaries [9, 63, 64]. PFG NMR diffusion measurements with beds of zeolites do in fact comply with this limiting case when the long-range diffusivity is much larger than the intracrystalline diffusivity (Figure 3.1.1, bottom left) and one is only analyzing the intracrystalline constituent of the propagator (viz., the narrow one).

The influence of the confining boundaries on the effective diffusivity, as reflected by Eqs. (3.1.16) and (3.1.17), has been applied repeatedly to determine the pore

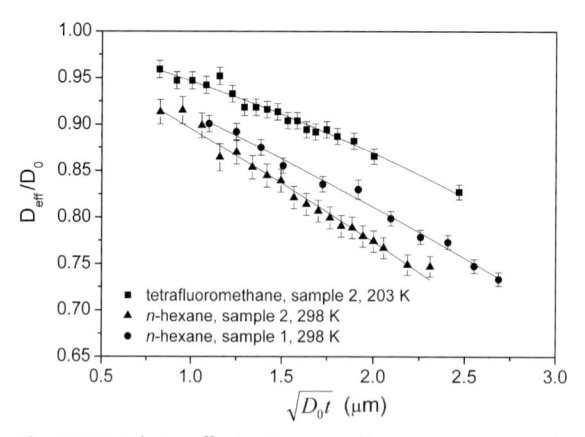

Fig. 3.1.11 Relative effective intracrystalline diffusivities $D(t)/D_0$ as function of $\sqrt{D_0 t}$ for n-hexane under single-component adsorption (circles) and for n-hexane (triangles) and tetrafluoromethane (rectangles) under two-component adsorption in zeolite NaX. The lines shown in the diagram represent the best fits of Eqs. (3.1.16) and (3.1.17) to the experimental data of $D(t)$ with D_0 and R as fitting parameters.

surface in rocks or beds of sand grains [64–71]. This concept has been successfully applied for the first time to beds of zeolites, as discussed in Ref. [72]. Figure 3.1.11 shows the results of these studies, which have been performed with two different samples of NaX, one loaded with n-hexane (two molecules per supercage), the other with n-hexane and tetrafluoromethane (one molecule of each per supercage). In both samples, for the n-hexane measurements a temperature of 298 K was chosen, where the n-hexane molecules where found to be totally confined so that data analysis could be based on Eq. (3.1.16). Measurement of the sample containing two different diffusants presented the options of operating with one and the same sample under the conditions of reflecting boundaries [Eq. (3.1.16), ^{1}H PFG NMR with n-hexane] and absorbing boundaries [Eq. (3.1.17), ^{19}F PFG NMR with CF$_4$]. The latter type of measurement had to be carried out at 203 K as only at these low temperatures the diffusivities were small enough to allow molecular displacements sufficiently small in comparison with the crystallite radii (as a supposition for the series expansion). Even at this low temperature the long-range diffusivity was found to be large enough to permit the limiting case of absorbing boundaries, i.e., of Eq. (3.1.17). It is significant that the good fit between the experimental data and the theoretical curves was only possible by involving the second-order terms in $(D_0 t)^{1/2}$.

It is worth noting that within a range of 20%, five different methods of analyzing the crystallite size, viz., (a) microscopic inspection, (b) application of Eq. (3.1.7) for restricted diffusion in the limit of large observation times, (c) application of Eq. (3.1.15) to the results of the PFG NMR tracer desorption technique, and, finally, consideration of the limit of short observation times for (d) reflecting boundaries [Eq. (3.1.16)] and (e) absorbing boundaries [Eq. (3.1.17)], have led to results for the size of the crystallites under study that coincide.

3.1.6
Conclusion

Owing to its ability to monitor the probability distribution of molecular displacements over microscopic scales from hundreds of nanometers up to several millimeters, PFG NMR is a most versatile technique for probing the internal structure of complex materials. As this probing is based on an analysis of the effect of the structural properties on molecular propagation, the properties of the material studied are those which are mainly of relevance for the transport processes inherent to their technical application.

The examples presented in this chapter refer to nanoporous materials. They have found widespread industrial application, in particular for the upgrading of substances by heterogeneous catalysis and separation. In most of these applications, molecular transport in these materials could be rate determining for the overall process. Thus, in addition to the information on structural properties, with the relevant parameters of the different regimes of molecular transport, including the regimes of intracrystalline diffusion, of long-range (intraparticle) diffusion and the penetration through surface (and other) barriers, PFG NMR is also able to provide key numbers characterizing the technical state of the materials under study.

In accord with the vivid development of this method over the last couple of years, the present chapter has presented both well-established methods, such as the evaluation of compacted zeolite particles with respect to their intracrystalline and intraparticle diffusivities, and examples of novel applications, such as the diffusion analysis of adsorption hysteresis, whose future significance cannot yet be predicted. In any case, owing to its exceptionally high sensitivity with respect to molecular displacements, PFG NMR will continue to be not only a valuable complement of all NMR imaging techniques, but it is also often combined with them. Already such a combination provides invaluable space-dependent information about molecular transport and an insight into the physical–chemical properties of matter on a macroscopic scale (e.g., by diffusion-contrasted imaging). Exciting examples are metabolic studies in medical applications and imaging of reactant transport in chemical reactors. Further development of these techniques will depend decisively on the attainability of higher resolutions, down to the submicrometer range.

References

1 B. Blümich **2000**, *NMR Imaging of Materials*, Clarendon Press, Oxford.
2 S. Stapf, S. I. Han, C. Heine, B. Blumich **2002**, *Concept Magn. Reson.* 14, 172–211.
3 P. T. Callaghan, A. Coy, D. MacGowan, K. J. Packer, F. O. Zelaya **1991**, *Nature* 351, 467–469.
4 P. T. Callaghan **1991**, *Principles of NMR Microscopy*, Clarendon Press, Oxford.
5 P. T. Callaghan **1990**, *J. Magn. Reson.* 88, 493–500.
6 R. M. Cotts, **1991** *Nature* 351, 443–444.
7 J. Kärger, W. Heink **1983**, *J. Magn. Reson.* 51, 1–7.
8 J. Kärger, H. Pfeifer, W. Heink **1988**, *Adv. Magn. Reson.* 12, 2–89.
9 P. P. Mitra, P. N. Sen, L. M. Schwartz, P. Ledoussal **1992**, *Phys. Rev. Lett.* 68, 3555–3558.
10 J. Kärger, D. M. Ruthven **1992**, *Diffusion in Zeolites and Other Microporous Solids*, Wiley & Sons, New York.
11 B. Newling, S. N. Batchelor **2003**, *J. Phys. Chem. B* 107, 12391–12397.
12 C. Meier, W. Dreher, D. Leibfritz **2003**, *Magn. Reson. Med.* 50, 510–514.
13 K. I. Momot, P. W. Kuchel **2003**, *Concept Magn. Reson. A* 19A, 51–64.
14 W. Heink, J. Kärger, H. Pfeifer **1978**, *Chem. Eng. Sci.* 33, 1019–1023.
15 R. M. Barrer **1978**, *Zeolites and Clay Minerals as Sorbents and Molecular Sieves*, Academic Press, London.
16 P. Galvosas, F. Stallmach, G. Seiffert, J. Kärger, U. Kaess, G. Majer **2001**, *J. Magn. Reson.* 151, 260–268.
17 F. Stallmach, J. Kärger **1999**, *Adsorption* 5, 117–133.
18 J. Kärger, C. M. Papadakis, F. Stallmach **2004**, (Structure-mobility relations of molecular diffusion in interface systems), in *Molecules in Interaction with Surfaces and Interfaces*, eds. R. Haberlandt, D. Michel, A. Pöppel, R. Stannarius, Springer, Heidelberg.
19 J. Kärger, J. Caro **1977**, *J. Chem. Soc., Faraday Trans. I* 73, 1363–1376.
20 J. Kärger, D. Freude **2002**, *Chem. Eng. Technol.* 25, 769–778.
21 J. Kärger **2002**, *Ind. Eng. Chem. Res.* 41, 3335–3340.
22 F. Keil **1999**, *Diffusion und Chemische Reaktion in der Gas/Feststoff-Katalyse*, Springer, Berlin.
23 F. J. Keil, R. Krishna, M. O. Coppens **2000**, *Rev. Chem. Eng.* 16, 71–197.
24 N. Y. Chen, T. F. Degnan, C. M. Smith **1994**, *Molecular Transport and Reaction in Zeolites*, VCH Publishers, New York.
25 D. Prinz, L. Riekert **1986**, *Ber. Bunsenges. Phys. Chem.*, 90, 413–420.

26 D. M. Ruthven, M. F. M. Post **2001**, (Diffusion in zeolite molecular sieves), in *Introduction to Zeolite Science and Practice*, eds. H. van Bekkum, E. M. Flanigen, J. C. Jansen, Elsevier, Amsterdam.
27 J. Kärger, S. Vasenkov, S. M. Auerbach **2003**, (Diffusion in zeolites), in *Handbook of Zeolite Science and Technology*, eds. S. M. Auerbach, K. A. Carrado, P. K. Dutta, Marcel Dekker, New York, Basel.
28 S. Vasenkov, J. Kärger **2002**, *Micropor. Mesopor. Mat.* 55, 139–145.
29 S. Vasenkov, W. Böhlmann, P. Galvosas, O. Geier, H. Liu, J. Kärger **2001**, *J. Phys. Chem. B.* 105, 5922–5927.
30 R. L. June, A. T. Bell, D. N. Theodorou **1992**, *J. Phys. Chem.* 96, 1051–1059.
31 F. Stallmach, A. Graser, J. Kärger, C. Krause, M. Jeschke, U. Oberhagemann, S. Spange **2001**, *Micropor. Mesopor. Mat.* 44, 745–753.
32 J. Kärger, M. Kocirik, A. Zikanova **1981**, *J. Colloid Interf. Sci.* 84, 240–24.
33 J. Kärger, H. Pfeifer **1994**, (NMR studies of molecular diffusion), in *NMR Techniques in Catalysis*, eds. a. Pines, A. Bell, Marcel Dekker, New York.
34 O. Geier, S. Vasenkov, J. Kärger **2002**, *J. Chem. Phys.* 117, 1935–1938.
35 F. Rittig, C. G. Coe, J. M. Zielinski **2002**, *J. Am. Chem. Soc.* 124, 5264–5265.
36 F. Rittig, C. G. Coe, J. M. Zielinski **2003**, *J. Phys. Chem. B* 107, 4560–4566.
37 F. Rittig, T. S. Farris, J. M. Zielinski **2004**, *AICHE J.* 50, 589–595.
38 J. A. Dunne, M. B. Rao, S. Sircar, R. J. Gorte, A. L. Myers **1996**, *Langmuir* 12, 5896–5904.
39 K. Malek, M. O. Coppens **2001**, *Phys. Rev. Lett.* 8712, 12550.
40 K. Malek, M. O. Coppens **2003**, *J. Chem. Phys.* 119, 2801–2811.
41 V. N. Burganos **1998**, *J. Chem. Phys.* 109, 6772–6779.
42 J. Weitkamp, L. Puppe **1999**, *Catalysis and Zeolites*, Springer, Berlin, Heidelberg.
43 P. Kortunov, S. Vasenkov, J. Kärger, M. Fé Elía, M. Perez, M. Stöcker, et al. **2005**, *Magn. Reson. Imag.* 23, 233–237.
44 S. Al-Khattaf, J. A. Atias, K. Jarosch, H. de Lasa **2002**, *Chem. Eng. Sci.* 57, 4909–4920.
45 C. M. Bidabehere, U. Sedran **2001**, *Ind. Eng. Chem. Res.* 40, 530–535.
46 P. J. Barrie, C. K. Lee, L. F. Gladden **2004**, *Chem. Eng. Sci.* 59, 1139–1151.
47 J. Kärger **1985**, *Adv. Colloid Interfac.*, 23, 129–148.
48 R. R. Valiullin, V. D. Skirda, S. Stapf, R. Kimmich **1997**, *Phys. Rev. E* 55, 2664–2671.
49 R. Valiullin, P. Kortunov, J. Kärger, V. Timoshenko **2004**, *J. Chem. Phys.* 120, 11804–11814.
50 V. Lehmann, R. Stengl, A. Luigart **2000**, *Mat. Sci. Eng. B – Solid* 69, 11–22.
51 R. Valiullin, P. Kortunov, J. Kärger, V. Timoshenko **2005**, *J. Phys. Chem. B* 109, 5746–5752.
52 J. Kärger, R. Valiullin, S. Vasenkov **2005**, *New J. Phys.* 7, 15.
53 K. W. Kehr, K. Mussawisade, T. Wichmann **1998**, (Diffusion of particles on lattices), in *Diffusion in Condensed Matter*, eds. J. Kärger, P. Heitjans, R. Haberlandt, Vieweg/Springer, Braunschweig, Berlin.
54 D. H. Everett **1967**, (Adsorption hysteresis), in *The Solid-gas Interface*, ed. E. Alison Flood, Marcel Dekker, New York.
55 F. Schüth, K. S. W. Sing, J. Weitkamp **2002**, *Handbook of Porous Solids*, Wiley-VCH, Weinheim.
56 L. Sarkisov, P. A. Monson **2000**, *Langmuir* 16, 9857–9860.
57 E. A. Ustinov, D. D. Do **2004**, *J. Chem. Phys.* 120, 9769–9781.
58 E. Kierlik, P. A. Monson, M. L. Rosinberg, L. Sarkisov, G. Tarjus **2001**, *Phys. Rev. Lett.* 87, 055701.
59 D. Wallacher, N. Kunzner, D. Kovalev, N. Knorr, K. Knorr **2004**, *Phys. Rev. Lett.* 92, 195704.
60 J. Kärger, W. Heink, H. Pfeifer, M. Rauscher, J. Hoffmann **1982**, *Zeolites* 2, 275–278.
61 J. Kärger **1982**, *AICHE J.* 28, 417–423.
62 J. Caro, M. Bülow, H. Jobic, J. Kärger, B. Zibrowius **1993**, *Adv. Catal.* 39, 351–414.
63 P. P. Mitra, P. N. Sen **1992**, *Phys. Rev. B* 45, 143–156.

64 P. P. Mitra, P. N. Sen, L. M. Schwartz **1993**, *Phys. Rev. B* 47, 8565–8574.
65 R. W. Mair, M. S. Rosen, R. Wang, D. G. Cory, R. L. Walsworth **2002**, *Magn. Reson. Chem.* 40, S29–S39.
66 R. W. Mair, M. N. Sen, M. D. Hurlimann, S. Patz, D. G. Cory, R. L. Walsworth **2002**, *J. Magn. Reson.* 156, 202–212.
67 R. W. Mair, R. L. Walsworth **2002**, *Appl. Magn. Reson.* 22, 159–173.
68 R. W. Mair, R. Wang, M. S. Rosen, D. Candela, D. G. Cory, R. L. Walsworth **2003**, *Magn. Reson. Imag.* 21, 287–292.
69 L. L. Latour, P. P. Mitra, R. L. Kleinberg, C. H. Sotak **1993**, *J. Magn. Res. A* 101, 342–346.
70 F. Stallmach, C. Vogt, J. Kärger, K. Helbig, F. Jacobs **2002**, *Phys. Rev. Lett.* 88, 105505.
71 F. Stallmach, J. Kärger **2003**, *Phys. Rev. Lett.* 90, 039602.
72 O. Geier, R. Q. Snurr, F. Stallmach, J. Kàrger **2004**, *J. Chem. Phys.* 120, 1–7.

3.2
Application of Magnetic Resonance Imaging to the Study of the Filtration Process
R. Reimert, E. H. Hardy, and A. von Garnier

3.2.1
Filtration Principles

Filtration is a unit operation by which particles suspended in a fluid are separated from this fluid when it flows through some means of filtration, the filter. The particles to be filtered can be solid or liquid (droplets), and the fluid carrying the particles can be either gaseous or liquid. However, only gas will be considered as the carrier fluid in the following discussions. In the filter the particles will accumulate due to obstacles that hinder their movement and due to adherence to the filter material and, later on, to the already accumulated particles. Therefore, the filter has to provide both the obstacles to the flow and the surface for adherence. These duties are concurrently well fulfilled by porous media, which can be arranged in either a flexible (filter bags, filter cloths) or an inflexible form (rigid filters). There is a distinctive different behavior depending on whether the separation of the particles takes place in the depth of the filter (in-bed filtration) or only on the surface of the filter. In the latter the particles are accumulated on the upstream side of the filter, which in this case is more or less a screen. The separated and accumulated (solid) particles form the filter cake, which itself acts as a porous filter medium. This type of filtration will not be considered in the course of this chapter.

A further distinction has to be made between solid and liquid particles as they behave very differently with respect to adhesion and in the accumulated state. Liquid particles adhere to the surface of the filter medium on coming into contact, provided they do not hit with too high a velocity. (If droplets hit an obstacle with a high impact they can disintegrate and the newly formed smaller droplets could rebound. Already separated and accumulated liquid might even be re-entrained by such an impact.) Rebounding is more likely for solid particles. This effect depends largely on the elasticity of the material of the solids but also on the adhesive forces between the particles and the collecting material of the filter. Accumulated separated liquid particles form a film on the surface of the collecting material and fill spaces, characterized by a gas velocity that is lower than average. Once thick enough, the film will be transported by the drag of the flowing gas and by gravity depending on the arrangement of the filter with respect to the earth's gravity field. The accumulated liquid fills part of the originally free gas volume lowering the porosity of the filter to a certain extent, but typically not totally due to the transport of the liquid, if the liquid is not too viscous. On the contrary, there is no such limitation for accumulated solid particles which can fill the voidage until the filter is clogged.

The filtration effect is related to the pressure loss of the fluid flowing through the filter. The pressure loss represents the flow conditions in the filter, i.e., the superficial flow velocity, the porosity and the diameter of the passages (pores or channels). In general, the higher the pressure loss the greater is the filtration efficiency.

The filtration efficiency is commonly expressed by the fractional collection efficiency $T(x)$, see Eq. (3.2.1), rather than by considering the total mass, as the filtration process depends strongly on the size x of the particles entrained in the carrier gas [1].

$$T(x) = \lim_{\Delta x \to 0} T(x + \Delta x) \quad (3.2.1)$$

$$T(x + \Delta x) = \frac{Q_{sep}(x + \Delta x)}{{}^m\Phi_{feed}(x + \Delta x)} \quad (3.2.2)$$

Here

$$Q_{sep}(x + \Delta x) = \frac{dm_{sep}(x + \Delta x)}{dt} \quad (3.2.3)$$

is the mass per time of separated particles with diameters between $x + \Delta x$ and

$${}^m\Phi_{feed}(x + \Delta x) = c_{L,G} \, u_0 \quad (3.2.4)$$

is the convective mass flow of particles entering the filter with diameters in the range of $x + \Delta x$, where $c_{L,G} = m_L/V_G$, m_L = mass of liquid dispersed in the carrier gas, u_0 = superficial gas velocity and V_G = volume occupied by the carrier gas.

3.2.2
In-bed Filtration

In an in-bed filter (also known as deep filtration) the particles are separated throughout the whole depth of the filter. The filter is either made of fibers (filter mats) or of collector bodies, see Figure 3.2.1. Collector bodies are applied when the operating conditions are extreme as per the temperature (up to 1400 °C) and the nature of the entrained particles (sticky and abrasive). For separating sticky particles or droplets of high viscosity a moving bed filter can be applied. Movement

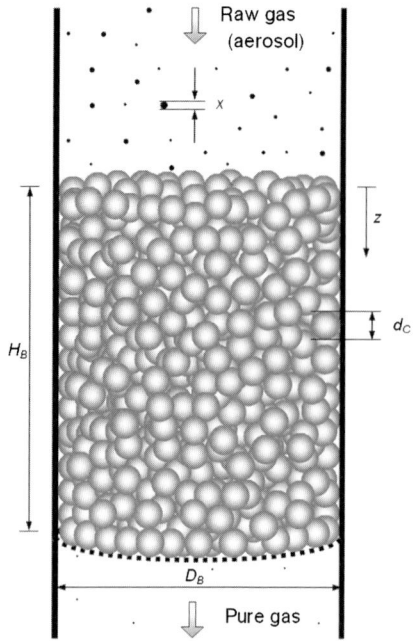

Fig. 3.2.1 Principles of in-bed filtration.

of the filter bed is caused by gravity. The collector bodies are withdrawn at the bottom, cleaned if necessary and re-fed to the top of the filter. For the sake of clarity only the separation of droplets in a filter bed made of spherical collector bodies will be considered further.

Calculation of $T(x,z)$ according to Eqs. (3.2.1) and (3.2.2) for in-bed filtration is based on the assumption that $Q_{sep}(x,z)$ is proportional to the mass flow of particles $^{m}\Phi(x,z)$ at the location z within the filter-bed ("first-order principle") [2]. Integration over the total bed height ($0 \leq z \leq H_B$) yields the overall separation efficiency.

$$T(x) = 1 - \exp\left[-1.5 \frac{1-\varepsilon}{\varepsilon} \frac{H_B}{d_c} \eta(x)\right] \qquad (3.2.5)$$

where ε = bed porosity, H_B = bed height, d_c = diameter of collector forming the filter bed and η = impact ratio, defined as

$$\eta = \frac{^{m}\Phi_{hit}}{^{m}\Phi_{cs}} \qquad (3.2.6)$$

where $^{m}\Phi_{hit}$ = mass flow of particles that hit the collector and are separated and $^{m}\Phi_{cs}$ = mass flow of incoming particles in the cross-sectional area of the collector perpendicular to the gas flow.

The assumption that the filtration can be treated by calculating the separation around one single collector body and integrating the result for the total bed underlies the scheme represented by Eqs. (3.2.5) and (3.2.6). For many decades research was dedicated to determining $\eta(x)$ [1]. An equation for calculating $\eta(x)$

presented by D'Ottavio and Goren [3] was recently adjusted by Schier and Reimert [4]:

$$\eta = \frac{St_{eff}^A}{B + St_{eff}^A} \qquad (3.2.7)$$

where St_{eff} = modified Stokes number [3], defined as

$$St = \frac{Cu \cdot \varrho_L \cdot x^2 \, u_{G,0}}{18 \, \mu_G \, d_c} \qquad (3.2.8)$$

where Cu = Cunningham correction [5], ϱ_L = liquid density (droplets) and μ_G = dynamic gas viscosity.

Whereas D'Ottavio and Goren and other workers proposed fixed values for A and B, Schier's experiments showed that the height H_B of the bed should be taken into account. This dependence needs to be considered especially when filters are optimized with regard to pressure loss [4].

$$A = 3.5 \, [1 - exp(-H_B/(5 \, d_c))] \qquad (3.2.9)$$

$$B = 8 + 26(d_c/H_B) \qquad (3.2.10)$$

Also, Schier's experiments revealed the necessity to consider the directional influence of the walls of the filter bed on the bed porosity ε. This effect is very important especially for small sized beds that are usually used in laboratories for investigations. For beds made of spherically shaped collectors several correlations exist describing the $\varepsilon(y/d_c)$ function where y denotes the distance from the wall in the radial direction. However, for relative large bed diameters D_B/d_c ranging from 5 to 25 it proved to be sufficient to use an averaged ε in Eq. (3.2.5), as proposed by Jeschar [6].

$$\varepsilon = 0.375 + 0.34 \, (d_c/D_B) \qquad (3.2.11)$$

In the course of separating the droplets some liquid will be stored within the bed. The accumulated liquid lowers the value of ε. However, in contrast to separation of solids the filter will not be blocked as the liquid can move out of the filter, as explained in Section 3.2.1. By introducing the saturation β the change in bed porosity due to accumulated liquid can be accounted for.

$$\beta = V_L/\varepsilon V_B \qquad (3.2.12)$$

$$\varepsilon_{ac} = \varepsilon \, (1 - \beta) \qquad (3.2.13)$$

where V_L = liquid filled part of the bed voidage and ε_{ac} = remaining bed porosity when liquid is accumulated.

3.2.3
Filtration Dynamics

According to the "first-order principle" the separation is assumed to occur [see explanation for Eq. (3.2.5)] with most liquid being separated at the bed entrance and accumulation starting there, resulting in the porosity ε_{ac} becoming a function of bed height z. As ε_{ac} influences the separation, according to Eqs. (3.2.5) and (3.2.13), the separation becomes a transient process. This could be modeled based on fundamental mass balances for the two liquid phases: liquid entrained in the gas and liquid accumulated in the bed. To verify the simulation results gained by the model applied to a laboratory size separator, the separation dynamics should be studied without interfering with the separation process itself. For this purpose NMR imaging (MRI) lends itself to being a non-invasive method for determining the accumulated liquid with temporal and spatial resolution.

3.2.3.1 Experimental Set-up

A vertically arranged NMR tomograph (Bruker Avance 200 SWB) with a static field of 4.7 T, field gradients of up to 0.14 T m^{-1} and an open-ended bird-cage probe head with a 64 mm inner diameter was used throughout the whole series of experiments. Droplet separators with a bed height H_B of 100 mm and diameters D_B of 38, 51 and 57 mm, respectively, were placed inside the measuring volume having a height of approximately 50 mm. Spherical non-porous glass beads with diameters d_c varying from 1.5 to 8 mm for different test runs served as collectors. In total the diameter ratio D_B/d_c, which influences the local and the total porosity as per Eq. (3.2.11), could be varied between 7 and 34. An air stream in which droplets of silicon oil (Goldschmidt, Essen, Germany, viscosity 0.97 Pa s) had been dispersed was routed through the separators. The complete set-up is shown in Figure 3.2.2.

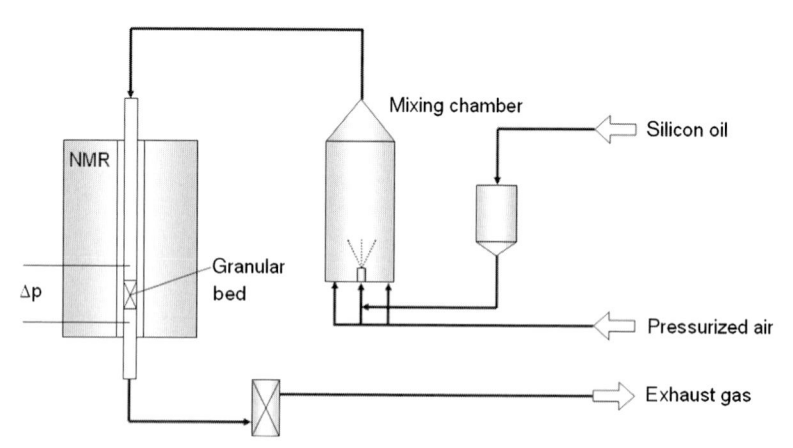

Fig. 3.2.2 Set-up for measuring in-bed filtration on-line in an NMR bird cage.

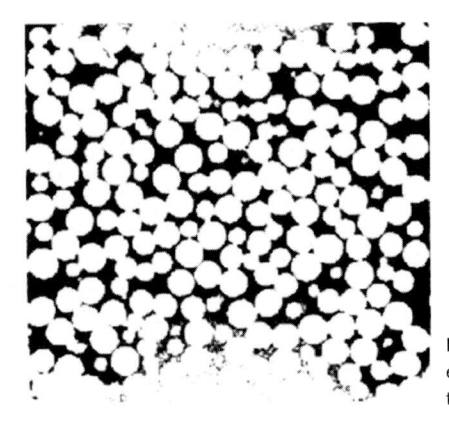

Fig. 3.2.3 Example of areas that are not exploitable due to the inhomogeneous field at the outer parts of the object.

3.2.3.2 Data Acquisition

Image matrices of 256^3 volume elements (voxels) and of 256^2 "picture elements" (pixels) were used for 3D and 2D measurements, respectively, corresponding to a spatial resolution of 250 μm. The use of larger matrices is limited by the increased acquisition time and the reduced signal-to-noise ratio. For 2D experiments, multi-slice versions with 11 slices perpendicular to the direction of superficial gas flow and two orthogonal slices in the direction of the gas flow were used and the slice thickness was set to 1 mm. At the upper and the lower ends of the bed and at its walls the signal intensity was rather distorted due to the inhomogeneous radio-frequency field, as exemplified by Figure 3.2.3 (upper and lower ends) and Figure 3.2.4 (walls). These parts of the bed were not taken into account in the data analysis. All measurements were performed with standard rapid spin-echo pulse sequences (RARE). The repetition time of typically 5 s and a RARE factor of 8 resulted in an acquisition time (for the 11 slices) of typically 160 s. Further details on the NMR methods are given in Chapter 5.2.1.2 by Gladden et al.

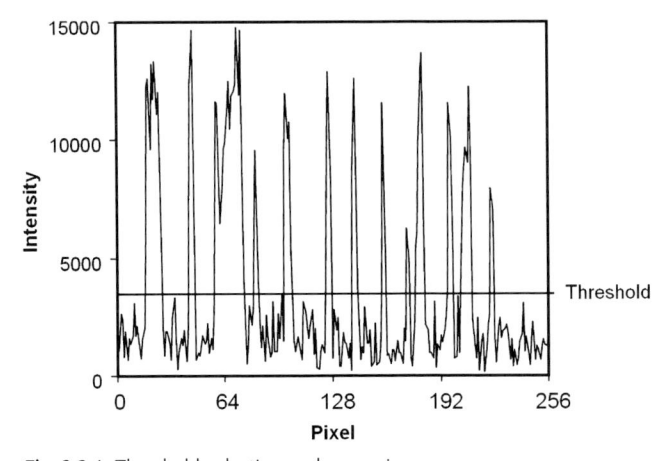

Fig. 3.2.4 Threshold selection and reasoning.

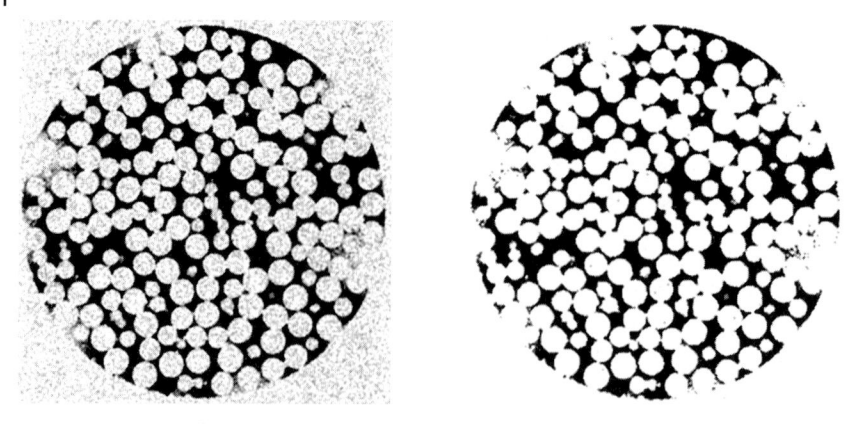

Fig. 3.2.5 Images before and after discriminating the signal intensity by the threshold.

Prior to further processing, the image intensities were binary gated using a threshold above the noise limit. For a filter filled with water, this is illustrated in Figure 3.2.5 for a pixel row in the middle of a slice, and in Figure 3.2.4 the result of binary gating for one slice is shown. Neither outside the bed nor in the image of the collector bodies does the signal (i.e., noise) remain. This type of data processing is important and necessary for generating quantitative data, for instance for wetting of collectors and for the liquid hold up (see Figure 3.2.6).

The software package PV-WAVE® from Visual Numerics was used for data processing.

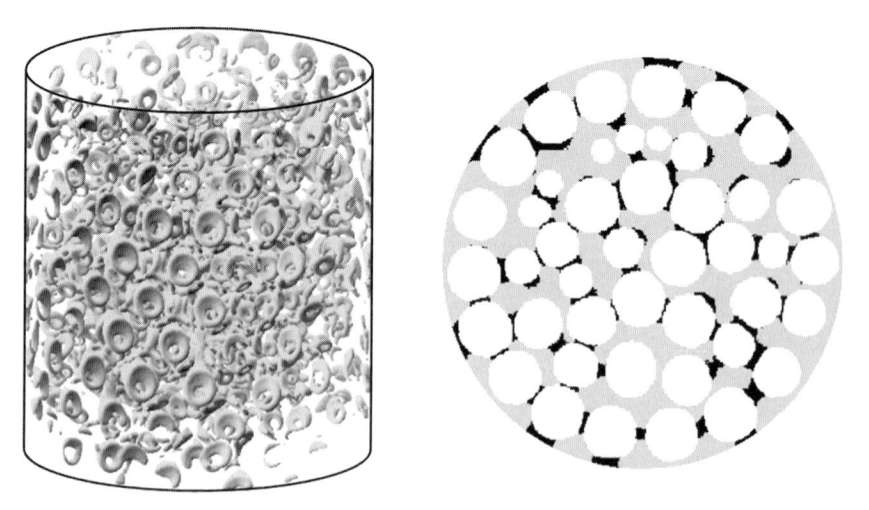

Fig. 3.2.6 Liquid filling the pores in the filter bed due to separation of the droplets.

3.2.3.3 MRI Results

The momentary conditions of liquid accumulated in the filter bed during uninterrupted filtration, in the following discussions denoted as liquid hold up, are well represented by Figure 3.2.6. This is possible as the time required to measure the 11 slices, 160 s, is small compared with the rather slow dynamics of the filtration process. In the surface projection on the left of Figure 3.2.6 the liquid surrounding the collector bodies is quite obvious. The 3D image was acquired with the static hold up. On the right side of the same figure the three types of pixels in a slice are shown. White areas represent the cuts through the collectors, identified from images of the filter filled with water. Black areas represent the liquid accumulated during filtration, the liquid hold up. The remaining large grey areas show the gas phase (with some liquid droplets in it, giving a signal below the threshold). From the distribution of the black spots the expected tendency of the liquid to accumulate where surfaces approach each other becomes obvious.

Likewise, Figure 3.2.7 shows how the liquid hold up increases with both the filtration time t_F and the height z in the bed. Of course, through data processing curves can be created which are more common for an engineering approach, as Figure 3.2.8 illustrates for the same type of filtration experiment. The liquid hold up which is defined by Eq. (3.2.12) can be obtained from the MRI data by dividing the number of liquid filled pixels (black area in Figure 3.2.6) by the number of void pixels (sum of grey and black pixels, Figure 3.2.6).

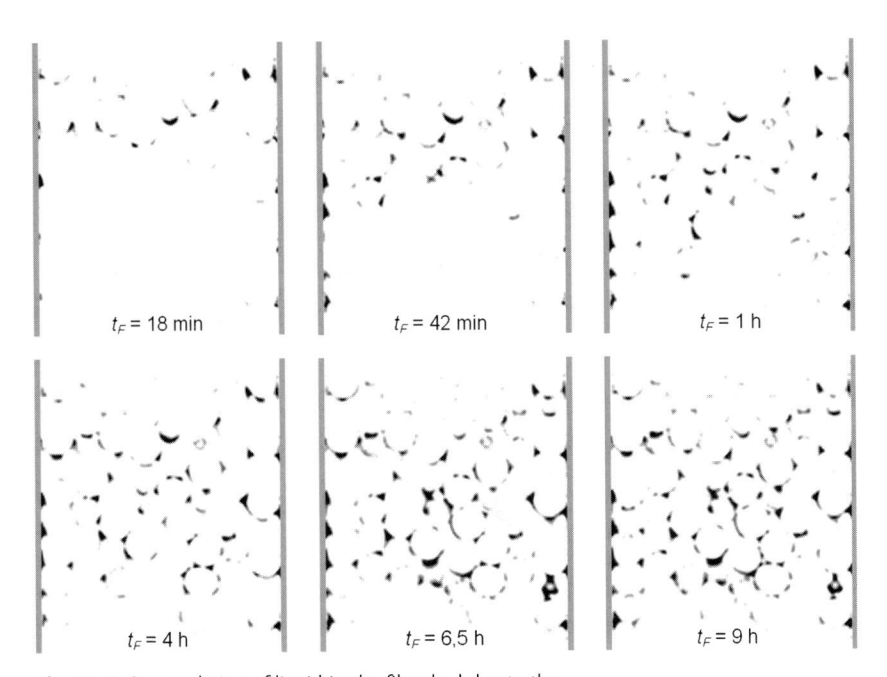

Fig. 3.2.7 Accumulation of liquid in the filter bed due to the separation of the droplets after data processing (d_c = 4 mm, $u_{G,0}$ = 1.0 m s^{-1}, Re = 261).

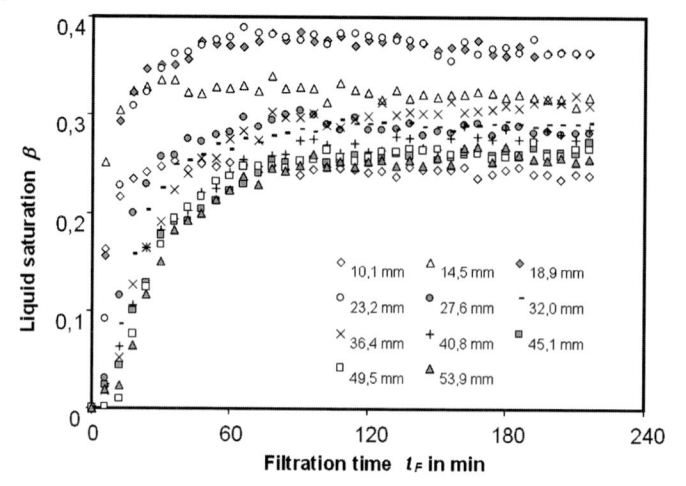

Fig. 3.2.8 Accumulation of liquid in the filter bed due to separation of the droplets after further data processing ($d_c = 4$ mm, $u_{G,0} = 1.0$ m s^{-1}, $Re = 261$).

It was mentioned earlier that the wall of the bed exerts a directional influence on the bed porosity. Numerous investigations have been devoted to establishing a relationship between porosity and distance from the wall. All these were based on integrating the porosity over the bed height resulting in a sole dependency function $\varepsilon(y/d_c)$. MRI offers the possibility to determine both $\varepsilon(y/d_c)$ and $\varepsilon(y/d_c, z)$ where z denotes the actual height within the bed. As an example Figure 3.2.9 shows the integral function $\varepsilon(y/d_c)$ and Figure 3.2.10 shows the relationship for a specific

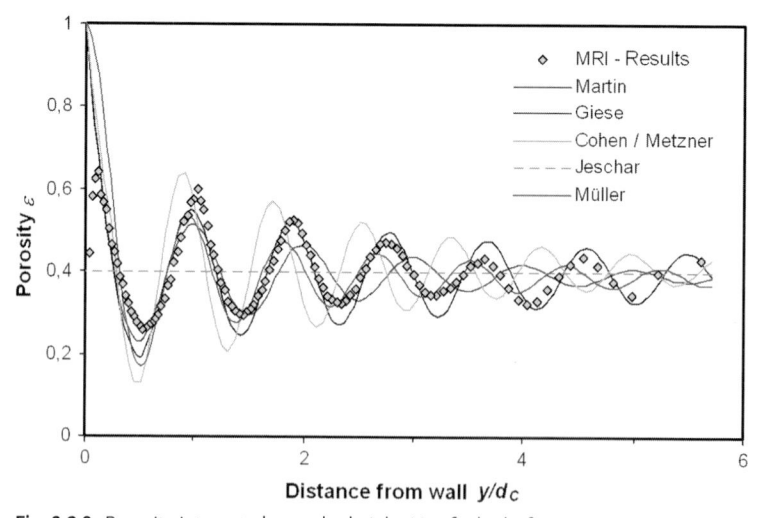

Fig. 3.2.9 Porosity integrated over the height H_B of a bed of glass beads ($d_c = 5$ mm, $D_B/d_c = 11.4$); measured for 180 slices, calculated with various correlations. [8]

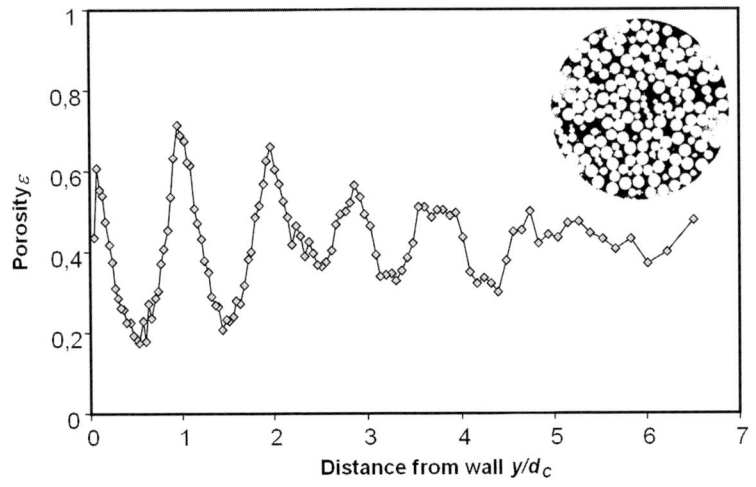

Fig.3.2.10 Local porosity in one slice z (d_c = 4 mm, D_B/d_c = 14.25).

height within the bed. As expected, the integration over the bed's height smoothes the curves. However, Figure 3.2.10 shows that locally the porosity might be quite different to the averaged value.

The averaged function $\varepsilon(y/d_c)$ can be well represented by a relationship established by Martin [7] consisting of a parabolic and an attenuated cosine part. The parameters for Eqs. (3.2.14) and (3.2.15) given by Martin were newly determined based on MRI results of the type shown in Figure 3.2.9. The differing curves shown in Figure 3.2.9 can be taken as proof of the difficulty of measuring ε with conventional means. Data

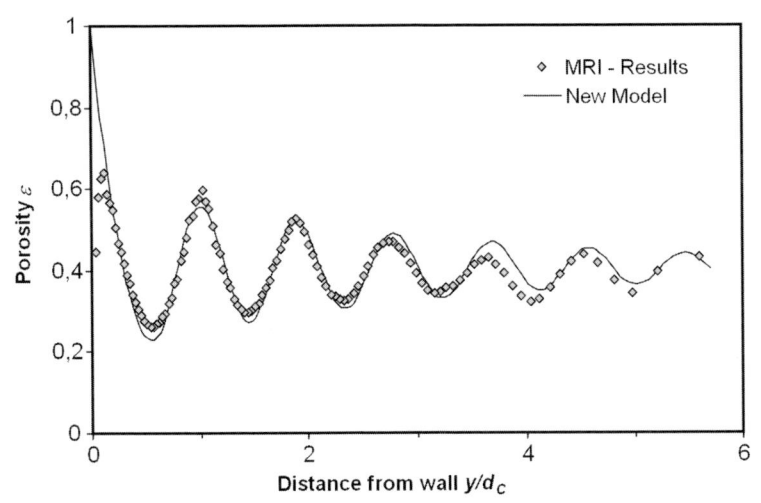

Fig. 3.2.11 Porosity integrated over the height H_B of a bed of glass beads (d_c = 5 mm, D_B/d_c = 11.4); measured for 180 slices, calculated with Eqs. (3.2.14) and (3.2.15).

relating to the curves in Figure 3.2.9 are given elsewhere [8]. Figure 3.2.11 shows that Eqs. (3.2.14) and (3.2.15) fit the measurements quite well when using the data set for the parameters A–D and ε, as cited below. However, to draw Figure 3.2.8, β [from Eq. (3.2.12)] was calculated with local values of ε making use of the unique possibility MRI offers to measure these parameters.

$$\varepsilon(y) = \varepsilon_{min} + (1 - \varepsilon_{min}) \left(\frac{y}{0.55\, d_c} - 1 \right)^2 \text{ for } \frac{y_c}{d_c} \leq 0.55 \quad (3.2.14)$$

$$\varepsilon(y) = \varepsilon_{\infty} + A\, exp\left(B\, \frac{y}{d_c}\right) \cos\left[2\pi \left(C\, \frac{y}{d_c} + D\right)\right] \text{ for } \frac{y_c}{d_c} > 0.55 \quad (3.2.15)$$

$A = 0.26$, $B = -0.32$, $C = 1.12$, $D = -0.14$, $\varepsilon_{min} = 0.23$ and ε_{∞} is from Eq. (3.2.11).

3.2.3.4 Modeling and Simulation

As mentioned at the beginning of Section 3.2.3 the separation process can be modeled by mass balances. Two mass balances have to be made as the liquid appears in two phases: dispersed in the gas phase (the droplets) and continuous in the accumulated state. Figure 3.2.12 describes the process and the conditions in addition to what was explained at the very beginning in Section 3.2.1.

Making use of some common simplifications and introducing the new properties described below, the mass balances can be written as:

$$\frac{\partial c_{L,G}}{\partial t} = -\nabla (c_{L,G}\, u_G) - q_{sep} \quad (3.2.16)$$

$$\frac{\partial \beta}{\partial t} = -\nabla (\beta\, u_L) + \frac{q_{sep}}{\varrho_L} (1 - \beta) \quad (3.2.17)$$

where

$$q_{sep} = \frac{Q_{sep}}{V_G} \quad (3.2.18)$$

is the volume specific separation per time.

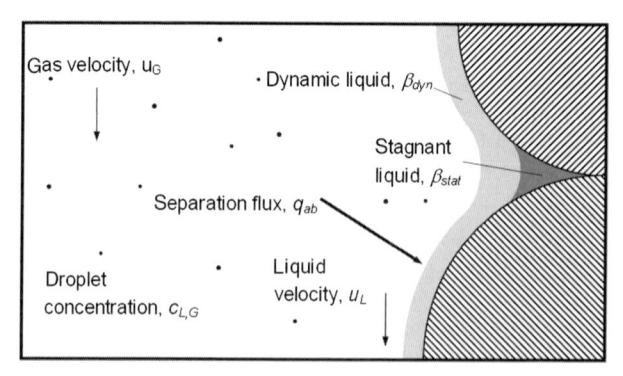

Fig. 3.2.12 Distribution and behavior of the liquid phases during in-bed filtration: the basis of modeling.

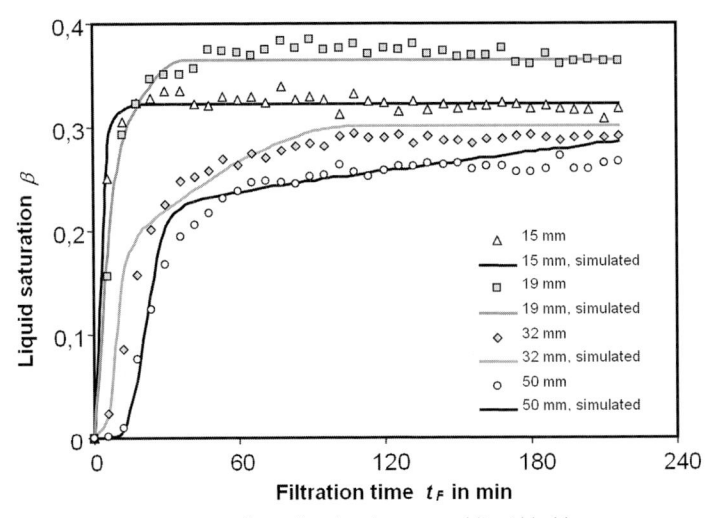

Fig. 3.2.13 Comparison of simulated and measured liquid hold up ($d_c = 4$ mm, $u_{G,0} = 1.0$ m s^{-1}, $Re = 261$), for 4 out of 11 slices.

$$u_G = \frac{u_{G,0}}{\varepsilon(1-\beta)} \qquad (3.2.19)$$

For simulation purposes, $c_{L,G}$, q_{sep} and hence β were replaced by fractional values, being functions of x. The procedure and the deduction of some of the equations mentioned have been described elsewhere [8]. Additionally, correlations and considerations were taken from the literature [9] for calculating the velocity u_L of the

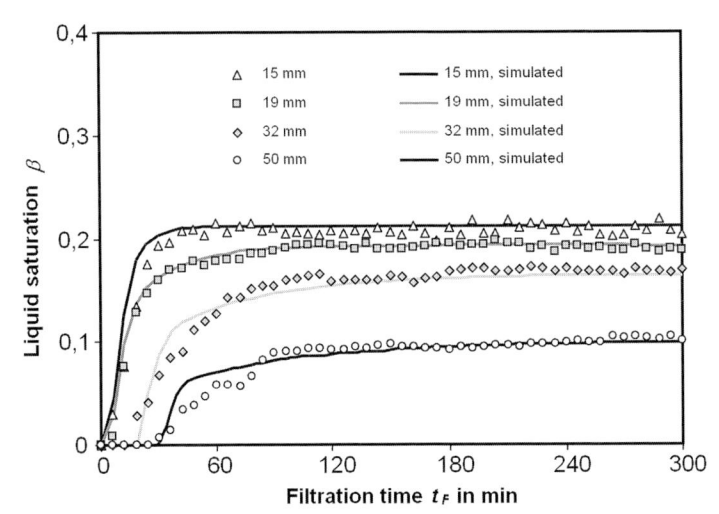

Fig. 3.2.14 Comparison of simulated and measured liquid hold up ($d_c = 4$ mm, $u_{G,0} = 2.5$ m s^{-1}, $Re = 1303$), for 4 out of 11 slices.

liquid film and an assumption was made about the distribution of accumulated liquid into a part β_{dyn}, moving with u_L, and a stationary part β_{stat}.

Figures 3.2.13 and 3.2.14 show sufficiently good agreement between MRI measurements and simulation results for various heights in the bed and for two quite different Re numbers. The higher values of β achieved for the lower Re number can be explained by a lower drag force exerted on the liquid film due to a lower gas velocity u_G. In addition, smaller diameters d_c of the collector bodies result in more interstitial pockets being formed per volume, giving more space for the liquid to accumulate therein.

3.2.4
Summary

To date, in-bed filtration was more or less a "black box" as any measurement within the bed was virtually impossible. The design of such filters was based on models that could be validated only by integral measurements. However, with the MRI method even the (slow) dynamics of the filtration process can be determined. The binary gated data obtained by standard MRI methods are sufficient for the quantitative description of the system. With spatially resolved measurements the applicability of basic mass balances based on improved models can be shown in detail.

In the course of the investigation detailed porosity measurements helped to improve the description of the influence of the wall on the porosity.

References

1 F. Löffler **2002**, (Dust separation), in *Ullmann's Encyclopedia of Industrial Chemistry*, (vol. 11), 6th edn, Wiley VCH, Weinheim.

2 M. Schier, R. Reimert **1999**, (Investigation in droplet separation in granular beds in the inertia dominated regime), in *High Temperature Gas Cleaning*, (vol. II), eds. A. Dittler, G. Hemmer, G. Kasper, Institut für Mechanische Verfahrenstechnik und Mechanik, Universität Karlsruhe (TH), Karlsruhe, 813–824.

3 G. D'Ottavio, S. Goren **1983**, (Aerosol capture in granular beds in the impaction dominated regime), *Aerosol Sci. Technol.* 2 (2), 91–108.

4 M. Schier, R. Reimert **2001**, (Tropfenabscheidung in Schüttschichtfiltern), *Chem.-Ing.-Tech.* 73 (8), 992–997.

5 P. Atkins **2001**, *Atkins' Physical Chemistry*, Oxford University Press, Oxford.

6 R. Jeschar **1964**, (Druckverlust in Mehrkornschüttungen aus Kugeln), *Arch. Eisenhüttenwesen* 35 (2), 91–108.

7 H. Martin **1978**, (Low Peclet number particle-to-fluid heat and mass transfer in packed beds), *Chem. Eng. Sci.* 33, 913–919.

8 V. van Buren **2004**, (Modellierung der Tropfenabscheidung im Anfahrbereich von Schüttschichtfiltern auf Basis von MRI-Untersuchungen), *Ph.D. Thesis*, Universität Karlsruhe, Aachen.

9 H. Brauer **1971**, *Grundlagen der Einphasen- und Mehrphasenströmungen*, Verlag Sauerländer, Aarau.

3.3
Multiscale Approach to Catalyst Design
Xiaohong Ren, Siegfried Stapf, and Bernhard Blümich

3.3.1
Introduction

Heterogeneous catalytic reactions are involved in many chemical processes ranging from refining to fine and specialty chemistry [1]. The rate constants of these reactions, and therefore the efficiency of the reactor, depend on: (a) the local environment of the catalyst surface including concentration and distribution of reagents and of possible deactivating substances in the vicinity of the active sites, (b) the rate of molecular transport as influenced by the topology and pore space of nanoporous cavities and (c) the concentration and temperature profiles as well as flow regimes as a function of the catalyst particle geometry on a macroscopic level and of the structure of the reactors. These various properties are not independent with respect to their effect on reactor performance and must be considered in an integrated approach. Therefore, the optimization of the catalyst ought to be conducted in conjunction with the reactor design, and is characterized according to the three levels mentioned. For the targeted development and optimization of the reaction process, leading to the production of any type of material, one requires powerful testing methods which are, in the best case, sufficiently flexible to quantify all the desired information separately, or even simultaneously. Among the methods that can be used to characterize catalyst materials and the relevant reactors, NMR is perhaps the ultimate technique because it provides a rich toolbox for the investigation of properties on all length scales of interest while remaining strictly non-invasive.

This chapter describes how NMR techniques can be used to quantify structural and dynamic properties on length scales from *nanometers* to *centimeters* for real, industrially relevant catalysts, and to derive the flow behavior on the macroscopic scale. We will focus on catalysts subject to deactivation caused by the formation of carbonaceous deposits, which is one of the major technological and economic problems in the chemical industry. Deactivation is a consequence of coke deposits being strongly adsorbed or bound to the surface, thereby blocking the active sites, the result being a decrease in the conversion efficiency from the reactants to the products of interest. At the same time and with the same consequences, coke deposits reduce the available pore space and hinder molecular diffusion of reactants and products. By burning off the coke with oxygen–nitrogen mixtures the catalyst is regenerated. Therefore, for an optimal catalyst design, the interest lies not only in achieving a high catalytic activity and selectivity, but also in the possibility of regenerating them several times so that their total "lifetime" is compatible with the cost of their production. Moreover, on a macroscopic level, these types of solid catalyzed processes are often carried out in fixed bed reactors, where an even distribution of the streamline is desired, so that the fluid can reach all pellets with similar probability and velocity, and remove the product from the reactor without a localized increase in concentration. An understanding of bypass effects and unbalanced feeding of the pellets is also required in order to avoid excessive heat generation by the usually exothermic reactions.

Some examples for NMR techniques that have been used for investigating catalytic reactors (most of which will be discussed in this chapter) and the corresponding length scales which they probe are as follows:

- 10^{-10}–10^{-9} m: *solid state* ^{1}H *and* ^{13}C *spectroscopy* allows the thickness of the coke deposit layer and its chemical composition to be determined (not discussed in this chapter, see e.g., Refs. [2, 3]);
- 10^{-10}–10^{-9} m: ^{129}Xe *spectroscopy* probes the pore size and the surface composition at the same time;
- 10^{-9}–10^{-6} m: *relaxometry* is sensitive to molecular reorientations due to rotation and translation and is affected by the pore surface curvature up to 1 μm;
- 10^{-9}–10^{-7} m: *cryoporometry* is an indirect method which exploits the relationship between freezing point and pore size via the NMR signal of the liquid;
- 10^{-6}–10^{-4} m: *diffusometry* determines the tortuosity of the pore space on scales of the molecular diffusion length during the encoding time (usually <1 s);
- 10^{-5}–10^{-1} m: *NMR imaging* provides information about packing density and order up to the size of the reactor itself;
- 10^{-5}–10^{-1} m: *velocity measurements and propagators* are extensions of diffusometry and imaging, and are able to quantify displacements from micrometers up to the size of the reactor itself.

From an economic point of view, it is important to stress that the majority of these experiments can be performed on a single standard NMR imaging spectrometer and even – at the cost of somewhat increased experimental times – employing the same probe head and hardware. In this study, most measurements have been carried out using standard Bruker Micro2.5 microimaging equipment with a 7 T (300 MHz proton resonance frequency) super wide bore magnet driven by a Bruker DMX 300 console. The exceptions are the relaxometry analyses, which were performed on a Stelar Fast Field Cycling Relaxometer (Stelar s.r.l., Mede, Italy).

The materials studied in our work are commercial, metal-containing catalyst pellets with the properties given in Table 3.2.1. All investigations on the nanometer scale, i.e., those which involved xenon spectroscopy, relaxometry, cryoporometry and diffusometry, were carried out on E-802 (Pt/Re doped, type "C") catalysts at different stages of coking and regeneration. In the coking process, coke was deposited in the catalyst by passing toluene (as a typical hydrocarbon) over a fixed bed at a total pressure of 6–10 bar (1 bar = 10^5 Pa) and a constant temperature of 540 °C. Decoking (regeneration) cycles of the catalyst were realized in a laboratory-built stainless-steel tubular flow reactor with a nitrogen–oxygen mixture with an oxygen content of 1–2%; the total pressure was 2 bar, and the temperatures were up to 550 °C. Samples of the same batch were removed from this reactor at regular intervals and up to three coking–regeneration cycles were performed. For a more detailed description of the process, see Ref. [2]. Imaging and velocity measurements were carried out with "clean" samples and distilled water as the flowing fluid, and were compared with beds of other catalyst pellets and glass beads of different diameters, all beds being generated by allowing pellets to sediment in an initially water-filled glass tube of 19 mm inner diameter.

Tab. 3.3.1 Physical properties of the packed porous particles. The relaxation times were determined at a Larmor frequency of 300 MHz for protons of water adsorbed into saturated catalyst pellets (average error 2%). The equivalent diameter is defined by 6 V_p/A_p where V_p and A_p are volume and external surface of the particles, respectively.

Material	Pt/Ni catalyst: G-43	Pt/Re catalyst: E-802
Manufacturer	Südchemie, Germany	Engelhard, USA
Shape	Cylindrical (type B)	Cylindrical (type C)
Pellet size	Ø 4.5 mm × 4.5 mm	Ø 1.5–2 mm × 3–8 mm
Equivalent diameter (mm)	4.5	2.2 (average)
T_1 (ms)	9.8	290
T_2 (ms)	4.4	13.0

3.3.2
^{129}Xe Spectroscopy

In the past two decades, ^{129}Xe NMR has been employed as a useful technique for the characterization of the internal void space of nanoporous materials. In particular, the xenon chemical shift has been demonstrated to be very sensitive to the local environment of the nuclei and to depend strongly on the pore size and also on the pressure [4–6]. Assuming a macroscopic inhomogeneity resulting from a distribution of adsorption site concentrations, ^{129}Xe NMR spectra of xenon in zeolites have been calculated, and properties such as line widths, shapes as well as their dependence on xenon pressure can be reproduced qualitatively. A fully quantitative analysis, however, remains difficult due to the different contributions to the xenon line shift. (See Chapter 5.3 for a more detailed description of Xe spectroscopy for the characterization of porous media.)

Catalyst pellets at different stages of the coking and regeneration process were subjected to thermally polarized ^{129}Xe gas at ambient temperature and a pressure of 20 bar inside a sealed sapphire cell of 10 mm outer diameter. Typically 4000 signals with a repetition time of 10 s were accumulated for each spectrum. Figure 3.3.1(a) shows the change in the Xe chemical shift during the regeneration process for the type C samples, where the numbers on the graphs indicate the percentage of coke removed relative to the maximum content of 16 wt-%. The average chemical shift decreases with decreasing coke content, and the linewidth is considerably broader if large amounts of coke are present. The broadening of the Xe line with coke content can, in principle, be an indication of the broader pore size distribution generated by the random deposition of coke in an otherwise fairly regular pore space. More likely, however, it can also be affected by a variation of the interaction strength of xenon with the available surface [7, 8]. This surface composition varies between pure Al_2O_3 to mostly or entirely coked, although the coke distribution on the surface is not known *a priori*. Figure 3.3.1(b) compares the dependence of chemical shift on the coke content at different stages of the deactivation–regeneration cycle. A correlation can clearly be identified but a hysteresis effect is observed, indicating that either pore space morphology or surface composition are affected by the cycle. For comparison, data obtained with a metal-free carrier material are shown. Owing to the absence of catalytically active sites, coking in this sample took considerably longer. A different coke composition could be responsible, and a generally slightly larger chemical shift is found while the hysteresis effect is much smaller than for the metal containing pellets, which hints at a more homogeneous coke distribution in the pore space.

As is shown in Figure 3.3.1(c) for the original catalyst pellets, a correlation between the Xe chemical shift and pore size can indeed be found when comparing the average ^{129}Xe chemical shift with pore sizes determined by BET (Brunauer–Emmet–Teller) adsorption. The systematic dependence is found despite the different accessibilities of the smallest pores by the N_2 molecule and the Xe atom. Although an influence of the surface coke content on the line shift certainly cannot be ruled out, the correlation between pore size and Xe chemical shift can still be used for characterization of the pore space, potentially also in combination with ^{129}Xe imaging employing hyperpolarized xenon.

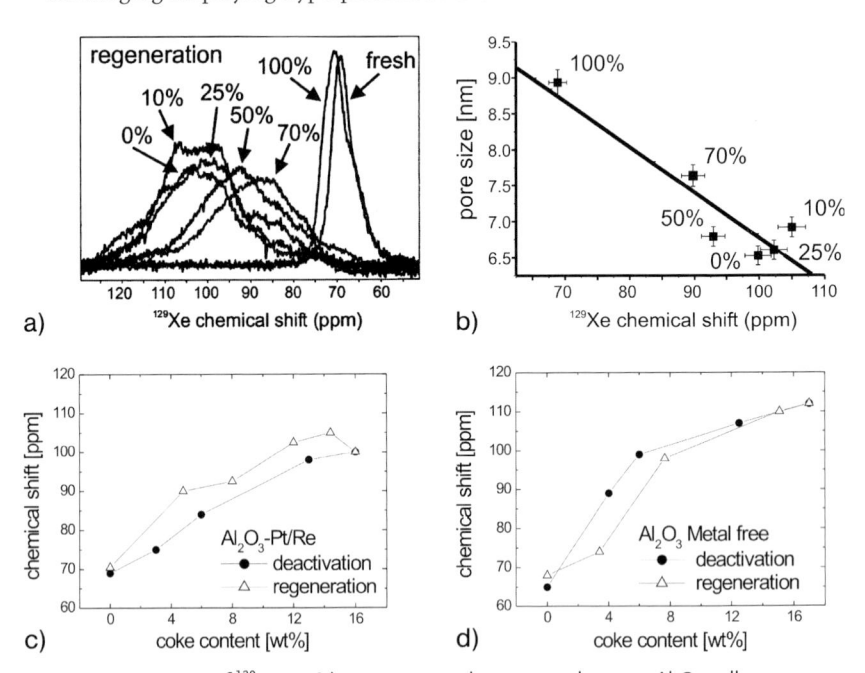

Fig. 3.3.1 (a) Spectra of ^{129}Xe at 20 bar in catalyst samples of various degrees of regeneration. (b) Correlation between ^{129}Xe chemical shift and coke content in catalyst samples compared with identical, but metal-free coked and regenerated porous Al_2O_3 pellets. (c) Average ^{129}Xe chemical shift for the catalyst samples com-pared with pore sizes obtained from BET measurements.

3.3.3
NMR Relaxometry

The term "relaxometry" is sometimes used in a general way describing all methods which determine relaxation times such as T_1 and T_2. Indeed, these relaxation times are strongly correlated to properties of the pore space and the confined fluid (see, e.g., Chapter 3.6). The relaxation times T_1 and, to a lesser degree also T_2, however, depend on the magnetic field strength, and the determination of this dependence – "relaxometry" in the stricter sense – provides additional information about surface roughness and adsorption characteristics of the fluid. From a more practical point of view, it allows one to choose an optimum field strength where T_1 becomes most sensitive for probing the desired material properties.

It has been shown that the relaxation dispersion of adsorbate molecules follows a characteristic behavior depending on the strength of the interaction between the molecule and the surface [9]. Essentially, the relaxation rate $T_1(\omega)^{-1}$ is proportional to the Fourier transform of the autocorrelation function, $G(\tau)$, of molecular reorientations in the time interval τ. It is the preferential orientation of a molecule at the surface that determines the evolution of $G(\tau)$. For weakly interacting molecules, such a preference might not exist, so that $G(\tau)$ is dominated by its initial decay due to molecular rotation (typically of the order 10^{-12} s), and contributions at low frequencies are small. Polar molecules, however, often experience a stronger effect because they are more likely to assume preferential orientations at a surface. The presence of surface-bound hydroxyl groups can increase this effect if hydrogen bonds become relevant. The distance traveled by a molecule during the reciprocal of the Larmor frequency determines the largest structure that can still influence the relaxation dispersion; for water at 10 kHz this amounts to several 100 nm. Non-Fickian diffusion phenomena have been observed under these circumstances, and results for liquids on surfaces of glass, TiO_2 and ZnO of different pore sizes have been reported [10]. In many cases, the frequency dependence of T_1 can be written as a power-law over a certain range of frequencies, $T_1(\nu) \propto \nu^\gamma$. The strongest frequency dependence is usually found at magnetic field strengths well below the typical laboratory fields of several Tesla, a range which is of particular interest to bench-top and mobile NMR scanners. In order to cover a frequency window as wide as possible, dedicated resistive magnets that are switched between polarization, relaxation and detection fields in intervals of a few ms are utilized to overcome the disadvantage of a low signal intensity at low fields; ^1H Larmor frequencies between a few kHz and several tens of MHz can be probed by these devices [11, 12].

In this study, NMR relaxometry has been applied to obtain information about the change in surface roughness and adsorbate properties for samples of catalyst type C for various levels of coke content. Figure 3.3.2 compares the change of $T_1(\nu)$ for n-heptane and water in deactivated and regenerated naphtha reforming Pt/Re–Al_2O_3 catalyst samples. For n-heptane, the frequency dependence is very weak for the fresh catalyst, indicating weak interactions with the surface. For water, however, the interaction is much stronger and a strong dispersion is visible for the fresh

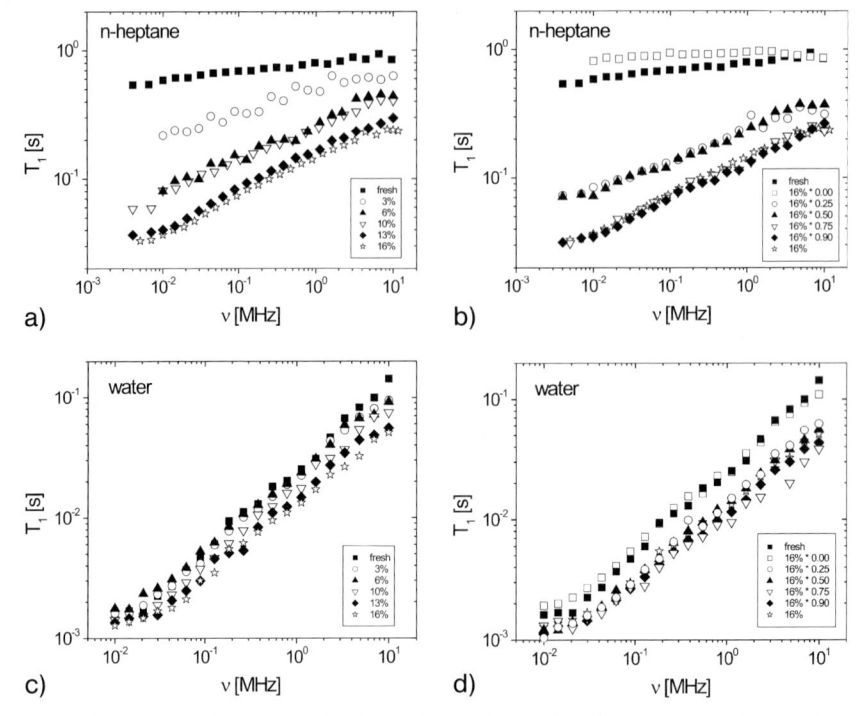

Fig. 3.3.2 Relaxation dispersion $T_1(\nu)$ for (a, b) n-heptane and (c, d) water at room temperature in catalyst pellets at various stages of coking and regeneration. Numbers indicate weight-percentages of coke (a, c) and residual coke content during regeneration (b, d).

catalyst ($T_1 \propto \nu^{0.7}$), a consequence of the preferential orientation of the water molecule relative to the surface.

Increasing the coke content leads to an increase in the frequency dependence for water and a reduction of T_1 at all frequencies. The same grouping effect is found during regeneration, the slope first becomes slightly smaller and then increases again. This trend follows exactly the change of the surface fractal dimension obtained from BET [2, 3] and is a consequence of the Lévy-walk type diffusion on a rugged surface [13], the roughness of which is evolving in the presence of coke deposits. The reduction of the absolute values of T_1 is only gradual but seems to follow the coke concentration. This weak dependence further corroborates the concept that the actual nature of the interaction is not a critical parameter and that changes in the pore space topology determine the relaxation behavior. However, the higher T_1 values for the fully regenerated sample again indicate a smaller interaction, which becomes obvious only in the last step of coke removal.

Contrary to the findings for water, the relative change of the absolute values of T_1 and of its frequency dependence is much more pronounced for n-heptane, which possesses a high affinity for the surface coke. One of the steps in the dispersion function can be interpreted as being related to the onset of full coverage of the surface. Because the maximum coke content corresponds to an average layer thickness of 0.5 nm – a value comparable to a monolayer coverage – this assumption is not unreasonable [14]. The results for different adsorbates give an indication of the high flexibility of the technique; while water relaxation senses the surface roughness, n-heptane can – after proper calibration – help in the understanding of the nanoscopic heterogeneity of coke distribution on the inner pore surface. For instance, it could be shown by relaxometry that the coke properties of metal-free pellets undergoing the same deactivation–regeneration cycles are considerably different from the catalytically grown coke [15].

3.3.4
Cryoporometry

NMR cryoporometry relies on the melting point depression, i.e., the difference in the melting point of crystals with a finite size d, $T_m(d)$, relative to the value of the bulk liquid T_m, which is given by the simplified Gibbs–Thomson equation [16]:

$$\Delta T_m = T_m - T_m(d) = k/d \tag{3.3.1}$$

Separation between solid and liquid fractions is achieved by taking advantage of the very short transverse relaxation time of the crystalline phase which can be filtered in an NMR echo experiment. By approximating the actual crystal size d by the pore diameter, the determination of pore sizes is possible provided the constant k is known; k is specific to the adsorbate, and one fluid with a particularly large value ($k = 182.5$ K nm) is cyclohexane. It has the further advantage of assuming a plastically crystalline state below the melting point, which reduces the risk of damaging the surrounding pore space [17]. This and its convenient bulk melting point of 279 K have made it a suitable choice for determining pore size distributions by monitoring its melting behavior with NMR, but also other substances have been applied to NMR cryoporometry [18, 19], in particular water [20, 21]. In principle, the method is capable of determining pore sizes in the range from approximately 1 nm to several 100 nm. It is obviously limited by the precision of the temperature measurements, but also by the minimum size below which crystallization does not occur (about 4 nm for cyclohexane [16]) and by the fact that the nanocrystal will usually be surrounded by a liquid-like transition zone near the pore wall [22]. Its behavior is indicative of the surface properties and it can also be investigated by NMR, an advantage over DSC techniques which are also used for exploiting the Gibbs–Thomson relationship but, unlike NMR, cannot be combined with imaging to reveal macroscopically heterogeneous pore size distributions [23].

For the experiments in type C catalysts, the pellets were overfilled with cyclohexane and initially cooled to 230 K. They were then reheated in steps of 1 K and allowed to equilibrate for 10 min before each measurement. The signal was determined from 32 accumulations with an echo sequence of 20 ms echo time to ensure that the signal from the plastically crystalline phase of cyclohexane had decayed fully. The typical heating curves of cyclohexane in the fresh and coked catalyst are displayed in Figure 3.3.3(a) As the temperature is increased, larger and larger crystals melt and contribute to the signal until at the bulk melting point (T_m = 279 K), all cyclohexane – including the excess outside of the pellets – is in the liquid state. Differentiating the total curve thus allows the reconstruction of the pore size distribution of the sample [17], which is presented in Figure 3.3.3(b). Fresh and coked catalyst pellets show a small but distinct difference in their pore size distribution (PSD) which is also observed in the BET (N_2 adsorption) measurements. The onset of melting occurs at lower temperatures for the coked sample, indicating a higher fraction of particularly small pores in the 5 nm range. Despite the different accessibilities of the pores by N_2 and the cyclohexane molecules, the results from both methods agree reasonably well. This finding supports the notion that the width of the PSD does not change significantly from the fresh to the fully coked catalyst, and that the apparent broadening found in the xenon spectral data is probably due to effects of interactions with the surface. A combined investigation of porous substances by spectroscopy, relaxometry and cryoporometry thus allows a separation of surface induced and pore size induced effects.

3.3.5
Diffusometry

The determination of self-diffusion coefficients as the simplest application of pulsed field gradient (PFG) NMR is a particularly suitable means of characterizing the pore space of catalyst pellets at different stages of coking and regeneration. As mentioned in the above section, a shrinkage of the average pore size of the type C catalysts from 9 to about 6.5 nm at maximum coking (16 wt-%) was found. This is partly explained by the equivalent coke surface layer thickness of 0.5 nm. However, while the pores are assumed to be cylindrical in shape, their exact geometry is not known. Small pores might be entirely blocked by coke plugs, thus isolating entire regions of the pore network and reducing the accessibility of the pore space significantly. For an understanding of the reaction process under conditions of partial coking, the geometrical properties that govern the transport of a molecule from outside the pellet to its interior need to be known. Under the conditions investigated in this study, a heterogeneous coke distribution within the pellet is not

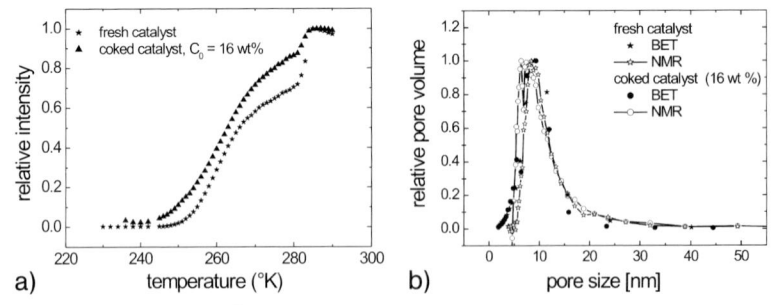

Fig. 3.3.3 (a) Hahn echo ^1H intensity during heating cycle of cyclohexane filling the pores of catalyst pellets. (b) Pore size distribution obtained from (a) in comparison with BET measurements.

expected (but see Section 3.3.7 for a different example), so that the *average* tortuosity of the pore space in the pellet is the only relevant information.

In order to verify the conditions of this averaging process, one has to relate the displacements during the encoding time – the interval Δ between two gradient pulses, set to typically 250 ms in these experiments – with the characteristic sizes of the system. Even in the bulk state with a diffusion coefficient D_0, the root mean square (rms) displacement $<Z^2>$ of n-heptane or, indeed, any liquid does not exceed several 10^{-5} m (given that $<Z^2> = 2D_0 \Delta$). This is much smaller than the smallest pellet diameter of 1.5 mm, so that intraparticle diffusion determines the measured diffusion coefficient (see Chapter 3.1). This intraparticle diffusion is hindered by the obstacles of the pore structure and is thus reduced relative to D_0; the ratio between the measured and the bulk diffusion coefficient is called the tortuosity τ. More precisely, the tortuosity τ is defined as the ratio of the mean-squared displacements in the bulk and inside the pore space over identical times:

$$\tau = \langle R_0^2(\Delta) \rangle / \langle R_i^2(\Delta) \rangle \tag{3.3.2}$$

This approach describes the porous medium well if the considered times Δ are such that a representative elementary volume of the entire pore space is covered by the moving particle. If this were not the case, a superposition of individual diffusion decays, originating from different regions inside the sample, manifests itself in a non-Gaussian shape of the distribution function of displacements. Because of partial mixing between these regions, the shape of these propagators, and therefore of the signal decay functions $S(q)$, will frequently also depend on time. In this study, propagators were measured for different encoding times Δ, and were found to be Gaussian and time-independent throughout, so that the diffusion coefficient is determined from

$$D = \frac{-1}{\Delta - \delta/3} \frac{\partial \ln E(q, \Delta)}{\partial q^2} \tag{3.3.3}$$

The intraporous self-diffusion coefficient D_i is then also time-independent and one can write $\tau = D_0 / D_i$.

It should be mentioned here that a different definition of the diffusion coefficient is often used in chemical engineering problems, which is more appropriate for the description of reactant or tracer transport. It takes into account the fact that the total fluid contained in a porous substance of porosity ε is reduced by this factor relative to the bulk, so that an effective diffusion coefficient D' of the reactants is defined such that

$$D' = \varepsilon \cdot D_i = \frac{\varepsilon D_0}{\tau} \tag{3.3.4}$$

The tortuosity for pore-filling liquids is ideally a purely geometric factor but can, in principle, depend on the fluid–surface interaction and the molecular size if very small pores are present such as in zeolites (see Chapter 3.1). To obtain a measure for a realistic situation, we have used n-heptane as a typical liquid and have computed τ

from the ratio of the diffusion coefficients in bulk and inside the saturated catalyst pellets at ambient temperature, both determined at $\Delta = 250$ ms using a stimulated echo sequence. The change of τ with increasing amounts of coke deposited on the catalyst type C, and the effect of regeneration, is shown in Figure 3.3.4. It can be seen that coating with coke reduces the connectivity of the pore space, leading to an increase of τ from 2.4 when fresh to 3.2 after 16 wt-% coke had been deposited on its surface. Regeneration, on the other hand, did not lead to an equivalent decrease of the tortuosity. Instead, a hysteresis in the tortuosity was observed: the tortuosity after regeneration was higher than that of the fresh catalyst sample. This is a finding very similar to the one reported above from Xe spectroscopy. The same coking–decoking cycle as discussed above has been performed three times for an identical set of samples and the variation of the tortuosity has been monitored under identical conditions [2]. In all three deactivation–regeneration processes, the change of the tortuosity in each cycle clearly shows the same tendency: The tortuosity increases with the increasing extent of deactivation and decreases with increasing degree of regeneration. After the third cycle, the tortuosity of the catalyst reaches almost that of the fresh catalyst, which means the diffusivity of the reactants or products in the regenerated catalyst is the same as in the fresh catalyst. Therefore, with appropriate care in semi-regenerative units, these bimetallic reforming catalysts can be used and regenerated several times before needing to be replaced [24].

Qualitatively similar results for partially coked samples have also been observed by other researchers. Garcia-Ochoa and Santos [25] studied the effect of coking on mass transport and morphology of Pt/Al_2O_3 and $NiMo/Al_2O_3$ catalysts by tracer pulse injection techniques. They found that when coking produces significant changes in the catalyst morphology, the tortuosity factor almost doubles for a catalyst containing 20 wt-% coke compared with a fresh catalyst. Wood and Gladden [26] confirmed the increase in tortuosity for deactivated hydroprocessing catalysts from a real industrial process using pentane and heptane probe molecules and PFG NMR.

3.3.6
Flow Propagators

While the previously described techniques were measuring the nanoscopic and microscopic properties of the catalyst pellets, respectively, fluid transport within

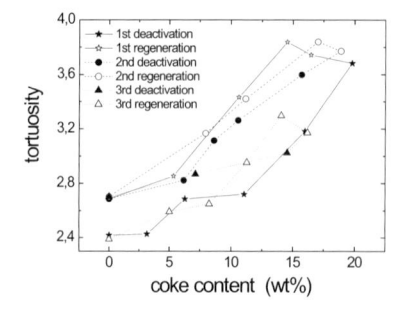

Fig. 3.3.4 Variation of the tortuosity τ inside the catalyst pellets during coking and regeneration, obtained by measuring the self-diffusion coefficient of n-heptane at room temperature.

the reactor itself is another important factor determining the reactor efficiency. A statistical description for the volume-averaged fluid displacement provides valuable information about the way that reactants are transported to, and products are transported from, the reaction sites. This is achieved by the propagator, or displacement probability density. A large number of propagator studies have been carried out on selected technically important systems, e.g., chromatographic columns [27, 28], packed bed reactors [29, 30], transport in geological media [31–34] and also of two-phase flow in reactors [35, 36]. Velocity measurements of gases have even become possible [37–40]. A common topic in the literature is the interpretation of the evolution of the propagator shape as a function of encoding time Δ.

In this study, propagators have been determined for the case of pure water flowing in packed beds of particles in a 19 mm id tube. In order to achieve a compensation for the possible background field gradient effects, the alternating gradient modification [41] of the standard PGSTE sequence was used in which each gradient pulse is split into two halves of duration $\delta/2$, being separated by a 180°-rf pulse. The encoding time Δ is defined as the interval between the additional 180°-pulses, and the echo intensity, normalized to zero gradient, is given as

$$E(\mathbf{q}, \Delta) = S(\mathbf{q}, \Delta)/S(0, \Delta) = \int \bar{P}(R, \Delta) \, exp[i2\pi \mathbf{q} R(\Delta)] dR \qquad (3.3.5)$$

employing the definition of the wave vector $\mathbf{q} = (2\pi)^{-1} \gamma \delta \mathbf{g}$.

The one-dimensional propagators $\bar{P}(Z, \Delta)$ and $\bar{P}(X, \Delta)$ of displacements parallel and perpendicular to the flow axis, respectively, were determined for a range of encoding times Δ and volume flow rates.

Figure 3.3.5(a) shows the evolution of the propagator for water flowing through a random array of cylindrical Pt/Al_2O_3 catalyst pellets of type C. The propagators are qualitatively similar to those obtained with flow among non-porous spherical particles of similar size [Figure 3.3.5(b)]. One observes a decrease in intensity of a peak near zero displacement and the development of a secondary maximum which becomes centered about the average flow displacement. The first peak originates from those molecules residing in quasi-stagnant regions of the bed, i.e., they are not immobile but remain near the surface of a pellet or within the free space between adjoining pellets. The second peak is related to the fluid molecules possessing a significant net displacement because of convection. The relative weight of both peaks at a given time has been described as being affected by the dimensionless Péclet number, $Pe = l<v>/D$, which expresses essentially the ratio of the time that a fluid element requires to diffuse across a unit of characteristic structural size l (e.g., a pore) to the time to cross this distance by flow with the average velocity, $<v>$. Higher Pe values give rise to a more persistent stagnant peak. Even without any knowledge about the actual shape and distribution of the flow streamlines, the observation of a decrease of the stagnant fraction with longer Δ can be interpreted as a mixing with the flowing part and is identified as an important feature in the characteristics of reactor performance where stable stagnant zones are to be avoided. To quantify the stagnant fraction, a decomposition into two Gaussian functions has been suggested to provide a practicable

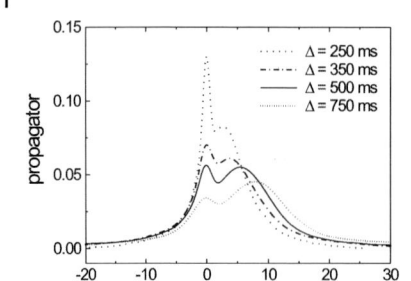

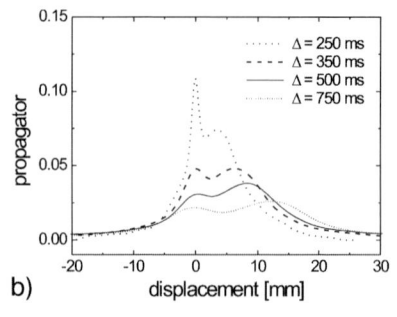

Fig. 3.3.5 Propagators for water flow at 2.93 mm³ s⁻¹ through the fixed-bed reactor: (a) spherical glass beads, pore diameter d_p = 2 mm and (b) cylindrical pellets with average equivalent diameter of 2.2 mm.

means of quantifying the two components [42]. Figure 3.3.6 shows the volume fractions of the stagnant and the flowing part for type C catalyst pellets in comparison with a packing of glass bead of similar particle size. In the catalyst bed, the mixing of water between stagnant pools and the fraction carrying the majority of the fluid transport is found to be similar to the small beads and faster than for beads of a size comparable to the catalyst pellets, suggesting that the cylindrical pellet shape has a positive effect on the dispersion properties of the flow. We note that for the timing chosen in these experiments, the intraporous fluid signal is fully suppressed due to its short transverse relaxation time (see Table 3.3.1), therefore does not give rise to a persistent stagnant peak which would otherwise be expected.

Apart from the separation into two fractions, the propagator can also be interpreted in terms of an average quantity, the second moment of displacements, which is proportional to the dispersion coefficient $D^*(\Delta)$. Rather than computing $D^*(\Delta)$ from the shape of the propagator directly, it is also possible to obtain it from the initial slope of the signal function $E(q,\Delta)$ in a 1D NMR experiment [43]:

$$D*(\Delta) \approx -(4\pi^2 \Delta)^{-1} \lim_{q \to 0} \frac{\partial \ln E(q, \Delta)}{\partial q^2} \quad (3.3.6)$$

The time-dependent dispersion coefficients for water flowing in a bed of catalyst type C are compared with flow in a packing of 2 mm glass beads at different flow

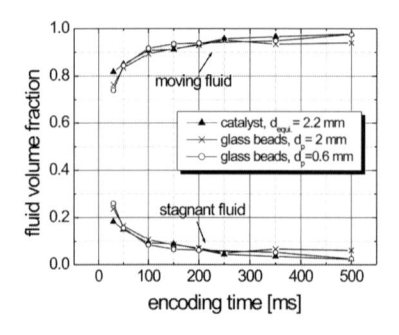

Fig. 3.3.6 Comparison of the relative fractions of stagnant and moving fluid obtained from the propagators for catalyst and glass beads.

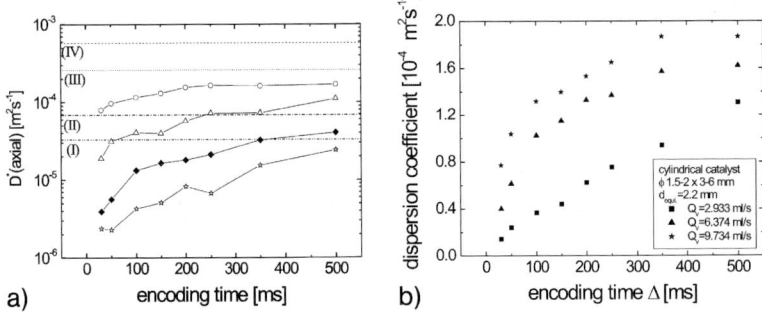

Fig. 3.3.7 Time dependence of the axial dispersion coefficients D^* for water flow determined by NMR; horizontal lines indicate the asymptotic values obtained from classical tracer measurements. (a) Water flow in packings of 2 mm glass beads at different flow rates; and (b) water flow in catalyst.

rates, respectively, in Figure 3.3.7. Owing to the outflow effect which would reduce the measured dispersion coefficient below its real value, measurements at longer Δ could not be performed. Most measurements were thus carried out in the pre-asymptotic regime, as is demonstrated by the horizontal lines in Figure 3.3.7(a), which indicate the long-time limit of D^* obtained with classical tracer techniques [44]. PFG NMR measurements without spatial resolution can thus be regarded as suitable methods for investigating the details of dispersion, reaction and mixing processes on time scales below 1 s, which are not routinely accessible by conventional tracer methods.

3.3.7
NMR Imaging

3.3.7.1 Spin Density Imaging and Velocimetry

The determination of propagators and dispersion coefficients provides an averaged description of transport in a fixed bed, but does not reveal the true nature of the flow process. First of all, conventional NMR imaging serves to determine the packing heterogeneity of the bed itself. For beds of low aspect ratio (defined as d_t/d_p where d_t and d_p are the diameter of the tube and the particles, respectively), radial ordering is observed which results in an oscillating radial fluid density function, as is demonstrated in Figure 3.3.8(a) for a cross section of a 1 mm thick slice through random packing of 2 mm glass beads ($d_t/d_p = 9.5$) in the absence of flow. The radial spin density is shown below the 2D image. Local maxima of fluid density are observed near the tube wall, and ring-like arrangements of the spherical pellets can be followed for up to five bead diameters, but even for the pellets with irregular shape and size distribution, a weaker ordering effect can also be seen [Figure 3.3.8(b)]. The less pronounced minima in the radial function for the catalyst pellets are a consequence of the residual intraporous water signal which is still partly visible in the image.

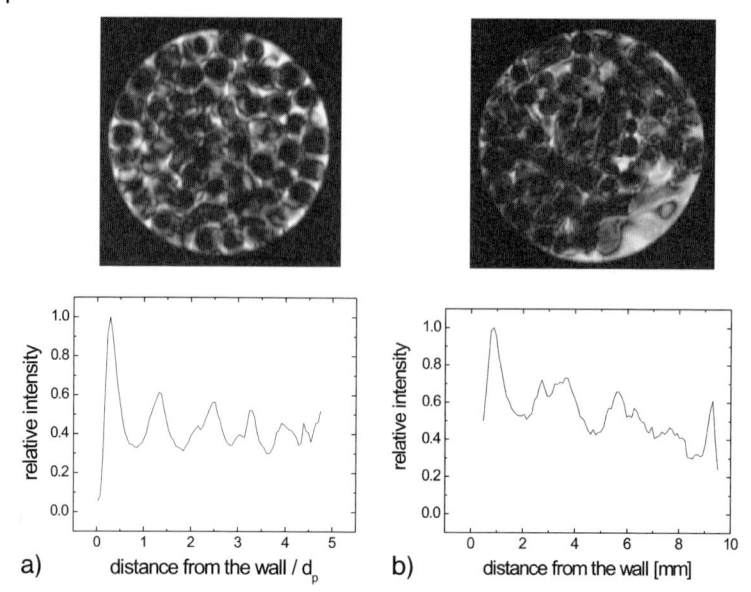

Fig. 3.3.8 Two-dimensional cross section through the fixed-bed reactor (upper) and radial spin-density distribution obtained from the images (lower) for (a) spherical glass beads of 2 mm in diameter and (b) cylindrical pellets with average equivalent diameter of 2.2 mm.

For the same samples, velocity images were obtained by inserting a velocity encoding gradient pair into the imaging sequence [14]. The encoding time was set to $\Delta = 6.9$ ms so that an instantaneous picture of the velocity distribution in the bed was obtained, as this time is insufficient to allow for significant velocity changes of the fluid elements. The results are shown in Figure 3.3.9 [45, 46]. At first sight it becomes obvious that the regions of highest spin density in Figure 3.3.8 do not necessarily correspond to regions of highest flow velocity; on the contrary, high velocities seem to occur in narrow to intermediate channels, while large voids, such as the region at the bottom right of Figure 3.3.9(b), do not show high velocities but might still contribute strongly to the total flow rate due to their size. In general, high velocities occur at distinct spots throughout the cross section. These localized regions represent cuts through a "flow backbone" [47], which have been successfully demonstrated to be deterministic in the sense that a close match between numerical simulations and measured velocities has been found [48, 49]; see also Chapter 2.9. An analysis of the radial distribution of velocities, as shown in the lower part of Figure 3.3.9, indicates that on average, a statistical prediction of the radial dependence of fluid transport can indeed be made. One noticeable fact is that large velocities do occur in the vicinity of the tube wall in agreement with observations of flow through beds of similar aspect ratio [50, 51].

One possibility to visualize fluid transport directly in the fixed bed by means of a series of time-encoded displacement images is given by the spin tagging technique

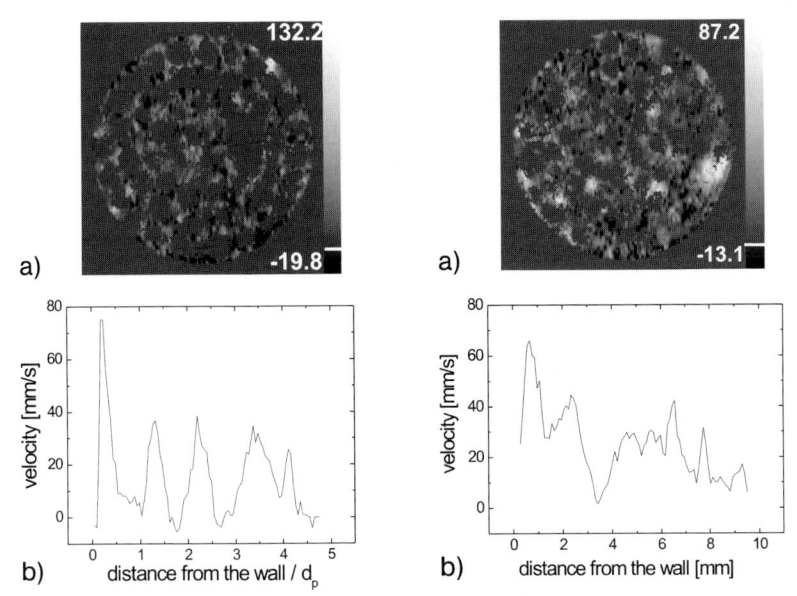

Fig. 3.3.9 Velocity encoded imaging (upper) and radial velocity distribution (lower) for a flow of water at 2.93 mm³ s⁻¹ through the fixed bed reactor, taken in the same slice as for Figure 3.3.8. (a) Spherical glass beads of 2 mm in diameter and (b) cylindrical pellets with average equivalent diameter of 2.2 mm.

[52, 53]. It consists of saturating an axial slice in the sample by applying a selective 90°-rf pulse, followed by an ordinary imaging module after a waiting time τ. Spins from this saturated region do not contribute to the signal in the image; by subtracting each image from a reference frame, a *positive* is formed which highlights the fluid from the excited slice. The flow pattern of water in a bed of 7 mm diameter glass beads is shown in Figure 3.3.10 with an average flow velocity of 19 mm s⁻¹ and evolution times in the range 1 ms ≤ τ ≤ 500 ms. The tagging experiment, demonstrating the deformation of the excited axial slice of 2 mm thickness, reveals the actual flow path of fluid elements around the beads up to displacements of almost 20 mm. It can be seen clearly that the fastest motions occur near the wall. Moreover, the structure of the packing can be identified. The flow conditions remained stationary throughout this period, i.e., no fluctuations or chaotic contributions were identified.

3.3.7.2 Parameter Imaging

The combination of microscopic and macroscopic information is made possible by what can be called "parameter imaging". In the general sense, it consists of the encoding of properties such as spectral line shifts, relaxation times, diffusion coefficients, etc., in the image by suitable combination of corresponding modules into one pulse sequence. Parameter images are to be distinguished from mere

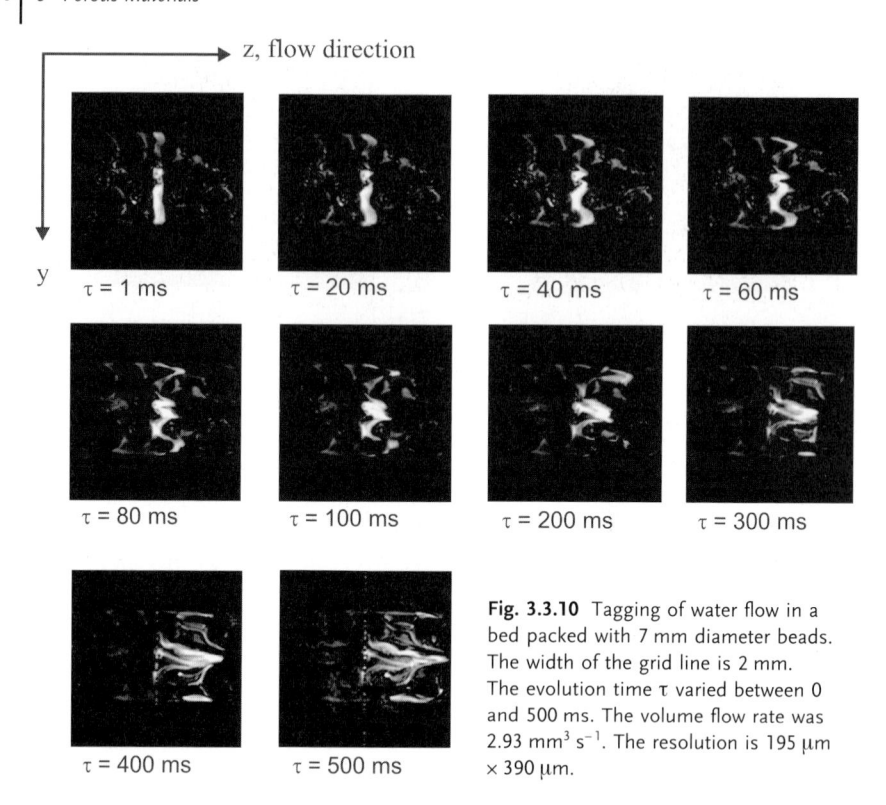

Fig. 3.3.10 Tagging of water flow in a bed packed with 7 mm diameter beads. The width of the grid line is 2 mm. The evolution time τ varied between 0 and 500 ms. The volume flow rate was 2.93 mm^3 s^{-1}. The resolution is 195 μm × 390 μm.

"weighted images", in that they allow a quantitative analysis and a correlation with other properties via separate calibration. Velocity imaging, mentioned above, is actually one application of parameter imaging as it provides both the spin density and the average velocity as information for each pixel. Similarly, the correlations of the coke contents, which have been established by the techniques discussed earlier in this chapter, can be exploited to visualize heterogeneities of the catalyst samples.

As a demonstration, we present the investigation of the coke distribution in a single catalyst particle subjected to different reaction conditions, which is actually of considerable relevance in the industrial environment. Using cylindrical type B pellets with an equivalent diameter of 4.5 mm, the assumption of heterogeneous coke formation and removal during the reaction–regeneration cycle may no longer be justified. Indeed, a dependence of the coke distribution on the regeneration temperature can even be seen optically by cutting the samples, as seen in Figure 3.3.11(a) – the left-hand and right-hand samples have undergone regeneration at 550 and 400 °C, respectively, to assume the same coke content of 7.65 wt-%. However, a quantification of the coke distribution cannot be achieved straightforwardly by optical means.

In Figure 3.3.2, the strong dependence of the ^{1}H relaxation time of n-heptane on coke content was shown for low magnetic field strengths; although less pronounced, this T_1 dependence still holds for high fields [2]. For large catalyst pellets

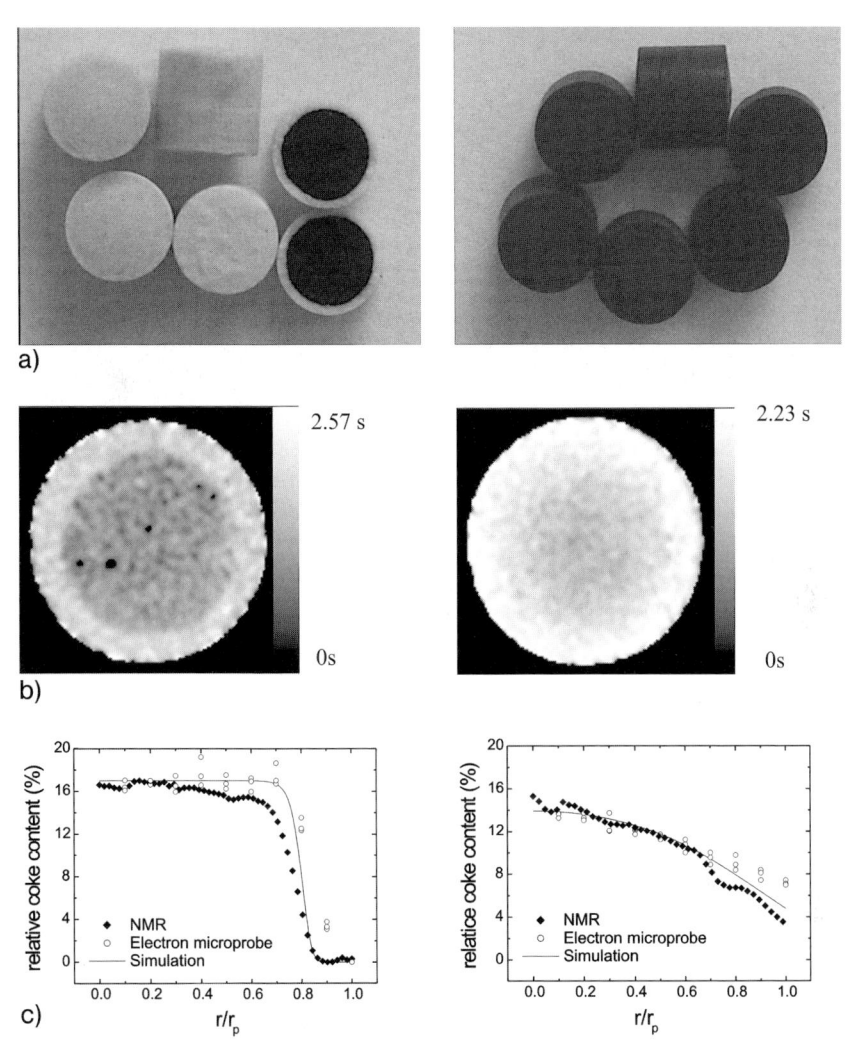

Fig. 3.3.11 Partially regenerated, coked Al$_2$O$_3$ catalyst samples with the same residual coke content of 7.65%: left, regenerated at 550 °C; right, regenerated at 400 °C. (a) Optical photographs of cut samples; (b) NMR images with the relaxation time T_1 as contrast parameter; (c) radial coke profiles determined from (b), compared with electron microprobe measurements and simulated distributions.

(4.5 mm diameter), the spatial distribution of coke becomes important. In particular, during reformation at high temperatures, complete coke removal occurs only at the outer edge of the pellet. Figure 3.3.11(b) shows the T_1 parameter maps obtained from series of images with varying repetition times. It clearly reveals radial heterogeneities in the sample also found from destructive classical techniques, and uneven coke distributions for optically uniform, spent catalyst pellets are identified that are inaccessible by other techniques. After calibration following known relationships, maps of the coke distribution in recycled catalyst grains are obtained in terms of coke concentration. The rotationally averaged coke contents are shown in Figure 3.3.11(c) and are compared with results from electron microscopy and with simulations of the regeneration process. They demonstrate the degree of efficiency of the regeneration reaction at different recycling temperatures: lower temperatures generally lead to a more even removal of coke from the internal surface of the carrier.

An alternative way to map the coke distribution is via the contrast of the tortuosity that was established in Figure 3.3.4. Figure 3.3.12(a) shows images acquired with several different diffusion encoding gradients, and the diffusion coefficient map [Figure 3.3.12(b)] computed from them. Because of the weaker dependence of the self-diffusion coefficient on the coke content, the image quality is not as good as in the relaxation map, but the two regions of low and high coke content (outer ring and central part, respectively) can still be distinguished. The fluctuations found in the image are at least partly due to actual heterogeneities of the coke distribution and also appear in the relaxation maps. In a similar fashion, the other probing techniques for nanoscopic behavior discussed above, ^{129}Xe spectroscopy and cryoporometry, can be combined to generate maps of the coke concentration inside the pellets.

3.3.8
Conclusions and Summary

A number of NMR techniques have been presented that probe crucial factors determining reactor efficiency over a wide range of length scales. These are, among others, the accessibility of the catalytically active metal sites on the pellet surface; the intraparticle tortuosity, which governs molecular transport of reactants and products; and the macroscopic flow profile being responsible for the fluid transport to the pellets themselves. On the microscopic level, cryoporometry is able to determine the pore size distribution directly, while ^{129}Xe spectra are influenced by pore sizes and surface compositions simultaneously. The latter, in conjunction with surface geometry information including the pore structure itself, is accessed by relaxometry investigations, which are made even more flexible by the appropriate choice of adsorbate molecules. Combining all three techniques with different solid-state encoding schemes that address the coke layer directly ultimately leads to an improved picture of the molecular-scale catalytic efficiency. The content, composition and spatial distribution of coke residues, and their respective influence on reactor performance, can be quantified separately. Fluid transport, on the other hand, is measured by PFG methods on scales of micrometers (diffusometry) to centimeters (velocity propagators and flow imaging). It is combined with imaging techniques that allow the degree of heterogeneity of both packing and transport in the bed to be displayed.

One can thus safely state that NMR techniques have become flexible and valuable tools for characterizing industrial porous catalysts in length scales from nanometers to centimeters, and to correlate information on these scales, for

3.3 *Multiscale Approach to Catalyst Design* | 281

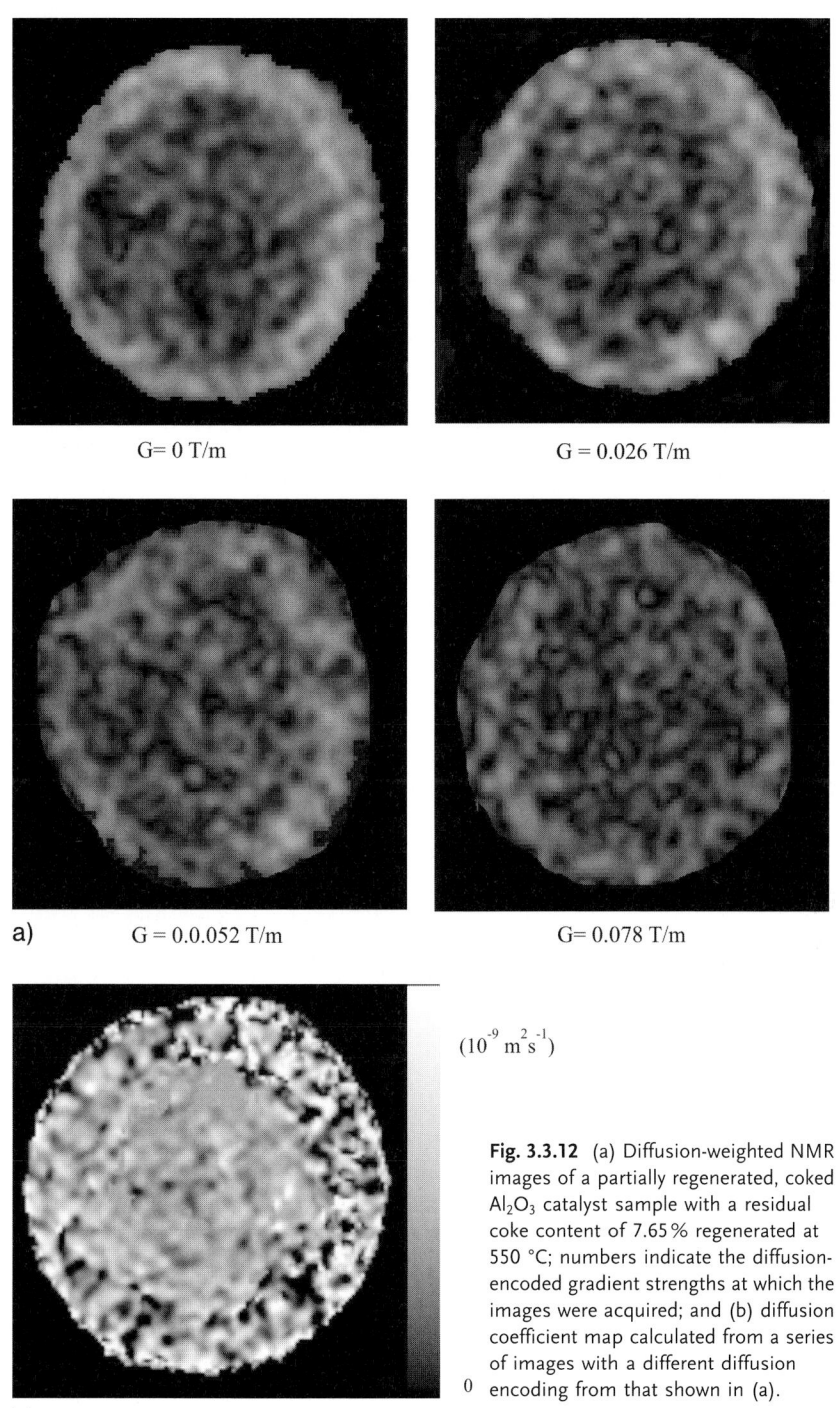

G= 0 T/m

G = 0.026 T/m

a) G = 0.0052 T/m

G= 0.078 T/m

b)

$(10^{-9} m^2 s^{-1})$

Fig. 3.3.12 (a) Diffusion-weighted NMR images of a partially regenerated, coked Al_2O_3 catalyst sample with a residual coke content of 7.65% regenerated at 550 °C; numbers indicate the diffusion-encoded gradient strengths at which the images were acquired; and (b) diffusion coefficient map calculated from a series of images with a different diffusion encoding from that shown in (a).

instance by generating parameter images. Independent calibration measurements are required to establish the correlation between the NMR parameter (such as chemical shift or relaxation times) with the material property of interest (such as coke content or local transport coefficients). The majority of these investigations can be carried out with one single spectrometer employing standard imaging hardware, which is a definite plus and puts the relatively high installation and running costs of the NMR equipment into a more favorable perspective. The next step of integrating all the available tools and of observing the reactor efficiency directly through *in situ* reaction monitoring has already been taken (see Chapters 5.3, 5.4, 5.5).

Acknowledgements

This project was supported by the Deutsche Forschungsgemeinschaft (BL 231/25–1) and by the German Academic Exchange Service (DAAD), which are thus gratefully acknowledged. We thank D.-H. Tang for the measurements with the conventional tracer technique, K. Kupferschläger, M. Adams and G. Schroeder for technical support, and A. Jess and M. Liauw for helpful discussions.

References

1. M. P. Dudukovic, F. Larachi, P. L. Mills **1999**, (Multiphase reactors – revisited), *Chem. Eng. Sci.* 54, 1975.
2. X. H. Ren, M. Bertmer, S. Stapf, D. E. Demco, B. Blümich, C. Kern, A. Jess **2002**, (Deactivation and regeneration of a naphtha reforming catalyst), *Appl. Catal. A: General* 228, 39.
3. X. H. Ren, M. Bertmer, H. Kühn, S. Stapf, D. E. Demco, B. Blümich, C. Kern, A. Jess **2002**, (^{1}H, ^{13}C and ^{129}Xe NMR study of changing pore size and tortuosity during deactivation and decoking of a naphtha reforming catalyst), *NATO Sci. Ser. II:Math., Phys. Chem.* 76, 603.
4. J. Fraissard, T. Ito, **1988** (Xe-129 NMR-study of adsorbed xenon – a new method for studying zeolites and metal-zeolites), *Zeolites* 8, 350.
5. C. I. Ratcliffe **1998**, *Annu. Rep. N. M. R. Spectrosc.* 36, 123.
6. J. L. Bonardet, J. Fraissard, A. Gedeon, M. A. Springuel-Huet **1999**, (Nuclear magnetic resonance of physisorbed Xe-129 used as a probe to investigate porous solids), *Catal. Rev.* 41, 115.
7. C. Tsiao, R. E. Botto **1991**, (Xe-129 investigation of coal micropores), *Energy Fuels* 5, 87.
8. M. A. Springuel-Huet, J. L. Bonardert, A. Gédéon, J. Fraissard **1997**, (Xe-129 NMR for studying surface heterogeneity: well-known facts and new findings), *Langmuir* 13, 1229.
9. R. Kimmich **1997**, *NMR Tomography, Diffusometry, Relaxometry*, Springer-Verlag, Berlin.
10. T. Zavada, R. Kimmich **1998**, (The anomalous adsorbate dynamics at surfaces in porous media studied by nuclear magnetic resonance methods. The orientational structure and Lévy walks), *J. Chem. Phys.* 109, 6929.
11. R. Kimmich, E. Anoardo **2004**, (Field-cycling NMR relaxometry), *Prog. Nucl. Magn. Reson. Spectrosc.* 44, 257.
12. E. Anoardo, G. Galli, G. Ferrante **2001**, (Fast-field-cycling NMR: Applications and instrumentation), *Appl. Magn. Reson.* 20, 365.

13 T. Zavada, R. Kimmich **1999**, (Surface fractal probed by adsorbate spin-lattice relaxation dispersion), *Phys. Rev. E* 59, 5848.

14 X. H. Ren, S. Stapf, H. Kühn, D. E. Demco, B. Blümich **2003**, (Molecular mobility in fixed bed reactors investigated by multiscale NMR techniques), *Magn. Reson. Imag.* 21, 261.

15 S. Stapf, X. Ren, E. Talnishnikh, B. Blümich **2005**, (Spatial distribution of coke residues in porous catalyst pellets analyzed by field-cycling relaxometry and parameter imaging), *Magn. Reson. Imag.* 23, 383.

16 C. L. Jackson, G. B. McKenna **1990**, (The melting behavior of organic material confined in porous solids), *J. Chem. Phys.* 93, 9002.

17 J. H. Strange, M. Rahman, E. G. Smith **1993**, (Characterization of porous solids by NMR), *Phys. Rev. Lett.* 71, 3589.

18 D.W. Aksnes, K. Forland, L. Kimtys **2001**, (Pore size distribution in mesoporous materials as studied by ^{1}H NMR), *Phys. Chem. Chem. Phys.* 3, 3203.

19 J. H. Strange, J. Mitchell, J. B. W. Webber **2003**, (Pore surface exploration by NMR), *Magn. Reson. Imag.* 21, 221.

20 J. B. W. Webber, J. H. Strange, J. C. Dore **2001**, (An evaluation of NMR cryoporometry, density measurement and neutron scattering methods of pore characterization), *Magn. Reson. Imag.* 19, 395.

21 R. M. E. Valckenborg, L. Pel, K. Kopinga **2002**, (Combined NMR cryoporometry and relaxometry), *J. Phys. D – Appl. Phys.* 35, 249.

22 S. Stapf, R. Kimmich **1997**, (Translational mobility in surface induced liquid layers investigated by NMR diffusometry), *Chem. Phys. Lett.* 275, 261.

23 J. H. Strange, J. B. W. Webber **1997**, (Spatially resolved pore size distributions by NMR), *Meas. Sci. Technol.* 8, 555.

24 B. E. Leach **1983**, *Applied Industrial Catalysis*, Academic Press, New York.

25 F. Garcia-Ochoa, A. Santos **1996**, (Coke effect in mass transport and morphology of Pt-Al$_2$O$_3$ and Ni-Mo-Al$_2$O$_3$ catalysts), *AIChE J.* 42, 524.

26 J. Wood, L. F. Gladden **2003**, (Effect of coke deposition upon pore structure and self-diffusion in deactivated industrial hydroprocessing catalysts), *Appl.Cat. A: General*, 249, 241.

27 U. Tallarek, E. Bayer, G. Guiochon **1998**, (Study of dispersion in packed chromatographic columns by pulsed field gradient nuclear magnetic resonance), *J. Am. Chem. Soc.* 120, 1494.

28 J. Park, S. J. Gibbs **1999**, (Mapping flow and dispersion in a packed column by MRI), *AIChE J.* 45, 655.

29 L. F. Gladden **1999**, (Applications of in situ magnetic resonance techniques in chemical reaction engineering), *Top. Catal.* 8, 87.

30 M. D. Mantle, A. J. Sederman, L. F. Gladden **2001**, (Single- and two-phase flow in fixed-bed reactors: MRI flow visualisation and lattice-Boltzmann simulations), *Chem. Eng. Sci.* 56, 523.

31 R. A. Waggoner, E. Fukushima **1996**, (Velocity distribution of slow fluid flows in Bentheimer sandstone: an NMRI and propagator study), *Magn. Reson. Imag.* 14, 1085.

32 K. J. Packer, J. J. Tessier **1996**, (The characterization of fluid transport in a porous solid by pulsed gradient stimulated echo NMR), *Mol. Phys.* 87, 267.

33 P. Mansfield, B. Issa **1996**, (Fluid transport and porous rocks I: EPI studies and a stochastic model of flow), *J. Magn. Reson. A* 122, 137.

34 S. Sheppard, M. D. Mantle, A. J. Sederman, M. L. Johns, L. F. Gladden **2003**, (Magnetic resonance imaging study of complex fluid flow in porous media: flow patterns and quantitative saturation profiling of amphiphilic fracturing fluid displacement in sandstone cores), *Magn. Reson. Imag.* 21, 365.

35 L. F. Gladden, M. D. Mantle, A. J. Sederman, E. H. L. Yuen **2002**, (Magnetic resonance imaging of single- and two-phase flow in fixed-bed reactors), *Appl. Magn. Reson.* 22, 201.

36 L. F. Gladden, M. H. M. Lim, M. D. Mantle, A. J. Sederman, E. H. Stitt **2003**, (MRI visualisation of two-phase flow in structured supports and trickle-bed reactors), *Catal. Today* 79, 203.

37 I. V. Koptyug, S. A. Altobelli, E. Fukushima, A. V. Matveev, R. Z. Sagdeev **2000**, (Thermally polarized ^{1}H NMR microimaging studies of liquid and gas flow in monolithic catalysts), *J. Magn. Reson.* 147, 36.

38 I. V. Koptyug, L. Y. Ilyina, A. V. Matveev, R. Z. Sagdeev, V. N. Parmon, S. A. Altobelli **2001**, (Liquid and gas flow and related phenomena in monolithic catalysts studied by H-1 NMR microimaging), *Catal. Today* 69, 385.

39 S. L. Codd, S. A. Altobelli **2003**, (A PGSE study of propane gas flow through model porous bead packs), *J. Magn. Reson.* 163, 16.

40 R. W. Mair, R. Wang, M. S. Rosen, D. Candela, D. G. Cory, R. L. Walsworth **2003**, (Applications of controlled-flow laser-polarized xenon gas to porous and granular media study), *Magn. Reson. Imag.* 21, 287.

41 R M. Cotts, M. J. R. Hoch, T. Sun, J. T. Markert **1989**, (Pulsed field stimulated echo methods for improved NMR diffusion measurements in heterogeneous systems), *J. Magn. Reson.* 83, 252.

42 U. Tallarek, D. van Dusschoten, H. Van As, E. Bayer, G. Guiochon **1998**, (Study of transport phenomena in chromatographic columns by pulsed field gradient NMR), *J. Phys. Chem. B* 102, 3486.

43 J. D. Seymour, P. T. Callaghan **1997**, (Generalized approach to NMR analysis of flow and dispersion in porous media), *AIChE J.* 43, 2096.

44 X. H. Ren, S. Stapf, B. Blümich **2005**, (NMR velocimetry of flow in model fixed-bed reactors of low aspect ratio), *AIChE J.* 51, 392.

45 X. H. Ren, S. Stapf, B. Blümich **2005**, (Magnetic resonance visualisation of flow and pore structure in packed beds with low aspect ratio), *Chem. Eng. Technol.* 28, 219.

46 D. Tang, A. Jess, X. Ren, B. Blümich, S. Stapf **2004**, (Axial dispersion and wall effects in narrow fixed bed reactors – a comparative study based on RTD- and NMR-measurements), *Chem. Eng. Technol.* 27, 866.

47 A. Klemm, H. P. Müller, R. Kimmich **1997**, (NMR microscopy of pore-space backbones in rock, sponge, and sand in comparison with random percolation model objects), *Phys. Rev. E* 55, 4413.

48 A. Klemm, R. Kimmich, M. Weber **2001**, (Flow through percolation clusters: NMR velocity mapping and numerical simulation study), *Phys. Rev. E* 63, 04514.

49 B. Manz, L. F. Gladden, P. B. Warren **1999**, (Flow and dispersion in porous media: Lattice–Boltzmann and NMR studies), *AIChE J.* 45, 1845.

50 J. C. Park, K. Raghavan, S. J. Gibbs **2002**, (Axial development and radial non-uniformity of flow in packed columns), *J. Chromatogr. A* 945, 65.

51 J. N. Papageorgiou, G. F. Froment **1995**, (Simulation models accounting for radial voidage profiles in fixed-bed reactors), *Chem. Eng. Sci.* 50, 3043.

52 E. A. Zerhouni, D. M. Parrish, W. J. Rodgers, A. Yang, E. P. Shapiro **1988**, (Human heart: Tagging with MR imaging – a method of noninvasive assessment of myocardial motion), *Radiology* 169, 59.

53 M. V. Icenogle, A. Caprihan, E. Fukushima **1992**, (Mapping flow streamlines by multistripe tagging), *J. Magn. Reson.* 100, 376.

3.4
Pure Phase Encode Magnetic Resonance Imaging of Concrete Building Materials
J. J. Young, T. W. Bremner, M. D. A. Thomas, and B. J. Balcom

3.4.1
Introduction

Understanding the interaction of water and ionic species with building materials is fundamental to producing durable structures. It is a curious dichotomy that these species, which are fundamental to the formation of the materials, are also their principal agents of decay. Concrete was first used as a construction material by the Ancient Greeks and Romans, and is still one of the most common building materials worldwide [1]. There is a critical need for research into concrete durability due to the significant and growing annual repair costs. In Canada alone, it has been estimated that $16 billion is spent on concrete structural repair each year [2].

Concrete is a composite material composed of cement paste with interspersed coarse and fine aggregates. Cement paste is a porous material with pore sizes ranging from nanometers to micrometers in size. The large pores are known as capillary pores and the smaller pores are gel pores (i.e., pores within the hydrated cement gel). These pores contain water and within the water are a wide variety of dissolved ions. The most common pore solution ions are OH^-, K^+ and Na^+ with minor amounts of SO_4^{2-} and Ca^{2+}. The microstructure of the cement paste is a controlling factor for durable concrete under set environmental exposure conditions.

There are many deterioration processes that influence the durability of concrete. Cycles of freezing and thawing, deicer salt scaling, chloride-induced corrosion of embedded reinforcing steel, the alkali-silica reaction (ASR) and sulfate attack are all common processes that affect the service life of concrete materials [1]. The presence of water plays a fundamental role in all of these deterioration mechanisms; without water the potential for damage would generally be minimal. The transport of chloride ions will control the damage induced by deicer salt scaling and corrosion of reinforcing steel. ASR is an expansive reaction involving siliceous minerals in some aggregates and the alkalis present in the cement paste. Additionally, lithium may play an important role in suppressing these expansions.

The sensitivity of Magnetic Resonance (MR) to the local concentration, molecular dynamics and molecular environment of these nuclei make it well suited for the study of deterioration processes in concrete materials. Hydrogen (water), lithium, sodium, chlorine and potassium are all MR sensitive nuclei and play an important role in cement chemistry. The ability of MRI to spatially resolve and non-destructively examine test samples as a function of treatment or exposure has the potential to provide new insight to better understand deterioration mechanisms and mass transport properties of concrete materials.

The non-destructive nature of MRI is critical to undertaking long duration experiments that require measurements to be made at discrete points in time. For example, measuring the ingress of ions due to diffusion is a very slow process (it can take many months to have appreciable penetration), and it is desirable to determine profiles at various times due to the potential interaction with the cement paste matrix (for example chlorine and lithium can become bound to the cement paste hydrates). Destructive testing methods require one sample for each time point measurement. As this introduces variability among test samples, *de facto*, it is necessary to study, and destroy, at each time point, a large number of samples. Fewer samples would be required if each sample could be non-destructively tested at all experimental time points allowing more specimen variables to be tested (for example different cement chemistry).

Several other research groups have employed MRI techniques to investigate the distribution of moisture in cement based and related materials with varying degrees of success. Pel et al. [3] employed a custom-built MR spectrometer with a spin-echo technique to study moisture in porous materials. The Stray Field Magnetic Resonance Imaging (STRAFI) technique was applied by Bohris [4] to image water penetrating into cement pastes. Leech et al. [5] used a one-dimensional Single-point Imaging (SPI) technique to measure the ingress of water into concrete specimens and Kaufmann et al. [6] used a three dimensional SPI technique to examine the ingress of water into concrete samples. Sodium in calcium silicate bricks has been measured by Pel et al. [7]. In our laboratory we have studied water [8–13], chlorine [14], sodium [14] and lithium [15] in cement based materials.

The focus of the UNB MRI Research Center for the last decade has been the development of a general quantitative laboratory-based MRI measurement technique for porous materials. The Single-point Ramped Imaging with T_1 Enhancement (SPRITE) [16–18] class of measurements developed at this laboratory are very well suited for rapid imaging of cement-based materials due their robustness in dealing with short signal lifetimes and the lack of image artifacts. Many of the concrete relevant processes studied require rapid time resolution in one, two or three dimensions, which the SPRITE technique has been able to provide. In addition, the SPRITE technique also allows for simplified image contrast to aid interpretation of the local image intensity. The following discussion focuses on the development of this technique and its application to water content measurements and more recently chlorine, sodium and lithium measurements in cement-based materials.

SPRITE is termed a pure phase encode technique because spatial encoding occurs through the application of variable amplitude magnetic field gradients (which yield spatially varying frequencies) applied for fixed periods of time. Variable frequency with a fixed evolution or encoding time yields a variable signal phase.

3.4.2
Single-Point Imaging – The SPRITE Techniques

MRI of concrete materials is challenging due to the short signal lifetimes and the low experimental sensitivity of magnetic resonance in general. The bulk T_1 and T_2 and T_2^* of ^1H from water in white Portland cement systems have been shown to be in the order of a few milliseconds and a few hundred microseconds, respectively [19]. It is important to note that as there is a distribution of pore sizes in the cement paste, there will be an associated distribution of T_1 and T_2 signal lifetimes [20].

The SPI [21, 22] method was a very promising imaging technique for the study of concrete materials due to its ability to measure short signal lifetimes with a local image intensity, which is readily understood in terms of the image acquisition parameters. However, the SPI technique is slow, noisy, causes large mechanical vibrations and has a demanding magnetic field gradient duty cycle. The SPI technique can be understood through examination of the pulse sequence diagram and the signal equation. We will consider this technique further because it is the basis of the more sophisticated SPRITE experiment.

Figure 3.4.1 shows the implementation of a three-dimensional SPI measurement. The position is encoded in reciprocal space, k-space, by adjusting the amplitude of the x, y and z phase encode magnetic field gradients as the MR signal is excited and detected with radio frequency (rf) pulse excitation. A radio-frequency pulse excites the sample while the magnetic field gradient is applied and a single point on the free induction decay (FID) is acquired after a chosen encoding time (t_p). A fast Fourier transform is applied to the k-space data to reconstruct an image. As each k-space point requires a radio frequency pulse, it is much slower than traditional spin-echo techniques that acquire multiple k-space points with a single excitation pulse. SPI acquires the signal while the magnetic field gradients are on. Therefore, it is able to image short MR signal lifetimes that are not possible with standard imaging techniques, such as spin-echo imaging, because echo times cannot be made short enough, nor can magnetic field gradients stabilize fast enough, to acquire high quality k-space data. The disadvantages of using SPI are mainly caused by the way gradient amplitudes are adjusted. Electrical currents are passed through the magnetic field gradient coil and high magnetic field gradients are required to spatially encode the data points at the periphery of k-space. Switching the magnetic field gradients on and off repeatedly causes large mechanical vibrations, noise and large duty cycles on the magnetic field gradient amplifiers

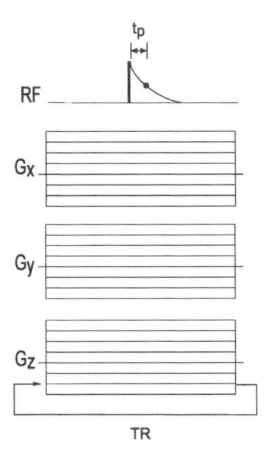

Fig. 3.4.1 Schematic description of the three-dimensional SPI technique. G_z, G_x and G_y are the phase encode magnetic field gradients and are amplitude cycled to locate each k-space point. A single data point is acquired at a fixed encoding time t_p after the rf excitation pulse from the free induction decay (FID). TR is the time between excitation (rf) pulses. Notice that the phase encode magnetic field gradients are turned on for the duration of the k-space point acquisition.

and coil. The switching also slows down the measurement due to the finite time required for the magnetic field gradient to stabilize.

Equation (3.4.1) is the SPI signal equation, where the signal intensity, S, at any point in an image is related to the local proton density, ϱ, weighted by the phase encode time, t_p, the repetition time TR, the flip angle α and the spin–lattice and spin–spin relaxation times T_1 and T_2^*. For density weighted imaging, ϱ must be determined spatially. Density weighted images can be obtained through the manipulation of t_p, TR and α coupled with an understanding of the sample signal lifetimes. Concrete materials generally have bulk T_1 values of a few milliseconds and TR is usually set to a minimum of a hundred milliseconds, causing the bracketed term to be approximately one. The measured signal is solely weighted by α (which is experimentally set) and the T_2^* decay. Ideally, if t_p could be made much less than T_2^* the image would be density weighted. However, this is not usually the case as T_2^* is in the order of a few hundred microseconds for concrete samples and t_p is limited by the spectrometer to tens of microseconds. T_2^* must therefore be measured in order to produce a density weighted image. T_2^* can potentially vary spatially, and therefore, a T_2^* mapping method should be implemented to determine its effect on the image intensity [23]. We emphasize that our goal in the majority of our concrete materials imaging is to acquire a quantitative map of the local concentration of the species of interest in the sample.

$$S = \varrho \cdot e^{\frac{-t_p}{T_2^*}} \cdot \left(\frac{1 - e^{\frac{-TR}{T_1}}}{1 - \cos\alpha \cdot e^{\frac{-TR}{T_1}}} \right) \cdot \sin\alpha \tag{3.4.1}$$

The SPRITE [16] technique was developed to reduce the dangerous mechanical vibrations and noise produced during an SPI experiment. In addition to the reduced noise and mechanical vibration, the experiment is much faster than SPI and the image intensity is still easily understood. However, one experimental shortcoming is a magnetic field gradient duty cycle which is even more demanding than an SPI experiment.

Figure 3.4.2 illustrates the basic implementation of the SPRITE method. While many aspects of SPRITE are the same as SPI (k-space is traversed in a similar manner and a single point is acquired at each point in k-space) it is the ramping of at least one phase encode magnetic field gradient that allows for the improved performance. For the SPRITE experiment, magnetic field gradients are turned on then stepped rapidly in small steps (not turned on then off for every acquisition point as is the case with SPI), ensuring that a large reduction in mechanical vibration and noise is realized. The SPRITE measurement is also faster due to the reduced TR (TR is reduced to a few milliseconds from a hundred milliseconds with SPI). The duty cycle is more demanding because magnetic field gradients are turned on for sustained periods during the measurement.

The SPRITE signal equation is identical to Eq. (3.4.1) and it is possible to acquire density weighted images in the same manner as an SPI experiment. The TR in a SPRITE experiment is only a few milliseconds, which is similar to the T_1 of a concrete specimen. Therefore to remove this influence from the image, α can be

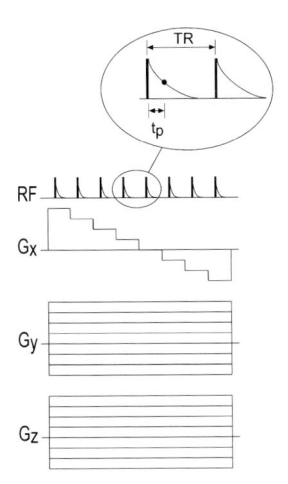

3.4 Pure Phase Encode Magnetic Resonance Imaging of Concrete Building Materials

Fig. 3.4.2 Schematic description of the three-dimensional SPRITE imaging technique. G_z, G_x and G_y are the phase encode magnetic field gradients and are amplitude cycled. A single data point is acquired at a fixed encoding time t_p after the rf excitation pulse from the free induction decay (FID). TR is the time between rf pulses. Notice that G_x is ramped ($+G_{x,\,max}$ to $-G_{x,\,max}$) and one k-space point is acquired for each value of the magnetic field gradient. G_y and G_z are on during the G_x magnetic field gradient ramp and turned off at the end.

made small so that the bracketed term is approximately 1. The influence of T_2^* decay should be removed as per the case of an SPI experiment in order to extract a density weighted image.

While it is true that the SPRITE technique is greatly improved compared with SPI experiments in terms of speed, reduced noise and reduced mechanical vibrations, neither technique is optimized for two-dimensional, three-dimensional or density weighted imaging. The SPRITE technique illustrated in Figure 3.4.2 is far from optimal. Both SPI and SPRITE commence data acquisition at the outer most k-space points and move through the center of k-space (zero magnetic field gradient) and proceed to the k-space extremities (see Figure 3.4.3). SPI and SPRITE both capture an FID at each k-space point; however, the center of k-space (zero magnetic field gradient) contains the most quantitative information about the sample (it is a bulk quantity measurement), yet it is acquired after many radiofrequency pulses which have induced a longitudinal magnetization steady state

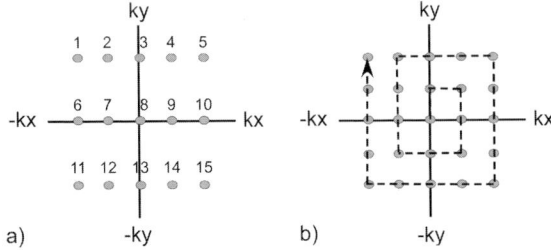

Fig. 3.4.3 (a) Two-dimensional k-space acquisition using SPI or SPRITE. The k-space data acquisition is indicated numerically. High magnetic field gradient amplitudes are applied at the extremities of k-space. (b) A generic two-dimensional centric scan SPRITE method. The first k-space point acquired is at the k-space origin where most of the quantitative information exists (before many excitation pulses have been applied), after which the acquisition moves outwards through the k-space.

[the bracketed term in Eq. (3.4.1)] This term greatly complicates the quantitative interpretation of first generation SPRITE images.

Three centric scan SPRITE methods have been proposed for imaging in one, two and three dimensions. They are the double half k [24], spiral [18] and conical [17] SPRITE techniques shown in Figures 3.4.4–3.4.6. Centric scan methods acquire the center point of k-space first (see Figure 3.4.3), and then acquire subsequent k-space points with linear, spiral or conical trajectories depending on the measurement type. These methods have been proven to be faster, and generally more quantitative, due to the simplified signal equation [Eq. (3.4.2) (symbols are as in Eq. 3.4.1)]. Spiral and conical SPRITE are faster not only due to short TR times, but also because they exclude data acquisition at the periphery of k-space. The signal equation is simplified because the k-space origin data points are acquired first in a data acquisition when the longitudinal sample magnetization is at equilibrium. This removes T_1 weighting from the image when acquired with appropriate measurement parameters [17].

$$S = \varrho \cdot e^{\frac{-t_p}{T_2^*}} \cdot \sin\alpha \qquad (3.4.2)$$

There are three principal reasons why pure phase encode, centric scan, SPRITE methods yield quantitative density weighted images. (a) The T_2, due to surface relaxation is too short lived in cement based materials to be observed with traditional frequency encode methods – let alone quantified. In addition, the pore size distribution inherent to most porous media yield multi-exponential T_1 and T_2 relaxation time constants. Traditional MRI methods cannot be quantified in a simple manner due to this multi-exponential signal decay. Centric scan SPRITE image intensities depend on the T_2^* time constant however, not on T_1, nor on T_2. The decay constant, T_2^* is single exponential for the vast majority of porous media – fully or partially saturated with minimal variation observed with the level of saturation. (b) The mismatch in magnetic susceptibility of the cement paste matrix and pore solution ensure that the magnetic field homogeneity is significantly perturbed in porous media leading to significant local distortions in the magnetic field homogeneity which can cause image distortion in traditional MRI methods [25]. A geometrically distorted image cannot be quantitative, by definition, as the local image intensity is unreliable. Locally distorted magnetic fields in pure phase encode SPRITE methods do not cause image distortion. (c) Significant magnetic field inhomogeneity leads to a linewidth restriction on resolution in traditional frequency encode MRI methods. The linewidth restriction is entirely absent in

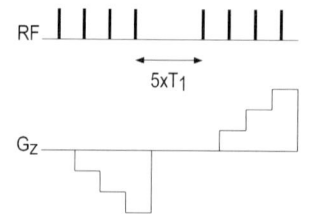

Fig. 3.4.4 Schematic description of the one-dimensional double half k (DHK) SPRITE technique. The phase encode magnetic field gradient, G_z, ramped through half of k-space beginning at the center and a single data point is acquired at a fixed time (t_p) after the rf excitation pulse. The second half of k-space is acquired after a $5T_1$ time delay. The time between rf pulses is TR.

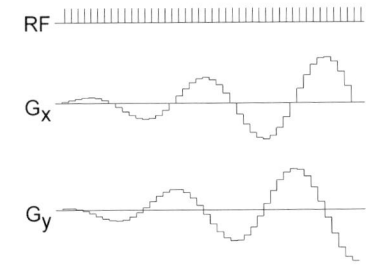

Fig. 3.4.5 Schematic description of the two-dimensional SPIRAL-SPRITE technique. G_x and G_y are the phase encode magnetic field gradients that are amplitude cycled to traverse k-space along an Archimedean Spiral. A single data point is acquired from the FID at a fixed encoding time t_p after an rf excitation pulse. TR is the time between rf pulses.

pure phase encode SPRITE imaging [25]. The minimal measurement dead time in SPRITE methods ensures that systems with T_2^* signal lifetimes as short as tens of microseconds can be readily imaged. This time window is three to four orders of magnitude reduced from the signal lifetimes encountered in routine clinical MRI. Quite literally SPRITE techniques are able to observe and quantify the fluid content in samples that are invisible to other MRI methods.

The SPI and the SPRITE techniques have been applied to numerous concrete material problems. Some of these applications are presented in the following discussion. It is very important, however, to point out that these measurement techniques are not limited to cement-based material problems, but are fairly general in application. They can be, and have been, applied to many different material science problems.

3.4.3
Hydrogen (Water) Measurements

3.4.3.1 Drying Profiles

Water is a fundamental component of concrete and is a principal determinant of durability. From the time of the initial contact of water and Portland cement, through the initial hardening and curing period and throughout the service life of the concrete, water enables the development of desirable properties of concrete. At the same time, water can be an integral component of concrete deterioration mechanisms. While in service, the change in moisture content of the concrete will influence concrete properties such as strength, creep, shrinkage and elasticity [26].

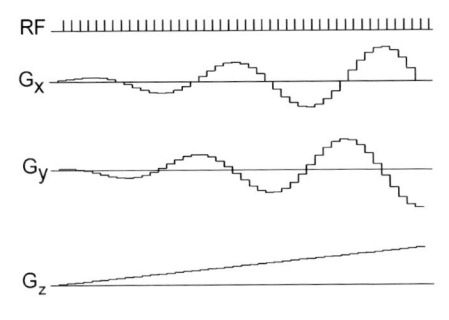

Fig. 3.4.6 Schematic description of the three-dimensional Conical-SPRITE technique. The G_x, G_y and G_z phase encode magnetic field gradients that are amplitude cycled to form a conical traverse through k-space. A single data point is acquired from the FID after an rf excitation pulse at a fixed encoding time t_p. TR is the time between rf pulses.

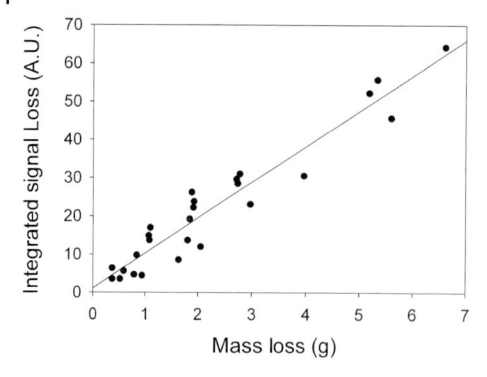

Fig. 3.4.7 The linear relationship between the integrated change in SPI image intensity and the gravimetric mass loss due to drying of concrete specimens [9].

The SPI/SPRITE techniques detect evaporable water and not water that is chemically combined in the cement paste [23]. Chemically combined water will have very short signal lifetimes (<20 μs [20]) and evaporable water will have longer signal lifetimes that can be measured using this technique. This has been verified by comparing the integrated signal loss of MRI drying profiles with the gravimetric results as shown in Figure 3.4.7. In addition, specimens have been measured using SPI after oven drying (oven drying at 105 °C to a constant mass is generally accepted as the method to remove evaporable water from concrete) and negligible image intensity is observed.

A one-dimensional concrete drying experiment is presented in Figure 3.4.8 to demonstrate the use of the SPI technique to determine the evaporable water content of a concrete specimen [9]. SPI measurement parameters TR and α were set at 100 ms and 6°, respectively, to produce an image that is only weighted by T_2^*. To remove the T_2^* weighting, T_2^* mapping was implemented using 8 different phase encode times (t_p) ranging from 55 to 300 μs. Gravimetric measurements

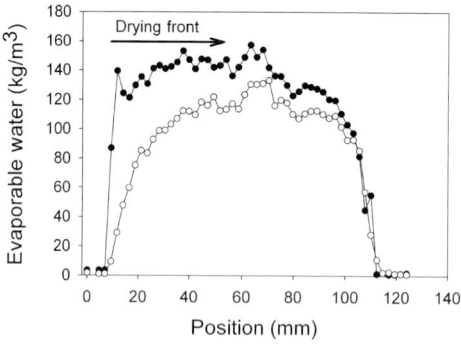

Fig. 3.4.8 One-dimensional SPI drying profiles of concrete moist-cured for 28 days and of a 0.6 water–cement ratio [9]. The specimen was sealed except for one face and exposed to a drying regime at 38 °C and 40% relative humidity for 28 days. The spatial moisture content after 28 days of moist curing and 28 days of drying are indicated by the symbols (●) and (○), respectively. The measurement parameters were: field of view (FOV) 150 mm, acquisition points 64, t_p = (55 – 300 μs, 8 values), α = 6°, TR = 100 ms, acquisition time 3.5 min per encoding time.

were made to determine the total mass of evaporable water so that an absolute scale could be applied to the density weighted image (gravimetric measurements are not required to assign absolute quantities to density weighted images – a reference standard may be included in an imaging experiment).

Concrete made with a water-to-cement ratio of $w/c = 0.6$, white Portland cement and quartz aggregates was cast in 100-mm long and 48-mm diameter cylinders. The specimen was then cured at 100% relative humidity for 28 days and allowed to dry from one exposed face for 28 days at 38 °C and 40% relative humidity. At the end of the moist curing period, and after the drying period, 1D MRI measurements were performed on the specimen. Figure 3.4.8 is the 1D MRI profile of evaporable water content in the concrete specimen. The profiles show that significant amounts of water can be removed with extended drying periods.

The density weighted images and their absolute quantification demonstrated through concrete drying experiments are very powerful tools for material science. This measurement methodology and data analysis is general and can be extended to other porous media. The results from MRI moisture profiles can also be used to measure moisture diffusivity that enable moisture transport models to be developed for a wide range of materials.

3.4.3.2 Water Uptake

The surface of concrete may be subjected to cycles of wetting and drying during normal service life. This process is a durability concern as harmful ions such as chlorides or sulfates can easily penetrate with the adsorbed water and there is the potential for saturation prior to freezing. Measuring the rate of water penetration into a dry concrete surface is thus important to the quality control of concrete mixes – hopefully yielding a durable product.

Cano Barrita [27] cast concrete specimens with w/c of 0.6, dried the specimens at 38 °C and 20% relative humidity, then measured the penetration of water in a capillary uptake type of experiment. A 3D centric scan SPRITE measurement was selected, as an image could be acquired in 150 s and the image would therefore be weighted only by the T_2^* decay. 3D images were acquired at various exposure times and the central 2D image slice was extracted from the data to measure the penetration depth with time.

The water penetration images are shown in Figure 3.4.9 at 0, 1, 2 and 4 h of contact with water. The bright dot at the top of each image is a doped water reference that can be used for image scaling. The penetrating water front is the bright region at the bottom of the images. The dark voids in the image are aggregates that contain no water. The outline of the specimen can also be observed in the image, due the fact that not all of the evaporable water is removed under the sample preparation drying scheme.

The penetration front was measured from the images in Figure 3.4.9 and plotted against the square root of time in Figure 3.4.10. The plot indicates that this relationship is linear and its slope is a measure of the sorptivity [28]. This type of experiment, coupled with gravimetric measurements, allow for the modeling of

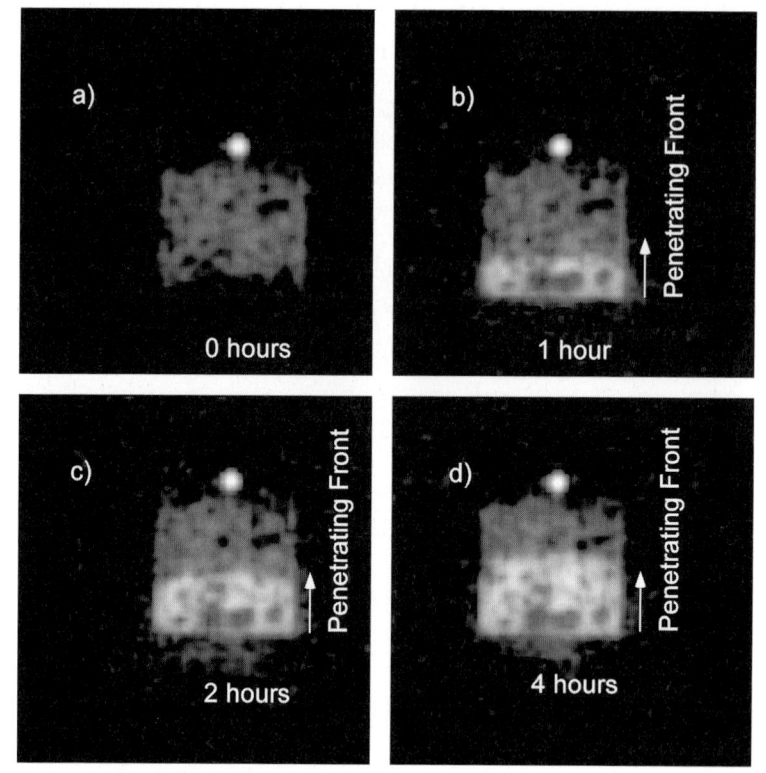

Fig. 3.4.9 Two-dimensional slice images taken from a three-dimensional Spiral-SPRITE water uptake experiment [27]. (a) Initially dry, (b) after 1 h, (c) after 2 h and (d) after 4 h of partial immersion in water. The measurement parameters were: FOV 100 × 100 × 100 mm, $t_p = 150$ µs, $\alpha = 20°$ and an acquisition time 2.5 min per image.

unsaturated flow into concrete. These results have been used, coupled with gravimetric results, to determine how various concrete parameters (moist curing time, mixture proportions and environmental exposure) can influence diffusivity. This information can be used to classify the quality of the cover concrete.

The application of conical SPRITE to three-dimensional capillary uptake problems would be advantageous. The increased speed would allow for T_2^* mapping experiments to be performed so that density weighted images could be measured. This would mean that complementary gravimetric measurements need not be undertaken. The experiment could be set up with the water reservoir and sample remaining in the magnet throughout the measurement. This would be beneficial as the measurement could be continuously performed, which is especially important at early exposure stages when the water invasion is most rapid.

This type of water uptake measurement is very general in application. It is applicable to most porous media, where moisture transport properties need to be measured.

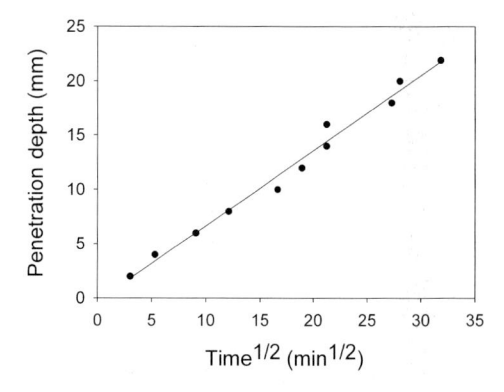

Fig. 3.4.10 Sorptivity determined from Figure 3.4.9 [27]. There is a linear relationship between penetration depth and the square root of time.

3.4.3.3 Freezing–Thawing

In northern climates freezing and thawing damage is a durability concern for concrete structures. Cycles of freezing and thawing are detrimental to concrete because of the cracking which results and the eventual disintegration of the cement paste. The development of air entrainment has greatly improved the durability of concrete subjected to cycles of freezing and thawing. The key to freezing and thawing damage is the distribution of moisture in concrete at the time of freezing and thawing. In addition to pure freezing and thawing damage, the use of deicer salts has engendered another form of deterioration – deicer salt scaling [29].

It is well known that the melting point of water confined to small pores is depressed [30, 31]. Therefore in concrete as the temperature decreases, the amount of frozen water will increase. Under normal temperature variation not all water in the pore structure will be frozen. The change from water in the liquid form to solid ice drastically reduces the T_2^* of hydrogen (T_2^* ice <9 µs [32]). Ice will not be observed in an image, even with the SPRITE techniques, and our experimental images will be maps of unfrozen water distribution.

The work of Prado et al. [12] and Choi et al. [11] has lead to a more general investigation of cyclical freezing and thawing deterioration in mortar. Figure 3.4.11 shows two-dimensional images extracted from a 3D centric scan SPRITE data set showing thawing and freezing waves moving into mortar. Bright regions are thawing regions while darker regions are partially frozen. Voids are water excluding large fine aggregates. Finite residual image intensity at all temperatures shows the presence of unfrozen water in small pores.

Damage to concrete is induced by cycles of freezing and thawing. The number of cycles to cause damage is highly variable. Laboratory testing can be accelerated through rapid freezing and thawing cycles; however a rapid method must be used to determine the amount of water freezing and thawing. MRI is probably the only measurement that can rapidly provide this type of information spatially resolved. MRI can also measure the pore size distribution under freezing conditions. The ability to measure and quantify all of these parameters should help answer the important questions related to understanding and preventing deicer salt scaling.

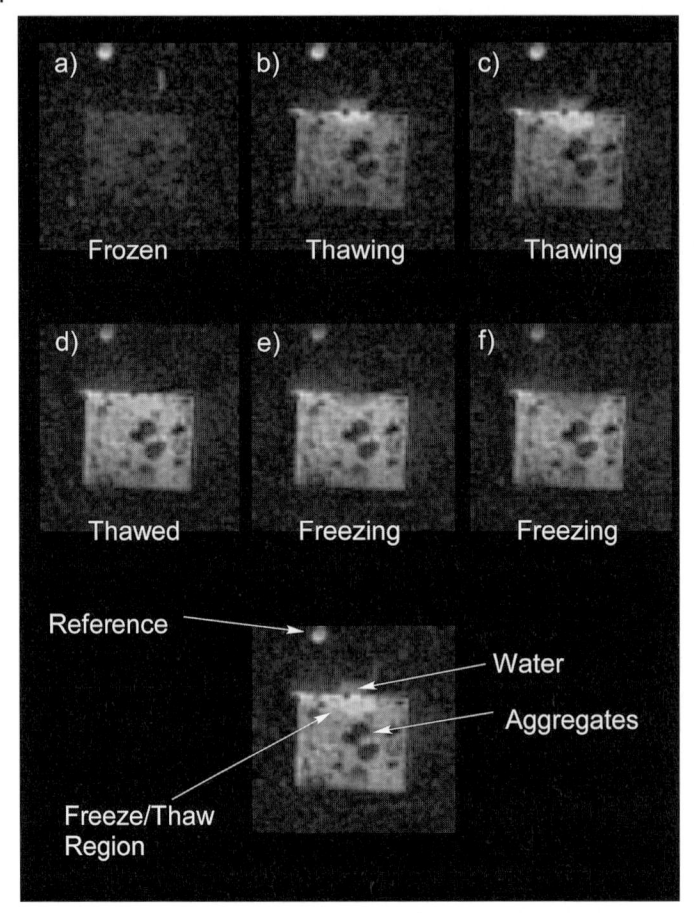

Fig. 3.4.11 Two-dimensional slice images taken from a three-dimensional Spiral-SPRITE cyclical freezing and thawing experiment. (a) The initially frozen specimen, (b) thawing front, (c) thawing front, (d) thawed specimen, (e) freezing front and (f) freezing front. The measurement parameters were: FOV 60 × 60 × 60 mm, acquisition points 64 × 64 × 32, t_p = 300 μs, α = 9°, acquisition time 3.0 min per image.

This methodology developed to observe water freeze–thaw in concrete materials, may be used quite generally to observe solid–liquid phase transitions in many different materials of industrial and technological interest. The method could be also applied to other problems involving freezing and thawing of water in confined pores.

3.4.3.4 Crack Detection

Cracking can be caused by numerous physical and chemical phenomena in Portland cement materials. For example, freeze–thaw damage and ASR manifest themselves through crack development that may be studied using MRI. In addition, the easy access that cracks provide for aggressive species to penetrate concrete is a durability concern and therefore methods need to be developed to study cracking in cement-based materials.

MR has been used to measure the pore size distribution by exploiting the dependence of T_1 and T_2 on the surface to volume ratio of water-filled pores [20, 31, 33]. The T_1 and T_2 dependence is described by Eqs. (3.4.3) and (3.4.4) [34] where ϱ_1 and ϱ_2 are spin–lattice and spin–spin surface relaxivity constants, and S/V is the surface-to-volume ratio of the pore. These equations provide the basis of a methodology for crack detection in cement paste specimens [13].

$$\frac{1}{T_1} = \varrho_1 \frac{S}{V} \qquad (3.4.3)$$

$$\frac{1}{T_2} = \varrho_2 \frac{S}{V} \qquad (3.4.4)$$

The SPRITE technique is ideal for determining the spatial distribution of water in cement-based materials because of the relatively short relaxation times. The technique is not well suited for imaging cracks in porous materials because of the relatively long relaxation times of water in cracks and low resolution of the technique. SPRITE measurements would suppress the signal from the water filled cracks due to long T_1 contrast and highlight water in the paste making it very difficult to detect cracking using this technique.

The distribution of short T_1 and T_2 in the paste, and longer T_1 and T_2 in the water filled cracks, makes the traditional frequency encode spin-echo imaging technique a natural choice for this research due to the natural suppression of short T_2 components in the paste. The spin-echo approach is necessary to image the water filled cracks because the cracks will produce high signal spin-echo image intensity without a large quantity of signal from the paste [13].

A cement paste phantom was prepared from white Portland cement. A thin sheet of copper was placed in the top of the specimen at the time of casting to produce a synthetic crack 300-µm wide, 8-mm long and 3–4-mm deep. The sheet was removed the following day. The specimens were 10 × 10 × 10 mm. At the time of demolding, the specimen was slightly damaged and two cracks formed on the corners of the specimen. The thicknesses of these cracks was estimated to be 100 µm and 60 µm. Prior to the MRI measurement the samples were stored in water to ensure saturation.

Figure 3.4.12 is a volume rendered 3D spin-echo image of the specimen. Spin-echo imaging observes the signal from the paste and the water filled cracks. The signal intensity from the cracks is much higher than from the paste. The signal from the paste comes from the distribution of long T_1 and T_2 components. Spin-echo imaging naturally suppresses the short T_2 components of the paste signal. This technique for crack detection relies on attaining a high signal component from the crack. As the surrounding paste produces a low signal relative to the cracks, detecting cracks smaller than the nominal resolution is possible.

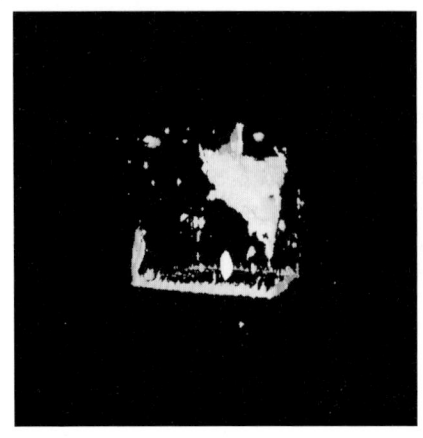

Fig. 3.4.12 Three-dimensional rendered spin-echo image of water filled cracks in a cement paste specimen [13]. Three cracks are visible in the image: a large triangular crack in the forefront, a smaller crack in the bottom left corner and a sheet-like structure at the top of the image. Water droplets can also be observed condensing on the cement paste surfaces. The measurement parameters were: FOV 20 × 20 × 20 mm, acquisition points 128 × 128 × 64, nominal resolution 156 × 156 × 312 µm, echo time 2.7 ms, repetition time 500 ms and acquisition time 270 min.

The geometrical requirements for crack detection can be estimated. A contrast-to-noise ratio (CNR) of $2\sqrt{2}$, defined by the difference in signal-to-noise ratio (SNR) of the structures of interest, provides a 95% confidence level that two neighboring voxels contain different structures [35]. Using a weighted average of SNR of the crack and paste and assuming that a crack is a sheet-like structure that extends through the depth and length of a voxel, the minimum crack width can be estimated. Using the CNR, the minimum detectable crack width in the 3D spin echo image would be 50 µm.

The spin-echo successfully imaged water held in the cracks of cement paste. This technique can be used to resolve cracks much smaller than the nominal resolution by relying on water saturation of the crack, the connectivity of the crack structure and the fact that it is relatively easy to detect a high intensity structure on a low intensity background.

This measurement methodology is potentially a very powerful tool to be used in studying concrete durability. Many durability problems manifest themselves as cracks (ASR, freeze–thaw damage, corrosion induced cracking) yet there are currently no good instrumental methods to investigate crack initiation and the influence of cracking on material transport properties. This methodology should assist in answering this question.

3.4.4
Chlorine and Sodium Measurements

Concrete exposed to deicer salts, or to a marine environment is subjected to chloride and sodium loading. The ability of concrete to resist the penetration of chlorides and sodium is a primary design consideration in marine or cold environments. The ingress of chlorides into concrete is a major problem due to chloride-induced corrosion of the reinforcing steel and deicer salt scaling [a process by which a thin layer (< 1 mm) of concrete deteriorates from the surface of the concrete]. The penetration of sodium from sea water or deicer salts is generally

3.4 Pure Phase Encode Magnetic Resonance Imaging of Concrete Building Materials

not considered to be such a problem, because sodium is naturally found in the pore solution of hydrated cement paste. However, sodium plays an important role in ASR, as certain siliceous aggregates in the presence of a high OH⁻ concentration will allow sodium and potassium to react with silica, forming an expansive gel.

Traditionally, the penetration of chlorides and sodium is measured destructively by grinding layers of concrete and chemically analyzing the powder samples. These data are used to calculate diffusion coefficients for the ions. This procedure is very slow, has low spatial resolution, and is destructive. The measured data are critically important for the development of service life models and therefore a rapid, high-resolution method to monitor the ingress of these ions is desirable.

The SPRITE class of measurements is required for MRI measurements of the low sensitivity, low concentration and short signal lifetimes of these nuclei. The absolute sensitivity of ^{35}Cl is only 0.4% and ^{23}Na is 9% compared with ^{1}H [36]. The short signal lifetimes of these nuclei (bulk T_2^* around a hundred μs and T_1 around a few ms) [14] require the use of pure phase encode methods. The low concentration of these nuclei coupled with the low absolute sensitivity requires a measurement technique that is rapid enough to permit numerous signal averages. It is important to note that sodium and chlorine are both quadrupolar nuclei and Eq. (3.4.1) is therefore not true for SPI/SPRITE measurement of these species; however, Eq. (3.4.2) is correct when centric scan SPRITE methods are employed. This provides one more illustration of the value of the centric scan SPRITE methods.

Quartz fine aggregates and normal Portland cement were used to prepare mortar at a w/c of 0.6. Mortar was cast into 30-mm diameter by 30-mm length cylinders and cured for 3 days. At the end of the curing period the specimens were oven dried at 105 °C until a constant mass was reached. Epoxy was then applied to the curved surface and the specimen was placed such that 1–2 mm of its height was immersed in a 20% NaCl solution.

Figures 3.4.13 and 3.4.14 are 1D SPRITE profiles of chloride and sodium penetrating into the mortar. The penetration of these ions is driven by capillary suction as the mortar is initially dry. While this study does not truly model field exposure conditions (the penetration of these ions is usually modeled using diffusion) there are some very significant findings. The signal magnitude for both chlorine and sodium decrease with time. The decrease in signal intensity is attributed to the binding of these nuclei to the cement paste matrix, which reduces signal lifetimes such that they are not detected. The most significant finding from this study is the fact that normal Portland cement (not just white Portland cement as used in hydrogen studies) can be measured. This implies that with sensitivity improvements, cores could be extracted from field concrete and the distribution of chlorine and sodium could be rapidly determined.

The time savings and improved spatial resolution of MRI measurements over traditional measurements are profound advantages of MRI. MRI can provide similar information in less than 1 h when literally it would otherwise require days. This would be of particular interest in diffusion type experiments where spatial concentration changes with time are measured. In addition, the same sample can be

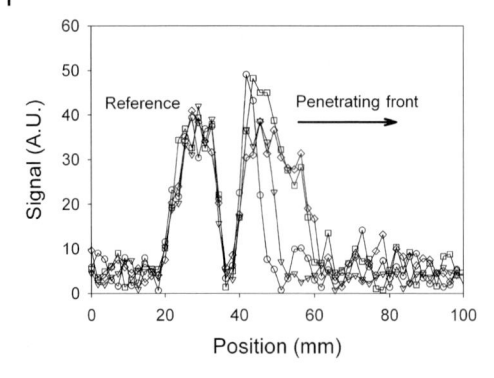

Fig. 3.4.13 One-dimensional SPRITE image of the chloride distribution in mortar specimens $w/c = 0.60$ at 3 h (○), 9 h (▽), 2 days (□) and 5 days (◇) [27]. The measurement parameters were: FOV 100 mm, $t_p = 300$ μs, $G_{max} = 25.6$ G cm^{-1}, $\alpha = 44°$ and acquisition time 20.7 min per encoding time.

measured with time, reducing uncertainties in the measured diffusion coefficient. Further work will include the development of the measurement to determine chlorine binding isotherms for various cement chemistries.

Once again these measurements are not limited to concrete materials, most MR sensitive ions can be measured in porous media using these techniques.

3.4.5
Lithium Measurements

ASR is an expansive reaction involving certain siliceous minerals found in some aggregates and alkali hydroxides present in cement paste pore water solution. The expansions produced by ASR cause cracking and serviceability problems in concrete structures. In addition to the direct damage caused by ASR, other processes such as ingress of chloride ions and freezing and thawing damage can be accelerated.

Lithium has been found to prevent ASR expansion [37]. It is used either to mitigate further distress in ASR-affected structures by topical application of lithium solutions or as a means of using ASR aggregates in new structures when other methods of ASR mitigation are not feasible. As a critical amount of lithium is needed in the pore solution of cement paste to arrest the expansion [38], a method to spatially resolve and quantify the lithium is desirable.

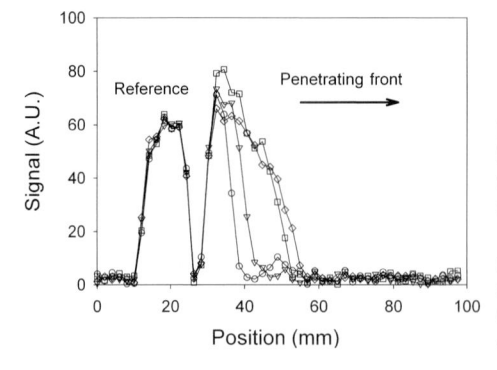

Fig. 3.4.14 One-dimensional SPRITE image of the sodium distribution in mortar specimens $w/c = 0.60$ at 2.5 h (○), 8 h (▽), 2 days (□) and 5 days (◇) [27]. The measurement parameters were: FOV 100 mm, $t_p = 300$ μs, $G_{max} = 9.5$ G cm^{-1}, $\alpha = 49°$ and acquisition time 7.4 min per encoding time.

The centric scan, one-dimensional, DHK SPRITE measurement was used to study the ingress of lithium. This measurement technique was selected due to the low absolute sensitivity of ^{7}Li (27% of ^{1}H [36]), the small amounts that are present and the short signal lifetimes (bulk T_1 of 10 ms and T_2^* of 120 μs). In addition to the robust, quantitative nature of this technique, lithium is a quadrupolar nucleus and interpretation of the image intensity is more complex than spin ½ nuclei. Once again Eq. (3.4.2) is quantitatively correct for even quadrupolar nuclei due to the fact the longitudinal steady state does not influence the image intensity.

Diamond [38] has shown that considerable amounts of lithium become bound during the hydration process. A study of the influence of hydration on the observable lithium content in cement paste samples was undertaken. In parallel, a traditional approach to lithium content in the pore solution was conducted (physical extraction of the pore solution and subsequent chemical analysis). The relationship between pore solution extraction and bulk MR is shown to be linear (see Figure 3.4.15). This result indicates that MR will be able to image lithium held in the pore solution but not lithium bound to the cement paste matrix. MR and pore solution extraction results also confirm that large amounts of lithium become bound during hydration.

Mortar specimens were prepared to determine the effectiveness of MRI in a time resolved lithium penetration experiment [15]. This work used a non-reactive aggregate and commercially available LiNO$_3$ solution to simulate topical treatments to concrete. These results will aid the development of a more general measurement of concrete core extracted from a lithium treated structure suffering from ASR.

Quartz fine aggregate and normal Portland cement were used to prepare mortar with a w/c of 0.5. A cylindrical specimen, 43 mm in diameter and 50 mm long, was cast and cured under sealed conditions for 3 days at 23 °C. The specimen was then oven dried at 105 °C for 1 day prior to exposure to lithium nitrate solution. The specimen was then placed such that the bottom of the cylinder was submerged approximately 1–2 mm into a lithium nitrate solution with Teflon tape applied to the curved surface.

Figure 3.4.16 shows one-dimensional DHK-SPRITE profiles of lithium penetrating into dry mortar. In addition to the determination of the penetration depth of lithium shown at various times, there is an overall increase in signal intensity at a given point, which is attributed to the filling of the various pore sizes in time. The

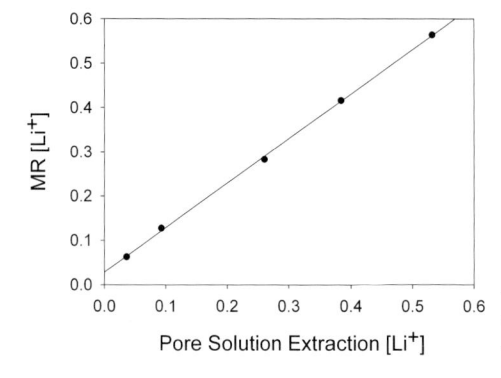

Fig. 3.4.15 Lithium concentrations in the pore solution determined by MR and the chemical analysis of the pore solution.

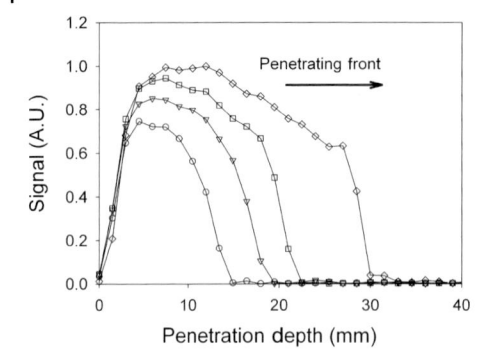

Fig. 3.4.16 Lithium distribution at times of 3 h (○), 6 h (▽), 9 h (□) and 22 h (◇) immersion in LiNO$_3$ for mortar [15]. DHK-SPRITE measurement parameters: FOV = 100 mm, 64 points, t_p = 30 μs, α = 5°, 5T_1 = 100 ms, scans = 1024 and acquisition time 6 min.

results presented are very promising for the measurement of lithium distributions in field concretes, as normal Portland cement was used in this study.

MRI lithium measurements are playing an important role in the development and understanding of the role of lithium in treating and preventing ASR. The spatial resolution and speed make this measurement technique very attractive for monitoring the removal of lithium from the pore solution during ASR and also in investigating how well various surface treatments will slow ASR.

3.4.7
Conclusion

The SPI and SPRITE class of measurements for imaging short MR signal lifetimes are quantitative and have signal equations that are readily understood. The centric SPRITE methods are much faster and feature signal equations that are easier to interpret. This feature makes these measurements more readily density weighted and better suited to imaging quadrupolar nuclei.

The measurement applications of density weighted drying profiles, water ingress, water phase transitions, crack detection, chlorine, sodium and lithium imaging applied to cement-based materials can be easily translated to other porous media. Density weighted MRI will no doubt prove to be a powerful tool in material science research.

References

1 S. Mindess, J. F. Young, D. Darwin **2003**, *Concrete*, 2nd edn, Prentice Hall, Englewood Cliffs, NJ.

2 NRC's Institute for Research in Construction-Building Envelope and Structure Research, *Durability and Repair of Concrete Structures*, http://irc.nrc-cnrc.gc.ca/bes/drcs.html. 10–1-2002, 1–5-2005.

3 L. Pel, K. Hazrati, K. Kopinga, J. Marchand **1998**, *Magn. Reson. Imag.* 16, 525–528.

4 A. J. Bohris, U. Goerke, P. J. McDonald, M. Mulheron, B. Newling, B. Le Page **1998**, *Magn. Reson. Imag.* 16, 455–461.

5 C. Leech, D. Lockington, P. Dux **2003**, *Mater. Struct./Mater. Construct.* 36, 413–418.

6 J. Kaufmann, W. Studer, J. Link, K. Schenker **1997**, *Magaz. Concrete Res.* 49, 157–165.

7 L. Pel, K. Kopinga, E. F. Kaasschieter **2000**, *J. Phys. D, Appl. Phys.* 33, 1380–1385.

8 S. D. Beyea, B. J. Balcom, T. W. Bremner, P. J. Prado, A. R. Cross, R. L. Armstrong, P. E. Grattan-Bellew **1998**, *Solid State Nucl. Magn. Reson.* 13, 93–100.

9 P. F. D. Cano-Barrita, B. J. Balcom, T. W. Bremner, M. B. MacMillan, W. S. Langley **2004**, *Mater. Struct./Mater. Construct.* 37, 522–531.

10 F. d. Cano Barrita, T. W. Bremner, B. J. Balcom **2003**, *Magaz. Concrete Res.* 55, 517–524.

11 C. Choi, B. J. Balcom, S. D. Beyea, T. W. Bremner, P. E. Grattan-Bellew, R. L. Armstrong **2000**, *J. Appl. Phys.* 88, 3578–3581.

12 P. J. Prado, B. J. Balcom, S. D. Beyea, R. L. Armstrong, T. W. Bremner **1997**, *Solid State Nucl. Magn. Reson.* 10, 1–8.

13 J. J. Young, P. Szomolanyi, T. W. Bremner, B. J. Balcom, *Cement Concrete Res.* **2004**, 34, 1459–1466.

14 F. D. J. Cano, T. W. Bremner, R. P. McGregor, B. J. Balcom **2002**, *Cement Concrete Res.* 32, 1067–1070.

15 J. J. Young, B. J. Balcom, T. W. Bremner, M. D. A. Thomas, K. Deka **2004**, *SP-222: Seventh CANMET/ACI International Conference on Recent Advances in Concrete Technology*, Las Vegas, NV, pp. 231–238.

16 B. J. Balcom, R. P. MacGregor, S. D. Beyea, D. P. Green **1996**, *J. Magn. Reson., Series A* 123, 131.

17 M. Halse, D. J. Goodyear, B. MacMillan, P. Szomolanyi, D. Matheson, B. J. Balcom **2003**, *J. Magn. Reson.* 165, 219–229.

18 P. Szomolanyi, D. Goodyear, B. Balcom, D. Matheson **2001**, *Magn. Reson. Imag.* 19, 423–428.

19 M. Bogdan, B. J. Balcom, T. W. Bremer, R. L Armstrong **1995**, *J. Magn. Reson., Series A* 116, 266–269.

20 W. P. Halperin, J. Y. Jehng, Y. Q. Song **1994**, *Magn. Reson. Imag* 12, 169–173.

21 S. Emid, J. H. N. Creyghton **1985**, *Physica B & C* 128, 81–83.

22 S. Gravina, D. G. Cory **1994**, *J. Magn. Reson., Series B* 104, 53–61.

23 S. D. Beyea, B. J. Balcom, P. J. Prado, A. R. Cross, C. B. Kennedy, R. L. Armstrong, T. W. Bremner **1998**, *J. Magn. Reson.* 135, 156–164.

24 K. Deka, M.B. MacMillan, A. V. Ouriadov, I. V. Mastikhin, J. J. Young, P. M. Glover, G. R. Ziegler, B. J. Balcom **2005**, *J. Magn. Reson.* in press.

25 S. D Beyea **2000**, University of New Brunswick PhD Thesis.

26 A. M. Neville **1995**, *Properties of Concrete*, 4th and final edn, Longman, Harlow.

27 F. d. Cano Barrita **2002**, University of New Brunswick PhD Thesis.

28 F. d. Cano Barrita, T. W. Bremner, B. J. Balcom **2003**, *Magaz. Concrete Res.* 55, 517–524.

29 M. Pigeon, R. Pleau **1995**, *Durability of Concrete in Cold Climates*, E & FN Spon, London.

30 J. H. Strange, M. Rahman, E. G. Smith **1993**, *Phys. Rev. Lett.* 71, 3589–3591.

31 J. Y. Jehng, D. T. Sprague, W. P. Halperin **1996**, *Magn. Reson. Imag.* 14, 785–791.

32 B. J. Balcom, J. C. Barrita, C. Choi, S. D. Beyea, D. J. Goodyear, T. W. Bremner **2003**, *Mater. Struct./Mater. Construct.* 36, 166–182.

33 A. Plassais, M. P. Pomies, N. Lequeux, P. Boch, J. P. Korb **2001**, *Magn. Reson. Imag.* 19, 493–495.

34 R. L. Kleinberg, W. E. Kenyon, P. P. Mitra **1994**, *J. Magn. Reson. Series A* 108, 206–214.

35 E. M. Haacke, W. R. Brown, R. M. Thompson, R. Venkatesan **1999**, *Magnetic Resonance Imaging Physical Principles and Sequence Design*, J. Wiley & Sons, New York.

36 *Bruker Almanac*, **1991**.

37 W. J. McCoy, A. G. Caldwell **1951**, *J. Am. Concrete Inst.* 22, 693–706.

38 S. Diamond **1999**, *Cement Concrete Res.* 29, 1271–1275.

3.5
NMR Imaging of Functionalized Ceramics
S. D. Beyea, D. O. Kuethe, A. McDowell, A. Caprihan, and S. J. Glass

3.5.1
Introduction

Improved characterization of the morphological/microstructural properties of porous solids, and the associated transport properties of fluids imbibed into these materials, is crucial to the development of new porous materials, such as ceramics. Of particular interest is the fabrication of so-called "functionalized" ceramics, which contain a pore structure tailored to a specific biomedical or industrial application (e.g., molecular filters, catalysts, gas storage cells, drug delivery devices, tissue scaffolds) [1–3]. Functionalization of ceramics can involve the use of graded or layered pore microstructure, morphology or chemical composition.

Fabrication processing of these materials is highly complex, particularly for materials created to have interfaces in morphology or a microstructure [4–5], for example in co-fired multi-layer ceramics. In addition, there is both a scientific and a practical interest in studying the influence of a particular pore microstructure on the motional behavior of fluids imbibed into these materials [6–9]. This is due to the fact that the actual use of functionalized ceramics in industrial and biomedical applications often involves the movement of one or more fluids through the material. Research in this area is therefore bi-directional: one must characterize both how the spatial microstructure (e.g., pore size, surface chemistry, surface area, connectivity) of the material evolves during processing, and how this microstructure affects the motional properties (e.g., molecular diffusion, adsorption coefficients, thermodynamic constants) of fluids contained within it.

Characterization of these materials requires a method capable of measuring non-destructively and non-invasively not only the parameters of interest, but also how these parameters vary with space: that is, an imaging method. While X-ray and neutron scattering and electron microscopy are highly valuable methods for characterizing the structure of materials on a microscopic and atomic level directly [10], it is difficult to determine how averages of the relevant properties vary with space. X-ray tomography permits the imaging of materials, but only provides maps of density.

Unlike electron microscopy, the ultimate resolution ($\approx$10 µm) of Nuclear Magnetic Resonance Imaging (NMRI) will never approach that of the pore scale. NMR signals do, however, contain a wealth of information from nanoscale phenomena that is ideal for characterizing macroscopic spatial variations in materials designed to exhibit nanoscale behavior. NMRI provides statistical averages of the underlying structure and properties, resolved on a spatial scale of tens to hundreds of micrometers. Furthermore, magnetic resonance simultaneously permits elucidation of the interactions between fluids and porous solid substrates [11–12].

Magnetic resonance has been used extensively to study fluids in porous solids [13–15], including ceramics [16, 17]. In recent years, we have applied NMR imaging and relaxometry of perfluorinated fluids (gases and liquids) to the study of ceramics and other porous materials [18–22]. This chapter will survey several such avenues of research, and their application to the experimental characterization of functionalized ceramics.

3.5.2
Experimental Background

3.5.2.1 How Does Microstructural Confinement Affect the Properties of Fluids?
The study of how fluids interact with porous solids is itself an important area of research [6]. The introduction of wall forces and the competition between fluid–fluid and fluid–wall forces, leads to interesting surface-driven phase changes, and the departure of the physical behavior of a fluid from the "normal" equation of state is often profound [6–9]. Studies of gas–liquid phase equilibria in restricted geometries provide information on finite-size effects and surface forces, as well as the thermodynamic behavior of constrained fluids (i.e., shifts in phase coexistence curves). Furthermore, improved understanding of changes in phase transitions and associated critical points in confined systems allow for material science studies of pore structure variables, such as pore size, surface area/chemistry and connectivity [6, 23–25].

In porous media, liquid–gas phase equilibrium depends upon the nature of the adsorbate and adsorbent, gas pressure and temperature [24]. Overlapping attractive potentials of the pore walls readily overcome the translational energy of the adsorbate, leading to enhanced adsorption of gas molecules at low pressures. In addition, condensation of gas in very small pores may occur at a lower pressure than that normally required on a plane surface, as expressed by the Kelvin equation, which relates the radius of a curved surface to the equilibrium vapor pressure [25].

Conventional bulk measurements of adsorption are performed by determining the amount of gas adsorbed at equilibrium as a function of pressure, at a constant temperature [23–25]. These bulk "adsorption isotherms" are commonly analyzed using a kinetic theory for multilayer adsorption developed in 1938 by Brunauer, Emmett and Teller (the BET Theory) [23]. BET adsorption isotherms are a common material science technique for surface area analysis of porous solids, and also permit calculation of adsorption energy and fractional surface coverage. While more advanced analysis methods, such as Density Functional Theory, have been developed in recent years, BET remains a mainstay of material science, and is the recommended method for the experimental measurement of pore surface area. This is largely due to the clear physical meaning of its principal assumptions, and its ability to handle the primary effects of adsorbate–adsorbate and adsorbate–substrate interactions.

The substantial literature on the bulk characterization of porous materials using conventional techniques provides a useful foundation as a starting point for overlapping the basic physics of traditional materials analysis with the parameters that can be measured using NMR.

3.5.2.2 How is NMR Sensitive to Pore Microstructure?

NMR signals are highly sensitive to the unusual behavior of pore fluids because of the characteristic effect of pore confinement on surface adsorption and molecular motion. Increased surface adsorption leads to modifications of the spin–lattice (T_1) and spin–spin (T_2) relaxation times, enhances NMR signal intensities and produces distinct chemical shifts for gaseous versus adsorbed phases [17–22]. Changes in molecular motions due to molecular collision frequencies and altered adsorbate residence times again modify the relaxation times [26], and also result in a time-dependence of the NMR measured molecular diffusion coefficient [26–27].

The inherent sensitivity of NMR signals to the fluid-substrate interactions via a large number of mechanisms provides a direct connection between the NMR measurables, the pore structure and the motional characteristics of the imbibed fluid. While the large number of potential NMR variables makes the experimental design and analysis complex, it also provides the potential for a measurement method capable of measuring and spatially resolving the parameters of interest to functionalized ceramics.

Accomplishing this requires a fundamental understanding of how the physical properties of the fluid and the porous solid affect the NMR measurables, which then permits the reverse study of how to interpret the NMR measured parameters in terms of the structure/transport parameters of interest.

3.5.3
NMR Relaxation Behavior of Perfluorinated Gases

3.5.3.1 Introduction to Gas Phase Relaxation

In our NMR studies of ceramics we have largely focused on the use of perfluorinated gases such as CF_4, SF_6 and c-C_4F_8 as molecular probes of the ceramic microstructure. These gases have been shown to have strong NMR signals (due to high fluorine spin density), convenient thermodynamic properties and, most importantly, they are highly sensitive to their physical environment via changes in their relaxation times and molecular diffusion coefficients [18–22]. While the use of protonated gases is also possible, the perfluorinated gases used in our experiments provided more convenient relaxation behavior (i.e., shorter T_1) and thermodynamic properties, and the increased chemical shift of fluorine relative to protons leaves open the possibility of combining spectroscopic measurements with our current experiments. In addition, porous solids typically have no ^{19}F background signal.

In comparison with water, perfluorinated gases are easy to imbibe along with the fluid into the materials, even in materials with low permeability (as long as the pores are not closed), and the increased molecular diffusion coefficient permits rapid exploration of the molecular probes of the pore structure. While hyperpolarized gases, such as ^{129}Xe and ^{3}He, have been shown to be tremendously successful in studying some porous systems such as packed beds [28], they are expensive and difficult to use, and the effect of surface impurities in "real" materials often serves to rapidly negate the advantage of hyperpolarization. By utilizing the shortened T_1 values and increased NMR sensitive nuclei per molecule of gases such as SF_6, it is possible to obtain high sensitivity without the need for hyperpolarization [21].

For liquids, the dominant relaxation mechanism is the nuclear–nuclear dipole interaction, in which simple motion of one nucleus with respect to the other is the most common source of relaxation [12, 27]. In the gas phase, however, the physical mechanism of relaxation is often quite different. For gases such as the ones listed above, the dominant mechanism is the spin–rotation interaction, in which molecular collisions alter the rotational state of the molecule, leading to rotation-induced magnetic fluctuations that cause relaxation [27]. The equation governing spin-rotation relaxation is given by

$$\frac{1}{T_1} = M \frac{\tau}{1 + (\omega_0 - \omega_j)^2 \tau^2} \tag{3.5.1}$$

where τ is the rotational correlation time, ω_0 and ω_j are the nuclear and molecular Larmor rotational frequencies and M is a constant, which is different for each molecular structure.

It has been established by Dong and Bloom [29] and Courtney and Armstrong [30] that this equation can be translated into one based upon the kinetic theory of gases using

$$\tau = \frac{1}{\varrho \sigma \bar{v}} = \frac{C}{\varrho} \tag{3.5.2}$$

where ϱ is the molecular number density, σ is the rotational state decorrelation cross section and $\bar{v}$ is the mean molecular speed. Indeed, Courtney and Armstrong [30] noted that Eq. (3.5.1) may be written as

$$\frac{T_1}{\varrho} = A + \frac{B}{\varrho^2} \tag{3.5.3}$$

where A and B are simply fit parameters. Equations (3.5.2) and (3.5.3) can then be combined to relate T_1 in terms of τ:

$$T_1 = \frac{A}{\tau} + B\tau \tag{3.5.4}$$

where $A' = AC$ and $B' = B/C$.

However, these parameters are temperature and Larmor frequency dependent. Such a model can be conveniently parameterized in terms of collision frequency (f), so that the rate of decorrelation is given by

$$\frac{1}{\tau} = \frac{f}{b} \tag{3.5.5}$$

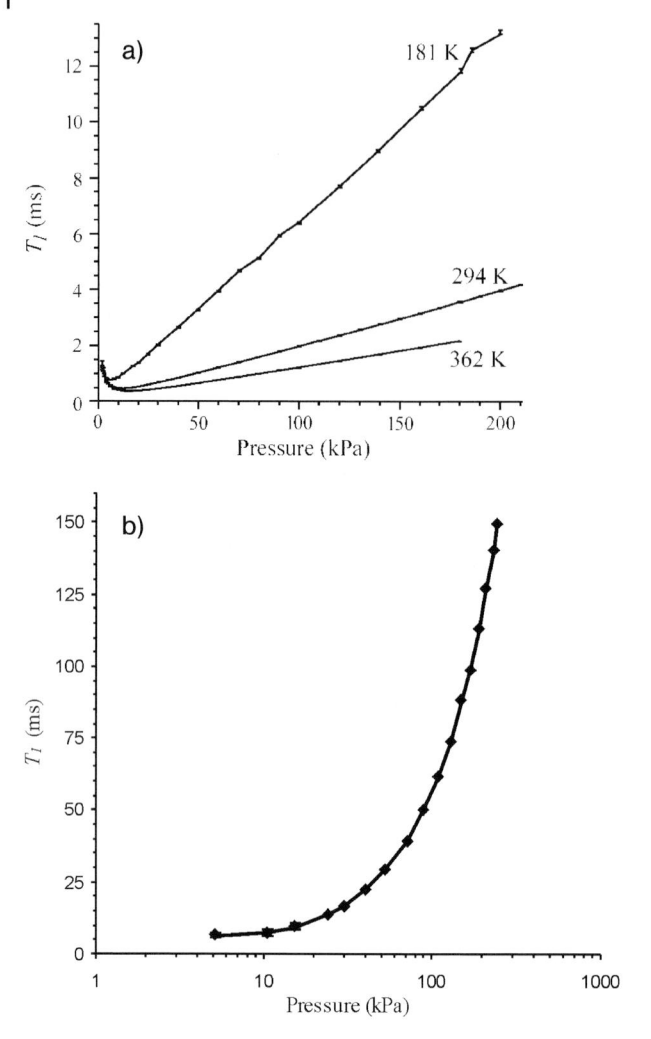

Fig. 3.5.1 Spin–lattice relaxation data for (a) CF$_4$ and (b) c-C$_4$F$_8$ gas as a function of pressure. The solid curve is the model prediction. Data for CF$_4$ were measured at 181, 294 and 362 K. Small temperature variations were measured for each T_1 data point, and were included in the model prediction, leading to the slight fluctuations observed in both the data and the model for CF$_4$ at 191 K. The x axis of the c-C$_4$F$_8$ data is shown logarithmically to highlight the agreement at low pressures.

where b is then simply the number of collisions required to randomize the rotational state. Using Lennard–Jones collision dynamics with formulaic temperature and Larmor frequency dependence, such a model can fit T_1 relaxation data as a function of gas pressure robustly for a variety of gases such as c-C$_4$F$_8$ and CF$_4$ (see Figure 3.5.1).

3.5.3.2 NMR Relaxation of Liquids Versus Gases in Porous Media

NMR relaxation of liquids such as water in porous solids has been studied extensively. In the fast exchange regime, the spin-lattice relaxation rate of water in pores is known to increase due to interactions with the solid matrix (so-called "surface relaxation"). In this case, T_1 can be described by Eq. (3.5.6):

$$\frac{1}{T_1} = \frac{1-\eta}{T_{1,\text{bulk}}} + \frac{\eta}{T_{1,\text{surf}}} \tag{3.5.6}$$

where η is the fraction of molecules at the surface, $T_{1,\text{bulk}}$ is the bulk fluid relaxation time and $T_{1,\text{surf}}$ is the relaxation of water molecules at the solid surface. For liquid phase relaxation, therefore, the relaxation rate increases as the pore size decreases (surface-to-volume ratio increases) [31].

For polyatomic gases in porous media, however, the relaxation rate commonly *decreases* as the pore size decreases [18–19]. Given that the relaxation mechanism is entirely different, this result is not surprising. If collision frequency determines the T_1, then in pores whose dimensions are in the order of the typical mean free path of a gas, the additional gas–wall collisions should drastically alter the T_1. For typical laboratory conditions, an increase in pressure (or collision frequency) causes a proportional lengthening of T_1, so the change in T_1 from additional wall collisions should be a good measure of pore size.

A parameterized model for gas relaxation as described can be easily modified to include the effect of wall collisions by utilizing the known additivity of correlation times

$$\frac{1}{\tau} = \frac{1}{\tau_{\text{gas}}} + \frac{1}{\tau_{\text{wall}}} \tag{3.5.7}$$

Combining Eqs. (3.5.5) and (3.5.7) gives us

$$\frac{1}{\tau} = \frac{f_{\text{gas}}}{b_{\text{gas}}} + \frac{f_{\text{wall}}}{b_{\text{wall}}} \tag{3.5.8}$$

The dominant relaxation mechanisms of polyatomic gases in porous solids are expected to be fairly different depending upon the phase of the fluid; i.e., collision mediated relaxation of gaseous molecules versus dipole–dipole surface relaxation of liquids. Depending upon the temperature and pressure, liquid and gaseous phases may coexist within the pores. Measurements of the gas T_1 values performed at temperatures above and below T_c will therefore exhibit dramatically different relaxation behavior as the equilibrium of gaseous and physisorbed species change. It is important to note, however, that while the measured relaxation times will be different above and below the critical temperature (i.e., the temperature at which bulk gas will not liquify at any pressure) due to the change in the phase equilibrium, the fundamental relaxation mechanisms remain the same.

As a first approximation, we can introduce wall collisions by simply introducing separate effectiveness parameters for molecule–molecule and molecule–wall collisions, in which f_{wall} can be described through kinetic theory as

$$f_{wall} = \frac{1}{4}\frac{S}{V}\bar{v} \tag{3.5.9}$$

where S/V is the surface-to-volume ratio. The unknown variable b is then simply a fit parameter that can be determined using materials with a known S/V.

3.5.4
Results and Discussion

3.5.4.1 Gas Relaxation
3.5.4.1.1 Temperatures Much Greater than the Critical Temperature

We have recently tested the T_1 model described above by obtaining T_1 measurements in powder samples with known S/V. Samples used were constructed from fumed silica (CAB-O-SIL M-5 and TS-500, Cabot Corp.), and were either hydrophilic (M-5) or treated by the manufacturer to be hydrophobic (TS-500). Powder of each type was pressed into a polycarbonate cylinder, with a degree of compression controlling the pore space volume of each sample. These materials have a very high specific surface area (200 m^2 g^{-1} for M-5, 212 m^2 g^{-1} for TS-500), which is not expected to change significantly even at the maximum compaction pressure used.

Samples were sealed and degassed using a standard vacuum station, and stored under vacuum conditions to prevent surface contamination. Inversion recovery measurements of T_1 as a function of pressure were obtained using semiconductor grade CF$_4$ gas (Air Products), with the pressures measured using an Omega PX303–100A5V absolute pressure transducer. Temperature was monitored using a copper–constantan thermocouple attached to the outside of the sample chamber, and the temperature varied between 292 and 298 K throughout the whole experiment (±0.3 K during an individual T_1 measurement). CF$_4$ has a critical temperature of 228 K. T_1 measurements were obtained in approximately 2 min using at least 50 delay times. Recovery curves were fit with a three-parameter single-exponential curve.

Figure 3.5.2 shows the results obtained using M-5 and TS-500 samples with S/V values of 3.03×10^7 and 3.28×10^7 m^{-1}, respectively, and porosities of 0.936 and 0.938, respectively. Note the significant deviation of the relaxation behavior from that of bulk CF$_4$ gas (dotted lines in Figure 3.5.2). The experimental data were first fitted to the model described above, assuming an increase in collision frequency due purely to the inclusion of gas–wall collisions, assuming normal bulk gas density. However, this model merely shifts the T_1 versus pressure curve to the left, whereas the data also have a steeper slope than bulk gas data. This pressure dependence can be empirically accounted for in the model via the inclusion of an additional fit parameter. Two possible physical mechanisms can explain the necessity of this parameter.

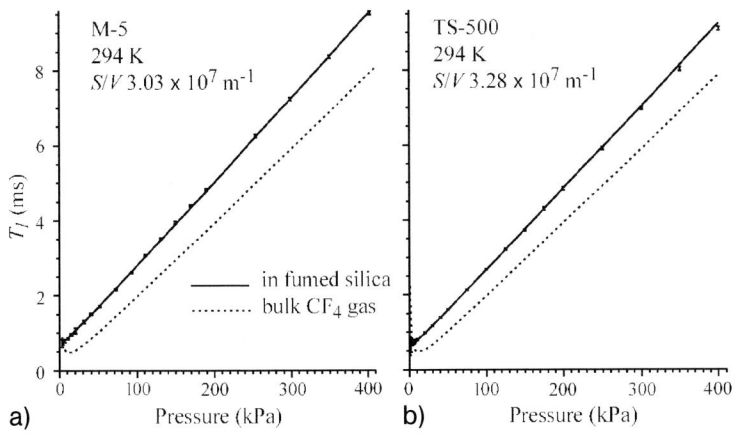

Fig. 3.5.2 T_1 of CF_4 in (a) fumed silica (M-5) and in (b) fumed silica with a hydrophobic surface coating (TS-500) as a function of pressure. The T_1 curves are not only shifted left because of the additional gas–wall collisions, but they also have steeper slopes than curves for bulk gas (dotted lines). Two physical models of gas molecules being adsorbed, either in the sense of surface adsorption or having high density within the attractive potentials of pore walls, predict the steeper slopes and fit the data equally well (solid lines). With calibration, the increase in T_1 above that of bulk gas is a measure of surface-to-volume ratio.

The first possibility is that the attractive potential associated with the solid surface leads to an increased gaseous molecular number density and molecular velocity. The resulting increase in both gas–gas and gas–wall collision frequencies increases the T_1. The second possibility is that although the measurements were obtained at a temperature significantly above the critical temperature of the bulk CF_4 gas, it is possible that gas molecules are adsorbed onto the surface of the silica. The surface relaxation is expected to be very slow compared with spin–rotation interactions in the gas phase. We can therefore account for the effect of adsorption by assuming that relaxation effectively stops while the gas molecules adhere to the wall, which will then act to increase the relaxation time by the fraction of molecules on the surface. Both models are in accord with a measurable increase in density above that of the bulk gas.

Regardless of its physical origin, Figure 3.5.2 demonstrates that the inclusion of an additional multiplicative parameter leads to good agreement between the model and the data. Whether or not either or both physical processes described above are present, the success of this empirical model permits the determination of S/V in materials.

3.5.4.1.2 Temperatures Much Less than the Critical Temperature

We have also obtained measurements for the relaxation time of octafluorocyclobutane (c-C_4F_8) gas in porous Vycor glass. Vycor (Corning Glass, ≈96.5 wt-% SiO_2) has a specific surface area of approximately 120 $m^2\,g^{-1}$, and was cleaned using H_2O_2 and evacuated to 10^{-8} Torr before all measurements. The Vycor was placed in

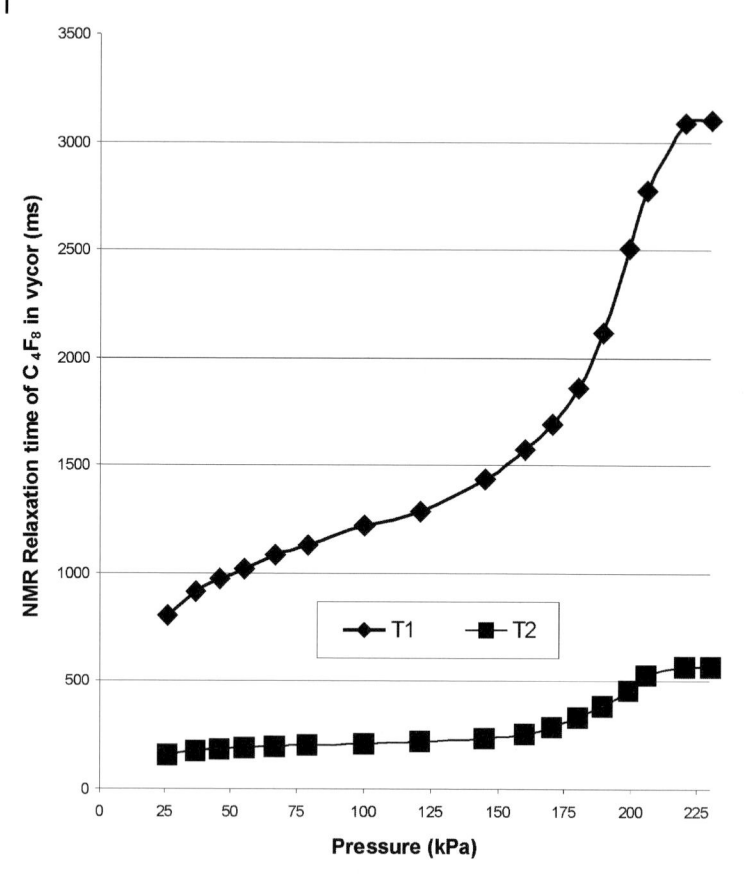

Fig. 3.5.3 Spin–lattice (T_1) and spin–spin (T_2) relaxation times as a function of pressure for c-C_4F_8 gas in porous Vycor glass at 291 K.

a sealed polycarbonate container that fitted tightly around the sample so as to occlude any bulk gas. This container was encased in a temperature controlled water jacket that maintained the temperature (as measured by a copper–constantan thermocouple attached to the container) at 291.2 ± 0.1 K. Given that the critical temperature of c-C_4F_8 is 388 K, measurements of relaxation time versus pressure should yield significantly different results, compared with CF_4, due to the fact that the phase equilibrium will now be driven towards the adsorbed phase.

As shown in Figure 3.5.3, the relaxation time versus pressure curves are dramatically different from those obtained using CF_4 at a temperature well above its critical point. Indeed, while the overall form of the T_1 curves for CF_4 in fumed silica was similar to that of the bulk gas, the shape of the T_1 plots for c-C_4F_8 in Vycor more closely resembles that of an adsorption isotherm (T_1 of CF_4 in Vycor is largely invariant with pressure, as gas–wall collisions in this material are more frequent than gas–gas collisions). This is not surprising given that we expect the behavior of this gas at 291 K to be shifted towards the adsorbed phase. The highest pressure measured was at the capillary condensation transition, at which point the gas completely condenses within the pores.

The measured NMR signal amplitude is directly proportional to the mass of adsorbate present, and the NMR signal versus pressure (measured at a fixed temperature) is then equivalent to the adsorption isotherm (mass of adsorbate versus pressure) [24–25]. As in conventional BET measurements, this assumes that the proportion of fluid in the adsorbed phase is significantly higher than the gaseous phase. It is therefore possible to correlate each relaxation time measurement with the calculated number of molecular layers of adsorbate, N (where $N = 1$ is monolayer coverage), also known as fractional surface coverage.

Figure 3.5.4 shows the T_1 for c-C_4F_8 in Vycor (normalized to the bulk gas T_1 at the same pressure) plotted as a function of fractional surface coverage. At low fractional surface coverage, the T_1 is significantly higher than the bulk T_1 at the same pressure. This is probably due to the fact that the behavior of c-C_4F_8 at low surface coverage is very similar to that of CF_4 in fumed silica; that is, the c-C_4F_8 is dominated by the gas phase and the relaxation is determined by collision mediated spin–rotation interactions. Given that the pore size of Vycor is 40 Å (approximately three orders of magnitude less than the mean free path), it is expected that the T_1 will be significantly higher than the bulk value at low surface coverage.

As the gas pressure (and therefore surface coverage) is increased, however, the relative T_1 difference steadily decreases with a minimum value at approximately 2–3 molecular layers. As the surface coverage exceeds a monolayer, we expect that the

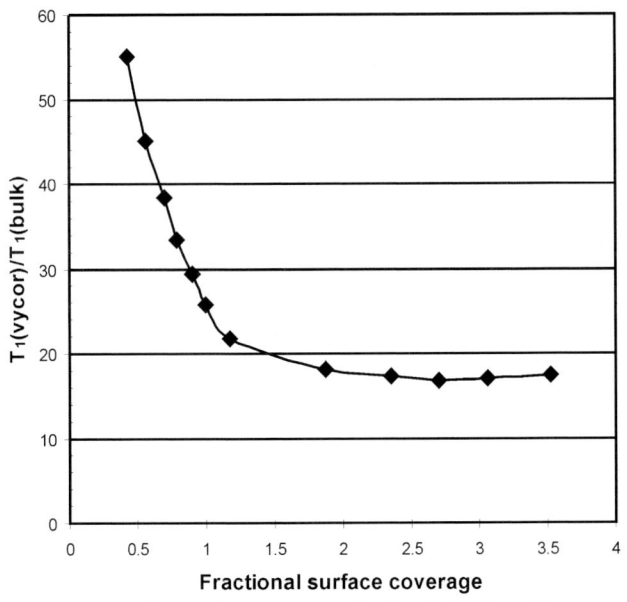

Fig. 3.5.4 Spin–lattice relaxation (normalized to the bulk gas T_1 at the same pressure) as a function of fractional surface coverage (i.e., the number of adsorbate monolayers).

residence time of the adsorbate will increase, and the behavior will shift from one that is dominated by gas phase relaxation (in fast exchange with surface molecules) to one that is dominated by liquid phase surface relaxation (in fast exchange with gas molecules). While the actual mechanisms remain the same, the difference lies in the relative weightings of the two relaxation mechanisms.

Whereas in CF_4 we can ignore the surface relaxation term, this term is significant for c-C_4F_8 at 291 K, with the relative weighting becoming increasingly important as we add additional molecular layers, as shown below. This is equivalent to Eq. (3.5.6), with the "bulk fluid" term set to the spin–rotation relaxation of the bulk gas. It is clear that in such a system, in comparison with CF_4 at 294 K, the effect of liquid phase surface relaxation cannot be ignored.

3.5.4.2 NMR Imaging
3.5.4.2.1 NMR Imaging of Functionalized Ceramics

Full evaluation of functionalized ceramics requires the ability to characterize the spatial variations in structure and morphology. Using NMRI, it is possible to map the underlying structure on a spatial scale of hundreds of microns.

As stated above, the NMR signal intensity can be directly proportional to the fluid density [11]. However, it is also clear that the fluid density may or may not be proportional to the underlying material density, depending upon the relative importance of gas adsorption. Fluid density images are only maps of material density if the pores are completely full of liquid, or if the effect of adsorption is minimized. Using a correctly chosen gas (e.g., CF_4, C_2F_6, SF_6) and temperature–pressure combination, combined with NMR acquisition parameters chosen to weight the images to density (i.e., long repetition times, TR, and short echo times, TE), the resulting images are direct maps of local porosity (and conversely, density). Density-weighted NMRI can, therefore, reveal underlying spatial variations in material density.

Indeed, even for materials designed to be homogeneous, residual density variations introduced can be detected. Figure 3.5.5 shows the radially averaged density map of a cylindrically shaped alumina (made from Sumitomo AKP-20 99.997% alumina powder) ceramic, obtained using NMR imaging of SF_6 gas. The experimental results were shown to agree with the corresponding X-ray computed tomography scans for the same materials, in which the materials were shown to

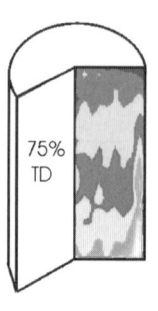

Fig. 3.5.5 Radially averaged density map of a "uniform" 75% Theoretically Dense (TD) Al_2O_3 ceramic cylinder, measured using NMR imaging of imbibed SF_6 gas. The maximum and minimum ceramic densities were 3.10–2.86 g cm^{-3}.

have a density gradation of 2.86–3.10 g cm^{-3}. The image was obtained at an ^{19}F Larmor frequency, using spin-echo volume imaging (5 ms echo time, repetition time of 50 ms, 16 averages), in a total imaging time of 1 h.

Results for C_2F_6 NMRI of co-axial cylinders of zinc oxide (ZnO) ceramic (from Aldrich Chemicals), are shown in Figure 3.5.6. This material was created using two separately pressed ceramic powder compacts. The inside and outside cylinders were compacted at different pressures, giving 0.57 relative density for the inner cylinder and 0.47 relative density for the outer cylinder. The two cylinders were then co-fired at 775 °C. Images of the individually fired components are also shown in Figure 3.5.6(a) and (b). The materials were placed in a Plexiglass holder contained within a polycarbonate cylinder, and were initially evacuated to 10^{-3} Torr before each experiment. The images were both obtained at an ^{19}F Larmor frequency using a gas pressure of 500 kPa, a 5-mm thick slice (5.4-ms echo time, repetition time of 400 ms) and 40 signal averages in a total imaging time of 17 min.

A small amount of bulk gas, which was not occluded by the Perspex holder, is visible in the images as a partial ring of elevated image intensity. Analysis of the images shows that the relative ceramic density (density$_{ring}$/density$_{center}$) of the inner and outer regions is different, depending upon whether they were individually fired or co-fired. Results showed a 4% increase in the relative density when the materials were co-fired. This clearly demonstrates that densification during sintering differs when the materials contain compositional or morphological interfaces [32].

NMRI of c-C_4F_8 in yttria-stabilized tetragonal zirconia polycrystal (Y-TZP) ceramics is shown in Figure 3.5.7. Following partial sintering these materials were dipped in a molten salt of aluminum nitrate nanohydrate for 10 min. The molten salt partially infiltrated the porous ceramic compact. The samples were then reheated to 600 °C (well below the sintering temperature) to decompose the infiltrated salt to an alumina precursor. The cylindrical samples were then cut in half axially. Figure 3.5.7 shows the NMR image of a longitudinal slice of two such

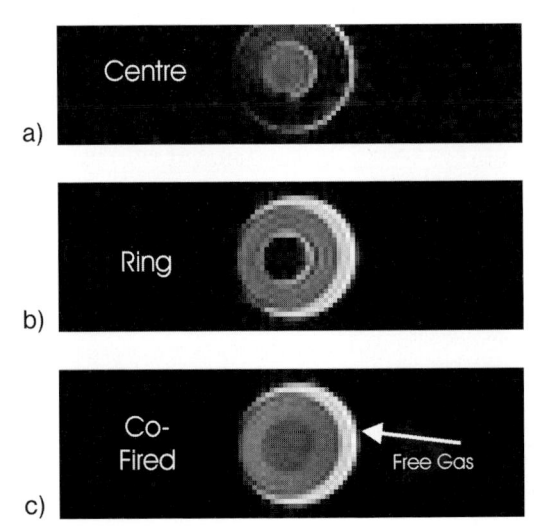

Fig. 3.5.6 (a) and (b) Density maps of two individually fired ZnO ceramics, and (c) a co-fired co-axial ceramic, obtained using NMR imaging of imbibed C_2F_6 gas.

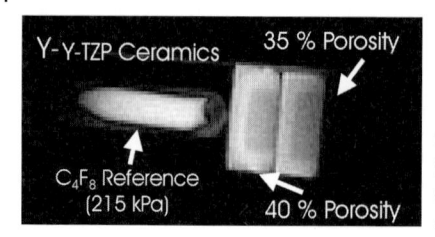

Fig. 3.5.7 NMR image of imbibed c-C$_4$F$_8$ gas in a 35 and 40% porosity Y-TZP ceramic, containing an alumina surface treatment. Adapted from Ref. [20].

samples with bulk porosities of 35 and 40%, held in a Plexiglass holder contained within a polycarbonate cylinder. The image was obtained at an ^{19}F Larmor frequency using a gas pressure of 175 kPa, a 7-mm thick slice (5.4-ms echo time, repetition time of 600-ms) and 16 signal averages in a total imaging time of 10 min, with an in-plane resolution of 0.75 × 0.75 mm.

The images clearly reveal the region infiltrated by the alumina salt. It is interesting to note, however, that the area containing the alumina exhibits an *increased* signal intensity, when imaged using c-C$_4$F$_8$ at 291 K, which was not present when imaged using C$_2$F$_6$ at similar pressures and temperatures [20]. One would not expect the presence of alumina particles to increase the material porosity. Images of c-C$_4$F$_8$ (i.e., of a gas near its condensation point) are expected to be maps of the adsorbed gas density, which is a function not only of porosity, but also of surface area and surface chemistry.

3.5.4.2.2 BET Adsorption Isotherm Imaging

The results given in Sections 3.5.4.1.1 and 3.5.4.1.2 clearly show that surface adsorption of gas can significantly affect the NMR results, depending upon the thermodynamic properties of the gas and the temperature and pressure at which the data were measured. Using NMR acquisition parameters chosen to weight the image to fluid density, the signal intensity is proportional to the mass of adsorbate contained within the pores. Therefore, the MR image at a specific pressure and temperature is a direct map of the local amount of gas adsorbed in each pixel. If we were to obtain MR images at a variety of continuously increasing and decreasing

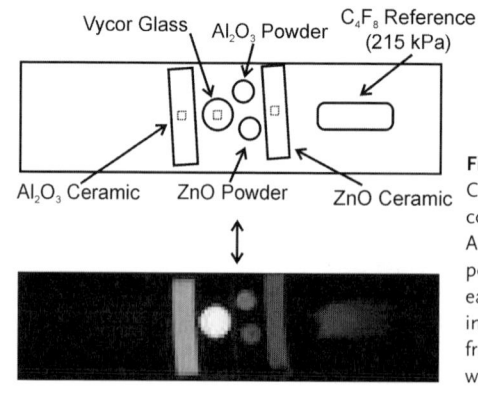

Fig. 3.5.8 Schematic and NMR image of C$_4$F$_8$ gas at 80 kPa in a hybrid phantom containing Vycor glass, a nanoparticulate Al$_2$O$_3$ powder, a nanoparticulate ZnO powder and sintered ceramics made from each of these powders. Dashed boxes indicate the regions of interest (ROIs) from which the isotherms in Figure 3.5.9 were extracted. Adapted from Ref. [21].

pressures, at a fixed temperature, the resulting data would be a pixel by pixel map of the amount of gas adsorbed as a function of pressure at a constant temperature: that is, the adsorption isotherm.

Recently we validated this concept using a compound material consisting of five individual materials with varying microstructural properties. The materials were Vycor glass, a nanoparticulate Al_2O_3 powder, a nanoparticulate ZnO powder and sintered ceramics made from each of these powders. The constituents were placed in a Plexiglass holder to occlude the bulk gas signal. When placed together in the imaging field-of-view these materials, while individually uniform, formed a mock heterogeneous phantom with a spatially varying microstructure. Figure 3.5.8 shows an NMR image of these materials at 80 kPa. A small sealed vial of C_4F_8 gas was used as a reference, and is used during post-processing of the data to correlate signal intensity with a fixed density of gas molecules.

The images were obtained at an ^{19}F Larmor frequency, using a slice thickness of 6 mm and in-plane resolution of 0.75×0.75 mm. The echo time in all images was 5.4 ms, and the TR was increased as a function of gas pressure so as to maintain a

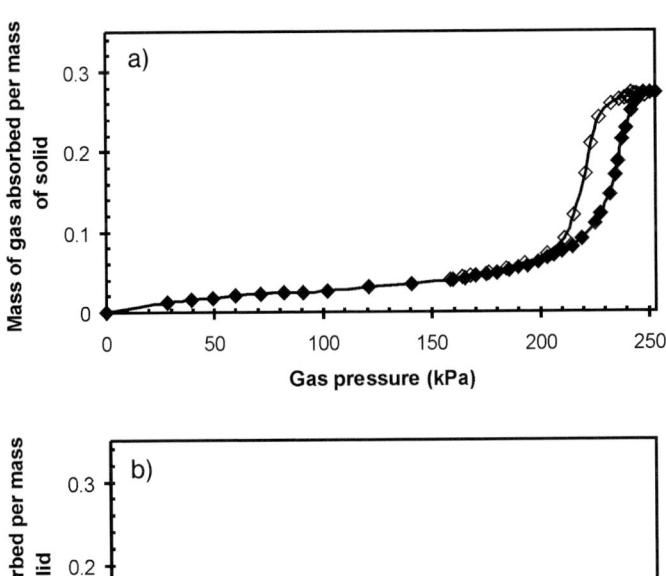

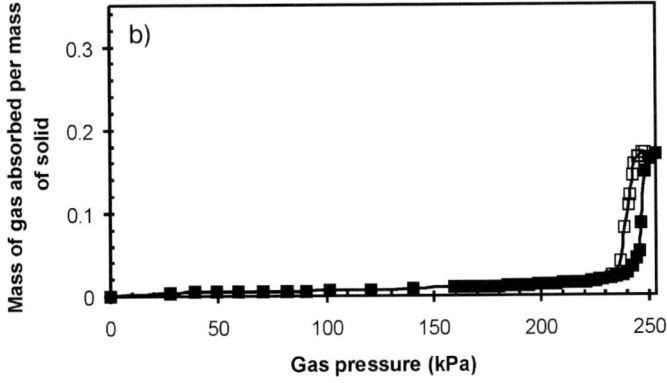

Fig. 3.5.9 NMR measured local adsorption/desorption isotherms extracted from ROIs within the (a) Al_2O_3 and (b) ZnO ceramics shown in Figure 3.5.8. Adapted from Ref. [21].

TR equal to five times larger than the measured bulk T_1. At low pressures, up to 32 signal averages were obtained, while at the highest pressures only two averages were necessary. Typical data acquisition times were approximately 20 min. Images were obtained at 38 increasing and 22 decreasing equilibrium pressures.

The adsorption/desorption isotherms measured by NMR (equivalent to conventionally measured isotherms), extracted from two different regions of the imaging field of view corresponding to the two ceramics, are shown in Figure 3.5.9. Once these local isotherms are extracted, they are simply the local adsorption for that point in space contained within the material, measured non-invasively and non-destructively. Conventional analysis techniques for adsorption isotherms (such as BET theory) can therefore be applied to the data, to determine the microstructural properties corresponding to that isotherm curve.

Table 3.5.1 shows the BET specific surface areas and adsorption energies measured by NMR [20–22] determined for the five materials, along with the conventionally measured N_2 BET specific surface areas. Adsorption energies using N_2 were not reported, as they are expected to be different because of variations in the gas properties. The NMR and conventionally measured results agree within error, over the full range of surface areas. Furthermore, the adsorption energies corresponding to each surface chemistry (i.e., Al_2O_3, ZnO and SiO_2) are significantly different for each of the three chemistries, while materials with the same chemistry have, within error, the same NMR measured adsorption energy. These results validate the accuracy of the isotherms extracted from the NMR images.

Tab. 3.5.1 BET results obtained using conventional bulk N_2 and NMR spatially resolved C_4F_8 adsorption isotherms.

Materials	Conventional N_2 specific surface area (m^2 g^{-1})[a]	NMR C_4F_8 specific surface area (m^2 g^{-1})[a]	NMR C_4F_8 net energy of adsorption (kJ mol^{-1})
ZnO powder	19[b]	17 ± 2[c]	8.7 ± 0.7
ZnO ceramic	7	7.5 ± 0.6	8.5 ± 0.5
γ-Al_2O_3 powder	41[b]	42 ± 1	5.8 ± 0.9
γ-Al_2O_3 ceramic	34	35 ± 2	5.6 ± 0.9
Vycor glass	112	119 ± 6	3.5 ± 0.4

a Data obtained at 77 K for N_2 and 291.2 K for c-C_4F_8.
b Data provided by manufacturer.
c Errors are propagated from the error values obtained for the slope and y intercept as determined by linear regression.

This method was also applied to the surface modified Y-TZP ceramics shown previously in Figure 3.5.7. Linearized BET plots [24] extracted from regions of interest within the 35% porosity ceramic corresponding to the outer ring (infiltrated by alumina) and the central region. Analysis of these data reveals that the relative surface area of the outer ring was 63% higher than the central region, due to the presence of the alumina particulate. The increased surface adsorption therefore explains the increased NMR image intensity of c-C_4F_8 in this region, in which the fine alumina particulate increased the surface area while occupying only a small pore volume fraction, leading to a decrease in porosity that was below the detection limits when performing porosity mapping using C_2F_6 NMR imaging.

3.5.5
Conclusions and Future Research

NMR signals are highly sensitive, via a number of different mechanisms, to the physical and chemical properties of porous materials. Using the set of NMR-based measurement methods that we have developed, it is possible to non-invasively and non-destructively characterize both the microstructural properties of the materials and relaxation properties of fluids imbibed into these materials.

Validation of these techniques has permitted the application of the methods to the study of a variety of functionalized ceramic materials. For example, research is currently underway into the study of the temporal evolution in microstructure and drug transport properties of resorbable calcium polyphosphate bioceramics. Such materials have been shown to hold great promise as implantable biomedical devices with the capacity for long-term continuous or variable rate drug delivery. MRI of non-hyperpolarized perfluorinated gases is also being applied to a variety of other porous systems, such as NMR imaging of lungs *in vivo*.

3.5.6
Description of NMR Equipment

All experiments were performed in a 1.9-T horizontal bore magnet (Oxford Instruments, Oxford, UK) with a clear bore diameter of 31 cm. Magnetic field gradients were produced by a 12-cm id water-cooled gradient set (Resonance Research, Billerica, MA, USA), capable of a maximum output of 300 mT m^{-1}, and were driven by Techron 7700 amplifiers (Techron Inc., Elkhart, IN, USA). Rf excitation was accomplished using either a quadrature driven birdcage coil (Morris Instruments, Ottawa, ON, Canada), or an 8-turn laboratory-built solenoid coil, driven by an ENI LPI-10 1000 W amplifier or a Matec Model 525 class-C amplifier.

Acknowledgements

The authors thank Kevin Ewsuk, Chris DiAntonio, Terry Garino, Dale Zschiesche and Denise Bencoe of Sandia National Laboratories for fabricating the ceramics and performing the N_2 adsorption measurements. a. F.M. and D. O. K. thank Dr. Doug Smith of Nanopore Inc. for fabrication of the fumed silica samples, and S. D. B. thanks Dr. Eiichi Fukushima of New Mexico Resonance and Dr. Mark

Filiaggi of Dalhousie University for useful discussions. Partial support for this work was provided by Sandia National Laboratories, a multiprogram laboratory operated by Sandia Corporation, a Lockheed–Martin Company, for the US Department of Energy under contract No. DE-AC04–94AL85000. D. O. K. acknowledges the support of the National Institutes of Health, and S. D. B. acknowledges the support of the Natural Sciences and Engineering Research Council of Canada.

References

1 T. Ishikawa, H. Yamaoka, Y. Harada, T. Fujii, T. Nagasawa **2002**, *Nature* 146, 64.
2 J. R. Groza **2002**, in *Nanostructured Materials – Processing, Properties and Potential Applications*, ed. C. C. Koch, William Andrew Publishing, New York, chap. 4.
3 N. Kossovsky, A. Gelman, E. Sponsler, H. Hnatyszyn **1994**, (Surface modified nanocrystalline ceramics for drug delivery application), *Biomaterials* 15, 1201.
4 S. J. Glass, D. J. Green **1999**, *J. Am. Ceram. Soc.* 82, 2745.
5 W. D. Kingery, H. K. Bowen, D. R. Uhlmann **1976**, *Introduction to Ceramics*, 2nd edn, John Wiley & Sons, New York.
6 L. D. Gelb, K. E. Gubbins, R. Radhakrishan, M. Sliwinska-Bartkowiak **1999**, *Rep. Prog. Phys.* 62, 1573.
7 M. E. Kainourgiakis, T. A. Steriotis, E. S. Kikkinides, G. Romanos, A. K. Stubos 2002, *Colloids Surf. A* 206, 321.
8 B.-H. Eom, I. Yu **1997**, *J. Kor. Phys. Soc.* 31, 833.
9 L. D. Gelb, K. E. Gubbins **1998**, *Langmuir* 14, 2097.
10 T. Garino, M. Mahoney, M. Readey, K. Ewsuk, J. Gieske, G. Stoker, S. Min **1995**, *Proceedings of the 27th International SAMPE Technical Conference*, Albuquerque, NM, USA, p. 610.
11 P. T. Callaghan **1991**, *Principles of NMR Microscopy*, Clarendon Press, Oxford.
12 B. Bluemich **2000**, *NMR Imaging of Materials*, Clarendon Press, Oxford.
13 T. W. Watson, C. T. P. Chang **1998**, (Characterizing porous media with NMR methods), *Prog. NMR Spectrosc.* 31, 343–386.
14 E. Fukushima **1999**, (Nuclear magnetic resonance as a tool to study flow), *Annu. Rev. Fluid. Mech.* 31, 95–123.
15 S. D. Beyea, B. J. Balcom, T. W. Bremner, R. L. Armstrong, P. E. Grattan-Bellew **2003**, (Detection of microcracking in cementious materials with space resolved 1H nuclear magnetic resonance relaxometry), *J. Am. Ceram. Soc.* 86 (5), 800–805.
16 W. A. Ellingson, J. L. Ackerman, L. Garrido, J. D. Weyand, R. A. Dimilia **1987**, (Characterization of porosity in green-state and partially densified Al_2O_3 by nuclear magnetic resonance imaging), *Ceram. Eng. Sci. Proc.* 8, 503–512.
17 M. J. Lizak, M. S. Conradi, C. G. Fry **1991**, (NMR imaging of gas imbibed into porous ceramic), *J. Magn. Reson.* 95, 548.
18 D. O. Kuethe, T. Pietrass, V. Behr **2005**, *J. Magn. Reson.* 177, 195–203.
19 A. Caprihan, C. F. M. Clewett, D. O. Kuethe, E. Fukushima, S. J. Glass **2001**, (Characterization of partially sintered ceramic powder compacts using fluorinated gas NMR imaging), *Magn. Reson. Imag.* 19, 311–317.
20 S. D. Beyea, A. Caprihan, C. F. M. Clewett, S. J. Glass **2002**, (Spatially resolved adsorption isotherms of thermally polarized perfluorinated gases in Y-TZP ceramic materials using NMR imaging), *Appl. Magn. Reson.* 22, 175–186.
21 S. D. Beyea, A. Caprihan, S. J. Glass, A. DiGiovanni **2003**, (Non-destructive characterization of nanopore microstructure: spatially resolved BET isotherms using nuclear magnetic resonance imaging), *J. Appl. Phys.* 94 (2), 935–41.
22 S. D. Beyea, S. L. Codd, D. O. Kuethe, E. Fukushima **2003**, (Studies of porous media by thermally polarized gas NMR: current status), *Mag. Reson. Imag.* 21 (3–4), 201–205.

23 S. Brunauer, P. H. Emmett, E. Teller **1938**, *J. Am. Chem. Soc.* 60, 309.
24 S. J. Gregg, K. S. W. Sing **1982**, *Adsorption, Surface Area and Porosity*, 2nd edn, (Academic Press Inc., New York.
25 S. Lowell, J. E. Shields **1991**, *Powder Surface Area and Porosity*, 3rd edn, Chapman & Hall, New York.
26 R. L. Armstrong **1987**, (Magnetic resonance relaxation effects in polyatomic gases), *Magn. Reson. Rev.* 12, 91–135.
27 R. Kimmich **1997**, *NMR Tomography, Diffusometry, Relaxometry*, Springer-Verlag, Berlin.
28 R. W. Mair, G. P. Wong, D. Hoffman, M. D. Hurliman, S. Patz, L. M. Schwartz, R. L. Walsworth **1999**, (Probing porous media with gas diffusion NMR), *Phys. Rev. Lett.* 83, 3324–3327.
29 R. Y. Dong, M. Bloom **1970**, (Determination of spin-rotation constants in fluorinated methane molecules by means of nuclear spin relaxation measurements), *Can. J. Phys.* 48, 793.
30 J. A. Courtney, R. L. Armstrong **1972**, (A nuclear spin relaxation study of the spin-rotation interaction in spherical top molecules), *Can. J. Phys.* 50, 1252.
31 R. L. Kleinberg, W. E. Kenyon, P. P. Mitra **1994**, (Mechanism of NMR relaxation in rock), *J. Magn. Reson. A* 108, 206–214.
32 S. J. Glass, T. J. Garino, J. G. Argüello, A. F. McDowell, S. D. Beyea, M. W. Reiterer, K. G. Ewsuk, (Sintering experiments, finite element modeling, and NMR characterization to assess the effects of green density and constrained sintering on ZnO densification and microstructure), in preparation.

3.6
NMR Applications in Petroleum Reservoir Studies
George J. Hirasaki

3.6.1
Introduction

NMR well logging and down-hole fluid analysis are tools to evaluate the properties of the geologic formations and their fluids during drilling or soon after drilling [1, 2]. The "porosity," or the fluid storage capacity of rock, is measured directly because the initial amplitude of ^{1}H NMR is proportional to the water and hydrocarbon contents of the rock. Minerals such as silica and calcite do not have hydrogen and thus do not contribute to the signal. The pore size distribution of water-filled pores can be estimated because the relaxation time is proportional to the volume to surface ratio of the pore. The "permeability" of the rock is estimated from the porosity and parameters of the estimated pore size distribution. When hydrocarbons accumulate in a rock formation, not all of the water is displaced but a fraction, known as the "irreducible water saturation," is retained by capillarity in the small pores, such as between clay flakes. This irreducible water saturation, or the "clay and capillary bound water", is estimated from the cumulative relaxation time distribution as a fraction of the pore space with a relaxation time shorter than a "cut-off" value of the relaxation time.

The response from the water and hydrocarbon can be distinguished by measuring the distributions of the diffusion coefficients simultaneously with the distributions of the relaxation times. The resulting distributions are displayed on a two-dimensional diffusion coefficient–relaxation time map and the distributions for brine, oil and gas can usually be distinguished. The relative magnitude of each of the distributions, corrected by the "hydrogen index" (the density of ^1H relative to that of water), determines the "saturation" (fraction of the pore space occupied by each fluid phase) occupied by brine, oil and free-gas. The oil viscosity and the amount of solution-gas in the "live-oil" are estimated from the position of the oil distribution relative to the correlation line for liquid hydrocarbons (dead-oil).

These rock and fluid properties are estimated based on several assumptions. The estimation of pore size from the relaxation time distribution of water-filled pores is based on the assumption that the fast-diffusion limit is valid and no diffusional-coupling between pores of different size is occurring. The estimation of irreducible water saturation at some specified capillary pressure assumes that only water occupies the smallest pores, which corresponds to the clay and capillary-bound water, and these pores have no diffusional-coupling with larger pores. The estimation of diffusion coefficients assumes that induced or internal field gradients due to magnetic susceptibility differences between the fluids and minerals are negligible. The estimation of oil viscosity and solution-gas from the oil diffusivity–relaxation time distribution assumes that the oil relaxes as the bulk fluid, that is, there is no contribution from the surface relaxation due to the oil wetting the pore walls.

3.6.2
NMR Well Logging and Fluid Analysis

3.6.2.1 NMR Wire-line Logging

Well logging is the acquisition of a continuous record of the geologic formation properties with well depth. A multitude of tools are used simultaneously and NMR is only one of the many measuring instruments on the logging tool. The NMR wire-line module is typically about 13-cm diameter and 7-m length. The measurement is usually made while lowering or raising the tool, but sometimes stationary measurements are made for T_1 at specific depths. Logging tools differ from laboratory spectrometers and MRI instruments as the tool is inside the bore hole and the sample is outside. Various vendors have different antenna designs. The measured volume of the formation can be either concentric cylindrical shells outside the bore hole or on one side of the bore-hole wall [1, 2]. A field gradient is associated with the magnetic field, as the strength of the magnetic field declines away from the magnet. The depth of investigation ranges from about 1 to 11 cm. More than one depth of investigation can be measured simultaneously by acquiring data at several discrete Larmor frequencies. This makes it possible to measures the profile due to flushing by invasion of the drilling fluid filtrate. NMR logging tools operate at a frequency of about 1 MHz. The measured frequency is changed to compensate for the change in the magnetic field strength with change in temperature of the magnet.

3.6.2.2 Measuring While Drilling

NMR measurements can be made while drilling with the tool placed behind the drill bit and information transmitted to the surface [3, 4]. Measurements of T_1 are more reliable because it is less sensitive to motion, but improved tool design has made measurements of T_2 possible [4].

3.6.2.3 Fluid Analysis Tool

In situ NMR measurements can be made in conjunction with down-hole fluid sampling [5, 6]. The NMR relaxation time and diffusivity can be measured under high-temperature, high-pressure reservoir conditions without loss of dissolved gases due to pressure depletion. In cases when the fluids may be contaminated by invasion of the filtrate from oil-based drilling fluids, the NMR analysis can determine when the fluid composition is approaching that of the formation [5, 6].

3.6.3
NMR Measurements

3.6.3.1 Inversion Recovery or Saturation Recovery for T_1

Data acquisition for T_1 distribution by inversion recovery or saturation recovery is slow (compared with T_2 measurements). Such measurements are made with the logging tool stationary (at a fixed depth). If the measurements were made at normal logging speed, the signal would be averaged over a range of depths and depth resolution would be poor. They are not frequently used in wire-line logging measurements. On the other hand, inversion recovery or saturation recovery measurements are commonly used with measuring-while-drilling or down-hole fluid analysis tools because these methods are not sensitive to tool motion or flowing fluids (compared with T_2 measurements). The rate of change in depth while drilling is slow compared with wire-line logging and fluid sampling is at a fixed depth.

3.6.3.2 Gradient CPMG for T_2 and Diffusion

The CPMG sequence is commonly used to measure the T_2 relaxation time distribution. It can also measure diffusion as the magnetic fields of logging tools are not homogeneous, so that the echo signal amplitude is sensitive to diffusion as a function of the echo spacing and the strength of the magnetic field gradient at the sensitive volume of the tool. The CPMG sequence can also be made sensitivity to T_1 by varying the polarization or "wait" time between CPMG sequences.

3.6.3.3 Diffusion-editing Sequences

When the measurement of a distribution of diffusivity is desired, improved signal-to-noise is needed. This can be achieved by using a diffusion-editing pulse sequence [7] rather than the CPMG sequence with varying echo spacing. This method varies the spacing between the first three pulses for sensitivity to diffusion, and the subsequent echoes are collected with short echo spacing for signal accumulation and relaxation information.

3.6.4
NMR Fluid Properties

Accurate interpretation of the formation properties (porosity, permeability and irreducible water saturation) requires reliable estimates of NMR fluid properties or the relationship between diffusivity and relaxation time. Estimation of oil viscosity and solution-gas content require their correlation with NMR measurable fluid properties. These include the hydrogen index, bulk fluid relaxation time and bulk fluid diffusivity [8].

3.6.4.1 Hydrogen Index

The initial amplitude (before attenuation by relaxation or diffusion) of the NMR response from a logging tool or laboratory spectrometer is calibrated by a measurement with water as a standard. The ratio of the initial amplitude of some other fluid to that of water is the Hydrogen Index or HI of the fluid. The hydrogen index of a fluid is defined as the ratio of the ^{1}H density of a fluid compared with that of water under standard conditions. Brine will have an HI of less than unity at high salinity [8]. The HI of gases can be calculated from gas composition and PVT (Pressure–Volume–Temperature) analysis. The light crude oils and refined oils (density less than 0.9 g cm^{-3} under standard conditions) at ambient conditions have an HI close to unity [8]. Heavier crude oils ($\varrho > 0.9$ g cm^{-3}) often have lower measured values for HI because the spectrometer or logging tool may not detect fast-relaxing components and the aromatic hydrocarbons have lower HI as a result of lower H/C ratios [9, 10]. Live oils with a significant amount of dissolved gas and elevated temperature may have an HI of significantly less than unity [10].

3.6.4.2 Relaxation Time

Degassed water under ambient conditions has a relaxation time (T_1 and T_2) of about 4 s at 30 °C [11, 12]. However, air-saturated brines may have a relaxation time of about 2–3 s. Light hydrocarbons are even more sensitive to dissolved oxygen [10]. For example, the relaxation time of deoxygenated pentane is 14 s while air-saturated pentane is about 3 s. The correlation for degassed alkanes between the relaxation time (T_1), viscosity (η) and temperature (T) is given by Eq. (3.6.1) [13].

$$T_1 = 0.0096 \frac{T}{\eta} \tag{3.6.1}$$

for T_1 in seconds, T in K and η in centipoise.

Morriss et al. [14] correlated the log mean (LM) T_2 relaxation times with viscosity for a number of crude oil samples and viscosity standards using Eq. (3.6.2), as is illustrated in Figure 3.6.1,

$$T_{2,\text{LM}} = \frac{1.2}{\eta^{0.9}} \tag{3.6.2}$$

for $T_{2,\text{LM}}$ in seconds and η in centipoise.

Live oil with dissolved methane does not follow the above correlations as methane relaxes by a spin–rotation mechanism, even when dissolved in liquid hydrocarbons [13]. The T_1 relaxation time as a function of η/T is illustrated in Figure 3.6.2 for different gas/oil ratios expressed in units of $m^3\ m^{-3}$ as a parameter. The solid line is the fit for zero gas/oil ratio and is given by Eq. (1).

The relationship between T_1 and T_2 was examined for a number of liquid alkanes and crude oils [15]. It was concluded that there is no difference for light oils, apparently because light oils satisfy the fast-motion condition (the correlation time is less than the Larmor period). However, viscous oils do not satisfy this condition as the departure between T_1 and T_2 correlates with an increasing viscosity and Larmor frequency.

The relaxation of gaseous methane, ethane and propane is by the spin–rotation mechanism and each pure component can be correlated with density and temperature [15]. However, the relaxation rate is also a function of the collision cross section of each component and this must be taken into account for mixtures [16]. This is in contrast to the liquid hydrocarbons and their mixtures that relax by dipole–dipole interactions and thus correlate with the viscosity/temperature ratio.

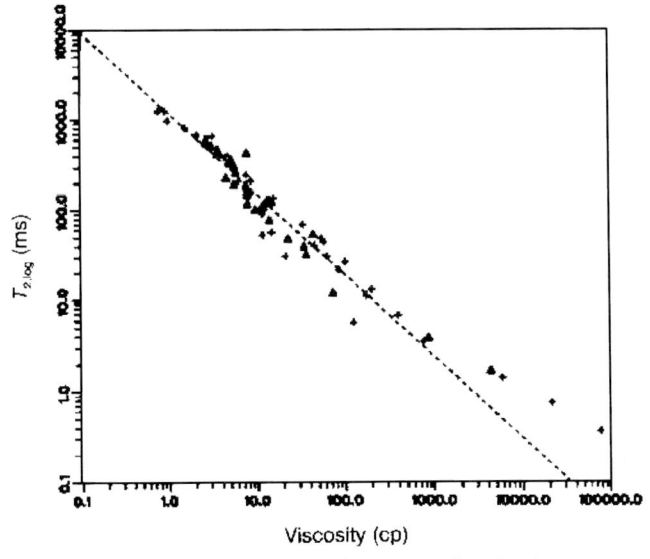

Fig. 3.6.1 Correlation of relaxation with viscosity of crude oils and viscosity standards [14].

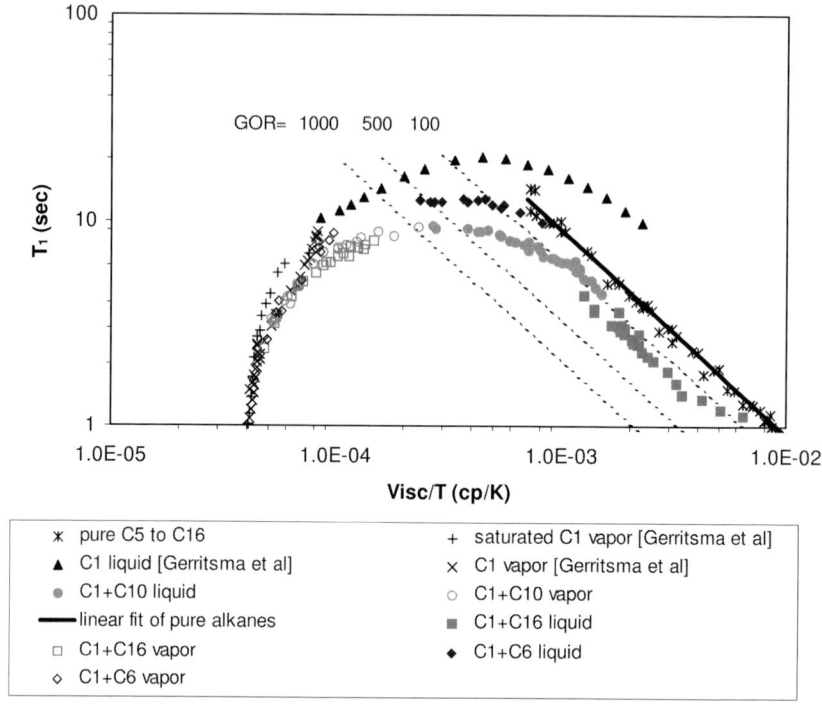

Fig. 3.6.2 Relaxation time of pure alkanes or methane saturated alkanes as a function of viscosity, temperature and gas/oil ratio (GOR, $m^3\ m^{-3}$) [13]. The solid line is for zero GOR. The dashed lines are for the indicated GOR.

3.6.4.3 Diffusivity

Diffusivity correlates linearly with the ratio of temperature and viscosity. Therefore the diffusivity can also be expected to correlate with relaxation time because the latter correlates with temperature and viscosity according to Eq. (3.6.1). Figure 3.6.3 illustrates the correlation between relaxation time and diffusivity with the gas/oil ratio as a parameter [13]. The correlation between diffusivity and relaxation time extends to hydrocarbon components in a mixture and there is a mapping between the distributions of diffusivity and relaxation time for crude oils [17].

3.6.5 Porosity

Porosity (ϕ) determination with NMR is a direct measurement as the response is from the fluid(s) in the pore space of the rock. The initial amplitude (before relaxation) of the NMR response of the fluid(s) saturated rock (corrected for hydrogen index) is compared with the amplitude of the response of bulk water having the same volume as the bulk volume of the rock sample. The 2 MHz NMR

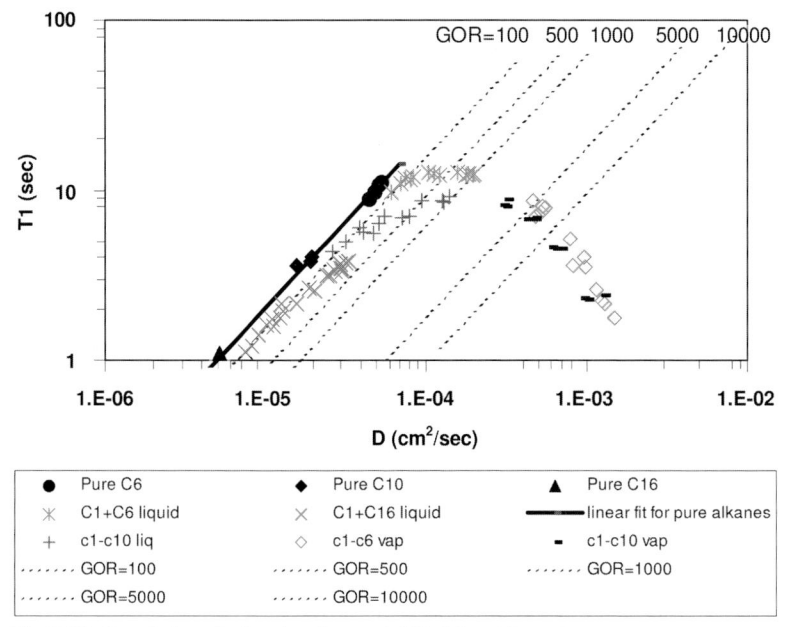

3.6 NMR Applications in Petroleum Reservoir Studies | 327

Fig. 3.6.3 Correlation of relaxation time and diffusivity with gas/oil ratio (GOR, $m^3\ m^{-3}$) as a parameter [13]. The solid line is for zero GOR. The dashed lines are for the indicated GOR.

porosity is compared with buoyancy porosity for 192 sandstone samples in Figure 3.6.4 [18]. The comparison between the NMR porosity and buoyancy porosity was less favorable when the CPMG sequence was used for NMR porosity in an 85 MHz spectrometer. The higher field strength results in higher internal field gradients due to paramagnetic material in the rock. These large induced gradients resulted in some portion of the sample relaxing faster than the CPMG echo spacing and could not be detected adequately. For this reason, high-field NMR is not usually used for NMR logging or laboratory rock measurements.

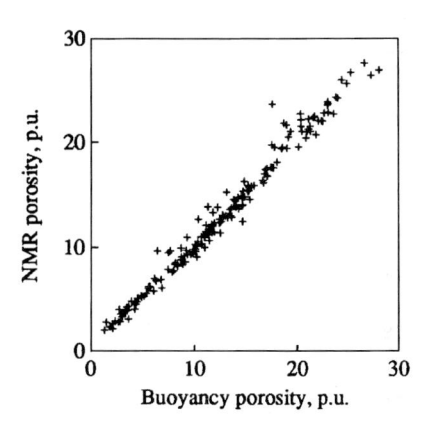

Fig. 3.6.4 Comparison of 2-MHz NMR porosity with buoyancy porosity for 192 sandstones [18].

3.6.6
Surface Relaxation and Pore Size Distribution

3.6.6.1 Fast Diffusion Limit

The key to obtaining pore size information from the NMR response is to have the response dominated by the surface relaxation rate [19–26]. Two steps are involved in surface relaxation. The first is the relaxation of the "spin" while in the proximity of the pore wall and the other is the diffusional exchange of molecules between the pore wall and the interior of the pore. These two processes are in series and when the latter dominates, the kinetics of the relaxation process is analogous to that of a stirred-tank reactor with first-order surface and bulk reactions. This condition is called the "fast-diffusion limit" [19] and the kinetics of relaxation are described by Eq. (3.6.3):

$$M(t) = M(0)\, e^{-t/T_{1,2}}$$

$$\frac{1}{T_1} = \frac{1}{T_{1,b}} + \varrho_1 \frac{S}{V}, \qquad \text{for } \varrho_1 \frac{a}{D} \ll 1 \qquad (3.6.3)$$

$$\frac{1}{T_2} = \frac{1}{T_{2,b}} + \varrho_2 \frac{S}{V} + \frac{(\gamma G t_E)^2 D}{12}, \qquad \text{for } \varrho_2 \frac{a}{D} \ll 1$$

where $M(t)$ is the magnetic response at time t; $T_{1,2}$ is T_1 or T_2; $T_{(1,2),b}$ is T_1 or T_2 of the bulk fluid; $\varrho_{(1,2)}$ is the surface relaxivity for T_1 or T_2; S/V is the surface to volume ratio of the pore; a is the characteristic length of the pore; D is the fluid diffusivity; γ is the gyromagnetic ratio; G is the magnitude of the gradient of the magnetic field; t_E is the echo spacing.

3.6.6.2 Measurement of Surface Relaxivity

The surface relaxivity is the parameter that correlates relaxation time with pore size [2, 21–23]. The surface relaxivity is generally larger for sandstone compared with carbonates and larger for water compared with oil. Wettability is an important factor in surface relaxation because a thin film of another phase on the pore walls can prevent surface relaxation of the non-wetting phase. The estimated value of the surface relaxivity is dependent on the method used to estimate the surface/volume ratio of the porous medium. Image analysis of pore sizes generally underestimates surface area because it may not be able to resolve the smaller pores that have a larger specific surface area [24]. Surface area measured by the BET method tends to overestimate the effective pore surface area because this method measures the total surface area, including that of microporous clay [25, 26]. Restricted diffusion measurements for estimation of the surface/volume ratio are weighted more to larger pore sizes [26]. These larger pores are usually representative of the characteristic pore size that correlate with formation permeability. The most common method of estimating surface relaxivity for consolidated reservoir rocks is by using mercury capillary pressure curves, as will be discussed later.

Temperature and measured surface relaxivity have shown only a weak dependence [21, 27] in some cases, whereas the opposite temperature dependence between silicate and carbonate surfaces has been shown in others [28].

3.6.6.3 Pore Size Distribution

Natural rocks seldom have a single pore size but rather a distribution of pore sizes. If all pores are in the fast-diffusion limit, have the same surface relaxivity and have no diffusional coupling, then the pores will relax in parallel with a distribution of relaxation times that corresponds to the distribution of the pore sizes. The magnetization will decay as a sum of the exponentials as described by Eq. (3.6.4).

$$M(t) = \sum_i f_i\, e^{-t/T_i}$$

$$M(0) = \sum_i f_i \qquad (3.6.4)$$

The relaxation time for each pore will still be expressed by Eq. (3.6.3) where each pore has a different surface/volume ratio. Calibration to estimate the surface relaxivity is more challenging because now a measurement is needed for a rock sample with a distribution of pore sizes or a distribution of surface/volume ratios. The mercury–air or water–air capillary pressure curve is usually used as an estimator of the cumulative pore size distribution. Assuming that all pores have the same surface relaxivity and ratio of pore body/pore throat radius, the surface relaxivity is estimated by overlaying the normalized cumulative relaxation time distribution on the capillary pressure curve [18, 25]. An example of this process is illustrated in Figure 3.6.5. The relationship between the capillary pressure curve and the relaxation time distribution with the pore radii, assuming cylindrical pores is expressed by Eq. (3.6.5).

$$P_c(S_w) = \frac{2\sigma\,\cos\theta}{r}$$

$$\frac{1}{T_{1,2}} - \frac{1}{T_{1,2,b}} = \varrho_{1,2}\frac{S}{V} = \varrho_{1,2}\frac{2}{r} \qquad (3.6.5)$$

where $P_c(S_w)$ is the capillary pressure; S_w is saturation of the wetting phase; σ is the surface tension; θ is the contact angle; and r is the pore radius.

The $T_{1,2}$ denotes T_1 or T_2, and if it is the latter, negligible diffusion effects are assumed. The surface relaxivity is estimated as the value that best overlays the two curves. The ratio of T_1/T_2, at a ^{1}H frequency of 2 MHz for 48 rocks had a median value of 1.6 with a standard deviation of 0.7 [29].

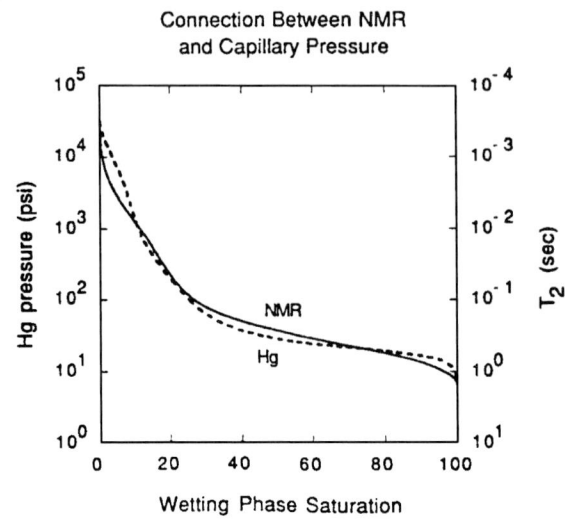

Fig. 3.6.5 Comparison of mercury injection porosimetry curve with NMR T_2 cumulative distribution [25].

3.6.7
Irreducible Water Saturation

3.6.7.1 FFI and BVI

If the NMR response is capable of estimating the pore size distribution, then it also has the potential to estimate the fraction of the pore space that is capable of being occupied by the hydrocarbon and the remaining fraction that will only be occupied by water. The Free Fluid Index (*FFI*) is an estimate of the amount of potential hydrocarbons in the rock when saturated to a given capillary pressure. It is expressed as a fraction of the rock *bulk* volume. The Bulk Volume Irreducible (*BVI*) is the fraction of the rock *bulk* volume that will be occupied by water at the same capillary pressure. The fraction of the rock *pore* volume that will only be occupied by water is called the irreducible water saturation ($S_{iwr} = BVI/\phi$). The amount of water that is irreducible is a function of the driving force to displace water, *i.e.*, the capillary pressure. Usually the specified driving force is an air–water capillary pressure of 0.69 MPa (100 psi).

Interpretation for irreducible water saturation assumes that the rock is water-wet or mixed-wet (water-wet during drainage but the pore surfaces contacted by oil becomes oil-wet upon imbibition). If a porous medium is water-wet and a non-wetting fluid displaces the water (drainage), then the non-wetting fluid will first occupy the larger pores and will enter the smaller pores only as the capillary pressure is increased. This process is similar to the accumulation of oil or gas in the pore space of a reservoir. Thus it is of interest to estimate the irreducible water saturation that is retained by capillarity after the hydrocarbon accumulates in an oil or gas reservoir. The *FFI* is an estimate of the amount of potential hydrocarbon in the rock when saturated to a given capillary pressure. Also, the actual *in situ* water saturation in the formation relative to the irreducible water saturation is an indicator of whether water will be produced. Water saturation that is greater than the irreducible water saturation is capable of being produced when the fluids are driven by a pressure gradient. The water saturation in the formation (undisturbed by drilling) is estimated from deep resistivity logs that measure beyond the filtrate invasion region.

The T_1 relaxation time distribution as water is displaced by air or kerosene (representing oil) is shown in Figures 3.6.6 and 3.6.7 [20]. Figure 3.6.6 shows the amplitude of the longer relaxation times going to zero as water is displaced progressively from the larger pores. Figure 3.6.7 also shows water (the peak on the left) being progressively displaced from larger pores by kerosene (the peak the on right) and the signal for kerosene becoming progressively greater in amplitude. This figure shows that the relaxation times of the kerosene peaks in the partially saturated cases corresponds to the relaxation time of the bulk kerosene (top curve). This is because a thin film of water shields the kerosene from the pore surfaces. The case of 0% water saturation is that of a core that was dried and fully saturated with kerosene. No water is present to shield the kerosene from the pore surfaces and kerosene has a contribution from the surface relaxation. Thus it relaxes faster (shifted to the left) than bulk kerosene or the kerosene in the partially saturated cases.

3.6.7.2 Relaxation Time Cut-off

The observation that the irreducible water occupies the pores with the shorter relaxation times suggests that it may be possible to determine a relaxation time

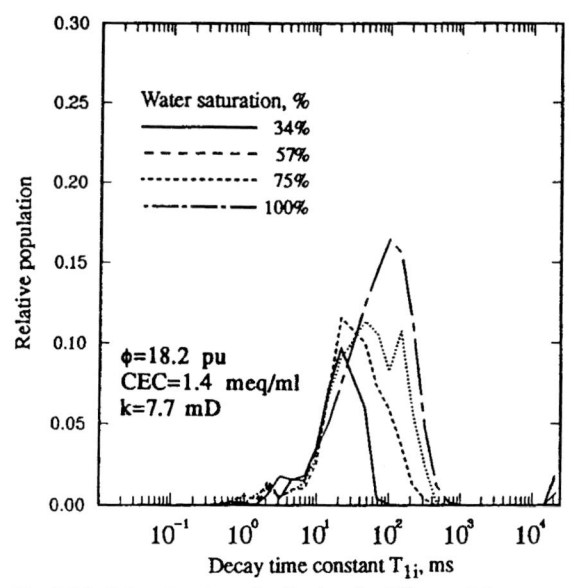

Fig. 3.6.6 Relaxation time distribution for different air/water saturations [20].

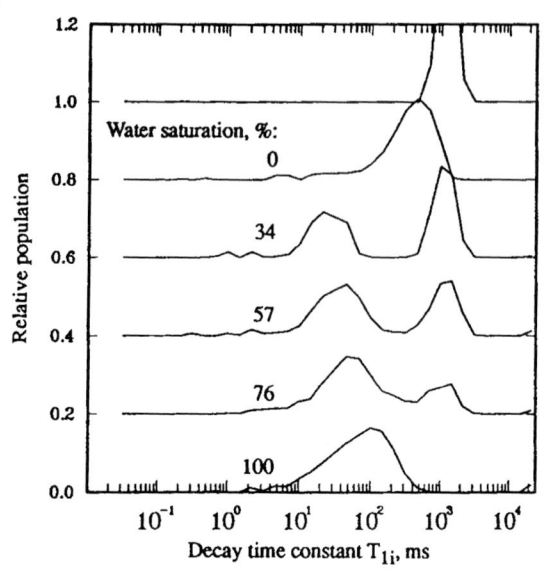

Fig. 3.6.7 Relaxation time distribution for different kerosene/water saturations. The uppermost curve is bulk kerosene [20].

"cut-off" that distinguishes the bound water from the mobile water. Analysis of many rock samples resulted in representative values of T_1 cut-off of 46 ms for sandstones [20], T_2 cut-off of 33 ms for sandstones and 92 ms for carbonates [18].

The oil and water response in Figure 3.6.7 could easily be distinguished in this example with kerosene as the oil. If the oil was a crude oil with a broad distribution of relaxation times, the oil may have non-zero response at relaxation times shorter than the T_1 cut-off. This could result in mistaking a part of the oil response as *BVI*. The correct approach in this case is to use diffusion measurements to distinguish between water and oil. This will be discussed under fluid identification (Section 3.6.9).

3.6.7.3 Diffusional Coupling

The concept of a T_2 cut-off that partitions the relaxation time distribution between the pores which can be displaced and those that cannot does not always apply. An exception is when there is significant diffusional coupling between the micropores that retain water at a high capillary pressure and the macropores in close proximity to the microporous system [26, 27]. A spectral BVI model or a forward model has been suggested to interpret these systems [30, 31, 53].

3.6.8
Permeability

Permeability (k) is the transport coefficient for the flow of fluids through a porous medium and has the units of length squared. NMR measures the porosity and the relaxation time distribution, from which the pore size distribution can be estimated. Three different models exist for estimating permeability. They are based on: (a) the log mean relaxation time of 100% brine saturated rock, (b) the ratio of *FFI/BVI* or (3) vuggy carbonates.

3.6.8.1 Log Mean Relaxation Time

This model (known as the SDR model) uses the log mean relaxation of 100% brine saturated rock as the estimator of the effective pore size [32]. This limits the method to rocks in the absence of hydrocarbons, which models permeability to be proportional to the porosity to the fourth power and log mean T_2 to the second power, Eq. (3.6.6). Variations of this method use different exponents or the formation resistivity factor rather than porosity [2]:

$$k = 4.6 \, \phi^4 \, T^2_{2,LM}, \quad 100\% \, S_w \tag{3.6.6}$$

3.6.8.2 Coats–Timur Equation

This model can be used in the presence of hydrocarbons because it is based in the ratio of *FFI/BVI* as an estimator of pore size, Eq. (3.6.7):

$$k = 10^4 \, \phi^4 \left(\frac{FFI}{BVI}\right)^2 \tag{3.6.7}$$

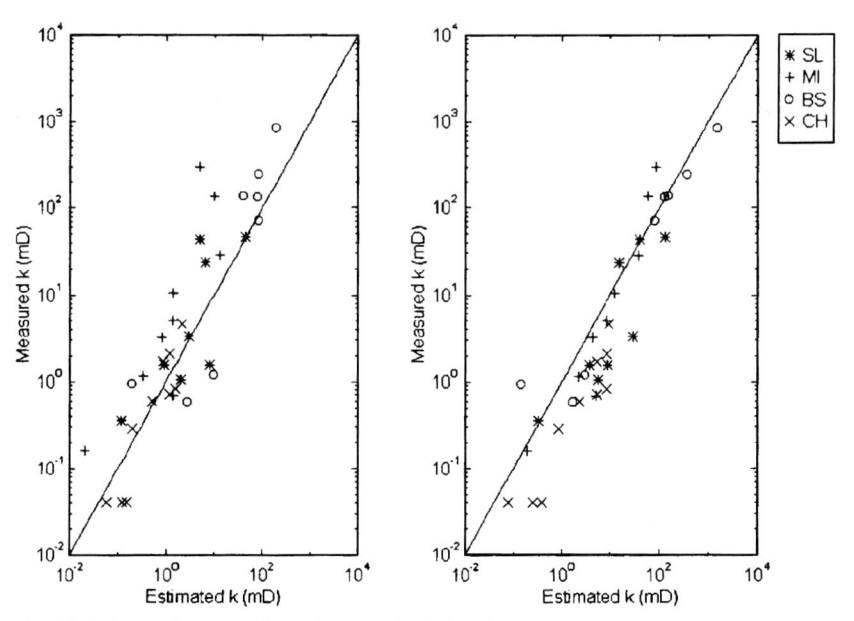

Fig. 3.6.8 Comparison of estimated permeability for 35 sandstones. Left: the log mean relaxation time model. Right: the Coats–Timur model [33].

3.6.8.3 Comparison of Correlations

NMR estimated permeabilities from Eqs. (3.6.6) or (3.6.7) are compared with measured permeabilities for 35 sandstones [33] in Figure 3.6.8. An investigation of a well characterized set of sandstone samples showed that Eq. (3.6.6) significantly underestimated the permeability when the rock has significant internal gradients and higher surface relaxivity due to the presence of paramagnetic minerals [33]. The *FFI* model, Eq. (3.6.7), was less sensitive to internal gradient effects. Thus the permeability correlation will be more accurate if calibrated for a specific formation. The relationship between these two models is analyzed in Ref. [34].

3.6.8.4 Vuggy Carbonates

Permeability estimation for vuggy carbonates is more difficult because NMR measures the pore body size while it is the pore throats connecting the pore bodies that dominate the permeability. One approach to estimating the permeability of vuggy carbonates has been to ignore the NMR response with T_2 values longer than 750 ms in the estimation of the porosity and mean relaxation time used in the permeability correlation [35]. This modified porosity and mean relaxation time is used in Eq. (3.6.6) with a modification of the prefactor. The resulting correlation for vuggy carbonates, called the Chang model, is given by Eq. (3.6.8) [35]:

$$k_{\text{vuggy carbonate}} = 4.75 \left(\phi_{\text{NMR},750} \right)^4 \left(T_{2,750} \right)^2 \quad (3.6.8)$$

The Chang model is good for low permeability samples but underestimates the permeabilities of the high permeability samples in a West Texas field, because it assumes that the vugs do not contribute to permeability. The correlation was significantly improved from below 0.1 md to greater than 1000 md by using a tortuosity dependent weighting between NMR parameters with and without the 750-ms cut-off [36].

$$k = 4.75 \left[\left(\frac{\phi}{\phi_{750}} \right)^a \phi_{750} \right]^4 \left[\left(\frac{T_{2LM}}{T_{2LM,750}} \right)^a T_{2LM,750} \right]^2 \quad (3.6.9)$$

$$a = 1 - \frac{\tau}{\tau_{max}}, \quad 0 \leq a \leq 1$$

The maximum value for the tortousity was 30 and the value for the tortusity of each sample was estimated from the formation factor. The formation factor was taken either from resistivity measurents or estimated from NMR measurements and the Myers' pore-combination model [37]. A comparison of the estimated permeability correlations is given in Figure 3.6.9. The SDR model overestimates the permeability of low permeability samples and the Chang model underestimates the permeability of high permeability samples. The modified Chang model, Eq. (3.6.9), improves the estimate of permeability for both low and high permeability samples.

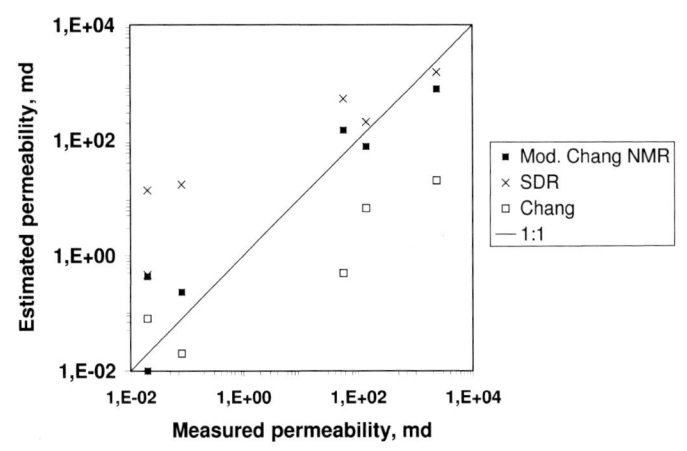

Fig. 3.6.9 Permeability estimated with the modified Chang model compared with the SDR and original Chang models [36].

3.6.9
Fluid Identification

Low field NMR is not capable of identifying fluids by chemical shift. The early approach to distinguishing oil and water was to dope the water with $MnCl_2$ in carbonates or chelated $MnCl_2$ in sandstones (chelating is to prevent loss due to ion exchange with clays). The role of $MnCl_2$ is to shorten the relaxation time of water such that it can be distinguished from oil. Current logging tools can distinguish between water, oil and gas by difference in diffusivity and/or T_1 and T_2. The early methods either utilized diffusivity contrast by changing the echo spacing in the CPMG sequence to distinguish fluids, which can be referred to as the "shifted spectrum" method [38, 39] or they utilized T_1 contrasts by varying the polarization time (also called "wait time") to distinguish fluids by the "differential spectrum" method [3, 39]. By using a combination of diffusion time and wait times and a model to correlate the relaxation time and diffusivity distributions of the crude oil, the potential of these approaches was further advanced [17, 40, 41]. This makes it possible to estimate the wettability, saturation and viscosity of oil from NMR measurements.

NMR interpretation has made significant advances with diffusion-editing pulse sequences and two-dimensional inversion of diffusivity and T_2 relaxation [7, 40–44]. The 2D inversion can also be used to compare T_1 and T_2 relaxation with each other [42]. Distributions of these two characteristic parameters can now be displayed on a 2D map and the relationship between them more easily visually interpreted. The 2D distribution map can be interpreted by comparing the measured distribution with the line for the bulk diffusivity of water and the correlation lines for the hydrocarbon components in crude oils, shown in Figure 3.6.10 as dashed lines [40–46]. Figure 3.6.10 shows water and oil having overlapping relaxation time distributions but when displayed in correlation with diffusivities, they are easily distinguished. The

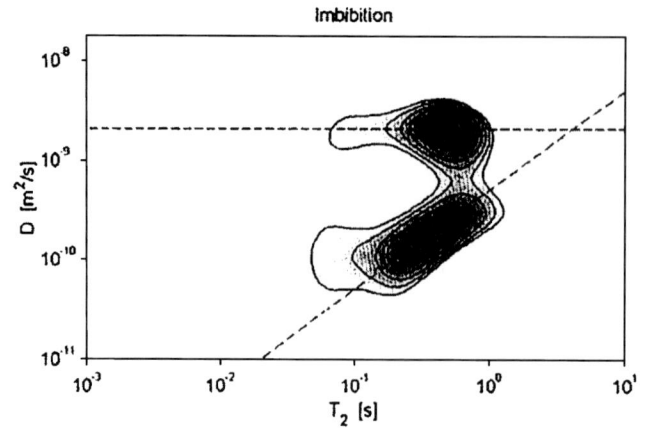

Fig. 3.6.10 Distributions of diffusivity and relaxation times for partially brine- and oil-saturated Bentheim sandstone [43].

saturation of each fluid is estimated from the integral over each corresponding distribution, which represents the volume of the respective fluid.

The oil viscosity is estimated from the log mean relaxation time and/or log mean diffusivity of the oil distribution and the gas/oil ratio [17, 40, 41, 44].

3.6.10
Exceptions to Default Assumptions

3.6.10.1 Wettability
Interpretation of NMR well logs is usually made with the assumption that the formation is water-wet such that water occupies the smaller pores and oil relaxes as the bulk fluid. Examination of crude oil, brine, rock systems show that a mixed-wet condition is more common than a water-wet condition, but the NMR interpretation may not be adversely affected [47]. Surfactants used in oil-based drilling fluids have a significant effect on wettability and the NMR response can be correlated with the Amott–Harvey wettability index [46]. These surfactants can have an effect on the estimation of the irreducible water saturation unless compensated by adjusting the T_2 cut-off [48].

3.6.10.2 Internal Field Gradients
The NMR response is usually interpreted assuming that the magnetic field gradient is equal to that designed by the tool. However, paramagnetic minerals can result in induced or internal gradients that are much larger than the gradient of the tool design [49]. The presence of internal gradients is identified by an apparent diffusivity greater than that of the fluids present [50, 51]. Internal gradients can be used to advantage in characterizing the pore structure [52]. This latter approach is the subject of the next chapter in this book.

3.6.10.3 Diffusional Coupling

Estimation of pore size distribution is based on the assumption that there is no diffusional coupling between pores of different sizes. Microporous carbonate grainstones have been identified as an example where this assumption is not valid [31].

3.6.11
Conclusions

NMR has proven to be a valuable tool for formation evaluation by well logging, downhole fluid analysis and laboratory rock characterization. It gives a direct measure of porosity as the response is only from the fluids in the pore space of the rock. The relaxation time distribution correlates with the pore size distribution. This correlation makes it possible to estimate permeability and irreducible water saturation. When more than one fluid is present in the rock, the fluids can be identified based on the difference in the fluid diffusivity in addition to relaxation times. Interpretation of NMR responses has been greatly advanced with the ability to display two distributions simultaneously.

Acknowledgments

The author acknowledges the contributions of many graduate students and colleagues and also the financial sponsorship of the industrial consortium on Process in Porous Media, NSF CTS-9321887, DOE DE-AC26–99BC15201 and DOE DE-PS26–04NT15515.

References

1 G. R. Coates, L. Xiao, M. G. Prammer **1999**, *NMR Logging; Principles & Applications*, Halliburton Energy Services, Houston.

2 K.-J. Dunn, D. J. Bergman, G. A. Latorraca **2002**, *Nuclear Magnetic Resonance Petrophysical and Logging Applications*, Pergamon, New York.

3 M. G. Prammer, E. Drack, G. Goodman, P. Masak, S. Menger, M. Morys, S. Zannoni, B. Suddarth, J. Dudley **2000**, *The Magnetic Resonance While-Drilling Tool: Theory and Operations*, SPE 62981, presented at the SPE ATC&E, Dallas, TX, 1–4 October, 2000.

4 J. Morley, R. Heidler, J. Horkowitz, B. Luong, C. Woodburn, M. Poizch **2002**, *Field Testing of a New Nuclear Magnetic Resonance Logging-While-Drilling Tool*, SPE 77477, presented at the SPE ATC&E, San Antonio, TX, 29 September–2 October, 2002.

5 J. Bouton, M. G. Prammer, P. Masak, S. Menger **2001**, *Assement of Sample Contamination by Downhole NMR Fluid Analysis*, SPE 71714, presented at the 2001 SPE ATC&E, New Orleans, 30 September–3 October, 2001.

6 R. Akkurt, C.-M. Fransson, J. M. Witkowsky, W. M. Langley, B. Sun, A. McCarty **2004**, *Fluid Sampling and Interpretation with the Downhole NMR Fluid Analyzer*, SPE 90971, presented at the SPE ATC&E, Houston, 26–29 September, 2004.

7 M. D. Hurlimann, L. Venkataramanan **2002**, (Quantitative Measurement of Two-Dimensional Distribution Func-

tions of Diffusion and Relaxation in Grossly Inhomogeneous Fields), *J. Magn. Reson.* 157, 31–42.

8 R. L. Kleinberg, H. J. Vinegar **1996**, (NMR Properties of Reservoir Fluids), *Log Analyst* 37 (6), 20–32.

9 G. A. LaTorraca, K.-J. Dunn, P. R. Webber, R. M. Carlson **1998**, (Low-Field NMR Determination of the Properties of Heavy Oils and Water-in-Oil Emulsions), *Magn. Reson. Imaging* 16 (5/6), 659–662.

10 Q. Zhang, S.-W. Lo, C. C. Huang, G. J. Hirasaki, R. Kobayashi, W. V. House **1998**, *Some Exceptions to Default Rock and Fluid Properties*, paper FF presented at the SPWLA 39th Annual Logging Symposium, Keystone, CO, May 26–29, 1998.

11 K. Krynicki **1966**, (Proton spin-lattice relaxation in pure water between 0 °C and 100 °C), *Physica* 32, 167–178.

12 J. C. Hindman, A. Svirmickas, M. Wood **1973**, (Relaxation processes in water. A study of the proton spin-lattice relaxation time), *J. Chem. Phys.* 59 (3), 1517–1522.

13 S.-W. Lo, G. J. Hirasaki, W. V. House, R. Kobayashi **2002**, *Mixing Rules and Correlations of NMR Relaxation with Viscosity, Diffusivity, and Gas/Oil Ratio of Methane/Hydrocarbon Mixtures*, presented at *SPEJ*, March 24–34, 2002.

14 C. E. Morriss, R. Freedman, C. Straley, M. Johnston, H. J. Vinegar, P. N. Tutunjian **1997**, (Hydrocarbon saturation and viscosity estimation from NMR logging in the Belridge diatomite), *Log Analyst* 38 (2), 44-59.

15 G. J. Hirasaki, S. W. Lo, Y. Zhang **2003**, (NMR properties of petroleum reservoir fluids), *Magn. Reson. Imaging* 21, 269–277.

16 Y. Zhang, G. J. Hirasaki, W. V. House, R. Kobayashi **2002**, *Oil and Gas NMR Properties: The Light and Heavy Ends*, paper HHH prepared for the SPWLA 43rd Annual Logging Symposium, Osio, Japan, June 2–5, 2002.

17 R. Freedman, S. Lo, M. Flaum, G. J. Hirasaki, A. Matteson, A. Sezginer **2001**, (A new NMR method of fluid characterization in reservoir rocks: Experimental confirmation and simulation results), *SPEJ* December, 452–464.

18 C. Straley, D. Rossini, H. Vinegar, P. Tutunjian, C. Morriss **1997**, (Core analysis by low-field NMR), *Log Analyst* 38 (2), 84–94.

19 K. R. Brownstein, C. E. Tarr **1979**, (Importance of classical diffusion in NMR studies of water in biological cells), *Phys. Rev. A* 19, 2446–2453.

20 C. Straley, C. E. Morriss, W. E. Kenyon, J. J. Howard **1995**, (NMR on partially saturated rocks: Laboratory insights on free fluid index and comparison with borehole logs), *Log Analyst* 36 (1), 40–56.

21 R. L Kleinberg, W. E. Kenyon, P. P. Mitra **1994**, (Mechanism of NMR relaxation of fluids in rock), *J. Magn. Reson., Ser. A* 108, 206–214.

22 I. Foley, S. A. Farooqui, R. L. Kleinberg **1996**, (Effect of paramagnetic ions on NMR relaxation of fluids at solid surfaces), *J. Magn. Reson., Ser. A* 123, 95–104.

23 W. E. Kenyon **1997**, (Petrophysical principles of applications of NMR logging), *Log Analyst* 38 (2), 21–43.

24 J. J. Howard, W. E. Kenyon, C. Straley **1993**, (Proton magnetic resonance and pore size variations in reservoir sandstones), *SPEFE* September, 194–200.

25 R. L. Kleinberg **1996**, (Utility of NMR T_2 distributions, connections with capillary pressure, clay effect, and determination of the surface relaxivity parameter ϱ_2), *Magn. Reson. Imaging* 14 (7/8), 761–767.

26 M. D. Hurlimann, K. G. Helmer, L. L. Lator, C. H. Sotak **1994**, (Restricted diffusion in sedimentary rocks. Determination of surface-area-to-volume ratio and surface relaxivity), *J. Magn. Reson., Ser. A* 111, 169–178.

27 L. L. Latour, R. L. Kleinberg, A. Sezginer **1992**, (Nuclear magnetic resonance properties of rocks at elevated temperatures), *JCIS* 150 (2), 535–548.

28 S. Godefroy, M. Fleury, F. Deflandre, J.-P. Korb **2002**, (Temperature effect on NMR surface relaxation in rocks for well logging applications), *J. Phys. Chem. B* 106, 11183–11190.

29 R. L. Kleinberg, S. A. Farooqui, M. A. Horsfield **1993**, (T_1/T_2 Ratio and frequency dependence of NMR relaxation

in porous sedimentary rocks), *JCIS* 158, 195–198.
30 G. R. Coats, D. Marschall, D. Mardon, J. Galford **1997**, *A New Characterization of Bulk-Volume Irreducible Using Magnetic Resonance*, paper QQ presented at the SPWLA 38 Annual Logging Symposium, Houston, TX, June 15–18, 1997.
31 T. S. Ramakrishnan, L. M. Schwartz, E. J. Fordham, W. E. Kenyon, D. J. Wilkinson **1999**, (Forward models for nuclear magnetic resonance in carbonate rocks), *Log Analyst* 40 (4), 260–270.
32 W. E. Kenyon, P. I. Day, C. Straley, J. F. Willemsen **1988**, (A three-part study of NMR longitudinal relaxation properties of water-saturated sandstones), *SPEFE* September, 622–636.
33 C.-C. Huang **1997**, (Estimation of rock properties by NMR relaxation methods), *MS Thesis*, Rice University, 108–113.
34 R. Sigal **2002**, (Coates and SDR permeability: Two variations on the same theme), *Petrophysics* 43 (1), 38–46.
35 D. Chang, H. Vinegar, C. Morriss, C. Straley **1997**, (Effective porosity, producible fluid, and permeability in carbonates from NMR logging), *Log Analyst* 38 (2), 60–72.
36 I. Hidajat, K. K. Mohanty, M. Flaum, G. Hirasaki **2004**, (Study of vuggy carbonates using NMR and X-ray CT scanning), *SPEFE&E* 7 (5), 365–377.
37 M. T. Myers **1991**, *Pore Combination Modeling: A Technique for Modeling the Permeability and Resistivity Properties of Complex Pore Systems*, paper SPE 22662 presented at the 1991 SPE ATC&E, Dallas, TX, 6–9 October, 1991.
38 R. Akkurt, H. J. Vinegar, P. N. Tutunjian, A. J. Guillory **1996**, (NMR logging of natural gas reservoirs), *Log Analyst* 37 (6), 33–42.
39 R. Akkurt, M. G. Prammer, M. A. Moore **1996**, (Selection of optimal acquisition parameters for MRIL logs), *Log Analyst* 37 (6), 43–52.
40 R. Freedman, N. Heaton, "Fluid Characterization using Nuclear Magnetic Resonance Logging," *Petrophysics*, **2004**, 45(3), 241–250.
41 R. Freedman, N. Heaton, M. Flaum, G. J. Hirasaki, C. Flaum, M. Hürlimann **2003**, (Wettability, saturation, and viscosity from NMR measurements), *SPEJ* December, 317–327.
42 B. Sun, K.-J. Dunn **2004**, (Methods and limitations of NMR data inversion for fluid typing), *J. Magn. Reson.* 169, 118–128.
43 M. D. Hurlimann, M. Flaum, L. Venkataramanan, C. Flaum, R. Freedman, G. J. Hirasaki **2003**, (Diffusion-relaxation distribution functions of sedimentary rocks in different saturation states), *Magn. Reson. Imaging* 21, 305–310.
44 N. J. Heaton, C. Cao Minh, J. Kovats, U. Guru **2004**, *Saturation and Viscosity from Multidimensional Nuclear Magnetic Resonance Logging*, SPE 90564 presented at SPE ATC&E, Houston, TX, 26–29 September, 2004.
45 M. Flaum, J. Chen, G. J. Hirasaki **2004**, *NMR Diffusion Editing For D-T_2 Maps: Application To Recognition Of Wettability Change*, *Petrophysics* 46 (2), 113–123.
46 J. Chen, G. J. Hirasaki, M. Flaum **2004**, *Study of Wettability Alteration from NMR: Effect of OBM on Wettability and NMR Responses*, paper presented at the 8th International Symposium on Reservoir Wettability, Houston, TX, May 16–18, 2004.
47 G. Q. Zhang, C.-C. Huang, G. J. Hirasaki **2000**, (Interpretation of wettability in sandstones with NMR analysis), *Petrophysics* 41 (3), 223–233.
48 J. Chen, G. J. Hirasaki, M. Flaum **2004**, *Effects of OBM Invasion on Irreducible Water Saturation: Mechanisms and Modifications of NMR Interpretation*, SPE 90141 presented at SPE ATC&E, Houston, TX, 26–29 September, 2004.
49 G. Q. Zhang, G. J. Hirasaki, W. V. House **2003**, (Internal field gradients in porous media), *Petrophysics* 44 (6), 422–434.
50 B. Sun, K.-J. Dunn **2002**, (Probing the internal field gradients of porous media), *Phys. Rev. E* 65, 051309-1-7.
51 M. D. Hurlimann, A. Matteson, J. E. Massey, D. F. Allen, E. J. Fordham, F. Antonsen, H. G. Rueslatten **2004**, (Application of NMR diffusion editing as chlorite indicator), *Petrophysics* 45 (5), 414–421.

52 Y.-Q. Song, S. Ryu, P. N. Sen **2000**, (Determining multiple length scales in rocks), *Nature* 406, July 13, 178–181.

53 V. Anand, G. J. Hirasaki **2005**, (Diffusional Coupling Between Micro and Macroporosity for NMR Relaxation in Sandstones and Grainstones), *SPWLA Transactions Annual Logging Symposium*, New Orleans, June 26–29.

3.7
NMR Pore Size Measurements Using an Internal Magnetic Field in Porous Media
Yi-Qiao Song, Eric E. Sigmund, and Natalia V. Lisitza

Abstract

This chapter reviews the recent development of an NMR method to obtain detailed pore size information in porous media. This method detects molecular diffusion of the fluid in the pore space subjected to the magnetic field inhomogeneity produced by the susceptibility contrast between the fluid and the solid materials. Inhomogeneous internal fields have been found in a broad range of materials, such as water-saturated rocks, blood cell solutions or trabecular bones. The spin-magnetization decay due to diffusion in the internal field (DDIF) is dependent on the microstructure of the medium. Thus, DDIF provides a probe of pore structure on the scale of the NMR diffusion length. In this chapter, the experimental execution, theoretical formalism and the numerical analysis of the DDIF technique are reviewed. Application of DDIF in several systems will be discussed.

3.7.1
Introduction

Heterogeneity is often the intrinsic aspect of many materials that determines their properties and utility. Examples are ubiquitous in our environment, from natural materials such as rocks, soil and packed snow, to man-made materials such as concretes and food products, and to biological tissues such as bones and lungs. Heterogeneity of the material is reflected in its porous structure, which in turn defines the material properties. For example, pore sizes and pore connectivity are essential for the flow of crude oil in rocks and water movement in aquifers. Bones in animals and humans are also porous, particularly the spongy bones. The internal structure of bones contributes to their mechanical strength and the loss of such structure and bone mass (as occurs in osteoporosis) may lead to bone fracture and deformation. The length scales of various structures in these materials fall in the range of 10^{-4}–1 mm, somewhere between the length scales determined by NMR spectroscopy at the low end and MR imaging at the high end.

Extensive experimental techniques have been developed for porous material characterization [1], including direct imaging [2–5] and bulk measurement techniques for the statistical properties of the pore space. NMR is one such bulk measurement that is both non-destructive and compatible with large samples.

Since it was proposed in the early 1980s [6, 7], spin-relaxation has been extensively used to determine the surface-to-volume ratio of porous materials [8–10]. Pore structure has been probed by the effect on the diffusion coefficient [11, 12] and the diffusion propagator [13, 14]. Self-diffusion coefficient measurements as a function of diffusion time provide surface-to-volume ratio information for the early times, and tortuosity for the long times. Recent techniques of two-dimensional NMR of relaxation and diffusion [15–21] have proven particularly interesting for several applications. The development of portable NMR sensors (e.g., NMR logging devices [22] and NMR-MOUSE [23]) and novel concepts for ex situ NMR [24, 25] demonstrate the potential to extend the NMR technology to a broad application of field material testing.

NMR diffusion measurements are usually performed by measuring the decoherence of the magnetization of the diffusing spins under the influence of applied magnetic field gradients. However, any medium for which a difference in magnetic susceptibility $\Delta\chi$ exists between the solid and liquid components will exhibit a spatially varying *internal* magnetic field when placed in a homogeneous external field B_0. The influence of such internal field inhomogeneity on relaxation or diffusion experiments has been studied both experimentally [26–30] and theoretically [31–35]. Several sequences have been designed to suppress, avoid, or cope with the effects of internal gradients for either relaxometry/diffusometry, e.g., [36, 37] or image quality issues [38]. Other techniques such as the BOLD effect in functional brain imaging [39] or some bone imaging techniques [40, 41] take advantage of the susceptibility effects for their operation.

The approach described in this chapter employs magnetic susceptibility effects in the following way. The diffusion of the spins through the internal field alone will generate a magnetization decay. As the magnitude and spatial variation of the internal field is controlled by the length scale of the porous medium, the spin magnetization decay can be a probe of the microstructure. As will be shown later, in some cases a distribution of pore sizes can be extracted from such an analysis. This general technique is termed "decay due to diffusion in the internal field", or DDIF [42, 43].

The outline of this chapter is as follows. Firstly, the DDIF concept and experiment are discussed both intuitively and quantitatively, including details of the pulse sequences and data processing algorithms. Secondly, a range of applications of DDIF are presented, including pore space measurements in rocks and microstructural measurements in blood cells. Finally, the advantages of the DDIF technique are summarized.

3.7.2
DDIF Concept

3.7.2.1 Qualitative Picture
Consider a porous medium with magnetic susceptibility difference $\Delta\chi$ between the confining solid and the permeating fluid (Figure 3.7.1). Magnetic field gradients will develop in the pore space. The spatial distribution of this internal magnetic

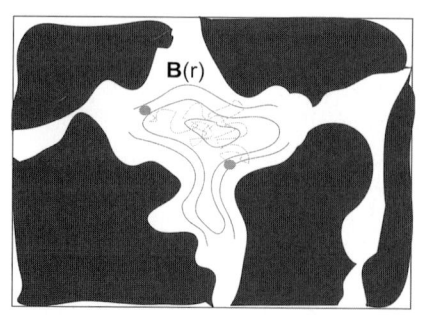

Fig. 3.7.1 Schematic of the DDIF effect in porous medium. The black areas are solid grains and the white areas are pore space. Diffusing spins in permeating fluid sample the locally variable magnetic field B(r) (solid contours sketched inside pore space) as it diffuses.

field will be determined by the structure of the underlying porous medium. While this assertion is qualitatively not surprising, quantitative connections actually exist between the pore structure and the magnetic field structure. For example, a numerical study of the internal magnetic field within the pore space of a random-dense pack of magnetized spheres [35] has demonstrated the near equivalence of two pair-correlation functions: that of the mass distribution of the grains and that of the internal magnetic field in the pore space. Intuitively, as the dipolar magnetic field decays monotonically, the pore size is the only length that determines the magnetic field variation due to susceptibility difference.

The essence of the DDIF method is to first establish a spin magnetization modulation that follows the spatial variation of the internal magnetic field within the individual pore. Such modulation is created by allowing spins to precess in the internal magnetic field. Then the diffusion-driven time-evolution (often decay) of such a modulation is monitored through a series of signal measurements at various evolution times t_D. The time constant of this decay corresponds to the diffusion time of a molecule (or spin) across the pore and thus is a direct measure of the pore size.

From an NMR perspective, the technique's sensitivity to structure is as follows. The experiment starts by letting spins precess in the internal field for a time period of t_e. During this encoding period, each spin acquires a phase that is proportional to its local magnetic field: $\Phi = \gamma B(r) t_e$. As a result, the internal field variation within a pore is encoded in the phase of the spins. After the encoding period, the spin magnetization is rotated to along the applied field direction to stop the precession and spins are allowed to diffuse for a time period of t_D. If the diffusion distance during t_D meets or exceeds the average length scale of this modulation, the amplitude of the modulation will deteriorate. As the whole sample is often comprised of a range of pore sizes and therefore gradient magnitudes, a distribution of decay times is expected. This distribution is characteristic of the geometry of the underlying distribution of the length scales and serves as a useful "fingerprint" of the material [43].

3.7.2.2 Theory

The magnetization evolution due to diffusion and relaxation can be studied theoretically using the Bloch-Torrey equation [44]:

$$\frac{\partial}{\partial t}m(r,t) = D\nabla^2 m(r,t) - \mu m(r,t) \tag{3.7.1}$$

Here $m(r,t)$ is the relative difference of the longitudinal magnetization M and its equilibrium value M_0; $m = (M - M_0)/M_0$, D is the bulk diffusion coefficient and μ is the bulk relaxation rate. The general solution to the Torrey-Bloch equation can be written as

$$m(r,t) = e^{-\mu t}\sum_{n=0}^{\infty} a_n\phi_n(r)e^{-t/\tau_n}, \tag{3.7.2}$$

where ϕ_n and τ_n are normalized eigenfunctions and eigenvalues, respectively; ϕ_n with different indices are orthogonal to each other.

If the confining geometry is known, the following boundary condition can be applied to determine the eigenvalues:

$$D\hat{n}\cdot\nabla\phi_n = \varrho\phi_n \tag{3.7.3}$$

where ϱ is the surface relaxivity. This condition defines the surface as a magnetization "sink" and its strength determines the slope of the eigenfunction at the surface. While we consider a uniform surface relaxivity here, in fact the DDIF method can accomodate some inhomogeneity in relaxivity. A spatially inhomogeneous relaxivity would lead to a distribution of relaxation times, just as occurs with a distribution of pore sizes in porous media. However, as will be described later, relaxation and diffusion information are carefully separated in the analysis process whether they are spatially distributed or not. Thus, the diffusion-based measurement is, to a first order, uncompromised by inhomogeneous relaxivity.

Once the specific geometry of the medium is known, the Torrey equation can be solved [45–47]. To date, solutions for simple porous structures such as packed cylinders and spheres are not yet available. However, a one-dimensional model of both the magnetic field distribution and the diffusion is particularly useful for being intuitive and for developing the interpretation.

In a one-dimensional pore, the Torrey-Bloch equation can be solved analytically and with fast diffusion and weak relaxation, it can be shown that the eigenvalues are [46]

$$\tau_0 \approx \frac{d}{\varrho} \tag{3.7.4}$$

$$\tau_n \approx \frac{d^2}{D\pi^2 n^2}, n = 1, 2, 3, \dots \tag{3.7.5}$$

where d is the pore size and ϱ the surface relaxivity. The eigenfunctions $\phi_n(r)$ for this case are readily calculated. The "ground state" eigenfunction $\phi_0(r)$ is approximately constant across the pore, while the higher order functions ($n > 0$) oscillate in space [48]. The weight of these "modes" in the full magnetization is given by the coefficients a_n, which can be selectively manipulated to some extent through the choice of pulse sequence. The more control that is exercised on which modes are "excited", the less error will be associated with the data analysis for extracting microstructural properties. For example, in an inversion recovery experiment, the only non zero coefficient is a_0 [46]. The ground state mode is sometimes referred to as the relaxation mode for this reason. The decay of the modes of $n > 0$ is primarily due to diffusion, hence the name "diffusion" modes. In the DDIF experiment, as will be seen later, great care is taken to separate the relaxation mode from the remaining diffusion modes. Furthermore, in the weak encoding limit, defined by small spin precession angles Φ due to the internal field, the primary non zero mode excited is a_1. In this case, an observed distribution of decay times can be mostly ascribed to a distribution of structural length scales.

It should be noted that the decomposition shown in Eq. 3.7.2 is not necessarily a subdivision of separate sets of spins, as all spins in general are subject to both relaxation and diffusion. Rather, it is a classification of different components of the overall decay according to their time constant. In particular cases, the spectrum of amplitudes a_n represents the populations of a set of pore types, each encoded with a modulation determined by its internal gradient. However, in the case of stronger encoding, the initial magnetization distribution within a single pore type may contain multiple modes ϕ_n. In this case the interpretation could become more complex [49].

Table 3.7.1 Rf pulse phases (in degrees) for the DDIF experiment.

Ph1	Ph2	Ph3	ACQ
90	0	0	+
270	0	0	−
90	180	0	−
270	180	0	+

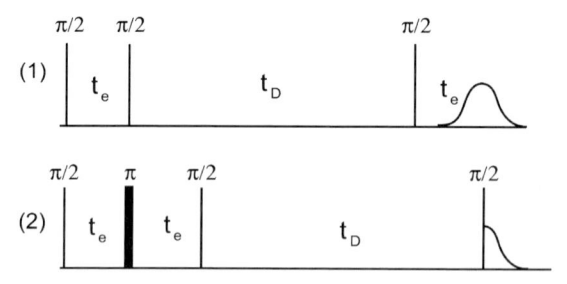

Fig. 3.7.2 DDIF (1) and reference (2) scan pulse sequences. The tipping angles of the pulses are marked as π and $\pi/2$. t_e and t_D are two time periods. An echo signal is detected for DDIF and an FID for the reference sequence. Phase cycling of rf pulses is shown in Tables 3.7.1 and 3.7.2.

3.7.2.3 Experiment

The DDIF experiment consists of a stimulated echo pulse sequence [50] and a reference scan to measure and separate the effect of spin-lattice relaxation. The pulse diagrams for these two are shown in Figure 3.7.2. Details of the experiments have been discussed in Ref. [51] and a brief description will be presented here.

The stimulated echo sequence includes an encoding period (t_e), followed by a diffusion period (t_D) and then a decoding period where an echo is formed and detected. During t_D, the magnetization is stored and the translational diffusion takes place. One of the advantages of such a sequence is to access long diffusion times and correspondingly long distances, as the limiting timescale on t_D is T_1 (often longer), not T_2. The phases of the rf pulses, which determine the direction of the nutation field B_1 in the rotating frame, determine which component of the encoded magnetization is stored; these components are usually denoted by "sine" or "cosine", referring to the component of a given isochromat that is longitudinally oriented after the second rf pulse. The DDIF experiment is optimized with the "sine" modulation [52, 51]. The 4-step phase cycling for this case is shown in Table 3.7.1 and includes a 2-step variation in the first and second pulse phases.

Table 3.7.2 Rf pulse phases (in degrees) for the reference experiment.

Ph1	Ph2	Ph3	Ph4	ACQ
0	90	0	0	+
180	90	0	0	−
0	90	180	0	−
180	90	180	0	+
0	270	0	0	+
180	270	0	0	−
0	270	180	0	−
180	270	180	0	+

The reference scan is to measure the decay due to spin-lattice relaxation. Compared with the corresponding stimulated echo sequence, the reference scan includes a π pulse between the first two $\pi/2$ pulses to refocus the dephasing due to the internal field and the second $\pi/2$ pulse stores the magnetization at the point of echo formation. Following the diffusion period t_D, the signal is read out with a final detection pulse. The phase cycling table for this sequence, including 2-step variation for the first three pulses, is shown in Table 3.7.2. The output from this pair of experiments are two sets of transients. A peak amplitude is extracted from each, and these two sets of amplitudes are analyzed as described below.

The two timescales in the pulse sequence, t_e and t_D, play different roles. The encoding time t_e determines the local phase winding, $\Phi = \gamma B(r)t_e$, and thus the complexity of its subsequent evolution. As mentioned earlier, two factors cause the magnetization decay to be distributed (i.e., multi-exponential): (a) a range of gradient environments and (b) a range of excited magnetization modes per environment. A useful regime in the case of porous media is $\alpha = \gamma \Delta B^i t_e \ll 1$, where (ΔB^i) is the variance of the internal field distribution. In this weak encoding limit, the eigenmode spectrum a_n satisfies $a_n \propto \alpha \propto t_e$. This linear regime is a useful signature, as it is found that the eigenmode spectrum is dominated by a_1, thus the decay time constant provides a direct measure of the pore size [49]:

$$d = \pi\sqrt{D\tau} \qquad (3.7.6)$$

The diffusion time t_D is the evolution time for the encoded magnetization pattern, and its useful range is ultimately limited by T_1 and the signal-to-noise ratio (SNR). For experiments with water at room temperature, t_D can often be as long as several seconds, probing pore sizes of up to several hundreds of microns.

Experimentally, an elegant property of the DDIF experiment is its simplicity. As the structural information is encoded by the internal field provided by the static uniform field and the material itself, the pulse sequence requires no applied field gradients and thus suffers from none of their complications, such as eddy currents, mechanical ringing, miscalibrations or non linearity.

3.7.2.4 Numerical Analysis

In the spirit of Eq. (3.7.2), the numerical model to fit the data is a superposition of decaying exponentials:

$$E(t_D) = \sum_j A_j exp(-t_D/\tau_j) \qquad (3.7.7)$$

where E is echo signal. Note that the notation A_j is used for the experimental coefficients to distinguish them from the eigenmode coefficients a_n, which are not necessarily equivalent. The simplest fitting procedure consists of performing a Laplace transform on the input magnitudes E. However, as mentioned earlier, the diffusion modes ($n > 1$) are of more interest than the relaxation mode ($n = 0$) for the DDIF experiment, so a common modification of the analysis is to subtract the relaxation mode from the data. This is done using the data from the reference scan, which probes the relaxation mode only. The model expression for this subtraction is

$$E_s = E - a_0 \times R \qquad (3.7.8)$$

where E is the DDIF amplitude data, R the reference amplitude data and E_s the subtracted data. The fraction of the relaxation mode a_0 can be determined in various ways. The simplest method is to extract the ratio $a_0 = E(t_{Dmax})/R(t_{Dmax})$ at the longest t_D point in the experiment, with the assumption that $t_{Dmax} \gg \tau_n$, so that all diffusive modes have decayed away, leaving only the relaxation mode. A method employing more of the data would be

$$a_0 \approx \frac{LI(E) \cdot LI(R)}{LI(R) \cdot LI(R)} \qquad (3.7.9)$$

where $LI(E)$ and $LI(R)$ denote the Laplace inversion spectra of E and R, respectively. A further refinement of the subtracted value can be gleaned from a series of trial Laplace inversions from a range of a_0 values near the above estimate(s), in which a square-deviation χ^2 is calculated between the fit and the data. The maximum value of the trial a_0 that is consistent with the known noise statistics of the original data is chosen as a_0. After a_0 is determined, the E_s data contains only contributions from the diffusion modes. These data are then passed to the Laplace inversion algorithm.

3.7.2.5 Laplace inversion

The Laplace inversion (LI) is the key mathematical tool of the DDIF experiment. The ability to convert the measured multi-exponential decay into a distribution of decay times is crucial to the DDIF pore size distribution application. However, unlike other mathematical operations, the Laplace inversion is an ill-conditioned problem in that its solution is not unique, and is fairly sensitive to the noise in the input data. In this light, significant research effort has been devoted to optimizing the transform and understanding its boundaries [17, 53, 54].

Firstly, it has been found that the estimation of all of the amplitudes of the LI spectrum cannot be made with a standard least-squares based fitting scheme for this ill-conditioned problem. One of the solutions to this problem is a numerical procedure called regularization [55]. In this method, the optimization criterion includes the misfit plus an extra term. Specifically in our implementation, the quantity to be minimized can be expressed as follows [53]:

$$||KA - E_s||^2 + \alpha ||A||^2 \qquad (3.7.10)$$

Here K is the kernel matrix determining the linear operator in the inversion, A is the resulting spectrum vector and E_s is the input data. The matrix element of K for Laplace inversion is $K_{ij} = exp(-t_i/\tau_j)$ where $\{t_i\}$ and $\{\tau_j\}$ are the lists of the values for t_D and decay time constant τ, respectively. The inclusion of the last term $\alpha ||A||^2$ penalizes extremely large spectral values and thus suppresses undesired "spikes" in the DDIF spectrum.

Secondly, although stable solutions covering the entire temporal range of interest are attainable, the spectra may not be well resolved: that is, for a given dataset and noise, a limit exists on the smallest resolvable structure (or separation of structures) in the Laplace inversion spectrum [54]. Estimates can be made on this resolution parameter based on a singular-value decomposition analysis of K and the signal-to-noise ratio of the data [56]. It is important to keep in mind the concept of the spectral resolution in order to interpret the LI results, such as DDIF, properly.

3.7.3
Applications

3.7.3.1 Pore Body and Throat

An example of DDIF data on a Berea rock sample is shown in Figure 3.7.1 illustrating the decay data (A), the pore size distribution after Laplace inversion (B) and the images of the pore space (C). The DDIF signal decays faster initially than the reference signal. At $t_D > 1$ s, the DDIF and reference signals show a similar rate of decay. The data at the long t_D were used to determine a_0 and the subtraction was performed as described in the previous section. The pore size distribution is illustrated in panel (B) with a dominant peak around 85 µm. This length is very close to the visual estimate of pore size from the microscopy in panel (C).

In addition, mercury intrusion porosimetry results are shown together with the pore size distribution in Figure 3.7.3(B). The overlay of the two sets of data provides a direct comparison of the two aspects of the pore geometry that are vital to fluid flow in porous media. In short, conventional mercury porosimetry measures the distribution of pore throat sizes. On the other hand, DDIF measures both the pore body and pore throat. The overlay of the two data sets immediately identify which part of the pore space is the pore body and which is the throat, thus obtaining a model of the pore space. In the case of Berea sandstone, it is clear from Figure 3.7.3(B) that the pore space consists of a large cavity of about 85 µm and they are connected via 15-µm channels or throats.

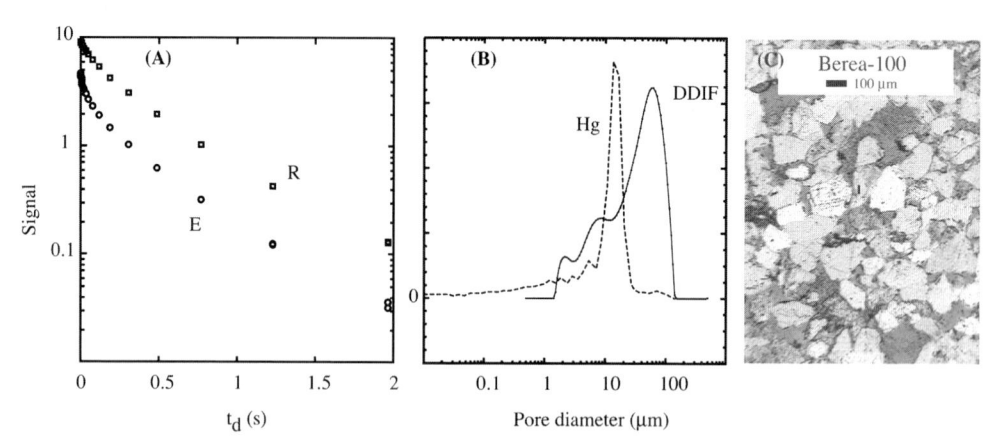

Fig. 3.7.3 (A) DDIF (circles) and reference (squares) data for a Berea sandstone sample. Measurements were performed at a proton Larmor frequency of 85.1 MHz. t_e = 100 µs. Signal-to-noise ratio is approximately 10^3. (B) Pore size distribution (solid line) obtained from DDIF, in comparison with mercury (Hg) porosimetry (dashed line). The peak in the Hg data indicates a pore throat size of 15 µm. The overlay of the two results identify the pore throat and pore body. (C) Optical microscopy of the 30-µm thin section of the Berea sample. The pore spaces are indicated by the blue regions, which were impregnated with blue epoxy prior to sectioning. Figure from Ref. [51] with permission.

3.7.3.2 Pore Shape

Pore shape is a characteristic of pore geometry, which is important for fluid flow and especially multi-phase flow. It can be studied by analyzing three-dimensional images of the pore space [2, 3]. Also, long time diffusion coefficient measurements on rocks have been used to argue that the shapes of pores in many rocks are sheet-like and tube-like [16]. It has been shown in a recent study [57] that a combination of DDIF, mercury intrusion porosimetry and a simple analysis of two-dimensional thin-section images provides a characterization of pore shape (described below) from just the geometric properties.

In this study, a mercury intrusion experiment was performed with a constant injection rate by regulating the intrusion pressure [58]. This is different from the conventional mercury intrusion experiment where the intrusion pressure is initially kept constant to record the mercury intrusion volume, then incremented to record the resultant incremental intrusion. In our experiment, the injection rate was kept extremely low so that the pressure loss due to flow was negligible compared with the capillary pressure. The data from this constant-rate mercury intrusion (CRMI) method, also called APEX [58], was collected through the pressure fluctuations as a function of intrusion volume, shown in Figure 3.7.4. The pressure fluctuations were analyzed to identify pore bodies and pore throats. For example, the sudden drop in pressure indicates the mercury meniscus entering a wide region (pore body) from a narrow region (pore throat). One of the parameters obtained from such an experiment is the distribution of the pore body volumes, shown in Figure 3.7.4 with a pronounced peak at around 20 nL. The

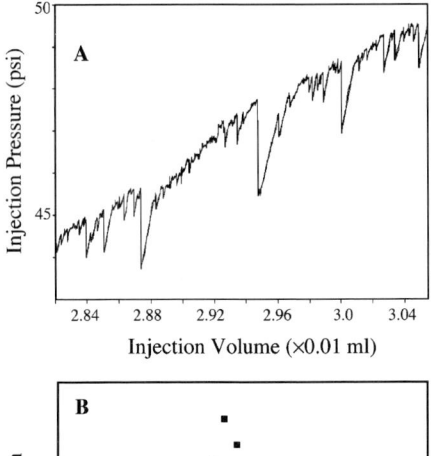

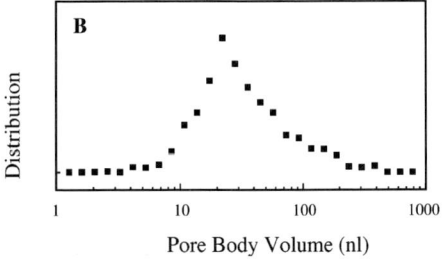

Fig. 3.7.4 (A) CRMI results of pressure versus volume on the Berea sandstone sample with a porosity of 20% and permeability 0.2 darcy. The two lines are raw CRMI data and the corrected data by a calibration run. Transducer noise was also filtered. The amount of the correction is fairly small and the two data sets overlap. (B) CRMI pore body volume distribution showing a predominant peak at around 20 nL. Figure from Ref. [57] with permission.

usual assumption of a spherical pore will give a pore diameter of 340 µm, which is inconsistent with the thin-section micrograph and the DDIF result that the pore body size is about 85 µm in this sample. This clearly indicates that the pore body is far from a spherical shape.

Given the pore body volume and one of the linear dimensions from DDIF, two shapes are possible: a pancake-like oblate or a tube-like oblong, with the short axis being 85 µm. These two shapes will have very different 2D cross-sectional images. Imagine that one cuts a long tube at random positions and orientations. The cross-section images would be dominated by ellipses of the size of the tube diameter, corresponding to sections perpendicular to the long axis of the tube. On the other hand, sections of an oblate would be dominated by the elongated ellipses, and there will be very few sections that approximate circles with a diameter of the short axis. The thin-section micrograph images from this sample showed that the majority of the pore cross sections are not extensively elongated and they are better described by a small aspect ratio. Thus, the combination of results from three techniques indicated that the pores are very much elongated and tube-like.

3.7.3.3 Pore Space Evolution

Carbonate rocks often contain pores of various sizes from sub-micron to centimeters or larger [59]. The pore sizes and their connectivity are the critical elements affecting the fluid transport properties. Dissolution or leaching by fresh water after the original deposition is considered one of the main pathways for the creation of secondary porosity in many carbonate formations. Rocks from the Bombay offshore basin in India [60] demonstrate the effects of dissolution on pore space. The varying levels of leaching and cementation throughout the geological history have created a great range of porosities at different depths and alternating zones of high (>100 mdarcy) and low (<1 mdarcy) permeabilities. By determining the pore size distribution, a consistent trend of the pore space evolution as a function of the degree of dissolution was found, leading to a dramatically altered new pore network [61, 62].

The observed DDIF spectra, as illustrated in Figure 3.7.5 for a few representative samples over the entire porosity range, show a clear trend as a function of porosity. For samples with porosity less than 10%, the DDIF spectra are dominated by the pores around 1–2 µm, with a shoulder extending to about 20 µm. The samples with porosity between 10 and 20% showed a DDIF spectrum still dominated by pores of a few microns; however, the porosity in the 10-µm region is much stronger. The high porosity samples, 20–36%, showed dramatically different DDIF spectra, which are dominated by the large pores. For a few samples of high porosity, the DDIF spectra extend to above 100 µm.

Detailed analysis of the DDIF spectra showed a constant microporosity for almost all samples. This lends strong support to the hypothesis that the original rocks after burial, compaction and the initial diagenesis had a common pore geometry. Such a base rock appears to be dominated by the micropores with approximately 2/3 micropores, 1/3 mesopores and very few macropores. The effect

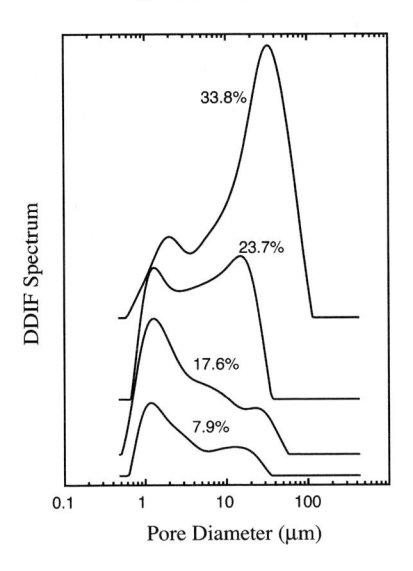

Fig. 3.7.5 Plot of the DDIF spectra for a several carbonate samples from an Indian off-shore basin over a range of porosities. The porosities of the samples are indicated next to the curves. The plots are shifted for clarity. The DDIF experiments were performed on a Bruker Biospec spectrometer at a magnetic field of 2.14 T (Nalorac Cryogenics, Inc.).

of leaching is to remove a portion of the base rock, adding macro- and mesoporosity. The large leached pores have formed a network with a different geometry, having a body-to-throat ratio substantially lower than that of the micropore network. This low body-to-throat ratio of about 5 for the leached pores is similar to that of many grain-supported sandstones [63]. Consequently, the leaching network may allow an efficient flow of oil compared with other samples with progressively higher body-to-throat ratios.

The technique of DDIF provides a quantitative characterization of the complex pore space of the rocks to supplement conventional mineralogy, chemistry and petrology analyses. A combination of DDIF, Hg intrusion, NMR T_2 and image analysis has become the new paradigm to characterize porous rocks for petroleum applications [62, 61].

3.7.3.4 Drainage and Imbibition

The pore geometry described in the above section plays a dominant role in the fluid transport through the media. For example, Katz and Thompson [64] reported a strong correlation between permeability and the size of the pore throat determined from Hg intrusion experiments. This is often understood in terms of a capillary model for porous media in which the main contribution to the single phase flow is the smallest restriction in the pore network, i.e., the pore throat. On the other hand, understanding multiphase flow in porous media requires a more complete picture of the pore network, including pore body and pore throat. For example, in a capillary model, complete displacement of both phases can be achieved. However, in real porous media, one finds that displacement of one or both phases can be hindered, giving rise to the concept of residue saturation. In the production of crude oil, this often dictates the fraction of oil that will not flow.

DDIF has been applied to understand two-phase flow (air and water) in a Berea sandstone sample and the relationship to the pore geometry [65]. Several different states of saturation were studied: full saturation and partial saturation by three methods, i.e., centrifugation, co-current imbibition and counter-current imbibition. Imbibition is a process in which a porous sample absorbs the wetting fluid through capillary force. In the case of co-current imbibition, the bottom of the rock sample was kept in contact with water, so the water is imbibed into the rock and the water and air flowed in the same direction. For counter-current imbibition, the whole sample was immersed and the water was drawn into the center of the rock as, the air was forced out; in this case, the water and air flowed in opposite directions.

DDIF spectra were obtained on the sample at the several saturation states [65]. DDIF spectra for full saturation and co-current imbibition with water saturation of 35.5 and 57.9% (Figure 3.7.6) were found to be of similar shape with a dominant peak at large pores and a shoulder extending to smaller pore sizes. This result

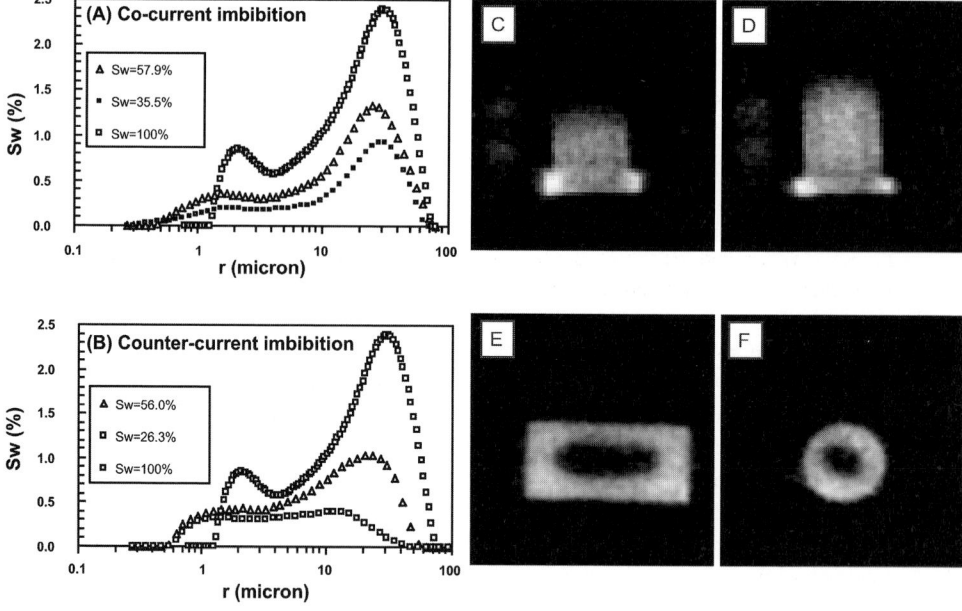

Fig. 3.7.6 DDIF spectra and SPRITE MRI images of Berea obtained in different saturation states. (A) The DDIF spectra during co-current imbibition at different water saturation (S_w) levels. Note the similar shape of DDIF spectra at different S_w. (B) The DDIF spectra during counter-current imbibition acquired at different water saturation levels. Note the change in the DDIF spectral shape for the different saturation levels. (C, D) A pair of images show 2D longitudinal slices from 3D Conical-SPRITE MRI data sets obtained during co-current imbibition. The time interval between the two images was 10.5 min. The images show a piston-like water penetration. (E, F) 2D slices from a 3D Conical-SPRITE MRI data set obtained during counter-current imbibition. The overall water saturation was 26.3%. The penetrating waterfronts have not reached the sample center. Figure from Ref. [65] with permission.

indicates that during co-current imbibition, water fills the pore bodies and the throats uniformly at the pore level. In addition, it was demonstrated by SPRITE magnetic resonance imaging (MRI) that the invading water showed approximately a rectangular shape, and exhibits piston-like water front movement. Behind the advancing waterfronts, no further increase of water saturation was observed and a uniform water saturation distribution was observed.

The DDIF data on the counter-current imbibition showed a different pattern (Figure 3.7.6). The MRI images at short imbibition times showed an invading front entering the sample from all directions and an initial lack of signal in the center of the sample. However, these images alone do not address the invasion at the pore level. DDIF data in Figure 3.7.6 obtained at short imbibition time shows a substantially flat spectrum and lacks the large peak at $r = 30 - 40$ μm, which is very different from the spectrum at the full saturation. The much reduced signal at the pore body size indicates that many of the large pores are not filled completely with water; instead, residual air occupies the center of these pores. At a longer imbibition time (water saturation $S_w = 56\%$), the DDIF spectrum shows a significant growth in the peak at $r = 30$ μm indicating that more pore bodies are being filled. Thus, the DDIF technique complements the macroscopic MRI of water imbibition with a microscopic probe of the selective filling of different pore sizes as a function of time. These data can inform on microstructural models dealing with the weakening of materials as a function of hydration or other uptake, such as the ingress of chlorine ions known to weaken concrete.

3.7.3.5 Water Permeation Through a Blood Cell Membrane

Transport of small molecules, ions and water through the cell membrane serves various biological functions important for the metabolism of the cell, maintenance of ion concentration and other cell activities [66, 67]. One of the important transport processes is the permeation of water molecules through the membrane of an erythrocyte in blood. Characterization of the diffusive water exchange between red blood cells (RBC) and plasma by NMR has generally been done on the basis of either relaxation in the presence of contrast agents [68–73] or diffusion in the presence of applied magnetic field gradients [70–77]. Both approaches proved to be very useful in probing tissue microstructure; in particular, a series of articles by Benga and coworkers [78, 79] showed the usefulness of the relaxation method to determine the water permeation time scale in blood under various disease conditions.

The relaxation method relies on the fact that the presence of hemoglobin in RBC causes the intracellular proton relaxation times ($T_{1c} = 570$ ms, $T_{2c} = 10$ ms) to be shorter than those in the plasma ($T_{1p} = 1700$ ms, $T_{2p} = 100$ ms) [73, 80]. The third time scale is the water residence time in RBC, T_r. The essence of the method is that the paramagnetic relaxation enhancing agent is added to the blood, so that T_{1p} is proportional to the inverse of the concentration of the agent. When the agent concentration is low and T_{1p} and T_{1c} are much longer then T_r, a proton will travel between the two compartments several times before its decay, thus, all protons will experience the same fraction of time for being inside and outside the cells. This gives rise to a uniform relaxation time for all protons and a single exponential decay in the relaxation experiments. In this regime, the residence time cannot be determined. When the concentration of the contrast agent is increased so that $T_{1p} < T_r$, the signal decay exhibits multi-exponential characteristics. Often the faster decay component corresponds to the relaxation in the plasma and the slower decay time constant is related to the water residence time in the blood cell.

The diffusion techniques rely on measurement of the diffusive decay of the NMR signal of water protons due to applied magnetic field gradients. The apparent diffusion coefficients (ADC) for water inside the RBC is smaller than that in plasma due to higher intracellular protein concentration and to a restriction of the water motion by cellular membranes. When the diffusion time is shorter than T_r, the diffusive decay reveals double exponential behavior indicating two different ADCs. When the diffusion time becomes comparable to T_r, the exchange of water molecules diminishes the difference in ADCs, resulting in an approximate single exponential decay. Measuring the magnetization decay as a function of the gradient strength for different times allows a determination of the exchange parameters.

Lisitza et. al. [81, 82] have demonstrated an alternative method to directly probing the decay of spin magnetization in native blood samples without either the use of a contrast agent or applied magnetic field gradients. The new method employs the essential concept of DDIF to use the internal magnetic field inhomogeneity to create magnetization modulation between RBC and plasma, so that water transport through the cell membrane can be directly measured from the magnetization decay. They demonstrated the presence of internal magnetic field variations due to blood cells by observing a significantly broader ^{1}H spectrum in a blood sample than those of a plasma sample and a blood sample with all the cells crushed [see Figure 3.7.7(A)]. In the crushed-cell sample, membrane was removed by centrifugation so that the proteins such as hemoglobin maintained their identical concentration, and were uniformly distributed within the sample. This observation is consistent with the fact that the susceptibility of the RBCs is noticeably different from that of the plasma due to the high concentration of hemoglobin. As a result, the magnetic field inside the cells is likely to be different from that in the plasma, setting up a spatially varying internal field.

The DDIF spectrum obtained [81, 82] in the packed erythrocyte is shown in Figure 3.7.7.(B), together with the distribution of the relaxation time T_1 measured using the inversion-recovery method. The relaxation experiment shows approximately single exponential decay with a single peak at 1 s in the LI spectrum. In the DDIF spectrum, the peak at 1 s remains with a slightly broader width. This is from the relaxation mode that corresponds to the peak in the T_1 spectrum. In addition to the relaxation, there are two fast decaying components in the DDIF spectrum. These components are missing in the relaxation spectrum and, thus, are related specifically to DDIF. The decay time of the fastest component ($\approx$ 1 ms) corresponds to a diffusion distance of about 1.5 μm, which is very close to the thickness of the erythrocyte (2 μm). Therefore, this peak can be interpreted as the diffusion modes that are either inside the RBC or in the plasma space. Given the concentration of RBC in the sample, the blood cells were closely packed and the spacing between the

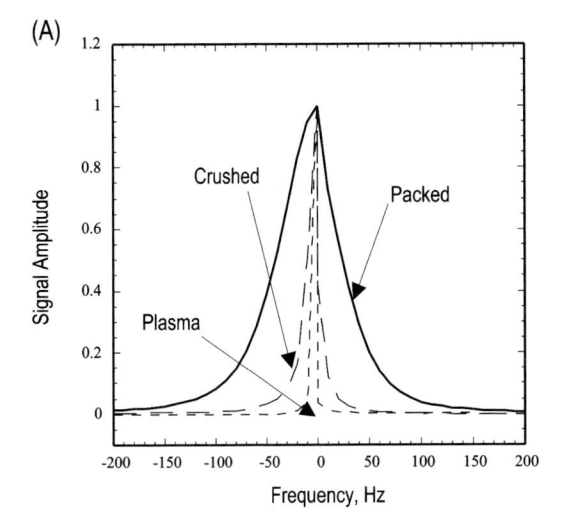

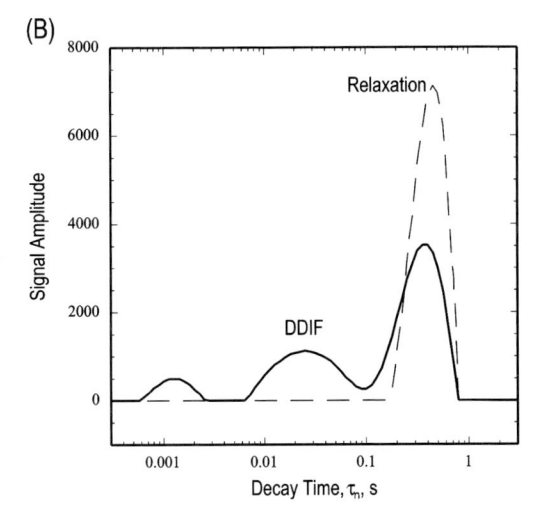

3.7 NMR Pore Size Measurements Using an Internal Magnetic Field in Porous Media | 355

Fig. 3.7.7 (A) NMR frequency spectra of different blood samples. The linewidths were found to be 70, 15 and 10 Hz for packed, crushed erythrocyte and plasma samples, respectively. (B) DDIF (solid line) and inversion recovery (dashed line) spectra obtained for the packed erythrocyte sample.

cells was similar to the cell size. The intermediate spectral component has an average value of 15 ms for its decay time, which agrees well with the values reported for the exchange times between the RBC and plasma [72, 73, 83]. Thus, this component has been interpreted as the exchange mode [84], which directly reflects the water diffusion across the membrane. Potentially, the DDIF experiment might provide more information about the dynamics of diffusion than the conventional techniques; for example, the width of the exchange mode could be related to the distribution of the exchange times over the sample, which, in turn, could reflect the local structure of the membrane.

3.7.4
Summary

The DDIF (decay due to diffusion in an internal field) technique appears to be a useful experimental approach to the study of porous media. Using the internal magnetic field gradients present in a medium of non-uniform magnetic susceptibility to encode diffusion information allows a direct connection to the underlying pore structure that generates the gradients. The distribution of decay times, in the weak encoding limit, reflects the distribution of internal gradients and thus of pore sizes within the material. This distribution is extracted via a Laplace inversion. Though the Laplace inversion is an ill-conditioned problem, the numerical techniques of relaxation-mode subtraction, regularization and resolution determination allow a stable and interpretable spectrum to be obtained.

Several applications of this technique have been demonstrated. These include measurements in rock samples that, in combination with other methods, measure static microstructural properties such as pore size and shape, as well as dynamic evolution of the pore distribution during the process of dissolution or water uptake. Other media containing inhomogeneous fields, such as blood cell solutions, also fall within the purview of the DDIF technique, and structural properties have been extracted from blood cell solution DDIF spectra. In summary, the DDIF technique has potential for the study of structure on the micron scale as a complement to existing methods, such as x-ray tomography or neutron scattering, to study porous media and other materials with internal structures that are important in many chemical engineering applications.

Acknowledgements

The authors thank L. Venkataramanan, S. Ryu, P. N. Sen, M. D. Hürlimann, D. F. Allen, W. Kenyon and A. Boyd for numerous discussions during the development of DDIF and its applications.

References

1 P. Z. Wong (1999), *Methods in the Physics of Porous Media*, Academic Press, London.
2 B. P. Flannery, H. W. Deckman, W. G. Roberge, K. L. D'Amico **1987**, *Science* 237, 1439.
3 J. T. Fredrich, B. Menendex, T. F. Wong **1995**, *Science* 268, 276.
4 F. M. Auzerias, J. Dunsmuir, B. B. Ferréol, N. Martys, J. Olson, T. S. Ramakrishnann, D. H. Rothman, L. M. Schwartz **1996**, *Geophys. Res. Lett.* 23, 705.
5 J. G. Berryman, S. C. Blair **1986**, *J. Appl. Phys.* 60, 1930.
6 M. H. Cohen, K. S. Mendelson **1982**, *J. Appl. Phys.* 53, 1127.
7 P. G. de Gennes **1982**, *C. R. Acad. Sci. Ser. II* 295, 1061.
8 R. L. Kleinberg, S. A. Farooqui, M. A. Horsfield **1993**, *J. Colloid Interface Sci.* 158, 195.
9 L. L. Latour, P. P. Mitra, R. L. Kleinberg, C. H. Sotak **1995**, *J. Magn. Reson. Ser. A* 112, 83.

10 W. P. Halperin, F. D'Orazio, S. Bhattacharja, J. C. Tarczon **1989**, in *Molecular Dynamics in Restricted Geometries*, eds. J. Klafter, J. Drake, John Wiley & Sons, New York.

11 D. E. Woessner **1963**, *J. Phys. Chem.* 67, 1365.

12 P. P. Mitra, P. N. Sen, L. M. Schwartz, P. Le Doussal **1992**, *Phys. Rev. Lett.* 68, 3555.

13 D. G. Cory, A. N. Garroway **1990**, *Magn. Reson. Med.* 14, 435.

14 P. T. Callaghan, A. Coy, D. MacGowan, K. J. Packer, F. O. Zelaya **1991**, *Nature (London)* 351, 467.

15 Y.-Q. Song, L. Venkataramanan, M. D. Hurlimann, M. Flaum, P. Frulla, C. Straley **2002**, *J. Magn. Reson.* 154 (2), 261.

16 R. L. Kleinberg **1994**, *Magn. Reson. Imaging* 12, 271.

17 R. M. Kroeker, R. M. Henkelman **1986**, *J. Magn. Reson.* 69, 218.

18 H. Peemoeller, R. K. Shenoy, M. M.. Pintar **1981**, *J. Magn. Reson.* 45, 193.

19 M. D. Hürlimann, L. Venkataramanan **2002**, *J. Magn. Reson.* 157, 31.

20 M. D. Hürlimann, L. Venkataramanan, C. Flaum **2002**, *J. Chem. Phys.* 117, 10223.

21 M. D. Hürlimann, M. Flaum, L. Venkataramanan, C. Flaum, R. Freedman, G. J. Hirasaki **2003**, *Magn. Reson. Imaging* 21, 305.

22 R. Kleinberg **1995**, in *Encyclopedia of Nuclear Magnetic Resonance*, eds. D. M. Grant, R. K. Harris, John Wiley & Sons, New York.

23 G. Eidmann, R. Savelsberg, P. Blümler, B. Blümich **1996**, *J. Magn. Reson. A* 122, 104.

24 C. A. Meriles, D. Sakellariou, H. Heise, A. J. Moulé, A. Pines **2001**, *Science*, 293, 82.

25 T. M. Brill, S. Ryu, R. Gaylor, J. Jundt, D. D. Griffin, Y.-Q. Song, P. N. Sen, M. D. Hürlimann **2002**, *Science* 297, 369.

26 M. Winkler, M. Zhou, M. Bernardo, B. Endeward, H. Thomann **2003**, *Magn. Reson. Imaging* 21, 311.

27 M. D. Hürlimann **1998**, *J. Magn. Reson.* 131, 232.

28 M. D. Hürlimann, K. G. Helmer, C. H. Sotak **1998**, *Magn. Reson. Imaging* 16, 535.

29 G. C. Borgia, R. J. S. Brown, P. Fantazzini **1995**, *Phys. Rev. E* 51, 2104.

30 G. C. Borgia, R. J. S. Brown, P. Fantazzini **1996**, *Magn. Reson. Imaging* 14, 731.

31 P. N. Sen, S. Axelrod **1999**, *J. Appl. Phys.* 86, 4548.

32 L. J. Zielinski, P. N. Sen **2000**, *J. Magn. Reson.* 147, 95.

33 R. M. Weisskoff, C. S. Zuo, J. L. Boxerman, B. R. Rosen **1994**, *Magn. Reson. Med.* 31, 601.

34 R. J. S. Brown **1961**, *Phys. Rev.* 121, 1379.

35 B. Audoly, P. N. Sen, S. Ryu, Y.-Q. Song **2003**, *J. Magn. Reson.* 164, 154.

36 W. D. Williams, E. F. W. Seymour, R. M. Cotts **1978**, *J. Magn. Reson.* 31, 271.

37 R. F. Karlicek Jr., I. J. Lowe **1980**, *J. Magn. Reson.* 37, 75.

38 W. R. Nitz **2002**, *Eur. Radiol.* 12 (12), 2866.

39 S. Ogawa, T. Lee, A. S. Nayak, P. Glynn **1990**, *Magn. Reson. Med.* 14 (1), 68.

40 J. Ma, F. W. Wehrli **1996**, *J. Magn. Reson.* 111, 61.

41 S. N. Hwang, F. W. Wehrli **1999**, *J. Magn. Reson.* 139, 35.

42 Y.-Q. Song **2000**, *J. Magn. Reson.* 143, 397.

43 Y.-Q. Song, S. Ryu, P. N. Sen **2000**, *Nature (London)* 406, 178.

44 H. C. Torrey **1956**, *Phys. Rev.* 104, 563.

45 M. Kac **1966**, *Am.. Math. Monthly* 73, 1.

46 K. R. Brownstein, C. E. Tarr **1979**, *Phys. Rev. A* 19, 2446.

47 S. Sridhar, A. Kudrolli **1992**, *Phys. Rev. Lett.* 72, 2175.

48 K. R. Brownstein, C.. E. Tarr **1977**, *J. Magn. Reson.* 26, 17.

49 N. V. Lisitza, Y.-Q. Song **2001**, *J. Chem. Phys.* 114, 9120.

50 E. L. Hahn **1950**, *Phys. Rev.* 80, 580.

51 Y.-Q. Song **2003**, *Concept Magn. Reson.* 18A (2), 97.

52 N. V. Lisitza, Y.-Q. Song **2002**, *Phys. Rev. B* 65, 1724061.

53 E. J. Fordham, A. Sezginer, L. D. Hall **1995**, *J. Magn. Reson. Ser. A* 113, 139.

54 G. C. Borgia, R. J. S. Brown, P. Fantazzini **1998**, *J. Magn. Reson.* 132, 65.

55 A. N. Tikhonov, V. Y. Arsenin **1977**, *Solutions of Ill-posed problems*, John Wiley & Sons, New York.
56 Y.-Q. Song, L. Venkataramanan, L.. Burcaw **2005**, *J. Chem. Phys.*, 122, 104104.
57 Q. Chen, Y.-Q. Song **2002**, *J. Chem. Phys.* 116, 8247.
58 H. H. Yuan, B. F. Swanson **1989**, *SPE Form. Eval.* March, 17.
59 M. E. Tucker **1991**, *Sedimentary Petrology*, 2nd ed., Blackwell Scientific Publications, Oxford.
60 K.. Satyanarayana, R.. R. Sharma, D. K. Dasgupta, K. K. Das **1999**, *Atlas of Carbonate Microfacies from the Reservoirs of Bombay Offshore Basin, India*, Oil and Natural Gas Corporation Ltd, Mumbai, India.
61 W. E. Kenyon, D. F. Allen, N. V. Lisitza, Y.-Q. Song **2002**, SPWLA 43rd Annual Meeting (Japan).
62 Y.-Q. Song, N. V. Lisitza, D. F. Allen, W. E. Kenyon **2002**, *Petrophysics* 43, 420.
63 Y.-Q. Song **2001**, *Magn. Reson. Imaging* 19, 417.
64 A. J. Katz, A. H. Thompson **1986**, *Phys. Rev. B* 34, 8179.
65 Q. Chen, M. Gingras, B. Balcom **2003**, *J. Chem. Phys.* 119, 479.
66 F. Crick **1970**, *Nature (London)* 225, 420.
67 J. A. Dix, A. K. Solomon **1984**, *Biochem. Biophys. Acta* 773, 219.
68 G. Santyr, I. Kay, R. Henkelman, M. Bronskill **1990**, *J. Magn. Reson.* 90, 500.
69 C. Kroeker, M. Stewart, M. Bronskill, R. Henkelman **1988**, *Magn. Reson. Med.* 6, 24.
70 S. Bradamante, E. Barchiest, S. Pilotti, G. Borasi **1988**, *Magn. Reson. Med.* 8, 440.
71 W. Spees, D. Yablonskiy, M. Oswood, J. J. H. Ackerman **2001**, *Magn. Reson. Med.* 45, 533.
72 J. Stanisz, J. G. Li, G. Wright, R. Henkelman **1998**, *Magn. Reson. Med.* 39, 223.
73 J. G. Li, J. Stanisz, R. Henkelman **1998**, *Magn. Reson. Med.* 40, 79.
74 Z. Luz, S. J. Meiboom **1963**, *Chem. Phys.* 39, 366.
75 J. Xie, A. Szafer, A. Anderson, K. M. Johnson, J. Gore **1996**, *Proc. ISMRM* 4, 1332.
76 L. L. Latour, K. Svoboda, P. Mitra, C. Sotak **1994**, *Proc. Natl. Acad. Sci.* 91, 1229.
77 G. Stanisz, A. Szafer, G. Wright, R. Henkelman **1997**, *Magn. Reson. Med.* 37, 103.
78 G. Benga, V. Borza, O. Popesku, V. I. Pop, A. Muresan **1986**, *J. Membrane Biol.* 89, 127.
79 G. Benga, O. Popesku, V. Borza, V. I. Pop, A. Hodarnau **1989**, *J. Membrane Biol.* 108, 105.
80 M. E. Fabry, M. Eisenstadt **1975**, *Biophys. J.* 15, 1101.
81 N. V. Lisitza, W. S. Warren, Y. Song **2002**, 43rd Experimental NMR Conference (Pacific Grove, CA).
82 N. V. Lisitza, W. S. Warren, Y.-Q. Song **2004**, *Biophys. J.*, submitted for publication.
83 M. Herbst, J. Goldstein **1984**, *Biochem. Biophys. Acta* 805, 123.
84 L. J. Zielinski, Y. Song, S. Ryu, P. N. Sen **2002**, *J. Chem. Phys.* 117, 5361.

4
Fluids and Flows

4.1
Modeling Fluid Flow in Permeable Media
Jinsoo Uh and A. Ted Watson

4.1.1
Introduction

Many processes of interest to industry and society involve the flow of fluids through permeable media. Industrial examples include filtration to remove solids, membranes to separate gases and catalyst supports within chemical reactors. There are many examples within the human body, including the transfer of gases to blood within the lung, the flow of blood and nutrients through bone and organs and the transfer of therapeutic agents as well as toxins through the dermis. Large-scale examples include the transport of petroleum or water through permeable underground geological structures.

In this chapter, we focus on the flow of one or two immiscible fluid phases through rigid permeable media, although our approach can be used to investigate many other physical processes. We determine the properties used to model flow in heterogeneous media at a scale, or with accuracy, not previously achieved. This work has a number of potential applications. The determination of spatially-resolved properties, such as porosity and permeability, can lead to a better understanding of properties of heterogeneous media and to methods to predict properties for other media based on more convenient or tractable experiments. We can improve our understanding and descriptions of complex processes within media, such as those that involve chemical and biological changes, by spatially resolving fluid states and properties within permeable media. Examples of areas of application include bioremediation and tissue engineering. Also, our general approach can be extended to investigate processes within deformable permeable media, such as bone and cartilage.

Our approach to determine the properties of heterogeneous media utilizes mathematical models of the measurement process and, as appropriate, the flow process itself. To determine the desired properties, we solve an associated system and parameter identification problem (also termed an inverse problem) to estimate the properties from the measured data.

In this chapter, we describe the approaches used to mathematically model the flow of immiscible fluid phases through permeable media. We summarize the elements of system and parameter identification, and then describe our methods for determining properties of heterogeneous permeable media.

4.1.2
Modeling Multiphase Flow in Porous Media

We mathematically model processes in order to understand them, predict responses to inputs or stimuli and to control them. The basic approach to model transport phenomena using continuum principles is well developed. Mathematical statements representing conservation principles (such as conservation of mass, momentum and energy) and constitutive relationships among various quantities that represent states of the associated process are used to model physical processes. To describe flow in permeable media, one could in principle use the well-established conservation and constitutive equations for fluids within the pore space, with associated boundary conditions at the solid–fluid interface. We refer to this as the "microscopic" approach. However, the actual geometry for virtually all naturally occurring media essentially defies precise description. Even if the actual geometry could be determined, the associated boundary value problems would hardly be tractable, except for possibly very idealized situations. Consequently, "macroscopic" models are typically used to describe flow within permeable media. The macroscopic equations represent a continuum representation at a coarser scale – representing averages over many "pores", as compared with the microscopic description, for which the continuum represents averages over many molecules. Many models, initially developed empirically, are firmly established using local volume-averaging of the associated molecular-scale continuum equations, or other homogenization methods (see, for example, Ref. [1]). These models establish the functional relationships among fluid states and the means for predicting the response to various inputs, given the fluid and media properties. However, the media properties represent local effective empiricisms, and their determination remains an outstanding problem.

The creeping flow of a single fluid phase through a rigid permeable medium is modeled with the continuity equation and Darcy's Law:

$$\frac{\partial [\phi(\mathbf{z})\varrho]}{\partial t} = -\nabla \cdot (\varrho \mathbf{v}) \tag{4.1.1}$$

$$\mathbf{v} = -\frac{k(\mathbf{z})}{\mu}(\nabla p - \varrho \mathbf{g}) \tag{4.1.2}$$

The system states (dependent variables) are the pressure, p, and the superficial (Darcy) velocity, $\mathbf{v}$. The density, ϱ, and viscosity, μ, are fluid properties, and $\mathbf{g}$ is the acceleration of gravity. The porosity, $\phi(\mathbf{z})$, and permeability, $k(\mathbf{z})$, represent the macroscopic properties of the media. Both are spatially dependent and are represented as continuous functions of position $\mathbf{z}$, as explicitly noted. While the permeability is in principle a second-order tensor, we take it to be isotropic here, as is commonly done, although the general case could be addressed. The macroscopic properties correspond formally to volume averages over an element centered at location **z**. To develop the macroscopic equations, it is not necessary to specify the size of the averaging volume, provided it is considerably larger than a "pore", and considerably smaller than the sample. However, it is useful to consider an *intrinsic* macroscopic property, which corresponds to the minimum size possible for the averaging volume. This size, also called the minimum representative elementary volume [2], is not precisely known, but it roughly corresponds to tens of pores. The intrinsic property thus corresponds to the maximum resolution possible. Clearly, coarser representations could be computed directly as integrals of the intrinsic property.

A generalized Darcy equation and equation of continuity for each fluid phase is used to describe the flow of multiple immiscible fluid phases:

$$\frac{\partial [\phi(\mathbf{z}) \varrho_i s_i]}{\partial t} = -\nabla \cdot (\varrho_i \mathbf{v}_i) \tag{4.1.3}$$

$$\mathbf{v}_i = -\frac{k(\mathbf{z}) k_{ri}(s_i)}{\mu_i} (\nabla p_i - \varrho \mathbf{g}) \tag{4.1.4}$$

The relative permeability to phase i, $k_{ri}(s_i)$, is taken to be a function of fluid saturation s_i, which is the fraction of the pore space occupied by phase i; it is supposed that any associated spatial variations are largely taken into account through the permeability. For two-phase flow, fluid saturations are related by

$$s_1 + s_2 = 1 \tag{4.1.5}$$

The capillary pressure function relates the local pressures of the two fluid phases:

$$p_c(s_1) = p_1 - p_2 \tag{4.1.6}$$

where phase 1 is taken as the non-wetting phase. The system states are the pressure and saturation of each fluid phase and the respective fluid velocities. The macroscopic media properties, in addition to the porosity and permeability which arise in the single-phase model, are the capillary pressure function and the two relative permeability functions.

Specific situations are simulated by solving the set of system equations [i.e., Eqs. (4.1.1 and 4.1.2) or (4.1.3–4.1.6)] with pertinent boundary and initial conditions, fluid properties and macroscopic properties. Fluid properties are generally readily obtained. Consider now the media properties, specifically the porosity and permeability, which are required for simulating all flows through permeable media.

The porosity and permeability depend solely on the structure of the medium. In principle, if the entire geometry of the solid surface were known, these properties could be calculated by local volume-averaging [1]. As mentioned previously, this is not the case, except possibly for certain idealized situations. Arguably, there should be some simplified representation for the medium so that, if the pertinent information were determined, the properties could be reliably calculated, but such a model representation has yet to be identified. Consequently, the properties must be determined experimentally.

The porosity can be determined by knowing locally the amount of fluid that saturates the pore space. Conventionally, an average value of the porosity is determined gravimetrically. X-ray CT scanning [3] or MRI [4] can be used to determine spatial distributions of porosity – the latter method is demonstrated in this chapter.

Properties defined through constitutive relationships, such as permeability, cannot be measured directly; they must be calculated through a mathematical model of the experiment with measurements of system states, or functions of these states. This is demonstrated for conventional permeability experiments based on flowing a fluid unidirectionally through a permeable sample, and recording the flow rate into the sample and the pressure drop across the sample. For a sample of length L with a constant cross-sectional area A_c and negligible gravity effects, the apparent permeability can be calculated by

$$k_a = -\mu \frac{Q}{A_c} \frac{L}{\Delta p} \tag{4.1.7}$$

where Q is the volumetric flow rate and Δp is the pressure drop measured over the sample length L. This equation represents a mathematical model of this flow experiment. It follows from the state equations, Eqs. (4.1.1 and 4.1.2), solved for time-independent conditions with pertinent boundary condition and the *assumption that the permeability is spatially uniform*. Note that if the material were *heterogeneous* – surely the expected situation for naturally occurring permeable media – the value of the permeability so calculated will not be a correct representation of the permeability for the sample. Consequently, we refer to the value so calculated to as an "apparent" value – i.e., it is calculated on the basis of an equation that may not suitably model the actual physical phenomena.

With the conventional experimental design, information about spatial variations of the permeability is not available. With MRI, we can obtain information *within* the sample, so that we may determine the *spatial distribution* of the permeability. Clearly, the computational procedure required to estimate the entire distribution will not be as simple as that reflected by Eq. 4.1.7. We will use the principles of system and parameter identification, discussed in the following section, to determine the various macroscopic properties from experiments.

4.1.3
System and Parameter Identification

We refer to system and parameter identification as the principles to determine the most appropriate equations, and properties within those equations, to describe physical phenomena. In particular, we refer to parameter identification as the estimation of properties within a specified model from observations of states or functions of states. This is also called the "inverse problem", so named in contrast to the "forward problem": the prediction of states (or outputs) from specified inputs and properties. System identification refers to the selection of the most appropriate equations. Here, we will consider system equations developed from first principles, i.e., statements of conservation principles and associated constitutive relationships. One might suppose that the equations are thus specified. However, a characteristic of the problems considered here is that the unknown properties are *functions*. System identification principles may be used to select appropriate finite-dimensional representations for the properties.

Suppose that we have a set of n measurements arranged within a vector $\mathbf{Y}$. Further suppose that the errors associated with the measurements are random, and additive, so that we can represent the measurements as

$$\mathbf{Y} = \mathbf{F}(\mathbf{X};\mathbf{t}) + \boldsymbol{\varepsilon} \tag{4.1.8}$$

where $\mathbf{F}$ represents the true response, which is a function of the system states $\mathbf{X}$ and perhaps also a set of independent variables $\mathbf{t}$ (e.g., the set of spatial coordinates and time). Suppose further that we can calculate $\mathbf{X}$ as follows:

$$\mathbf{X} = \mathbf{f}(\mathbf{X}, \mathbf{t}; \mathbf{P}) \tag{4.1.9}$$

where $\mathbf{f}$ denotes the computational procedure or mapping, based on the mathematical model of the experiment, which relates the properties $\mathbf{P}$ and independent variables to the states; for a nonlinear process, this procedure may be a function of the state, as denoted here, where $\mathbf{P}$ denotes the true (although unknown) properties.

Statistical theory provides for the construction of a function incorporating differences between the measured and corresponding calculated values, with the best estimates being the properties that minimize that function. This procedure is shown in detail in the following section. Generally, the implementation is significantly impacted by the functionality of the properties. Three classes of problems are identified:

1. The properties are represented by a set of constants, $\mathbf{P} = \mathbf{C}$, a vector with m elements.
2. The properties are a set of functions of independent variables, $\mathbf{P}(\mathbf{t})$.
3. The properties are a set of functions of state (dependent) variables, $\mathbf{P}(\mathbf{X})$.

The parameter identification problem associated with the conventional permeability experiments is within the first class (with $m = 1$). By contrast, the problems we consider here are within the second and third classes; these are *functional* estimation problems. Ultimately, however, these are solved with finite-dimensional representations, although an essential aspect of the solution of these infinite-dimensional (function) estimation problems is the selection of the appropriate representations.

In Section 4.1.4.1, we develop the estimation of the relaxation distribution functions from NMR data. These are used to determine porosity and saturation distributions. In Section 4.1.4.2, we develop the estimation of permeability distributions. Both of these problems are within the second class, as the models of the measured responses are based on properties that are functions of independent variables within the models. The multiphase flow properties, considered in Section 4.1.4.3, are functions of a state variable (i.e., fluid saturation), and thus the estimation of these properties falls within the third class.

4.1.4
Determination of Properties

In this section, we describe our experimental and analysis methods to determine spatially dependent porosity and saturation distributions, permeability functions and saturation-dependent multiphase flow properties: the relative permeability and capillary pressure functions.

4.1.4.1 Porosity and Saturation Distributions

If the amount of fluid within a fully saturated permeable medium is known as a function of position, the spatially resolved porosity distribution can be determined. If the medium is saturated with two fluids, and the signal from one can be distinguished, the fluid saturation can be determined. In this section, we will develop a method to determine the amount of a single observed fluid using MRI, and demonstrate the determination of porosity. In Section 4.1.4.3, we will demonstrate the determination of saturation distributions for use in estimating multiphase flow functions.

The intrinsic NMR magnetization intensity is proportional to the number of spins, so that with a suitable calibration [5] the amount of fluid corresponding to each voxel can, in principle, be determined by spin-density imaging. However, the intrinsic magnetization intensity is an equilibrium quantity that is not measured directly. For fluids in porous media, transverse relaxation is often significantly enhanced due to the fluid–solid interactions and self-diffusion in the presence of magnetic-susceptibility induced field gradients. Consequently, spin-density is not accurately determined directly from images as significant relaxation occurs before the signal is acquired. The intrinsic magnetization intensity can be estimated from the observed signal if the associated relaxation rate is known. However, this is complicated by the fact that the relaxation of fluids in permeable media is affected by the local properties of the media, including the microscopic structure of the pore space and the composition of the solid matrix constituents, both of which can vary spatially.

We can perform spatially resolved Carr–Purcell–Meiboom–Gill (CPMG) experiments, and then, for each voxel, use magnetization intensities at the echo times to estimate the corresponding number density function, $P(\tau)$, which represents the amount of fluid associated with the characteristic relaxation time τ. The corresponding intrinsic magnetization for the voxel, M_0, is calculated by

$$M_0 = \int_0^\infty P(\tau)d\tau \qquad (4.1.10)$$

One can consider this procedure as that of modeling the relaxation process within each voxel, and extrapolating the signal to a zero echo time [6]. This basic element – the estimation of the relaxation distribution within a heterogeneous system – has many applications, including the determination of fluid saturations [7], microscopic structures as described by the pore-size distribution [8] and compartmentalization of biological tissues [9]. While a number of different methods have been proposed for estimating relaxation distributions, our particular approach is unique, in that we develop it as a functional estimation problem using nonparametric regression theory [10]. Our method is presented in the following sub-section.

4.1.4.1.1 Determination of NMR Relaxation Distribution

We represent the NMR relaxation distribution by the continuous number density function, $P(\tau)$, of characteristic relaxation time τ. Our measurements correspond to a series of CPMG echoes, represented by

$$Y(t_j) = \int_0^\infty P(\tau) K(t_j, \tau) d\tau + \varepsilon_j, \, j = 1, \ldots, n \qquad (4.1.11)$$

where ε_j is a random measurement error and the kernel function is given by $K(t_j, \tau) = \exp(-t_j/\tau)$ [6, 11].

The CPMG pulse sequence is composed of the initial 90° radiofrequency (rf) excitation pulse followed by a series of 180° rf pulses spaced to allow a train of

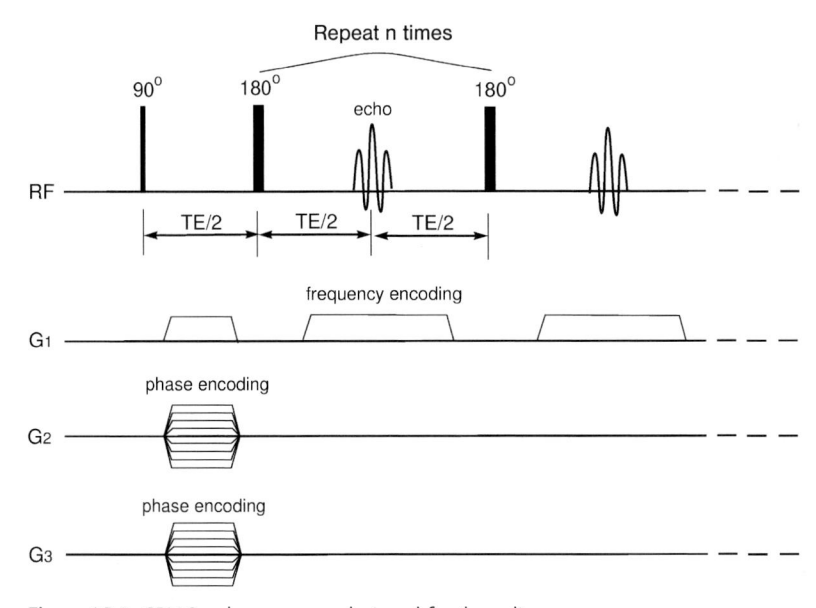

Figure 4.1.1 CPMG pulse sequence designed for three-dimensional imaging. TE is echo time, and G_1, G_2 and G_3 represent the gradient magnetic fields along the directions of z_1, z_2 and z_3, respectively.

echoes to be observed (Figure 4.1.1). The spatial information is revolved by frequency-encoding in the axial direction (z_1) and phase-encodings in the transverse directions (z_2, z_3).

Our goal is to estimate the function $P(\tau)$ from the set of discrete observations $Y(t_j)$. We use a nonparametric approach, whereby we seek to estimate the function without supposing a particular functional form or parameterization. We require that our estimated function be relatively smooth, yet consistent with the measured data. These competing properties are satisfied by selecting the function that minimizes, for an appropriate value of the regularization parameter λ, the performance index:

$$J = [\mathbf{Y} - \mathbf{F}]^T \mathbf{W}[\mathbf{Y} - \mathbf{F}] + \lambda \int_0^\infty \left[\frac{d^2 P(\tau)}{d\tau^2}\right]^2 d\tau \qquad (4.1.12)$$

The data vector $\mathbf{Y}$ is comprised of the measurements, and the vector $\mathbf{F}$ represents the corresponding values calculated by the integral equation [Eq. (4.1.11)] with a specified distribution function $P(\tau)$. The first term on the right-hand side of Eq. (4.1.12) reflects the goodness of fit to the data by the corresponding calculated values for a given estimate of the distribution function. The weighting matrix $\mathbf{W}$ is chosen on the basis of maximum likelihood principles [11]. The second term is the regularization function. It serves to smooth the solution by penalizing the magnitude of its second derivative. The regularization term is weighted relative to the data-fitting term by the regularization parameter, λ, providing a trade-off between the smoothness of the solution and the goodness of fit to the data.

We employ a B-spline basis representation for the distribution function:

$$P(\tau) = \sum_{i=1}^{n_s} c_i B_i^m(\tau, \mathbf{x}) \qquad (4.1.13)$$

where m is the order of the spline, $\mathbf{x}$ is the extended partition (the location of the knot) and n_s is the spline dimension, which is given by the sum of the number of interior knots and the order of the spline. B-splines are a superior way of representing the distribution function as they can accurately approximate any continuous function on a finite domain [12]. Thus, with sufficient numbers of knots, we will be able to represent the true, but unknown, function.

Using Eq. (4.1.13), the estimation problem is cast in discrete form as the determination of the coefficients $\mathbf{c} = [c_1, \ldots, c_{n_s}]$ that minimize the quadratic performance index:

$$J = [\mathbf{Y} - \mathbf{Ac}]^T \mathbf{W}[\mathbf{Y} - \mathbf{Ac}] + \lambda \mathbf{c}^T \mathbf{M}^T \mathbf{Mc} \qquad (4.1.14)$$

where the matrices $\mathbf{A}$ and $\mathbf{M}$ are derived from Eqs. (4.1.11–4.1.13). We incorporate linear equality and inequality constraints that ensure the estimated distribution is non-negative and vanishes outside the range of relaxation times that includes the relaxation time for the bulk fluid and the minimum relaxation time detectable by the experiment. For a given value of regularization parameter, λ, the unique set of

B-spline coefficients that globally minimize the performance index is determined.

We use a method that implements the Unbiased Prediction Risk criterion [13] to provide a data-driven approach for the selection of the regularization parameter. The equality constraints are handled with LQ factorization [14] and an iterative method suggested by Villalobos and Wahba [15] is used to incorporate the inequality constraints [10]. The method is well suited for the relatively large-scale problem associated with analyzing each image voxel as no user intervention is required and all the voxels can be analyzed in parallel.

4.1.4.1.2 Experimental Results and Discussion

Our method is demonstrated with experiments on a Bentheimer sandstone sample. The sample was prepared to be cylindrically shaped with a diameter of 2.5 cm and a length of 2.0 cm. The sample was fully saturated with de-ionized water under vacuum. We performed the CPMG imaging experiment described in the previous section to measure the magnetization intensity at 50 echoes spaced by 4.6 ms for each of $32 \times 16 \times 16$ voxels within the field of view of $3.0 \text{ cm} \times 3.0 \text{ cm} \times 3.0 \text{ cm}$. The corresponding voxel size is $0.938 \text{ mm} \times 1.88 \text{ mm} \times 1.88 \text{ mm}$. We used 1 s of repetition time (TR) and the total imaging time was 4.3 min.

Figure 4.1.2 shows the estimated normalized relaxation distribution, $P(\tau)/\int_0^\infty P(\tau)d\tau$, for a selected voxel. Using this function, the magnetization intensity is calculated as a function of relaxation time, and is presented in Figure 4.1.3 together with the relaxation data. Note that the calculated magnetization intensity is consistent with the measured values, indicating a precise fit of the data. The magnetization datum at the first echo is 88% of the value of the estimated intrinsic magnetization, indicating that significant relaxation has occurred before the first

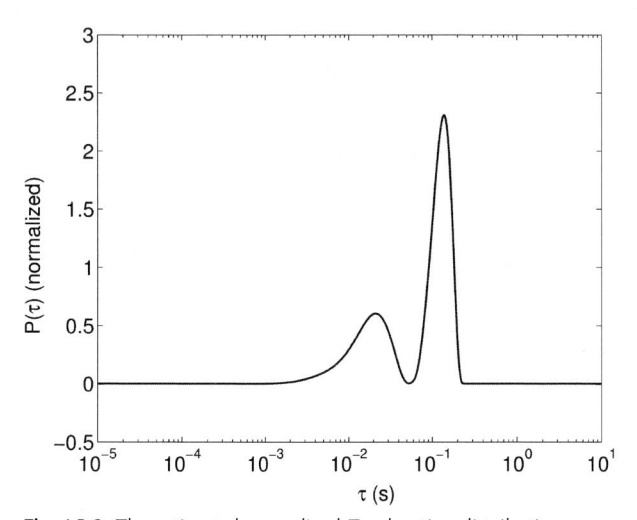

Fig. 4.1.2 The estimated normalized T_2 relaxation distribution for the selected voxel in the Bentheimer sample.

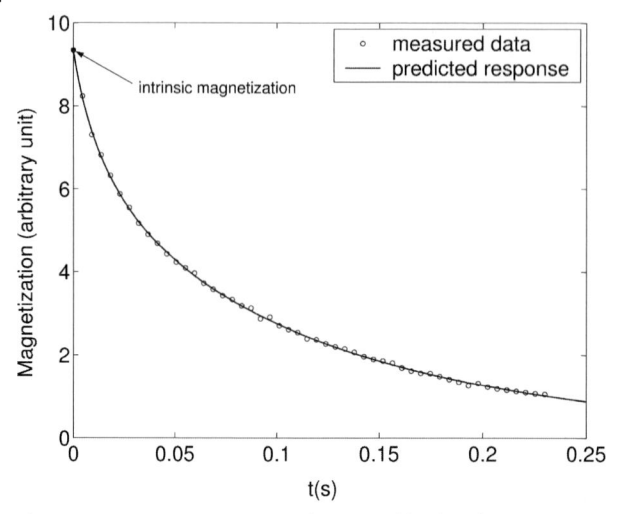

Fig. 4.1.3 Intrinsic magnetization determined by the relaxation data corresponding to the selected voxel in the Bentheimer sample.

observed echo. A porosity image of a horizontal layer (the seventh one from the bottom among the 16 layers of voxels) of the sample is presented with a grayscale in Figure 4.1.4. The porosity in that layer ranges from 0.16 to 0.34, indicating that the sample is macroscopically heterogeneous.

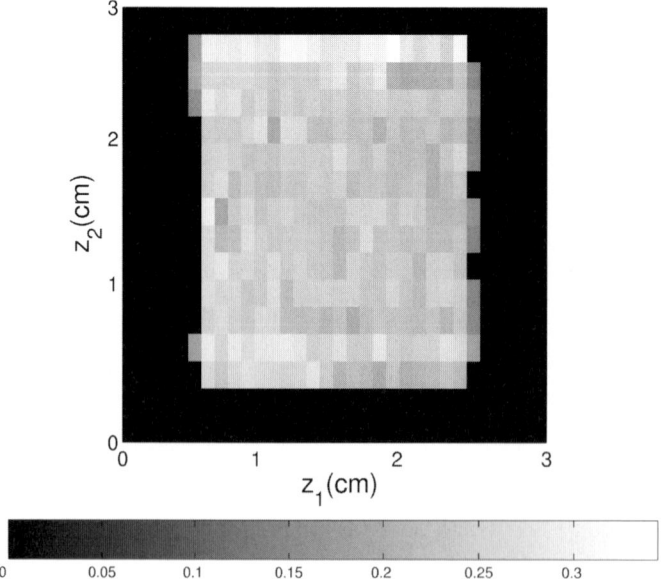

Fig. 4.1.4 Porosity distribution within a horizontal layer of the Bentheimer sample. Axis z_1 is parallel with the static magnetic field.

4.1.4.2 Permeability Distribution

The lack of a method to determine the spatial distributions of permeability has severely limited our ability to understand and mathematically describe complex processes within permeable media. Even the degree of variation of intrinsic permeability that might be encountered in naturally occurring permeable media is unknown. Samples with permeability variations will exhibit spatial variations in fluid velocity. Such variations may significantly affect associated physical phenomena, such as biological activity, dispersion and colloidal transport. Spatial variations in the porosity and permeability, if not taken into account, can adversely affect the determination of any associated properties, including multiphase flow functions [16].

We have developed a method to spatially resolve permeability distributions. We use MRI to determine spatially resolved velocity distributions, and solve an associated system and parameter identification problem to determine the permeability distribution. Not only is such information essential for investigating complex processes within permeable media, it can provide the means for determining improved correlations for predicting permeability from other measurements, such as porosity and NMR relaxation [17–19].

We use a conventional experimental design, in that fluid is flowed at a constant flow rate through a sample. We measure the pressure drop and the distribution of velocity within the sample, as described in the following section. We then estimate the permeability distribution from the measured data, as described in Section 4.1.4.2.2.

4.1.4.2.1. NMR Velocity Imaging

While NMR has been used to image flow for decades [20, 21], the reliable determination of fluid velocity within permeable media is problematic. For flow in conduits, the velocity can be related to the averaged phase shift [22], but this may not hold for fluids within permeable media [23]. Instead, we determine a velocity density distribution for each volume element and calculate the corresponding superficial average velocity from the density distribution. The imaging time is increased by a factor that is the number of velocity encoding steps. As the flow is maintained at a constant flow rate, the relatively long imaging times do not compromise the experiment.

Our velocity encoding protocol uses a pulsed-field-gradient-stimulated echo (PFGSTE). An important feature of the PFGSTE NMR method is that nuclear spins moving in the presence of a magnetic field gradient, $\mathbf{G}(t')$, exhibit a phase shift in the observed echo. For a time-dependent gradient, the phase shift of a spin is

$$\Delta \varphi(t) = \gamma \int_0^t \mathbf{G}(t) \cdot \mathbf{z}(t) dt \qquad (4.1.15)$$

We use intense gradient pulses, i.e., the pulse widths are much shorter than the time interval between them. The phase shift during the gradient pulse can then be written as

$$\Delta \varphi = \mathbf{q} \cdot \mathbf{z} \qquad (4.1.16)$$

Here, **q** is defined as

$$\mathbf{q} = \gamma \int_{\text{pulse}} G(t) dt \qquad (4.1.17)$$

where γ is the gyromagnetic ratio. The intense gradient pulses are advantageous for measuring molecular displacements particularly when there is a broad range of displacements in the sample [24]. An example of the stimulated-echo sequence used in this work is presented in Figure 4.1.5. The sequence detects spin displacements along the z_1 direction. The displacements along other directions can be measured with the corresponding directions of the gradient pulses.

The molecular translations are spatially resolved by combining the velocity encoding sequence with the conventional spatial imaging encodings. In this "velocity imaging", the phase shift of the spins reflects the information about their displacement as well as the spatial position. In this work, a flow-compensation [25] is implemented in the velocity imaging pulse sequence to eliminate the actifacts due to fluid flow.

We define the spin–displacement density function, $\varsigma(\mathbf{z}, \mathbf{Z})$, so that the density of spins – the number of spins divided by the voxel volume – that have displacements between $\mathbf{Z}$ and $\mathbf{Z} + d\mathbf{Z}$ in a voxel at $\mathbf{z}$ is $\varsigma(\mathbf{z}, \mathbf{Z}) d\mathbf{Z}$. The density function $\varsigma(\mathbf{z}, \mathbf{Z})$ can be expressed in terms of local spin density $\varrho(\mathbf{z})$ and the normalized displacement distribution function $P(\mathbf{z}, \mathbf{Z})$:

$$\varsigma(\mathbf{z}, \mathbf{Z}) = \varrho(\mathbf{z}) P(\mathbf{z}, \mathbf{Z}) \qquad (4.1.18)$$

The integral of $\varsigma(\mathbf{z}, \mathbf{Z})$ with respect to $\mathbf{Z}$ over its whole domain provides the spin density at the voxel:

$$\int \varsigma(\mathbf{z}, \mathbf{Z}) d\mathbf{Z} = \varrho(\mathbf{z}) \qquad (4.1.19)$$

It follows that

$$\int P(\mathbf{z}, \mathbf{Z}) d\mathbf{Z} = 1 \qquad (4.1.20)$$

The observed NMR signal, $S(\mathbf{k}, \mathbf{q})$, is modulated by the two wave vectors: the first, $\mathbf{k}$, is related to the spatial density of the spins and the second, $\mathbf{q}$, is related to a spatial displacement of spins:

$$S(\mathbf{k}, \mathbf{q}) = \iint \varsigma(\mathbf{z}, \mathbf{Z}) \exp(i\mathbf{k} \cdot \mathbf{z}) \exp(i\mathbf{q} \cdot \mathbf{Z}) d\mathbf{z} d\mathbf{Z} \qquad (4.1.21)$$

The joint density function for each voxel can be reconstructed by taking inverse Fourier transforms with respect to each of the wave vectors:

$$\varsigma(\mathbf{k}, \mathbf{q}) = \iint S(\mathbf{z}, \mathbf{Z}) \exp(-i\mathbf{k} \cdot \mathbf{z}) \exp(-i\mathbf{q} \cdot \mathbf{Z}) d\mathbf{k} d\mathbf{q} \qquad (4.1.22)$$

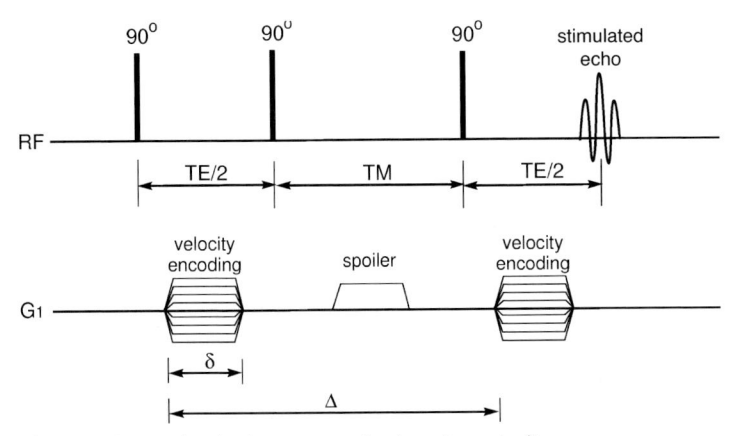

Fig. 4.1.5 A stimulated-echo sequence for detecting spin displacements. This sequence detects spin displacements along the z_1 direction. TM is mixing time.

The spin–displacement density function, $\varsigma(\mathbf{z}, \mathbf{Z})$, and the normalized displacement distribution function, $P(\mathbf{z}, \mathbf{Z})$, can be converted readily into the joint spin–velocity density function, $\varsigma(\mathbf{z}, \mathbf{v}_n)$, and the normalized velocity distribution function, $P(\mathbf{z}, \mathbf{v}_n)$, respectively, with the net velocity $\mathbf{v}_n$ defined as $\mathbf{v}_n = \mathbf{Z}/\Delta$. Once the velocity density function is determined for each of the volume elements, the superficial average velocity, $\mathbf{v}$, is calculated by [23]:

$$\mathbf{v}(\mathbf{z}) = \phi(\mathbf{z}) \frac{\int \mathbf{v}_n \varsigma(\mathbf{z}, \mathbf{v}_n) d\mathbf{v}_n}{\int \varsigma(\mathbf{z}, \mathbf{v}_n) d\mathbf{v}_n} \tag{4.1.23}$$

The porosity $\phi(\mathbf{z})$ for the voxel at $\mathbf{z}$, is determined as described in Section 4.1.4.1. Note that the observed spin–displacement density function, $\varsigma(\mathbf{z}, \mathbf{v}_n)$, is not actually associated with the intrinsic value of spin density, $\varrho(\mathbf{z})$, due to the transverse and longitudinal relaxation. However, this does not affect the calculated average velocity given in Eq. (4.1.23) because the spin density terms in the denominator and nominator cancel each other.

An example of the spin–velocity density function is demonstrated in Figure 4.1.6. A velocity imaging experiment was performed on water flowing through a 6-mm diameter tube. The velocity density function was spatially resolved along the axial direction of the tube, denoted by z_1 in the figure. It is observed that the velocity density function has a steep peak at zero velocity when the fluid is not flowing, but is shifted to a positive velocity when the flow rate was increased to 2.5 mL min^{-1}.

We demonstrate the procedure with an experiment conducted on a Bentheimer sandstone sample. For simplicity, we use a relatively thin sample and resolve only the two in-plane spatial coordinates. The sample is a rectangular parallelepiped shape having a length of 50 mm extending in the z_1 direction, width 25 mm along the z_2 direction and thickness 5 mm in the z_3 direction. The sample was sealed laterally with epoxy and mounted in Plexiglass end-plates with O-rings and tube

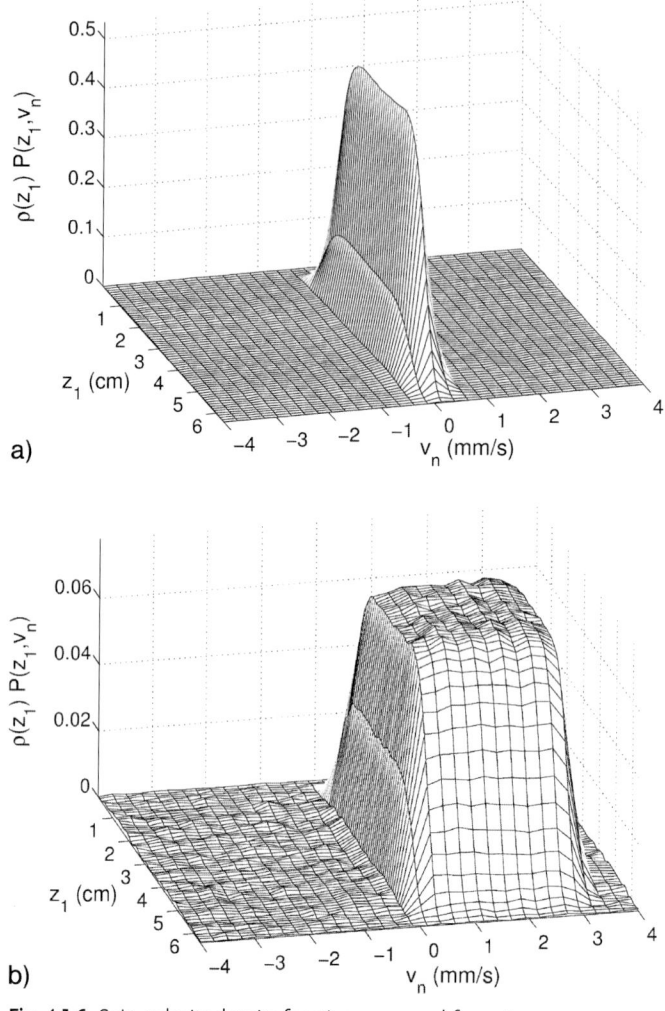

Fig. 4.1.6 Spin–velocity density function measured for water flowing through a 6-mm diameter tube at: 0.1 mL min^{-1} (left) and 2.5 mL min^{-1} (right). The velocity density function is spatially resolved along the z_1 direction.

fittings, with flow along the z_1 direction. NMR velocity imaging was carried out using the proton resonance signal from brine (3% NaCl and 0.03% NaN$_3$ by weight in distilled water). A stable brine flow through the sample was retained and the pressure drop between the inlet and outlet ports was monitored using a differential pressure transducer. The measured total flow rate was 1.5 mL min^{-1} and the pressure drop was 0.1 atm. The apparent average permeability for the entire sample calculated from the integrated form of Darcy's law is 1.0 darcy. In Figure 4.1.7, the arrows at each point represent the direction and relative magni-

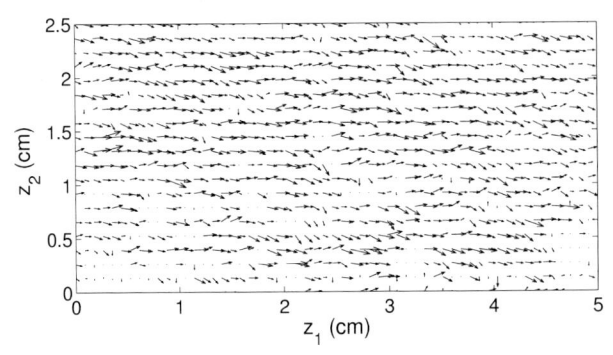

Fig. 4.1.7 Superficial average velocity data for the flow experiment with the thin Bentheimer sandstone sample. Each arrow represents the direction and relative magnitude of the superficial average velocity at the corresponding voxel. The velocities are measured for 58 × 20 voxels.

tude of the superficial average velocity at the corresponding voxel. The variations in the velocity clearly indicate the permeability is not uniform, and the sample is not (macroscopically) homogeneous.

4.1.4.2.2 Estimation of Permeability Distribution

We formulate an identification problem to determine the permeability distribution from the measured superficial velocity distribution. In this section, we first develop the model for our experiment, and present the estimation method.

For a steady-state, incompressible flow, we use Eq. (4.1.2) to eliminate the velocity in Eq. (4.1.1) and obtain the following equation for pressure:

$$\nabla \cdot \left[\frac{k}{\mu}(\nabla p - \varrho \mathbf{g}) \right] = 0 \tag{4.1.24}$$

For the conventional experimental design with fluid entering into the sample at $z_1 = 0$ and exiting from the sample at $z_1 = L$, and with gravity acting opposite to the z_3 direction, the boundary conditions are given by

$$\begin{aligned} p(0, z_2, z_3) &= p_{\text{in}} + \varrho g(z_3 - z_3^{\text{in}}) & \text{on } z_1 = 0 \\ p(L, z_2, z_3) &= p_{\text{out}} + \varrho g(z_3 - z_3^{\text{out}}) & \text{on } z_1 = L \\ \mathbf{v} \cdot \mathbf{n} &= 0 & \text{on } S_{\text{nf}} \end{aligned} \tag{4.1.25}$$

where $z_{3\text{in}}$ and $z_{3\text{out}}$ represent reference points at which p_{in} and p_{out} are specified, respectively. The unit vector n is normal to S_{nf}, the sealed periphery surface. Once the pressure distribution is determined, the velocity can be computed using the Darcy equation [Eq. (4.1.1)]. In our current formulation, the state equations [Eqs. (4.1.24 and 4.1.25)] are solved by using finite differences [26]. We solve these equations on a grid which is finer than the resolution of data, and then compute the corresponding average values of velocity for comparison with the measured values.

We now formulate the associated identification problem by arranging all the velocity measurements in a vector **Y** and, in the vector **F**, the corresponding values are calculated from the solution to Eqs. (4.1.24 and 4.1.25) using an estimate of the permeability function. The performance index is:

$$J = [\mathbf{Y} - \mathbf{F}]^T \mathbf{W}[\mathbf{Y} - \mathbf{F}] + \lambda \int \sum_{i=1}^{3} \left(\frac{\partial^2 k(\mathbf{z})}{\partial z_i^2}\right)^2 d\mathbf{z} \qquad (4.1.26)$$

The unknown permeability is represented using tensor product B-splines, which are given by the product of univariate B-splines:

$$k(\mathbf{z}) = \sum_{i=1}^{N_1} \sum_{j=1}^{N_2} \sum_{k=1}^{N_3} c_{i,j,k} B_i^m(z_1; \mathbf{x}_1) B_j^m(z_2; \mathbf{x}_2) B_k^m(z_3; \mathbf{x}_3) \qquad (4.1.27)$$

Here the B-spline $B_{im}(z_j; \mathbf{x}_j)$ is the ith B-spline basis function on the extended partition $\mathbf{x}_j$ (which contains locations of the knots in the z_j direction), and $c_{i,j,k}$ is a coefficient. We use cubic splines and sufficient numbers of uniformly spaced knots so that the estimation problem is not affected by the partition. The estimation problem now involves determining the set of B-spline coefficients that minimizes Eq. (4.1.26), subject to the state equations [Eqs. (4.1.24 and 4.1.25)], for a suitable value of the regularization parameter. At this point, the minimization problem corresponds to a nonlinear programming problem.

For a given experimental design, the keys to obtaining reliable estimates of permeability distributions are the successful solution to the minimization problem and suitable selection of the regularization parameter. Currently, we are determining the regularization parameter, λ, by examining the degree of fit [the first term in Eq. (4.1.26)] using several different values for λ, and estimating the maximum value of λ which does not compromise the fit to the data [27]. The minimization problem is solved by a global optimization procedure as there are probably many local minima in this problem. The global minimum is supposed to represent the minimum among the local minima. However, conventional global optimization algorithms generally do not provide a very precise determination of the minimum, particularly when a large number of parameters are involved. As our problem is ill-posed, the estimate associated with a sub-optimal solution can be significantly different from that of the actual minimum.

In our procedure, we use a combination of global and local optimization methods to determine the global minimum accurately. We use simulated annealing [28, 29] to provide a good starting point for the local algorithm. Then, a quasi-Newton method, Broyden–Fletcher–Goldfarb–Shanno (BFGS), is used to determine the B-spline coefficients that minimize Eq. (4.1.26). The gradient is efficiently calculated by the method of adjoint states [30, 31].

The estimate for the permeability distribution obtained using the velocity data shown in Figure 4.1.7 is presented in Figure 4.1.8. For this example, we used 30

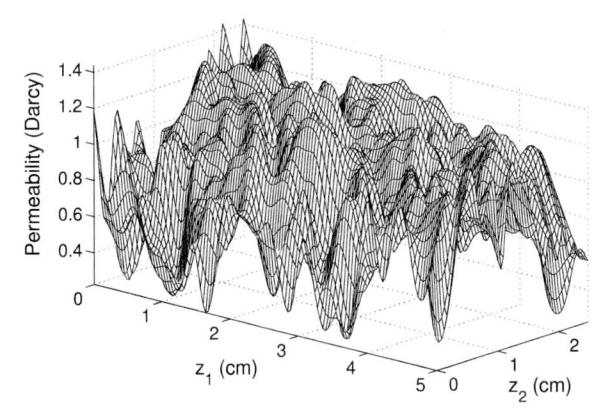

Fig. 4.1.8 Determined permeability distribution for the thin Bentheimer sample. The vertical axis represents the permeability value for the corresponding point.

equally spaced knots in each direction of z_1 and z_2. The spline dimension (number of coefficients estimated) was 1156. The permeability values ranged from 0.2 to 1.4 darcy, indicating a fairly large variation within the sample.

4.1.4.3 Multiphase Flow Properties

Relative permeability and capillary pressure functions, collectively called multiphase flow functions, are required to describe the flow of two or more fluid phases through permeable media. These functions primarily depend on fluid saturation, although they also depend on the direction of saturation change, and in the case of relative permeabilities, the capillary number (or ratio of capillary forces to viscous forces). Dynamic experiments are used to determine these properties [32].

Conventionally, the sample is initially saturated with one fluid phase, perhaps including the other phase at the irreducible saturation. The second fluid phase is injected at a constant flow rate. The pressure drop and cumulative production are measured. A relatively high flow velocity is used to try to negate capillary pressure effects, so as to simplify the associated estimation problem. However, as relative permeability functions depend on capillary number, these functions should be determined under the conditions characteristic of reservoir or aquifer conditions [33]. Under these conditions, capillary pressure effects are important, and should be included within the mathematical model of the experiment used to obtain property estimates.

While capillary pressure can be determined independently through experiments implementing a series of equilibrium states, this can be very time consuming, particularly if the entire capillary pressure function is to be reconciled. Furthermore, as there can be difficulties in re-establishing identical states of initial saturation, it is most desirable to determine capillary pressure and relative permeability functions simultaneously, from the same experiment.

Reliable determination of all three functions depends on the information content associated with the experiments. The conventional experimental design does not provide sufficient information to determine all three functions accurately [34]. Another consideration is that conventional analyses are all based on the assumption that the sample is uniform, and use an average value for porosity and an apparent value for permeability. Clearly, these properties vary spatially, and failure to account for the effects of spatial variations in the properties will lead to errors in the estimates of the functions [16].

We present a general approach for estimating relative permeability and capillary pressure functions from displacement experiments. The accuracy with which these functions are estimated will depend on the information content of the measurements, and hence on the experimental design. We determine measures of the accuracy with which the functions are estimated, and use these measures to evaluate different experimental designs. In addition to data measured during conventional displacement experiments, we show that the use of multiple injection rates and saturation distributions measured with MRI can substantially increase the accuracy of estimates of multiphase flow functions.

4.1.4.3.1 Mathematical Model and Property Estimation

We use the generalized Darcy expression [Eq. (4.1.4)] to eliminate the velocity in Eq. (4.1.3), obtaining the following equation for each fluid phase:

$$\frac{\partial}{\partial t}(\phi \varrho_i s_i) = \nabla \cdot \left(\frac{k k_{ri} \varrho_i}{\mu_i} (\nabla p_i - \varrho_i \mathbf{g}) \right) \tag{4.1.28}$$

The fluid properties and porosity and permeability are determined independently. Boundary and initial conditions are specified for the particular experiment to be considered. With specified multiphase flow functions, the state equations, Eqs. (4.1.28, 4.1.5 and 4.1.6), can be solved for the transient pressure and saturation distributions, $p_i(z,t)$ and $s_i(z,t)$, $i = 1, 2$. The values for $\mathbf{F}$ can then be calculated, which correspond to the measured data $\mathbf{Y}$.

The solution of these dynamic nonlinear differential equations is considerably more complex than the previous systems considered. In particular, stable solution methods are based on physically realistic multiphase flow functions that have the following properties: relative permeability functions are non-negative, monotonically increasing with their respective saturation, and are zero at vanishing saturations, and capillary pressure is monotonically increasing with respect to the saturation of the non-wetting phase. It is necessary that any iterative scheme for estimating the multiphase flow functions retain these characteristics at each step.

In order to ensure successful minimization of the performance index and to enhance our ability to determine the global optimum, we select the corresponding finite-dimensional representation in a different manner than before. We again use B-splines to represent the unknown functions:

$$k_{ri}(s_1) = \sum_{j=1}^{N_i} c_j^i B_j^m(s_1, \mathbf{x}^i), i = 1, 2 \qquad (4.1.29)$$

$$p_c(s_1) = \sum_{j=1}^{N_c} c_j^c B_j^m(s_1, \mathbf{x}^c) \qquad (4.1.30)$$

Using maximum likelihood theory, the estimates for the collection of spline coefficients c are those that minimize the performance index

$$J = [\mathbf{Y} - \mathbf{F}]^T \mathbf{W}[\mathbf{Y} - \mathbf{F}] \qquad (4.1.31)$$

subject to the constraints

$$\mathbf{Gc} \geq \mathbf{g}_0 \qquad (4.1.32)$$

The inequality constraints are chosen to ensure that physically realistic estimates are obtained at each step of the minimization procedure. We use a trust-region based, linear-inequality constrained Levenberg–Marquardt algorithm [35] to solve this minimization problem. In order to choose a suitable partition – numbers and location of knots – for the multiphase functions, we solve a sequence of minimization problems with increasing values of the spline dimensions. We generally look for the first solution for which satisfactory residuals – differences between measured and calculated values – are obtained. By starting with very few degrees of freedom, and beginning each successive problem with the prior estimates, our method is robust and relatively efficient. For further details see Watson and coworkers [32, 34].

4.1.4.3.2 Experimental Design

Displacement experiments can be relatively complex and time-consuming, so the experimental design can be a critical issue. Using suitable system and parameter identification methods, we obtain the best estimates of properties from the available data. It is most desirable to have some measures of the accuracy with which the properties are estimated. If that level of accuracy is less than desired, one can consider other ways of conducting the experiments so that additional information about the properties may be obtained.

In this case, the conventional experimental design is insufficient to ensure accurate estimates of all three multiphase flow functions [34]. We have considered two different modifications in the experimental design that can provide for improved estimates. These modifications can be incorporated separately, or together, thus representing a total of three different candidate experimental designs.

In the first we use a sequence of different injection rates, rather than simply a single rate, as done conventionally. The larger flow rates provide for greater fluid displacements, thus providing information for a greater range of saturation. Furthermore, the different flow rates result in relatively different emphases for capillary and relative permeability effects, resulting in greater information content about those properties.

In the second we include measurements of the saturation distributions obtained by MRI. These data reflect state quantities at spatial locations throughout the sample, whereas the conventional measurements made outside of the samples are sensitive only to an integral of the state quantities.

We also use a linearized covariance analysis [34, 36] to evaluate the accuracy of estimates and take the measurement errors to be normally distributed with a zero mean and covariance matrix $\mathbf{\Sigma}$ Assuming that the mathematical model is correct and that our selected partitions can represent the true multiphase flow functions, the mean of the error in the estimates is zero and the parameter covariance matrix of the errors in the parameter estimates is:

$$\mathbf{C} = (\mathbf{A}^T \mathbf{\Sigma}^{-1} \mathbf{A})^{-1} \tag{4.1.33}$$

where $\mathbf{A} = \partial \mathbf{F}/\partial \mathbf{c}$ is the sensitivity matrix of the model function with respect to the parameters. The covariance matrix of the estimated flow functions is then written as

$$\mathbf{V} = (\mathbf{D}^T \mathbf{C}^{-1} \mathbf{D})^{-1} \tag{4.1.34}$$

where the matrix $\mathbf{D}$ is the sensitivity of the flow functions with respect to the parameters at a specified saturation value. Using the diagonal of $\mathbf{V}$, v_{ii}, the upper and lower bounds of a point-wise confidence interval, f_{ci}, can be constructed for a flow function $f = [k_{r1}, k_{r2}, p_c]$ at a point i:

$$f_{ci} = f_i \pm \alpha \sqrt{v_{ii}} \tag{4.1.35}$$

where α is the appropriate quantile for the given confidence level and distribution.

The analysis is based on a chosen set of properties. While the accuracy will depend on those particular properties, global features can often be readily identified. We demonstrate the assessment of experimental design in the next subsection.

4.1.4.3.3 Results and Discussion

We consider several different experimental designs for determining multiphase flow functions from displacement experiments. Firstly, we consider the traditional experimental design, stated in the previous section. We refer to this as "single rate, conventional data." Secondly, we consider additional experimental data which are given by saturation distributions, resolved along the flow direction, at various times. We refer to this case as "single rate, with saturation data." The data are obtained by CPMG imaging, using the protocols presented in Section 4.1.4.1. By limiting the resolution to a single spatial dimension along the length of the sample, the saturation distribution can be obtained in a matter of seconds. This is sufficiently fast so that any changes in the saturation distribution during imaging will be negligible. Two other cases are based on using a few sequential, constant flow rates, of increasing size, rather than the single injection rate. "Multi-rate, conventional data" is based on measurements of the pressure drop and production, and "multi-rate, with saturation data" includes additional measurements of the saturation distribution. The complete details of these experimental designs, including flow rates and properties, are reported elsewhere [34].

Results of the covariance analysis for the accuracy of estimates of the relative permeability of water and capillary pressure functions, along with the specified "true" functions, are shown in Figures 4.1.9 and 4.1.10 (the results for the relative permeability of oil are not included here). The accuracy measures are presented as 95% confidence intervals.

The minimum value for water saturation attained in the sample for the single-rate cases is about 0.3. The data, therefore, can contain information about the flow function only over a range of water saturation above 0.3, and the confidence intervals for the single-rate experiments are finite only over that range. For the multi-rate cases, however, lower water saturation values are achieved as larger flow rates are used, and the confidence intervals are finite over a wider range of water saturation. Compared with the traditional experimental design, substantial improvements in the accuracy of the estimates are obtained by using multiple flow rates. This is quite remarkable as no additional equipment, and little extra effort, is actually required. The additional saturation measurements do improve the accuracy of the estimates further.

The estimation of flow functions from an actual experiment is reported next. A multi-rate primary drainage experiment was conducted on a Texas Cream limestone sample. Hexadecane was used as the oleic phase and deuterium oxide (D_2O) was used as the aqueous phase. Protons are imaged, so only the oil phase is observed. The pressure drop data, production data and saturation data are shown in Figures 4.1.11–

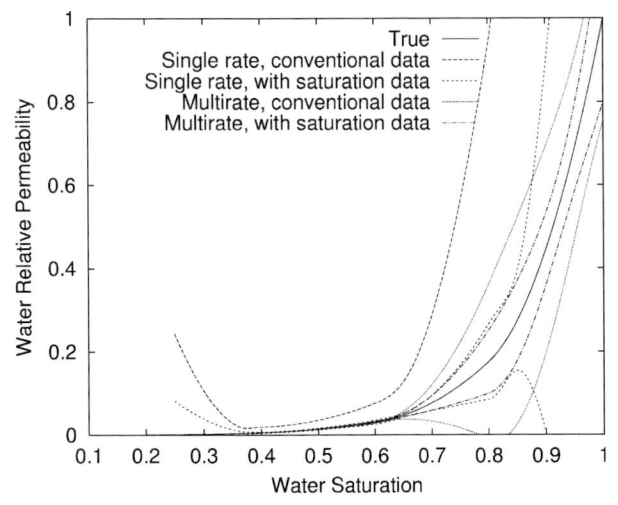

Fig. 4.1.9 Relative permeability of water and 95% confidence intervals for different experimental designs. For each case, 95% confidence intervals are shown with a pair of curves (Reprinted with permission for [34]).

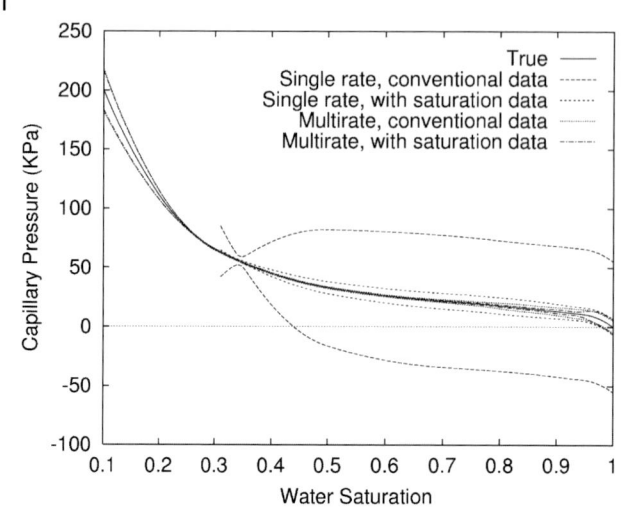

Fig. 4.1.10 Capillary pressure and 95% confidence intervals for different experimental designs (see Figure 4.1.9) (Reprinted with permission for [34]).

4.1.13, along with the corresponding values calculated using the estimated relative permeability and capillary pressure functions. Conventional data are fairly well matched, as are the saturation data, but not as precisely, particularly for the larger values of saturation. This is probably due to the fact that estimates of the permeability distribution were not available, so a single apparent value for the permeability was used. Also, it is recognized that the production data, which represent an integral of the saturation distribution, are easier to reconcile than the distribution itself.

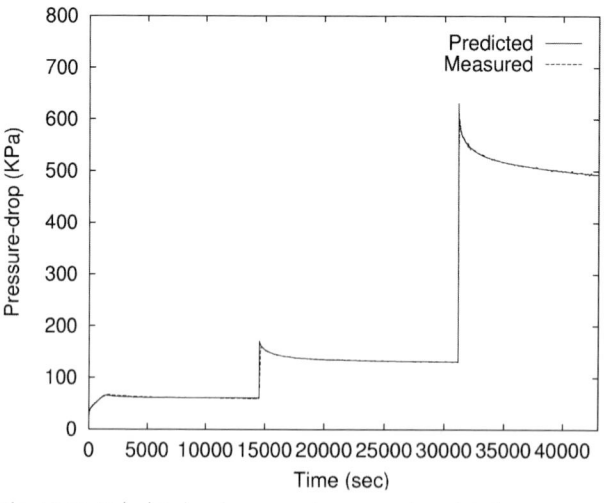

Fig. 4.1.11 Calculated and measured pressure-drop data in a multi-rate primary drainage experiment on the Texas Cream limestone sample (Reprinted with permission for [34]).

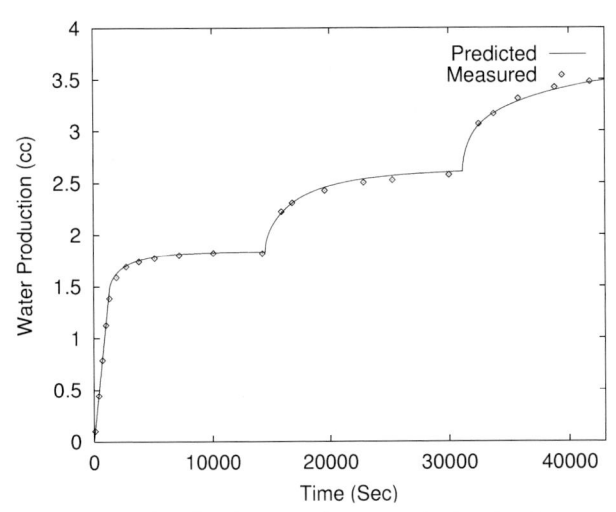

Fig. 4.1.12 Predicted and measured water production data in a multi-rate primary drainage experiment on the Texas Cream limestone sample (Reprinted with permission for [34]).

4.1.5
Summary

Spatially resolved, non-invasive NMR measurements within permeable media provide an unprecedented opportunity to extend the means for describing and

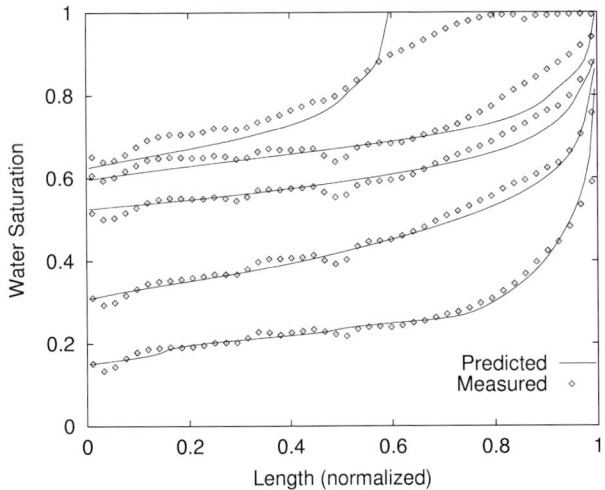

Fig. 4.1.13 Calculated and measured water saturation profiles for a multi-rate primary drainage experiment on the Texas Cream limestone sample. Starting from the upper left, profiles correspond to times of 12, 22, 238, 500 and 696 min. (Reprinted with permission for [34]).

understanding a heterogeneous media and physical processes occurring within such media. In order to utilize NMR measurements for such purposes, we model the observed physical phenomena and use system and parameter identification methods to extract properties within those models.

We demonstrated our approach by investigating the flow of single and multiple immiscible fluid phases through rigid permeable media. We developed a non-parametric approach for estimating relaxation distributions so that the intrinsic magnetization can be determined from MRI spin–density images. The intrinsic magnetization is then used to determine spatially resolved porosity and saturation distributions.

We presented a novel method to determine spatially resolved permeability distributions. We used MRI to measure spatially resolved flow velocities, and estimated the permeability from the solution of an associated system and parameter identification problem.

Our approach was demonstrated by determining multiphase flow functions from displacement experiments. Spatially resolved porosity and permeability distributions can be incorporated to mitigate errors encountered by assuming that the properties are uniform. We developed measures of the accuracy of the estimates and demonstrated improved experimental designs for obtaining more accurate estimates of the flow functions. One of the candidate experimental designs incorporated MRI measurements of saturation distributions conducted during the dynamic experiments.

These methods provide unprecedented resolution of porosity and permeability for heterogeneous media, and substantially advance the reliability of estimates of multiphase flow functions. This work provides the foundation for further studies of heterogeneous media and important processes that occur within such media, including those with chemical and biological changes that are intimately affected by media structures.

References

1 S. Whitaker **1999**, *The Method of Volume Averaging*, Dluwer Academic Publishers, Dordrecht, The Netherlands.
2 J. Bear **1972**, *Dynamics of Fluids in Porous Media*, American Elsevier, New York.
3 E. M. Withjack **1988**, *SPEEE*, December, 696.
4 A. T. Watson, J. T. Hollenshead, J. Uh, C. T. P. Chang **2002**, *Annu. Rep. N. M. R. Spectrosc.* 48, 113.
5 S. Chen, F. Qin, K-H Kim, A. T. Watson **1993**, *AIChE J.* 39, 925.
6 R. Kulkarni, A. T. Watson **1997**, *AIChE J.* 43, 2137.
7 S. Chen, F. Qin, A. T. Watson **1994**, *AIChE J.* 40, 1238.
8 J. Uh, A. T. Watson **2004**, *Ind. Eng. Chem. Res.* 43, 3026.
9 P. S. Belton, R. G. Ratcliffe **1985**, *Prog. Nucl. Magn. Reson. Spectrosc.* 17, 241.
10 J. Uh, A. T. Watson. *J. Magn. Reson.* submitted for publication.
11 H.-K. Liaw, R. Kulkarni, S. Chen., A. T. Watson **1996**, *AIChE J.* 42, 538.
12 L. L. Schumaker **1981**, *Spline Functions: Basic Theory*, John Wiley and Sons, New York.
13 R. L. Eubank **1999**, *Nonparametric Regression and Spline Smoothing*, Marcel Dekker, New York.

14 E. Anderson et al. **1994**, *LAPCK Users' Guide*, 2nd edn, The Society for Industrial and Applied Mathematics, Philadelphia.
15 M. Villalobos, G. Wahba **1987**, *J. Am. Stat. Assoc.* 82, 239.
16 R. Valestrand, A.-A. Grimstad, K. Kolltveit, J.-E. Nordtvedt, J. Phan, A. T. Watson **2003**, *Inverse Prob. Eng.* 11, 289.
17 A. Timur **1969**, *J. Petrol. Technol.* 21, 775.
18 W. E. Kenyon, P. I. Day, C. Straley, J. F. Willemsen **1988**, *SPE Form. Eval.* 3, 622.
19 P. N. Sen, C. Straley, W. E. Kenyon, M. S. Whittingham **1990**, *Geophysics* 55, 61.
20 A. N. Garroway **1974**, *J. Phys. D* 7, L159.
21 P. R. Moran **1982**, *Magn. Reson. Imag.* 1, 197.
22 A. Caprihan, E. Fukushima **1990**, *Phys. Rep.* 198, 195.
23 C. T. P. Chang, A. T. Watson **1999**, *AIChE J.* 45, 437.
24 C. M. Edwards, C. T. Chang, S. Sarkar **1993**, *Proceeding of the SCA Annual Technical Conference*, No. 9310.
25 J. M. Pope, S. Yao **1993**, *Concept. Magn. Reson.* 5, 281.
26 K. Seto, J. T. Hollenshead, A. T. Watson, C. T. P. Chang, J. C. Slattery **2001**, *Transport Porous Media*, 42, 351.
27 P. H. Yang, A. T. Watson **1988**, *SPE Res. Eng.* 3, 995.
28 S. Kirkpatric, C. D. Gelatt, Jr., M. P. Vecchi **1983**, *Science* 13, 671.
29 A. Corana, M. Marchesi, C. Martini, S. Ridella **1987**, *ACM Trans. Math. Software* 13, 262.
30 W. Chen, G. Gavalas, J. Seinfeld, M. Wasserman **1974**, *SPE J.* 14, 593.
31 D. Cacuci, C. Weber, E. Oblow, J. Marable **1980**, *Nucl. Sci. Eng.* 75, 88.
32 A. T. Watson, R. Kulkarni, J.-E. Nordtvedt, A. Sylte, H. Urkeda **1998**, *Meas. Sci. Technol.* 9, 898.
33 P. C. Richmond, A. T. Watson **1990**, *SPE Res. Eng.* 5, 121.
34 R. Kulkarni, A. T. Watson, J.-E. Nordtvedt, A. Sylte **1998**, *AIChE J.* 44, 2337.
35 P. C. Richmond **1988**, (Estimating Multiphase Flow Functions From Displacement Experiment), *Ph.D. Thesis*, Texas A&M University.
36 P. D. Kerig, A. T. Watson **1987**, *Soc. Pet. Eng. J.* 2, 103.

4.2
MRI Viscometer
Robert L. Powell

4.2.1
Introduction

In many industries, the rheological properties of process streams are indicative of other properties of the materials or have a direct role on the quality of the end product. For example, the flow properties of a polymer depend upon its molecular weight distribution while those of a slurry reflect the size, shape and concentration of the particles. The rheology of materials such as these are not easily characterized. It is well know that polymeric fluids exhibit a wide range of complex properties including a viscosity that varies with the shear rate, normal stresses in shearing flow, viscoelastic properties and extensional flow properties. Among these, the shear rate dependent shear viscosity is the most used property in characterizing materials.

Drawing samples at certain points of the process, taking them to a laboratory and measuring the shear viscosity over a range of shear rates with a conventional rheometer has been the main method of monitoring the viscosity of process streams. This method is labor and time intensive and in many systems laboratory results may not represent the *in situ* sample properties appropriately. Materials such as food stuffs or synthetic polymers that are undergoing a reaction during processing may change once they are withdrawn from the process. Characterizing such materials using in-line or on-line measurements can eliminate the problems associated with off-line measurements.

There are commercially available in-line or on-line viscometer devices. In-line devices are installed directly in the process while on-line devices are used to analyze a side stream of the process. Most devices are based on measuring the pressure drop and flow rate through a capillary. The viscosity is either determined at a single shear rate or, at most, a few shear rates. Complex fluids, on the other hand, exhibit a viscosity that cannot be so easily characterized. In order to capture enough information that allows, for example, a molecular weight distribution to be inferred, it is necessary to determine the shear viscosity over reasonably wide ranges of shear rates.

Many materials are conveyed within a process facility by means of pumping and flow in a circular pipe. From a conceptual standpoint, such a flow offers an excellent opportunity for rheological measurement. In pipe flow, the velocity profile for a fluid that shows shear thinning behavior deviates dramatically from that found for a Newtonian fluid, which is characterized by a single shear viscosity. This is easily illustrated for a power-law fluid, which is a simple model for shear thinning [1]. The relationship between the shear stress, σ, and the shear rate, $\dot{\gamma}$, of such a fluid is characterized by two parameters, a power-law exponent, n, and a constant, m, through

$$\sigma(\dot{\gamma}) = m\dot{\gamma}^n \qquad (4.2.1)$$

Figure 4.2.1 shows the dimensionless velocity profiles for flow in a pipe for four different values of the power-law exponent, ranging from 0.25, which is highly shear thinning, to 1, which is a linearly viscous or Newtonian fluid. The latter case shows the typical parabolic velocity profile. As the power-law exponent is decreased, the profile becomes progressively blunter. Near the center of the pipe, the velocity is almost independent of the radius. Most of the shearing occurs near the pipe wall. This is shown more dramatically in Figure 4.2.2, which gives the dimensionless shear rate as function of dimensionless radial position. The shear rate is simply the derivative of the velocity with respect to radius,

$$\dot{\gamma}(r) = \frac{dw(r)}{dz} \qquad (4.2.2)$$

Here, w is the axial velocity, r is the radial position, with $r = 0$ being the axis of the circular pipe, and the absolute value sign is employed to maintain a positive value of the shear rate. Figure 4.2.2 shows that the shear rate varies dramatically with radius, being zero at the center and a maximum at the tube wall. Furthermore, as the power-law exponent decreases, the shearing is confined to a narrower and narrower region near the pipe wall. As indicated by this simple model, shear thinning materials in pipe flow show both low shear rate and high shear rate behavior simultaneously. This is also seen in Figure 4.2.2 where the dimensionless viscosity is shown as a

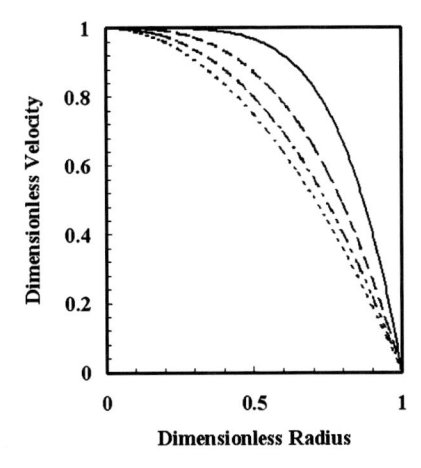

Fig. 4.2.1 Dimensionless velocity profiles for steady flow in a circular pipe for a power-law fluid for values of the power-law exponent $n = 0.25$ (solid line), through to 0.5, 0.75 and 1 (small dashed line), respectively. As the power-law exponent decreases, the profile becomes progressively more blunt with most of the shearing being confined to the region near the wall.

function of dimensionless radial position. For a Newtonian fluid, the viscosity is constant and in this system, has a value of 1. As n decreases, the viscosity is seen to vary across the pipe radius. This effect is dramatic. For $n = 0.25$, the viscosity varies by three orders of magnitude between the pipe wall (lowest) and a point that is one tenth away from the center of the pipe. The range of materials used for the tests described in this chapter is given in Table 4.2.1. The particular polymer melt used in these studies had a power-law exponent of almost 1. This is atypical [1] and reflects the low range of shear rates that could be accessed by our pumping equipment. Furthermore, materials with yield stresses can mimic power-law behavior with an exponent approaching zero. Here we describe a technique that exploits this variation in viscosity across the tube using magnetic resonance imaging (MRI).

Tab. 4.2.1 Materials and conditions for the studies discussed.

Fluid	Consistency (Pa s)[a]	Flow index	$\dot{\gamma}$ (s^{-1})	Yield stress (Pa)	Flow rates (mL s^{-1})
1% carboxymethylcellulose	2.8	0.74	0.1–31	–	13
					22
0.5% polyacrylamide	0.45	0.64	0.1–68	–	12
					43
1% poly(ethylene oxide)	0.52	0.7	0.1–124	–	26
0.34% xanthan gum	1.1	0.3	0.1–90	–	18
					81
0.53% microfibrous cellulose	3.2	0.34	0.2–100	3.15 ± 0.08	11
					15
					30
Low density polyethylene	110.3[b]	≈1	1–12	–	2.4

a See Eq. (4.2.1).
b Average value of rotational rheometer data over the shear rate range 0.1–12 s^{-1}.

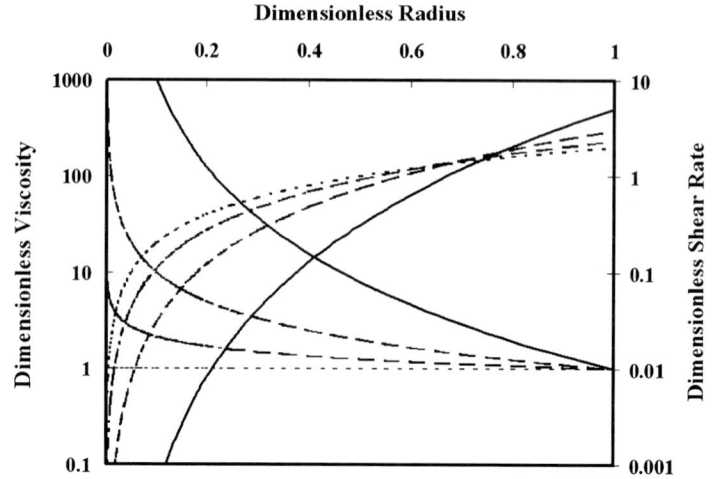

Fig. 4.2.2 Dimensionless shear rate and viscosity as a function of radius for a power-law fluid under the conditions shown in Figure 4.2.1. For a highly shear thinning material, the shear rate is large near the wall and close to zero near the center. The viscosity can vary by several orders of magnitude in the pipe.

MRI is one among many techniques that can be employed to determine the local velocity. Its advantage is that it is both non-invasive and can work with opaque systems. The latter is especially important in many industrial settings involving slurries and emulsions. Other systems for determining velocity have drawbacks. Pitot tubes and hot film anemometers are invasive and do not function well, if at all, in multiphase systems. Laser Doppler velocimetry and particle image velocimetry require the fluids to be transparent. The single technique that approaches MRI in its versatility is ultrasonic pulsed Doppler velocimetry (UPDV) [2–4]. However, MRI does have advantages over UPDV as the data over the entire sampling volume can be obtained regardless of the concentration of solids and because of the potential of MRI to allow a wide range of additional information, such as chemical and physical compositions, to be obtained.

From an experimental standpoint, the questions we address are what measurements must be made and how to make those measurements in order to ascertain the shear viscosity variation in the pipe. Our goal is to obtain spatially resolved data and to show that from a single set of such data, the viscosity of a complex fluid can be obtained over the range of shear rates in the pipe. The capability of measuring the velocity and the pressure drop simultaneously enables this hypothesis to be made. The velocity is used to determine the shear rates and the pressure drop allows the variation of the local value of shear stress to be measured. The measurements can be obtained in a standard process flow, without the need for additional pumping capacity or inducing flow in a side stream. We call this approach "pointwise" viscosity measurement. The shear rates are obtained from non-invasive flow measurements, using MRI, while the stress distribution can be readily calculated using standard transducers for pressure drop measurements. Therefore, this method can potentially be employed as an in-line or on-line viscosity probe for process control purposes to improve both process efficiency and end-product quality.

4.2.2
Theory

The theoretical basis for spatially resolved rheological measurements rests with the traditional theory of viscometric flows [2, 5, 6]. Such flows are kinematically equivalent to unidirectional steady simple shearing flow between two parallel plates. For a general complex liquid, three functions are necessary to describe the properties of the material fully: two normal stress functions, N_1 and N_2 and one shear stress function, σ. All three of these depend upon the shear rate. In general, the functional form of this dependency is not known *a priori*. However, there are many accepted models that can be used to approximate the behavior, one of which is the power-law model described above.

Viscometric flows used for measurements include well known flows, such as flow in a narrow gap concentric cylinder device and between a small angle cone and a flat plate. In both of these cases the flows established in these devices approximate almost exactly simple shearing flow. There are other viscometric flows in which the shear rate is not constant throughout, these include the wide gap concentric cylinder flow and flow in a circular pipe, discussed above.

Viscometric flow theories describe how to extract material properties from macroscopic measurements, which are integrated quantities such as the torque or volume flow rate. For example, in pipe flow, the standard measurements are the volume flow rate and the pressure drop. The fundamental difference with spatially resolved measurements is that the local characteristics of the flows are exploited. Here we focus on one such example, steady, pressure driven flow through a tube of circular cross section. The standard assumptions are made, namely, that the flow is uni-directional and axisymmetric, with the axial component of velocity depending on the radius only. The conservation of mass is satisfied exactly and the z component of the conservation of linear momentum reduces to

$$\frac{d r \sigma(r)}{dr} = r \frac{dP}{dz} \tag{4.2.3}$$

where P is the pressure. The pressure gradient is assumed to be constant and is characterized by a pressure drop, ΔP over a length L. Along with the requirement that the stress cannot be singular at $r = 0$, this allows Eq. (4.2.3) to be integrated as

$$\sigma(r) = -\frac{\Delta P}{2L} r \tag{4.2.4}$$

Here $\Delta P = P_2 - P_1$, where P_2 is the pressure downstream of P_1 and $\Delta P < 0$. Equation (4.2.4) shows that the local shear stress in a tube is determined by the pressure drop. Recalling the definition of the shear rate, Eq. (4.2.2), it is seen that both the shear stress and the shear rate are functions of the radius. To obtain the shear stress as a function of shear rate, in principle we can solve for the radius as a function of the shear rate and substitute this into the shear expression for the shear stress versus radius. From an experimental standpoint, we measure $w(r)$ in order to calculate $\dot{\gamma}(r)$, and then at each value of r, determine the shear stress from Eq. (4.2.4). The shear stress versus shear rate is obtained by choosing different radial positions and finding each corresponding value. The shear viscosity, $\eta\dot{\gamma}$, is defined through

$$\sigma(\dot{\gamma}) = \eta(\dot{\gamma})\dot{\gamma} \qquad (4.2.5)$$

and can be obtained by dividing $\sigma(\dot{\gamma})$ by $\dot{\gamma}$.

The power of this technique is two-fold. Firstly, the viscosity can be measured over a wide range of shear rates. At the tube center, symmetry considerations require that the velocity gradient be zero and hence the shear rate. The shear rate increases as r increases until a maximum is reached at the tube wall. On a theoretical basis alone, the viscosity variation with shear rate can be determined from very low shear rates, theoretically zero, to a maximum shear rate at the wall, $\dot{\gamma}_w$. The corresponding variation in the viscosity was described above for the power-law model, where it was shown that over the tube radius, the viscosity can vary by several orders of magnitude. The wall shear rate can be found using the Weissenberg–Rabinowitsch equation:

$$\dot{\gamma}_w = \frac{Q}{\pi R^3}\left(3 + \frac{d\ln Q}{d\ln(\Delta P)}\right) \qquad (4.2.6)$$

where R is the tube radius and Q is the volume flow rate. As Eq. (4.2.6) requires the measurement of Q for different ΔP, an estimate can be obtained by assuming that fluid is Newtonian with $\frac{d\ln Q}{d\ln(\Delta P)} = 1$ and

$$\dot{\gamma}_w = \frac{4Q}{\pi R^3} \qquad (4.2.7)$$

The second important feature of this technique is that it is independent of the constitutive relationship of the material. This is a direct reflection of its rigorous foundation in viscometric flow theory.

In addition to the measurement of the viscosity, this technique also allows the yield stress to be estimated. For a typical yield stress type material, there is a critical shear stress below which the material does not deform and above which it flows. In pipe flow, the shear stress is linear with the radius, being zero at the center and a maximum at the wall. Hence, the material would be expected to yield at some intermediate position, where the stress exceeds the yield stress. The difficulty with this method is in the determination of the point at which yielding occurs and, indeed, whether the material is appropriately modeled as having a yield stress or is better considered as having a highly shear thinning viscosity. As is seen in Figure 4.2.1, for a power-law exponent of 0.25, the velocity profile from $\frac{r}{R} = 0$ to $\frac{r}{R} = 0.4$ varies by 1%. From a purely experimental standpoint, data showing such an effect could either be interpreted through a low power-law exponent or through a yield stress. The latter interpretation can result in robust estimates of this difficult to measure parameter.

The discussion above that led to Eqs. (4.2.6 and 4.2.7) assumes that the no-slip condition at the wall of the pipe holds. There is no such assumption in the theory for the spatially resolved measurements. We have recently used a different technique for spatially resolved measurements, ultrasonic pulsed Doppler velocimetry, to determine both the viscosity and wall slip velocity in a food suspension [2]. From a rheological standpoint, the theoretical underpinnings of the ultrasonic technique are the same as for the MRI technique. Hence, there is no reason in principle why MRI can not be used for similar measurements.

As we shall see, the limitation on this technique stems from two related sources. Firstly, as Eq. (4.2.2) shows, the velocity data must be differentiated. If these are not sufficiently smooth, large errors can result. Secondly, near the tube center, the velocity gradient is almost zero. The precision of the measurement is particularly critical if meaningful data are to be obtained in this region.

4.2.3
Experimental Techniques

4.2.3.1 MRI

The foundations of pointwise rheological measurements by MRI derive from flow imaging experiments using a number of non-Newtonian solutions. In these studies, the measured velocity profiles compared well with the analytical solutions [7–9]. The constants of the models used in the analytical solutions were obtained by independent, conventional viscosity measurements. The next step was to consider the inverse problem, namely, given the velocity profile and the pressure drop, to determine the local viscosity. This was first achieved by Powell et al. [10, 11]. Subsequent work has sought to increase the accuracy of the technique as well as to define the design parameters for a practical instrument. The most direct scheme for analyzing the data calculates the shear rates without resorting to an *a priori* assumption with respect to the constitutive equation. Such work at UC Davis, CA, USA [10, 12–15] and elsewhere [16, 17] smoothes the velocity data using a polynomial fit prior to differentiation, or directly differentiates the discrete data [18] to obtain the shear rate. In some cases, the velocity data are fit to a functional form suggested by a constitutive model [19]. In all studies the MRI viscosity results agree well with conventional rheometer measurements. Its non-invasive character and applicability to opaque systems as well as its ability to provide shear viscosity data over a wide range of shear rates in a single measurement has made this technique promising as an on-line viscometer.

For some materials, the measured velocity profiles can be subject to different interpretations. Highly blunted profiles may indicate a complex fluid with a highly shear thinning viscosity, as shown in Figure 4.2.1. An alternative interpretation is that the material has a yield stress. This is especially attractive in those cases where previous studies have shown that there is good reason for such a constitutive assumption. For example, in the case of paper pulp suspensions, it is well known that the fibers form networks that have yield behavior [20]. Magnetic resonance imaging has been used to measure velocity profiles in such systems, and others, and to estimate their yield stresses [13, 21].

4.2.3.2 Procedure

The NMR velocity profile images presented here were obtained using a General Electric CSI–II/TecMag Libra spectrometer connected to a 0.6-T Oxford superconducting magnet (corresponding to 25.96 MHz ^{1}H resonance frequency) with a PowerMac/MacNMR user interface (TecMag, Houston, TX, USA). The horizontally oriented magnet has an inner bore diameter of 330 mm. A set of unshielded gradient coils driven by an Oxford-2339 water cooled gradient power amplifier produce three orthogonal gradients G_x, G_y and G_z, where G_z is parallel to the magnet bore.

For suspensions and polymer solutions, steady flow in a pipe was established using a closed flow loop system consisting of poly(vinyl chloride) (PVC) tubing (id = 26.2 mm) with 9 m of straight tubing upstream of the magnet bore (length/diameter $\cong$ 340) to ensure fully developed flow. More details of this system are given in Refs. [13–15]. In developing this system, we were focusing on the ultimate application to materials of a practical nature. For example, much of our work considered suspensions. In some cases particle dimensions could be of the order of 1 mm, when the pipe must be sufficiently large that the properties that are measured are not affected by walls of the pipe. Otherwise, a true measure of the rheological properties cannot be obtained. Similarly, in order to obtain fully developed pipe flow, the length to diameter ratio must be sufficiently large. For laminar flow, the ratio of the entrance length, that is the length required to reach within 1% of the final value, to the diameter, is given by 0.058 N_{Re} [22]. The Reynolds number, N_{Re}, is defined in the usual way in terms of density and average velocity, pipe diameter and viscosity. For non-Newtonian fluids, the expression for the Reynolds number is more complex [1], however, it can be estimated. In the work discussed here, the Reynolds numbers are 100 or less, implying that we are in the fully developed regime. For measurements with a polymer melt, a single screw extruder was used for pumping through the magnet with the extrudate exiting the die into the atmosphere at the end of the magnet.

This technique has been tested with a variety of systems, as given in Table 4.2.1. These include several aqueous solutions that were used to span a wide range of non-Newtonian behaviors, including shear thinning, a low shear rate Newtonian plateau, a high shear rate Newtonian regime and an apparent yield stress. We have also tested this with a melt of low density polyethylene (Sclair, Canada) with a melting temperature of 108 °C, melt index of 51.5 and a weight averaged molecular weight of 52 000. In all cases, benchmark data were provided using a conventional rotational rheometer, usually with a cone and plate but sometimes with a concentric cylinder geometry. There are several advantages to using the cone and plate geometry. The shear rate experienced by the sample during an experiment is the same everywhere. Even for complex liquids, loading the sample is relatively easy, but for polymer solutions and melts it is particularly straightforward.

The MRI system used for the measurements to date uses unshielded gradients that suffer from strong eddy currents. Sinusoidal shaped G_z gradient pulses are used to both ensure that the finite gradient rise time (1.8 ms) is not exceeded and that each pulsed gradient step size is equal in gradient magnitude. The spin-echo refocusing technique used to form the signal allows for improved image quality for quantitative assessment compared with the use of a gradient-echo. Gradient-echoes suffer from T_2^* relaxation and possible attenuation or phase shift effects that result from inhomogeneities in the static magnetic field. The advantages of using the NMR sequence of Figure 4.2.3 are that it is a robust, simplified approach that allows for quantitative, spatially resolved velocity data even in the presence of severe NMR hardware limitations. The primary disadvantage is that it is not as fast as other velocity profile measurement techniques, such as Real time ACquisition and Evaluation of motion (RACE) [23] and Spatial Modulation of Magnetization (SPAMM) [34]. However, both SPAMM and RACE provide less quantitative velocity/displacement data than the pulse sequence described above.

The two-dimensional, time-domain NMR signal resulting from Figure 4.2.3 is composed of spatial image information that is frequency- and phase-encoded to generate data in k- and q-space (or Fourier space). The reciprocal space variables k and q are conjugate variables of position and displacement, respectively. The fluid position is denoted by x and displacement along the flow direction by Δz. The reciprocal space variable k is k_x and is given by $(2\pi)^{-1}\gamma G_x t$, where t is the acquisition time variable and γ is the gyromagnetic ratio. The reciprocal space

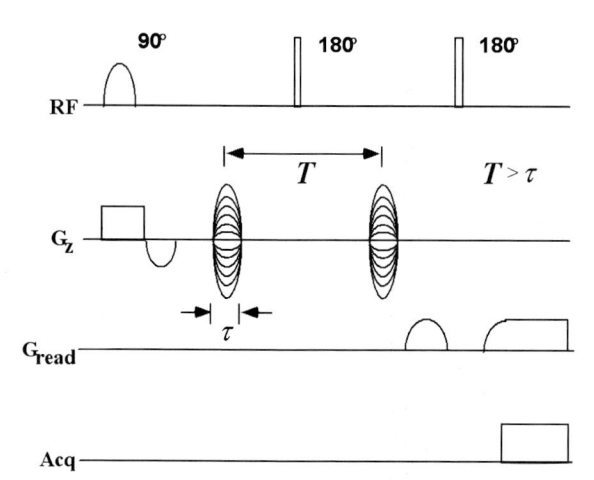

Fig. 4.2.3 PGSE timing diagram where G_z denotes both the slice select and the pulsed, sinusoidal shaped displacement encoding gradient and G_{read} displays the transverse imaging gradient.

variable q is q_z and is given by $(2\pi)^{-1}\gamma G_z \tau$ where τ is the duration of the pulsed gradient. Here, k-space is traversed during signal acquisition while G_z is stepped between separate acquisitions to obtain data in q-space.

4.2.4
Results

Equations (4.2.2 and 4.2.4) clearly show that the two required measurements for pointwise rheological determinations are the velocity profile and the pressure drop. In the present discussion we focus on the former. Pressure and pressure drop measurements are used in many applications. To perform these adequately, it is usually sufficient to know the range of pressures to be measured and to consult with the vendor literature. The more difficult and less standard measurement is the velocity profile and the corresponding velocity gradient. Of particular interest is a technique that has potential applications in real processes with materials that can be opaque. As described in the previous section, the one technique that has shown the most promise for velocity measurements under such conditions is MRI. Our discussion deals with the following areas: (a) obtaining MRI viscosity data; (b) showing the range of applications; (c) describing the limitations; (d) enhancing the quality of the data; (e) designing a system; and (f) measuring yield stress. There are other issues that have been addressed, such as methods to reduce data acquisition time [15]. These are discussed only to the extent that they directly impact the data quality. Finally, we also describe some of the outstanding issues for MRI.

The first task is to measure the velocity accurately so that the shear rate can be determined. For MRI, this can be accomplished in a variety of ways [11, 14, 15, 23–27]. We choose to discuss velocity profiles obtained using a displacement sensitive pulsed gradient with one-dimensional imaging that spatially resolves the displacements. Detailed descriptions of this methodology can be found in Arola et al. [13, 14]. A typical velocity profile is given in Figure 4.2.4. The fluid is a 0.6% by weight aqueous carboxymethylcellulose (CMC) solution. In obtaining this image, we have analyzed our data according to the procedures of Arola et al. [15]. We applied a Fourier transform along the flow direction to determine the velocity. The spatial information is provided in terms of the radial coordinate. These are calculated by applying a Hankel

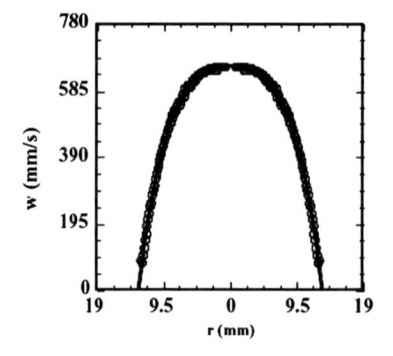

Fig. 4.2.4 Velocity profile for a 0.6% aqueous carboxymethylcellulose solution obtained by MRI. The data points are the MRI data. The line connecting these is the result of an eighth-order even polynomial fit.

transform in the spatially encoded direction, or, in this case, the x direction. This technique allows information obtained in the "natural" coordinate system of the magnetic field gradients (x, y, z) to be transformed into axial velocities with respect to a cylindrical coordinate system. It is based on the reciprocal relationships

$$S(k_r) = 2\pi \int_0^\infty \varrho(r) J_0(2\pi r k_r) r dr$$

$$\varrho(r) = 2\pi \int_0^\infty S(k_r) J_0(2\pi r k_r) k_r dr$$

(4.2.8)

Once the velocity profile has been obtained, the shear rate is calculated. This is the most difficult step. To ensure that the viscosity is determined without any bias, no assumption is made regarding the constitutive behavior of the material. Every effort is made to obtain smooth, robust values of the shear rate without any bias towards a particular model of the flow behavior. Particularly near the tube center, the velocity profiles are distorted by the discrete nature of the information. The size of a pixel is defined by the velocity and spatial resolutions. These are given by

$$\delta w = \frac{2\pi}{\gamma(\Delta G_z) N_z T \tau}$$

(4.2.9 a)

and

$$\delta r = \frac{2\pi}{\gamma G_x N_x D_w}$$

(4.2.9 b)

respectively, where, ΔG_z is the velocity phase encoding gradient stepping size, N_z is the number of samples (or steps) in the z direction, T is the time between the two G_z gradient pulses used in our PGSE sequence, N_x is the number of samples along that axis and D_w is the acquisition dwell time.

Our approach has two stages. Firstly, we implemented a technique known as velocity aliasing to increase the resolution of the velocity data [15]. This is achieved by setting the velocity encode gradients so large that the maximum phase evolution of a given fluid element for a pulsed gradient step exceeds $\pm\pi$ or 2π. The resulting image consists of a series of lines that must be unwrapped to represent the velocity profile. An actual aliased image from MRI is given in Figure 4.2.5. Figure 4.2.5(A) shows the velocity image obtained using a standard Fourier transform technique which provides the velocity in Cartesian coordinates for a 0.6% aqueous CMC solution. Figure 4.2.5(B) gives the aliased image. Near the pipe wall, the quality of the image is diminished as a result of using the Cartesian system, whereas near the pipe center, the intensity is fairly high, as it is a projection of three-dimensional information onto a plane. The unwrapping procedure in effect extends the velocity axis by splicing together the parts of the velocity profile. The effectiveness of this is most readily seen near the center of the pipe, where the velocity gradient is small and the velocity resolution is critical in providing a precise calculation of the shear

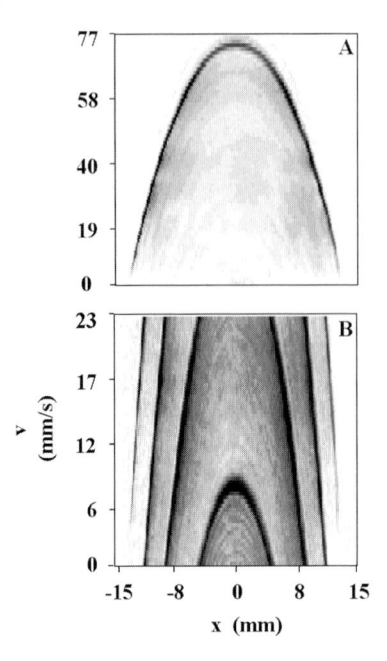

Fig. 4.2.5 Unaliased (A) and aliased (B) velocity images for a 0.6% aqueous carboxymethylcellulose solution obtained by MRI. The vertical axes represent the data obtained in the velocity encoded direction and the horizontal axis represents data obtained in the spatially encoded direction. Aliasing is achieved by setting the velocity encode gradients so large that the maximum phase evolution of a given fluid element for a pulsed gradient step exceeds $\pm\pi$ or 2π

rate. Figure 4.2.6 shows the velocity profiles for the aliased and unaliased data for the region near the center of the pipe where the difference is most apparent. For the unaliased data in this instance, the small velocity gradient near the pipe center results in 6–7 radial positions registering the same value of the velocity. This effect is greatly diminished when aliasing is used. In fact, for this particular case, it is

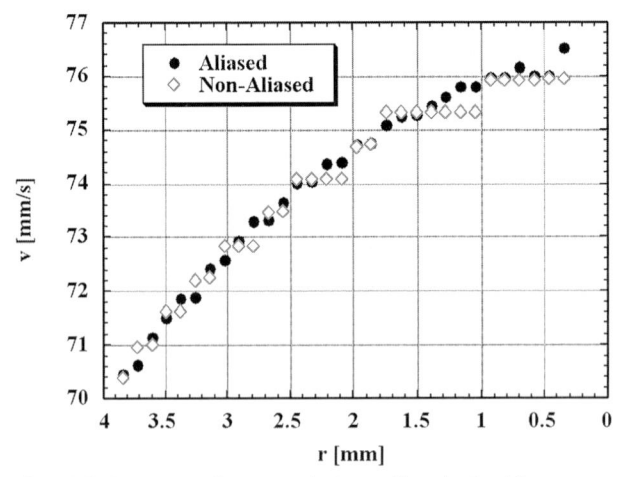

Fig. 4.2.6 Comparison between velocity profiles obtained from aliased (circles) and unaliased (diamonds) MRI data. The data near the pipe center are shown where the difference between the two methods is most apparent.

possible to improve the resolution by a factor of 3.4 using aliasing without adding to the experiment time. This last feature is critical when attempting to optimize a device for both cost and flexibility.

Secondly, the velocity profile is fit globally to an even order polynomial and the shear rate is determined by analytically differentiating this expression. The choice of an even order polynomial allows smoothing of the data without unnecessarily biasing the result towards a particular constitutive equation. The use of an even order polynomial can be justified in terms of a general construct of a hierarchical constitutive theory, such as that for the Rivlin–Ericksen fluids [28]. It might also be justified from a purely experimental standpoint in that if the velocity profile is to be symmetrical, the velocity is the same at a value of the angular coordinate θ of 0° and of 180°. Using an even order polynomial forces this to be the case. The result of such a fit is shown in Figure 4.2.4. The solid line represents the fit and the symbols are the data, as discussed above. The use of an even order polynomial is generally found to work fairly well.

Other schemes have been proposed in which data are fit to a lower, even order polynomial [19] or to specific rheological models and the parameters in those models calculated [29]. This second approach can be justified in those cases when the range of behavior expected for the shear viscosity is limited. For example, if it is clear that power-law fluid behavior is expected over the shear rate range of interest, then it would be possible to calculate the power-law parameters directly from the velocity profile and pressure drop measurement using the theoretical velocity profile

$$w = \left(\frac{-\Delta P R}{2mL}\right)^{1/n} \frac{R}{(1/n) + 1} \left(1 - \left[\frac{r}{R}\right]^{(1/n)+1}\right) \qquad (4.2.10)$$

and finding the values of the m and n that provide the best fit. However, in research applications that seek to determine the shear viscosity unambiguously without the prejudice of a particular constitutive relationship, the alternative that we propose which uses the even order polynomial should be implemented.

The central principle of pointwise rheological measurements, as symbolized by Eqs. (4.2.2 and 4.2.4), is that the local viscosity can be determined from the local shear stress and shear rate. The local shear rate is calculated from experimental data as described in the preceding paragraph. The pressure drop is used in Eq. (4.2.4) to calculate the local shear stress. The result of such an analysis is shown in Figure 4.2.7, which shows the shear stress and the shear rate as a function of the radius for a 0.6% by weight aqueous carboxymethylcellulose solution. In this figure, the data are represented by lines. The data for the shear stress are continuous over the domain of the pipe as they are calculated from Eq. (4.2.4). For the shear rate, we are using an analytical differentiation based on the fitted polynomial. As a result, we plot an analytical expression for the shear rate. From these data, the shear stress as a function of the shear rate can be determined at every point in the pipe by choosing a position and finding the corresponding $\dot{\gamma}$ and $\sigma(\dot{\gamma})$. The shear viscosity is calculated using Eq. (4.2.5) at every corresponding shear rate. The net effect is that from a single set of data, comprised of a velocity profile and a pressure drop, it is possible to calculate the viscosity over a wide range of shear rates, as shown in Figure 4.2.8. A conventional cone and plate viscometer was used to obtain the benchmark data. The data from the MRI correspond fairly well with those data over a range of nearly a decade and a half of shear rates. At the lowest shear rates, the discrepancy between the two techniques becomes apparent. In this regime, the data correspond to radial positions closest to the center of the tube where the velocity profile is nearly flat. The calculated shear rate provides values of viscosity that are consistently lower than the benchmark data. This implies that at a fixed value of shear stress, the shear rate that is being calculated is greater than the actual shear rate that corresponds to that shear stress. This difficulty in obtaining reliable values of the shear rate near the pipe center was one of the principle reasons that we developed the velocity aliasing technique. At the same time, the behavior shown in Figure 4.2.8 reflects all of the data that we have obtained to date, with deviations between the MRI data and the benchmark data occurring at low shear rates. Hence, we have used considerable effort to try to understand how to predict when this deviation will occur.

It is possible to alter the range of shear rates over which data can be obtained by changing the volume flow rate, as indicated by Eqs. (4.2.6 and 4.2.7). In practice, if the volume flow rate is decreased, the velocity profile near the center of the pipe usually has a larger velocity gradient. It is therefore possible to accurately determine the viscosity at lower values of the shear rate. Figure 4.2.9 compares the viscosities obtained by a cone and plate viscometer with those measured by MRI at two different volume flow rates. The MRI viscosity results agree well with each other and with the conventional viscometer except at the low shear rate regions. The differences between velocities of the successive pixels at the regions close to tube center become smaller than the velocity resolution, leading to almost flat velocity profiles. Although a general polynomial fit is used to calculate the shear rates, the velocity resolution limits the accurate fitting of the velocity data, especially for the regions close to the pipe center. The extent of this problem can be reduced by increasing velocity resolution (making δw smaller) or reducing the volumetric flow rate so that the regions where the successive velocity points can be identified separately extend closer to the center. Figure 4.2.9 shows that the viscosity results for the low flow rate case is more accurate than the high flow

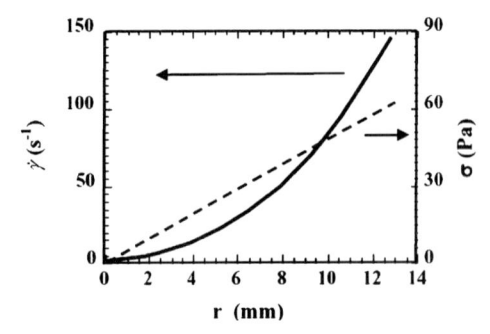

Fig. 4.2.7 Shear rate and shear stress as a function of radius for a 0.6% by weight aqueous carboxymethylcellulose solution. The shear stress as a function of shear rate is obtained by choosing values of the radius and finding the corresponding values of σ and $\dot{\gamma}$.

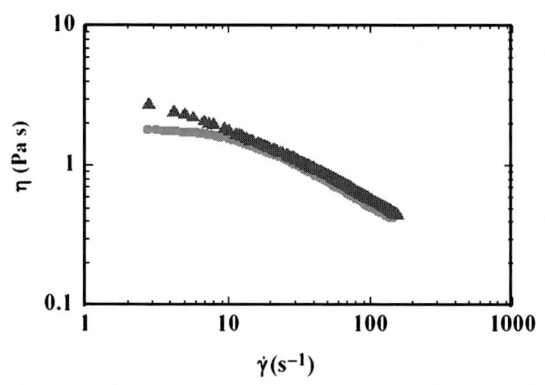

Fig. 4.2.8 Shear viscosity versus shear rate data for a 0.6% by weight aqueous carboxymethylcellulose solution. Data from MRI were obtained from one combined measurement of a velocity profile and a pressure drop. (▲) Cone and plate; (●) MRI.

rate case in the low shear rate regions. As the shear rate is increased, the deviation tends towards a constant minimum value. There is a slight offset from the benchmark data at the higher shear rates, which could be due to calibration errors in either the rotational viscometer or the pressure transducers. Either of these sources would lead to such a systematic error in the data. It is also important that the use of the Hankel transform results in a loss of information from the pixels close to the tube center. In general, this effect is not important as the velocity gradients are so small that it is difficult to obtain accurate shear rate data at values of the radius larger than those affected by his artifact.

Figure 4.2.10 shows the power of using MRI to determine viscosity. In this figure, the radial resolution of the MRI has been varied from 200 μm to 1 mm while the velocity resolution has remained constant at $\delta w = 2.2$ mm s^{-1}. The most important conclusions to draw from this are two-fold. Firstly, the effect of the radial resolution is small. On the scale shown in Figure 4.2.10, it is almost impossible to discern any effect whatsoever. Qualitatively, the net effect of changing the radial

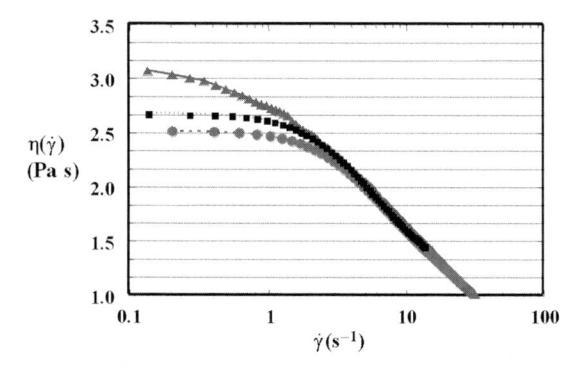

Fig. 4.2.9 Viscosity versus shear rate obtained using a cone and plate viscometer (▲) and MRI: (■) 13 mL s^{-1}, (●) 22 mL s^{-1}.

resolution is to increase the change in velocity that can be expected as the sequential data points that are compared are larger. On the other hand, this enhancement of the change in velocity is not offset by the reduced radial resolution. This is shown in Figure 4.2.11, which magnifies the data in Figure 4.2.10 at the lowest shear rates. Again, the comparison is made with the rotational viscometer data. Included in this are "error bars" that show two standard deviations in the rotational data. These were estimated by calculating the variation in the viscosity over three experimental trials. In all cases the MRI data are within the two standard deviation limit of the rotational data. Furthermore, all of the data are within 3% of the standard. In obtaining the data in Figures 4.2.10 and 4.2.11 the entire velocity profile was included, except for the data at $r = 0$. Under these circumstances, it is possible to find a correlation between the minimum shear rate at which data can be obtained and the radial resolution. We found that that there was a linear correlation of the form $\dot{\gamma}_{r,min} = 0.87\delta r + 0.01$. As we will see, this minimum value of the shear rate is a less conservative estimate than is found by considering the effect of the velocity resolution. It is in fact δw that identifies the lowest shear rate at which data can be reliably obtained.

We have applied the MRI technique to a wide range of fluids, as shown in Table 4.2.1. Many of these are systems that were chosen in order to have a wide range of properties that represent the full spectrum of behavior observed for complex liquids. The data shown in Figure 4.2.10 provide an excellent case in point. At high shear rates, the data are clearly seen to be shear thinning, and in fact are behaving as a power law. In this instance the MRI and the rotational rheometry data show very good agreement. The data also show a regime at low shear rates where Newtonian fluid behavior is observed. The measurements with MRI are also able to track this behavior, particularly if careful attention is paid to the low shear rate limitations of the system.

The MRI technique has also been used with systems of more practical importance, such as a polymer melt [19]. Here a low density polyethylene melt was

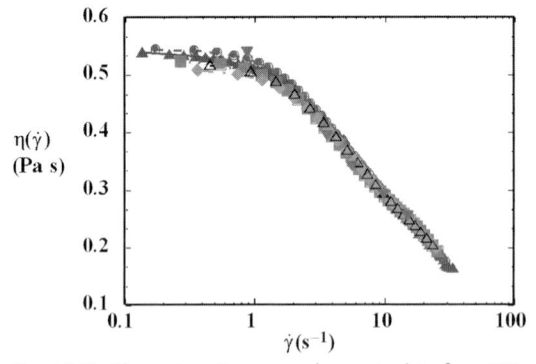

Fig. 4.2.10 Shear viscosity versus shear rate data for a 1% aqueous poly(ethylene oxide) solution. Rotational rheometry: (▲). MRI: $\delta r = 200$ μm (●), $\delta r = 300$ μm (■), $\delta r = 400$ μm (◆), $\delta r = 500$ μm (△), $\delta r = 1000$ μm (▼).

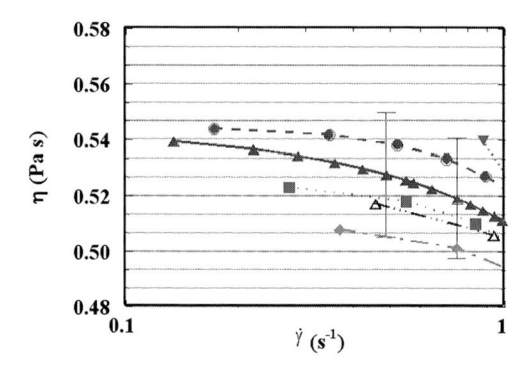

Fig. 4.2.11 Shear viscosity versus shear rate data for a 1% aqueous poly(ethylene oxide) solution. Symbols same as in Figure 4.2.10.

pumped though a 1.2-cm diameter quartz pipe placed in 0.6 T imaging magnet. The pipe was insulated and hot air was blown across it in order to maintain melt conditions and also to provide some degree of temperature control. The pressure was measured at one point in the die leading up to the quartz pipe and the pressure drop was then inferred by assuming atmospheric conditions at the outlet and applying standard end corrections to the exit pressure. Typical data obtained by this technique are shown in Figure 4.2.12, where we are simply providing a plot of the signal intensity at every pixel. The horizontal pixels represent the velocity encoded direction and the vertical pixels are the spatially encoded direction. While the data clearly indicated the velocity profile, their quality is degraded as a result of using a metal die in the imaging magnet and also placing the extruder near the magnet. In order to analyze images such as Figure 4.2.12, we consider the signal intensity as a function of the number of the pixel in the velocity encoded direction and carefully find the maximum signal intensity at each radial position. A typical velocity profile obtained by this technique is shown in Figure 4.2.13. These data clearly show that it is possible to obtain the velocity profile for the low density polyethylene melt, but that precision of the data does not approach that found previously, such is in Figure 4.2.10. For this reason, it is difficult to fit the data to an even order polynomial and determine the shear rate as a function of radius. In fact, from rotational rheometer data, it was found that the melt exhibited very slight shear thinning behavior up to a shear rate of about $10\ s^{-1}$, which is our estimate of the maximum shear rate associated with this velocity profile. As a result, we fit these data to a parabolic curve that reflects Newtonian behavior, which is shown in Figure 4.2.13 and represented by $w = 3.6\ (1-2.78r^2)$. Using Eq. (4.2.10) and making it specific for a Newtonian fluid, the maximum velocity, $w_{max} = 3.6$ cm s^{-1} = $(-\Delta PR^2/4\mu L)$. As ΔP is measured, it is possible to ascertain μ to be 117.7 Pa s over the shear rate range of from 1 to 12 s^{-1}. This compared very favorably with the result from rotational rheometry data of 114.2 to 106.4 Pa s over the shear rate range of from 0.1 to 12 s^{-1}.

Although the results from the polymer melt demonstrate the applicability of this technique to a system of great importance, it is also clear that there are design issues that would need to be addressed. The data in Figure 4.2.12 are of much lower quality than those obtained using polymer solutions at room temperature. There are many possibilities for this. The proximity of the extruder to the magnet and the

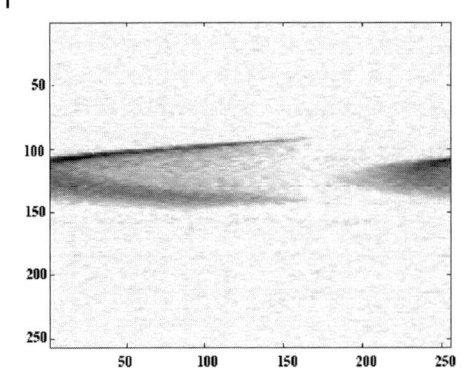

Fig. 4.2.12 Image showing the velocity in the horizontal direction versus position in the vertical direction for a low density polyethylene melt.

presence of metal piping near the gradient coil and the large temperature gradients introduce artifacts. The image was not analyzed with a Hankel transform, which contributed to the shading. Also, the spin–spin relaxation time for the polymer was 44 ms, which was less than the echo time of 50 ms.

The enabling feature of the MRI-based method is the ability to know *a priori* the validity of the measurement so that unknown samples can be measured with a high degree of reliability. The most critical parameter is the minimum shear rate at which the viscosity can be measured. As shown in Figure 4.2.9, MRI data can provide a false indication of the transition from shear thinning behavior to Newtonian behavior when using data near the pipe center where the shear rates are low. The critical parameter in determining this minimum shear rate is the velocity resolution, δw. The specific correlation between $\dot{\gamma}_{min}$ and δw is shown in Figure 4.2.14. Here we have identified $\dot{\gamma}_{min}$ to be that shear rate at which the data from the MRI deviates from the benchmark data by 6%. The velocity resolution and the minimum shear rate are made dimensionless by the mean velocity, $\bar{w}$, and $\bar{w}/R$, respectively. These data were obtained using a polyacrylamide solution and both the volume flow rate and δw were varied. The correlation implied by Figure 4.2.14 provides the means of the finding $\dot{\gamma}_{min}$ for a specified velocity resolution, mean

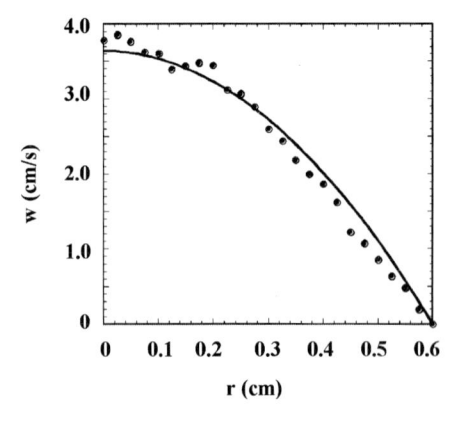

Fig. 4.2.13 Velocity profile for low density polyethylene melt. The line refers to a fit of the data based on a parabolic velocity profile.

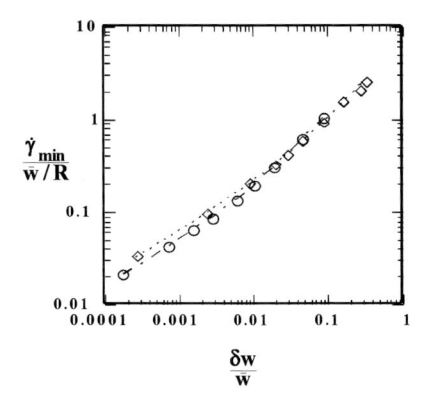

Fig. 4.2.14 Dimensionless correlation between the minimum shear rate at which the shear viscosity can be obtained via MRI, $\dot{\gamma}_{min}$ and the velocity resolution, δw. These data were obtained using an aqueous polyacrylamide solution at two different volume flow rates, or mean velocities, $\bar{w}$.

velocity and pipe radius. Such a correlation not only holds for MRI-based pointwise viscosity measurements but also for those using ultrasonic-pulsed Doppler velocimetry [2–4]. As such it appears to provide a universal estimate of the low shear rate limitations of spatially resolved viscosity measurement techniques. There are factors that may affect this, such as the radial resolution, however, to a first approximation, the correlation, as it stands, is robust.

The MRI-based technique can also provide measurements of the yield stress. Figure 4.2.15 shows the velocity profiles for a 0.53% by weight microfibrous cellulose (MFC) suspension. At all flow rates, the velocity profiles show a very flat region near the pipe center. The size of this region decreases as the flow rate is increased. As discussed above, there are various interpretations to such velocity profiles, including that the suspension is behaving as a power-law fluid with a small exponent. The alternative is that the suspension has a yield stress. In this case, it is possible to measure the yield stress by identifying the point in the pipe next to the plug flow region where the suspension is undergoing continuous shearing. Figure

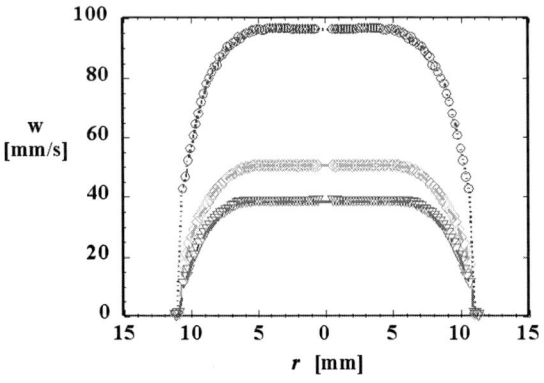

Fig. 4.2.15 Velocity profiles for a 0.53% MCC suspension at three different volume flow rates. The region near the center of the pipe shows a blunt velocity profile that can be interpreted in terms of a yield stress.

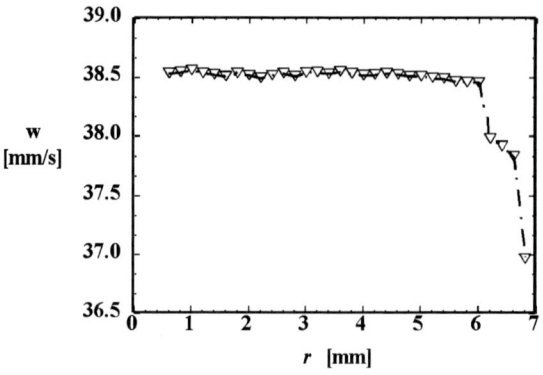

Fig. 4.2.16 Expanded velocity profile for 0.53 % MCC suspension at the lowest volume flow rate shown in Figure 4.2.15. The point at which there is a transition from the blunt velocity profile to the region of high shearing can be interpreted in terms of a yield stress.

4.2.16 shows this more explicitly by expanding the data for the lowest velocity. The plug flow region is clearly visible out to $r \approx 6$ mm. Using Eq. (4.2.4) the stress at that point can be calculated from the pressure drop. The value of 3.42 ± 0.34 Pa compares very favorably with the yield stress obtained using a conventional technique, 3.15 ± 0.08 Pa. The advantage of the MRI technique is, firstly, that it can be measured in a process stream. Secondly, many materials that might be viewed as having a yield stress may exhibit time dependent properties. The properties that would be measured off-line may not be indicative of those in the process stream. Finally, while the sizes of the fibers in the MFC suspensions are small relative to the typical dimensions of a viscometer, this is not necessarily the case for many industrial complex fluids such as pulp suspensions and tomato products. In such cases, the capability to make a measurement in industrial scale devices is of great benefit.

4.2.5
Conclusion

Magnetic resonance imaging has enabled the development of a completely novel type of viscometer. This technique is based on the capacity of MRI to accurately measure velocity profiles in opaque liquids. Its potential applications include many systems of industrial relevance, such as polymer melts and slurries. The data presented here clearly show that a wide range of fluid behaviors can be measured.

The intention of this work was not to consider whether the specific experimental system used here would be applicable to an actual industrial setting. Rather, the work has set the stage to move beyond the question of *whether* the technique can be applied, but *when*. The barriers to its implementation are two-fold. Firstly, the specific technology must be developed for a particular process line. Secondly, the cost of the MRI is prohibitive, although with the use of electromagnets and permanent magnets for MRI, this is becoming less of a factor. Furthermore, when compared with other on-line instruments that promise far less in terms of their ability to characterize a complex fluid comprehensively, the cost for a small MRI unit is not out of line. Essentially the same equipment that provides the velocity profiles can be used to determine chemical composition, concentration of a solid phase and even estimates of temperature. It seems likely, therefore, that an MRI-based rheometer either as a process or a research tool will at some point be commercialized. The more obvious choices for the initial implementation of such instruments are those facilities producing high value products that must meet critical rheological specifications, such as manufacturing plants producing engineering polymers, personal care products and processed foods. In all cases, the added benefit of being able to characterize other factors (chemical composition, etc.) would also be important.

References

1 R. B. Bird, R. C. Armstrong, O. Hassager **1987**, *Dynamics of Polymeric Liquids, Vol. 1: Fluid Mechanics*, John Wiley, New York.
2 N. Dogan, M. J. McCarthy, R. L. Powell **2002**, *J. Food Sci.* 67, 2235–2240.
3 N. Dogan, M. J. McCarthy, R. L. Powell **2003**, *J. Food Process. Eng.* 25, 571–587.
4 N. Dogan, M. J. McCarthy, R. L. Powell **2005**, *J. Text. Studies* in the press.
5 B. D. Coleman, H. Markovitz, W. Noll **1966**, *Viscometric Flows of Non-Newtonian Fluids*, Springer, Berlin.
6 K. Walters **1975**, *Rheometry* Chapman and Hall, London.
7 S. Sinton, A. Chow **1991**, *J. Rheol.* 35, 735–772.
8 M. J. McCarthy, J. E. Maneval, R. L. Powell **1992**, in *Advances in Food Engineering*, eds. R. P. Singh, A. Wirakartakamasuma, CRC Press, Boca Raton, (pp.) 87–99.
9 J. D. Seymour, J. E. Maneval, K. L. McCarthy, M. J. McCarthy, R. L. Powell **1993**, *Phys. Fluids A* 5, 3010–3012.
10 R. L. Powell, J. E. Maneval, J. D. Seymour, K. L. McCarthy, M. J. McCarthy **1994**, *J. Rheol.* 38, 1465–1470.
11 J. E. Maneval, K. L. McCarthy, M. J. McCarthy, R. L. Powell **1994**, (Nuclear Magnetic Resonance Imaging Rheometer), *US Patent Ser.* 08/146,497.
12 T.-Q. Li, M. Weldon, L. Odberg, M. J. McCarthy, R. L. Powell **1995**, *J. Pulp Paper Sci.* 21, J408–J414.
13 D. F. Arola, G. A. Barrall R. L. Powell, M. J. McCarthy **1999**, *J. Rheol.* 43, 9–30.
14 D. F. Arola, G. A. Barrall, R. L. Powell, K. L. McCarthy, M. J. McCarthy **1997**, *Chem. Eng. Sci.* 52, 2049–2057.
15 D. F. Arola, G. A. Barrall, R. L. Powell, M. J. McCarthy **1997**, *J. Mag. Reson. Anal.* 3, 175–184.
16 S. J. Gibbs, D. E. Haycock, W. J. Frith, S. Ablett, L. D. Hall **1997**, *J. Magn. Reson.* 125, 43–51.
17 L. Sun, G. Amin, L. D. Hall, W. Bolf, W. J. Frith, S. Ablett **1999**, *Meas. Sci. Technol.* 10, 1272–1278.
18 S. J. Gibbs, K. L. James, L. D. Hall **1996**, *J. Rheol.* 40, 425–440.
19 Y. Uludag, G. A. Barrall, D. F. Arola, M. J. McCarthy, R. L. Powell **2001**, *Macromolecules* 34, 5520–5524.
20 A. Swerin, R. L. Powell, L. Ödberg **1992**, *Nordic Pulp Paper Res. J.* 7, 126–132.
21 T.-Q. Li, M. J. McCarthy, K. L. McCarthy, J. D. Seymour, L. Ödberg, R. L. Powell **1994**, *AIChE J.* 40,1408–1411.
22 S. Whitaker **1992**, *Introduction to Fluid Mechanics*, Kreiger, Melbourne, FL.
23 T. Kahn, E. Muller, J. S. Lewin, U. Modder **1992**, *J. Comput. Assist. Tomogr.* 16, 54–61.
24 L. Axel, L Dougherty **1989**, *Radiology* 171, 841–845.
25 Y. Xia, P. T. Callaghan **1991**, *Maromolecules* 24, 4777–4786.
26 P. T. Callaghan **1991**, *Principles of Nuclear Magnetic Resonance Microscopy*, Clarendon Press, Oxford.
27 E. Fukushima **1999**, *Annu. Rev. Fluid Mech.* 95–123.

28 C. Truesdell, W. Noll **2004**, *The Non-Linear Field Theories of Mechanics*, Springer, Berlin.

29 M. K. Cheung, R. L. Powell, M. J. McCarthy **1997**, *AIChE J.* 43, 2596–2600.

4.3
Imaging Complex Fluids in Complex Geometries
Y. Xia and P. T. Callaghan

4.3.1
Introduction

Complex (polymer) fluids can exhibit fascinating rheological properties that depend upon the history of the deformation, rather than the instantaneous state of deformation as in Newtonian fluids. As early as AD 100, the great Roman historian Cornelius Tacitus described the harvest of bitumen near a lake in *The Histories V*. He wrote [1]:

"*Those whose business it is take it with the hand, and draw it on to the deck of the boat; it then continues of itself to flow in and lade the vessel till the stream is cut off.*"

In modern literature this unusual flow phenomenon is commonly termed the tubeless siphon, open siphon or Fano flow [2–4], where a fluid column can be drawn upward out of a container by a tube, even when the tube is completely above the fluid (Figure 4.3.1). While the gravitational force on a Newtonian fluid column is responsible for the pulling of additional fluid out of a container through a tube *inserted inside* the fluid, the driving mechanism for a tubeless siphon in a polymeric fluid is the entanglements within the polymer molecules. The stretching of the macromolecules in the fluid produces a restoring force along the flow direction, which acts to pull additional fluid out of the container. In addition to Fano flow, other extraordinary flow phenomena exist in rheology when the fluid involved contains very large polymeric molecules. We shall see that NMR imaging at microscopic resolution can be a unique tool in the studies of complex fluids in complex geometries. Although a full review of polymeric rheology and NMR imaging is beyond the scope of this chapter, it should be helpful to introduce some of the fundamentals in these two areas. We will then turn our attention to a recent NMR velocity imaging experiment of Fano flow. The final section describes two equally fascinating flow patterns in the category of polymeric rheology.

4.3.2
Rheological Properties of Polymeric Flow

Rheology concerns the study of the deformation and flow of soft materials when they respond to external stress or strain. If the ratio of its shear stress and shear rate is a straight line, the material is termed Newtonian; otherwise, it is termed non-Newtonian (Figure 4.3.2(a)). As the slope of the curve is the viscosity η, a shear-thinning fluid exhibits a reduced viscosity as the shear stress increases, whereas a shear-

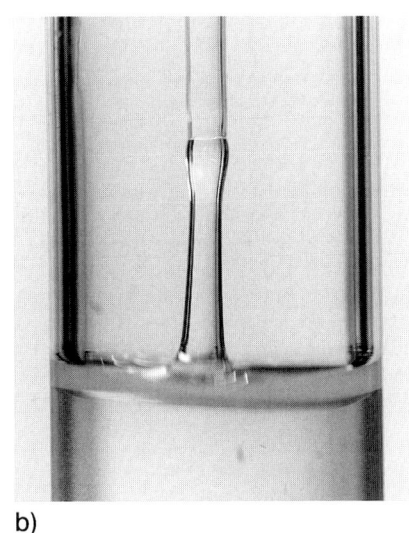

Fig. 4.3.1 (a) Photographs of a tubeless siphon formed by dissolving 0.5 % w/v poly(ethylene oxide) powder in tap water, where a Fano column can be seen between the tip of the glass pipette at the top and fluid reservoir at the bottom. (b) Excess fluid can be seen just below the fluid entrance. (c) A large amount of excess fluid eventually flows downwards outside and along the Fano column, which can disturb the vertical location of the column. These figures illustrate the fact that there is an optimum volume flow rate for a particular flow system.

thickening fluid exhibits an increased viscosity as the shear stress increases. These non-linear viscosities are caused by some shear-induced conformational changes at the molecular level in the material. For example, some semi-dilute polymer solutions become shear thinning because the shearing process reduces the molecular entanglements, while some suspensions of particles can become shear thickening because the shearing process induces particle aggregation. Some fluids can also show complex behavior of viscosities, for example, a region of shear thinning followed by another region of shear thickening as a function of shear rate, or shear thinning/thickening as a function of particle volume fraction in suspensions [5].

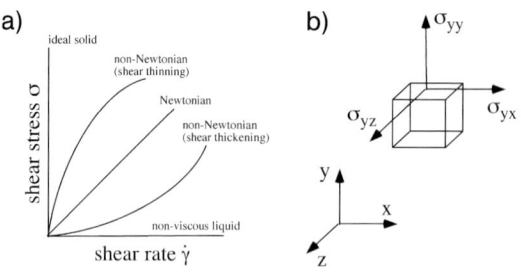

Fig. 4.3.2 (a) The relationships of shear stress and shear rate for several types of common materials, where the slope is the viscosity η. A non-Newtonian fluid is characterized by its non-linear viscosity. (A real solid will have a very large but finite viscosity.) (b) The definition of the stress components in a stress tensor in rheology.

To comprehend the physics of the fluid being deformed under external forces, we will consider the elements of a stress tensor for a fluid volume, as

$$\underline{\underline{\sigma}} = \begin{bmatrix} \sigma_{xx} & \sigma_{yx} & 0 \\ \sigma_{xy} & \sigma_{yy} & 0 \\ 0 & 0 & \sigma_{zz} \end{bmatrix} \quad (4.3.1)$$

where the first index refers to the orientation of the plane surface (the normal to the surface) and the second index refers to the direction of the stress (Figure 4.3.2(b)) [6]. In addition, for one-dimensional flow at a velocity v, the x is the direction of the fluid velocity and the y is the direction of velocity variation in the fluid. On the top surface of the fluid element in Figure 4.3.2(b), for example, three stress elements can, in general, be written as σ_{yx}, σ_{yy} and σ_{yz}, respectively. In this notation, the diagonal elements of the stress tensor are termed "normal stresses" and the off-diagonal elements are termed "shear stresses". The quantities that are important in polymeric rheology are the two differences between the normal stress elements ($N_1 = \sigma_{xx} - \sigma_{yy}$, $N_2 = \sigma_{yy} - \sigma_{zz}$).

For a simple planar shear flow with a no-slip boundary condition at all fluid–solid interfaces, the fluid can be viewed as being sheared by a tangential force F to give successive imaginary layers of velocity that increase linearly as the distance from the stationary plate increases, for which the shear viscosity is used to quantify the resistance of the fluid to flow (Figure 4.3.3(a)). With this configuration where both v_y and v_z equal zero, the shear stress σ_{yx} is F_x/A_y and the shear rate is the velocity gradient $\dot{\gamma} = \partial v_x/\partial y$. Among a number of constitutive equations that describe the relationship between the shear stress upon the shear rate, one of the better known is the "power law" fluid

$$\sigma_{xy} = K\dot{\gamma}^n \quad (4.3.2)$$

where K and n are constants for a particular fluid. The power law exponent, n, is unity for a Newtonian fluid and is less than unity for a shear-thinning fluid. Clearly the power law constitutive equation is phenomenological and suffers from a

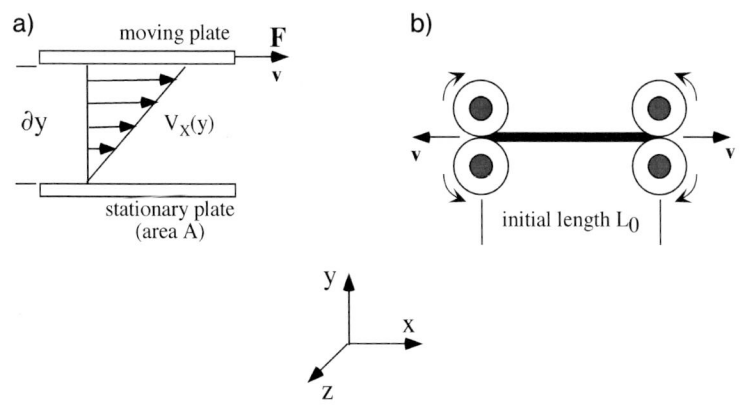

Fig. 4.3.3 (a) Shear flow of a Newtonian fluid trapped between the two plates (each with a large area of A). The shear stress (σ) is defined as F/A, while the shear rate ($\dot{\gamma}$) is the velocity gradient, $\partial v_x/\partial y$. The shear viscosity (η_s) is defined as the ratio of the shear stress and shear rate. (b) A polymeric material is being stretched at both ends at a speed of v. The material has an initial length of L_0 and an (instantaneous) cross-sectional area of A.

number of defects, including the divergence of the viscosity at zero shear in the case of shear-thinning behavior. Nonetheless, it does provide a fairly good empirical description in a number of experiments.

If the fluid is Newtonian, the relevant elements in the stress tensor are

$$\sigma_{yx} = \eta_s \dot{\gamma}, \quad \sigma_{xx} - \sigma_{yy} = N_1 = 0, \quad \sigma_{yy} - \sigma_{zz} = N_2 = 0 \tag{4.3.3}$$

where η_s is the shear viscosity. Water and glycerol are two common examples of Newtonian fluids, with shear viscosities of 1×10^{-3} Pa s (Pascal seconds) for water and 1×10^0 Pa s for glycerol. If the fluid is non-Newtonian, η_s becomes a function of the shear rate, and the normal stress differences N_1 and N_2 are no longer zero. This non-equality in the normal stress elements is due to the generation of anisotropic forces among/between/within macromolecules in an anisotropically deformed fluid.

Extensional flow describes the situation where the large molecules in the fluid are being stretched without rotation or shearing [5]. Figure 4.3.3(b) illustrates a hypothetical situation where a polymer material is being stretched *uniaxially* with a velocity of v at both ends. Given the extensional strain rate $\dot{\varepsilon}$ ($= 2v/L_0$) for this configuration, the instantaneous extensional viscosity η_e is related to the extensional stress difference ($\sigma_{xx} - \sigma_{yy}$), as

$$\sigma_{xx} - \sigma_{yy} = \eta_e \dot{\varepsilon} \tag{4.3.4}$$

The ratio of extensional viscosity η_e to shear viscosity η_s is known as the Trouton ratio, which is three for Newtonian fluids in uniaxial extension and larger than three for non-Newtonian fluids. For a viscoelastic fluid such as a polymer in solution, the uniaxial extensional viscosity characterizes the resistance of the fluid to stretching deformations. This extensional property, both time dependent and strain-rate dependent, is of great importance in rheological research and industrial processing such as the formation and spinning of biological and synthetic fibers.

4.3.3
NMR Microscopy of Velocity

The basic principles of NMR and NMR imaging are well understood [7–10]. At the heart of the technique is the linear proportionality between the nuclear Larmor frequency (ω) and the magnitude of the external magnetic field (B_0) in which the nuclei are immersed. By placing a sample in a magnetic field that has been deliberately made non-uniform, the Larmor frequency of the nuclear spins will differ from one location to another across the sample. The spatial position of the spin is, therefore, encoded by a shift of the precessional frequency. The normal practice in imaging experiments is to superimpose a linearly varying gradient onto the $\mathbf{B}_0$, so that the precessional frequency becomes

$$\omega(r) = \gamma\, \mathbf{B}_0 + \gamma \mathbf{G} \cdot \mathbf{r} \tag{4.3.5}$$

where γ is the gyromagnetic ratio of the nucleus (a constant) and $\mathbf{G}$ is the applied field gradient vector. By defining a quantity $\mathbf{k}$ that is the reciprocal space vector conjugate to the nuclear spin coordinate $\mathbf{r}$, NMR signal can be acquired in $\mathbf{k}$-space and an inverse Fourier transformation will generate a nuclear spin density image in $\mathbf{r}$-space. Although NMR imaging experiments can be performed in a three-dimensional manner, it is common to image only two dimensions by *electronically* selecting a two-dimensional slice through the sample (Figure 4.3.4(a)). This strategy reduces the size of data array from N^3 to N^2 where N is the number of discrete points to be sampled, and the result is consistent with the final two-dimensional visual display on a computer screen. By using a large imaging gradient $\mathbf{G}$ and scaling down the receiver coil, one can improve the imaging resolution to as fine as tens of microns, where NMR imaging has been termed NMR microscopy [10, 11].

This section concerns the macroscopically ordered motion, where a molecule with a velocity v moves to a new position over a specific time interval. In order to probe the motion of nuclear spins using NMR imaging, the usual imaging sequence for a spin location needs to be made motion-sensitive. The common strategy is to embed a Pulsed Gradient Spin Echo (PGSE) pair [12] into the imaging sequence. Because each spatial location is encoded with a unique set of gradient values, any nucleus that has not moved during a predetermined time interval in the PGSE will acquire no additional phase shift, while any motion of the nucleus will result in a change in phase shift. This phase shift (frequency change) can be detected by the PGSE pair and becomes a contrast factor in imaging, as

$$E_c(v) = \exp(i\gamma g v \Delta \delta) \tag{4.3.6}$$

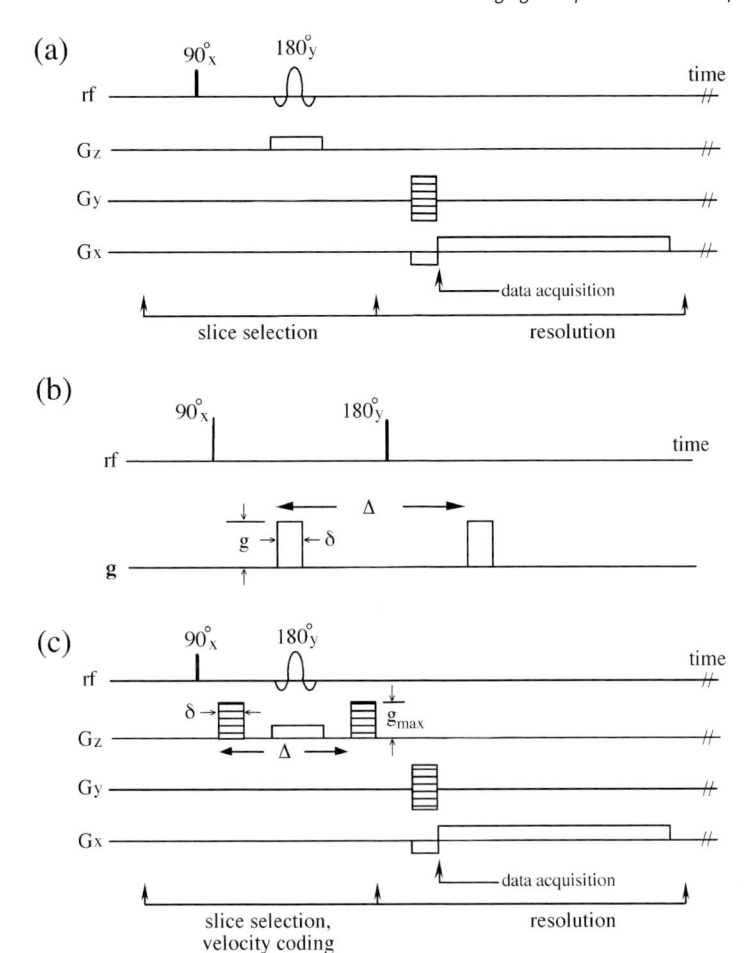

Fig. 4.3.4 (a) A 2D imaging pulse sequence using a Cartesian raster sampling in k-space (the read gradient G_x, the phase gradient G_y, the slice gradient G_z). The simultaneous application of a shaped pulse and a gradient field determines the profile of an imaging slice. The position of the imaging slice (z) is determined by the frequency (ω) of the excitation rf field with respect to the polarizing frequency (ω_0) of B_0. The k-space traverse in the 2D imaging plane is determined by the direction/duration/ magnitude of the phase and read gradients. (b) The PGSE sequence that records the motion of the spins. In the presence of the first gradient pulse, any phase shift depending upon the position of each nucleus is recorded. Between the pulses the molecules containing the nuclei keep changing their positions due to translational flow. Following the 180° rf pulse, all prior phase shifts have been inverted. Any nucleus which has not moved between the two gradient pulses will be refocused perfectly, and any motion of the nucleus will result in the incomplete refocusing. (c) The pulse sequence for the NMR velocity-imaging experiments. This imaging sequence constructs a 2D image in the xy plane, with the sensitivity to motion along the z direction.

where v is the velocity in the direction of the motion-sensitive gradient g, Δ and δ are the two time intervals specified by the PGSE pair (Figure 4.3.4(b)) [13]. By defining a quantity **q** that is the reciprocal vector conjugate to the displacement **R** of the nuclear spins, a series of motion-weighted NMR images can be acquired and a second Fourier transformation will construct an image of the velocity [14]. To avoid excessively long experiments, it is customary to limit the spatial imaging space (k-space) to two-dimensional and velocity-imaging space (q-space) to one-dimensional (Figure 4.3.4(c)). Velocity imaging experiments can therefore be carried out with the g gradient being successively stepped in a number of discrete steps to some maximum value g_{max}. Each g step corresponds to a "slice" in q-space [13], where one pair of complex images (one in-phase and one quadrature-phase) is reconstructed using the normal k-space reconstruction algorithm. The complex images are weighted progressively by $E_c(v)$ as the stepping up of g. Successful applications of NMR imaging of velocity include the imaging of vascular flows in plants [15–17], viscous polymer flows in capillary tubes [18] and water flow in sudden expansion/contraction geometries [19].

4.3.4
NMR Velocity Imaging of Fano Flow

Now let us take a look at a recent NMR imaging experiment of Fano flow, in which the local velocities in the tubeless column were mapped out quantitatively and nondestructively [20]. For such a set-up, the weight force of the column is balanced by the extensional stress difference $\sigma_{zz} - \sigma_{xx}$ associated with the vertical velocity gradient $(\partial v_z/\partial z)$, as

$$\sigma_{zz} - \sigma_{xx} = \eta_e(\partial v_z/\partial z) \tag{4.3.7}$$

Fano columns are most easily formed in viscoelastic solutions where extensional viscosities exceed shear viscosities by several orders of magnitude.

In this experiment, the fluid was about 600 mL of the high molar mass (8×10^6 dalton) poly(ethylene oxide) powder (Aldrich Cat. No. 37,283–8) mixed with tap water to form a 0.5% w/v viscous solution. The experimental arrangement for the Fano siphon is shown schematically in Figure 4.3.5, where the flow path was sealed except at the tip of the glass pipette and the top of the glass reservoir. Before the onset of the siphon, the tip of the pipette was lowered into the polymer fluid and then a pump started. Once a steady flow was established, the pipette tip was gradually lifted up to several millimeters above the surface of the fluid in the reservoir, resulting a tubeless column inside a 15-mm rf coil that is located in a 7-T/89-mm superconducting magnet. Although the Fano column was entirely inside the magnet and could not be inspected visually, the tuning indicator for the receiver coil on the front panel of the NMR instrument console proved to be a reliable and sensitive indicator for the steadiness of the column. This was because any unsteadiness of the Fano column would change the fluid level in the glass reservoir, which would in turn change the tuning of the NMR receiver. For this particular fluid system, a tubeless column of up to 6 mm could be maintained steady for several hours, at a measured flow rate of about 125 mL h^{-1}.

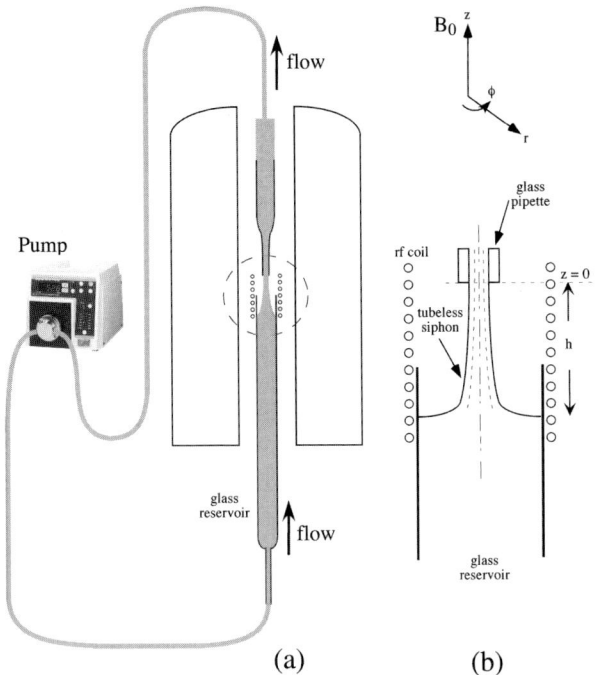

Fig. 4.3.5 (a) Schematic diagram showing the arrangement of Fano flow – the pipette and reservoir are inside the bore of the 7-T vertical superconducting magnet. (b) Geometry of Fano flow showing the tubeless siphon of height, h. Reprinted from Ref. [20], with permission from Elsevier.

Two-dimensional NMR velocity-imaging experiments were performed at 16 different heights along the tubeless column, at a vertical spacing of 0.5 mm apart and using the sequence shown in Figure 4.3.4. At each height, eight pairs of 2D complex images were obtained for eight different values of the g gradient, which was directed along the direction of siphon column (z). The PGSE pair had a motion-sensitive time interval Δ of 5 ms and a duration δ of 1 ms, with a maximum g gradient (g_{max}) of 0.21 T m^{-1}. The 2D imaging matrix had a digital size of 128 × 128 and a field of view of 4 × 4 mm, which resulted in a transverse resolution of 31.25 μm. The slice thickness of the 2D images was 1 mm, which was transverse to the column axis (z). The fluid was doped with 0.1 % CuSO$_4$ to reduce the spin–lattice relaxation time (T_1) of the fluid so that a repetition time of 400 ms could be used in the experiments, resulting an acquisition time of 15 min for each velocity image. Other experimental details can be found elsewhere [20].

Velocity images and profiles at several selected heights are shown in Figure 4.3.6, where the noisy points in the images indicate the air space where a liquid signal was not detected. When the fluid is inside the glass pipette, the velocity profile is nearly Poiseuille and a non-slip boundary condition is almost achieved. This is consistent with one of the early tube flow reports that the 0.5 % w/v solution of

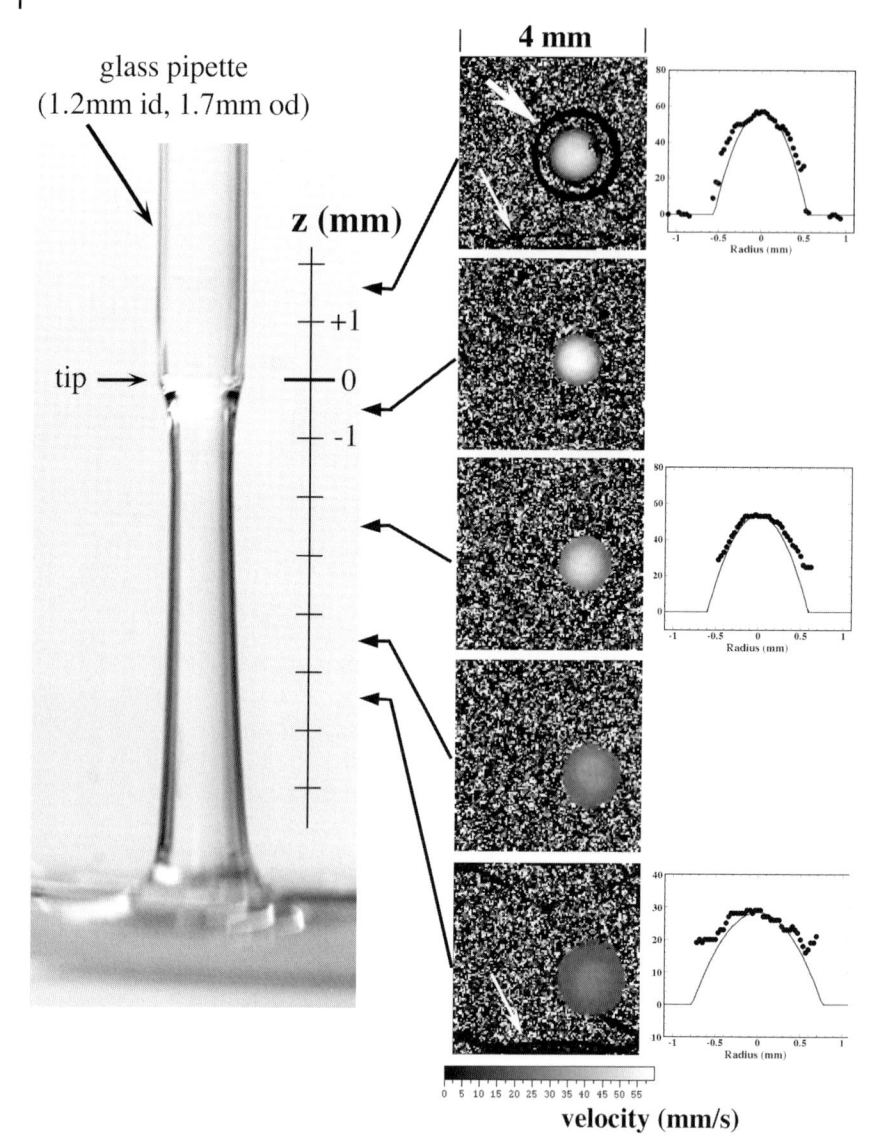

Fig. 4.3.6 Velocity maps and profiles at different heights of the Fano column. The dark ring surrounding the pipe at $z = 1.5$ mm (larger white arrow) is due to a layer of stationary fluid adhering to the pipe exterior following the dipping of the pipe into the reservoir at the start of the experiment. The small white arrows mark the NMR "foldbacks" from the stationary fluid at the inner surface of the fluid reservoir. In the velocity profiles, the solid curves are the calculated Poiseuille profiles in tube flow. Velocity images are reprinted from Ref. [20], with permission from Elsevier.

poly(ethylene oxide) has a power law exponent of 0.85 [18]. (A Poiseuille profile from a Newtonian fluid has a power law exponent of 1.) Below the pipette, the free surface of the tubeless column moves at a finite velocity. Figure 4.3.7 examines the relationship between the radius of the column and the velocity values of the column at different heights along the z axis. It is clear that the column radius increased significantly at increasing distance below the pipette entrance. At the same time, the velocity gradient at different locations inside the column decreased significantly at increasing distance below the pipette entrance. The extensional viscosity of this experiment was calculated to be 60 Pa s, when the average extensional strain rate was about 10 s^{-1} [21]. Given the zero-shear viscosity of 0.09 Pa s for this fluid, the Trouton ratio of the experiment was over 600.

The extensional viscosity characterizes the stretching of the fluid. Although it is a unique property of polymeric materials, it is difficult to measure directly. In this regard, the tubeless siphon method offers two unique advantages. (a) The fluid receives less severe treatment prior to testing. (b) The Fano column has no direct contact with any solid surface hence there is little shearing effect arising as a consequence of the "no-slip" boundary condition, which could complicate any unambiguous measurement. The successful application of non-invasive NMR velocimetry to the study of Fano flow at high resolution demonstrates the effectiveness of using NMR microscopic imaging in the measurement of extensional properties of elastic liquids.

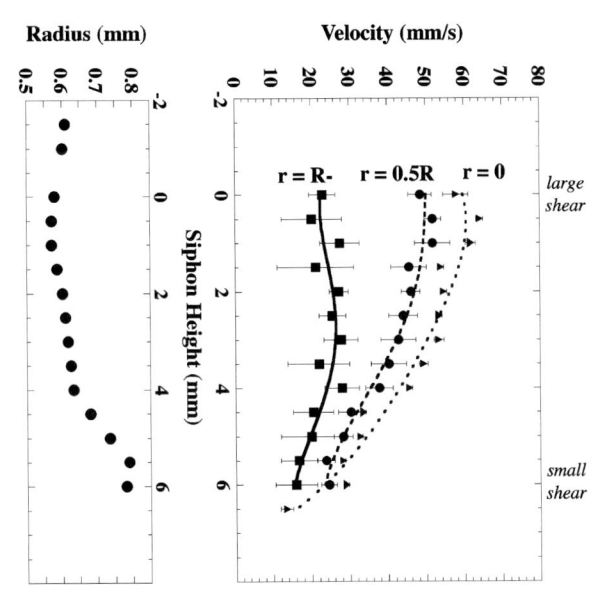

Fig. 4.3.7 The left figure shows the radius of the Fano column as a function of the column height (z), which was enlarged towards the column base. The right figure shows the local velocity values at three different locations (z) of the Fano column as a function of the column height. The three sets of velocity values were measured from the velocity profiles shown in Figure 4.3.6, at the free surface of the Fano column ($r = R^-$), one half radius of the column ($r = 0.5R$) and the center of the column ($r = 0$). The figure clearly shows that there was a large velocity shear near the fluid entrance, and the velocity at the free surface was almost constant.

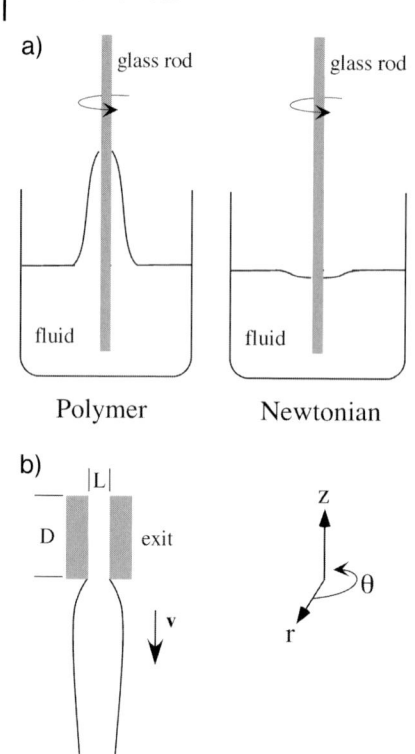

Fig. 4.3.8 Schematics of two extensional flows: (a) rod-climbing and (b) die-swelling.

4.3.5
Other Examples of Viscoelastic Flows

In addition to Fano flow, there are other fascinating flows that belong to the realm of the viscoelastic flow. Two examples are briefly described here, rod-climbing and die-swell. The rod-climbing (Weissenberg) effect describes the elevation or building-up of the fluid near a rotating rod (Figure 4.3.8(a)), in striking contrast to a Newtonian fluid in similar geometry where a slight dip (vortex) will occur in the region near the rod due to the centrifuge forces. A simple proposition for the mechanism of the rod-climbing effect is to consider the stretching of macromolecules around the circular streamlines near the rotating rod. The non-zero effect of the normal stress differences ($\sigma_{\theta\theta} - \sigma_{rr}$, $\sigma_{rr} - \sigma_{zz}$) are such that there will be an inward force on the molecules against the centrifuge forces and also a upward force on the molecules against the gravitational force.

The die-swell (extrudate swell) effect describes the significant expansion of the diameter of the fluid column after exiting from a small pipe (Figure 4.3.8(b)). Some polymer fluids can have a swelling of up to two or three times the exit diameter. A simple proposition for the mechanism of the die-swell phenomenon is that while the fluid is inside the exit pipe, it is subject to a velocity shear, similar to the pipe flow with a maximum shear stress at the wall [18]. This velocity shear stretches macromolecules by different amounts along its streamlines, depending upon the distance of the molecules from the wall. After exiting from the exit pipe, the macromolecules are suddenly free to relax vertically, consequently the diameter of its fluid column increases. An interesting feature of the die-swell phenomenon is the influence of the length-to-diameter ratio (L/D) of the exit pipe on the die-swelling: a large ratio can "delay" the occurrence of the swelling [22]. This feature clearly indicates the relaxations of macromolecular chains in die-swelling. As the die-swell expansion is a steady-state situation, the instrumental set-up similar to the tubeless siphon experiment described in Figure 4.3.5 can be readily adapted to image the amount of the expansion, the curvature of the extruding die-swell column and other viscoelastic properties of the fluid.

In summary, we have commented briefly on the microscopic applications of NMR velocity imaging in complex polymer flows in complex geometries, where these applications have been termed "Rheo-NMR" [23]. As some of these complex geometries can be easily established in small scales, NMR velocimetry and viscometry at microscopic resolution can provide an effective means to image the entire Eulerian velocity field experimentally and to measure extensional properties in elastic liquids non-invasively.

Acknowledgments

Y.X. acknowledges support from the Research Excellence Fund in Biotechnology from Oakland University and R01 grant (AR 45172) from NIH while P.T.C. acknowledges financial support from the Royal Society of New Zealand Marsden Fund and Centres of Research Excellence Fund.

References

1 A. J. Church, W. J. Brodribb (eds.) **1942**, *The Complete Works of Tacitus* (translated), Modern Library Edition for Random House, New York.

2 G. Fano **1908**, (Contributo allo studio dei corpi filanti), *Archivio di fisiologia* 5, 365–370.

3 K. K. K. Chao, M. C. Williams **1983**, (The ductless siphon: A useful test for evaluating dilute polymer solution elongational behavior. Consistency with molecular theory and parameters), *J. Rheol.* 27 (5), 451–474.

4 E. F. Matthys **1988**, (Measurement of velocity for polymeric fluids by a photochromic flow visualization technique: the tubeless siphon), *J. Rheol.* 32 (8), 773–788.

5 R. G. Larson **1999**, (The structure and rheology of complex fluids), *Topics in Chemical Engineering*, ed. K. E. Gubbins, Oxford University Press, New York.

6 H. A. Barnes, J. F. Hutton, K. Walters **1989**, *An Introduction to Rheology*, Elsevier, Amsterdam.

7 A. Abragam **1960**, *The Principles of Nuclear Magnetism*, Clarendon Press, Oxford.

8 E. Fukushima, S. B. W. Roeder **1981**, *Experimental Pulse NMR: A Nuts and Bolts Approach*, Addison-Wesley, Reading, MA.

9 C. P. Slichter **1992**, *Principles of Magnetic Resonance*, 3rd edn, eds. M. Cardona, P. Fulde, K. von Klitzing, H. J. Queisser, Springer Series in Solid-State Sciences, Springer-Verlag, Berlin.

10 P. T. Callaghan **1991**, *Principles of Nuclear Magnetic Resonance Microscopy*, Oxford University Press, Oxford.

11 B. Blümich, W. Kuhn (eds.) **1992**, *Magnetic Resonance Microscopy, Methods and Application in Materials Science, Agriculture and Biomedicine*, Weinheim, VCH.

12 E. O. Stejskal, J. E. Tanner **1965**, (Spin diffusion measurements: Spin echoes in the presence of a time-dependent field gradient), *J. Chem. Phys.* 42 (1), 288–292.

13 Y. Xia **1996**, (Contrast in NMR imaging and microscopy. Concepts in magnetic resonance), 8 (3), 205–225.

14 P. T. Callaghan, Y. Xia **1991**, (Velocity and diffusion imaging in dynamic NMR microscopy), *J. Magn. Reson.* 91, 326–352.

15 C. F. Jenner, Y. Xia. C. D. Eccles, P. T. Callaghan **1988**, (Circulation of water within wheat grain revealed by nuclear magnetic resonance micro-imaging), *Nature (London)* 336, 399.

16 Y. Xia, P. T. Callaghan **1992**, ("One-shot" velocity microscopy: NMR imaging of motion using a single phase-encoding step), *Magn. Reson. Med.* 23 (1), 138–153.

17 W. Köckenberger, J. M. Pope, Y. Xia, K. R. Jeff, E. Komor, P. T. Callaghan **1997**, (A non-invasive measurement of phloem and xylem water flow in castor bean seedlings by nuclear magnetic resonance microimaging), *Planta* 201, 53–63.

18 Y. Xia, P. T. Callaghan **1991**, (Study of shear thinning in high polymer solution using dynamic NMR microscopy), *Macro. Mol.* 24 (17), 4777–4786.

19 Y. Xia, K. R. Jeffrey, P. T. Callaghan **1992**, (Imaging velocity profiles: flow through an abrupt contraction and expansion), *AIChE J.* 38 (9), 1408–1420.

20 Y. Xia, P. T. Callaghan **2003**, (Imaging the velocity profiles in tubeless siphon flow by NMR microscopy), *J. Magn. Reson.* 164 (2), 365–8.

21 P. T. Callaghan, Y. Xia **2004**, (Nuclear magnetic resonance imaging and velocimetry of Fano flow), *J. Phys. Condens. Matter* 16, 4177–4192.

22 R. B. Bird, R. C. Armstrong, O. Hassager **1987**, *Dynamics of Polymer Liquids*, 2nd edn, John Wiley & Sons, New York.

23 P. Callaghan **1999**, (Rheo-NMR: nuclear magnetic resonance and the rheology of complex fluids), *Rep. Prog. Phys.* 62, 599–668.

4.4
Quantitative Visualization of Taylor–Couette–Poiseuille Flows with MRI[+]
John G. Georgiadis, L. Guy Raguin, and Kevin W. Moser

4.4.1
Introduction

When a viscous fluid is placed in an annulus between a rotating inner cylinder and a fixed outer cylinder with a zero externally imposed axial pressure gradient, and when the rotation speed of the inner cylinder exceeds a critical value, the flow reorganizes spontaneously in toroidal counter-rotating Taylor vortices (Figure 4.4.1). This is known as the Taylor–Couette problem, which is of archival significance because it marks the first instance when the predictions of the hydrodynamic stability theory matched the experiment [1]. The previous becomes the Taylor–Couette–Poiseuille flow [2] when an axial flow is superimposed. The name is due to the persistence of Taylor vortices and to the underlying stable flow that consists of a superposition of Couette and Poiseuille flow in the annular gap, which is naturally termed Couette–Poiseuille flow. We will refer collectively to all swirling annular flows involving a superposed axial flow as *Taylor–Couette–Poiseuille (TCP)* flows. Such flows are ubiquitous in many solid–liquid and liquid–liquid mixing processes (stirring, blending, dissolution, agitation and emulsification) and play an important role in a variety of engineering fields, including polymer processing, chemical separations, bioreactors, oil extraction and food engineering [3]. Given that the generation of turbulent flow fields in such systems results in excessive energy dissipation and possible damage to fluids consisting of long molecular chains, laminar flows provide a viable alternative for mixing enhancement. In laminar flow in a straight tube, mixing is primarily induced by the relatively slow molecular diffusion process [4]. The secondary flows in Taylor–Couette systems increase transverse mixing and produce more uniform concentration distributions in cross-sectional planes.

Laminar mixing can be classified in terms of two regimes: regular or chaotic [5]. In regular mixing, fluid particle trajectories are integrable over the entire geometry. For example, the streamlines associated with the secondary flow generated by Taylor vortices constitute an integrable system (using dynamical systems nomenclature), meaning that a particle remains on a unique streamline unless its trajectory is altered by molecular diffusion. In chaotic mixing, a number of pathlines become non-integrable and this behavior is associated with chaotic trajectories of certain fluid particles. The flow regime associated with chaotic mixing is often called chaotic advection, or *Lagrangian* turbulence, referring to coordinates which move with the tracked particle. The mixing properties of chaotic advection are often compared with those associated with *Eulerian* turbulence, the latter characterized by chaotic behavior relative to coordinates fixed in space. Recent studies of chaotic mixing in spatially periodic systems include continuous flow systems, such as the partitioned pipe mixer [6] and twisted pipe [7]. The partitioned pipe mixer consists of a series of rectangular plates held stationary inside a rotating tube, while the twisted pipe is a sequence of half-tori, each rotated by a pitch angle with respect to its neighbors. Laminar chaotic mixing has received a lot of attention in the chemical engineering literature [8].

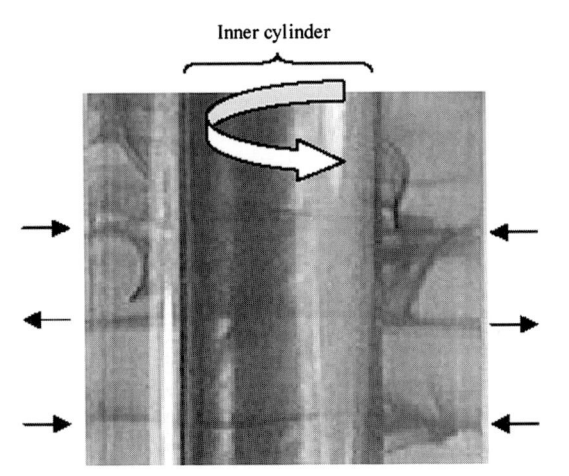

Fig. 4.4.1 Visualization of Taylor vortices by injecting dye. Arrows denote the flow direction of fluid jets formed between the toroidal recirculating cells stacked along the axial direction.

With the above discussion as motivation, we hope to demonstrate that the further study of the kinematics of Taylor–Couette flows offers a number of benefits for the scientist and engineer. Firstly, it provides the opportunity to introduce geometric (or topologic [9]) methods in fluid mechanics and transport phenomena. In transport phenomena such methods have not yet enjoyed in the attention they continue to receive in the field of dynamical systems, owing in part to the preference for a Eulerian point of view in fluid mechanics, as opposed to the Lagrangian point of view, which is standard in dynamics. Both the numerical simulation and the extraction of correct Lagrangian properties of flows rely on the reconstruction of kinematically admissible velocity fields, that is, fields that conserve volume. Secondly, TCP flows provide an excellent canonical problem for the study of the interplay between chaos and order in solid–liquid or liquid-liquid two-phase flows. The review manuscript by Ottino et al. [10] and recent studies of specialized flows [11–12] inspired us to introduce a new question: can we use chaotic phenomena by modifying the TCP flow to enhance the segregation of a dispersed particulate phase, that is, to "unstir"?

The objective of this chapter is to demonstrate the marriage of Magnetic Resonance Imaging (MRI) with nonlinear dynamics in order to address the above question. Specifically, we apply MRI to study the hydrodynamics of a *complex* Taylor–Couette–Poiseuille flow mode and then use dynamical methods to investigate the effect of flow *perturbation* on the passive transport of dispersed particulate matter. Henceforth, MRI refers to ^{1}H-MRI, a very reasonable assumption given the preponderance of the proton nucleus in most applications involving liquids. The majority of classical quantitative methods used to probe fluids rely on optical techniques, such as Particle Image Velocimetry [13], which gives "planar" (*en face*) information. Despite their excellent spatial and temporal resolution, the ability to measure velocity in topologically complex geometries remains limited as such methods rely on matching the index of refraction in the field of view. Additionally, the extraction of mass concentration information requires the introduction of an exogenous contrast agent, for example, through dye injection [6] or electrolysis [14]. Unlike MRI, such methods do not allow the estimation of local mass flux and the tensorial characteristics of the diffusion or dispersion coefficients. MRI is a superior methodology because, in addition to being completely non-invasive, it affords a plethora of contrast mechanisms [15–16]. Although this is the scope of earlier chapters in this volume, it is useful to briefly review some MRI methodology here for completeness. Classical Fourier Transform MRI methods collect one signal (echo) for each radiofrequency (rf) excitation pulse (spin-echo and gradient-echo imaging). These methods can be speeded up by using small tip angle, e.g., the radiofrequency-spoiled Fast Low-Angle SHot technique (FLASH). A number of fast (subsecond) MRI methods acquire the whole image after a single RF excitation by using multiple echoes. Such single-shot methods include echo planar imaging (EPI), spiral trajectory imaging (Spiral EPI), Rapid Acquisition with Relaxation Enhancement (RARE) and GRadient-and-Spin-Echo imaging (GRASE). Fukushima [17] reviewed the application of such methods in fluid mechanics up to 1999, and Mantle and Sederman [18] provided an excellent review of dynamic MRI methods up to 2003, with emphasis on chemical engineering applications. It is important to note at this juncture that we are discussing fast MRI in terms of the speed of resolving a dynamic "object" (such as an arbitrarily varying scalar field or an unsteady velocity field), and not a periodic or quasi-periodic object, by completing the image acquisition phase with requisite temporal resolution. Many of the so-called real-time (or cine-) MRI methods rely on acquiring parts of the image over several periods [19] and therefore can only resolve time-periodic or quasi-periodic flows.

Several previous studies have applied MRI velocimetry techniques to steady cylindrical Couette flows. For example, Hopkins et al. [20] used a stroboscopic sampling technique to study the stable, two-dimensional flow generated between a fixed inner cylinder and a rotating outer cylinder. Pulsed field gradient motion encoding techniques (q-space MRI) were later used by Hanlon et al. [21] to measure fluid velocities in their narrow-gap, temperature-controlled Couette flow rheometer. Seymour et al. [22] provided a detailed study of steady Taylor vortices with a combination of k-space and q-space MRI. While these studies provided important insight into stable Couette flows, there has been little application of MRI flow imaging techniques in complex TCP flows. To do so, fast imaging methods must be utilized so that hydrodynamic instabilities can be probed. Kose [23] used EPI to achieve acquisition times of 41 ms in order to acquire velocity information in Taylor–Couette flows at rotation speeds up to eight times the critical Taylor number, which corresponds to the unsteady wavy vortex regime. Sederman et al. [24] used a rotationally-compensated RARE technique to image the deformation of a water droplet in silicone oil sheared in a wide-gap Couette cell rotating at less than 120-revolutions per minute. Our group has used MRI to probe instabilities in Taylor–Couette flows [25–27] and quantified the accuracy of MRI velocimetry in a number of complex or rotating phantoms: tubes with stenosis [28–30], Rushton-type mixers [31] and packed beds [32].

In the following, we highlight the application of MRI spin tagging methods in the study of a specific TCP flow and explore the consequences of flow perturbation in stirring or segregating dispersed particles. By exploiting the spatial symmetry of the flow field, a combination of MRI and numerical methods is used to extract vectorial information about the flow. As spatial symmetries do not occur very often, the search continues for better spatial and temporal resolution MRI techniques for the quantification of complex flows. We will conclude this chapter with the preview of a couple of promising new techniques aimed at accelerating MRI velocity acquisition in rotating flows.

4.4.2
Taylor–Couette–Poiseuille Flow

4.4.2.1 Fundamental Hydrodynamics
The study of laminar vortical structures has been well documented in the Taylor–Couette (TC) flow for an extensive range of gap sizes [33]. Of particular interest are the experimentally determined boundaries in the map of flow regimes [34], and specifically the flow regime where the propagating helical vortex and stationary toroidal vortex modes become stable simultaneously and interact in the axisym-

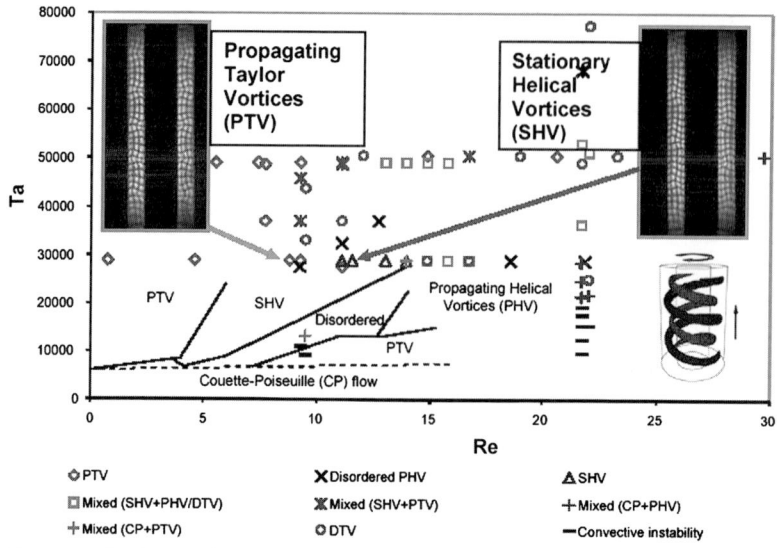

Fig. 4.4.2 The discrete data points represent Taylor–Couette–Poiseuille flow regimes observed with MRI for $\eta = 0.5$ [41]. The curved boundaries were obtained for $\eta = 0.77$ with optical techniques [38]. The two inserts show MRI spin-tagging FLASH images of the SHV and PTV hydrodynamic modes.

metric high-aspect ratio TC apparatus. The bifurcation diagrams in the vicinity of this so-called codimension-2 point [9] indicate the possibility for coexistence of these two modes for wide-gap annuli [35, 36]. Considering the analogy with TC flows presents fresh opportunities to study complex nonlinear interactions between various hydrodynamic modes at relatively low rotation rates (i.e., in the neighborhood of the primary bifurcation) in the TCP problem.

The flow domain of TCP can be described by two dimensionless hydrodynamic parameters, corresponding to the rotational speed of the inner cylinder and the imposed axial flow rate: the Taylor number, Ta, and the axial Reynolds number, Re, respectively:

$$Ta = \frac{4\eta^2}{1-\eta^2} \left(\frac{\Omega_1}{\nu}\right)^2 (R_2 - R_1)^4$$

$$Re = \frac{(R_2 - R_1)U}{\nu}$$

(4.4.1)

where R_1 and R_2 are the inner and outer cylinder radii, respectively, Ω is the inner cylinder angular velocity, U denotes the mean axial velocity imposed by the through-flow and ν is the fluid kinematic viscosity. TCP is characterized by two additional geometrical parameters: the ratio of the inner cylinder radius to the outer one, $\eta = R_1/R_2$, and the ratio of the length L of the cylinders to the annulus gap width $(d = R_2 - R_1)$, termed the aspect ratio $\Gamma = L/d$.

4.4 Quantitative Visualization of Taylor–Couette–Poiseuille Flows with MRI† | 421

This section focuses on steady and unsteady hydrodynamic modes that emerge as the rotational speed of the inner cylinder (expressed by Ta) and pressure-driven axial flow rate (scaled by Re) are varied, while the outer cylinder is kept fixed. These modes constitute primary, secondary and higher order bifurcations, which break the symmetry of the base helical Couette–Poiseuille (CP) flow and represent drastic changes in flow structure. Figure 4.4.2 presents a map of observed hydrodynamic modes in the (Ta, Re) space, and marks the domain where all of the hydrodynamic modes that interest us appear. We will return to this figure shortly.

4.4.2.2 MRI Velocimetry and Data Reduction

The logistics of conducting MRI velocity measurements in TCP are somewhat complicated due to the need to minimize magnetic susceptibility gradients in the experimental apparatus and to avoid interferences from the torque-producing system. Figure 4.4.3 gives the experimental set-up used for all the MRI experiments discussed here. The measurements are performed on a 4.7-T horizontal scanner with gradient capabilities up to 6.5 G cm^{-1} and rise time of 500 μs. The rf coil used in the experiments is a 6 cm id laboratory-built birdcage coil. The TCP test section is fabricated from MRI-compatible materials with $\eta = 0.5$ (inner radius $R_1 = 0.9525$ cm) and aspect ratio $\Gamma = 16$ (length $L = 15.24$ cm). Both cylinders are made from poly(methyl methacrylate), while the inlet/outlet ports and strainers were fabricated as a single piece from a stereolithography resin. Rubber lip seals and glass ball bearings are chosen to ensure smooth rotation of the inner cylinder and no leakage. Water enters the annulus with an imposed flow rate through perforated disks that are fastened to the outer stationary cylinder, while the inner cylinder rotates with a specified angular velocity. The axial flow is maintained by

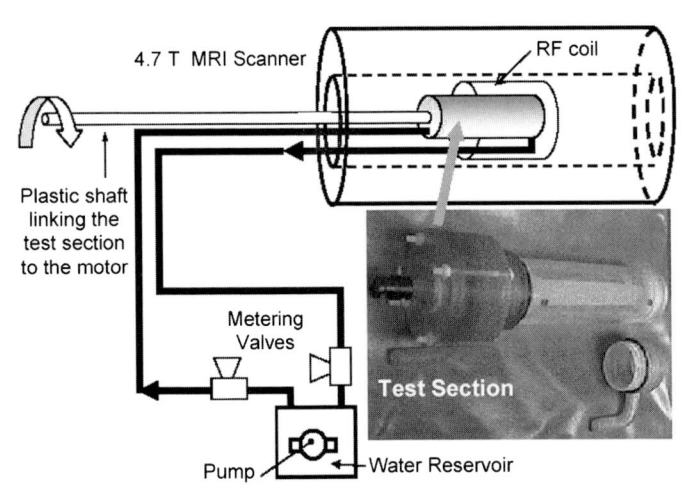

Fig. 4.4.3 Experimental set-up for the MRI investigation of Taylor–Couette–Poiseuille flow. The electrical motor driving the shaft, as well as the pump, are placed 7.3 m away from the scanner to avoid interference with the magnetic fringe field.

pumping water in a closed loop, while the inner cylinder is rotated via a stepper motor placed at a distance of 7.3 m from the MRI magnet, behind the 5-G line of the fringe field. Three 2.44-m long, 6.35-mm diameter plastic shafts linked together with universal plastic joints deliver power to the rotating shafts. The shafts are aligned with the scanner axis (approximately 0.9 m above the floor) and supported by six stands to prevent excessive bending. This distance has proven to be adequate to eliminate all interference between the motor and pump and the MRI scanner during imaging.

A spin-tagging technique was used to visualize the different flow regimes encountered as Ta and Re were varied. This MRI technique involves tagging nuclear spins by modulating the longitudinal spin polarization in two orthogonal directions, so that a Cartesian grid can be formed. In an image acquired immediately after the tagging, the grid appears undistorted. However, the grid deforms as the tags evolve with the flow. Fluid motion between the first and subsequent acquired images is extracted from the displacement and distortion of the grid. This allows a time-of-flight approach to the estimation of local velocity. By acquiring images every 0.5–1 s, direct visualization of flow patterns can be obtained in the form of cinematography [28]. Using this pulse sequence, tagged images of the TCP flow are acquired in both longitudinal and transverse slices. The rf coil length and T_1 relaxation time generally determine the upper limit on the flow evolution time. To observe the distorted grid, the tagged spins must remain within the rf coil over the total experiment time (flow evolution plus imaging times). In addition, the magnetization of the tagged spins returns exponentially to its rest state according to the T_1 relaxation times. With pure water, the flow evolution time can be as high as 3 s if the tagged spins remain in the rf coil.

The imaging of the tagged grids is performed via one of two imaging techniques on a number of longitudinal (meridian plane) and transverse (plane normal to axis) slices of the TCP test section. A classic spin-echo imaging sequence is used mainly for steady flows as it requires an acquisition time of the order of 6 min for a slice with a 256×256 pixel resolution. The snapshot FLASH (Fast Low Angle SHot) imaging technique (described in Figure 4.4.4) is used for the unsteady flows, with an acquisition time of 0.25 s for a 128×96 resolution [28]. Both sequences produce high contrast snapshots of material lines evolving from the originally straight Cartesian gridlines, but the second has clearly lower signal-to-noise ratios. Consecutive MRI scans are obtained for set values of the rotation rate (Ta) and flow rate (Re), and are then repeated by moving to the next set point by increasing the axial flow rate through the apparatus. A map summarizing all our results is given in Figure 4.4.2, where the nomenclature used to characterize the various flow regimes is consistent with Ref. [38]. For appropriate Re values, the base helical flow (Couette–Poiseuille) undergoes a Hopf-bifurcation to a state characterized by propagating toroidal vortices (PTV), which are translating along the cylinder axis with constant speed [14]. As the axial flow rate increases, the PTV regime is followed by the stationary helical vortices (SHV) regime and mixed regimes featuring disordered moving helices or PTVs.

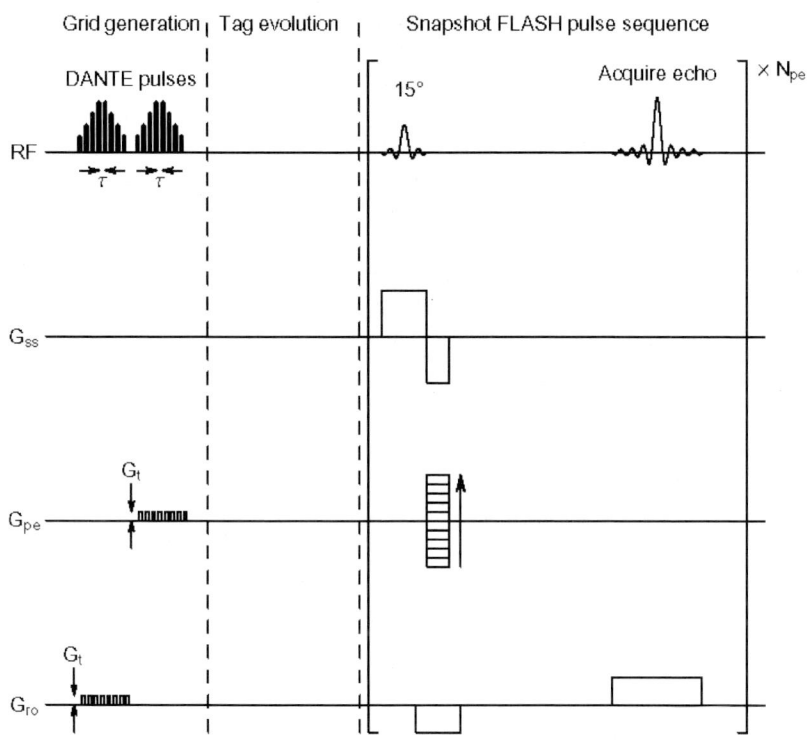

Fig. 4.4.4 MRI pulse sequence for spin-tagging velocimetry with tagging gradient G_t and snapshot FLASH imaging. A Cartesian grid of tagged fluid is generated by synchronizing two series of hard rf pulses (DANTE combs) with the tagging gradients applied along two orthogonal directions (pe = phase encoding, ro = readout) along the imaging slice. The separation of grid lines is $2\pi/(\gamma G_t \tau)$, where γ is the gyromagnetic ratio and τ is the delay between individual DANTE pulses. Following a judiciously chosen evolution period, the tags (now displaced by the flow) are imaged on the selected slice by the FLASH sequence [25, 26, 37]. The sequence shown in brackets needs to be repeated N_{pe} times, where N_{pe} is the resolution in the phase encoding direction.

The focus of this section is limited to the flow regimes observed for one particular Taylor number ($Ta^{1/2} = 170$) as Re is increased. As the arrows in Figure 4.4.2 indicate, the flow evolves from PTV, to a pure SHV regime, and finally to a spatially mixed state of SHV and PTV. The spatially mixed state of SHV and PTV corresponds to SHV filling the annular cavity starting from the inlet, and gradually converting to PTV downstream. Using the FLASH imaging sequence described in Figure 4.4.4, an oscillating instability growing in the downstream direction is observed in the out-flowing boundary jets forming between the counter-rotating of the SHV. Figure 4.4.5 demonstrates the gradual "blurring" of spin-tagging images acquired with the spin-echo sequence as the axial velocity is increased and SHV transitions to a mixed state of SHV and PTV.

The SHV flow regime has been studied systematically using the spin-echo and FLASH imaging sequences [39]. Figure 4.4.6 describes the steps for extracting velocity from the displacement of the material lines on a transverse slice obtained with classic spin-echo for the SHV mode. We start by comparing the image of Figure 4.4.6(a) with the undistorted (Cartesian) grid lines. The undistorted grid is easily recreated by using the corresponding pairs of stationary tags on the outer cylinder as markers. The in-plane velocity components are then estimated by computing the displacement of the grid intersection points (Figure 4.4.6(b)).

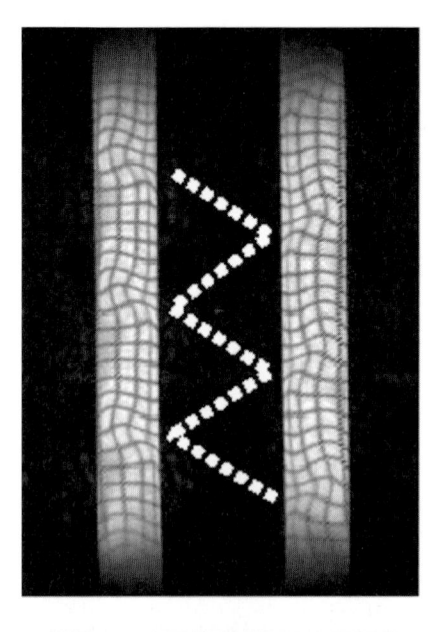

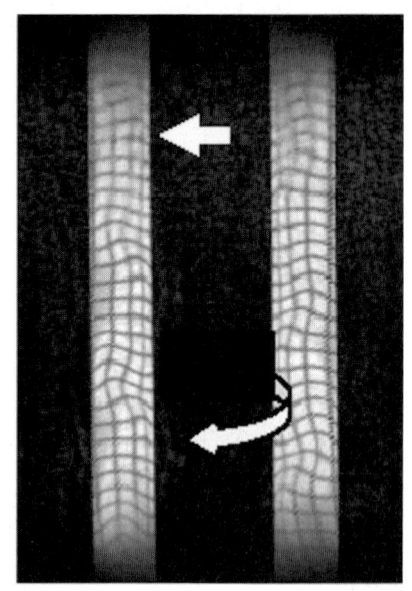

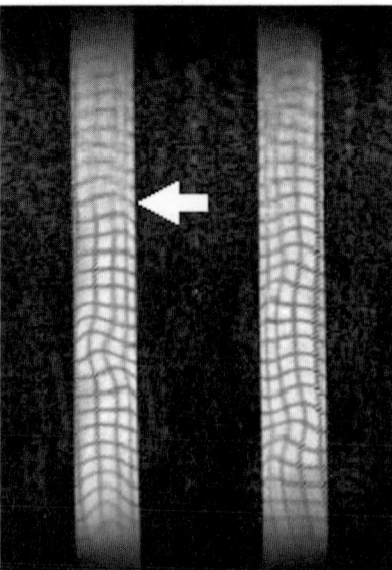

Fig. 4.4.5 Gradual blurring (staring on locations marked by arrow) of MRI spin-tagging spin-echo images of Taylor–Couette–Poiseuille flow as the axial flow is increased (from left to right). The images correspond to longitudinal sections of the flow and the axial flow is upwards. The dashed line marks the location of one of the stationary helical vortices which characterize the SHV mode. This flow regime corresponds to the transition from the SHV (steady) to partial PTV (unsteady) regimes as Re increases, as shown in Figure 4.4.2.

4.4 Quantitative Visualization of Taylor–Couette–Poiseuille Flows with MRI

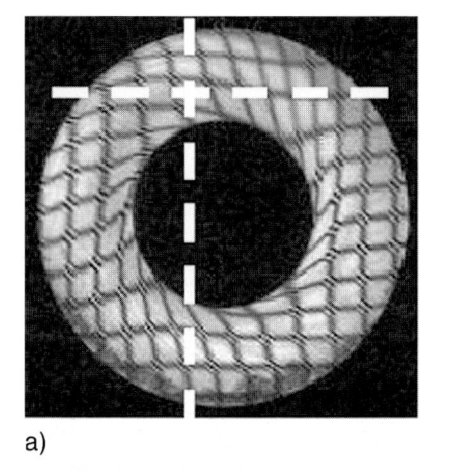

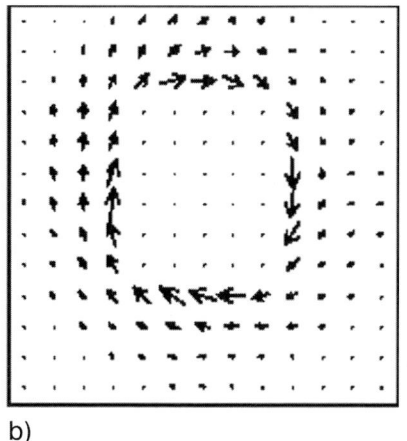

Fig. 4.4.6 Fluid velocity extraction from MRI spin-tagging in Taylor–Couette–Poiseuille flow (SHV mode). (a) Spin-echo image of deformed spin-tagged grid on a transverse slice. The crossed dash lines correspond to a pair of undistorted grid lines connecting stationary tags on the outer cylinder. (b) Evaluation of in-plane velocity on a transverse slice by computing the displacement of the grid line intersections. (c) Reconstruction of spin-tagged grid after the full 3D SHV field is reconstructed. Note the excellent agreement with (a).

Upon the reconstruction of the full 3D velocity field, the deformation of the Cartesian grid is recreated on selected slices (Figure 4.4.6(c)) in order to verify that the algorithm is properly implemented.

Longitudinal images reveal the helical structure of the SHV flow as outflow boundaries on one side match the inflow boundaries on the other, and are used to measure the axial wavelength λ (defined as the axial length of a pair of counter-rotating vortices). Transverse images taken at five different axial positions are nearly identical within the proper "screw" symmetry group. We have essentially verified that the flow field remains invariant as it is mathematically rotated around and translated along the axis with the proper "pitch" λ [39]. The 3D-velocity field for SHV flow is then reconstructed from the grid deformation in transverse and longitudinal sections, as explained in Ref. [27] and is represented in Figure 4.4.7(a) by volumetrically rendering the two key streamtubes. Given the helical symmetry, one slice can provide all the topological information about the SHV mode. Figure 4.4.7(b) gives the SHV *flow portrait* in terms of streamlines on the longitudinal slice. An important characteristic of the SHV flow portrait is the existence of periodic orbits of fluid particles with trajectories covering the surfaces of the family of nested helical tubes contained in the two counter-rotating recirculation cells depicted in Figure 4.4.7(a). It is important to note that the reconstructed SHV velocity field has been validated numerically by solving the Navier–Stokes equations in TCP for the appropriate parameters: $\eta = 0.5$, $Ta^{1/2} = 170$ and $Re = 11.14$ [27, 41]. The computed streamlines are superimposed on the longitudinal planes depicted in Figure 4.4.7(b). We are now ready to model a perturbation of the flow in the vicinity of the SHV mode, which is consistent with the experimental observations of SHV–PTV transitions, and to study its effect on the dispersion of any particulates placed in the flow.

4.4.2.3 Particle Dispersion in Oscillatory Taylor–Couette–Poiseuille Flow

Let us consider the SHV mode of the TCP flow as the base state of a dynamical system described by three velocity components V_r, $V\theta$ and V_z relative to the fixed cylindrical co-ordinate system depicted in Figure 4.4.7(b). This dynamical system is described mathematically by the equations of motion of the particle trajectories in the 3D (r, θ, z) coordinate system:

$$\begin{cases} \dfrac{dr}{dt} = V_r(r, \theta, z) \\ \dfrac{d\theta}{dt} = \dfrac{1}{r} V_\theta(r, \theta, z) \\ \dfrac{dz}{dt} = V_z(r, \theta, z) \end{cases} \quad (4.4.2)$$

The helical symmetry of the velocity field makes it possible to consider Poincaré maps for the equivalent dynamical system. Indeed the velocity field can be fully described by giving the velocity vector in a longitudinal half section, for example, the $r > 0$ part of the (r, z)-plane at $\theta = 0$. As the reconstructed flow field is periodic in the axial direction with wavelength λ, the construction of a Poincaré map can be accomplished by collecting the points at which fluid trajectories intersect a set of longitudinal (r, z)-subdomains at $\theta = 0$, spaced λ distance apart along z, as shown in Figure 4.4.7(b). For example, starting with a particle P located on the longitudinal subdomain $[r, 0 < z < \lambda]$, the next intersection R of its trajectory with at $\theta = 0$ occurs on the longitudinal sub-domain centered at $z = (1 + 1/2)\lambda$, while the nth intersection occurs at the subdomain centered at $z = (n + 1/2)\lambda$, where n is an integer. The Poincaré map can be built by projecting all such intersection points back onto the longitudinal subdomain $[r, 0 < z < \lambda]$ at $\theta = 0$, and then omitting the initial location. If a cloud of particles are placed at various initial positions, construction of the Poincaré map can be used to traverse the 2D phase space of the dynamical system given by Eq. (4.4.2). The symmetry of the flow guarantees [40] that there exists a local change of variables $(r, \theta, z) \rightarrow (\varrho, \zeta, z)$ so that system (4.4.2) becomes

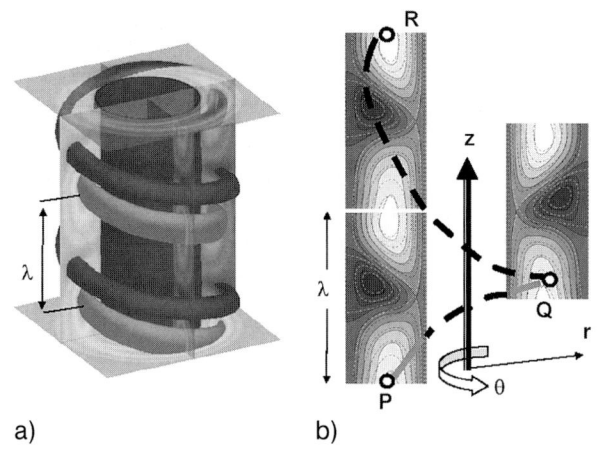

Fig. 4.4.7 (a) Reconstruction of the Stationary Helical Vortex (SHV) mode from MRI data acquired with the spin-tagging spin-echo sequence [27]. The axial flow is upwards and the inner cylinder is rotating clockwise. The two helices represent the counter-rotating vortex streamtubes. (b) Construction of Poincaré map for SHV [41]. The orbit of a typical particle is represented by the curve PQR wrapped around the z axis. The streamlines of the counter-rotating vortices are represented by streamfunction contours (solid lines) on each longitudinal plane (r, z). The superimposed dashed line contours show the numerical predictions for SHV [27].

$$\begin{cases} \varrho = \dfrac{r^2}{2}, & \dfrac{d\varrho}{dt} = \dfrac{\partial H(\varrho,\varsigma)}{\partial \varsigma} \\ \varsigma = z + \dfrac{\lambda}{2\pi}\theta, & \dfrac{d\varsigma}{dt} = -\dfrac{\partial H(\varrho,\varsigma)}{\partial \varrho} \\ & \dfrac{dz}{dt} = V_z\left(\sqrt{2\varrho},\varsigma\right) \end{cases} \quad (4.4.3)$$

where H is a Hamiltonian function, which is related to the SHV streamfunction Ψ by a simple coordinate transformation: $H(\varrho,\varsigma) = \Psi(r,\varsigma)$.

As a first step towards the analytical study of the dynamics of the mixed PTV–SHV regime in the vicinity of the SHV base flow, we consider a non-autonomous perturbation of system (4.4.3) as follows:

$$\begin{cases} \dfrac{d\varrho}{dt} = \dfrac{\partial H(\varrho,\varsigma,\phi)}{\partial \varsigma} \\ \dfrac{d\varsigma}{dt} = -\dfrac{\partial H(\varrho,\varsigma,\phi)}{\partial \varrho} \\ \dfrac{d\phi}{dt} = f_\varepsilon \end{cases} \quad (4.4.4)$$

This corresponds to a Hamiltonian system which is characterized by a weak oscillatory perturbation of the SHV streamfunction $\Psi(r, \varsigma) \rightarrow \Psi(r, \varsigma) + \varepsilon\Psi_1(r, \varsigma) \times \sin(f_\varepsilon t)$. The equations of fluid motion (4.4.4) are used to compute the inertial and viscous forces on particles placed in the flow. Newton's law of motion is then integrated in 3D for a neutral particle (which is identical to a fluid particle) and a particle with twice the density of the fluid, both subjected to a Stokesian viscous drag as they interact with the flow. Figure 4.4.8 shows the Poincaré maps of these particles obtained from the computed particle trajectories in a flow with a perturbation amplitude of $\varepsilon = 0.5$ and two frequencies, $f_\varepsilon = 0.16$ Hz and $f_\varepsilon = 0.80$ Hz, for a specific initial particle placement. We should note in passing that both these excitation frequencies are outside the range of the "natural" frequencies for the periodic orbits in the unperturbed SHV mode, and that the dimensionless amplitude $\varepsilon = 0.5$ corresponds to a perturbation energy that is only 3.5% of the average kinetic energy of the SHV flow.

The "speckle" patterns of Figure 4.4.8(a) and (c) indicate that the oscillatory flow perturbation induces chaotic trajectories for both the neutral and heavy particles, and this can be verified by examining carefully the detailed trajectories of each particle [41]. This behavior of the particles placed in the deterministic flow given by Eq. (4.4.4) is a manifestation of Lagrangian chaos. Figure 4.4.8(a) and (c) shows that by increasing the frequency, the chaotic regions become limited and the phase space contracts. The Poincaré sections for the neutral particle driven by the higher frequency show a mixture of regular and chaotic regions, while for the heavy particle, these regions become confined to the centers of the counter-rotating vortices. By considering a large sample of initial particle locations, similar results are obtained for a cloud of particles [41]. Such observations demonstrate that by controlling the excitation frequency, the perturbed SHV system can be "tuned" to stir or to unstir particles which are dispersed in the flow.

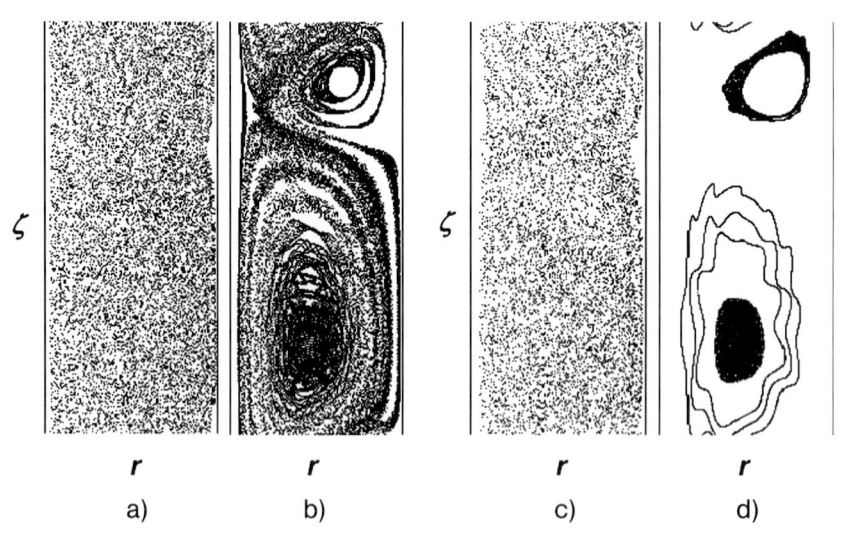

Fig. 4.4.8 Poincaré sections on the (r, ζ) plane for $\varepsilon = 0.5$. (a) Neutrally buoyant particle with $f_\varepsilon = 0.16$ Hz. (b) Neutrally buoyant particle with $f_\varepsilon = 0.80$ Hz. (c) Heavy particle with $f_\varepsilon = 0.16$ Hz. (d) Heavy particle with $f_\varepsilon = 0.80$ Hz. Heavy particles have twice the density of the fluid.

4.4.3
Future Directions

In this section, we review some emerging techniques for the imaging of rotating fluid systems in terms of both software and hardware innovations in MRI. Our group has developed the framework of a procedure to speed up the MRI velocity acquisition in TCP flows by using "prior knowledge" to reduce the amount of flow encoding [42]. Essentially, we augment the MRI acquisition phase by a velocity reconstruction step, which is based on fluid mechanical constraints, that is, on equations governing the fluid motion (Figure 4.4.9(a)). Let us demonstrate the procedure by decomposing the Taylor–Couette velocity field near the first bifurcation in a series of spatial basis functions that obey the flow kinematics (boundary conditions and divergence-free velocity field). Exploiting the z periodicity, we can represent the axial velocity component as follows:

$$u_z = \sum_{i=1}^{N'} \sum_{k=1}^{M'} \left[F'_{i,k}(t)\cos\left(\frac{2\pi k}{\lambda}\varsigma\right) + G'_{i,k}(t)\sin\left(\frac{2\pi k}{\lambda}\varsigma\right) \right] \frac{1}{R_2 r} T'_i(\hat{r}) \quad (4.4.5)$$

As the radial functions T'_i are chosen to satisfy the imposed kinematic constraints, one needs only determine the time-dependent prefactor F' and G' values and therefore less k-space acquisition is needed. The left panel of Figure 4.4.9(b) shows the axial velocity component on a longitudinal slice of the classical Taylor–Couette flow obtained by phase-contrast MRI velocimetry. The measurements justify the use of the z symmetry and axisymmetry [implied by Eq. (4.4.5)]. In the right panel of Figure 4.4.9(b), the arrows represent the 2D velocity measurements with MRI and the contours show the reconstructed streamlines. We have shown [42] that by using velocity decompositions like that given by Eq. (4.4.5), one can reduce the data acquisition by a factor of 4 with only 1% loss of velocity accuracy – which is higher than the accuracy of MRI velocimetry *per se* [28, 29]. It is worthwhile noting that kinematically constrained MRI methods also help with the faster imaging of non-Newtonian flows in TCP, as so far the constraints do not involve any assumptions about the fluid rheology. In addition to accelerating MRI data acquisition, this approach creates a framework for order reduction via modal decomposition. This reduction is extremely helpful in the study of hydrodynamic instabilities, such as the one described in the previous section. With respect to reducing the MRI encoding requirements for flows in more complex geometries, we have developed methods that incorporate coarse velocity data into the computational reconstruction of the velocity field in a way that it obeys both flow kinematics and dynamics [43].

We conclude with a discussion of hardware innovations that show promise. Conventional "ultrafast" MRI obtains spatial resolution through repeated applications of magnetic field gradients of different strengths during successive repetitions of the basic MRI experiment, successive echo acquisitions, or some combination of the two. Advances in gradient systems and pulse sequence design have enabled much faster imaging techniques, greatly improving the ability of MRI to image dynamic processes. However the use of recalled echoes for phase encoding

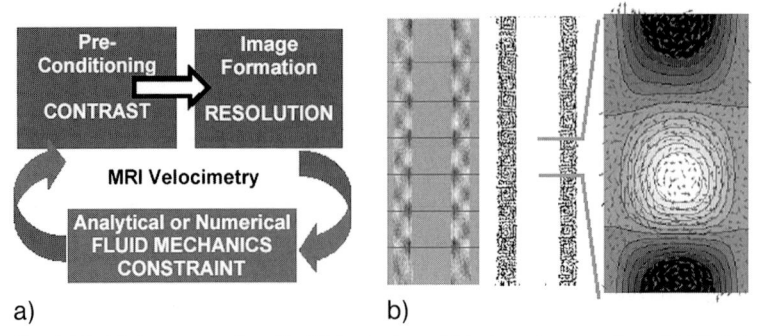

Fig. 4.4.9 (a) A new paradigm for MRI velocimetry based on prior knowledge. The introduction of mathematical constraints imposed by physical laws governing the flow allows the reduction of the encoding phase as well as more efficient post-processing of MRI data. (b) Efficient reconstruction of Taylor–Couette cells on the longitudinal plane (r, z) starting from coarse MRI velocity measurements obtained with phase-contrast velocimetry.

carries the penalty of temporal blurring, as the data from which the image is formed are obtained over a period of acquisition of multiple echoes. *Partially parallel imaging methods such as SMASH and SENSE have been very effective in reducing imaging times without increasing the rate of gradient switching [44–49].* Instead, these methods reduce imaging time by simultaneously receiving MR signals from *multiple elements* in an array of sensors. Using arrays of four to eight coils, acceleration factors of two to three have become common in clinical applications.

Striving to push partially parallel imaging to its natural limit, Wright et al. [50–51] have demonstrated the acquisition of an entire image in a *single echo*. Because an entire image is acquired in a single signal acquisition, the Single-echo Acquisition MRI technique provides an extremely fast "shutter speed" for snapshot flow imaging. Spatial localization in one direction is accomplished by using a surface coil consisting of a 64-element array of parallel, narrow (strip) elements. The second direction and slice selection are achieved using conventional gradients. A prototype 64-channel MRI receiver and a 64-element receiver coil array were constructed to demonstrate and evaluate the method. A single echo corresponding to a projection along the long axis of each element is obtained from each array element simultaneously, using the 64-channel receiver. The research group has recently achieved *continuous* single-echo acquisition of the velocity in a rotating gel phantom for approximately 13 s, at 100 frames per second.

4.4.4
Summary

MRI has been used as a non-invasive quantitative visualization technique to investigate a class of complex Taylor–Couette–Poiseuille (TCP) flows, which constitute a prototype of many mixing or fractionation processes. Here we focused on the vicinity of the Stationary Helical Vortex (SHV) regime characterized by a spatially mixed state involving spatiotemporal oscillations. The helical symmetry of the SHV flow allowed the reconstruction of the 3D flow field by using MRI spin-tagging with spin-echo. Qualitative information about hydrodynamic mode interactions was additionally obtained by imaging on longitudinal and transverse slices with spin-tagging FLASH techniques for a wide range of flow conditions. Given the annular topology and aspect ratio of the TCP flow domain, only MRI techniques can provide such spatiotemporal information. The SHV velocity field extracted via MRI was then described as a dynamical system and was perturbed by superimposing a weak oscillatory mode in order to study the fate of particles dispersed in the flow. Neutral particles exhibited chaotic behavior in the perturbed field. As the excitation frequency and the density of the particle were increased, the phase space contracted, thus indicating that laminar chaotic flows can lead to particle segregation. This observation that weak parametric excitation of solid–liquid systems can lead to "unstirring" constitutes a promising first step in the search for better laminar separation methods. We have additionally discussed two promising research directions which aim at improving the speed of MRI velocimetry in rotating flows by at least an order of magnitude. The first involves the use of "prior-knowledge" to minimize acquisition, while the second relies on a phased-array surface coil to sample all k-space within a single echo.

Acknowledgement

This chapter is dedicated to P. C. Lauterbur who continues to inspire us.

References

1 G. I. Taylor **1923**, *Phil. Trans. R. Soc.* A233, 289.
2 J. Legrand, F. Coeuret **1986**, *Chem. Eng. Sci.* 41, 47.
3 *Perry's Chemical Engineering Handbook*, McGraw-Hill, New York, **1999**.
4 G. I. Taylor **1953**, (Dispersion of soluble matter in solvent flowing slowly through a tube), *Proc. R. Soc. Lond. A* 219, 186.
5 C. Castelain, A. Mokrani, P. Legentilhomme, H. Peerhossaini **1997**, *Expts. Fluids* 22, 359.
6 H. A. Kusch, J. M. Ottino **1992**, *J. Fluid Mech.* 236, 319.
7 S. W. Jones, O. M. Thomas, H. Aref **1989**, *J. Fluid Mech.* 209, 335.
8 J. M. Ottino **1989**, *The Kinematics of Mixing: Stretching, Chaos and Transport*, Cambridge University Press, Cambridge.
9 R. Abraham, J. E. Marsden, T. Ratiu **1988**, *Manifolds, Tensor Analysis, and Applications*, 2nd edn, Springer Verlag, New York.
10 J. M. Ottino, F. J. Muzzio, M. Tjahjadi, J. G. Franjione, S. C. Jana, H. A. Kusch **1992**, *Science* 257, 754.
11 T. Shinbrot, M. M. Alvarez, J. M. Zalc, F. J. Muzzio **2001**, *Phys. Rev. Lett.* 86, 1207.
12 I. J. Benczik, Z. Toroczkai, T. Tel **2002**, *Phys. Rev. Lett.* 89, 164501.
13 R. J. Adrian **1991**, (Particle-imaging techniques for experimental fluid mechanics), *Annu. Rev. Fluid Mech.* 23, 261–304.
14 L. G. Raguin, M. Shannon, J. G. Georgiadis **2001**, *Int. J. Heat Mass Transfer* 44 (17), 3295.

15 P. T. Callaghan **1991**, *Principles of Nuclear Magnetic Resonance Microscopy*, Clarendon Press, Oxford.

16 B. Blumich **2000**, *NMR Imaging of Materials*, Clarendon Press, Oxford.

17 E. Fukushima **1999**, *Annu. Rev. Fluid Mech.* 31, 95.

18 M. D. Mantle, A. J. Sederman **2003**, *Prog. Nucl. Magn. Reson. Spectrosc.* 43, 3.

19 S. B. Reeder, E. Atalar, A. Z. Faranesh, E. R. McVeigh **1999**, *Magn Reson. Med.* 42, 375.

20 J. A. Hopkins, R. E. Santini, J. B. Grutzner **1995**, *J. Magn. Reson.* 117, 150.

21 A. D. Hanlon, S. J. Gibbs, L. D. Hall, D. E. Haycock, W. J. Frith, S. Ablett **1998**, *Magn. Reson. Imag.* 16 (8), 953.

22 J. D. Seymour, B. Manz, P. T. Callaghan **1999**, *Phys. Fluids* 11, 1104.

23 K. Kose **1994**, *Phys. Rev. Lett.* 72 (10), 1467.

24 A. J. Sederman, M. D. Hollingsworth, M. L. Johns, L. F. Gladden **2004**, *J. Magn. Reson.* 171, 118.

25 K. W. Moser, L. G. Raguin, A. Harris, H. D. Morris, J. G. Georgiadis, M. Shannon, M. Philpott **2000**, *Magn. Reson. Imag.* 18 (2), 199.

26 K. W. Moser, L. G. Raguin, J. G. Georgiadis **2001**, *Phys. Rev. E*, 64:016319.

27 L. G. Raguin, J. G. Georgiadis **2004**, *J. Fluid Mech.* 516, 125.

28 K. W. Moser, E. C. Kutter, J. G. Georgiadis, R. O. Buckius, H. D. Morris, J. R. Torczynski **2000**, *Expts. Fluids* 29 (5), 438.

29 K. W. Moser, J. G. Georgiadis, R. O. Buckius **2000**, *Magn. Reson. Imag.* 18 (9), 1115.

30 K. W. Moser, J. G. Georgiadis, R. O. Buckius **2001**, *Ann. Biomed. Eng.* 29 (1), 9.

31 K. W. Moser, J. G. Georgiadis **2003**, *Magn. Reson. Imag.* 21 (1), 127.

32 K. W. Moser, J. G. Georgiadis **2004**, *Magn. Reson. Imag.* 22 (2), 257.

33 S. T. Wereley, R. M. Lueptow **1998**, *J. Fluid Mech.* 364, 59–80.

34 C. D. Andereck, S. S. Liu, H. L. Swinney **1986**, *J. Fluid Mech.* 164, 155–183.

35 W. F. Langford, R. Tagg, E. J. Kostelich, H. L. Swinney, M. Golubitsky **1988**, *Phys. Fluids* 31, 776–785.

36 M. Golubitsky, W. F. Langford **1988**, *Physica D* 32, 362–392.

37 T. J. Mosher, M. B. Smith **1990**, *Magn. Reson. Med.* 15, 334.

38 A. Tsameret, V. Steinberg **1994**, *Phys. Rev. E* 49, 4077.

39 K. W. Moser **2001**, (Quantitative Measurement of Velocity and Dispersion via Magnetic Resonance Imaging), *Ph.D. Thesis*, University of Illinois at Urbana-Champaign.

40 I. Mezic, S. Wiggins **1994**, *J. Nonlinear Sci.* 4, 157.

41 L. G. Raguin **2004**, (Theoretical and Experimental Study of a Continuous Hydrodynamically-enhanced Separation System Paradigm), *Ph.D. Thesis*, University of Illinois at Urbana-Champaign.

42 L. G. Raguin, J. G. Georgiadis **2005**, *Expts. Fluids*, submitted for publication.

43 L. G. Raguin, A. K. Kodali, D. V. Rovas, J. G. Georgiadis **2004**, *Proc. 2004 IEEE Eng. Med. Biol.*

44 J. S. Hyde, A. Jesmanowicz, W. Froncisz, J. B. Kneeland, T. M. Grist **1987**, *J. Magn. Reson.* 70, 512.

45 M. Hutchinson, U. Raff **1988**, *Magn. Reson. Med,* 6 (1), 87.

46 J. R. Kelton, R. L. Magin, S. M. Wright **1989**, *Proc. (Works-in-Progress) Eighth Annual Meeting, SMRM*, Amsterdam, The Netherlands, p. 1172.

47 D. K. Sodickson, W. J. Manning **1997**, *Magn. Reson. Med.* 38, 591.

48 K. P. Pruessmann, M. Weiger, M. B. Scheidegger, P. Boesiger **1999**, *Magn. Reson. Med.* 42, 952.

49 W. E. Kyriakos, L. P. Panych, D. F. Kacher, C. F. Westin, S. M. Bao, R. V. Mulkern, F. A. Jolesz **2000**, *Magn. Reson. Med.* 44, 301.

50 S. M. Wright, M. P. McDougall, D. G. Brown **2002**, *Proc. IEEE Eng. Med. Biol.* 1181.

51 S. M. Wright, M. P. McDougall **2004**, *Proc. Intl. Soc. Magn. Reson. Med.* 12, 533.

4.5
Two Phase Flow of Emulsions
Nina C. Shapley and Marcos A. d'Ávila

4.5.1
Introduction

Emulsions are physical systems composed of two liquid phases where one phase is dispersed into the other in the form of droplets [1]. These systems are of great scientific and technological interest and applications are widely found in the pharmaceutical, food and polymer processing industries. Emulsions are not in thermodynamic equilibrium; therefore, surfactants are usually added to the system in order to stabilize the dispersion kinetically. Typical droplet sizes range from 0.1 to 10 μm and the size distribution is usually polydisperse, exhibiting a lognormal distribution [2]. Other distributions such as Gaussian [3], bimodal [4] and monodisperse [5, 6] can be obtained depending on the component properties and preparation methods. Despite the widespread use of emulsions in technological applications, their flow behavior is not well understood, especially in concentrated systems [7].

The action of a flow field can affect the emulsion properties. In turn, the emulsion can affect the overall flow and rheological properties. For example, a shear flow can induce or suppress emulsion coalescence [8, 9], a process by which multiple droplets fuse together to form larger droplets. A shear flow may also overcome the destabilization of an emulsion due to creaming, which is the gravitational separation of the droplets and the continuous phase due to density differences. Finally, a shear flow may induce droplet breakup [10]. On the other hand, droplet deformation and the droplet volume fraction can affect flow due to changes in rheological properties [11, 12] and can induce droplet migration [13]. In a wide variety of applications ranging from food [14] to polymer processing [15], the quality of the product relies on the degree of control of the droplet size distribution and the emulsion flow properties. Hence, interest has developed to gain an improved understanding of the behavior of emulsions undergoing flow processes.

Direct observations of emulsion flows are usually limited to dilute systems, in part due to limitations on experimental techniques. Conventional techniques such as optical microscopy, video imaging and laser-Doppler velocimetry require sample transparency, and therefore can only be utilized in dilute emulsions (droplet volume fractions of less than 5%), refractive index-matched systems or highly confined systems. Static and dynamic light scattering [16], phase Doppler methods [8] and laser-Doppler velocimetry [17] rely on single-scattering of laser light by the droplets, and video microscopy consists of following the concentration or trajectory of tracer droplets in an otherwise transparent system [18, 19]. As concentrated emulsions are most often the focus of current fundamental studies and practical applications, the use of optical methods for such materials generally involves sample removal and extreme dilution, which is invasive, or the development of a model, transparent system.

Nuclear magnetic resonance imaging (NMRI) has been recognized to have great potential as an experimental tool for engineering science research due to its versatility and accuracy and as an R&D tool in product and process development [20, 21]. Applications identified in different fields of chemical engineering range from chemical reaction to food and to biochemical engineering [20]. For example, NMRI can be used to quantify the effectiveness of food processing operations such as mixing, or to map the moisture content or temperature profile inside various model foods [20]. Although these techniques have been used mostly in laboratory research due to their high operational costs, low field NMRI devices have been shown to be able to perform routine process and quality control tasks in industry [20, 22, 23] and the potential for the use of these techniques to solve industrial problems is widely recognized [24].

The capability of NMRI-based methods to perform noninvasive, spatially-resolved flow measurements positions this technique as an important tool for the study of flow and microstructure of emulsions, especially for concentrated systems, which are opaque. In previous studies of emulsion flow, the capabilities of various NMRI methods to study these systems have not been fully explored [25, 26]. On the other hand, recent results demonstrate the flexibility of this technique to characterize flow [27] and rheology [28] of emulsions and other dispersed systems [29, 30], foods [31], polymer melts [32], stratified fluids [33], Taylor instabilities[34], viscometric flows of complex fluids [35], turbulent flows [36], flow in porous media [37, 38] and mixing [39, 40]. The study of emulsions under static conditions with NMRI has also been utilized to gather important information such as spatially-resolved droplet size distributions [41, 42]. Also, it has been shown that destabilization mechanisms such as creaming [43], coalescence [4] and Ostwald ripening [44] can be studied using magnetic resonance techniques. A unique aspect of NMRI is that intensity (or phase angle) values in each pixel of an image can represent local amounts of flow or diffusion, while the NMR frequency spectrum reflects the chemical composition or indicates the presence of the droplet phase or continuous phase.

Although NMRI is a very well-suited experimental technique for quantifying emulsion properties such as velocity profiles, droplet concentration distributions and microstructural information, several alternative techniques can provide similar or complementary information to that obtained by NMRI. Two such techniques, ultrasonic spectroscopy and diffusing wave spectroscopy, can be employed in the characterization of concentrated emulsions *in situ* and without dilution [45].

After NMRI, ultrasonic spectroscopy is probably the most versatile technique for characterizing emulsions. Ultrasonic spectroscopy has been used to measure droplet concentration [46] and size [47] distributions, and velocity profiles can be obtained through ultrasound Doppler velocimetry [48–50]. When ultrasound waves are pulsed through the emulsion, measurements of the wave speed and attenuation of the pulses over a range of frequencies can be related to droplet volume fraction and size distributions along the path of propagation by equations derived from multiple scattering theory [51] with effective medium approximations [52]. Coefficients in the model equations depending on droplet size and concentration are then fit to the data, after assuming a known droplet size distribution function, such as a lognormal distribution [53].

In terms of measuring emulsion microstructure, ultrasonics is complementary to NMRI in that it is sensitive to droplet flocculation [54], which is the aggregation of droplets into clusters, or "flocs," without the occurrence of droplet fusion, or coalescence, as described earlier. Flocculation is an emulsion destabilization mechanism because it disrupts the uniform dispersion of discrete droplets. Furthermore, flocculation promotes creaming in the emulsion, as large clusters of droplets separate rapidly from the continuous phase, and also promotes coalescence, because droplets inside the clusters are in close contact for long periods of time. Ideally, a full characterization of an emulsion would include NMRI measurements of droplet size distributions, which only depend on the interior dimensions of the droplets and therefore are independent of flocculation, and also ultrasonic spectroscopy, which can characterize flocculation properties.

For a flocculated emulsion, it is supposed that at low ultrasonic wave frequencies, (e.g., 1–10 MHz), the attenuation due to thermal dissipation at droplet interfaces is reduced and at high wave frequencies, (e.g., 10–150 MHz), the attenuation due to multiple scattering effects is enhanced, compared with a well-dispersed emulsion [54]. Although velocity and attenuation measurements can be obtained for highly concentrated systems, the theoretical correlations only contain terms that are at most quadratic in droplet volume fraction. The absence of higher order terms suggests that the models may be most appropriate for moderately concentrated systems, and indeed typical bulk volume fraction values in ultrasonic experiments range from 0.1 to 0.3. For highly concentrated emulsions, alternative relationships may be needed. Finally, ultrasonic spectroscopy is a point-wise measurement method. A scan of the source and detector over a typical sample yields a concentration and size distribution measurement in approximately several minutes, resulting in comparable time resolution to NMRI concentration and size distribution measurements [55].

Diffusing wave spectroscopy is another technique utilized to measure emulsion microstructure in nearly optically opaque systems such as concentrated emulsions. In this technique, the intensity of multiple-scattered light from the sample is recorded over time. The intensity autocorrelation function indicates a characteristic time scale (which can range from 10^{-8} to 10^5 s) [56] for temporal fluctuations in intensity. The time scale is then related to the mean square displacement of the droplets through the Stokes–Einstein diffusivity for Brownian motion. Assuming a known displacement, the product of the droplet radius and the local viscosity is calculated [57]. In a flocculated system, the floc radius and the effective floc viscosity are obtained instead of the individual droplet properties [57]. Alternatively, if the droplet size and viscosity are known, the droplet mean square displacement can be calculated, and then the technique can detect the difference between a series of separate flocs and a connected network of droplets [58]. Either point-wise or volume-averaged measurements can be made [57]. The technique is sensitive to local rearrangements of droplets during shear, even to fluctuations on length scales much smaller than the wavelength of light [56], and can be used in an "echo" form in oscillatory shear that recalls NMR spin-echo and PGSE methods described in the next section [59, 60]. Hence, the greatest appeal of alternative techniques such as ultrasonic spectroscopy and diffusing wave spectroscopy is likely to be the ability to capture droplet cluster size distributions and the local connectivity of droplets, which is complementary to NMRI characterization of emulsion microstructure.

The purpose of this chapter is to describe the methodology used to determine flow, spatial homogeneity and microstructure of model oil-in-water emulsions through NMRI methods, and to show the capability of the NMRI technique to perform such measurements accurately. A summary of the research carried out by the authors on flow and mixing of emulsions in horizontal concentric cylinders is presented. Specifically, the research emphasizes gravitational effects on flow and mixing of oil-in-water emulsions. As the flows are run at relatively low velocities, the capability of NMRI to measure slow flow processes of two-phase systems is demonstrated. Section 4.5.2 describes the NMRI set-up and methods used to perform the work on flow and mixing of emulsions. Some results obtained using the apparatus and the methods described in Section 4.5.2 are presented in Sections 4.5.3 and 4.5.4. In Section 4.5.5, future perspectives on the subject are presented.

4.5.2
NMRI Set-up and Methods

4.5.2.1 NMRI Set-up

Our NMRI apparatus was designed to perform measurements in a 7-T superconducting magnet with a horizontal bore (Oxford), where a Bruker microgradient system with inner diameter of 53 mm was attached to it, providing gradient intensities up to 95 G cm^{-1}. The flow cell consisted of two concentric cylinders made of glass with outer cylinder radius $R_o = 1.1$ cm, inner cylinder radius $R_i = 0.4$ cm and length $L = 28$ cm. Housing machined from an Ultem rod and a laboratory-built Alderman-Grant radiofrequency coil [61] with a resonant frequency of 300 MHz were attached to the system and provided satisfactory results for our purposes. A stepper motor was attached to the system and was used to generate a steady flow where the outer cylinder rotated at a constant velocity V and the inner cylinder remained stationary. The angular velocity of the outer cylinder can be expressed as (V/R_o). Data were acquired through a Bruker Biospec system and the raw data were available as a binary file that was accessed by a Silicon Graphics workstation. The Bruker software Paravision (v. 2.1) was used to set the experimental conditions such as pulse sequences and defining parameters, but not all the data could be processed using Paravision. Therefore, computer codes for data processing were written in Matlab 5.3 (Mathworks) for the various experiments involved in this work. Figure 4.5.1 shows a photograph of the probe and Figure 4.5.2 shows the rf coil design. All measurements shown here were performed in the cross section located in the center of the probe in order to avoid end effects in the measurements (see Figure 4.5.3).

As indicated in Figure 4.5.3, gravitational effects are especially significant in the system due to the horizontal orientation of the cylinders. Gravity acts in the plane $(x-y)$

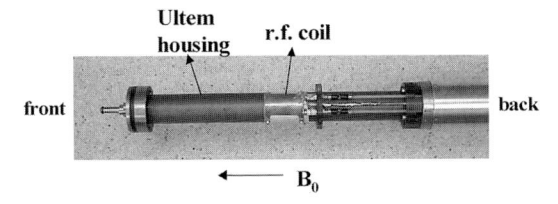

Fig. 4.5.1 Photograph of the NMRI probe designed to generate a concentric cylinder shear flow. The cylinder axis is parallel to the direction of the static magnetic field B_0. The front of the probe is inserted into the magnet and the stepper motor is attached to the back of the probe. As the stepper motor has magnetic components, it was kept about 5 m away from the magnet for safety reasons.

of the main emulsion motion. Therefore, density differences can cause droplets to cross flow streamlines and move into a non-uniform configuration within the plane of the rotational motion. As non-uniform droplet distributions in turn influence local emulsion properties, the flow and droplet concentration fields become coupled together. Such interplay between the flow and droplet concentration distributions would probably not be as apparent for vertically oriented cylinders, where gravity is directed perpendicular to the plane of the main rotational motion.

4.5.2.2 Velocity Measurements

There are two main methods to measure velocity fields and profiles using NMRI: time-of-flight velocimetry (TOF) and phase encoding velocimetry. In this section these methods are briefly described with a discussion of how they were used to perform accurate velocity measurements in oil-in-water emulsions. These methods

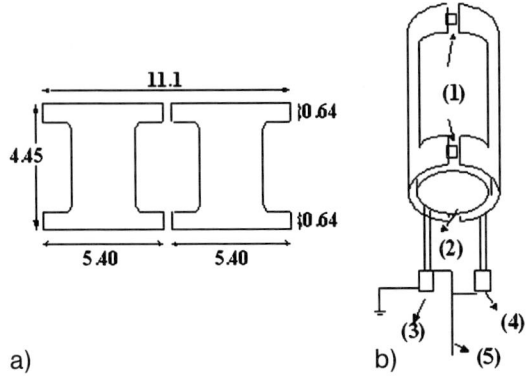

Fig. 4.5.2 Schematic drawing of an Alderman-Grant rf coil. (a) Plane view of the coil showing the dimensions in centimeters. The smoothed edges are 90° arches with radius of 0.22 cm. (b) Coil in "wrapped" view, indicating the main components: (1) chip capacitors of 5.6 pF, (2) guard ring, (3) tuning and (4) matching variable capacitors of 40 pF and (5) rf cable. Guard rings were used to prevent electric field propagation through the sample and they were separated from the coil by a thin layer of Teflon ($\approx$0.1 mm).

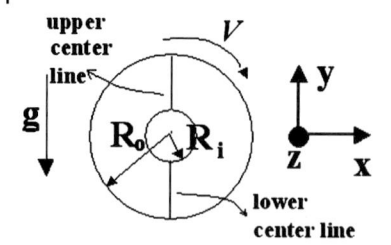

Fig. 4.5.3 Cross section of a concentric cylinder flow cell. The directions z and y correspond, respectively, to the static magnatic field B_0 and the gravitational field directions.

were used to measure the bulk velocity; however, it is not difficult to measure velocities for both phases separately, as methods for water and oil signal suppression through selective excitation can be utilized.

4.5.2.2.1 Time-of-flight Velocimetry

Time-of-flight velocimetry (TOF) consists of saturating a predetermined region of a flowing system by applying an rf pulse to null the signal of this region. A slice is taken in the direction perpendicular to the main flow. The displacement of the tagged nuclei is obtained after a known period of time, and the velocity is determined from the ratio of the displacement to time. The advantage of this technique is that it is easy to implement and it gives accurate velocity measurements at a steady state. However, it cannot resolve flows at low velocities (<0.1 mm s^{-1}) and it is limited to unidirectional flow. Other methods based on tagging, such as the DANTE pulse sequence [62] have been used in order to obtain steady-state flow fields in two dimensions [33]. Figure 4.5.4(a) shows a TOF pulse sequence. In this figure, the first rf pulse corresponds to the saturation pulse, which nulls the signal of the region specified by the gradient, in this case the x direction gradient G_x. The remaining pulse sequence corresponds to the usual spin-echo imaging pulse sequence. The magnetic field gradients G_x, G_y and G_z are the frequency-encoding, phase-encoding and slice-selection gradients, respectively. Figure 4.5.4(b) shows a TOF image for an isooctane-in-water emulsion with a 0.4 oil volume fraction flowing at $V = 2.0$ cm s^{-1}.

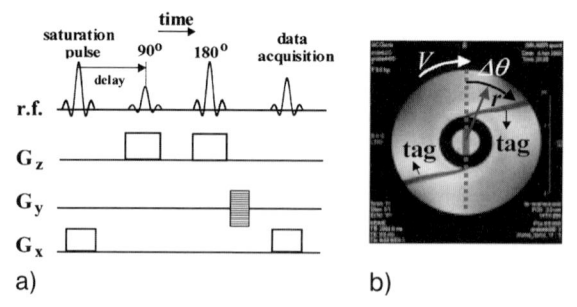

Fig. 4.5.4 (a) TOF pulse sequence: the first rf pulse corresponds to the saturation pulse that nulls the tag signal and the subscripts x, y and z correspond to the coordinate directions shown in Figure 4.5.3. (b) TOF image obtained using the pulse sequence shown in (a) for an isooctane-in-water emulsion with a 40% oil volume fraction and $V = 2.0$ cm s^{-1}.

The tags correspond to the displacement after the delay time from the initial state shown by the dotted line at the center line. Here, the delay time was 0.5 s and the image was obtained in approximately 5 minutes.

4.5.2.2.2 Phase Encoding Velocimetry

Dynamic displacements of a nuclear spin ensemble can be measured using pulsed-field-gradients (PFG) incorporated into either spin-echo (pulsed-field gradient spin-echo, PGSE) or stimulated-echo (pulsed-field gradient stimulated-echo, PGSTE) pulse sequences. The PGSTE pulse sequence has the advantage of avoiding signal loss due to spin–spin relaxation (T_2) and it consists of the application of three 90° rf pulses and two pulsed gradients with magnitude G and duration δ separated by a time interval Δ (Figure 4.5.5). In liquids, molecular coherent motion is detected through the phase component of the signal in a PFG sequence, whereas the molecular random motion is detected through the signal intensity. Therefore, simultaneous flow and diffusion measurements can be performed using PFG methods and they are widely used to obtain both macroscopic velocity fields and self-diffusion coefficients of liquids. Diffusive motion will be discussed in the next section, where the random motion of molecules confined in a spherical geometry is the basis of droplet size distribution measurements in emulsions.

In the case of the macroscopic flow of emulsions, velocity fields can be measured by applying a PFG pulse sequence together with imaging gradients and this method is known as phase encoding imaging. This method produces direct images of velocity profile distributions of the system [63–66]. The principle of velocity phase imaging relies on the behavior of the phase $\phi(t)$ of the spin system. The phase component of the NMR signal for an arbitrary volume (or voxel) V with spin density $\varrho(r, t)$ is given by

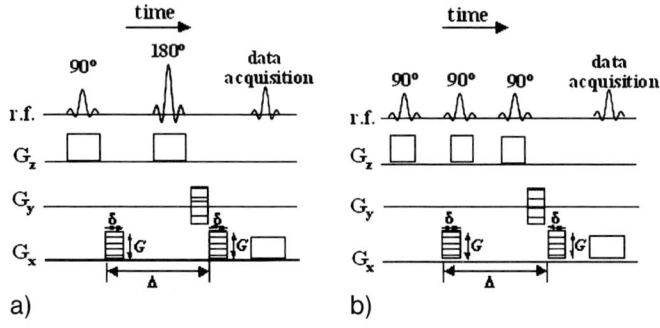

$$S = \int_V \varrho(r, t)\exp(i\phi)dV = \int_V \varrho(r, t)[\cos\phi + i\sin\phi]dV \quad (4.5.1)$$

Fig. 4.5.5 Pulsed field gradient sequences to obtain velocity and diffusion data: (a) spin-echo (PGSE) and (b) stimulated-echo (PGSTE). The application of imaging gradients G_x, G_y and G_z allows the measurement of velcocity maps and spatially-resolved diffusion coefficients and size distribution in emulsions.

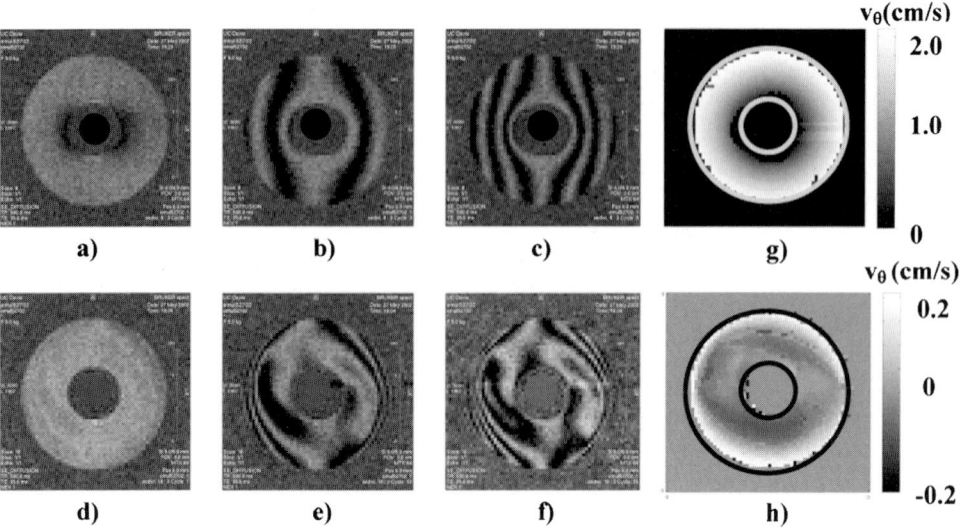

Fig. 4.5.6 Images of the real component of the signal at different gradient intensities G for an isooctane-in-water emulsion with 10% droplet volume fraction at: (a) $V = 2$ cm s^{-1} and $G = 0$; (b) $V = 2$ cm s^{-1} and $G = 10$ mT m^{-1}; (c) $V = 2$ cm s^{-1} and $G = 20$ mT m^{-1}; (d) $V = 0.2$ cm s^{-1} and $G = 0$; (e) $V = 0.2$ cm s^{-1} and $G = 150$ mT m^{-1}; and (f) $V = 0.2$ cm s^{-1} and $G = 200$ mT m^{-1}. Gradients were oriented in the vertical (y) direction. Note that this is a variation of the pulse sequence shown in Figure 4.5.5, where the flow gradients were oriented in the horizontal (x) direction. After combining data such as (a–f) from the x and y directions, velocity maps of the calculated tangential velocity component v_θ are shown in (g–h): (g) $V = 2$ cm s^{-1} (co-rotating flow); and (h) $V = 0.2$ cm s^{-1} (counter rotating flow).

where the phase $\phi(t) = \gamma_G \int_0^t G \cdot r \, dt$. A Taylor series expansion of r with respect to time, where the second and higher order terms are neglected leads to

$$\phi(t) = \gamma_G (r_0 \cdot p_0 + v \cdot p_1) \tag{4.5.2}$$

where $p_0 = \int_0^t G \, dt$ and $p_1 = \int_0^t G t \, dt$ are the zeroth and first moments of the gradient, respectively; r_0 is the position of the spin at time $t = 0$ and v is the average velocity during the application of the pulse sequence. In the imaging sequence shown in Figure 4.5.5(a), the phase of the NMR signal is proportional to the gradient intensity, the gradient duration, the evolution time and the velocity.

A series of images can be obtained with gradient intensity values evenly spaced between zero and a maximum value. As the gradient intensity increases, so does the accumulated phase angle ϕ as shown in Eq. (4.5.2). Two sets of images are reconstructed from the recorded NMR signal: the real and imaginary parts of the complex values in each pixel. In each pixel of the real or imaginary image, the intensity value oscillates as successively stronger gradients are applied. The frequency of the intensity value oscillation in each pixel is proportional to the velocity in the pixel. Figure 4.5.6 shows images of the real component of the signal at different gradient intensities G applied in the y direction for an isooctane-in-water emulsion with 10% droplet volume fraction flowing at (a–c) 2 cm s^{-1} and (d–f) 0.2 cm s^{-1}. The relationship between intensity oscillation frequency and velocity is reflected in Figure 4.5.6(a–c), where the intensity values vary rapidly near the edges of the flow cell (large y component of velocity) and slowly near the vertical center line (small y component of velocity). Also, the contrasting behavior of phase accumulation between images (a–c) and (d–f) shows that the flow patterns in these situations are markedly different.

Velocity maps such as Figure 4.5.6(g–h) can be obtained by processing the data presented in Figure 4.5.6(a–f). The details of this procedure have been described by d'Avila [26] and Shapley et al. [27]. In brief, the magnitude of the velocity is calculated from the set of real images, by taking the Fourier transform of the intensity oscillation in each pixel. The location of the peak along the frequency spectrum is then related to the velocity magnitude through the known gradient strength and timing parameters in the NMRI pulse sequence. Meanwhile, the sign of the velocity is calculated from the set of imaginary images. The advantages of calculating the velocity from multiple gradients rather than from a single value are that the multi-gradient method allows the resolution of a wide range of velocities and is insensitive to many artifacts. After the sequential measurement of velocity components, the components can be combined to form vector maps of the flow field [26, 27].

4.5.2.3 Concentration Measurements

The difference in resonant frequencies between nuclei in molecules or chemical groups of distinct chemical structure is called the "chemical shift". The chemical shift arises due to the difference between the main magnetic field and the local magnetic field experienced by the nuclei of a molecule, depending on the electronic environment of the nuclei. Therefore, it is possible to associate a particular molecule or molecular group with its chemical shift. Spectroscopy techniques used to identify and assign chemical structures are largely based on chemical shifts. In the case of emulsions, the continuous and dispersed phases are identified by the presence of two distinct peaks. Peak intensity in the spectrum is proportional to the number of spins in the sample; i.e. it is proportional to the concentration. When magnetic field gradients are applied, spatially-resolved NMR spectra can be obtained. With this information, a quantitative map of the volume fractions can be obtained. The application of imaging gradients to obtain an NMR spectrum at each pixel is known as chemical shift imaging (CSI) [67]. Figure 4.5.7(a) shows a CSI pulse sequence used to study mixing of concentrated emulsions, which consists of the application of a spin-echo pulse sequence together with phase-encoding gradients G_x and G_y, and a slice-selection gradient G_z. A CSI image of a partially creamed emulsion for the horizontal concentric cylinder geometry (Section 4.5.2.1), showing the NMR spectrum in two different regions in the flow cell is presented in Figure 4.5.7(b). Other methods based on signal intensity can be used to determine concentration distributions in systems of solid particles dispersed in liquids [39, 68, 69], as the signal from the solids decays rapidly to an undetectable

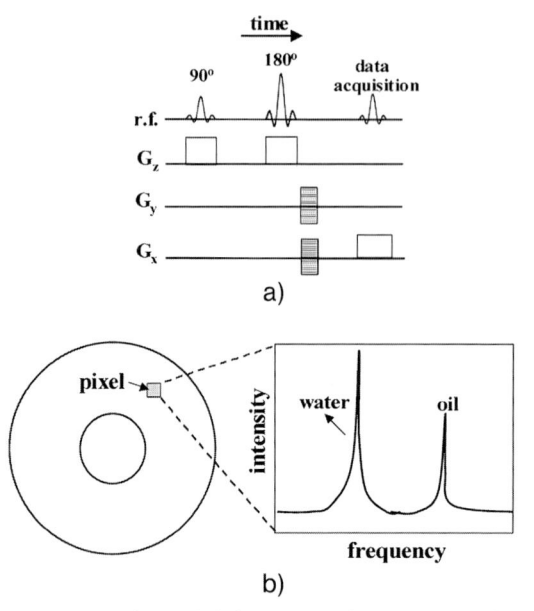

Fig.4.5.7 (a) Chemical shift imaging pulse sequence and (b) schematic drawing of CSI data for a given pixel of an oil-in-water emulsion inside the horizontal concentric cylinders geometry.

value, but in the case of emulsions, where both dispersed and continuous phases contribute to the total signal, concentration measurements must rely on signal suppression such as selective excitation [70].

4.5.2.4 Droplet Size Distribution Measurements

Pulsed-field gradient methods can be used to measure emulsion droplet sizes, where typical droplet diameters range from 1 to 20 μm. Methods such as PGSE and PGSTE (see Section 4.5.2.2) with imaging gradients can be used to measure spatially-resolved droplet size distributions. The choice to use either PGSE or PGSTE depends on the system to be studied. We have used PGSTE, which has the advantage of avoiding signal loss due to T_2 relaxation.

In a PFG pulse sequence, if the molecules are free to move, it can be shown that the normalized signal S/S_0 can be described by an exponential decay as follows,

$$S/S_0 = \exp(-Db) \tag{4.5.3}$$

where $b = \gamma_G^2 G^2 \delta^2 (\Delta - \delta/3)$ and S_0 is the signal S when $G = 0$. The slope of the curve $\ln(R = S/S_0)$ versus b yields the molecular self-diffusion coefficient D.

In the case of emulsions, the oil phase is confined in a spherical geometry (droplets), and the random motion of the oil molecules is restricted to the droplet boundary. In this case, the signal decay function S/S_0, assuming a Gaussian phase distribution of spins for random motion of molecules confined in a sphere of radius a, is given by the following expression [71]:

$$S/S_0 = R(g, \delta, \Delta, D, a) = \exp\left[-2\gamma_G^2 G^2 \sum_{m=1}^{\infty} f(\alpha_m)\right] \qquad (4.5.4)$$

and

$$f(\alpha_m) = \left[\alpha_m^2(\alpha_m^2 a^2 - 2)\right]^{-1}\left[2\delta/\alpha_m^2 D - [2 + \exp(-\alpha_m^2 D(\Delta - \delta)) - 2\exp(-\alpha_m^2 D\delta)]\big/(\alpha_m^2 D)^2 - [2\exp(-\alpha_m^2 D\Delta) + \exp(-\alpha_m^2 D(\Delta + \delta))]\big/(\alpha_m^2 D)^2\right] \qquad (4.5.5)$$

Here, S_0 is the signal when $G = 0$, D is the self-diffusion coefficient, γ_G is the gyromagnetic ratio and α_m are roots of the Bessel function equation $\alpha_m a J'_{3/2}(\alpha_m a) - (1/2)J_{3/2}(\alpha_m a) = 0$. If the system is polydisperse, the signal decay is due to contributions from droplets of different sizes. Then, the signal attenuation is given by the volume average over all sizes as

$$(S/S_0)_{obs} = \int_0^{\infty} a^3 P(a) R(D, g, \delta, \Delta, a) da \bigg/ \int_0^{\infty} a^3 P(a) da \qquad (4.5.6)$$

Here $P(a) = 1/(8a^2\sigma^2\pi)^{1/2}\exp[-(\ln 2a - \ln 2\bar{a})^2/(2\sigma^2)]$ is a lognormal size distribution function with mean radius $\bar{a}$ and standard deviation σ. Fitting Eqs. (4.5.4–4.5.6) to experimental values of (S/S_0) gives $\bar{a}$ and σ for the droplet size lognormal distribution.

The assumption that the spin phase distribution is Gaussian is valid only for free diffusion, but it is an approximation for the case of restricted diffusion, which could lead to inaccuracies in droplet size measurements when using Eq. (4.5.4) to determine droplet sizes. However, Balinov et al. [72], by comparing predictions of signal attenuation using Eq. (4.5.4) with molecular dynamics simulations, has demonstrated that this equation accurately describes the NMR signal attenuation for a wide range of experimental conditions and droplet sizes. According to Balinov, for a system with the emulsion and NMR conditions described here, droplets with radii smaller than about 14 μm have attenuation curves described well by Eq. (4.5.4) [72]. Also, droplet size distributions in emulsions determined using Eq. (4.5.4) showed good agreement when compared with other techniques, such as the Coulter counter [3], optical microscopy [73] and static light scattering [42] and provide a reliable method for emulsion droplet size distribution measurements even at low magnetic fields (0.5 T) [22]. The limitation of this method is that it is necessary to assume an *a priori* droplet size distribution. In this case, we assume a lognormal distribution, which is the typical droplet size distribution exhibited by an emulsion. Experimentally, a lognormal distribution of droplet sizes in the emulsion was confirmed by static light scattering [26]. Data processing efforts based on regularization methods in order to determine size distributions without assuming a known distribution have appeared in the literature [74]. Such analysis is a promising area for the improvement of pulsed-field gradient methods.

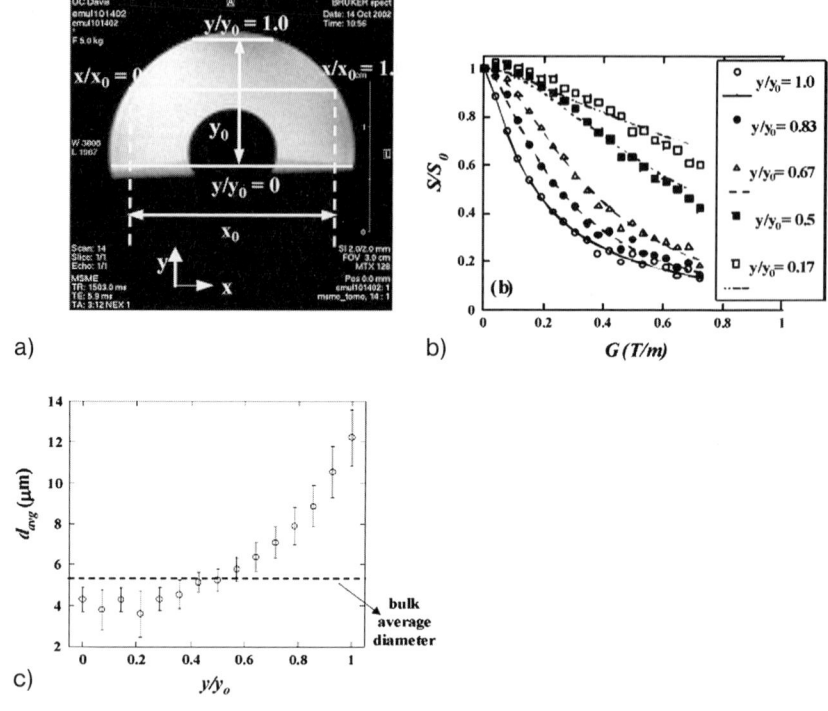

Fig. 4.5.8 (a) Spin-echo image showing the droplet phase in a fully creamed emulsion; (b) PGSTE data for the emulsion shown in (a) at different locations along the y direction in the creamed layer; and (c) average droplet size obtained using Eqs. (4.5.4–4.5.6) as a function of the position. Here the signal from the water phase was suppressed by adding a small amount of manganese chloride, an agent which accelerates the decay of the NMR signal [75].

Figure 4.5.8 shows the spatially-resolved droplet size distribution of a fully creamed isooctane-in-water emulsion obtained using the PGSTE pulse sequence shown in Figure 4.5.5(b). It can be seen that this method is able to provide droplet size distributions with spatial resolution.

4.5.3
Complex Flows of Homogeneous Emulsions

Homogeneous emulsions sheared in the horizontal concentric-cylinder geometry, depending on the characteristics of the emulsion and the shearing, can exhibit complex flow behavior. Here a homogeneous emulsion means a system where the dispersed phase appears to be uniformly distributed inside the flow cell. Complex flow transitions in oil-in-water emulsions were studied by our group using NMRI. It was found that this transition occurs when the outer cylinder velocity is sufficiently small such that buoyancy effects due to the density difference between the continuous and dispersed phases are comparable to the viscous effects due to rotation of the outer cylinder. Emulsions of isooctane-in-water with several droplet volume fractions were prepared by blending of the components and homogenizing. The resulting systems were non-buoyant. Results presented here are for an isooctane-in-water emulsion stabilized with 0.5 wt%. of Tween 20, which is a nonionic surfactant. The emulsion is polydisperse with an average droplet diameter of 3.5 µm. Densities of water ϱ_{water} and isooctane ϱ_{oil} were 1.0 and 0.69 g cm^{-3}, respectively. The bulk viscosity of an emulsion with bulk oil volume fraction ϕ_{bulk} = 0.4 is approximately 2.0 cP, so the emulsions studied here do not have high viscosities, in contrast to high viscosity emulsions, which exhibit viscoelastic behavior. Here we will show some results obtained using NMRI in this study. The experimental parameters and a detailed discussion of this buoyancy-driven flow transition can be found in Shapley et al. [27].

Figure 4.5.9 shows TOF images of an emulsion with bulk oil volume fraction ϕ_{bulk} = 0.4 flowing at two different outer cylinder velocities. The outer cylinder rotates in the clockwise direction. When V = 2.0 cm s^{-1} (Figure 4.5.9(a)), it can be observed by the tag displacement that the flow is co-rotating. When the outer cylinder velocity is decreased to a critical velocity $V = V_c$ = 0.1 cm s^{-1} (Figure 4.5.9(b)), the flow exhibits a partially counter-rotating behavior, where flow in the opposite direction to the outer cylinder motion is observed. The co-rotating flow presented in Figure 4.5.9(a) is well described by a Newtonian profile in this geometry, showing that the emulsion flow is similar to a single-phase flow. The counter-rotating flow profile was observed for emulsions with several volume fractions. Figure 4.5.10 shows TOF velocity profiles for emulsions with volume fractions of 0.3, 0.4 and 0.5. It can be seen that the counter-rotating behavior is observed and that the velocity profiles are similar.

The hypothesis that the transition to complex flow is buoyancy driven is supported by a scaling analysis to estimate the bulk density difference $\Delta \varrho$ required to reverse a viscous flow. For the viscous and gravitational contributions to the flow to be comparable, one needs,

$$\Delta \varrho \, g \sim \frac{\mu \, V}{R_o^2} \tag{4.5.7}$$

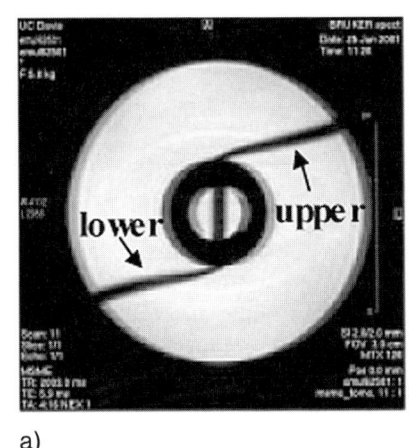

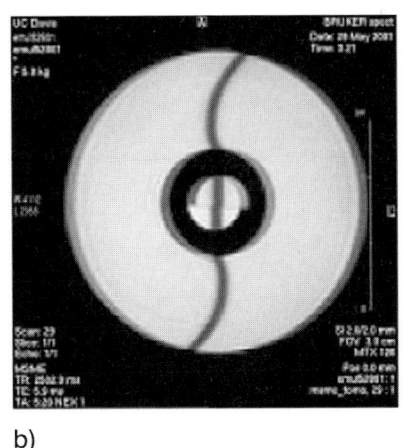

Fig. 4.5.9 TOF images for (a) V = 2 cm s^{-1} and (b) V = 0.1 cm s^{-1}. The outer cylinder rotates in the clockwise direction.

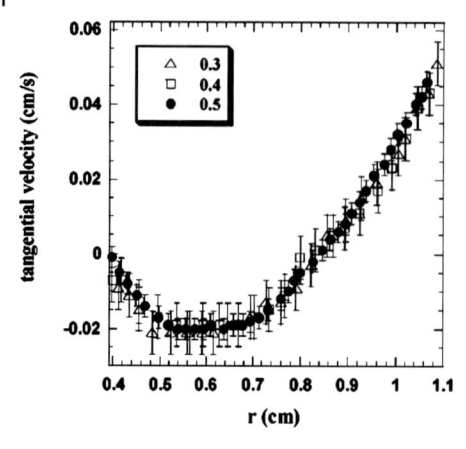

Fig. 4.5.10 Velocity profiles along the upper center line (see Figure 4.5.3) measured by TOF for isooctane-in-water emulsions with several oil volume fractions at $V = 0.05$ cm s^{-1}. Here $r = 0.4$ and $r = 1.1$ cm indicate the inner and outer cylinder walls, respectively.

where $g = 980$ cm s^{-2} is the gravitational constant, μ is the emulsion viscosity, V is the outer cylinder rotation speed and R_o is the outer cylinder radius. By choosing characteristic values of $\mu \approx 2$ cP, $V \approx 0.1$ cm s^{-1} and $R_o \approx 1.1$ cm, one finds that $(\Delta \varrho) \approx 10^{-6}$ g cm^{-3}. This result shows that even virtually undetectable fluctuations in the density can lead to a considerable effect on the flow under many conditions of interest. A perturbation analysis of the continuity and Navier–Stokes equations allowed us to obtain an expression for the velocity component in the angular θ-direction and it is given by

$$\hat{v}_\theta^{(0)} = \frac{\hat{\Gamma}}{2\hat{r}} \left(\frac{\hat{r}^2 \ln \hat{r} - (R_i/R_o)^2 \ln(R_i/R_o) + \hat{r}^2 (R_i/R_o)^2 \ln(R_i/(\hat{r} R_o))}{1 - (R_i/R_o)^2} \right)$$
$$+ \frac{1}{\hat{r}} \frac{\hat{r}^2 - (R_i/R_o)^2}{1 - (R_i/R_o)^2}$$
(4.5.8)

where $\hat{r} = r/R_o$ and $\hat{v}_\theta = v_\theta/V$. The derivation of the equation above is described in detail in Shapley et al. [27]. The second term on the RHS of Eq. (4.5.8) corresponds to the Newtonian fluid solution. The first term on the RHS accounts for the bulk density fluctuations due to buoyancy effects in the system. The dimensionless parameter $\hat{\Gamma}$ can be used as a means to characterize the transition to the counter-rotating flow. Low values of $\hat{\Gamma}$ correspond to the Newtonian flow, whereas the counter-rotating behavior is well described for higher values of $\hat{\Gamma}$. Figure 4.5.11 shows velocity profiles obtained for both TOF and phase encoding methods at the vertical center line (see Figure 4.5.3) for an emulsion with $\phi_{\text{bulk}} = 0.4$. The figure confirms that velocity measurements made from the two methods agree within the limit of experimental uncertainty. Also, in Figure 4.5.11, the lines correspond to the fitting of Eq. (4.5.8), where the parameter $\hat{\Gamma}$ is adjustable. It can be seen that Eq. (4.5.8) describes the velocity profile very well for both counter-rotating (Figure 4.5.11(a)) and co-rotating flows (Figure 4.5.11(b)).

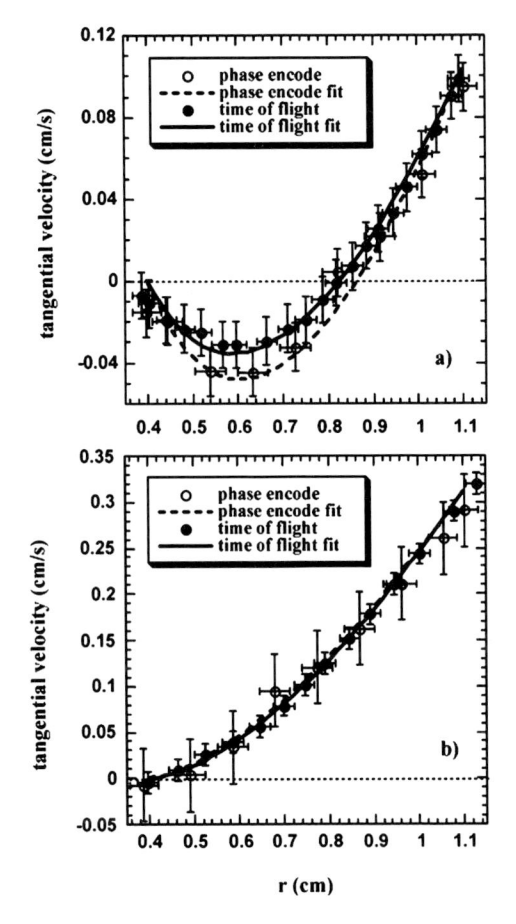

Fig. 4.5.11 Measured TOF and phase encoding velocity profiles from the lower, vertical center line and corresponding fits using Eq. (4.5.8). Outer cylinder rotation speed is: (a) 0.1 cm s^{-1} (counter-rotating); (b) 0.31 cm s^{-1} (co-rotating).

Figure 4.5.12 shows a plot of $\hat{\Gamma}$ as a function of the outer cylinder velocity V. In this figure, the error bars correspond to the uncertainty of the fitted values of $\hat{\Gamma}$ obtained by TOF and phase encoding. It can be seen that the counter-rotating flow is characterized by high values of $\hat{\Gamma}$, ranging between 8 and 12. Based on this figure, $\hat{\Gamma}$ clearly distinguishes the flow transition between the co-rotating and counter-rotating flow states. The analysis therefore captures the impact of buoyancy effects on the flow pattern of the system.

4.5.4
Mixing of Concentrated Emulsions

When droplets are initially not uniformly dispersed inside the flow cell, i.e. the emulsion is not homogeneous, the presence of a shear flow will induce mixing and the flow behavior of the system will be dependent on the spatial distribution of both phases. Therefore, in order to study the flow and mixing of an initially non-homogeneous emulsion it is necessary to obtain information on how both phases

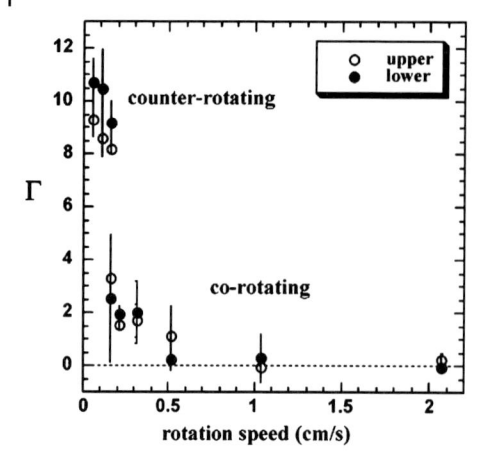

Fig. 4.5.12 The dimensionless parameter $\hat{\Gamma}$ plotted against outer cylinder rotation speed V. Values of $\hat{\Gamma}$ were fit from the lower (●) or upper (○) vertical center line (see Figure 4.5.3) velocity profile.

are distributed inside the flow cell. Spatial distribution of droplet sizes in polydisperse emulsions can also be an important factor in the mixing behavior of emulsions at low rotation velocities, as discussed by d'Avila et al. [40]. Here we will present the application of NMRI methods in the study of mixing in a horizontal concentric-cylinder geometry. The emulsion has the same properties as the one shown in the previous section, with bulk volume fraction $\phi_{bulk} = 0.4$. In the case of an initially creamed oil-in-water emulsion, if the outer cylinder velocity is sufficiently small, the mixing process occurs in a distinct pattern, where buoyancy effects influence the kinetics and also the pattern of mixing, which is markedly different from the mixing of bulk immiscible single phase fluids, where the initial stages of mixing involve stretching and folding of fluid regions to form interleaved layers [76].

Chemical shift imaging (CSI) was used to monitor the oil volume fraction during the mixing process. Figure 4.5.13 shows normalized volume fraction profiles along the vertical center-line (see Figure 4.5.3) at different times. The mixing time is expressed in strain units as $\gamma = tV/(R_o - R_i)$, where t is the time. One revolution of the outer cylinder corresponds to 9.83 strain units. The initial condition ($\gamma = 0$)

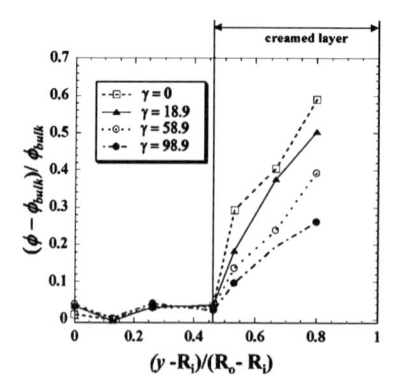

Fig. 4.5.13 Concentration profiles at different strains along the upper vertical center line (see Figure 4.5.3) for $\phi_{bulk} = 0.4$ and $V = 0.025$ cm s^{-1}.

corresponds to the system after 3 h of creaming. The mixing process in this figure consisted of the rotation of the outer cylinder at $V = 0.025$ cm s^{-1}. As shown in Figure 4.5.13, the creamed layer corresponds to the high oil volume fraction region, where the highest oil concentration is located in the top of the creamed layer. It can be seen in the figure that mixing proceeds through the depletion of the creamed layer, where the layer thickness seems to remain constant during the mixing process.

Velocity profiles obtained by TOF in the upper region of the probe during mixing are shown in Figure 4.5.14 for $V = 0.05$ cm s^{-1}. It can be seen that for $\gamma < 290$, the emulsion is flowing only near the outer cylinder wall. At higher values of γ, the flow exhibits counter-rotating profiles, which evolve to steady state at large γ (see Section 4.5.3). The progress of mixing and flow development over the entire cross section can be presented in terms of two averaged parameters, the mixing intensity I_ϕ [77] and the velocity difference intensity I_v [78]:

$$I_\phi \equiv \sum_{i=1}^{N} \frac{(\phi_i - \phi_{bulk})^2}{(\phi_{i\ t=0} - \phi_{bulk})^2};$$

$$I_v \equiv \frac{\int_{R_i}^{R_o} [v_\theta(r) - v_{\theta,steady}(r)]^2\, dr}{\int_{R_i}^{R_o} [v_\theta(r)_{t=0} - v_{\theta,steady}(r)]^2\, dr}$$

(4.5.9)

Here v_θ is the measured tangential velocity profile at time t and (v_θ,steady) is the value at steady-state. Both intensity indices have a value of unity at $t = 0$, and approach zero as t approaches infinity. Figure 4.5.15 shows the variation of the intensity indices with average strain, for an outer cylinder velocity of 0.05 cm s^{-1}. These plots indicate that the mixing process occurs in two stages, where the velocity profile develops only after the droplet concentration profile is essentially uniform. It can be seen that I_ϕ decays to zero at approximately 100 strain units, whereas I_v shows that the steady-state velocity profile is reached only when $\gamma \approx 400$. From Figure 4.5.14 it can be seen that when $\gamma = 115$, flow is detected

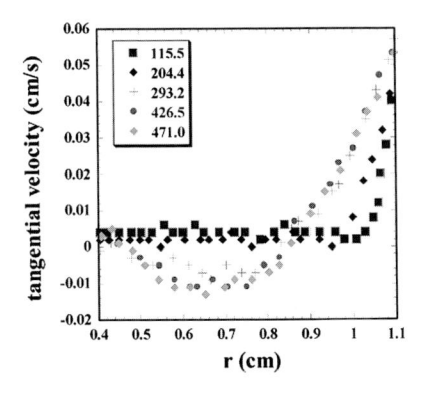

Fig. 4.5.14 Velocity profiles on the upper vertical center line, for a series of strain values for $\phi_{bulk} = 0.4$ and $V = 0.05$ cm s^{-1}; $r = 0.4$ and 1.1 cm indicate the inner and outer cylinder walls, respectively.

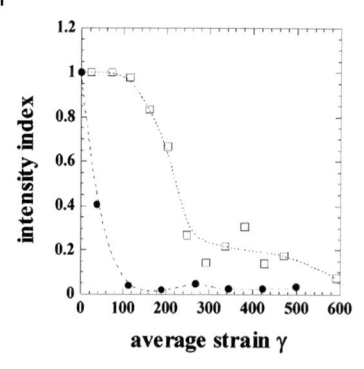

Fig. 4.5.15 Variation of the velocity difference intensity index (□) and concentration mixing intensity index (●) with average strain, for bulk droplet volume fraction of 0.4. Lines are shown to guide the eye.

only in the region close to the outer cylinder wall. Therefore, most of the mixing process occurs when most of the fluid inside the flow cell is stationary. This led us to propose that the mixing process in this case occurs through the transport of the droplets out of the creamed layer through a boundary layer as shown in Figure 4.5.16. According to the proposed model, a simple mass balance on the droplets in the creamed layer leads to a concentration decay rate in the creamed layer and a boundary layer thickness that are consistent with experimental observations of concentration and velocity profiles over time.

The droplet size at the top of the creamed layer should be larger than in the bottom, as larger droplets reach the top wall during the creaming process; therefore, during mixing, the droplet size at the top of the creamed layer should decrease. Figure 4.5.17 shows a plot of the average droplet size as a function of time at the top of the creamed layer ($y/h = 1$) for flow at $V = 0.05$ cm s^{-1}. It can be seen that the average droplet diameter decreases during the mixing. Droplet size distributions were found using a PGSTE pulse sequence (Figure 4.5.5(b)) and Eqs. (4.5.4–4.5.6) applied to the pixel located at the top of the creamed layer on the center line. In order to avoid flow artifacts, the flow was stopped for 5 min before starting the measurements. However, even in this case, flow artifacts for $t > 0$ were

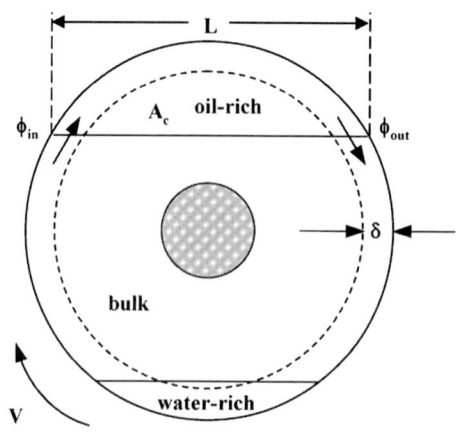

Fig. 4.5.16 Schematic drawing of a boundary layer mixing mechanism. It is proposed that a thin layer with thickness δ has a linear velocity profile with average velocity $V/2$. Material with bulk droplet volume fraction ϕ_{in} is drawn into the creamed layer (area A_c) and material with average creamed layer volume fraction ϕ_{out} is swept out. The remainder of the emulsion (inside the dashed circle) is stagnant.

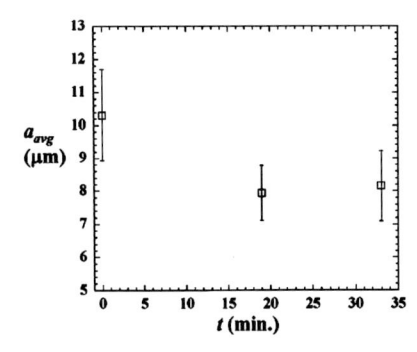

Fig. 4.5.17 Plot of average droplet diameter as a function of time at the top of the creamed layer ($y/h = 1$).

observed from the application of the diffusion gradients G in any direction. It has been recognized that such artifacts would tend to cause over-prediction of the droplet size; therefore the decrease in droplet size should be more pronounced than that shown in Figure 4.5.17 and the result shown in this figure does not have quantitative accuracy. On the other hand, the decrease in the droplet size at the top of the creamed layer during the mixing process agrees qualitatively with the proposed mixing mechanism that a flowing boundary layer transports the largest droplets out of the creamed layer. A more detailed discussion of the mixing mechanism can be found in d'Avila et al. [40]. Details of the experimental procedure used to determine droplet sizes can be found in d'Avila et al. [42] and d'Avila [26].

The flow artifacts detected in the droplet size measurements are similar to those reported by Goux et al. [79] and Mohoric and Stepisnik [80]. In their work natural convection effects led to an increase in the decay of signal attenuation curves, causing over-prediction in the self-diffusion coefficient of pure liquids. In order to avoid flow effects in droplet size distributions, flow compensating pulse sequences such as the double PGSTE should be used. It has been demonstrated recently that this sequence facilitates droplet size measurements in pipe flows [81].

4.5.5
Future Directions

This chapter presents a methodology based on NMRI to study the flow of emulsions. Methods to measure velocity, concentration and droplet size with spatial resolution were successfully implemented by our group and we believe that the methods described here can be extended to other studies on the flow and stability of emulsions. These methods have great potential, as they non-invasively provide spatially and temporally detailed information about the flow and microstructure of emulsions. The research summary presented here on flow and mixing of emulsions has shown novel features of flowing emulsions, involving the interplay between buoyancy and viscous effects. In addition, the mixing behavior of emulsions presents characteristics that are different from the mixing of immiscible single fluids. Our research has also shown the possibility of accurately measuring droplet size distributions with spatial resolution in creamed emulsions, which led to important information regarding the droplet size inside the creamed layer during mixing. However, it was shown that under certain conditions, flow artifacts might lead to errors in droplet size measurements. In such cases, velocity compensated pulse sequences, for example, the double PGSTE should be used. We believe that the extension of the methodology presented here has great potential to lead to further improvements in our understanding of flow, mixing and destabilization of emulsions. Here are some topics that we consider important for future research in this subject:

1. *Studies of flow-induced coalescence* are possible with the methods described here. Effects of flow conditions and emulsion properties, such as shear rate, initial droplet size, viscosity and type of surfactant can be investigated in detail. Recently developed, fast (3–10 s) [82, 83] PFG NMR methods of measuring droplet size distributions have provided nearly "real-time" droplet distribution curves during evolving flows such as emulsification [83]. Studies of other destabilization mechanisms in emulsions such as creaming and flocculation can also be performed.
2. *Flow effects on non-neutrally buoyant emulsions and suspensions* can be studied in various geometries. For example, flow in rotating cylinder and narrow gap concentric cylinder geometries in both horizontal and vertical orientations can be studied. Flow instabilities in settling suspensions in a horizontal rotating cylinder have recently been reported [84]. Measurements of velocity fields have not been reported in the literature, but can be performed by using the methods presented in this work.
3. *Studies of shear-induced droplet migration* are rare in non-dilute emulsion systems but are achievable with the various methods described here [85].
4. *Spatially-resolved measurement of the droplet size distribution* can be accomplished by the implementation of velocity compensated pulse sequences, such as the double PGSTE [81] in a spatially resolved imaging sequence. Accurate measurements of spatially resolved droplet size distributions during flow and mixing of emulsions would provide truly unique information regarding flow effects on the spatial distribution of droplets.
5. *Rheo-NMR* [86] methods have been shown to be well-suited to emulsion rheology studies [28] and could be combined with any of the topics described above. The combination of structural and rheological measurements is a promising area for further research.

Acknowledgements

The authors gratefully acknowledge the supervision of Professors Ronald J. Phillips, Stephanie R. Dungan and Robert L. Powell and Dr. Jeffrey H. Walton at the University of California, Davis, and financial support of this work from NASA (NRA- 96 – HEDS – 01), CNPq (Brazil) – (200382/97–7), FAPESP (Brazil) – 03/01892–0 and NSF (CTS-98126).

References

1 B. P. Binks **1998**, (Emulsions – recent advances and understanding), in *Modern Aspects of Emulsion Science*, ed. B. P. Binks, The Royal Society of Chemistry, Cambridge.

2 R. J. Hunter **1995**, *Foundations of Colloid Science – Volume I*, Clarendon Press, Oxford.

3 P. T. Callaghan, K. W. Jolley, R. S. J. Humphrey **1983**, (Diffusion of fat and water in cheese as studied by pulsed field gradient nuclear magnetic-resonance), *Colloid Interface Sci.* 93, 521.

4 H.-Y. Lee, M. J. McCarthy, S. R. Dungan **1998**, (Experimental characterization of emulsion formation and coalescence by nuclear magnetic resonance restricted diffusion techniques), *J. Am. Oil Chem. Soc.* 75, 463.

5 T. G. Mason, J. Bibette **1996**, (Emulsification in viscoelastic media), *Phys. Rev. Lett.* 77, 3481.

6 J. Bibette, **1991**, (Depletion interactions and fractionated crystallization for polydisperse emulsion purification), *J. Colloid Interface Sci.* 147, 474.

7 A. Z. Zinchenko, R. H. Davis **2002**, (Shear flow of highly concentrated emulsions of deformable drops by numerical simulations), *J. Fluid Mech.* 455, 21.

8 V. Mishra, S. M. Kresta, J. H. Masliyah **1998**, (Self-preservation of the drop size distribution function and variation in the stability ratio for rapid coalescence of a polydisperse emulsion in a simple shear field), *J. Colloid Interface Sci.* 197, 57.

9 A. Nandi, A. Mehra, D. V. Khakhar **1999**, (Suppression of coalescence in surfactant stabilized emulsions by shear flow), *Phys. Rev. Lett.* 83, 2461.

10 H. A. Stone **1994**, (Dynamics of drop deformation and breakup in viscous fluids), *Annu. Rev. Fluid Mech.* 26, 65.

11 R. G. Larson **1999**, *The Structure and Rheology of Complex Fluids*, Oxford University Press, New York.

12 X.-F. Yuan, M. Doi **1998**, (A general approach for modelling complex fluids: its application to concentrated emulsions under shear), *Colloid Surf. A* 144, 305.

13 P. C.-H. Chan, L. G. Leal **1979**, (Motion of a deformable drop in a 2nd-order fluid), *J. Fluid Mech.* 92, 131.

14 P. Walkenstrom, A. M. Hermansson **2002**, (Microstructure in relation to flow processing), *Curr. Opin. Colloid Interface Sci.* 7, 413.

15 S. Velankar, P. Van Puyvelde, J. Mewis, P. Moldenaers **2001**, (Effect of compatibilization on the breakup of polymeric drops in shear flow), *J. Rheol.* 45, 1007.

16 Y. A. Antonov, P. Van Puyvelde, P. Moldenaers **2004**, (Effect of shear flow on the phase behavior of an aqueous gelatin-dextran emulsion), *Biomacromolecules* 5, 276.

17 J.-B. Salmon, L. Becu, S. Manneville, A. Colin **2003**, (Towards local rheology of emulsions under Couette flow using dynamic light scattering), *Eur. Phys. J. E* 10, 209.

18 M. R. King, D. T. Leighton **2001**, (Measurement of shear-induced dispersion in a dilute emulsion), *Phys. Fluids* 13, 397.

19 J. A. Pathak, M. C. Davis, S. D. Hudson, K. B. Migler **2002**, (Layered droplet microstructures in sheared emulsions: finite-size effects), *J. Colloid Interface Sci.* 255, 391.

20 L. F. Gladden **2003**, (Magnetic resonance: ongoing and future role in chemical engineering research), *AIChE J.* 49, 2.

21 M. D. Mantle, A. J. Sederman **2003**, (Dynamic MRI in chemical process and reaction engineering), *Prog. Nucl. Mag. Reson. Spec.* 43, 3.

22 G. J. W. Goudappel, J. P. M. van Duynhoven, M. M. W. Mooren **2001**, (Measurement of oil droplet size distributions in food oil/water emulsions by time domain pulsed field gradient NMR), *J. Colloid Interface Sci.* 239, 535.

23 D. F. Arola, G. A. Barrall, R. L. Powell, K. L. McCarthy, M. J. McCarthy **1997**,

(Use of nuclear magnetic resonance imaging as a viscometer for process monitoring), *Chem. Eng. Sci.* 52, 2049.

24 L. D. Hall, M. H. G. Amin, S. Evans, K. P. Nott, L. Sun **2001**, (Magnetic resonance imaging for industrial process tomography), *J. Electron. Imag.* 10, 601.

25 B. Newling, S. J. Gibbs, L. D. Hall, D. E. Haycock, W. J. Frith, S. Ablet **1997**, (Chemically resolved NMR velocimetry), *Chem. Eng. Sci.* 52, 2059.

26 M. A. d'Avila **2003**, *Flow and Characterization of Emulsions by Nuclear Magnetic Resonance*, Ph.D. Dissertation, University of California at Davis.

27 N. C. Shapley, M. A. d'Avila, J. H. Walton, R. L. Powell, S. R. Dungan, R. J. Phillips **2003**, (Complex flow transitions in a homogeneous, concentrated emulsion), *Phys. Fluids* 15, 881.

28 K. G. Hollingsworth, M. L. Johns **2004**, (Rheo-nuclear magnetic resonance of emulsion systems), *J. Rheol.* 48, 787.

29 A. W. Chow, S. W. Sinton, J. H. Iwamiya, T. S. Stephens **1994**, (Shear-induced particle migration in Couette and parallel-plate viscometers: NMR imaging and stress measurements. *Phys. Fluids* 6, 2561.

30 J. R. Abbott, N. Tetlow, A. L. Graham, S. A. Altobelli, E. Fukushima, L. A. Mondy, T. S. Stephens **1991**, (Experimental observations of particle migration in concentrated suspensions – Couette flow), *J. Rheol.* 35, 773.

31 Y. J. Choi, K. L. McCarthy, M. J. McCarthy **2002**, (Tomographic techniques for measuring fluid flow properties), *J. Food Sci.* 67, 2718.

32 Y. Uludag, G. A. Barrall, D. F. Arola, M. J. McCarthy, R. L. Powell **2001**, (Polymer melt rheology by magnetic resonance imaging), *Macromolecules* 34, 5520.

33 E. K. Jeong, S. A. Altobelli, E. Fukushima **1994**, (NMR imaging studies of stratified flows in a horizontal rotating cylinder), *Phys. Fluids* 6, 2901.

34 J. D. Seymour, B. Manz, P. T. Callaghan **1999**, (Pulsed gradient spin echo nuclear magnetic resonance measurements of hydrodynamic instabilities with coherent structure: Taylor vortices), *Phys. Fluids* 11, 1104.

35 M. M. Britton, R. W. Mair, R. K. Lambert, P. T. Callaghan **1999**, (Transition to shear banding in pipe and Couette flow of wormlike micellar solutions), *J. Rheol.* 43, 897.

36 T.-Q. Li, J. D. Seymour, M. J. McCarthy, K. L. McCarthy, L. Ödberg, R. L. Powell **1994**, (Turbulent pipe-flow studied by time-averaged NMR imaging – measurements of velocity profile and turbulent intensity), *Magn. Reson. Imag.* 12, 923.

37 R. W. Mair, M. D. Hurlimann, P. N. Sen, L. M. Schwartz, R. L. Walswothry **2001**, (Tortuosity measurement and the effects of finite pulse widths on xenon gas diffusion NMR studies of porous media), *Mag. Reson. Imag.* 19, 345.

38 K. J. Packer **2003**, (Magnetic resonance in porous media: forty years on), *Mag. Reson. Imag.* 21, 163.

39 K. L. McCarthy, Y. Lee, J. Green, M. J. McCarthy **2002**, (Magnetic resonance imaging as a sensor system for multiphase mixing), *Appl. Mag. Reson.* 22, 213.

40 M. A. d'Avila, N. C. Shapley, J. H. Walton, S. R. Dungan, R. J. Phillips, R. L. Powell **2003**, (Mixing of concentrated emulsions measured by nuclear magnetic resonance imaging), *Phys. Fluids* 15, 2499.

41 P. J. McDonald, E. Ciampi, J. L. Keddie, M. Heidenreich, R. Kimmich **1999**, (Magnetic-resonance determination of the spatial dependence of the droplet size distribution in the cream layer of oil-in-water emulsions: Evidence for the effects of depletion flocculation) *Phys. Rev. E* 59, 874.

42 M. A. d'Avila, R. L. Powell, R. J. Phillips, N. C. Shapley, J. H. Walton, S. R. Dungan **2005**, (Magnetic resonance imaging: A technique to study flow and microstructure of emulsions), *Braz. J. Chem. Eng.* 22, 49.

43 B. Newling, P. M. Glover, J. L. Keddie, D. M. Lane, P. J. McDonald **1997**, (Concentration profiles in creaming oil-in-

water emulsion layers determined with stray field magnetic resonance imaging), *Langmuir* 13, 3621.
44 N. Hedin, I. Furo **2001**, (Ostwald ripening of an emulsion monitored by PGSE NMR), *Langmuir* 17, 4746.
45 M. M. Robins, A. D. Watson, P. J. Wilde **2002**, (Emulsions – creaming and rheology), *Curr. Opin. Colloid Interface Sci.* 7, 419.
46 P. V. Nelson, M. J. W. Povey, Y. Wang **2001**, (An ultrasound velocity and attenuation scanner for viewing the temporal evolution of a dispersed phase in fluids), *Rev. Sci. Instrum.* 72, 4234.
47 R. Chanamai, N. Herrmann, D. J. McClements **2000**, (Probing floc structure by ultrasonic spectroscopy, viscometry, and creaming measurements), *Langmuir* 16, 5884.
48 J. Bouillard, B. Alban, P. Jacques, C. Xuereb **2001**, (Liquid flow velocity measurements in stirred tanks by ultrasound doppler velocimetry), *Chem. Eng. Sci.* 56, 747.
49 S. Manneville, J.-B. Salmon, A. Colin **2004**, (A spatio-temporal study of rheo-oscillations in a sheared lamellar phase using ultrasound), *Eur. Phys. J. E* 13, 197.
50 N. Dogan, M. J. McCarthy, R. L. Powell **2003**, (Comparison of in-line consistency measurement of tomato concentrates using ultrasonics and capillary methods), *J. Food Proc. Eng.* 25, 571.
51 V. J. Pinfield, M. J. W. Povey, E. Dickinson **1996**, (Interpretation of ultrasound velocity creaming profiles), *Ultrasonics* 34, 695.
52 D. J. McClements, N. Herrmann, Y. Hemar **1998**, (Influence of flocculation on the ultrasonic properties of emulsions: theory), *J. Phys. D: Appl. Phys.* 31, 2950.
53 R. Chanamai, N. Herrmann, D. J. McClements **1998**, (Influence of flocculation on the ultrasonic properties of emulsions: experiment), *J. Phys. D: Appl. Phys.* 31, 2956.
54 R. Chanamai, N. Herrmann, D. J. McClements **2000**, (Probing floc structure by ultrasonic spectroscopy, viscometry, and creaming measurements), *Langmuir* 16, 5884.
55 P. V. Nelson, M. J. W. Povey, Y. Wang **2001**, (An ultrasound velocity and attenuation scanner for viewing the temporal evolution of a dispersed phase in fluids), *Rev. Sci. Instrum.* 72, 4234.
56 J. L. Harden, V. Viasnoff **2001**, (Recent advances in DWS-based micro-rheology), *Curr. Opin. Colloid Interface Sci.* 6, 438.
57 Y. Hemar, D. N. Pinder, R. J. Hunter, H. Singh, P. Hebraud, D. S. Horne **2003**, (Monitoring of flocculation and creaming of sodium-caseinate-stabilized emulsions using diffusing-wave spectroscopy), *J. Colloid Interface Sci.* 264, 502.
58 E. ten Grotenhuis, M. Paques, G. A. van Aken **2000**, (The application of diffusing-wave spectroscopy to monitor the phase behavior of emulsion-polysaccharide systems), *J. Colloid Interface Sci.* 227, 495.
59 P. Hebraud, F. Lequeux, J. P. Munch, D. J. Pine **1997**, (Yielding and rearrangements in disordered emulsions), *Phys. Rev. Lett.* 78 (24), 4657.
60 Y. Nicolas, M. Paques, A. Knaebel, A. Steyer, J.-P. Munch, T. B. J. Blijdenstein, G. A. van Aken **2003**, (Microrheology: structural evolution under static and dynamic conditions by simultaneous analysis of confocal microscopy and diffusing wave spectroscopy), *Rev. Sci. Instrum.* 74, 3838.
61 D. W. Alderman, D. M. Grant **1979**, (Efficient decoupler coil design which reduces heating in conductive samples in superconducting spectrometers), *J. Magn. Reson.* 36, 447.
62 P. T. Callaghan **1991**, (Principles of nuclear magnetic resonance microscopy), Clarendon Press, Oxford.
63 P. T. Callaghan, Y. Xia **1991**, (Velocity and diffusion imaging in dynamic NMR microscopy), *J. Magn. Reson.* 91, 326.
64 A. Caprihan, E. Fukushima **1990**, (Flow measurements by NMR), *Phys. Rep.* 198, 195.

65 J. M. Pope, S. Yao **1993**, (Quantitative NMR imaging of flow), *Conc. Magn. Reson.* 5, 281.

66 E. Fukushima **1999**, (Nuclear magnetic resonance as a tool to study flow), *Annu. Rev. Fluid Mech.* 31, 95.

67 H. Rumpel, J. M. Pope **1993**, (Chemical shift imaging in nuclear magnetic resonance: a comparison of methods), *Conc. Magn. Reson.* 5, 43.

68 A. M. Corbett, R. J. Phillips, R. J. Kauten, K. L. McCarthy **1995**, Magnetic resonance imaging of concentration and velocity profiles of pure fluids and solid suspensions in rotating geometries), *J. Rheol.* 39, 907.

69 S. Bobroff, R. J. Phillips **1998**, (Nuclear magnetic resonance imaging investigation of sedimentation of concentrated suspensions in non-Newtonian fluids), *J. Rheol.* 42, 1419.

70 S. Yao, A. G. Fane, J. M. Pope **1997**, (An investigation of the fluidity of concentration polarisation layers in crossflow membrane filtration of an oil-water emulsion using chemical shift selective flow imaging), *Mag. Reson. Imag.* 15, 235.

71 K. J. Packer, C. Rees **1972**, (Pulsed NMR studies of restricted diffusion. 1. Droplet size distributions in emulsions), *J. Colloid Interface Sci.* 40, 206.

72 B. Balinov, B. Jonsson, P. Linse, O. Soderman **1993**, (The NMR self-diffusion method applied to restricted diffusion – simulation of echo attenuation from molecules in spheres and between planes), *J. Mag. Reson. A* 104, 17.

73 X. Li, J. C. Cox, R. W. Flumerfelt **1992**, (Determination of emulsion size distribution by NMR restricted diffusion measurement), *AIChE J.* 38, 1671.

74 K. G. Hollingsworth, M. L. Johns **2003**, (Measurement of emulsion droplet sizes using PFG NMR and regularization methods), *J. Colloid Interface Sci.* 258, 383.

75 R. M. Kroeker, R. M. Henkelman **1986**, (Analysis of biological NMR relaxation data with continuous distributions of relaxation-times), *J. Mag. Reson.* 69, 218.

76 P. DeRoussel, D. V. Khakhar, J. M. Ottino **2001**, (Mixing of viscous immiscible liquids. Part 1: Computational models for strong-weak and continuous flow systems), *Chem. Eng. Sci.* 56, 5511.

77 Y. Lee, M. J. McCarthy, K. L. McCarthy **2001**, (Extent of mixing in a two-component batch system measured using MRI), *J. Food Eng.* 50, 167.

78 R. E. Hampton, A. A. Mammoli, A. L. Graham, N. Tetlow, S. A. Altobelli **1997**, (Migration of particles undergoing pressure-driven flow in a circular conduit), *J. Rheol.* 41, 621.

79 W. J. Goux, L. A. Verkruyse, S. J. Salter **1990**, (The impact of Rayleigh-Benard convection on NMR pulsed-field-gradient diffusion measurements), *J. Mag. Reson.* 88, 609.

80 A. Mohoric, J. Stepisnik **2000**, (Effect of natural convection in a horizontally oriented cylinder on NMR imaging of the distribution of diffusivity), *Phys. Rev. E* 62, 6628.

81 M. L. Johns, L. F. Gladden **2002**, (Sizing of emulsion droplets under flow using flow-compensating NMR-PFG techniques), *J. Mag. Reson.* 154, 142.

82 C. Buckley, K. G. Hollingsworth, A. J. Sederman, D. J. Holland, M. L Johns, L. F. Gladden **2003**, (Applications of fast diffusion measurement using Difftrain), *J. Mag. Reson.* 161, 112.

83 K. G. Hollingsworth, A. J. Sederman, C. Buckley, L. F. Gladden, M. L. Johns **2004**, (Fast emulsion droplet sizing using NMR self-diffusion measurements), *J. Colloid Interface Sci.* 274, 244.

84 W. R. Matson, B. J. Ackerson, P. Tong **2003**, (Pattern formation in a rotating suspension of non-Brownian settling particles), *Phys. Rev. E* 67 (5), art. 050301, part 1.

85 K. G. Hollingsworth, J. P. Hindmarsh, A. J. Sederman, L. F. Gladden, M. L. Johns **2003**, *Fast Magnetic Resonance Characterisation of Multi-Phase Liquids*, paper presented at the 7th International Conference on Magnetic Resonance Microscopy, Snowbird, UT, September, 2003.

86 P. T. Callaghan **1999**, (Rheo-NMR: nuclear magnetic resonance and the rheology of complex fluids), *Rep. Prog. Phys.* 62, 599–670.

4.6
Fluid Flow and Trans-membrane Exchange in a Hemodialyzer Module
Song-I Han and Siegfried Stapf

4.6.1
Objective

The purpose of a hemodialyzer is to support or replace the function of the human kidney, that is, to clean used blood from various organic contaminants and waste products. Frequently, the malfunction of the kidneys also requires the mechanical removal of water from the blood, a process that is called ultrafiltration [1, 2]. Common to the wide range of specialized products in dialysis is the aim of optimizing the filtration process in order to reduce the time required for a dialysis session – typically 4 hours with modern technology – which is, at the same time, a relevant cost factor. In order to meet this aim, molecular exchange is induced between the human blood and a second fluid phase through a semipermeable membrane of large surface area. This is achieved by pumping blood inside narrow (typically a few 100 µm) hollow-fiber membranes, between 10 000 and 20 000 that are bundled inside a cylindrical container, and generating a counterflow of the cleaning fluid in the surrounding space of this container (Figure 4.6.1 (a, b)). The membrane is typically made of cuprophane, cellulose acetate, polyacrylonitrile, polysulfone or poly(methyl methacrylate) [3–5]. Exchange through the membrane is induced by a combination of diffusion, mechanical dispersion, convection, pressure differences and osmotic pressure. The total efficiency of the dialysis process depends on an optimal filtration rate but also on avoiding backfiltration because the contaminants are able to pass through the membrane in either direction [4–6]. The main interest for a better understanding of the filtration process thus lies in the quantification of the contribution of diffusion and trans-membrane mass flow to the net flow distribution. The general flow behavior is influenced by the membrane material itself, by how evenly and densely the capillary membranes are packed, and also by the operating parameters of the dialysis process (e.g., pressure, flow rate) [7, 8]. Magnetic resonance flow imaging methodologies provide ideal and unique tools to visualize and quantify counterflow characteristics in an operating hemodialyzer module, among others because the system (membrane material, blood) is opaque, which makes it inaccessible to other imaging techniques [9–12]. The properties of counterflow through miniaturized ($\varnothing$ = 9 mm) model hemodialyzer modules (Figure 4.6.1 (c, d)) [13, 14] with two different membrane materials were investigated by various magnetic resonance flow imaging tools and are presented in this chapter.

4.6.2
Methods

Velocity-encoding 2D NMR imaging methods characterize general patterns of spatial velocity distributions and directly visualize different characteristics of flow behavior depending on the properties of the materials and operating parameters of the dialysis process. The spatial distribution of flow (z) velocities in a cross section (x, y) of the dialyzer is displayed via a 2D parameter map $v_z(x, y)$ where velocity is encoded into the signal phase by a PFG pair within the imaging sequence (see Chapter 1, Introduction). Another variant of a two-dimensional display is the plot of downstream velocity as a function of the transverse (x) coordinate in a projection image, $P(v_z, x)$; here, the propagator, i.e., the distribution of flow velocities v_z, is obtained from a stepwise variation of the encoding PFG pair. Furthermore, velocity distribution functions averaged over an extended volume (the entire sample or the sample volume of interest) can be obtained by Pulsed Gradient Spin-Echo (PGSE) measurements in longitudinal (z) and transverse (x) directions relative to the pressure gradient. A 2D Velocity Exchange SpectroscopY (VEXSY) [15] experiment autocorrelates the velocity distribution along one spatial direction, here the mean flow direction (z), at two different times separated by a mixing (or observation) time t_m. The method consists of applying two PFG pairs independently of each other (see Chapter 1, Introduction). In the 2D VEXSY map, the time evolution of the velocity distribution of an ensemble of flowing molecules is visualized. By investigating the projections on the two diagonals of the plot, either average velocities or velocity *changes* during t_m can be quantified; in particular, it is this latter information that allows one to gain insight into the peculiarities of the filtration process [16–18].

4.6.3
Materials

The flow behavior in miniaturized hemodialyzer modules with two types of biocompatible membrane materials, SMC and SPAN, was investigated by using doubly distilled water as the flowing fluid in both compartments, subsequently termed "membrane side (M)" and "dialysate side" (D), respectively (Figure 4.6.1 (c, d)) [12]. SMC stands for Synthetically Modified Cellulose and SPAN for Special PolyAcryloNitrile-based copolymer (Akzo Nobel, Membrana GmbH), both types representing standard membrane material. The capillaries made from this hollow

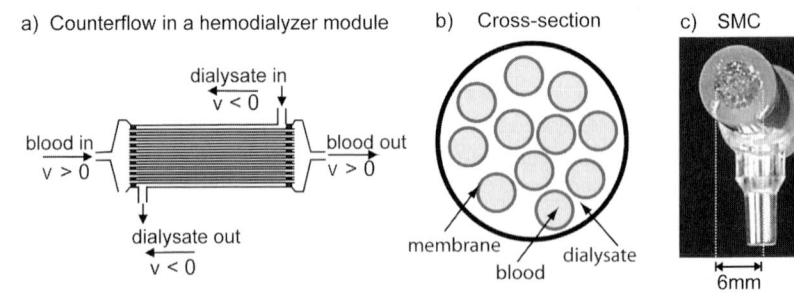

Fig. 4.6.1 (a) Diagram of the counterflow principle of membrane and dialysate flow in a hemodialyzer module. (b) Schematic depiction of the module cross section. (c–d) Photographs of the cross sections of the mini-hemodialyzer modules used in this study.

fiber membrane have to fulfill high standards of excellent hemocompatibility, optimized bundle construction and have to sustain various clinical sterilizing processes. The miniaturized SMC module consists of 174 capillaries of 200 µm inner diameter with a wall thickness of 8 µm and its membrane material is suitable for low flux hemodialysis, whereas the miniaturized SPAN module consists of about 310 capillaries of the same inner diameter but with a wall thickness of 30 µm and its material is suitable for high flux hemodialysis. SMC sustains a wider range of sterilizing processes while SPAN exhibits outstanding hemocompatibility.

4.6.4
Results and Discussion

4.6.4.1 Velocity Encoded Images of SMC and SPAN

A simple but illustrative approach for the visualization of flow properties and heterogeneities is to combine the measurement of the velocity distribution with imaging along one spatial direction. In Figure 4.6.2, the longitudinal velocity component v_z is plotted versus the transverse position coordinate x.

Flow inside the capillary membranes, depicted in the lower half of the plot and indicated by positive velocities, shows a regular pattern. The single capillaries are resolved, and flow inside each capillary possesses almost identical maximum velocities. Flow outside the membranes (upper half, negative velocities) reveals a different pattern. Here, the different flow characteristics between the SMC and SPAN modules become distinct. Obviously, the capillaries in the SMC module are not packed in a regular manner. Large spaces in between the capillaries cause an irregular flow pattern in the dialysate-side with a maximum velocity of about -15 mm s^{-1} (Figure 4.6.2(a)), which is comparable to the maximum velocity in

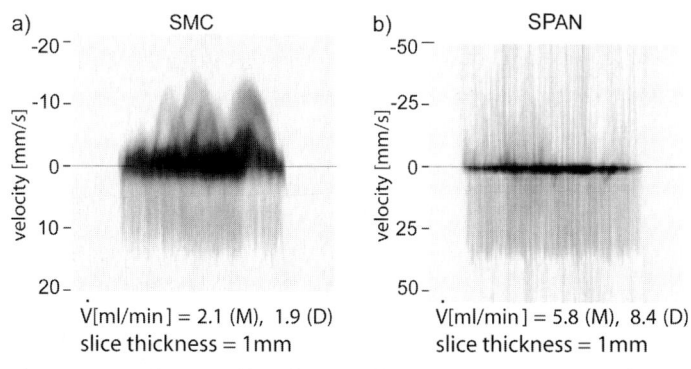

Fig. 4.6.2 1D velocity profiles of counterflow through the mini-hemodialyzer modules of type SMC (a) and SPAN (b), where velocity along z is plotted versus the x position axis (total width 9 mm), with z representing the flow direction. Positive velocities correspond to membrane-side flow (M) and negative velocities to dialysate-side flow (D). The flow rate $\dot{V}$ is labeled in the figure. Slice selection along z was employed, and a gradient pulse duration of $\delta = 2$ ms, and an encoding time $\Delta = 5$ ms were used. The spatial resolution is 31 µm and the velocity resolution 0.63 mm s^{-1}.

the membrane side of about +13 mm s^{-1}. Also noticeable is the high intensity around zero and small negative velocities in the flow image of the SMC module (black region in the center). This means that a large portion of the dialysate is trapped or slowed down in the inter-membrane space. This could be due to hold-up caused by entangled membranes so that some portion of the flowing dialysate has to find its flow path around irregular inter-membrane spacing. If this interpretation is correct, it also means that considerable transverse velocity components must be present; this aspect will be discussed later. In the SPAN module (Figure 4.6.2(b)), a much higher maximum velocity in excess of –50 mm s^{-1} in the dialysate-side is present compared with the maximum velocity of 35 mm s^{-1} in its membrane-side. Also, a very regular packing of the membrane bundle becomes obvious, being reflected in a homogeneous spatial pattern resolving single capillary membranes in both compartments, which contain regular intra-capillary and inter-capillary flow patterns. This confirms the higher packing homogeneity of the capillaries in the SPAN modules, which results in coherent and unhindered flow. Note that very narrow spaces between the capillaries seem to exist due to compact packing as opposed to the characteristics in the SMC module, causing partially very high flow velocities reaching outside the measurable range of the experimental setting used in Figure 4.6.2. One has to keep in mind that "too narrow" spaces can cause an increased pressure drop and result in lower velocities. However, here the direct visualization of the regular narrower spacing together with the presence of higher flow velocities makes the origin of this behavior unambiguous. This experiment gives one an idea about the packing characteristics of the capillary membranes and the resulting flow pattern inside the hemodialyzer modules. Judging by the flow pattern, a better exchange efficiency is to be expected for the SPAN module due to more efficient contact between the two compartments.

In the following step, 2D velocity maps were generated which represent 2D image slices along the xy plane encoded by the longitudinal velocity v_z utilizing the NMR phase information of each image pixel. Two sets of 2D spin density images (a, c) together with the corresponding velocity maps (b, d) obtained from the SMC and the SPAN module are presented in Figure 4.6.3.

In the spin density images of the SMC module (Figure 4.6.3(a)), the single capillaries are clearly resolved. Artifacts due to motional blurring (during ≈10 ms observation time) along the transverse directions x and y can be discerned. This means that noticeable transverse dynamics are present, as predicted previously, either caused by non-straight or moving capillary membranes, or by a trans-membrane filtration process. The corresponding velocity map (Figure 4.6.3(b)) obtained from the NMR phase distribution shows a different contrast which is given by velocity instead of spin density, but the correlation of the spatial pattern between these image pairs is clearly recognizable. Very low spin density spots in Figure 4.6.3(a) that are obviously due to large spacing between the loosely packed membranes correlate with a rather large negative velocity (two black spots in Figure 4.6.3(b)). This seems contradictory, but if we consider that a large portion of dialysate is significantly trapped or hindered causing small flow velocities, large

magnitude image
a) resolution 40 µm, slice=1mm

SMC

0   position [mm]   7

velocity map
b) ⊓ 0.12T/m, δ=2ms, Δ=4ms

SMC

velocity [mm/s]: 132, 88, 44, 0, -44, -88

0   position [mm]   7

c) resolution 40 µm, slice=1mm

SPAN

0   position [mm]   7

d) ⊓ 0.08T/m, δ=2ms, Δ=4ms

SPAN

velocity [mm/s]: 60, 30, 0, -30, -60

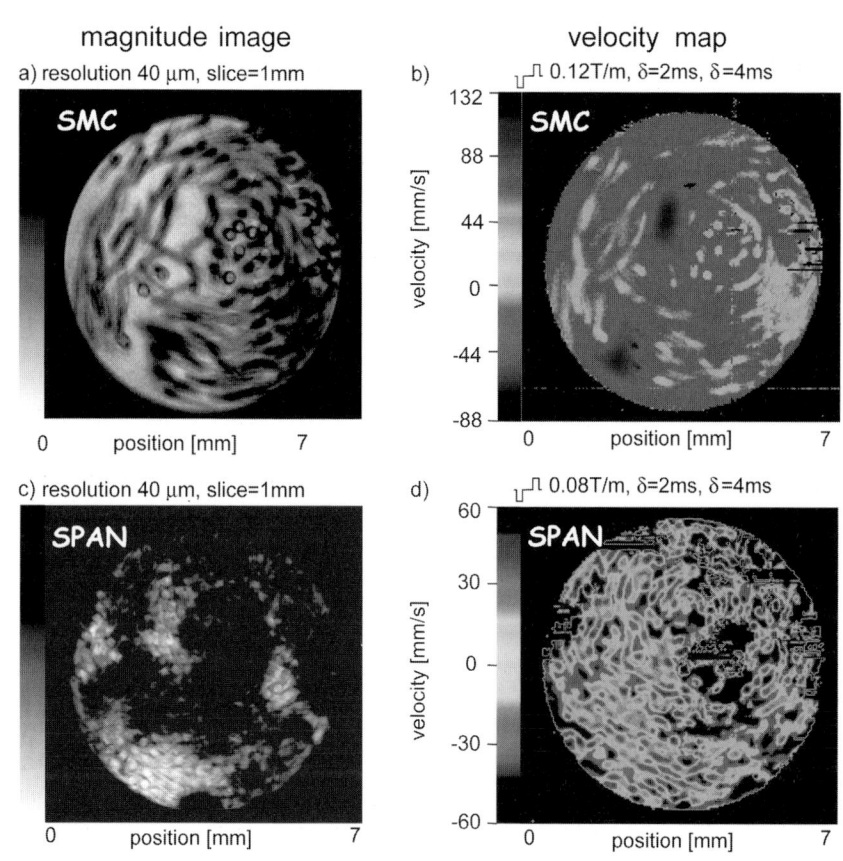

0   position [mm]   7

Fig. 4.6.3 2D spin density images (a, c) and corresponding 2D velocity maps (b, d) along the cross section of miniaturized hemodialyzer modules of the type SMC and SPAN. The applied flow rate on the SMC module is 2.1 mL min^{-1} for the membrane side and 1.9 mL min^{-1} for the dialysate side, and for the SPAN module 2.6 mL min^{-1} for the membrane side and 3.7 mL min^{-1} for the dialysate side.

inter-membrane spacing should result in unhindered flow, and therefore considerably large flow velocities. This can be confirmed by the magnitude of flow of the above mentioned black spots in Figure 4.6.3(b) (corresponding to about –60 mm s^{-1}) presenting the dialysate flow. One can confirm that the membranes in SMC are not well packed causing an uneven velocity distribution in the dialysate-side, and that most of the velocities are represented in the range of ± 15 mm s^{-1}, as can be readily recognized in Figure 4.6.2(a) and embedded in Figure 4.6.3(b) as the grey color scale distribution.

The images obtained from the SPAN module (Figure 4.6.3(c and d)) show completely different characteristics compared with those from the SMC module. Noticeable features are the dense and evenly packed capillary membranes and the lower quality and inhomogeneities in the images. As already discussed for the 1D profile (Figure 4.6.2(b)), both intra- and inter-membrane flow seems unhindered and fast, and particularly narrow spacing between some capillary membranes causes velocities higher than the average value. Such high velocities result in signal loss due to outflow from the sensitive volume of the resonator, in artifacts in the spin density image due to motional blurring which indicates the presence of large transverse motion and, finally, in high fluctuations in velocity. Because the velocities are calculated from the NMR phase value, those ϕ exceeding the continuity range of $-\pi < \phi < \pi$ cause phase wrapping, making the unambiguous determination of velocities impossible. Phase wrapping in cases of steady and moderate spatial variation of velocities can be corrected for by a suitable routine (see, for instance, Chapters 2.9, 4.2), but not in the case of sudden phase changes from one position to another. Black spots in Figure 4.6.3(d) showing a rupture in the velocity image from one pixel to the next are due to this "irreparable" phase wrapping. Most of the presented velocities lie in the range of ± 35 mm s^{-1}. Those resolved capillaries, containing large positive intra-membrane flow velocities (dark grey), are first surrounded by a layer of low velocities (light grey) and subsequently surrounded by large negative velocities of the dialysate flow (dark grey). Thus counterflow is indeed present around each single capillary membrane, which should result in very good filtration efficiency [see Figure 4.6.1(b) for illustration]. The observation that the distribution of low and zero velocities around each capillary membrane is relatively broad proves the presence of permeation across the porous hollow fiber membranes indicating considerable trans-membrane flow. Thus, from the point of view of investigating filtration, the SPAN module seems to be more promising.

The 2D velocity images confirmed the observation obtained from the 1D velocity profiles, and gave further insights by adding one more spatial dimension. The visualization of the spatial distribution of the capillary membranes and the assignment to the resulting flow pattern became reliable by means of 2D velocity maps. However, the even distributions of the membrane-side flow and the very different dialysate flow patterns between the SMC and SPAN module is more readily distinguishable from the 1D profiles (Figure 4.6.2). Yet, the most important question about the physical mechanism of exchange, i.e., whether diffusion or convection dominates trans-membrane flow, cannot be answered by these measurements.

4.6.4.2 Velocity Distribution in SMC and SPAN

After the investigation of the spatial velocity distribution and characterizing the different flow patterns, probability densities of velocities averaged over the cross section of the samples measured by PGSE techniques are discussed. Figure 4.6.4(a–d) presents such averaged velocity distributions along the x, y and z directions, for a given flow rate inside the modules of type SPAN and of SMC, respectively. The averaged z velocity propagator shows a fairly uniform distribution for the membrane-side, representing ordered laminar flow inside the hollow fiber membranes on the left-hand side of the distribution (positive velocities). This indicates that laminar flow with identical condition (e.g., velocity, capillary diameter and pressure difference) is present in almost all capillaries. The dialysate-side shows a "fast decaying" distribution representing negative velocities outside of the membrane on the right-hand side of this distribution. One reason for this latter non-uniform decay function is the non-spherical geometry of the space between the hollow fibers for which the flow behavior is not as simple as for pipe flow. Another reason is the non-perfectly packed fibers, which causes different sizes of spacing between them, resulting in an overlap of various velocity distribution patterns. The velocity distributions along x and y are symmetrical about zero and have much narrower Gaussian distributions compared with z (Figure 4.6.4(b versus d)). In case of the SPAN module, the transverse velocity distribution is clearly broadened compared with the distribution due to self-diffusion, which is superimposed on the figures and labeled as "static" (Figure 4.6.4(b)). This indicates the presence of stronger trans-membrane motion than in case of pure diffusive motion within the SPAN membranes. The SMC module shows no noticeably

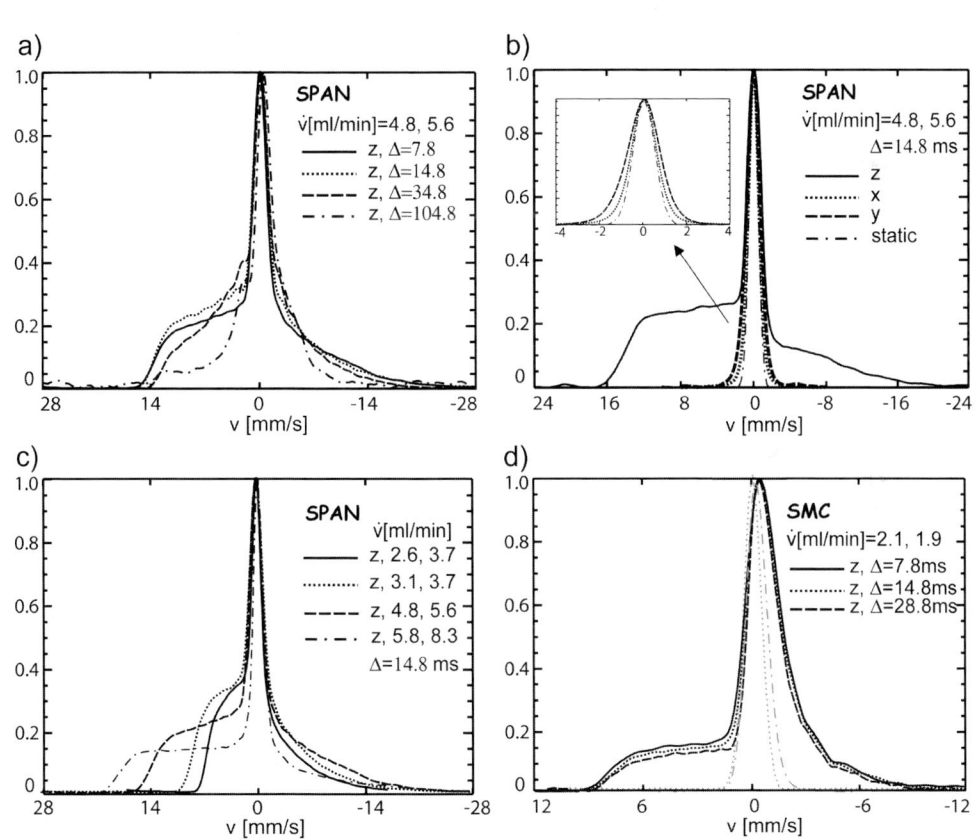

Fig. 4.6.4 Statistical velocity distributions. (a) SPAN. Longitudinal (z) velocity distribution at varying observation times Δ. (b) SPAN. Velocity distribution along x, y and z compared with the computed distribution caused by self-diffusion ("static"). Flow rates: 4.8 mL min^{-1} (M), 5.6 mL min^{-1} (D). (c) SPAN. Longitudinal (z) velocity distribution at varying flow rates. (d) SMC. Longitudinal (z) velocity distribution at varying Δ of between 7.8 and 28.8 ms together with the x velocity distribution, and the velocity distribution caused by self-diffusion ("static") is displayed.

broader transverse velocity distribution than expected from pure self-diffusion (Figure 4.6.4(d)). Note that the influence of bent capillaries is largely covered in the range of translational displacement due to self-diffusion for the given experimental parameters. For example, a bending of 5° relative to the pressure axis imposes a transverse velocity of 1 mm s^{-1} on those molecules traveling with v_z = 10 mm s^{-1}, which is well within the range of transverse velocities caused by self-diffusion for the given observation time.

The effect of varying observation times, Δ, on the statistical z velocity distribution has been investigated (Figure 4.6.4(a, d)). While it was found that the z velocity distributions of the SMC module hardly differ for different Δ values (Figure 4.6.4(d)), there is a clear dependence of the shape of the velocity distribution function of the SPAN module on the Δ value (Figure 4.6.4(a)). This supports the assumption that inside each of the capillaries in the SMC module, it is mainly pure laminar flow that is present, which is only broadened by diffusion. For observation times Δ of between 9 and 30 ms over which the velocity is measured, the velocity profile is therefore practically independent of Δ (Figure 4.6.4(d)). The sensitive dependence of the velocity distribution pattern on the encoding time Δ in the SPAN module indicates the presence of non-laminar flow and strong transverse velocity components caused by transverse mass flow such as convection. No attempt at a more detailed discussion will be made here.

4.6.4.3 VEXSY on SMC and SPAN

Finally, the information gained by obtaining the two-dimensional autocorrelation of two z velocity distribution functions at subsequent time intervals of duration Δ, separated by t_m from the 2D VEXSY experiment will be discussed. This is the direct, and therefore most straightforward, approach to investigating the trans-membrane flow characteristics, because this 2D correlation map represents nothing other than the change of the z velocity distribution during a given mixing time t_m. Therefore, the 2D VEXSY maps contain time-invariant velocities on the main diagonal corresponding to the averaged velocity propagator along z as obtained by a 1D PGSE sequence. For an initial velocity v_1, the resulting velocity change to v_2 (which can be positive or negative) after the mixing time t_m is reflected as a broadening along the off-diagonal axis. More generally, any additional signature that appears off the main diagonal is due to a change in velocity, which can be caused by isotropic self-diffusion, mechanical dispersion and non-laminar mass flow such as convection and inversional flow. Therefore this 2D autocorrelation map contains the fingerprint of trans-membrane flow, and in conjunction with simulated results, delivers insight into the physics of filtration in an operating hemodialyzer module under given experimental conditions.

2D VEXSY experiments on miniaturized hemodialyzer modules of type SMC and SPAN at differing mixing times t_m and various operating flow rates are presented in Figures 4.6.5 and 4.6.6. In addition, Figure 4.6.6(f) presents experimental results on flow through a bundle of non-permeable glass capillaries of 500 µm inner diameter for comparison. These glass capillaries have been prepared

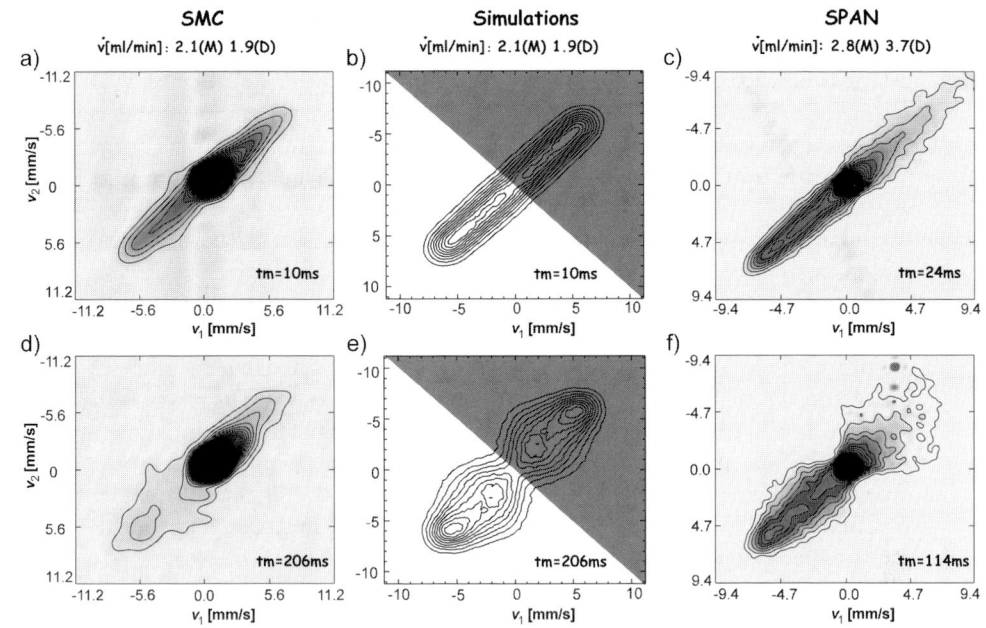

Fig. 4.6.5 2D VEXSY experiments performed on axial counterflow of water in mini-hemodialyzer modules of type SMC and SPAN. Two different mixing times t_m for two typess of modules are displayed. The applied flow rate for the SMC module is 2.1 mL min^{-1} (M) and 1.9 mL min^{-1} (D). The flow rate for the SPAN module is 2.8 mL min^{-1} (M) and 3.7 mL min^{-1} (D). Results from simulations of the 2D VEXSY map with parameters taken from the experiments with the SMC modules are displayed for comparison (b, e). Here, only the half-space to the lower left of the map is to be considered.

in such a way that water can only flow through the inside of the capillaries. Also, 2D VEXSY maps have been simulated for counterflow through a bundle of circular capillaries of 200 μm inner diameter, where a porosity of 20% (equal to the nominal porosity of the modules) and laminar flow conditions were assumed. For velocity exchange, only the mechanism of random self-diffusion in addition to laminar flow was accounted for. The porosity of the hollow fiber membranes is simulated by giving those particles colliding with the wall a certain probability to permeate through the wall. Because in the simulations, only intra-capillary flow is calculated and displayed, only the comparison with experimental results on the membrane-side (lower left half-space of the map) is possible. Selected simulation results are presented in Figures 4.6.5 and 4.6.6 for comparison with experimental results.

2D VEXSY experiments have been performed on SMC and SPAN modules at comparable operating flow rates, by systematically varying the mixing time t_m. Figure 4.6.5 shows one set of experimental results in comparison with simulated data. SMC and SPAN modules reveal markedly different features. The 2D VEXSY maps for the SMC module show a more steady and smooth broadening perpendicular to the main diagonal for all velocities. The broadening is found to be smallest at the maximum velocity because the velocity gradient, dv_z/dr, is minimal in the center of the pipes where the velocity maximum is located. The 2D VEXSY maps for the SPAN modules show larger and distinctly increasing broadening perpendicular to the main diagonal for velocities along the diagonal larger than zero up to a given value of moderate velocity, both for the membrane and dialysate side. The characteristic pattern that the largest broadening occurs at ≈1/3 of the maximum velocity is because of the combined effect of the presence of large velocities and velocity gradients. In the center of the tube, particles possessing maximum velocities experiencing minimum velocity gradients are flowing. Near the tube wall, the low velocities (momentum) do not translate into maximum diffusive broadening despite the highest velocity gradients present. This characteristic pattern is more distinct for the inter-membrane flow compared with intra-membrane flow, and becomes more pronounced for longer mixing times t_m (see the pattern change from top to bottom in Figures 4.6.5 and 4.6.6). The comparison between simulation and experiment results in good agreement in terms of broadening. Also, the pattern of the 2D VEXSY maps of the simulations and the experiments on the SMC module are comparable. Because the simulation assumes 20% wall porosity, and only accounts for laminar flow and self-diffusion, this confirms the characteristics concluded from various experimental results of the

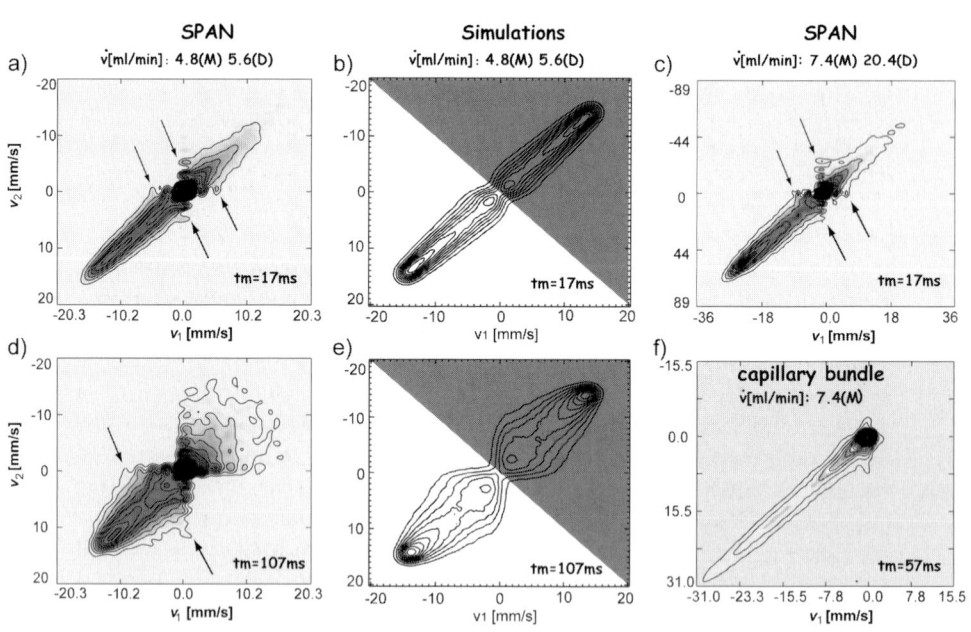

Fig. 4.6.6 2D VEXSY experiments performed on an axial counterflow of water in mini-hemodialyzer modules of type SPAN. Results for the same mixing times t_m and different flow rates are displayed in (a) and (c). The applied flow rates are indicated above the figures for the membrane side (M) and the dialysate side (D). Results from simulations of the 2D VEXSY map with comparable conditions as for (a) and (d) on the SPAN module are displayed for comparison. (f) Experimentally obtained VEXSY map on unidirectional flow through a bundle of non-permeable capillaries of 500 μm inner diameter and 1 mm outer diameter.

NMR flow imaging toolbox, that diffusion rather than mass flow such as convection dominates the trans-membrane filtration in the SMC hemodialyzer module under the given operating condition. It is also worth noting the comparison between the experimental 2D VEXSY pattern on the intra-membrane flow side obtained from flow through permeable capillaries (Figure 4.6.6(c)) and non-permeable capillaries in (Figure 4.6.6(f)). Allowing a considerable mixing time t_m of about 60 ms, not much broadening can be seen if only self-diffusion is present in addition to laminar flow through non-permeable capillaries. The stronger broadening of the experimental 2D VEXSY pattern for the SMC module (Figure 4.6.5(a)) and SPAN module (Figure 4.6.6(a)) for even lower flow rates and lower mixing times demonstrates how efficiently pure Brownian self-diffusion can serve as the trans-membrane filtration across a porous membrane of 20% porosity [19].

When comparing the simulation and experiment for the SPAN module displayed in Figure 4.6.6(a, b, d and e), the agreement in terms of quantity of broadening is reasonable. However, if one takes a closer look at the characteristic VEXSY pattern, there is a distinct difference between experiment and simulation. The experimental 2D VEXSY map shows characteristics "lappets" patterns (see arrows indicating these in Figure 4.6.6) around low or near-zero velocities, which are different from the smooth transition of broadening without edges. This off-diagonal "lappets" pattern is present in the first and the third quadrant of the 2D map with a clear tendency to "grow" into the second and fourth quadrant. This can be interpreted in such a way that molecules reaching the membrane walls, i.e., possessing relatively low velocities, have the tendency to change their flow direction when permeating through the porous membranes and getting carried along by the counterflow of the dialysate – or *vice versa*. The reverse process actually represents backfiltration, which is to be avoided in the real clinical process but is expected under these experimental conditions where water is used as the transported fluid on both sides. In order to obtain such a well defined pattern, mass flow and convection at the porous interface has to contribute to the filtration mechanism, besides pure statistical diffusion. Their contribution within the experimentally given mixing time, however, was not coherent and large enough to manifest itself as inversion of flow – as it would appear as signal patterns in the second and fourth quadrants – on the 2D VEXSY map. Note that this "lappets" pattern is more distinctive (but less broadened) for shorter t_m and almost disappears for long t_m [see this tendency in Figure 4.6.6(a–d)]. This is because a short observation time takes a snapshot at coherent mass flow, but the effect of statistical motion is accumulated for longer mixing times. The slightly stronger broadening on a closer look at the experimental results, as compared with the simulations, can be attributed to the contribution of convection, which was not accounted for in the simulation.

In order to delve further into the discussion on the velocity change and its physical origin, the broadening perpendicular to the main diagonal was more closely examined. For this purpose, projections onto the secondary diagonal of the 2D VEXSY maps, representing the distribution of velocity changes, were studied, as presented in Figure 4.6.7. Whereas the velocity change distribution of the SMC module appears to be Gaussian (a), the pattern for the SPAN module clearly differs from Gaussian (b, c). This characteristic can be more easily understood if the logarithm of the distribution is displayed as a function of the square of the velocity-change axis square as in (d). In this plot, a Gaussian distribution that represents pure diffusive broadening should result in a straight correlation, as can be estimated for curve (1). The curves derived from the SPAN module (2, 3) clearly deviate from a linear behavior. This proves again the presence of transverse mass flow other than just diffusive motion. Such effective filtration due to a strong contribution of the transverse mass flow may arise from the stronger capillary force of the membrane material, geometric features or rougher surface texture of the

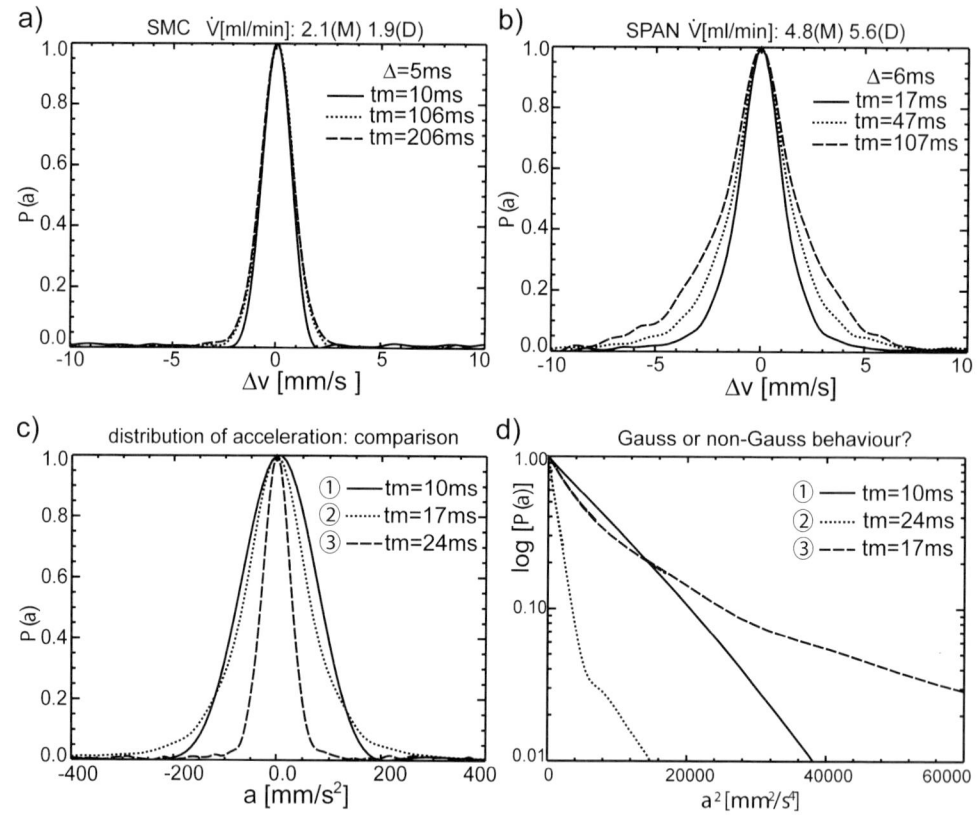

Fig. 4.6.7 Projections along the secondary diagonal from the 2D VEXSY experiments presented partly in Figure 4.6.5 and 4.6.6. (a) Distribution of velocity change obtained among others from Figure 4.6.5(a, d) of the SMC module. (b) Distribution of velocity change obtained among others from Figure 4.6.6(a, d) of the SPAN module. (c) Three out of six distributions presented in (a) and (b) are displayed as the distribution of acceleration, which is obtained by dividing the velocity change (Δv) by the mixing time t_m. Curve (1) is from the SMC module showing a Gaussian distribution, and curves (2) and (3) are from the SPAN module showing non-Gaussian distributions. (d) Alternative representation of the data shown in (c). The logarithm of the probability density of acceleration is plotted against the square of the acceleration. A Gaussian type of distribution should result in a linear dependence as is approximated by dataset 1.

SPAN module. It was mentioned at the beginning that the special polyacrylonitrile fibers of SPAN have a wall thickness of 30 µm, which is considerably thicker than the 8 µm wall thickness of the SMC modules [19]. As a consequence, the presence of stronger capillary effects from the special porous fiber material of the SPAN module would be a reasonable conclusion. Furthermore, the texture of the special polyacrylonitrile fibers is expected to have better surface properties, supporting the permeation of molecules as compared with synthetically modified cellulose. In conclusion, both convection and diffusion effectively contribute to the filtration efficiency in a SPAN module, whereas for the SMC membrane, diffusion is the driving force for molecular exchange, the efficiency of which is also considerable and benefits from the large surface-to-volume ratio.

4.6.5
Conclusion

This chapter demonstrates the versatility and usefulness of the dynamic NMR microscopy toolbox for the characterization of counterflow characteristics through a bundle of porous hollow fiber membranes. Two different types of miniaturized hemodialyzer modules with different fiber materials have been studied. In particular, the 2D VEXSY experiment, which autocorrelates velocity distribution functions at two different times, is custom-made to study filtration mechanisms of counterflow, as the velocity change is directly recorded in a 2D VEXSY map. Whether pure diffusion or mass flow dominates the filtration process can be understood from their characteristic patterns. A comparison between experimental and simulated 2D VEXSY maps supported the interpretation. Although a 2D VEXSY map contains the direct fingerprint of the velocity change characteristics, such measurement needs to be accompanied by dynamic NMR imaging experiments. 1D velocity images give valuable information about the generic difference in flow pattern between the dialysate and membrane compartment. However, only a velocity-encoded 2D map with slice selection or a 3D map can allocate the characteristic flow patterns in a spatially resolved manner, so that the flow profiles can be correlated, e.g., with the packing of the capillary membranes. In conclusion, the characterization of complex fluid flow necessitates the use of various methods out of the NMR imaging toolbox.

Acknowledgements

We would like to thank Peter Blümler and Simone Laukemper-Ostendorf who initially established the collaboration between the company Membrana and the RWTH. They started NMR flow imaging studies to characterize filtration in hemodialyzer modules, and Volker Göbbels measured the first 2D VEXSY data of counterflow in such applications. All experimental work has been accomplished in the Magnetic Resonance Center (MARC) directed by Bernhard Blümich, whose support and leadership is greatly acknowledged.

References

1 R. W. Baker **2004**, *Membrane Technology and Applications*, 2nd edn, Wiley, Chichester.
2 W. H. Hörl, K. M. Koch, R. M. Lindsay, C. Ronco, J. F. Winchester (eds.) **2004**, *Replacement of Renal Function by Dialysis*, 5th edn, Springer, Berlin.
3 A. K. Cheung, J. K. Leypoldt **1997**, (The hemodialysis membranes: a historical perspective, current state and future prospect), *Sem. Nephrol.*, 17, 196–213.
4 J. T. Daugirdas, P. G. Blake, T. S. Ing (eds.) **2003**, *Handbook of Dialysis*, 3rd edn, Lippincott Williams & Wilkins, Philadelphia, PA.
5 W. L. Henrich (ed.) **2003**, *Principles and Practice of Dialysis*, 3rd edn, Lippincott Williams & Wilkins, Philadelphia, PA.
6 A. R. Nissenson, R. N. Fine (eds.) **2005**, *Clinical Dialysis*, 4th edn, McGraw-Hill Professional, New York.
7 (a) C. Ronco, M. Scabardi, M. Goldoni, A. Brendolan, C. La Crepaldi, G. Greca **1997**, (Impact of spacing filaments external to hollow fibers on dialysate flow distribution and dialyzer performance), *Int. J. Artif. Organs*, 20, 261–266; (b) J. Botella, P. M. Ghezzi, C. Sanz-Moreno **2000**, (Adsorption in hemodialysis), *Kidney Int.* 58 (76), 60; (c) B. J. Pangrle, E. G. Walsh, S. Moore, D. Dibiasio **1989**, (Investigation of fluid flow patterns in a hollow fiber module), *Biotech. Tech.* 3, 67–72.
8 A. Frank, G. G. Lipscomb, M. Dennis **2000**, (Visualization of concentration fields in hemodialyzers by computed tomography), *J. Membrane Sci.* 175, 239–251.
9 P. A. Hardy, C. K. Pohb, Z. Liao, W. R. Clark, D. Gaob *2002*, (The use of magnetic resonance imaging to measure the local ultrafiltration rate in hemodialyzers), *J. Membrane Sci.* 204, 195–205.
10 (a) T. Osuga, T. Obata, H. Ikehira **2004**, (Detection of small degree of nonuniformity in dialysate flow in hollow fiber dialyzer using proton magnetic resonance imaging), *Magn. Reson. Imag.* 22, 417–420; (b) J. Zhang, D. L. Parker, J. K. Leypoldt **1995**, (Flow distributions in hollow fiber hemodialyzers using magnetic resonance Fourier velocity imaging), *Am. Soc. Artif. Intern. Organs* 41, M678–82.
11 K. Rombach, S. Laukemper-Ostendorf, P. Bluemler **1998**, (Applications of NMR flow imaging in materials science), in *Spatially Resolved Magnetic Resonance*, eds. P. Bluemler, B. Bluemich, R. Botto, E. Fukushima, Wiley-VCH, New York.
12 S. Han **2002**, *Correlation of Position and Motion by NMR: Pipe Flow, Falling Drop, and Salt Water Ice*, Dissertation, Shaker Verlag.
13 B. Blümich, P. Blümler, L. Gasper, A. Guthausen, V. Göbbels, S. Laukemper-Ostendorf, K. Unseld, G. Zimmer **1999**, *Macromolec. Symp.* 141, 83.
14 V. Göbbels **1999**, *Zweidimensionale Magnetische Resonanz an porösen Medien*, Dissertation, RWTH, Aachen, Germanay.
15 P. T. Callaghan, B. Manz **1994**, (Velocity exchange spectroscopy), *J. Magn. Reson. A*, 106, 260.
16 B. Blümich **2000**, *NMR Imaging of Materials*, Oxford Science Publications, Oxford.
17 S. Han, B. Blümich **2000**, *Appl. Magn. Reson.* 18, 101.
18 S. Han, S. Stapf, B. Blümich **2000**, (Two-dimensional PFG-NMR for encoding correlations of position, velocity and acceleration in fluid transport), *J. Magn. Reson.* 146, 169.
19 W. R. Clark, D. Y. Gao *2002*, (Properties of membranes used for hemodialysis therapy), *Sem. Dialysis*, 15, 191–195.

4.7
NMR for Food Quality Control

Michael J. McCarthy, Prem N. Gambhir, and Artem G. Goloshevsky

4.7.1
Introduction

Quality of food products and the ability to guarantee the quality of a food product is becoming increasingly important in a global economy where there are multiple sources for the food product. This need to measure, control and guarantee quality has resulted in an emphasis to develop more analytical techniques/sensors to measure a product for both external and internal quality. Consider quality evaluation of fresh fruits and vegetables.

Quality evaluation of fresh fruits and vegetables is critical for grading, processing, sorting and marketing. While modern mechanical harvesting of products reduces production costs, it also increases the need for proper sorting of agricultural products. Most fruits and vegetables are not harvested at the same stage of maturity due to biological and environmental differences. Even hand-picked fresh market product grading would benefit from additional quality sorting, especially based on internal quality features.

Automated sorting machines sort fruits into several categories according to specified quality evaluation rules. The quality evaluation can roughly be divided into two categories: external and internal. External quality factors include size, shape, surface color and surface defects; internal factors include voids, solids content, disorders and composition. Detecting external quality factors is easier than detecting internal factors. Automated sorting light reflection and image processing systems are commercially used for external quality evaluation. There are few sensor systems commercially available for internal quality evaluation of agricultural products, but those internal quality assessment devices that are available are based on X-ray and infrared (IR) technology. The X-ray devices are efficient, however they are limited in that only defects that relate to density difference are observed and hence this limits their application. An additional concern for consumers is the use of X-ray radiation on foods. IR based methods can primarily provide information on composition of the product, such as sugar content, water content or oil content. A few newer near infrared-based devices can provide limited information on internal quality defects, however these will be limited in application to particular fruits that have appropriate transmission characteristics, primarily apples. These techniques do not generally penetrate throughout the entire product and provide insufficient information on structural defects (e.g., holes, bruises, freeze damage) in the product. Because of the need for internal quality evaluation, researchers have been working to find methods for evaluating internal quality attributes of agricultural and food products nondestructively by measuring their physical, acoustical, electrical, optical, X-ray and nuclear magnetic resonance (NMR) properties. NMR based sensors have the potential to detect multiple quality factors simultaneously. These factors include composition and structural defects.

Most importantly the MR based systems acquire information from the internal parts of the fruit. This complements the primarily surface information currently obtained in the handling of fresh fruits and vegetables.

Internal defects in oranges include freeze damage, dry regions, over mature (graining), puff and crisp, mold damage, seeds (in seedless varieties) and hollow centers. Shown in Figure 4.7.1 is a magnetic resonance image of two Clementine citrus fruit with seeds and hollow centers. Detection of these internal defects is not possible with external inspection in most cases. Each of these defects can have important economic consequences.

Freeze damage occurs in oranges when temperatures drop below freezing. The extent of the damage to individual oranges and to the orchard is difficult to determine as the damage develops slowly over time. The external appearance of the fruit is not noticeably changed. The internal damage develops slowly over time. Internal damage results in water-soaked areas on the segment membranes and injured juice sacs and vesicles. The injured juice sacs and vesicles eventually become dry and collapse over time. This loss of juice occurs over a period of several weeks. During this time a fermented smell may develop. Ideally early detection of the extent of freeze damage would assist in determining the appropriate action for the orchard, either pick as fast as possible and send to juice production, or in the case of only slight damage use the fruit for export. While freeze damage does not occur every year in citrus production it is a common event. Consider two recent events in California, the freeze of 1990 and that of 1998, each of which cost California citrus producers over 700 million dollars [1].

Other internal defects, while not as spectacular as freeze damage make a significant impact on economics. Mold growth and damage in citrus fruit will develop over a week to 10 days. Initially there is a wound in the fruit and the mold typically will grow inside the fruit (the surface having been treated to eliminate mold). If the orange with the internal mold damage is not sorted out of a shipment, the mold can infect a significant portion of the lot and usually the entire shipment is lost. Defects such as seeds in seedless varieties are more of an eating quality issue, as shown in Figure 4.7.1. Detection of seeds would provide a definite

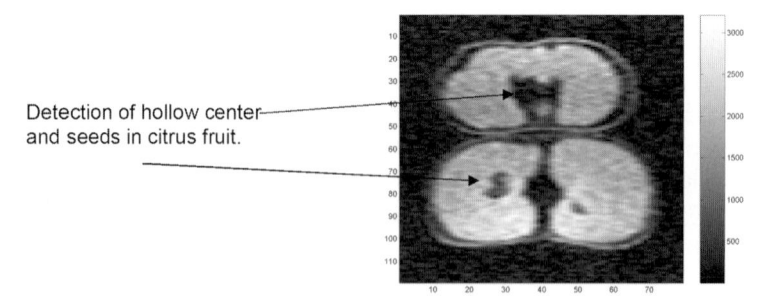

Fig. 4.7.1 Internal defects in citrus fruits. Proton MRI from the central plane of two Clementine citrus fruits. The plane is 5 mm thick and taken at a 0.1 T field strength. The upper fruit shows a hollow center defect, which leads to deformation of the fruit during shipment, and the bottom fruit displays seeds, an eating quality defect.

advantage to the shipper, as the product could be guaranteed seedless. Similarly, in processed foods there is a need to monitor quality and ensure product specifications are met.

This chapter will describe the use of nuclear magnetic resonance and magnetic resonance imaging to characterize the quality attributes of foods and for use in process optimization, shelf-life determination and component migration.

4.7.2
Relationship of NMR Properties to Food Quality

The consumer is interested in how foods taste, smell, feel and look. These qualitative descriptors are often highly correlated to the physical/chemical attributes of the food product. The ability of NMR/MRI to measure and quantify physical and chemical properties directly and indirectly provides a powerful tool for quality assessment [2]. Direct measurements of composition are often possible using simple one-dimensional spectra or just initial signal intensity in the time domain. Indirect measurements of composition may often be achieved through correlations with relaxation times. These decay times for the NMR signal may be strong functions of moisture content, textural properties and/or physical phases. Diffusion coefficient measurements have a strong dependence on food material structure, such as particle size and other structural features in porous food matrices. MRI measurements of component spatial distribution give an insight into the rates of drying, freezing and other transient processes, which impact on the final product quality and process efficiency. Collectively NMR and MRI measurement protocols permit a wide range of quality measurements in food systems.

4.7.3
Applications of NMR in Food Science and Technology

The wide range of applications of NMR in food science and technology can broadly be divided into the following categories and subcategories. The flow diagram in Figure 4.7.2 gives a brief outline of the various groups of applications with a few examples. In natural products several internal quality parameters such as fat, moisture and sugar content are important in fair pricing for marketing and determining the stage of maturity and ripeness. NMR has been widely used for the determination of oil content and/or moisture content in a variety of intact grains, such as oil seeds and cereal seeds [3]. NMR is also utilized for the nondestructive evaluation of the internal quality of fruits and vegetables [2]. In processed foods NMR has found applications in the characterization of food products, quality control, authentication of fruit juices and wines, measurement of solid/fat ratio and moisture content in various bakery and dairy products. The application of NMR to the characterization of food systems will be described using examples from high-resolution and low-resolution NMR and MRI experiments.

4.7.3.1 High Resolution NMR

High resolution solution state NMR spectroscopy is being used increasingly to observe signals from highly mobile molecules for the characterization of complex samples. The advantages of the high-field spectrometers are the good dispersion, very high signal-to-noise ratio and high dynamic range, enabling analysis of molecules with both strong and weak signals. This technique has enormous potential as it can provide compositional information about a wide range of different metabolites in a rapid and non-destructive manner. NMR is very useful in identifying marker compounds when monitoring chemical composition of solutions, which can be exploited as fingerprinting methods. This method has several applications in the fields of food authenticity and quality assurance. Some of the applications are the detailed and direct characterization of fruit juices [4–6], coffee [7, 8], wine [9, 10] and beer [11, 12].

In liquid food systems water tends to dominate composition as well as the NMR signal. It is customary in NMR to acquire repetitive responses of rf pulses to improve the signal-to-noise ratio and resolution to overcome the effects of instrumental imperfections and for handling of low concentration and low abundant nuclei. This results in the collection of large unwanted noise signals and solvent signals. In food products, it is often necessary to eliminate the unwanted intense water signal that overshadows the weak signal from the desired constituent of the sample. Over the last few years NMR technology, including digital filtering, gradient system and built in pulse sequences, has been developed to suppress the noise and large unwanted solvent peaks.

One of the oldest methods for water signal suppression is WEFT (Water Elimination Fourier Transform). This method takes advantage of the fact that compared with water, a macromolecule usually has a shorter value for proton T_1 and a much lower diffusion coefficient (Becker 2000, www.cis.rit.edu/htbooks/nmr/chap-11/chap-11.htm). In this method an inversion recovery sequence is applied with the pulse interval time τ chosen in such a way that the water signal goes through zero (T_1 ln 2). Another simple and effective approach is the Jump and Return (JR) technique. This uses a 90° pulse followed by a short precession period τ and then a second 90° of opposite phase. The solvent line on resonance is restored to the z axis, whereas all off-resonance magnetizations retain most components in the xy plane, thus giving rise to the FID without solvent resonance [13]. A number of other approaches for solvent suppression are also available. Solvent signal suppression is efficiently performed by the following techniques/methodologies [14].

1. *WET pre-saturation:* The WET sequence (Water Suppression Enhanced through T_1 Effects) employing a z gradient, uses a number of pulses of variable lengths for multiple suppression. Each selective rf pulse is followed by a dephasing field gradient pulse and the tip angle of the selective rf pulse is varied to optimize the WET sequence. Multiple solvent frequencies can be saturated in a fast and efficient way by such an approach. In combination with ^{13}C decoupling, it is easy to remove the ^{13}C satellites from the solvent.

Flow Diagram

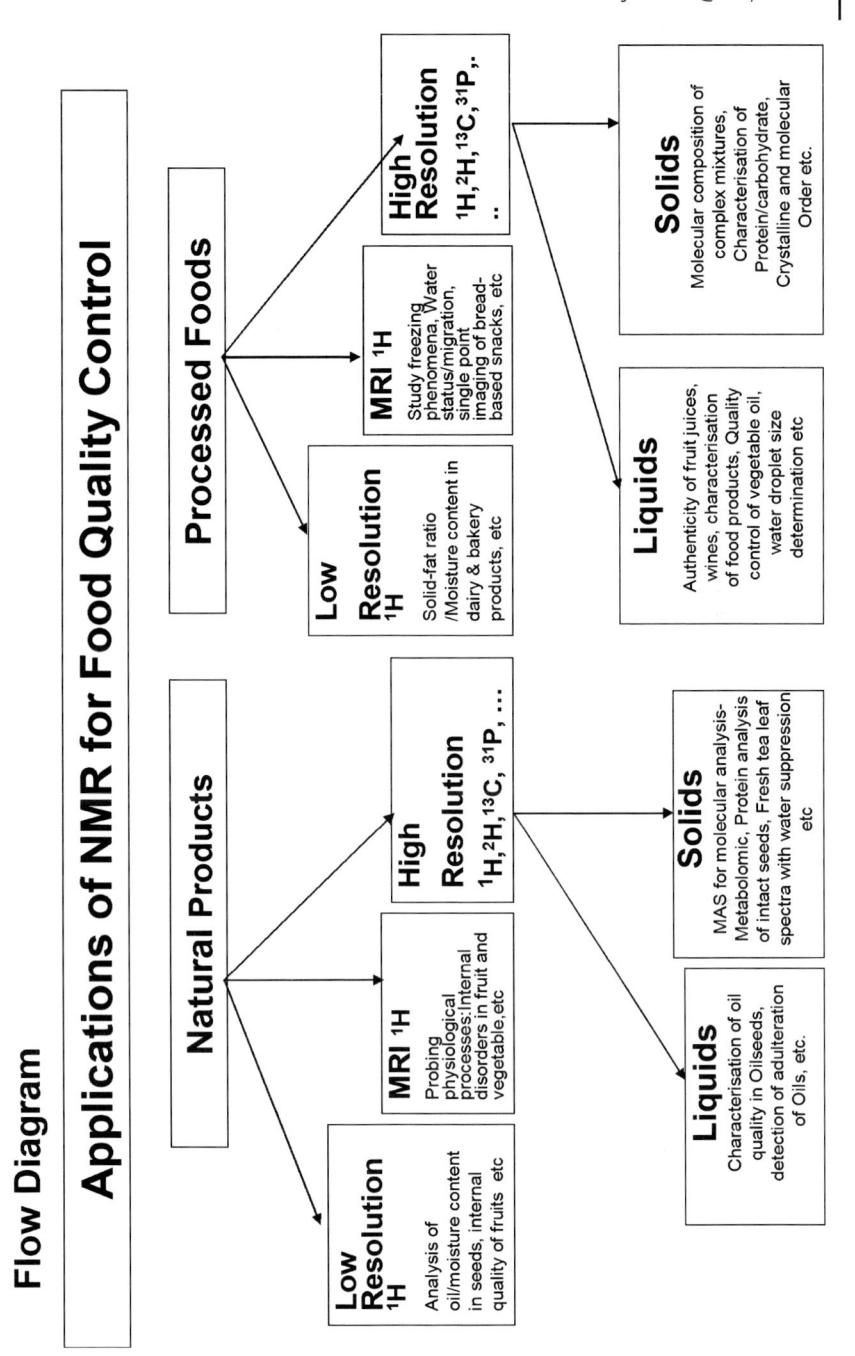

Applications of NMR for Food Quality Control

Processed Foods

Low Resolution ¹H
Solid-fat ratio/Moisture content in dairy & bakery products, etc

MRI ¹H
Study freezing phenomena, Water status/migration, single point imaging of bread-based snacks, etc

High Resolution
¹H, ²H, ¹³C, ³¹P, ..

Solids
Molecular composition of complex mixtures, Characterisation of Protein/carbohydrate, Crystalline and molecular Order etc.

Liquids
Authenticity of fruit juices, wines, characterisation of food products, Quality control of vegetable oil, water droplet size determination etc

Natural Products

Low Resolution ¹H
Analysis of oil/moisture content in seeds, internal quality of fruits etc

MRI ¹H
Probing physiological processes: Internal disorders in fruit and vegetable, etc

High Resolution
¹H, ²H, ¹³C, ³¹P, ...

Solids
MAS for molecular analysis-Metabolomic, Protein analysis of intact seeds, Fresh tea leaf spectra with water suppression etc

Liquids
Characterisation of oil quality in Oilseeds, detection of adulteration of Oils, etc.

Fig. 4.7.2 Application of NMR for the measurement of quality factors in foods.

2. *Pre-saturation*: In this technique prior to data acquisition, a highly selective low-power rf pulse irradiates the solvent signals for 0.5 to 2 s to saturate them. No irradiation should occur during the data acquisition. This method relies on the phenomenon that nuclei which have equal populations in the ground and excited states are unable to relax and do not contribute to the FID after pulse irradiation. This is an effective pulse sequence of NOESY-type pre-saturation that consists of three 90° pulses: RD – 90° – t_1 – 90° – t_m – 90° – FID, where RD is the relaxation delay and t_1 and t_m are the pre-saturation times.
3. *Soft-pulse multiple irradiation*: In this method, pre-saturation is done using shaped pulses having a broader excitation profile. Therefore, it is a more suitable method for the suppression of multiplets. This technique is very effective, easy to apply and easy to implement within most NMR experiments. In aqueous solutions, however, slowly exchanging protons would be detectable due to the occurrence of transfer of saturation. In addition, the spins with resonances close to the solvent frequency will also be saturated.

However, all these suppression techniques have a common disadvantage in that the desirable signals of interest lying under the solvent signal are also suppressed.

4.7.3.1.1 Apple Juice

The high field proton NMR spectra of apple juices have been shown to be sensitive to a number of factors including cultivar type, microbiological activity and enzymatic activity [4]. The significant spectral differences between cultivars may be utilized in identifying the origins of apple juices. The study showed that in addition to speciation and authentication of juices, NMR will be a valuable tool in the analysis of biochemical changes occurring in fruits and their juices. The incomplete spectral assignment reported in this study could be improved by using multidimensional methods. In a recent communication [12], diffusion ordered spectroscopy (DOSY) was described as being applied to fruit juices as a complementary aid to spectral assignment. The proton NMR spectra of apple and grape juice samples (Figure 4.7.3) show that the aliphatic region is dominated by many typical amino acid patterns, superimposed with aliphatic organic acids. As expected, the sugar region in the grape juice proton spectrum is dominated by glucose and fructose. Most importantly, in grape juice the rich aromatic composition of aromatic/phenolic compounds makes it an interesting sample for identification of phenolic compounds that are nutritionally important.

4.7.3.1.2 Tomato Juice

A detailed analysis of the proton high field NMR spectra of tomato juice and pulp has recently been acquired [15]. The combination of suitable selective and two-dimensional techniques (J-resolved, COSY, TOCSY, DOSY, etc.) was used for assignment of each spin system and resolved the complex pattern in the 1D overlapped spectra. A High-Resolution Magic Angle Spinning (HR-MAS) technique was used to obtain well resolved proton spectra of tomato pulp with resolution comparable to the corresponding juice. The comparison between the juice and pulp spectra shows that essentially all the water-soluble substances present in the pulp are extracted in the juice, but HR-MAS NMR is also able to detect some insoluble compounds such as lipids.

4.7.3.1.3 Beer

Beer contains a complex mixture of, mainly, water, ethanol, carbohydrates and fermentable sugars. The proton NMR spectrum of beer (Figure 4.7.4) shows strongly overlapped peaks arising from several carbohydrates and some minor components in both the aliphatic and the aromatic regions. The spectral assignment was carried out with the aid of two-dimensional methods and nearly 30 compounds were identified [11]. However, owing to strong signal overlap, HR-

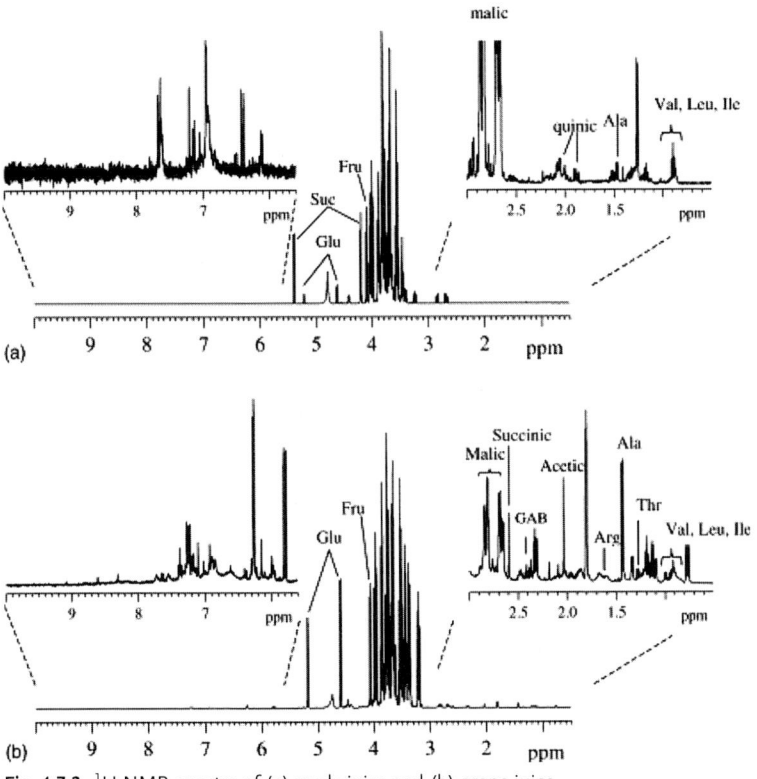

Fig. 4.7.3 ^{1}H NMR spectra of (a) apple juice and (b) grape juice samples. Vertical expansions are shown for aliphatic and aromatic regions and some assignments are indicated. (Permission granted to reprint this figure from Ref. [12].)

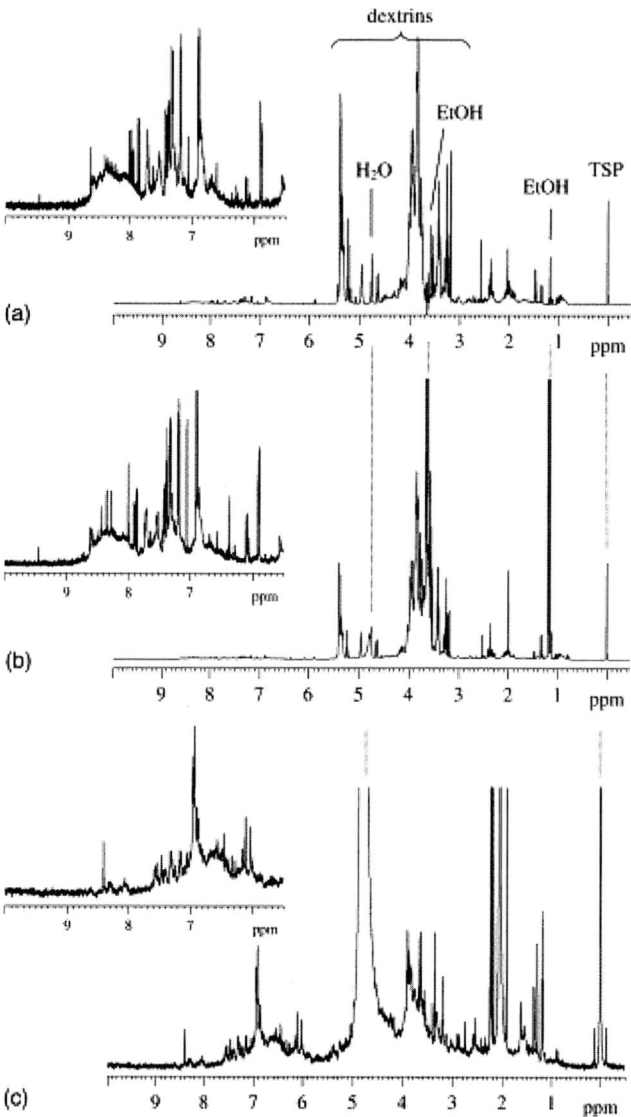

Fig. 4.7.4 ^{1}H NMR spectra of (a) beer 1 (a lager) (b) beer 2 (an ale) and (c) aqueous phenolic extract of beer 2. The vertical inserts show the expansions of the aromatic regions. (Permission granted to reprint this figure from Ref. [12].)

NMR alone is insufficient to enable the full assignment of the beer spectra to be made. Application of Principal Component Analysis (PCA) to the spectral profiles of beers of differing type (ales and lagers) showed some distinction on the basis of the aliphatic and sugar compositions, whereas the PCA of the aromatic profiles gave a clearer separation. The application shows the potential of HR-NMR in the rapid identification of the nature and origin of the beer samples. The possibility of further improvement in the spectral assignment of beer has recently been shown [12] through application of Diffusion Ordered Spectroscopy (DOSY). However, the spectral information could be further improved through the use of more than one dimension or ^{13}C DOSY NMR.

4.7.3.1.4 Coffee

The NMR spectra using PCA and Linear Discriminant Analysis (LDA) obtained for instant spray dried coffees from a number of different manufacturers demonstrated [8] that the concentration of the extracted molecules is generally high enough for clear detection. The compound 5-(hydroxymethy)-2-furaldehyde was identified as the primary marker of differentiation between two groups of coffees. This method may be used to determine whether a fraudulent retailer is selling an inferior quality product marked as being from a reputable manufacturer [8].

4.7.3.1.5 Olive Oil

The application of ^{13}C NMR for the rapid analysis of the oil composition of oil seeds is well known [16]. ^{13}C NMR has recently been applied to the quantitative analysis of the most abundant fatty acids in olive oil [17]. The values obtained by this method differed by only up to 5% compared with GLC analysis. The quantitative analysis was applied to the olefinic region of the high resolution ^{13}C NMR spectrum of virgin olive oil to detect adulteration by other oils which differed significantly in their fatty acid composition. The application of the methodology for the detection of adulteration of olive oil by hazelnut oil is more challenging as both oils have similar chemical profiles and further experiments are in progress.

In the case of the low abundance of some compounds, there are difficulties with signal overlap. To overcome these difficulties, there have been developments involving NMR hyphenation with techniques such as HPLC and mass spectrometry. In LC/NMR methods of analysis, NMR is used as the detector following LC separation and this technique is capable of detecting low concentrations in the nanogram range. This technique has been reported for the detection and identification of flavanoids in fruit juices and the characterization of sugars in wine [17].

4.7.3.2 Solid Food

Many components of food are in the solid state and possess very short T_2. The linewidths from solid components are generally too wide to be observed directly by solution state NMR methods. However, these components can be detected by the special techniques of solid state NMR. These techniques involve the use of cross polarization excitation (from ^{1}H to ^{13}C), high power ^{1}H decoupling (to inhibit relaxation pathways) and magic angle spinning (to reduce chemical shift anisotropy effects) and are known as CPMAS. The samples are spun at an angle of 54° relative to the magnetic axis at the high speeds between 3 and 15 kHz (compared with 10–20 Hz used for liquid samples), which reduces or eliminates the broadening and permits the acquisition of liquid-like spectra [18, 19]. This method of ^{13}C CPMAS has been applied to study common solid food components including wheat proteins [20], starch [21], and oils [22, 23].

Solid state ^{13}C CPMAS NMR spectra of Wheat High Molecular Weight (W.HMW) subunits show well resolved resonances identical with spectra of dry protein and peptide samples [24]. Most of the amino acids side-chain resonances are found in the 0–35 ppm region followed by the alpha resonances of the most abundant amino acids: glycine, glutamine and proline at chemical shifts of 42, 52 and 60 ppm, respectively, and the carbonyl carbons show a broad peak in 172–177 ppm region. The CPMAS spectra of hydrated whole HMW provides important information on the structural characteristics.

4.7.3.3 Low Resolution/Low Field

Low resolution spectrometers utilize a low-field permanent magnet to generate the polarizing magnetic field. The range of magnetic field strengths is from 0.01 to 0.65 T. Analysis of the signal has primarily been performed in the time-domain although some applications are beginning to appear using frequency-domain techniques. The main features of the time-domain signal that are used for analysis are:

1. The initial amplitude of the signal is directly proportional to the total number of protons in the sample.
2. The signals from different components or sample environments are separated based on differences in relaxation rates.
3. The proton signal acquired is the superposition of the signals from all proton-containing components in the sample.

Low Resolution Nuclear Magnetic Resonance (LR-NMR) systems are routinely used for food quality assurance in laboratory settings [25]. NMR based techniques are standardized and approved by the American Oil Chemist's Society (AOCS) (AOCSd 16b-93, AOCS AK 4–95), the International Union of Pure and Applied Chemistry (IUPAC) (solid fat content, IUPAC Norm 2.150) and the International Standards Organization (ISO) (oil seeds, ISO Dis/10565, ISO CD 10632). In addition to these standardized tests, low resolution NMR is used to measure moisture content, oil content and the state (solid or liquid) of fats in food. Table 4.7.1 summarizes common food products that are analyzed by low-resolution NMR for component concentration.

Tab. 4.7.1 Applications of low resolution NMR in quality assurance testing.

Moisture content	Oil content, fat solid/liquid ratio
peanuts	chocolate
flour: rice, wheat and corn	marzipan
lentils	rice
biscuits and cookies	milk powders
oil seeds	baby foods
milk powders	peanuts
cornstarch	cheese
baby foods	oil seeds
confections	biscuits and cookies

The measurement of solid fat content is one of the most common and successful quality measurements using LR-NMR. This technique relies upon the dramatic difference in spin–spin relaxation times between solid and liquid fat (often three orders of magnitude). Two data points from the free induction decay are used to measure the amount of solid fat. The sample is first tempered and then placed in the LR-NMR. The sample reaches equilibrium in around 2–6 s and a single rf pulse is applied to the sample. The signal from the sample is recorded. This signal is the superposition of signal from both the solid and liquid fat. The intensity of the signal immediately after the pulse is proportional to the total amount of liquid plus solid fat. The magnetization of the solid fat decays rapidly and at 70 μs is less than 1% of the initial value. The signal from liquid fat decays with a spin–spin relaxation time of $\approx$100 ms and the intensity at 70 μs is essentially unchanged. This measurement is simulated for a sample of oil that is 80% liquid in Figure 4.7.5. The initial rapid decay is a result of the solid component of the spin–spin relaxation and the almost horizontal line portion is from the liquid fraction with a much slower relaxation rate. In the actual experiment the signal cannot be recorded immediately after the rf pulse due to instrumental limitations. The time between the pulse and recording the signal is called the dead time and is approximately 5 μs. During this delay the solid component signal decays significantly and so standards are used to correct for any unrecorded signal. This method is commonly referred to as the direct method. The reproducibility of the direct method is 0.4% and it is the fastest and simplest of all the NMR based methods. Other methods for analysis of fats and oils are either a slight variation of the direct method or are a complete mathematical deconvolution of the entire signal decay [26].

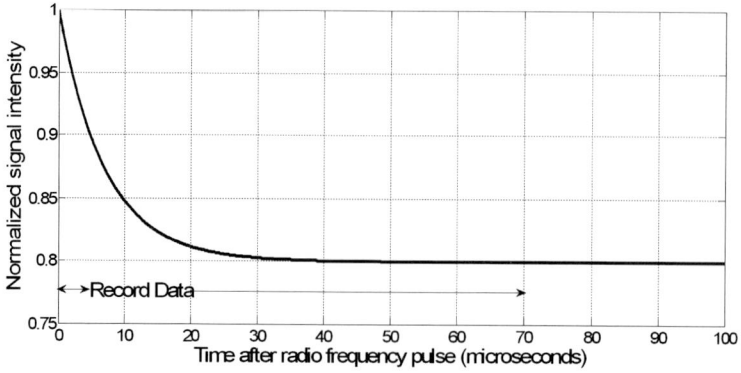

Fig. 4.7.5 Simulated proton signal decay from a low resolution NMR measurement of the solid/liquid ratio from an oil with 20% solid and 80% liquid (%-w/w).

4.7.3.4 On-line Sensors

The simplicity and speed of low-resolution NMR measurements has resulted in many efforts to utilize NMR as a process sensor. In addition to measurement speed, NMR has additional advantages as a process sensor, including, the spectroscopy is a non-contact, non-invasive technique, it has a linear signal from the detection limits ≈10 ppt to 100% and multiple quality factors can be measured using the same device [27]. Initial efforts at process control focused on using low-field strength magnets. A large number of applications have been on food and agricultural products. Two groups, Tri-Valley Research and The Southwest Research Institute made many important contributions to these early systems [28]. The applications of these early systems were primarily to measure moisture content in wheat, corn, dry soup mixes and agricultural products or for the measurement of fluid viscosity.

The measurement of viscosity is important for many food products as the flow properties of the material relate directly to how the product will perform or be perceived by the consumer. Measurements of fluid viscosity were based on a correlation between relaxation times and fluid viscosity. The dependence of relaxation times on fluid viscosity was predicted and demonstrated in the late 1940s [29]. This type of correlation has been found to hold for a large number of simple fluid foods including molten hard candies, concentrated coffee and concentrated milk. Shown in Figure 4.7.6 are the relaxation times measured at 10 MHz for solutions of rehydrated instant coffee compared with measured Newtonian viscosities of the solution. The correlations and the measurement provide an accurate estimate of viscosity at a specific shear rate.

Several commercial companies now offer process compatible NMR systems. These systems are either low resolution based on relaxation time measurements or high resolution Fourier Transform spectral measurements. The low resolution systems are manufactured by Process Control Technologies (www.pctnmr.com) and Progression, Inc. (www.progression-systems.com). Progression's systems

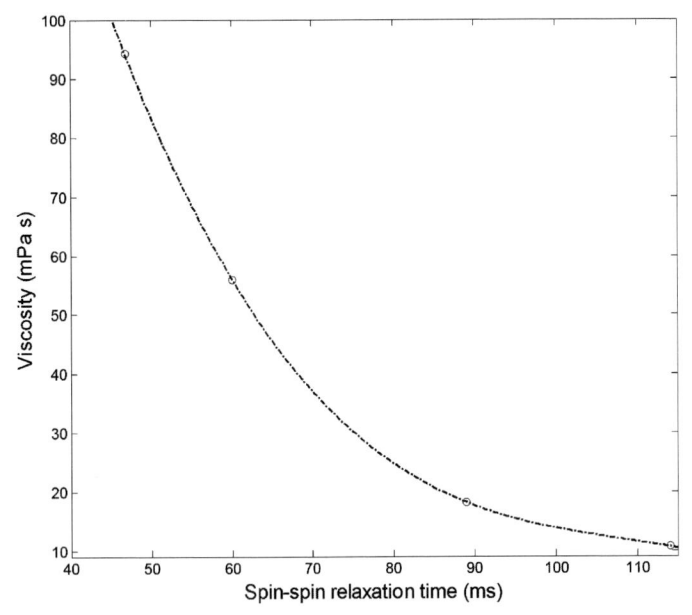

Fig. 4.7.6 The relationship between spin–spin relaxation time and viscosity for a concentrated coffee solution (32–46%-w/w). The coffee solution is rehydrated from commercial spray dried coffee.

have been applied primarily to controlling polyolefin and/or thermoplastic production. The high resolution based sensors have been used to measure component concentrations in industrial petroleum refineries. Recently this system has been used to measure quality factors in fluid foods, for example alcohol content in beer and water and fat content in dairy products (www.process-nmr.com).

4.7.3.5 Portable NMR Systems

Many samples are difficult or impossible to bring to the laboratory for analysis in an NMR spectrometer. Historically these types of samples, for example fruit on trees, or food in large packages, have rarely been studied with statistically relevant sample sizes. Achieving measurements on sufficient sample to understand natural variations is most easily achieved by taking the system to the samples. Recent work in magnet design, spectrometer design and analysis of NMR signals acquired in non-uniform magnetic fields have resulted in a range of portable NMR systems that can address taking the system to the samples. The most versatile portable NMR system is called the NMR Mouse (Mobile Universal Surface Explorer) ([30] and www.nmr-mouse.de). This device has been applied to measure properties of polymers, automobile tires, foods and human subjects (www.nmr-mouse.de). The NMR Mouse is based on a small magnet and rf coil assembly that achieves an active volume external to the magnet. This active volume is of the order of mm away from the surface of the NMR Mouse [30]. As the magnet assembly can be easily lifted and moved with one hand the NMR signal can be acquired from small regions of very large objects. Small regions can even be imaged using this assembly. The main magnetic field is shaped to produce a linearly decreasing field orthogonal to the magnet–rf coil surface. The frequency of excitation is used to select a plane above the surface and phase encoding is used to spatially resolve signals in the other two dimensions. The main limitation of the NMR Mouse is the small active region for measurements.

Achieving a larger active volume than that of the NMR Mouse requires an alternative magnet design. The most common alternative designs are the Halbach cylinder and a C-shaped magnet. Figure 4.7.7 shows a Halbach magnet designed to be used to measure spin–spin relaxation rates of whole Navel oranges. (Quantum Magnetics, San Diego, CA, USA). This is a very similar system to one recently constructed to measure relaxation decays for rock cores [31]. Other examples include a triangular-shaped magnet system to measure moisture content of wood chips [32], an array of cylindrical magnets to create a large NMR Mouse like system [33] and an Earth's field system for use in Antarctica [34].

4.7.3.6 Magnetic Resonance Imaging

Magnetic resonance imaging provides engineers with a unique tool for estimation of material transport coefficients. MRI has been applied to measure transport phenomena in many types of food processing and for food property characterization [35]. Books by Hills [36], McCarthy [2] and Ruan and Chen [37] describe many of these applications in detail, including dehydration, freezing, freeze-drying, heat transfer and moisture migration. The following example demonstrates a means to calculate the effective diffusivity, from a well-defined measurement of the concentration profile as a function of time.

If the flux (J_0) and the concentration profile [$C(x)$] are measured, the diffusivity can be obtained through Fick's first law:

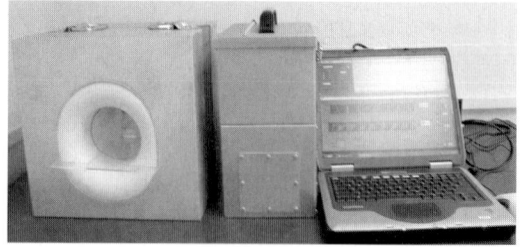

Fig. 4.7.7 A portable NMR system for measuring freeze damage in Navel oranges. The Halbach magnet is shown above the completed system with two Navel oranges for comparison. The complete system has the battery powered spectrometer electronics housed in a metal box and the system is run by a portable computer.

$$D(x) = D(C(x)) = -J_0/(dC/dx) \qquad (4.7.11)$$

The work of Crank [38] provides a review of the mathematical analysis of well defined component transport in homogeneous systems. These mathematical models and measured concentration profile data may be used to estimate diffusivities in homogenized samples. The use of MRI measurements in this way will generate diffusivities applicable to models of large-scale transport processes and will thereby be of value in engineering analysis of these processes and equipment.

Consider moisture migration in pasta post-cooking. This will occur in a food service environment when pasta is held at serving temperature for an extended period of time. Pasta should be served with a firm *al dente* texture, however the texture of pasta held at serving temperature degrades over time. The mechanism for the degradation of texture is the redistribution of moisture in the pasta after cooking. The cooking process establishes a moisture gradient in the pasta from the surface to the center, where the surface has a high moisture content and the center has a lower moisture content [39]. The decrease in firmness of the pasta correlates well with the equilibration of the moisture in the pasta. To estimate the shelf-life of the pasta an effective moisture diffusivity is needed.

Figure 4.7.8 illustrates a proton spin-echo MRI of lasagna noodles cooked for 9.5 min and concentration profiles calculated from the data. The signal intensity is bright at the surfaces of the sheets and decreases as the moisture content becomes lower towards the center of the sheet. Prior to calculating the diffusivities the MRI data are converted from signal intensities into moisture concentrations. This is needed because the signal depends not only upon the number of nuclei per unit volume but also upon both the spin–spin and spin–lattice relaxation times. These two relaxation times vary as a function of moisture content in the pasta [40]. An analytical mass transfer model based on Fick's second law is used to describe the movement of the moisture. The moisture redistribution data were used in combination with a model for moisture migration to estimate effective moisture diffusivities. Effective diffusivities as a function of cooking time were calculated and ranged from $1.5\text{--}4.0 \times 10^{-7}$ cm^2 s^{-1}. The fitting procedure and model have been described in detail by McCarthy and coworkers [40].

Magnetic resonance imaging of physical properties is also applicable to process control. Viscosity is a physical property affecting the final product quality for a wide range of fluids in the food industry. Viscosity measurements are critical to process control of unit operations and final end use [41]. Current in-line viscometers typically measure viscosity only at one or a few shear rates [42], for instance, conventional tube viscometry yields only one viscosity data point from a single measurement. Control of the process and product quality would be improved significantly if viscosity over a wide range of shear rates could be measured. MRI can be used as a viscometer based on analysis of a measured velocity profile of the material flowing in a tube coupled with a simultaneous measurement of the pressure drop driving the flow [43, 44].

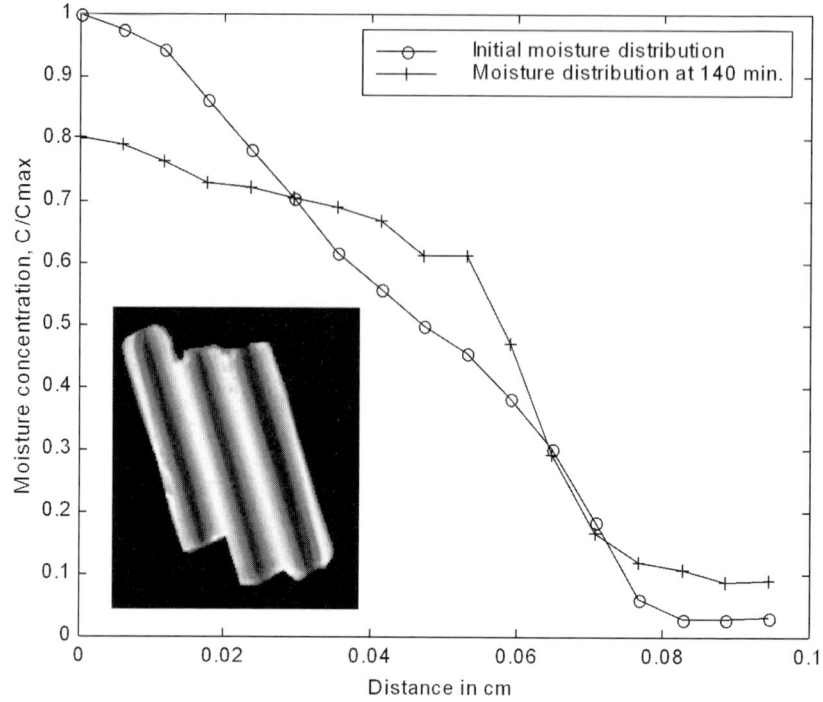

Fig. 4.7.8 Moisture distribution in lasagna pasta as a function of distance from the surface to the center (0.1 cm). The moisture profiles are extracted from the central sheet of a set of three sheets. The image data at one time point is shown by the insert. The lighter gray is directly proportional to moisture content and occurs at the edge of the pasta sheet.

A velocity profile $u(r)$ is obtained using MRI flow imaging and, with respect to radial position r, the values of shear rate $\dot{\gamma}(r)$, ranging from zero at the tube center to a maximum at the tube wall, can be calculated from the velocity profile as local velocity gradients:

$$\dot{\gamma}(r) = -\frac{du(r)}{dr} \tag{4.7.12}$$

The shear rate is computed locally using a global curve fit to the velocity profile. If $\Delta P/L$ is the pressure drop driving the flow across the pipe length L, the shear stress $\sigma(r)$ is expressed as

$$\sigma(r) = -\frac{\Delta P}{2L} r \tag{4.7.13}$$

As the shear rate and shear stress are radially dependent, each radial position provides a viscosity data point $\eta(\dot{\gamma})$ at a different shear rate:

$$\eta(\dot{\gamma}) = \frac{\sigma(r)}{\dot{\gamma}(r)} \tag{4.7.14}$$

These measurements have been demonstrated to achieve results comparable to laboratory research grade rheometers. Unfortunately, current commercially available MRI systems are almost all built around large, superconducting, high-field magnets. The costs of these systems have precluded widespread utilization of MRI based sensors in industry.

Microfabrication techniques with the features of integration, reproduction and precision are particularly suitable for the implementation of miniaturized designs and significant reduction in product costs. A miniaturized magnetic resonance probe for on-line/in-line flow studies can be microfabricated by combining an rf coil [45, 46] with microfabricated gradients [47] and electronics [48] around a small tube/capillary.

Utilization of a microfabricated rf coil and gradient set for viscosity measurements has recently been demonstrated [49]. Shown in Figure 4.7.9 is the apparent viscosity of aqueous CMC (carboxymethyl cellulose, sodium salt) solutions with different concentrations and polymer molecular weights as a function of shear rate. These viscosity measurements were made using a microfabricated rf coil and a tube with id = 1.02 mm. The shear stress gradient, established with the flow rate of 1.99 ± 0.03 μL s^{-1} was sufficient to observe shear thinning behavior of the fluids.

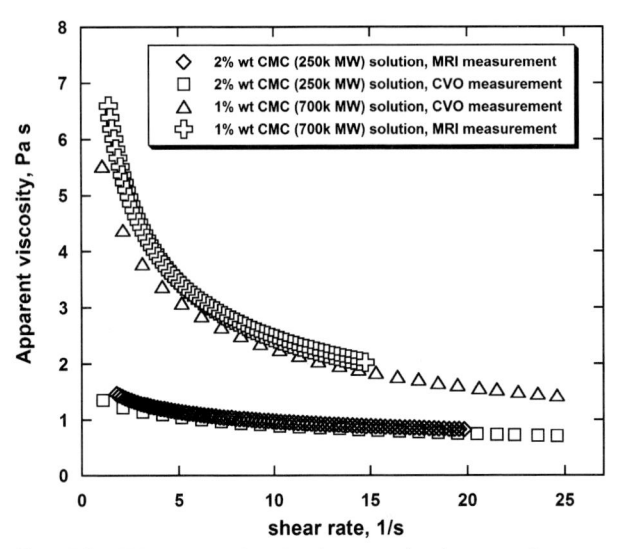

Fig. 4.7.9 MRI apparent viscosity-shear rate data in comparison with a conventional rotational viscometer shear viscosity-shear rate data. (Permission granted to reprint Figure 4 on page 517 in Ref. [49].)

4.7.4
Summary

The range of structures and compositions in food materials is extremely large. Macroscopic (cm) to molecular (nm) length scales are observed in foods. NMR is unique in that the measurement is capable of probing structure and composition on all the length scales of importance in food systems. The compositional and structural complexity and diversity account directly for many of the sensory and processing properties that are observed. These observed properties relate directly to the consumers concept of quality in addition to manufacturing quality standards. A variety of NMR experimental techniques are used to characterize and measure the structure and composition of foods. The information obtained from NMR can be used effectively to quantify quality parameters as well as provide vital information for understanding transport processes in production, storage and consumption.

References

1 J. P. Tiefenbacher, R. R. Hagelman, R. J. Cecora **2000**, *California Citrus Freeze of December 1998: Place, Perception and Choice – Developing a Disaster Reconstruction Model*, Quick Response Report #125, James and Marilyn Lovell Center for Environmental Geography, San Marcos, TX.

2 M. J., McCarthy **1994**, *Magnetic Resonance Imaging in Foods*, Chapman & Hall, New York.

3 P. N., Gambhir **1992**, (Applications of low-resolution pulsed NMR to the determination of oil and moisture in oilseeds), *Trends Food Sci. Technol.* 3, 191–196.

4 P. S. Belton, I. Delgadillo, A. M. Gil, P. Roma, F. Casuscelli, I. J. Colquhoun, M. J. Dennis, M. Spraul **1997**, (High field NMR studies of apple juices), *Magn. Reson. Chem.* 35, S52-S60.

5 Y. Lu, L.Y Foo **1999**, (The polyphenol constituents of grape pomace), *Food Chem.* 65, 1–8.

6 A. M. Gil, I. F. Duarte, I. Delgadillo, I. J. Colquhoun, F. Casuscelli, E. Humpfer, M. Spraul **2000**, (Study of compositional changes of mango during ripening by use of nuclear magnetic resonance spectroscopy), *J. Agric. Food Chem.* 48, 1524–1536.

7 M. Bosco, R. Toffanin, D. Palo, L. Zatti, A. Segre **1999**, (High resolution ^{1}H NMR investigation of coffee), *J. Sci. Food Agric.* 79, 869–878.

8 A. J. Charlton, W. H. Farrington, P. Brereton **2002**, (Application of 1H NMR and multivariate statistics for screening complex mixtures: quality control and authenticity of instant coffee), *J. Agric. Food Chem.* 50, 3098–3103.

9 A. Ramos, H. Santos **1999**, (NMR studies of wine chemistry and wine bacteria), *Annu. Rep. N. M. R. Spectrosc.* 37, 179–202.

10 I. J. Kosir, J. Kidric **2001**, (Identification of amino acids in wines by one- and two-dimensional nuclear magnetic resonance spectroscopy), *J. Agric. Food Chem.* 49, 50–56.

11 I. Duarte, A. Barros, P. S. Belton, R. Righelato, M. Spraul, E. Humpfer, A. M. Gil **2002**, (High-resolution nuclear magnetic resonance spectroscopy and multivariate analysis for the characterisation of beer), *J. Agric. Food Chem.* 50, 2475–2481.

12 A. M. Gil, I. Duarte, E. Cabrita, B. J. Goodfellow, M. Spraul, R. Kerssebaum **2004**, (Exploratory applications of diffusion ordered spectroscopy to liquid food: an aid towards spectral assignment), *Anal. Chim. Acta* 506 (2), 215–223.

13 E. D. Becker **2000**, *High Resolution NMR: Theory and Chemical Applications*, Academic Press, New York, (pp.) 241–42.

14 K. Albert **2002**, (LC-NMR: theory and experiment), in *On –Line LC-NMR and Related Techniques*, ed. K. Albert, John Wiley & Sons, Chichester, (pp.) 13–17.

15 A. P. Sobolev, A. Segre, R. Lamanna **2003**, (Proton high-field NMR study of tomato juice), *Magn. Reson. Chem.* 41, 237–245.

16 P. N. Gambhir **1994**, (^{13}C NMR Spectroscopy of intact oilseeds – A rapid method for measurement of fatty acid composition), *Proc. Indian Acad. Sci. (Chem. Sci.)* 106 (7), 1583–1594.

17 I. Kyrikou, M. Zervou, P. Petrakis, T. Mavromoustakos **2003**, (An effort to develop an analytical method to detect adulteration of olive oil by hazelnut oil), in *Magnetic Resonance in Food Science*, eds. P. S. Belton, A. M. Gil, G. A. Web, D. Rutledge, The Royal Society of Chemistry, Cambridge, (pp.) 223–230.

18 P. A. Keifer **1998**, (New methods for obtaining high resolution NMR spectra of solid-phase-synthesis resins, natural products, and solution-state combinatorial chemistry libraries), *Drugs Future* 23, 301–307.

19 E. Humpfer, M. Spraul **1998**, (The use of high resolution NMR in food analysis), *Semin. Food Anal.* 3, 287–302.

20 J. N. Shoolery **1995**, (The development of experimental and analytical high resolution NMR), *Proc. Nucl. Magn. Reson. Spectrosc.* 28, 37–52.

21 M. J. Gidley, S. M. Bociek **1985**, (Molecular-organization in starches – A C-13 CP MAS NMR Study), *J. Am. Chem Soc.* 107, 7040.

22 T. M. Eads, W. R. Croasmun **1998**, (NMR applications to fats and oils), *J. Am. Oil Chem. Soc.* 65 (1), 78–83.

23 S. M. Bociek, S. Ablett, I. T. Norton **1985**, (A C-13-NMR study of the crystal polymorphism and internal mobilities of the triglycerides, tripalmitin and tristearin), *J. Am. Oil Chem. Soc.* 62 (8), 1261–1266.

24 A. M. Gil, E. Alberti, A. Naito, K. Okuda, H. Saito, A. S. Tatham, S. Gilbert **1998**, (Solid state ^{13}C NMR studies of wheat high molecular weight subunits), in *Advances in Magnetic Resonance in Food Science*, eds. P. S. Belton, B. P. Hills, G. A. Webb, The Royal Society of Chemistry, Cambridge, (pp.) 127–134.

25 P. J. Barker, H. J. Stronks **1990**, (Application of the low resolution pulsed NMR "MINISPEC" to analytical problems in the food and agriculture industries), in *NMR Applications in Biopolymers*, eds. S. J. W. Finley, A. S. Serianni, Plenum Press, New York, (pp.) 481–498.

26 M. C. M. Gribnau **1992**, (Determination of solid/liquid ratios of fats and oils by low-resolution pulsed NMR), *Trends Food Sci. Technol.* 3, 186–190.

27 T. W. Skloss, A. J. Kim, J. F. Haw **1994**, (High resolution NMR process analyzer for oxygenates in gasoline), *Anal. Chem.* 66 (4), 536–542.

28 P. J. McDonald **1995**, (The use of nuclear magnetic resonance for on line process control and quality assurance), in *Food Processing Recent Developments*, ed. A. G. Gaonkar, Elsevier, Oxford, (pp.) 23–36.

29 N. Bloembergen, E. M. Purcell, R. V. Pound **1948**, (Relaxation effects in nuclear magnetic resonance absorption), *Phys. Rev.* 73, 679–718.

30 B. Blümich, V. Anferov, S. Anferova, M. Klein, R. Fechete, M. Adams, F. Casanova **2002**, (Simple NMR-Mouse with a bar magnet), *Concepts Magn. Reson.* 15 (4), 255–261.

31 S. Anferova, V. Anferov, D. G. Rata, B. Blümich, J. Arnold, C. Clauser, P. Blümler, H. Raich **2004**, (A mobile NMR device for measurements of porosity and pore size distribution of drilled core samples), *Concepts Magn. Reson. Part B* 23B (1), 26–32.

32 P. J. Barale, C. G. Fong, M. A. Green, P. A. Luft, A. D. McInturff, J. A. Reimer, M. Yahnke **2002**, (The use of a permanent magnet for water content measurements of wood chips.), *IEEE Trans Appl. Superconductivity* 12 (1), 975–978.

33 M. W. Hunter, P. T. Callaghan, R. Dykstra, C. E. Eccles **2003**, (Design and construction of a portable one-sided access NMR probe), in *Book of Abstracts 7th International Conference on Magnetic*

Resonance Microscopy, Snowbird, UT, USA, September 20–23, The University of Utah, P-7.
34 R. Dykstra, P. T. Callaghan, C. D. Eccles, M. W. Hunter **2003**, (A portable NMR for remote measurements), in *Book of Abstracts 7th International Conference on Magnetic Resonance Microscopy*, Snowbird, UT, USA, September 20–23, The University of Utah, P-8.
35 F. Mariette **2004**, (Relaxation RMN et IRM : un couplage indispensable pour l'étude des produits alimentaires), *C. R. Chim.* 7, 221–232.
36 B. P. Hills **1998**, *Magnetic Resonance Imaging in Food Science*, Wiley Interscience, New York.
37 R. R. Ruan, P. L. Chen **1998**, *Water in Foods and Biological Materials: A Nuclear Magnetic Resonance Approach*, Technomic, Lancaster, PA, USA.
38 J. Crank **1983**, *Mathematics of Diffusion*, Oxford University Press, Oxford.
39 J. J. Gonzalez, K. L. McCarthy, M. J. McCarthy **2000**, (Textural and structural changes in lasagna after cooking), *J. Texture Stud.* 31, 93–108.
40 K. L. McCarthy, J. J. Gonzalez, M. J. McCarthy **2002**, (Change in moisture distribution in lasagna pasta post cooking), *J Food Sci.* 67 (5), 1785–1789.
41 J. F. Steffe **1996**, *Rheological methods in food process engineering* 2nd edn, Freeman Press, East Lansing, MI, USA.
42 P. J. Cullen, A. P. Duffy, C. P. O'Donnel, D. J. O'Callaghan **2000**, (Process viscometry for the food industry), *Trends Food Sci. Technol.* 11, 451–457.

43 J. E. Maneval, K. L. McCarthy, M. J. McCarthy, R. L. Powell, July 2, **1996** *U. S. Patent*, No. 5532593.
44 R. L. Powell, J. E. Maneval, J. D. Seymour, K. L. McCarthy, M. J. McCarthy **1994**, (Nuclear magnetic resonance imaging for viscosity measurements), *J. Rheol.* 38 (5), 1465–1470.
45 A. G. Webb **1997**, (Radiofrequency microcoils in magnetic resonance), *Prog. Nucl. Magn. Reson. Spectrosc.* 31, 1–42.
46 A. G. Goloshevsky, J. H. Walton, M. V. Shutov, J. S. de Ropp, S. D. Collins, M. J. McCarthy **2005**, (Development of low field NMR microcoils), *Rev. Sci. Instrum.* 76, 24101–24106.
47 A. G. Goloshevsky, J. H. Walton, M. V. Shutov, J. S. de Ropp, S. D. Collins, M. J. McCarthy **2005**, (Integration of biaxial planar gradient coils and an rf microcoil for NMR flow imaging), *Meas. Sci. Technol.* 16, 505–512.
48 J. Dechow, T. Lanz, M. Stumber, A. Forchel, A. Haase **2003**, (Preamplified planar microcoil on GaAs substrates for microspectroscopy), *Rev. Sci. Instrum.* 74 (11), 4855–4857.
49 A. G., Goloshevsky, J. H. Walton, M. V. Shutov, J. S. de Ropp, S. D. Collins, M. J. McCarthy **2005**, (Nuclear magnetic resonance imaging for viscosity measurements of non-Newtonian fluids using a miniaturized rf coil), *Meas. Sci. Technol.* 16, 513–518.

4.8
Granular Flow
Eiichi Fukushima

4.8.1
Introduction

4.8.1.1 The Need for NMR Measurements of Granular Matter

Granular matter is all around us. It ranges from natural materials such as sand and asteroids to artificial materials such as pharmaceutical tablets and dry cereal. There is great and practical interest in static granular matter from the standpoint, for example, of how stable a slope is against avalanches or how much force the wall of a silo might have to withstand when it is full of grain.

Besides the static properties, granular matter has the unique property that even though it can withstand stress in its static state, it flows more like a liquid when the conditions are appropriate. There are cases in which we wish to prevent such flows, for example, avalanches. On the other hand, the fact that grains can flow assists us immeasurably in aiding transport of the material, for example, coal or grains flowing out of silos. On a more sophisticated level, we are interested in mixing dissimilar materials as often happens in food processing and pharmaceutical industries.

The fact that granular flow studies represents a major scientific challenge is shown by the fact that there is, as yet, no known equation of motion that describes the flowing motion of grains from first principles, such as the Navier–Stokes equation for Newtonian liquids. Thus, we are not able to predict the motion of the grains from a set of boundary conditions together with material properties. Of course, one thing that is known is that each particle collision satisfies conservation of momentum and energy (including the inelastic component that ends up as heat) and the subsequent movement obeys Newton's law of motion. This has spawned a large number of numerical simulations starting from specifications of parameters such as size, mass and coefficients of friction and restitution. Indeed, such simulation studies form the majority of current granular flow studies, overshadowing, at least for the moment, experimental investigations of granular motion.

Because of page length limitations, this chapter is merely a survey of past and current activities in the field without detailed descriptions. For a representative sample of typical studies on this subject, the reader is referred to some general reviews [1–3]. Presentations at granular flow meetings will also give more specific information. *Powders and Grains* is a quadrennial international meeting that brings together granular researchers from both the engineering and physics communities. Its proceedings [4] give an excellent overview of the field. *Granular Matter* published by Springer-Verlag is a journal devoted solely to granular studies. Other journals with regular contributions from this field include *Chemical and Engineering Science, Physical Review E* and *Chaos*. Finally, most relevant to this chapter, Duran has written a compact volume summarizing the state of experimental granular state work as of approximately 1999 [5].

Experiments in granular matter research, especially in granular flows, are difficult and represent a small fraction of all granular studies, being overshadowed by the innumerable computer simulation studies. (Some of the experiments are bordering on the heroic, when they involve avalanches. There is at least one group that have studied snow avalanches from inside a chamber over which the avalanches pass [6] and another group that studied 250 000 ping-pong balls avalanching down a ski jump [7]). Optical methods of measurement are still the most common experimental method, so far, though they are severely limited by the opacity of the materials. (The exception is diffusing wave spectroscopy [8] which measures the result of light scattering off (shiny) particles near the surface of the sample.) Traditional non-invasive methods for measuring flows either observe the surface of granular matter or use bulk methods to measure overall properties such as its average density. There is a need to seek non-invasive experimental methods to study the structure and motion within the bulk.

Ultrasound methods are not well suited for imaging granular flows in the bulk because the sound transmission is attenuated greatly by the innumerable air–solid boundaries in typical granular matter. At present, X-ray and other radiation methods can obtain static structures of granular matter but they are not well suited for velocity measurements because of lack of speed. Parker and coworkers have challenged this barrier with positron emission particle tracking of a (very large) rotating drum that has a 40 cm inner diameter and is 100 cm long [9].

Nuclear magnetic resonance (NMR) and its imaging modality magnetic resonance imaging (MRI) are non-invasive techniques, originally developed in physics, that are finding increasing use in a wide variety of disciplines from low-temperature physics to analytical chemistry to clinical imaging. It is a good candidate for granular matter because, in addition to being non-invasive, it can study a wide variety of parameters including particle density and velocity. A review article has been written on the use of NMR for studies of flowing matter including granular flow [10] and a bibliography of NMR experiments up to about 2000 is available [11].

NMR/MRI studies of granular matter are related to studies of concentrated suspensions by the same method, which is gaining acceptance [10]. However, there are major differences in the logistics of the experiments. The vast majority of concentrated suspension studies image the fluid carrying the particles rather than the particles, which are usually glass or plastic. It is a rare experiment that looks at the particles themselves rather than detecting the particles as voids in the image of protons in the liquid.

With the rare exception of xenon gas NMR of fluidized beds, which we discuss later, granular flow studies by NMR detect signals from the particles and not the surrounding medium. Because it is technically easier to obtain NMR signals from liquids rather than solids, the majority of granular NMR studies so far use solid particles containing liquids.

4.8.1.2 NMR Parameters of Relevance to Granular Matter
4.8.1.2.1 Density

By far the most prolific application of MRI is to medical imaging, mostly involving some form of anatomical imaging related to tissue density. Ironically, the quantitative measurement of signal amplitude, which is proportional to the tissue density, is more difficult than measurements of higher order parameters such as velocity. (This is because the higher order parameters are usually deduced from the phase of the signal, which is more robust against noise than the amplitude of the signal.)

The beauty of MRI in medical imaging is its sensitivity to different chemical or physical properties that give useful contrasts between volumes having different properties, for example, cancerous tissue versus normal tissue, both having similar density. This greatly reduces the need for accurate density measurements because it shifts the burden of image contrast to a parameter that is more sensitive than density.

The relative resolution of an MRI does not depend particularly on the size of the imaged object and usually is of the order of one part in 10^2. Thus, the absolute resolution obtainable by MRI depends on the size of the system being imaged. The density resolution depends on many factors, but is usually of the order of a few percent.

4.8.1.2.2 Velocity

There are two broad categories of velocity measurements by MRI. The conceptually simpler method is to use non-invasive tags on the images to monitor the time evolution of the sample. The most common tags are grids of lines that are superimposed on a 2D image, as discussed in Section 4.8.2.5. The image is repeatedly observed at different times after the tagging takes place to visualize the flow based on the evolution of the grid.

A more quantitative method is the so-called phase method, the phase being one of two parameters that an MRI image yields, the other being signal amplitude. We show below (Section 4.8.2.6) that the phase is correlated with the velocity of the sample, so a spatially resolved image of the signal phase can yield a velocity image.

4.8.1.2.3 Velocity fluctuations

Velocity fluctuations or correlations $<u(0), u(t)>$, where $u(t)$ stands for the fluctuating part of the velocity at time t, is an important concept in granular flow. It is correlated with granular temperature, which is difficult to measure by most techniques. The beauty of NMR is that the velocity fluctuations, averaged over adjustable time and space, is a directly measurable quantity, rather than a quantity that is calculated from monitoring the motion of specific particles, as would be done by optical measurements. Velocity fluctuations are determined by the attenuation of the NMR signal in the presence of a magnetic field gradient. Thus, it is basically an NMR diffusion or displacement measurement, as will be described in Section 4.8.2.8.

4.8.2
NMR Strategies

4.8.2.1 Special Requirements for NMR/MRI Experiments

Magnetic metals should not be used in any part of the sample holder or other components that will be in the vicinity of the magnet where the stray magnetic field can be considerable. As a rule of thumb, the stray magnetic field outside the magnet is fairly strong for distances approximately equal to the dimension of the magnet in that direction. Electric motors are composed of magnetic components and, in addition, do not work well in strong magnetic fields. Thus, motors, for example, for turning a horizontal cylinder should be located well away from the magnet and connected to the sample via a non-magnetic rod such as nylon, aluminum or brass. This is not only a logistics issue but also a safety issue of considerable magnitude.

Common non-magnetic metals that can be used around the magnet include aluminum and brass, as already mentioned, plus copper and some of the stainless steels. Alloys such as brass and stainless steel can have inclusions of magnetic impurities so they need to be checked, at least with strong magnets. Although dc and low frequency magnetic fields pass through conducting metals, ac (and rf) magnetic fields do not. Therefore, the use of non-magnetic metals near/in the rf probe should be limited to geometries that present a small cross section to the magnetic field direction.

4.8.2.2 Choice of Sample Materials and Magnetic Field Strength

NMR is a notoriously insensitive experimental method, because the nuclear magnetic moment is such a weak quantity, giving rise to magnetic fields that are weaker than most other sources of magnetic fields, with the possible exception of the fields due to the currents in the brain which give rise to signals detected by magnetoencephalography. When nuclear spins precess in the static field of the magnet, they give rise to very weak oscillating magnetic fields that are difficult to detect. The largest magnetic moment for common atoms belongs to hydrogen, which is fortunate because hydrogen is present in many naturally occurring objects. The other parameters under some control of the experimenters include the strength of the magnetic field of the magnet used in the NMR/MRI experiment and the density of the spins.

The NMR/MRI signal per spin depends steeply on the strength of the static field of the magnet. Therefore, one might think that the experiment should be carried out in the strongest magnetic field available. In chemical NMR, there is a constant push to use stronger magnetic fields for this and other reasons. Unfortunately (or fortunately – there are obvious downsides to using very strong magnets), granular studies cannot be carried out in very strong magnetic fields because of the inhomogeneity in the magnetic susceptibility of the system distorting the otherwise uniform static magnetic field.

Consider a single spherical particle in a uniform magnetic field. Because the magnetic susceptibility of the particle is different from that of air, the magnetic field lines must obey certain conditions at the boundary between the particle and air. This causes the magnetic field lines to curve near the surface of the particle, an effect that is magnified for smaller particles and for stronger magnetic fields. Thus, from a practical point of view, it is difficult to do NMR/MRI of many granular particles as small as 1 mm diameter in magnetic fields of 4.7 T or stronger.

Magnets with strong field strengths AND a large volume of the field are difficult (and expensive) to make. Granular particles need to be large enough to avoid electrostatic effects, and the sample needs to contain a sufficient number of particles to avoid wall effects and to achieve statistical homogeneity. Thus, magnets with moderately large useable volume with not so strong magnetic fields are well suited for granular studies. Much of the NMR/MRI experiments to date have been performed in 2 T magnets with a uniform field region of approximately 8 cm diameter spherical volume. The size of the uniform field region defines the size of the region to be studied. However, the sample assembly can exceed this size if permitted by the probe geometry.

It is well known that solid materials are characterized by lack of relative motion between the constituent atoms compared with liquids, which results in NMR linewidths much broader than for liquids. Also, it often happens that solids have much longer spin–lattice relaxation T_1 than in liquid samples. Thus, studying NMR of solids is more difficult than liquids and studying flows of solid material would seem to be quite impractical.

In order to avoid the problems of weak signals from typical solid samples, all NMR/MRI signals from granular particles to date have been obtained with solid particles containing liquids. The majority of experiments have been done either with seeds that naturally contain oils, predominantly mustard seeds, or pharmaceutical pills that contain liquid cores. In some cases, however, solid beads were impregnated with liquid to make them NMR-visible. In one case, it was oil-soaked sugar particles [12], while in another, acetone was imbibed into catalytic materials [13].

4.8.2.3 Density Imaging

Medical MRI results primarily as density images with some physical or chemical contrast. On the contrary, relatively few velocity imaging measurements are made, mostly in cardiology. (There are hybrid methods such as MRI angiography whereby only the flowing liquid is imaged, or other flow visualization applications, but these can still be considered to be anatomical images with flow as the contrasting property.)

What is not commonly known outside the NMR community is the relative difficulty in making accurate relative density measurements by NMR/MRI compared, for example, with making velocity measurements by NMR/MRI. This is because density measurements rely on the signal strength, which can be affected by any number of factors. In contrast, the most common way to measure flow velocity by NMR/MRI is to use the phase of the NMR signal rather than the amplitude, which vastly reduces its sensitivity to amplitude variations.

The other fact that is not commonly realized outside of the NMR/MRI community is that the most common MRI sequence, the spin-warp, obtains data from all parts of the sample in every repetition of the data acquisition step. In other words, if it takes 128 "scans" to make a 2D MR image, each of the 128 scans is getting information from all parts of the image rather than obtaining information from $1/128^{th}$ of the sample. Therefore, any change in the shape of the sample during data acquisition leads to placing the intensity in a wrong location of the image. This makes shaking and vibration experiments much more difficult than smooth flow experiments, such as a rotating drum experiment. Please refer to the section that discusses vibrating samples for more information.

The absolute strength of the MRI signal depends on many factors including the number of nuclear spins in an imaging voxel. Thus, there is a tradeoff between image resolution and signal-to-noise ratio, S/N. For the commonly used granular particles such as mustard seeds and pharmaceutical pills, the overall density of spins in the solid sample depends on the fraction of the particle occupied by the NMR-visible spins, in other words, the liquid portion. For typical particles, the fraction occupied by the liquid decreases with the overall size of the particle, so the particles at the small end of the scale are less sensitive. (The exception to this rule is liquid imbibed particles that are uniform in structure [12, 13].)

A practical ramification of this rule is utilized in segregation studies. The most obvious way to study segregation by NMR/MRI is to use two different types of beads, one that is NMR-visible and the other not. However, size segregation studies can be performed by simply using two different sizes of NMR-visible particles because they have different overall spin densities, for the reason stated above. Therefore, the image intensity will depend on the relative concentration of each particle in a voxel.

4.8.2.4 Problems of Imaging Static Structures

Measuring static granular structures by MRI incurs some complications. The majority of NMR/MRI experiments to date use heterogeneous particles, specifically, solid particles with liquid cores. Thus, a close packed array of such particles would not image as close packed but would show gaps because the solid shells would yield weak or no NMR signal.

There are several ways to deal with this problem. The simplest in concept is to use liquid imbibed particles that are uniform in structure, such as sugar cubes [12, 13] as already mentioned above. To my knowledge, this has never been done for structure studies. The second is to use NMR-invisible particles, for example, glass, and fill the spaces between the particles with a liquid such as water that can be imaged. The third is to use the usual radially heterogeneous particles and then use software to calculate the parameters. One first determines the centers of each particle and then decides whether the particles are in contact with each other or not by the distances between the centers [14].

4.8.2.5 Tagging for Flow Visualization

Non-invasive tagging can be performed in MRI by selectively saturating the spin magnetization at certain positions and watching the evolution of the image as the "tagged" spins move with time. The most common scheme is to tag equally spaced parallel thin slices oriented in a direction perpendicular to the gradient direction, rotate the gradient 90°, and then repeat the procedure orthogonal to the first set, thus creating a grid in a 2D plane.

There are many specific ways to generate equally spaced tags but they are all based on the same principle of manipulating the rf pulses to generate equally spaced bands of rf radiation in the frequency domain. It is well known that under ordinary conditions, meaning normal levels of nuclear spin excitation, the frequency spectrum of the rf excitation pulse(s) is approximately the Fourier transform of the pulses in the time domain. Thus, a single slice can be generated in the presence of a magnetic field gradient from a pulse (or pulses) that has a narrow "line" as its Fourier transform.

The most effective slice that we can imagine is a narrow square band of width d such that the intensity is constant across d but is zero outside. If that is to be the resulting frequency domain "spectrum", its Fourier transform is $\sin(\omega t)/(\omega t)$ $\equiv \text{sinc}(\omega t)$ [15], where ω is proportional to $1/d$. Thus, playing out $\text{sinc}(\omega t)$ in the presence of a linear magnetic field gradient will excite the spins in a rectangular slice orthogonal to the gradient direction. If the excitation is chosen to be 90°, then there will be a signal void in a subsequent image corresponding to the slice, because the spins will have been saturated only in that slice by the first pulse.

In order to replicate the slice (so that the tagging is suggestive of a "picket fence"), it is sufficient to sample the $\text{sinc}(\omega t)$ function, which can be done by taking the array of numbers that make up $\text{sinc}(\omega t)$ in memory and setting most of the data to zero so the remaining non-zero points are periodic. It has been demonstrated [16] that this is an easy way to generate multiple parallel tagging slices.

Tagging is not the method of choice for determining velocities because the calculation of the velocity is complex and the spatial resolution of the velocity information is only as good as the size of the grids. The best velocity images are made with the phase method, described below. However, tagging is a superb method to visualize flow.

4.8.2.6 Velocity Imaging by the Phase Method

A moving spin, undergoing precession, in a magnetic field gradient experiences a change in the Larmor precession frequency, resulting in a phase shift compared with a spin that does not move. The NMR/MRI experiment yields spatially resolved magnitude and phase of the nuclear spins from the sample, with the magnitude being proportional to the spin density whereas the phase is related to the velocity.

The instantaneous precession frequency ω in a magnetic field gradient G for a moving spin with coordinate $x = x_0 + vt + at^2/2 + \ldots$ is

$$\omega = \gamma B = \gamma G(x_0 + vt + at^2/2 + \ldots) \tag{4.8.1}$$

so that the phase, being an integral of that expression, develops with time as

$$\begin{aligned}\phi &= \gamma \int G(x_0 + vt + at^2/2 + \ldots) dt \\ &= \gamma x_0 \int G dt + \gamma v \int t G dt + \gamma(a/2) \int t^2 G dt + \ldots\end{aligned} \tag{4.8.2}$$

The integrals in this expansion are the temporal moments of G, usually denoted m_i for the ith moment and under the control of the experimenter, so the expression for phase becomes

$$\phi = \gamma x_0 m_0 + \gamma v m_1 + \gamma(a/2) m_2 + \ldots \tag{4.8.3}$$

The first term is the starting phase that depends on the starting position x_0. The second term is linear in v, which is exactly what we are looking for. The third term is linear in the acceleration a, and subsequent terms depend on higher order derivatives of the coordinate.

In an NMR/MRI flow experiment, we would like to measure parameters such as velocity without regard to the starting position of the particle. Thus, m_0 is always set to zero. The moments m_i are under the control of the experimenter in that they are manipulated by the choice of the time dependence of the gradient G. Thus, it is easy to see that m_0 can be set to zero by simply making sure that the time integral of the gradient is zero. The easiest way to accomplish this is to have a bipolar gradient of equal absolute amplitude and duration.

With $m_0 = 0$, the first surviving term is $\gamma v m_1$. So the equation is now

$$\phi = \gamma v m_1 + \ldots \tag{4.8.4}$$

If the acceleration is taken to be constant during the NMR/MRI measurement, the velocity can be obtained by repeatedly measuring the phase ϕ while the first moment m_1 is varied. Because MRI yields spatially resolved values for ϕ, a pixel by pixel determination of the velocity v can be made.

Typically, at least two different values of m_1 (besides $m_1 = 0$) are used because there are invariably phase shifts that arise from various factors that do not depend on the gradient moments, resulting in a non-zero intercept of ϕ versus m_1. Thus, a velocity image is time consuming because each set of measurements with a value of m_1 is an image in itself.

If we extend this logic, choosing the time variation of G to null m_1 would yield information about the acceleration a. (It is easy to show that nulling any moment m_i would null all lower moments m_j, $j < i$.) However, the higher order terms are increasingly noisier so the measurement of motional parameters of order higher than velocity is difficult.

4.8.2.7 Propagators

NMR can obtain the distribution of velocities without spatially resolving the velocity profile and such a distribution is called a propagator [17, 18]. By trading off the spatial information, we gain speed. In Section 4.8.3.5, we describe the formation of gas bubbles in a fluidized bed that happens too fast to be imaged at present. However, it is possible to make NMR measurements without spatial resolution, specifically propagators, velocity fluctuations, chemical shifts and linewidths, of rapid granular flows as we might find in fluidized beds with bubbles.

The requirement for such spatially averaged measurements is that the granular assembly be statistically stable, that is, the probability of finding grains in a particular distribution of situations, such as having certain velocities or of it being in a certain set of neighbor distances, is independent of time.

The propagator $P(R,t)$ is defined by the expression for the signal

$$S(q) = \int P(R,t)\exp(i2\pi q \cdot R)dR \tag{4.8.5}$$

and represents the probability that a particle starting at $t = 0$ will make a displacement R in time t. q is a vector $\gamma\delta g/2\pi$ where g is the amplitude of the gradient pulse that is applied to the spin system for duration δ. Wang et al. [19], as further described in Section 4.8.3.5, have measured the propagators for ^{129}Xe gas flowing through a fluidized bed. They showed that for low average velocities the distribution of velocities remain the same and just move to higher average velocities as the average gas velocity increases, showing that bubbles have not, yet, formed. Above a certain threshold value of flow, however, the propagator becomes broader, indicating the formation of bubbles.

4.8.2.8 Velocity Fluctuations and Diffusion

It has been known for 40 years, since the initial PGSE experiment of Stejskal and Tanner [20], that NMR is an excellent tool for measuring the statistical displacements of molecules with time, usually called diffusion. The exponential attenuation α of the PGSE signal as a function of time between the PGSE pulses is given by [21]

$$\alpha(t) = (1/\pi)\int D_{zz}(\omega) |F(\omega,t)|^2 d\omega \tag{4.8.6}$$

where

$$D_{zz}(\omega) = (1/\pi)\int <u(t)u(0)> \exp(i\omega t)dt \tag{4.8.7}$$

is the zz component of the self-diffusion tensor,

$$F(\omega,t) = \int [g\int g(t'')dt'']\exp(i\omega t')dt' \tag{4.8.8}$$

is the spectrum of the motion-encoding gradient that is under the control of the experimenter, and $u(t)$ is the randomly varying portion of the velocity. If an exponential velocity fluctuation autocorrelation function with a correlation time is assumed, it will be possible to extract the correlation time from the measured diffusion coefficient by suitably manipulating the magnetic field gradient $g(t)$.

Although this formalism had been worked out and applied to random molecular diffusion in liquids, it had not been applied to motions of granular particles until recently. Based on the work of Seymour and coworkers [22, 23], the formalism applies to granular flows, at least in the parameter space used. This is a fascinating finding because it will allow access to the correlation of velocity fluctuations of particles, a parameter of significance to granular studies. It seems reasonable that the correlation time for this process may be related to the correlation time of particulate collisions although it is not obvious what the proportionality factor should be. Furthermore, the velocity fluctuation intensity that defines the particulate diffusion is related to the granular temperature [24].

In a recent study, Huan et al. [25] performed NMR experiments in vibrofluidized beds of mustard seeds in which the small sample volume allowed pulses short enough that displacements in the ballistic phase were distinguishable from those in the diffusion phase. In this case, the average collision frequency is measured directly, bypassing the uncertainty of the multiplicative factor mentioned above. These workers also measured the height dependence of the granular temperature profile.

4.8.2.9 Chemical Shift and Linewidth

So far, the majority of granular matter studies by NMR/MRI have used liquid state proton measurements in solid materials. Because proton signals are relatively insensitive to chemical environment through the chemical shift effect and because the physical environments are relatively similar in all liquids, the resonance frequency and the NMR linewidth are not good indicators of granular parameters such as particle density and velocity.

An exception is the use of ^{129}Xe gas NMR to study fluidization of granular particles by Wang et al. [19]. ^{129}Xe has a large chemical shift range, which means the resonance frequency is very sensitive to the chemical environment. Thus, ^{129}Xe is very sensitive to any variation of magnetic field strength compared with protons.

This fact has been used to obtain a contrast between gas molecule spins that are in the bubbles in a fluidized bed and those in the emulsion phase. The gas molecules in the emulsion phase see a wide range of magnetic fields because of the magnetic susceptibility heterogeneity caused by the many solid–gas interfaces that distort the otherwise uniform magnetic field. The motion of the gas molecules between closely packed particles, therefore, modulates the magnetic field intensity seen by the spins and cause lifetime broadening of the NMR line. On the other hand, the spins in bubbles see more uniform magnetic fields because there are no particles in the bubbles to disturb the field. Furthermore, bubbles tend to be more spherical in shape, guaranteeing that the field inside the bubbles must be fairly uniform.

Thus, it is possible to distinguish the source of the NMR signal, whether it is from the bubbles or the emulsion. Because NMR is a fairly slow method, it is not possible to spatially resolve this information, at least at present. (Spatially resolving the NMR signal divides the signal intensity into the large number of voxels, incurring serious penalties in the S/N.) So, Wang and his collaborators obtained NMR spectra and deduced the fraction of gas that is in the bubble versus that in the emulsion in the entire sample region, as a function of fluidization from the ratio of the spectral areas. In this way, they were able to detect the onset of bubbly flow with increasing inlet gas flow.

4.8.3
Systems Studied

4.8.3.1 Static Structure

As an example of static structures studied by MRI, the distribution of contact angles was obtained for an assembly of 640 spheres of 3-mm diameter packed in a cylinder of aspect ratio ≈1 [14]. The structure was determined from the raw data by locating the centers of the spheres and then recreating the full spheres. Contact was deemed to be made when the centers were separated by the sphere diameter (within some uncertainty) and the orientation of the lines joining the centers yielded the contact angles. Because this stack was fairly short, there is a dominant peak at $\phi = 0$ where ϕ is the angle with respect to the horizontal, that is there were well defined horizontal layers. This work performed measurements as a function of compaction but the effects were minimal because the entire assembly was being studied. In a later study [26], the column was spatially resolved to yield interesting differences as a function of height.

4.8.3.2 Shaking

Shaking and vibrating experiments are more difficult to perform by NMR/MRI compared with flow experiments. This is because of the way in which MRI is most commonly performed, namely, by spin-warp imaging. As already mentioned in Section 4.8.2.3, spin-warp imaging repeatedly acquires data from all parts of the sample and reconstructs images as a Fourier transform of time *and* of the repetition just mentioned. Thus, if the sample configuration is different from acquisition to acquisition, it will appear that there is an extra component of the signal in the frequency domain. We refer the reader to standard references for spin-warp imaging [27].

Fluidized bed experiments can be realized either by shaking the sample (vibrofluidized) or by passing gas through the system. We will combine the discussion of both types of experiments under fluidized bed in Section 4.8.3.5.

4.8.3.2.1 Single Shake

Despite the difficulty of measuring vibrating granular material by MRI, one of the earliest studies of grains by MRI was of a vibrating box [28]. The group at Chicago chose to make the MRI measurements in the stationary sample *between* shaking to avoid the aforementioned problems. Initially, the static sample is tagged (see Section 4.8.2.5) with a "picket fence" of horizontal tags. The sample is immediately given one vertical shake and the evolution of the tags is measured as soon as the particles return to their static state. This results in the initially straight tags being bent by the convection pattern. In addition, indications of particulate diffusion could also be obtained from the broadening of the initially sharp tags upon shaking [29].

4.8.3.2.2 Vibrating Box

The group in Albuquerque made the first observations of an actively vibrating sample, not limited to single shakes. In order to overcome the difficulties of applying spin-warp imaging, a line-scan method was used in its place [30]. Line-scan, as the name implies, makes images by acquiring data along a line that is scanned across the sample, rather than sampling the signal from the entire sample all the time as is done with spin-warp. Therefore, the sample blurring is limited to the location where the blurring actually occurs rather than all over the image. The trade-off is poorer signal-to-noise ratio per time because information is acquired only where the line is at any given time.

The box of particles was shaken in the frequency-doubling mode so two halves of the sample could vibrate 180° out of phase. An interesting feature of these images was the highly sheared region between the parts moving out of phase, which appeared darker because of the enhanced incoherent motion due to collisions. Rather than being stationary in the center of the box, this region of maximum shear oscillates horizontally with the vertical motion of the particles.

4.8.3.3 Shear in Couette

Mueth et al. performed an experiment in which seeds were sheared in a vertical Couette having inner and outer diameters of 51 and 82 mm, and azimuthal velocity profiles were made by an MRI tagging technique [31]. Because MRI tags are always perpendicular to the magnetic field gradient direction, they are almost always straight and parallel to each other, making them less straightforward for measuring the azimuthal component of velocity. They solved the problem by analyzing the distortion of the parallel stripes around the azimuthal direction at each radial distance. This work lead to the interesting conclusion that the velocity profile is dominated by a Gaussian decay as a function of the radial distance from the inner or rotating wall.

4.8.3.4 Rotating Drum

The partially filled horizontal rotating drum is the configuration that is most studied by NMR/MRI. The smoothness of flow makes it ideal for spin-warp imaging, as already discussed in Section 4.8.3.2.2 above and the horizontal geometry is tailor-made for the common horizontal bore magnets of approximately 30 cm diameter bore. After the initial studies that showed the feasibility of using NMR/MRI for granular flow [32], there have been follow-up works that quantified the velocity profile [33] and extended the NMR/MRI capability to measuring fluctuations of velocity [22, 23]. In addition, the question of axial segregation has been addressed in several works.

4.8.3.4.1 Uniform Particles

Velocity profiles of particles in a partially filled horizontal drum were measured in the initial work [32] as well as in several subsequent works. For velocity images the phase method, described above, is the method of choice unless flow visualization is the main aim. In these cylindrical systems, there is solid body rotation of a large fraction of the beads that transports them to the top of the flowing layer. The flow takes place in a much thinner lenticular layer along the free surface, being thickest about half way along the free surface. On the surface, the velocity increases smoothly to near the center of the free surface and decreases smoothly to the bottom of the slope. As a function of depth, the velocity obeys a quadratic dependence on the depth close to the center of the slope, which is consistent with the notion that the bulk velocity profile adjusts itself to minimize rapid changes in the velocity [33]. It was found that this quadratic dependence is obeyed for a remarkably large range of depths, except quite close to the free surface where the particles flow slower than the quadratic dependence would predict. This tendency for the particles at the free surface not to keep up with the rotation rate for the faster rotation rates compared with particles deeper down in the flowing layer was also noted in the earlier work [33]. When the rotation rate is increased sufficiently, the topmost particles will flow even slower than some particles deeper in the layer.

Most, if not all, velocity measurements in the bulk, other than NMR/MRI, make measurements through clear end-walls of either long or short cylinders. End-walls, because of friction with the particles, change the dynamic angle of repose near the end-wall and can also cause convections in long cylinders with components of velocity in the axial direction [34, 35]. NMR/MRI experiments can avoid the end-wall effects by making measurements far from the end-walls in a long cylinder.

Measurements of velocity fluctuations were made near the center of a long, partially filled, horizontal cylinder by Seymour and coworkers [22, 23]. From these measurements, these workers derived the spatial dependence of the fluctuation correlation amplitude, which is what is commonly taken to be the granular temperature. Furthermore, the spatial dependence of the correlation time offers interesting insights into granular flow. It varies relatively little even in this extremely heterogeneous flow system, from particles deep down in the flowing layer that barely move to those at the free surface that can move quite rapidly. The correlation time varies only by a factor of 2 as a function of depth along an axis through the center of the cylinder and perpendicular to the free surface, whereas it varies only by approximately 50% along the free surface. Thus, the particles that barely move, either deep in the flowing layer or at the extreme ends of the free surface, have correlation times that are of the same order of magnitude as for the faster particles. The vast difference in velocities between the slowest and fastest particles must be due not to the correlation times but to the velocity between collisions and the average interparticle distance that increases with granular temperature.

4.8.3.4.2 Mixed Systems – Segregation

Segregation phenomena in partially filled horizontal rotating cylinders have been known for many decades. However, the details of how segregation takes place in the bulk have been slow to emerge, largely because of our inability to see into the bulk. Nakagawa et al. performed several MRI studies of radial segregation as well as how axial transport takes place, starting from the axially unmixed state in this geometry [36]. Metcalf and collaborators have further looked into these questions with MRI [37].

Studies of relatively slow segregation in closed sample configurations, such as the rotating drum, are perhaps the easiest granular experiments to perform by MRI. Measurements can be made with the sample stopped; the process can continue to evolve between the measurements. As mentioned above, MRI is much easier to perform on stationary systems as opposed to systems undergoing motion.

Hill, in her thesis work, used MRI to study what happens in the bulk during axial segregation by size in a partially filled horizontal drum [38]. She monitored the evolution of the distribution of particles as a function of the number of rotations by making MRI images of suitable areas of the drum with the rotation stopped. Only the smaller particles were NMR visible in her experiment, so she obtained time course images of the core. A slice containing the cylinder axis, shows bulges in the core that represent incipient bands of the smaller particles that could grow sufficiently to appear on the surface as bands of small particles. It is as if one could squeeze a part of the core and the core responds by bulging elsewhere. Thus, this study suggests that axial segregation is not simply driven by surface effects but involves some mechanism in the bulk.

Maneval et al. [39] measured the transverse velocity profile of a partially filled rotating horizontal cylinder as a function of mixture composition of two different sized beads. The shape of the profile depends critically on the mixture because the larger particles tend to flow very easily over the core of the smaller particles that form because of radial segregation. The presence of axial gradients of not just velocity but of shear suggests the latter as a possible driving force for axial segregation, in analogy with shear-induced migration in concentrated suspensions [40].

4.8.3.5 Fluidized Bed

Savelsberg et al. obtained proton NMR signals from fluidized seeds and catalyst particles containing liquids [13]. They found step changes in the signal intensity as gas flow increased, indicating phase changes such as fluidization, bubbling and slug flow. They also obtained q-space data from which particulate diffusion coefficients were calculated as a function of gas flow. Finally, propagators were obtained showing that displacements are small, of the order of one particle diameter, even in the reduced density slug flow phase for the measurement duration of 2.1 ms.

Huan et al. [41] measured the behavior of a small fluidized bed consisting of 45–80 mustard seeds in a small-bore vertical magnet. The small sample size allowed short pulses, and spatial distribution of collision correlation times and granular temperature were measured directly and compared with the hydrodynamic theory of Garzo and Dufty [42]. This paper [41] contains an excellent survey of previous experiments on fluidized beds.

Wang et al. [19], as already mentioned in Section 4.8.2.9, take advantage of the large chemical shift difference between ^{129}Xe gas nuclei that are in the emulsion versus those in bubbles to study the inception of the bubbling phase of a fluidized bed. They also measured the propagator, which shows the distribution of gas velocities in the sample. This is the only work that the author knows of that uses gas NMR to study granular flow of any type. Under normal circumstances, gas NMR is less sensitive than liquid NMR by the ratio of the spin densities, roughly three orders of magnitude, which is a serious handicap. These authors overcame this handicap by using hyperpolarized ^{129}Xe, which regains the S/N shortage. The use of hyperpolarized gas for NMR/MRI is not prohibitive but adds a significant complication and cost to an experiment that is already at above-average levels in complexity.

4.8.4
Future Outlook

4.8.4.1 Technical Capabilities and Possible Applications

Here are some possible directions in which the field could develop. It is certain that the methodology will improve for measurements of faster flows. Also, various higher order parameters will be measured. For example, it is already technically possible to measure the anisotropy of the velocity fluctuations. Furthermore, it is, in principle, possible to measure what we might call the spectrum of collisions to determine the deviations from the exponential correlation model by the use of a time-dependent magnetic field gradient [43]. It might become possible to image solids themselves rather than just the liquid component of the particle or to obtain signals simultaneously from both the solid and gas components of fluidized beds. More realistically, it should be possible to study the liquid and solid phases of wet granular flows. Improved methodology might allow measurements of some transient phenomena rather than steady state flows. This may open the way for NMR/MRI to contribute to the studies of pattern formation. Finally, the author is hopeful that there will be a way to measure 3D force chains by an NMR related method which includes NQR (nuclear quadrupole resonance, which produces signals that can depend on the strain of the solid material [44]). The trick is to figure out a suitable magnetic resonance parameter that depends on pressure, not an inconceivable task. In summary, the future is bright for the application of NMR to granular problems. After a somewhat tentative decade of seeing what measurements could be made, there are signs of branching out into more sophisticated measurements, such as granular temperature and studies of bubble formation in fluidized beds.

4.8.4.2 Practical Considerations of Cost and Complexity

Traditionally, experimental studies of granular media used fairly basic methodology. The early experiments were almost entirely visual; for example, the early segregation work of Oyama [45] used visual observations and sketches to describe axial segregation. With the advent of high speed photography, surface velocity studies were performed both in the laboratory [46] and in the field [6, 7]. NMR/MRI is among the group of more sophisticated methods, such as high intensity X-ray [47], diffusing wave spectroscopy [8] and positron emission tomography [9], that can be brought to bear on experimental investigations of granular systems.

NMR/MRI sets itself apart from many technologies in being extremely versatile. It is able to non-invasively measure not only particulate structure and density but also higher order parameters, such as velocity and velocity fluctuations. It has the option of spatially resolving such parameters (in which case the name MRI applies) or, instead, measuring distributions, that is, histograms, of such parameters without any spatial resolution, but with the tradeoff of significantly higher S/N.

The instrumentation is also as varied as the variety of parameters to be measured. However, the choice is severely limited if the instrumentation needs to be of a turnkey variety. An obvious example is MRI instruments for clinical medicine that are technologically well suited for particle density imaging. MRI technicians in clinical MRI facilities are well trained to make standard images but, with rare exceptions, would not be able to make the more complex measurements needed, for example, for velocity imaging. Unfortunately, clinical MRI instruments are very expensive and also require expensive physical facilities to house them. Furthermore, medical facilities are usually not willing to divert the use of their expensive MRI instrument for purposes other than their needs, especially if instrument settings need to be changed.

Because the major cost of a commercial MRI is the magnet, there have been attempts to develop portable MRI consoles, that is, the requisite instrumentation for MRI minus the magnet, that could be transported to a medical MRI facility so that a scientific experiment could be set up independently of the medical MRI console (see Chapter 2.2). There are also efforts to develop compact NMR/MRI instruments for non-traditional uses. Granular studies would easily fit into this category.

For many measurements specific to granular flows, such as velocity profiles, acceleration or of vibrating samples, a laboratory with NMR/MRI expertise is desirable. Therefore, much of the future progress in NMR/MRI studies of granular systems will come from collaborations of granular matter experts with such laboratories that have the expertise in NMR/MRI.

Acknowlegments

This work was supported by the Engineering Science Program, Basic Energy Sciences, US Department of Energy, via grant DE-FG03–98ER14912, but this sponsorship does not constitute an endorsement by the DOE of the views expressed in this work. I acknowledge the help of Ms. Cathy Clewett in the preparation of this manuscript.

References

1. H. M. Jaeger, S. R. Nagel, R. P. Behringer, **1996**, *Phys. Today*, April, 32–38.
2. H. M. Jaeger, S. R. Nagel, R. P. Behringer, **1996**, *Rev. Mod. Phys.* 68, 1259–1273.
3. J. Kakalios, *Am. J. Phys.* **2005**, 73, 8–22.
4. R. Garcia-Rojo, H. J. Herrmann, S. McNamara (eds.) **2005**, *Powder and Grains 2005: Proceedings of the 5th International Conference on Micromechanics of Granular Media*, Stuttgart, 18–22 July 2005, A. A. Balkema, Leiden.
5. J. Duran **2000**, *Sand, Powders, and Grains*, Springer, New York.
6. J. D. Dent, K. J. Burrell, D. S. Schmidt, M. Y. Louge, E. E. Adams, T. G. Jazbutis, *Ann. Glaciol.* **1998**, 26, 247–252.
7. J. N. McElwaine **2001**, K. Nishimura, *Ann. Glaciol.* 32, 241–250.
8. N. Menon, D. J. Durian **1997**, *Phys. Rev. Lett.* 79, 3407–3410.
9. Y. L. Ding, J. P. K. Seville, R. Forster, D. J. Parker **2001**, *Chem. Eng. Sci.* 56, 1769–1780.
10. E. Fukushima **1999**, *Annu. Rev. Fluid Mech.* 31, 95–123.
11. E. Fukushima **2001**, *Adv. Complex Systems* 4, 503–507.
12. P. Porion, N. Sommier, P. Evesque **2000**, *Europhys. Lett.* 50, 319–325.
13. R. Savelsberg, D. E. Demco, B. Bluemich, S. Stapf **2002**, *Phys. Rev. E* 65, 020301.
14. R. A. Waggoner, M. Nakagawa, S. J. Glass, M. Reese, E. Fukushima **1998**, (Particle compaction as observed by MRI), in *Spatially Resolved Magnetic Resonance*, eds. P. Blümler, B. Blümich, B. Botto, E. Fukushima, Wiley-VCH, Weinheim, (pp.) 299–304.
15. R. N. Bracewell **1965**, *The Fourier Transform and its Applications*, McGraw-Hill, New York.
16. M. V. Icenogle, A. Caprihan, E. Fukushima **1992**, *J. Magn. Reson.* 100, 376–381.
17. J. Kaerger, W. Heink **1983**, *J. Magn. Reson.* 51, 1–7.
18. P. T. Callaghan, S. L. Codd, J. D. Seymour **1999**, *Concepts Magn. Reson.* 11, 181–202.
19. (a) R. Wang, M. S. Rosen, D. Candela, R. W. Mair, R. L. Walsworth 2005, (Study of gas-fluidization dynamics with laser-polarized ^{129}Xe), *Magn. Reson. Imag.* 23, 203–207; (b) R. Wang **2005**, (Study of gas flow dynamics in porous and granular media with laser-polarized 129Xe NMR), *Ph.D. Thesis*, Department of Nuclear Engineering, Massachusetts Institute of Technology, February 2005.
20. E. O. Stejskal, J. E. Tanner **1965**, *J. Chem. Phys.* 42, 288–292.
21. J. Stepisnik **1985**, *Prog. Nucl. Magn. Reson. Spectrosc.* 17, 187–209.
22. J. D. Seymour, A. Caprihan, S. A. Altobelli, E. Fukushima **2000**, *Phys. Rev. Lett.* 84, 266–269.
23. A. Caprihan, J. D. Seymour **2000**, *J. Magn. Reson.* 144, 96–107.
24. S. B. Savage, R. Dai **1993**, *Mech. Mater.* 16, 225–238.
25. C. Huan, X. Yang, D. Candela, R. W. Mair, R. L. Walsworth **2004**, *Phys. Rev. E* 69, 041302.
26. M. Nakagawa, R. A. Waggoner, E. Fukushima **1999**, (Non-invasive measurement of fabric of particle packing by MRI), in *Mechanics of Granular Materials*, eds. M. Oda, K. Iwashita, A. A. Balkema, Rotterdam, Section 4.3.1, (pp.) 240–250.
27. P. T. Callaghan **1991**, *Principles of Nuclear Magnetic Resonance Microscopy*, Clarendon Press, Oxford.
28. E. E. Ehrichs, H. M. Jaeger, G. S. Karczmar, J. B. Knight, V. Yu. Kuperman, S. R. Nagel **1995**, *Science*, 267, 1632–1634.
29. V. Yu. Kuperman **1996**, *Phys. Rev. Lett.* 77, 1178–1181.

30 A. Caprihan, E. Fukushima, A. D. Rosato, M. Kos **1997**, *Rev. Sci. Instrum.* 68, 4217–4220.

31 D. M. Mueth, G. F. Debregeas, G. S. Karczmar, P. J. Eng, S. R. Nagel, H. M. Jaeger **2000**, *Nature (London)*, 406, 385–389.

32 M. Nakagawa, S. A. Altobelli, A. Caprihan, E. Fukushima, E.-K. Jeong **1993**, *Exp. Fluids* 16, 54–60.

33 M. Nakagawa, S. A. Altobelli, A. Caprihan, E. Fukushima **1997**, (NMR measurement and approximate derivation of the velocity depth-profile of granular flow in a rotating, partially filled, horizontal cylinder), in *Powders and Grains 97, Proceedings of the Third International Conference on Powders & Grains*, eds. R. P. Behringer, J. T. Jenkins, A. A. Balkema, Rotterdam, (pp.) 447–450.

34 A. Santomaso, M. Olivi, P. Canu **2004**, *Chem. Eng. Sci.* 59, 3269–3280.

35 A. Caprihan, B. E. Smith, J. E. Maneval, K. M. Hill, E. Fukushima **2005**, (Effects of end-wall friction on granular flow in rotating 2D and 3D cylinders), in *Powders and Grains 2005, Proceedings of the 5th International Conference on Micromechanics of Granular Media*, eds. R. Garcia-Rojo, H. J. Herrmann, S. McNamara, A. A. Balkema, Leiden, (pp.) 877–880.

36 M. Nakagawa, S. A. Altobelli, A. Caprihan, E. Fukushima **1997**, *Chem. Eng. Sci.* 52, 4423–4428.

37 G. Metcalf, L. Graham, J. Zhou, K. Liffman **1999**, *Chaos* 9, 581–593.

38 K. M. Hill, A. Caprihan, J. Kakalios **1997**, *Phys. Rev. Lett.* 78, 50–53.

39 J. E. Maneval, B. E. Smith, A. Caprihan, E. Fukushima **2005**, (Flow measurement inrotational flow of mixed-size granular systems), in *Powders and Grains 2005, Proceedings of the 5th International Conference on Micromechanics of Granular Media*, eds. R. Garcia-Rojo, H. J. Herrmann, S. McNamara, A. A. Balkema, Leiden, (pp.) 841–844.

40 D. Leighton, A. Acrivos **1987**, *J. Fluid Mech.* 181, 415–439.

41 C. Huan, X. Yang, D. Candela **2004**, *Phys. Rev. E* 69, 041302.

42 V. Garzo, J. W. Dufty **1999**, *Phys. Rev. E* 59, 5895–5911.

43 P. T. Callaghan and J. Stepisnik, *Adv. Magn. Opt. Reson.* **1996**, 19, 325–388.

44 (a) T. P. Das, E. L. Hahn **1958**, *Nuclear Quadrupole Resonance Spectroscopy*, Academic Press, New York; (b) J. B. Miller, G. A. Barrall **2005**, *Am. Sci.* 93, 50–57.

45 Y. Oyama **1939**, *Bull. Inst. Phys. Chem. Res. Rep.* 5, 600–639.

46 J. Rajchenbach **2002**, *Phys. Rev. Lett.* 88, 014301.

47 G. T. Seidler, L. J. Atkins, E. A. Behne, U. Noomnarm, S. A. Koehler, R. R. Gustafson, W. T. McKean **2001**, *Adv. Complex Syst.* 4, 481–490.

5
Reactors and Reactions

5.1
Magnetic Resonance Microscopy of Biofilm and Bioreactor Transport
Sarah L. Codd, Joseph D. Seymour, Erica L. Gjersing, Justin P. Gage, and Jennifer R. Brown

5.1.1
Introduction

Bioreactors are important in both industrial and natural settings. Industrial applications in the biotechnological production of chemicals and materials for uses in pharmaceuticals and foods utilize both eukaryotic cells in animals, plants, fungi and algae and prokaryotic cells in bacteria. Bioreactor conditions relating to cell transport and activity are found in many natural and medical systems. Medical implants such as catheters serve as capillary bioreactors. In the Earth's subsurface microbial bacteria play a large role in geological and chemical processes in both fractured and porous media. Cellular organisms are present in bioreactors in two broadly defined "states", as free floating cells (planktonic) and as surface attached microcolonies (biofilms) [1]. Transport phenomena in bioreactors are typically dependent on advection dominated mixing which determines the flux of mass through the reactor. Mass nutrient transport from the bulk fluid to the biofilm or free floating cells, and byproduct transport from the biofilm or free cells are controlled by the mixing within the bioreactor. Two classical methods for quantification of transport in systems with convection are the mass transfer coefficient [2] and the Residence Time Distribution (RTD) [3]. Magnetic Resonance Microscopy (MRM) can generate maps of the spatial velocity distribution combined with non-spatially resolved propagator data to provide information to determine the efficacy of the mass transfer coefficient approach and to quantify the RTD.

The characterization of mixing in reactors has a long history in chemical engineering and a classic paper on the topic is that of Danckwerts, in which the concept of a residence time distribution was developed [4, 5]. This statistical characterization of the mean times spent within a reactor of length L, allows a quantification of the transport in complex reactor flows. A significant strength of the RTD approach is the ability to diagnose effects in packed beds, such as channeling [3]. In partitioned pipe mixers [6] and capillary based microfluidic

NMR Imaging in Chemical Engineering
Edited by Siegfried Stapf and Song-I Han
Copyright © 2006 WILEY-VCH Verlag GmbH & Co. KGaA, Weinheim
ISBN: 3-527-31234-X

reactors [7, 8] chaotic advection [9] is responsible for mixing, and RTDs provide data on the complex mixing dynamics. MRM measurements of the bulk fluid flow in biofilm impacted capillary bioreactors show helical secondary flows qualitatively similar to those generated in curved microfluidic reactors [10]. MRM also provides for direct calculation of a modified scale dependent RTD, which is not identical to the RTD measured in the classical fashion but provides analogous information. The propagator contains all the information on the system dynamics and directly quantifies mixing. Calculating the RTDs from the propagators presents the data in a form familiar to the larger chemical engineering community and demonstrates the information that can be provided from MRM measured propagators [11, 12].

This chapter presents MRM data for capillary, packed bed and Couette Vortex Flow Reactor (VFR) types. A brief overview of the basic elements of the MRM techniques applied to study reactor flows in this chapter is given for completeness, but is in no way comprehensive. The RTD calculation from the propagator is discussed first, using the classic non-ideal capillary reactor undergoing Poiseuille flow with Taylor dispersion [13]. MRM velocity maps of secondary flows [14, 15] in a *Staphylococcus epidermidis* biofilm impacted capillary reactor are shown to demonstrate the limitations of mass transfer coefficients. The impact of the biofilm on the transport is quantified by the calculation of RTDs [15]. MRM characterization of the impact of bioactivity on porous media or packed bed reactors for the growth of *Pseudomonas aeruginosa* (FRD1 containing the plasmid pAB1) [16] is demonstrated using velocity maps and propagators [17] and the RTDs are calculated. Finally, the impact on the velocity field of a suspended *Fusobacterium nucleatum*, a rod like bacterium, in a Couette VFR is discussed.

5.1.2
Theory

5.1.2.1 Magnetic Resonance Microscopy (MRM)

MRM [18] involves the use of strong magnetic field gradients to investigate flow, diffusion and magnetic relaxation in matter on scales of the order of less than 100 μm spatial resolution. Information obtained from molecular motions over the entire sample or within the spatial resolution provides data on spatial scales down to 10 nm over timescales ranging from 10 μs to 1 s. Much of the potential of MRM lies not in its high resolution spin-density maps, but rather in the non-invasive, non-destructive spatially resolved flexibilities utilizing a range of physical and chemical molecular scale contrast mechanisms [18]. The most obvious contrast mechanism is spin density, or molecular density, to which chemical shift techniques can be applied to excite certain molecular species selectively, leading to the spatial mapping of a specific molecule. Alternatively, three-dimensional maps may be obtained where the pixel signal intensity is directly proportional to nuclear spin relaxation times (T_1, T_2 or $T_{1\rho}$). Relaxation times are related to the rotational and translational freedom of the molecule and are most often used to distinguish more solid-like materials from more liquid materials, or to provide soft matter contrast when spin density is similar. The magnetic resonance (MR) signal may also be encoded for translational molecular motion and quantitative spatial maps of self-diffusion and velocity fields can be obtained. A technique for combined velocity and T_2 relaxation maps has been introduced and rapid simultaneous transport and structure measurements will have significant application to bioreactors [19].

5.1.2.1.1 Transverse Magnetic Relaxation Maps

The MRM pulse sequence used to "image" the biofilm is an adaptation of a basic spin-warp slice selection sequence where the spins in the slice of interest ($z = -s/2$ to $s/2$) are excited at an initial time with a 90° radiofrequency (rf) pulse, Figure 5.1.1(a) [18]. Gradients in the x and y directions are stepped through such that for each gradient value the spins acquire a magnetization phase shift dependent on their location within that gradient. The signal is dephased and refocused at effective echo times of $1t_e$, $2t_e$, $3t_e$ ...$8t_e$ [20]. The MR signal, $S(k_x,k_y,t_e)$ obtained at each echo time is the Fourier inversion of the spin density, $\varrho(x,y,z)$ weighted by the average relaxation time parameter $T_2(x,y,z)$ at each image voxel.

$$S(k_x, k_y, t_e) = \int_{-s/2}^{s/2} \left[\int_{-\infty}^{\infty}\int_{-\infty}^{\infty} [\exp\{-t_e/T_2(x,y,z)\}\varrho(x,y,z)] \exp\{i2\pi(k_x x + k_y y)\} dx dy \right] dz \quad (5.1.1)$$

where $k_i = (2\pi)^{-1}\gamma G_i \tau_i$, and τ_i is the duration of the gradient in the i-direction, G_i. Fourier transformation of the MR signal provides the relaxation weighted spin density $\varrho(x,y)$ spatially averaged over the slice thickness in z and the gradients can be applied along any coordinate axis.

5.1.2.1.2 Velocity Maps

Velocity maps are obtained using another adaptation of the basic spin-warp slice selection sequence, with a pair of gradient pulses added to encode for molecular motion [18]. The pulses are located either side of the refocusing 180° pulse, Figure 5.1.1(b). The first gradient pulse encodes the spins with a phase in magnetization dependent on their location at that point in time, and the second gradient pulse reverses this phase encoding. If a spin moves in the time period Δ, the spin retains a residual phase shift directly proportional to the displacement that has occurred in the time interval Δ. The MR spin echo signal normalized to eliminate relaxation effects, $E(k_x,k_y,q) = S(k_x,k_y,q)/S(k_x,k_y,q=0)$, is the Fourier inversion of the density of the spins weighted by the coherent motion

$$E(k_x, k_y, q_i) = \int_{-s/2}^{s/2} \left[\int_{-\infty}^{\infty}\int_{-\infty}^{\infty} [\exp\{-i2\pi\Delta q_i v_i(x,y,z)\}\varrho(x,y,z)] \exp\{i2\pi(k_x x + k_y y)\} dx dy \right] dz \quad (5.1.2)$$

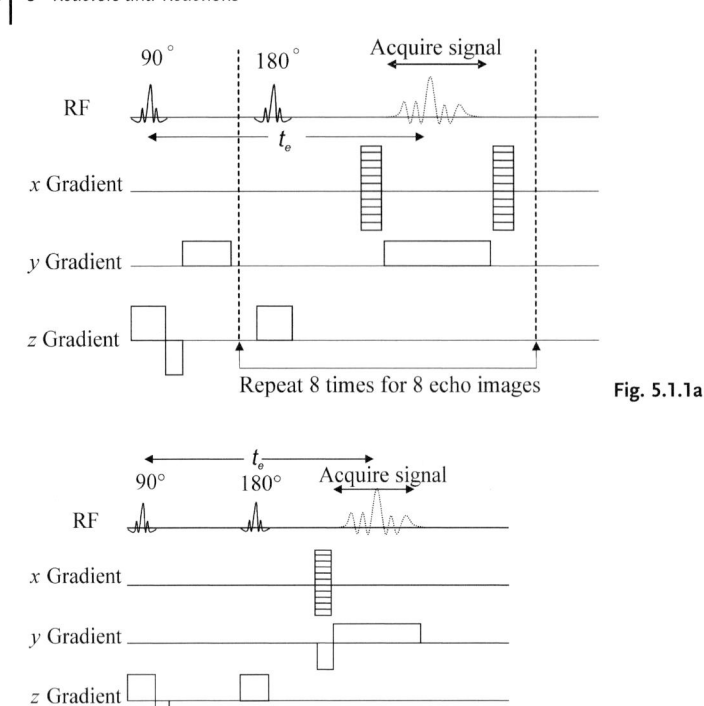

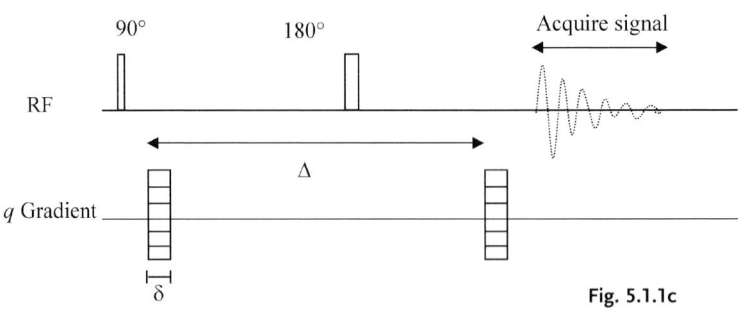

Fig. 5.1.1 (a) Pulse sequence or timing diagram indicating the application of radiofrequency excitation pulses and magnetic field gradients to generate a series of T_2 relaxation weighted images. To make a T_2 map the echo image is refocused eight times at different values of t_e. (b) Velocity map pulse sequence. To make a velocity map the echo image is repeated with two different values of the q gradient pulse. A velocity map is obtained by determining the phase shift at each image voxel due to the motion sensitizing bipolar q-gradient and hence calculating the average displacement that occurred for those spins over the encoding time Δ. (c) Pulsed Gradient Spin Echo (PGSE) pulse sequence for measurement of the propagator, or displacement probability $\bar{P}_s(Z, \Delta)$. The above sequence samples all of q-space with positive and negative q-values in order to obtain all the data points needed to construct artifact free propagators.

where $q_i = (2\pi)^{-1}\gamma g_i \delta$, and δ is the duration of the velocity encoding gradient applied in the i-direction, g_i. A map of the residual phase shift at each spatial location can then be interpreted as an image of the average velocity over time Δ in each image pixel. Additional time averaging occurs as the velocity maps take approximately 10 min to acquire, so flow changes on a time scale shorter than this are averaged for each pixel providing the stationary velocity [21].

5.1.2.1.3 Propagator [Probability Density of Displacements P(Z,Δ)]

In the pulsed gradient spin-echo sequence, Figure 5.1.1(c), only a pair of pulsed magnetic field gradients are applied so as to encode in the phase of the magnetization the location of all MR active spins at an initial time and then unwind that phase at a set time Δ later, generating phase shifts dependent on molecular dynamics over time Δ from the entire sample. No spatially localized image data are obtained. The measured echo signal $E(q,\Delta)$ is the Fourier inversion of the propagator averaged over the initial spin density $\varrho(z)$,

$$E(q) = \int_{-\infty}^{\infty} \bar{P}(Z, \Delta) \exp\{i 2\pi q Z\} dZ \qquad (5.1.3)$$

where $q = (2\pi)^{-1}\gamma g \delta$ is the Fourier reciprocal wavelength to displacement $Z = z' - z$ and $\bar{P}(Z, \Delta)$ the averaged propagator [18]. This allows a statistical measurement of the details of the dynamics over the entire sample to be obtained. The function can also be discussed in terms of the probability distribution of velocities for times Δ that are short relative to the time scale over which variations in velocity occur $\bar{P}(Z, \Delta) \equiv \bar{P}(v_z) = \bar{P}(Z/\Delta)$ [22, 23], an interpretation which provides the direct connection to the RTD.

5.1.2.2 Transport in Capillary and Vortex Flow Bioreactors

5.1.2.2.1 Mass Transfer Coefficients

Biofilms adhere to surfaces, hence in nearly all systems of interest, whether a medical device or geological media, transport of mass from bulk fluid to the biofilm–fluid interface is impacted by the velocity field [24, 25]. Coupling of the velocity field to mass transport is a fundamental aspect of mass conservation [2]. The concentration of a species $c(\mathbf{r},t)$ satisfies the advection diffusion equation

$$\underbrace{\frac{\partial c}{\partial t}}_{\text{rate of change of mass}} + \underbrace{\nabla \cdot (\mathbf{v}c)}_{\text{flux by advection}} = \underbrace{\nabla \cdot \mathbf{D} \cdot \nabla c}_{\text{flux by diffusion}} + \underbrace{R}_{\substack{\text{rate of mass} \\ \text{consumption} \\ \text{or production}}} \qquad (5.1.4)$$

The spatial and temporal evolution of the concentration field is dependent on the velocity field vector $\mathbf{v}(\mathbf{r},t)$, the diffusion tensor $\mathbf{D}(\mathbf{r},t)$ and any reactions occurring in the system $R(\mathbf{r},t)$. Non-dimensionalization of Eqn. (5.1.4) generates the Peclet number, $Pe = v_{z,max}l/D$, which represents the ratio of advective to diffusive mass transport [2]. The velocity field is determined by the conservation of momentum, which for an incompressible Newtonian fluid is the Navier–Stokes equation

$$\left(\frac{\partial \mathbf{v}}{\partial t} + \mathbf{v} \cdot \nabla \mathbf{v}\right) = -\frac{1}{\varrho}\nabla p + \nu \nabla^2 \mathbf{v} + \mathbf{g} \tag{5.1.5}$$

For axial capillary flow in the z direction the Reynolds number, $Re = v_{z,max}l/\nu$ = "inertial force/viscous force", characterizes the flow in terms of the kinematic viscosity ν the average axial velocity, $v_{z,max}$, and capillary cross sectional length scale l by indicating the magnitude of the inertial terms on the left-hand side of Eq. (5.1.5). In capillary systems for $Re < 2000$, flow is laminar, only the axial component of the velocity vector is present and the velocity is rectilinear, i.e., depends only on the cross sectional coordinates not the axial position, $\mathbf{v} = [0,0, v_z(x,y)]$. In turbulent flow with $Re > 2000$ or flows which exhibit hydrodynamic instabilities, the non-linear inertial term generates complexity in the flow such that in a steady state $\mathbf{v} = [v_x(x,y,z), v_y(x,y,z), v_z(x,y,z)]$.

The modeling of mass transport from the bulk fluid to the interface in capillary flow typically applies an empirical mass transfer coefficient approach. The mass transfer coefficient is defined in terms of the flux and driving force $J = k_c(c_{bulk}-c)$. For non-reactive steady state laminar flow in a square conduit with constant molecular diffusion D, the mass balance in the fluid takes the form

$$0 = -v_z(x,y)\frac{\partial c(x,y,z)}{\partial x^2} + D\left(\frac{\partial^2 c(x,y,z)}{\partial x^2} + \frac{\partial^2 c(x,y,z)}{\partial y^2}\right) \tag{5.1.6}$$

indicating that the mechanism for transport to the capillary surface is molecular diffusion not advection [26]. Non-dimensionalization with axial length L and non-axial length l leads to, $Pe_L l^2/L^2$, multiplying the advective term with the diffusive term of order 1. This leads to the natural introduction of the axial mass transfer Peclet number $Pe_L = v_{z,max}L/D$. The axial flow impacts the diffusive transport from the bulk fluid to the surface in two ways; through the diffusive sampling of the streamlines (Taylor dispersion) and the inflow of fluid at bulk concentration. The mass transfer coefficient can be derived for rectilinear capillary flow and the dimensionless mass transfer coefficient, the Sherwood number (Sh), scales linearly with the Reynolds number and the Schmidt number (Sc), $Sc = \nu/D$, $Sh = k_c L/D \sim Re\, Sc = Pe_L$. Analytical solutions for boundary layer flows generate correlations of the form $Sh \sim Re^n Sc^m$, with $n = m = 1/2$ for fluid–fluid interfaces with no interfacial velocity gradient and $n = m = 1/3$ for fluid–solid interfaces with a velocity gradient at the interface [2]. Following the analytical results, empirical values of $n = 7/8$ and $m = 1/3$ are determined for turbulent flows and for capillary flow with $Re > 2000$ indicating a stronger dependence of the mass transfer coefficient on Re due to turbulent eddy mixing. Mass transfer coefficient scaling relationships for biofilm systems have been developed from both experiment [25, 27] and simulation [24, 28, 29]. Correlations of the form $Sh \sim Re^{1/2}Sc^{1/2}$ [28] and $Sh \sim Re^{1/3}Sc^{1/3}$ [24] have been fit to simulation data.

The classical mass transfer coefficient approach fails to incorporate the finer details of the advection field into the correlation in a quantitative fashion. Of significant issue is the fact that in many low Re systems, secondary flows with coherent structure, e.g., vortices, occur due to boundary induced inertial effects and the mass transport is altered by the presence of bulk advection in the x and y directions [29], an issue studied extensively in the analysis of fluid mixing [6]. The concepts of dynamical systems theory have been applied to characterize complex advection in mixing devices [6, 30]. Qualitatively two mass transport effects can occur. The secondary flows can increase homogenization, so the concentration field is uniform and all points in the bulk fluid are near to the average bulk concentration, leading to enhanced mass transport to the biofilm interface relative to a rectilinear flow. Regions of diffusion limited transport and depleted concentration, i.e., hydrodynamic and concentration boundary layers, at the biofilm–fluid interface are decreased. Alternatively, secondary flows can generate a heterogeneous concentration distribution due to fluid regions where regular motions inhibit exchange with the bulk fluid. Mass transport is then dependent on the interaction between the biofilm structure and the structure of the advection field as to whether regions of high or low concentration are in contact with the interface. Lattice–Boltzmann simulations have clearly elucidated these variations in local Sherwood numbers in heterogeneous biofilms [29].

In non-linear mixing flows, e.g., a partitioned pipe mixer, dynamical systems simulations lead to a picture of the flow given by the Poincare section, the time repetition surface of fluid particle trajectories. The advection field exhibits regions of low mixing and high flow regularity surrounded by stronger mixing regions of less ordered flow, referred to as islands of regularity within a chaotic sea [31]. A mixing strength parameter based on the ratio of secondary to axial advection rate is used to characterize the flow, $\beta \sim (v_{x,max}L)/(v_{z,max}l)$ and the amount of chaotic motion increases with β [6]. It is convenient to introduce a non-axial Peclet number, $Pe_l = v_{x(y),max}l/D$, and an effective mixing strength can be expressed in terms of a ratio of mass transfer Peclet numbers as $\beta \sim (Pe_l/Pe_L)(L/l)^2$.

5.1.2.3 Transport in Packed Bed Bioreactors

The modeling of mass transport in packed bed reactors applies the theory of dispersion [32]. The conservation of mass for the average concentration <c> is derived

$$\underbrace{\frac{\partial \langle c \rangle}{\partial t}}_{\substack{\text{rate of change} \\ \text{of average mass}}} + \underbrace{\langle \mathbf{v} \rangle \cdot \nabla \langle c \rangle}_{\substack{\text{flux by average} \\ \text{advection}}} = \underbrace{\nabla \cdot \mathbf{D}^* \cdot \nabla \langle c \rangle}_{\substack{\text{flux by} \\ \text{dispersion}}} + \underbrace{R}_{\substack{\text{rate of average} \\ \text{mass consumption} \\ \text{or production}}} \quad (5.1.7)$$

in terms of the average velocity <v> and the dispersion coefficient $\mathbf{D}^*(\mathbf{r},t)$. The form of the dispersion coefficient arises from the averaging process and is related to the fluctuations in velocity about the mean. Several mechanisms of dispersion, mechanical, hold-up and Taylor dispersion, may be identified and scaling with Pe developed [32]. RTDs are often used to characterize macroscale behavior of packed beds [3].

5.1.2.4 Residence Time Distribution (RTD)

The RTD quantifies the number of fluid particles which spend different durations in a reactor and is dependent upon the distribution of axial velocities and the reactor length [3]. The impact of advection field structures such as vortices on the molecular transit time in a reactor are manifest in the RTD [6, 33]. MRM measurement of the propagator of the motion provides the velocity probability distribution over the experimental observation time Δ. The residence time is a primary means of characterizing the mixing in reactor flow systems and is provided directly by the propagator if the velocity distribution is invariant with respect to the observation time. In this case an exact relationship between the propagator and the RTD, $N(t)$, exists

$$N(L,t)dt = \bar{P}(v_z, \Delta)dv_z \tag{5.1.8}$$

where the reactor length L is considered to be the mean distance traveled by the spins over the observation time, i.e., $L = v_{z,av}\Delta$, and the residence time t of the spin is related to the velocity of the spin, v_z, by $t = L/v_z$ or $t = v_{z,av}\Delta/v_z$.

In most real systems the propagator transitions from the instantaneous velocity distribution, measured using short observation times, to an invariant form, measured for times greater than the asymptotic time. The asymptotic time is typically that at which all spins have experienced all velocity streamlines via mixing and dispersion mechanisms. A calculation of the RTD from a propagator measured in the pre-asymptotic region is most accurate at times around the residence time of the spins traveling at the mean velocity. If the RTD is displayed on a normalized time axis t/Δ then spins residing in the reactor for a time $t/\Delta = 1$ are traveling at the mean velocity, and spins residing for half or double this time are accurately represented by the RTD calculated from this propagator only if the velocity distribution does not change significantly over this observation time window. As demonstrated below, scale dependent RTDs calculated from propagators of velocity distributions that vary over Δ provide details of the mixing, even though they are not exactly the RTD of classical theory.

5.1.3
Reactors

5.1.3.1 Pure Fluid in a Non-ideal Circular Capillary Reactor

Two template examples based on a capillary geometry are the plug flow ideal reactor and the non-ideal Poiseuille flow reactor [3]. Because in the plug flow reactor there is a single velocity, v_0, with a velocity probability distribution $P(v) = v_0^{-1}\delta(v-v_0)$ the residence time distribution for capillary of length L is the normalized delta function $\text{RTD}(t) = \tau^{-1}\delta(t-\tau)$, where $\tau = L/v_0$. The non-ideal reactor with the parabolic spatial velocity distribution has a velocity probability distribution which is a hat function and the corresponding RTD indicates the large amount of fluid eluted rapidly by the maximum velocities. In the measurement of RTDs for a specified length of the Poiseuille flow capillary reactor; the radius, the average velocity and molecular diffusivity of the reactor fluid, determine the impact of Taylor dispersion as a mixing mechanism. The propagator data for Taylor dispersion of octane flowing in a 150 μm circular capillary [13] analyzed as a velocity probability distribution are shown in Figure 5.1.2(a). The data can be converted into the scale dependent RTDs shown in Figure 5.1.2(b). The natural scaling for the RTD is the

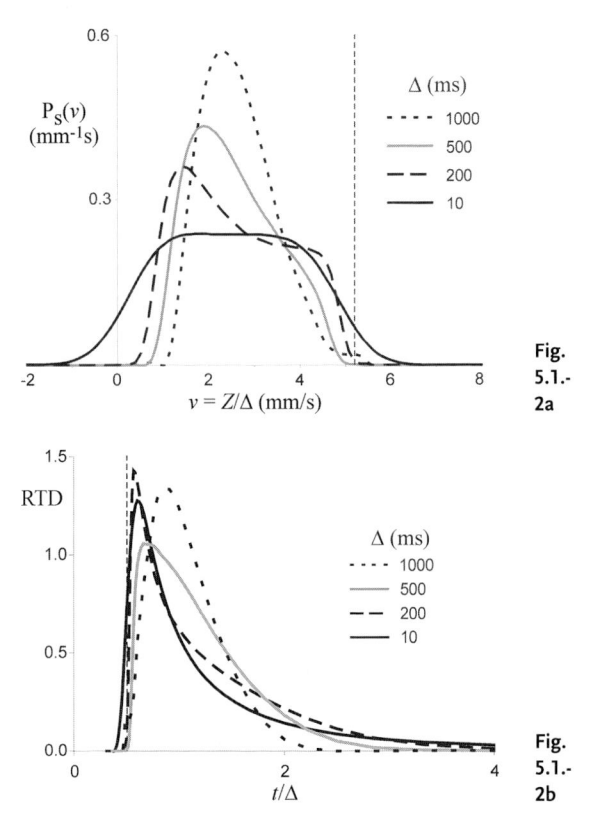

Fig. 5.1.2 Non-ideal capillary flow reactor (a) propagators [13] and (b) corresponding RTDs calculated from the propagator data. (a) The propagators indicate the distribution of average velocities over each observation time (Δ) ranging from 50 ms to 1 s. As the observation time increases the spins exhibit a narrowing distribution of average velocities due to the motional narrowing effect of molecular diffusion across the streamlines. The dashed vertical line represents the maximum velocity that would be present in the absence of molecular diffusion. (b) The RTDs correspond to varying conduit lengths L related to the observation time for the propagator and the mean velocity of the spins, i.e., $L = v_{z,av} \Delta$ is the mean distance traveled by the spins in the observation time. The time axis of the RTD is normalized to the observation time so that a time $t/\Delta = 1$ indicates when spins traveling at the mean velocity will exit the conduit. The dashed vertical line represents the time at which the first spins would exit in the absence of molecular diffusion.

observation time Δ, so as to provide direct comparison between the RTDs calculated from propagators measured at varying displacement times. The impact of Taylor dispersion mixing on the propagator is manifested as a transition from the uniform velocity distribution represented by a hat function at an MR displacement observation time of 10 ms to the Gaussian distribution at 1000 ms [Figure 5.1.2(a)] [13]. Correspondingly the RTD is seen to transition from the 10 ms measurement, where a large amount of fluid is eluted at short times and a significant tail is present at long times due to the slow velocities near the capillary wall, to an RTD where the majority of material is eluted at an intermediate time with no significant tail at long times due to the mixing by Taylor dispersion. The data in Figure 5.1.2 clearly demonstrate that the ability to measure propagators by MR fully characterizes reactor performance in terms of the RTD.

It is important to note that an RTD depends on the length of the conduit analyzed, whereas the measured propagator depends on the observation time Δ of spin displacement evolution. The MRM measured RTDs are normalized about the average distance of conduit L sampled by the spins in time Δ, and scale dependent analysis of the variation of the RTD with L is possible. The RTD generated from propagators measured over longer observation times form the best representations of the true RTDs for longer conduit lengths. Of course the spin displacement is limited to the length of the rf coil L_{rf} so the MR displacement observation time is limited by the maximum velocity $\Delta = L_{rf}/v_{z,max}$. Figure 5.1.3 provides a schematic diagram indicating the relationship between the high velocity tails of the MR measured propagator, displayed as a probability distribution of velocity, and the short time region of the RTD. Analogously the low velocity short displacement region of the propagator corresponds to the long time region of the RTD. The RTD can be used to characterize or troubleshoot reactor performance and can be calculated non-invasively from the MR measured propagator if the most relevant observation time is chosen relative to the conduit length of interest.

5.1.3.2 Biofilm Impacted Square Capillary Reactor

S. epidermidis is implicated in many medical implant infections. Mechanical heart valves, shunts, catheters and orthopedic devices are examples of implanted devices

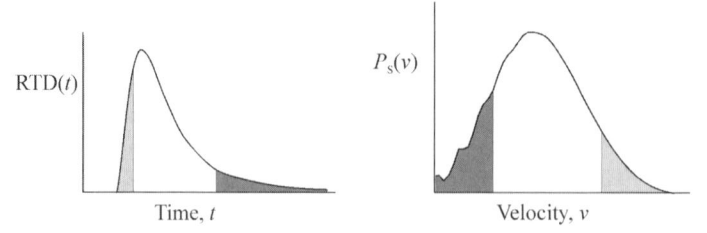

Fig. 5.1.3 Schematic of the relationship between the propagator and the RTD. The dark gray shaded area indicates the 25% of spins traveling with the slowest velocities and the light grey shaded area indicates the 25% of spins traveling at the fastest velocities.

which are commonly infected by *S. epidermidis* biofilms [34]. These biofilms are often continuously subjected to high flow rates which cause the biofilm structure to develop high tensile strength [35]. Quantifying how biofilms interact with bulk fluid flow, or advection, is critical to modeling mass transport in biofilm systems [27], shear stress influences on the structure of biofilms [36] and detachment rates [29].

Measuring spatial distributions of velocity within bioreactors is difficult in part due to biofilm opacity. Particle tracking, where microscopic fluorescent spheres are tracked with confocal laser microscopy, is time and labor intensive and measures a tracer response to the flow [37, 38]. In systems where the velocity is of higher spatial dimension and/or time dependent, streamlines everywhere tangent to the Eulerian velocity and streaklines followed by particles can diverge, and a tracer measurement provides particle size dependent Lagrangian tracking, as opposed to the spatial distribution of the molecules composing the bulk flow field [39]. MRM accurately and non-invasively images spatial distributions of velocity fields in biofilm systems [40]. Recently, images of fluid flowing through a circular biofilm fouled capillary demonstrated the ability of MRM to image the axial flow and calculate shear force on the biofilm surface [41]. MRM measurements of 3D velocity components in a square glass capillary fouled with biofilm indicate significant non-axial flows are generated by the biofilm, and correlation between T_2 magnetic relaxation maps displaying biofilm structure with the corresponding velocity patterns provide data on the impact of biofilm growth on advective mass transport [14, 15]. These issues are of significant importance in biosensor and bioseparation applications based on microfluidics or "lab on a chip" technology [42].

5.1.3.2.1 Biofilm Capillary Reactor Materials

The bacterial species *S. epidermidis* strain 35984 is studied in these experiments, and is chosen for its medical importance and the fact that it forms thick opaque biofilms, hundreds of microns thick, in a relatively short amount of time. The biofilms cultured for these experiments were grown in 1 mm square glass capillaries with gravity driven nutrient flow [15]. The nutrient feed is $1/10^{th}$ strength tryptic soy broth (DIFCO, Beckton Dickinson and Company) at a volumetric flow rate of 0.028 mL s^{-1}, corresponding to an average velocity in a clean capillary of 28 mm s^{-1} with $Re = 90$. Note Re is calculated with $l = 1$ mm, as for a 1 mm square duct the hydraulic radius is 1 mm. After a no-flow inoculation period of 4 h to allow the suspended bacteria to settle and attach to the glass, the biofilm is allowed to grow for 48 h under these flow conditions. The biofilm sample is incubated at 37 °C during all stages of the growth process to simulate human body temperature. After the 48 h growth period, the biofilm sample was then loaded into the magnet for study. The time dependent growth data shown in Figure 5.1.4 are obtained using the same temperature and flow rate but with the bioreactor in the magnet 4 h after inoculation so that growth could be monitored [15].

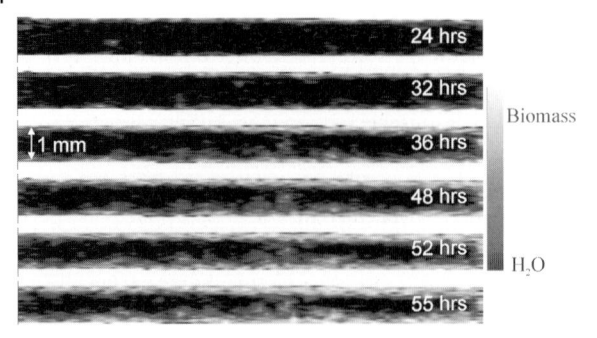

Fig. 5.1.4 Time lapse T_2 maps of biofilm structures as a function of growth. The T_2 maps are collected at different stages in the growth of a biofilm at the time shown after initial inoculation. Light colors indicate water restricted within a biofilm and black is the bulk unrestricted water. Images are acquired in the absence of flow, however the flow direction for growth periods between images is left to right [15].

5.1.3.2.2 Biofilm Capillary Reactor MRM Experiment Details

A Bruker Avance DRX spectrometer networked to a 250-MHz superconducting magnet was used to conduct the MRM experiments. Microimaging was accomplished using a Bruker Micro5 probe and gradient amplifiers to produce magnetic field gradients up to 2 T m^{-1}. An rf coil with a internal diameter of 5 mm was used for all the MRM experiments. The medical contrast agent, Magnevist® (Berlex Laboratories), was added to the feed at a concentration of 0.6 mL per liter of nutrient feed in order to decrease the experiment times. Low concentrations of Magnevist have been shown to have no effect on the growth of the biofilms [43]. The MRM measurements are of the ^{1}H protons on the H$_2$O molecules in the bulk fluid and within the biomass. T_2 maps are generated using a slice-selection 2D multi-spin echo imaging sequence [Figure 5.1.1(a)] with the following parameter values: T_R = 500 ms, t_e = 10 ms, 8 echo images. Total acquisition time for each map was 5 min during which time the flow is stopped to enhance image resolution and avoid in-flow–out-flow artifacts. Velocity data are acquired for flow at Re = 32, using a spin-echo sequence with bipolar velocity sensitizing gradients [Figure 5.1.1(b)] with parameters of: T_R = 2000 ms, t_e = 20 ms, g = 0 and 100 mT m^{-1}, δ = 1 ms, Δ = 7 ms. Total acquisition time for the velocity maps is 17 min. The T_2 and velocity maps are averaged over a slice of thickness of 0.3 mm with a field of view (FOV) of 2.5 mm × 20 mm with 64 × 128 pixels, for a spatial resolution of 39 × 156 µm per pixel.

5.1.3.2.3 Biofilm Capillary Reactor Structure Evolution

MRM non-invasive monitoring of biofilms during growth provides the potential for extended studies of nutrient and biocide effects on biofilm structure and transport [15]. Figure 5.1.4 demonstrates the monitoring of biofilm structure in a capillary bioreactor over a 30-h period. The restricted water molecules within the biofilm are displayed in the light grey to white color range associated with low T_2 values, and the free water is displayed as black [14, 15, 41, 44]. Note in the first image at 24 h the biofilm forms an approximately 100-μm thick layer of fairly uniform height. The biofilm contains hollowed regions near the capillary wall several of which are clearly visible slightly left of center at the bottom of the image. At 32 h after inoculation the biofilm thickness increases in a heterogeneous fashion with bumps of greater thickness. By 36 h after inoculation the biofilm exhibits significant tower like structures and heterogeneity. Calculation of roughness and surface enlargement from such images [41] can be made to quantify variation over time. Regions of high molecular mobility, shown dark, and low molecular mobility, shown bright, are heterogeneously dispersed throughout the biofilm for times of up to 55 h. The data clearly indicate the structural heterogeneity and further study is ongoing to determine if regions of low mobility due to cell clusters can be differentiated from dense EPS regions. An interesting point from the perspective of transport within the biofilm is the fact that what is directly measured is molecular mobility, from which biomass presence is inferred. It may be possible to analyze the data from a perspective such as volume averaging and attempt to correlate average material transport properties [45] to the measured spatial distribution of molecular mobility.

5.1.3.2.4 Biofilm Capillary Reactor Structure and Advective Transport

The cross sectional depth profile of a biofilm can be seen in the T_2 map for a slice perpendicular to the flow axis, Figure 5.1.5, showing the heterogeneity of the biofilm structure [15]. Quantitative velocity data $v_i(z)$ are plotted as a function of the axial position for two positions in the x direction. The lines A and B drawn through the images indicate the x position of the data presented in the graphs to the right, and are chosen in the bulk fluid and at the biofilm–fluid interface. To construct a biofilm indicator function the T_2 data are masked, setting values associated with free water equal to zero and normalized by the maximum T_2 value. This provides a biofilm signal to be displayed as a value above zero where the biomass is present, with 1 the maximum within the image at the region of lowest T_2, and zero in areas with no biomass [15]. All three components of velocity are normalized by the maximum z direction velocity so that the x and y components are displayed as a percentage of the axial velocity and are indicative of the mixing intensity [6].

Data for the bulk fluid, line A, indicate that v_z varies as a function of z but maintains a value near 0.75 of maximum velocity. The periodicity of v_x and v_y is clearly evident in the graph of line A and a 180° out of phase coupling of the components is seen with one positive when the other is negative. This indicates a preferred orientation to the plane of the oscillatory flow and this feature was seen in all the biofilms grown throughout this study. The secondary flow components are 0.1–0.2 of the maximum axial velocity and are spatially oscillatory. The significant non-axial velocities indicate non-axial mass transport has gone from diffusion dominated, $Pe_l = 0$, in the clean capillary, to advection dominated, $Pe_l \sim 2 \times 10^3$, due to the impact of the biofilm. For comparison, the axial Peclet number is $Pe_L \sim 2 \times 10^5$. Line B intersects areas covered by biomass and areas of only bulk

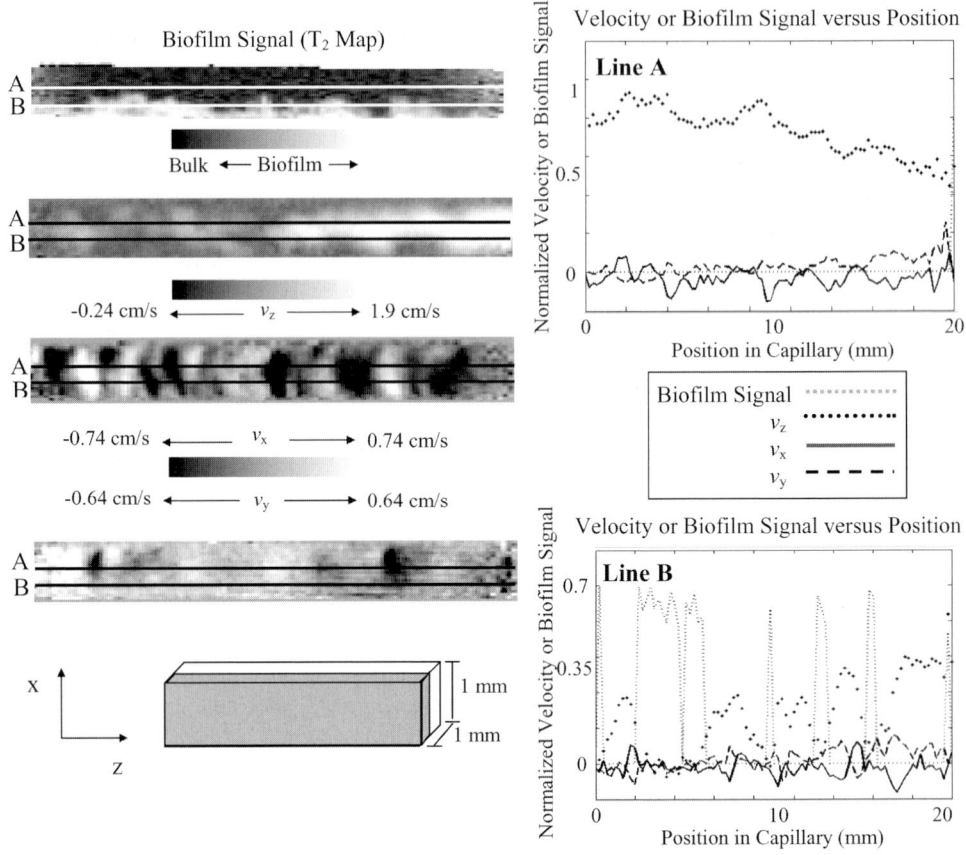

Fig. 5.1.5 Quantitative data on the correlation of biofilm and velocity for a slice perpendicular to the flow axis. The images on the left are from top to bottom: T_2 map, z velocity component, x velocity component and y velocity component. One dimensional profiles through lines A, in bulk fluid, and B, intersecting biofilm fluid interface, are shown on the right. The biofilm signal indicator, dotted grey line, has been normalized so that zero corresponds to no biomass and 1 corresponds to the highest density of biomass. The velocity components are normalized by the maximum axial (z) velocity. Non-axial velocity components are significant at both heights A and B with maximum non-axial velocity ≈20% of the axial maximum. In line A the 180° out of phase coupling of v_x, black line, and v_y, dashed line can be clearly seen. Line B indicates the strong correlation between the biomass and axial (z) velocity near the interface [15].

fluid as indicated in the graph of the biofilm signal. Comparing the T_2 map and the graph, line B is seen to intersect a large biofilm clump from an axial position of 3–7 mm at a depth approximately 120 μm below the biofilm–fluid interface and is near the biofilm surface along the entire observed length [15]. This graph demonstrates that in areas where there is biomass, indicated by a value above zero on the grey dotted line, v_z, indicated by the black dotted line drops down to values less than 0.1 of the maximum. The axial velocity is negative at sufficient depth within the biomass, as in the region from 3 to 7 mm, as is evidenced by the axial velocity image with regions of small negative velocity within the biofilm. Negative velocities have been demonstrated in simulations [29] and intimated in lower resolution MRI experiments [46]. The secondary velocity components v_x solid black line and v_y dashed line, are small within the biofilm and at the interface, and while not as regular as in line A, exhibit an out of phase spatial oscillation and provide amplitudes of 0.1 of maximum axial velocity. The spatial variations in the velocity field are significant over mm scales along the biofilm.

A perturbed helical flow structure of the advection field is indicated by the data. This is evidenced by velocity maps measured for consecutive slices perpendicular to the flow, Figure 5.1.6 [15]. Of greatest interest here are the secondary flow components. Note the onset of a vortex from slices 1 and 2 to a fully developed clockwise vortex flow in slice 3. In slices 4 and 5 the coherent vortex structure is rotated and decaying with axial position. It is important to remember that the velocity measured is for a displacement time Δ of 4 ms but, time averaging over the image acquisition time of 10 min occurs due to encoding of information in each acquired signal, so that the spatial distribution velocity components measured are stationary [21]. This is clear visualization of the complex spatial distribution of flow structures which generates a corresponding variation in mass transport [29]. Recent simulations of the structure of secondary flows, Dean vortices, in curved micro-channels suggest that "switching" between flow patterns with varying numbers of vortices during axial transit generates a chaotic flow and has an impact on mixing [10]. Figure 5.1.6 indicates a "switching" type of event with strong negative v_x and positive v_y components in slices 1 and 2, transitioning to the coherent vortex in slice 3 and an inverted configuration in slices 4 and 5 relative to the flow upstream of slice 3, with positive v_x and negative v_y [15]. The complexity of the

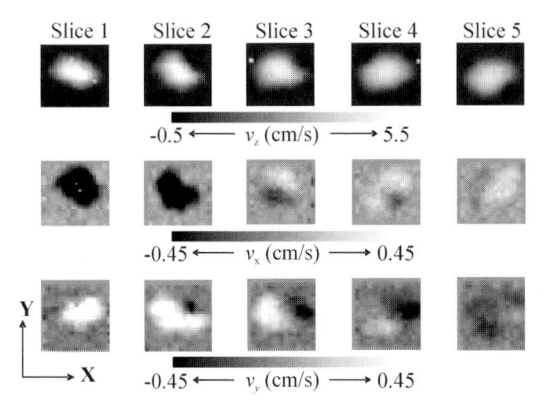

Fig. 5.1.6 These images display the three components of velocity in five consecutive slices through the transverse cross section of the capillary. The direction of bulk flow is from slice 1 to 5 with each slice 300-μm thick and contiguous. Positive axial velocity is out of the page. A negative x component represents flow to the left. A negative y component represents flow down the page. The helical nature of the flow is clear with non-axial flow changing directions, through the sequence of slices and going through a coherent clockwise vortex flow in slice 3 [15].

5.1.3.2.5 Biofilm Capillary Reactor Propagators and Residence Time Distributions

Figure 5.1.7 shows the propagator of the motion measured for a clean and a biofilm impacted capillary [14, 15] and the residence time distributions calculated for each from these velocity distributions. The clean capillary gives an experimental propagator equal to the theoretical velocity distribution convolved with a Gaussian diffusion curve [14], as shown in Figure 5.1.2. For the flow around the biofilm structure note the appearance of a high velocity tail indicating higher probability of large displacements relative to the clean capillary. The slow flow peak near zero displacement results from the protons trapped within the EPS gel matrix where the

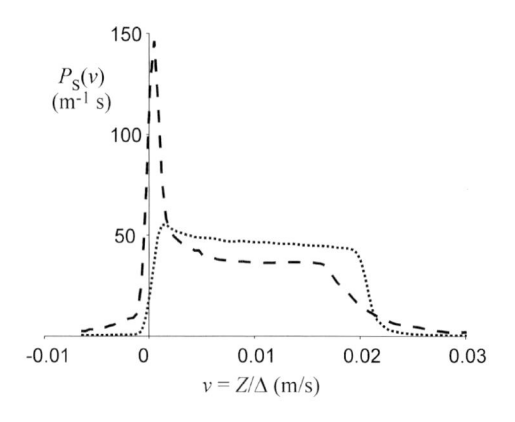

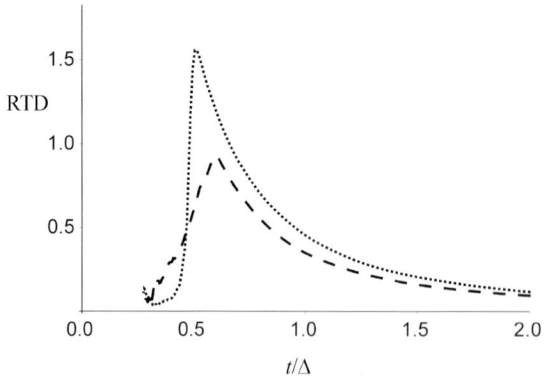

Fig. 5.1.7 (a) Propagators in both a clean square capillary (dotted) and for flow around a biofilm structure (dashed) for an observation time, $\Delta = 15$ ms. (b) Residence time distribution functions calculated from the propagator data shown in (a). The induction of a high velocity tail due to biofilm fouling is indicated by an increase of spins exiting the conduit ahead of the mean velocity. The trapping of spins in the EPS matrix induces an increase in the zero velocity peak and is indicated by an increase in the tail of the RTD.

primary transport mechanism is diffusive. The residence time distribution integrated over time $\int_0^{t'} N(L, t)dt$ gives the fraction of fluid that has been in a reactor of length L for less than time t'. In Figure 5.1.7 L is the average length traveled by the fluid in the time $\Delta = 15$ ms, i.e., $L = v_{z,av} \Delta$. The long tail in the biofilm propagator manifests itself as a feature at short residence times, whereas the large zero displacement peak in the biofilm propagator manifests itself as a long tail on the RTD [15].

5.1.3.3 Packed Bed Reactors

Packed bed bioreactors occur in a multitude of industrial and natural settings [40]. Industrial applications range from trickle bed reactors used in water filtration systems to hydrogel based enzymatic reactors for production of monoclonal antibodies from mammalian cells. The recovery of oil from or the remediation of contaminants within porous earth formations are strongly impacted by bioactivity and are in essence large packed bed bioreactors. In medical applications membrane and mechanical filtration devices experience biofouling and are variations of packed bed bioreactors. MR methods have found significant application in the study of transport in model and natural porous media due in large part to the ability to measure molecular dynamics in opaque systems [11, 40]. The characterization of porous media structure by molecular diffusion [47] and bulk flow [48, 49] are well established, as are the ability to study the displacement and time dependence of the hydrodynamic dispersion [48, 50–53]. Recently the impact of bioactivity on scale dependent dispersion has been quantified and the theory of Continuous Time Random Walks (CTRWs) applied to describe the transition in dynamics from normal to anomalous fluid transport [17].

5.1.3.3.1 Packed Bed Reactor Materials

A 5-mm id porous media column packed bed reactor was constructed using 241-µm diameter monodisperse polystyrene beads (Duke Scientific 4324A). Experiments utilized a dual syringe pump (Pharmacia P-500), which allows controlled volumetric flow rates. The column, tubing and pump were sterilized by repeated treatments with an ethanol solution (70%-wt). For bacterial experiments, the column was then inoculated with a strain of P. aeruginosa (FRD1 containing the plasmid PAB1) [16] and allowed to sit for approximately 10 h under very low flow (10 mL h^{-1}) of a 1/10 dilution of tryptic soy broth (TSB) media (Becton Dickenson) to allow initial bacterial attachment to the beads. The column was then aseptically transferred to the MR system. Either deionized water for the non-bacterial experiments or 1/10 TSB media under constant stirring to ensure oxygenation was pumped into the column from bottom to top with the effluent collected in a waste carboy. Magnevist® (Berlex Laboratories) was added to the deionized water and 1/10 TSB media at 0.05% v/v. The temperature of the column was held constant at 20 °C for all data collection. The temperature was increased daily for several hours in the bacterial experiments to 37 °C to encourage biofilm formation and growth.

5.1.3.3.2 Packed Bed Reactor MRM Experiment Details

The sample was placed in an 8-mm gradient coil in the Bruker Avance DRX250 spectrometer described previously. A sequence of MR pulse sequences were repeated over a 7–10 day period to monitor the effects of biofilm growth on velocity distributions, T_2 relaxation and averaged propagators [17, 54]. The MRM measurements are of the ^{1}H protons on the H_2O molecules in the bulk fluid and within the biomass. T_2 maps are generated using a slice-selection 2D multi-spin echo imaging sequence [Figure 5.1.1(a)] with the following parameter values: T_R = 500 ms, t_e = 10 ms, 8 echo images. Total acquisition time for each map was 40 min during which time the flow is stopped to enhance image resolution and avoid in-flow outflow artifacts. Velocity data are acquired for flow at Re = 0.9, using a spin echo sequence with bipolar velocity sensitizing gradients [Figure 5.1.1(b)] with parameters of: T_R = 2000 ms, t_e = 18 ms, g = 0 and 80 mT m^{-1}, δ = 1 ms, Δ = 6 ms. Total acquisition time for the velocity maps is 70 min. The T_2 maps are averaged over a slice of thickness of 0.2 mm while velocity maps are averaged over a slice of thickness of 1.0 mm. Both T_2 and velocity maps were taken with a field of view (FOV) of 7.0 mm × 7.0 mm with 128 × 128 pixels, for a spatial resolution of 55 × 55 μm per pixel. The PGSE pulse sequence [Figure 5.1.1(c)] was used to measure the echo signal $E(q,\Delta)$ and to compute the averaged propagators $P(Z/\Delta)$. The experimental parameters used for these experiments were T_R = 2 s and δ = 1 ms. 128 q sampling points were used within the range of gradient values of $-g_{max}$ to $+g_{max}$ in order to detect both positive and negative displacements. The maximum gradient value g_{max} was chosen in order to prevent data wrap and varied depending on the value of Δ ranging from 250 mT m^{-1} for Δ = 300 ms to 500 mT m^{-1} for Δ = 20 ms. The total acquisition time for 128 point sampling was approximately 35 min.

5.1.3.3.3 Packed Bed Reactor Structure and Advective Transport

MRM images of the impact of biofilm growth on the pore scale velocity and magnetic relaxation time in the packed bed reactor for fixed volumetric flow rate are shown in Figure 5.1.8. The pore scale velocity images indicate greater spatial variability of the velocity due to biofilm growth 3 and 6 days after inoculation with *P. aeruginosa*. Increased regions of no flow due to biomass blocked pores generate high velocity in unclogged pores to conserve mass for the fixed flow rate. The transverse magnetic relaxation time images indicate the spatial distribution of the biomass growth by measuring enhanced relaxation due to the restricted motion of water molecules within the biomass [44]. On day 3 the impact on pore scale velocity is more significant than on the magnetic relaxation, demonstrating the sensitivity of transport processes to small amounts of biofilm growth.

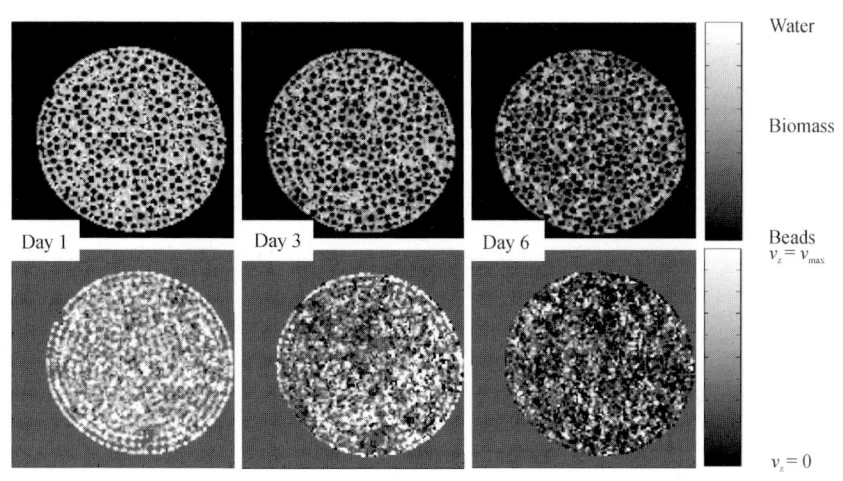

Fig. 5.1.8 MRM maps of velocity (bottom row) and T_2 magnetic relaxation (top row) as a function of biofilm growth time (left to right). Day 1 shows the clean porous media. Spatial resolution is 54.7 μm per pixel (128 × 128 - pixels) in plane, so the data reflect pore scale spatial distributions of velocity over a 1000-μm slice and biomass over a 200-μm slice.

5.1.3.3.4 Packed Bed Reactor Propagators and Residence Time Distributions

The propagators for both the clean reactor and a biofilm impacted reactor were measured at the same fixed volumetric flow rates of 1.67 mL min^{-1} corresponding to $Pe \sim 200$ based on the bead diameter. Figure 5.1.9 shows the propagator and RTD calculated from these data for the clean packed bed reactor. The propagators in Figure 5.1.9(a) show the temporal transition in dynamics. A Cauchy [55] like distribution is seen at $\Delta = 20$ ms, indicative of the velocity distribution within single pore spaces, as during the observation time spins are primarily localized within single pores [17]. The dynamics transition to Gaussian at $\Delta = 300$ ms, due to the effect of spins sampling multiple pore spaces, indicating the impact of dispersion mechanisms on the dynamics [17]. The corresponding RTDs in Figure 5.1.9(b) show the transition from early to later primary efflux times and a decrease in the long time elution tails due to mixing as the observation time increases.

Figure 5.1.10 shows the impact of biofilm growth on the packed bed dynamics. The propagators, Figure 5.1.10(a), for a displacement observation time of $\Delta = 300$ ms for the clean bead pack and biofilm growth on days 3 and 6 show a strong transition from normal transport with Gaussian statistics to anomalous transport with non-Gaussian statistics [17]. The growth of the biofilm, day 3 and day 6, alters the dynamics with increased probability of small displacements due to fluid that is entrained in the biomass and trapped in clogged dead end pores, and long tails in the distribution due to the higher probability of large displacements in high permeability unclogged channels [17]. These features are also evident in the RTDs, Figure 5.1.10(b), with the trapped fluid in the biomass and dead end pores generating long time tails and the rapid flow on the backbone channels causing

Fig. 5.1.9 (a) MR measured propagators and (b) corresponding calculated RTDs for flow in a model packed bed reactor composed of 241-µm monodisperse beads in a 5-mm id circular column for observation times Δ ranging from 20 to 300 ms. As the observation time increases the velocity distribution narrows due to the dispersion mechanisms of the porous media; this effect is observed in the RTDs as a narrowing of the time window during which spins will reside relative to the mean residence time as the conduit length is increased.

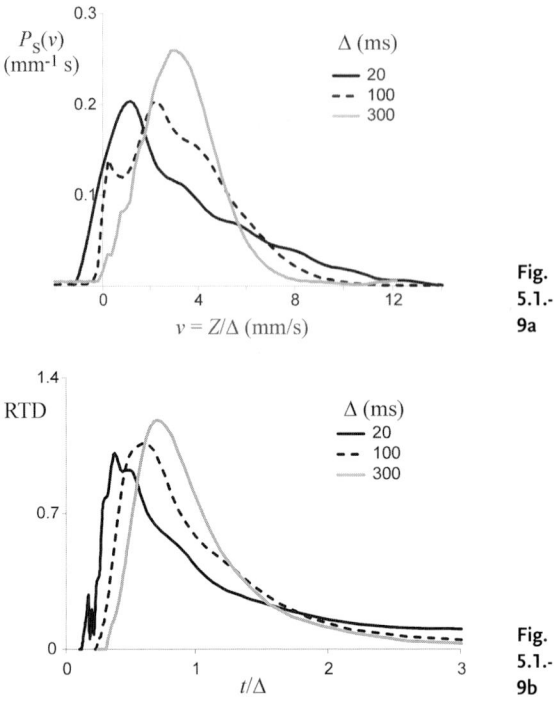

early time elution relative to the clean bead pack. This clearly demonstrates the classic channeling effect, for which RTDs have long been used to diagnose in packed bed reactors [3].

5.1.3.4 Vortex Flow Reactors (VFR)

Reactors which generate vortex flows (VFs) are common in both planktonic cellular and biofilm reactor applications due to the mixing provided by the VF. The generation of Taylor vortices in Couette cells has been studied by MRM to characterize the dynamics of hydrodynamic instabilities [56]. The presence of the coherent flow structures renders the mass transfer coefficient approaches of limited utility, as in the biofilm capillary reactor, due to the inability to incorporate microscale details of the advection field into the mass transfer coefficient model.

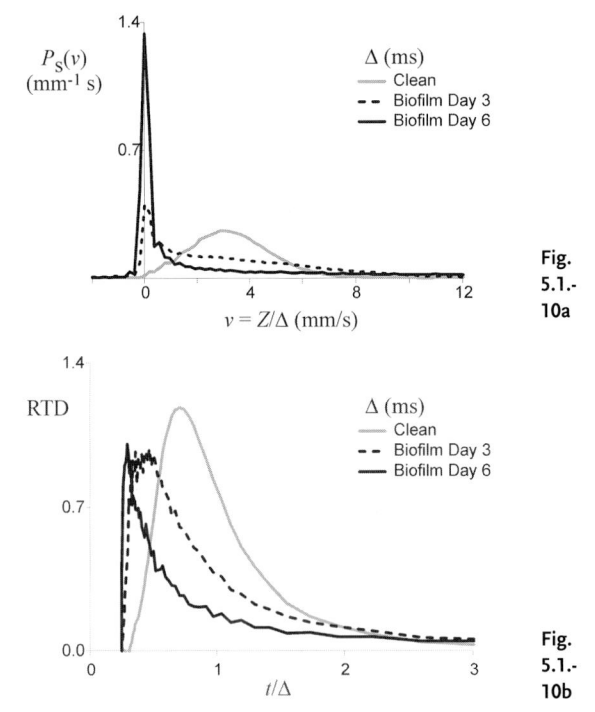

Fig. 5.1.10 (a) MR measured propagators and (b) corresponding calculated RTDs for flow in a model packed bed reactor composed of 241-μm monodisperse beads in a 5-mm id circular column for a fixed observation time of 300 ms and as a function of biofilm fouling. As the porous media becomes biofouled, a high velocity tail appears due to the formation of strong backbone flow paths and a zero velocity peak forms due to entrapment of spins in the EPS gel matrix and dead-end pores. The RTD exhibits related features as the distribution of residence times increases with biofilm growth.

5.1.3.4.1 Vortex Flow Reactor Materials

The vortex flow reactor was a glass Couette cell driven by a Bruker RheoNMR system. The cell consisted of a stationary outer glass tube with an id of 9 mm and a rotating inner glass tube with an od of 5 mm, giving a gap of 2 mm. The Couette was filled with cylindrical bacterial cells, F. nucleatum ($\approx 2 \times 20$ μm), suspended in water at a concentration of $\approx 10^{11}$ cells mL^{-1}.

5.1.3.4.2 Vortex Flow Reactor MRM Experiment Details

The sample was placed in the Couette designed to fit in the 10 mm gradient coil of the Bruker Avance DRX250 spectrometer described previously. Velocity data are acquired for a rotation rate of 6 rps corresponding to a shear rate of 47 s^{-1}, using a spin echo sequence with bipolar velocity sensitizing gradients [Figure 5.1.1(b)] with parameters of: $T_R = 1000$ ms, $t_e = 20$ ms, $g = 0$ and 25 mT m^{-1}, $\delta = 0.5$ ms, $\Delta = 10$ ms. Total acquisition time for the velocity maps is 17 min. The velocity maps are averaged over a slice of thickness of 1 mm with a field of view (FOV) of 15 mm × 20 mm with 64 × 128 pixels, for a spatial resolution of 23 × 156 μm per pixel.

5.1.3.4.3 Vortex Flow Reactor Advective Transport

Detailed studies of VFRs under varying flow throughput conditions are ongoing and here we present preliminary data on the impact of a cell suspension on the

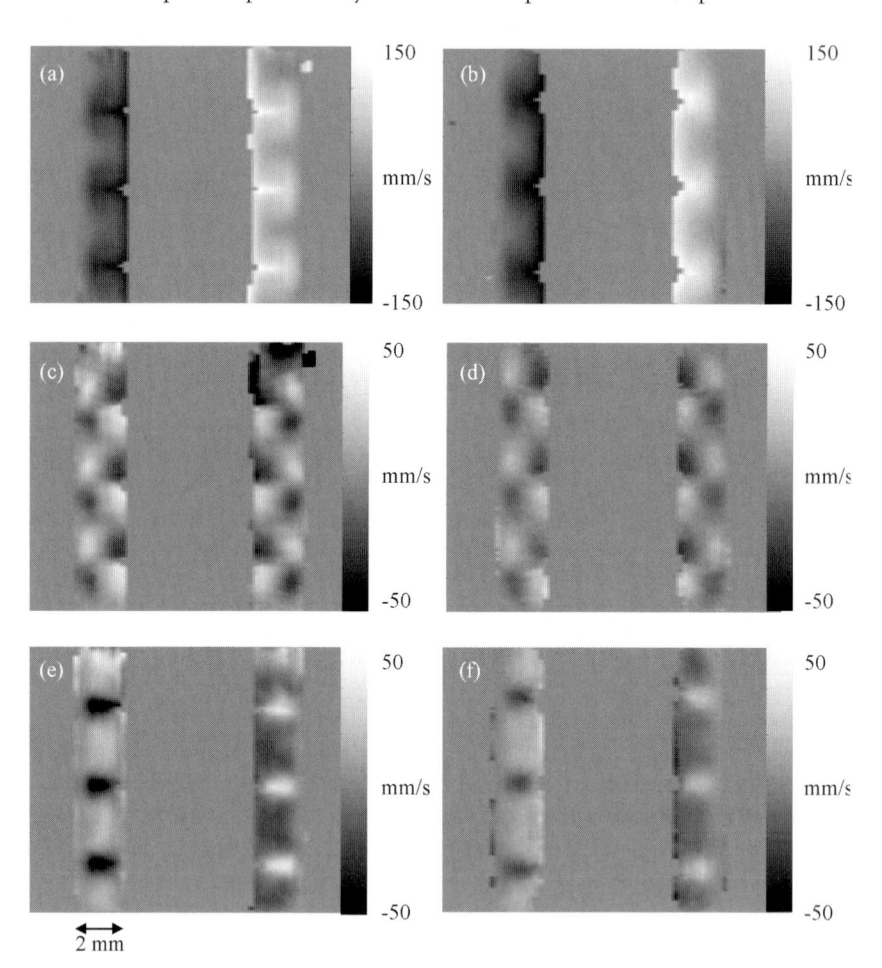

Fig. 5.1.11 MR velocity maps of a cross section through the center of a Couette cell with a 2 mm gap. The inner cylinder is rotating. (a) y component of velocity v_y corresponding to the tangential direction with flow out of the plane (black) and into the plane (white). (c) z component of velocity v_z corresponding to the axial direction with flow up (black) and down (white). (e) x component of velocity, v_x, corresponding to the radial direction with flow left (black) and right (white). (b), (d) and (f) are the x, y and z components of the velocity, respectively, for cylindrical bacterial cells, *F. nucleatum* (≈2 × 20 μm), suspended in water at a concentration of ≈10^{11} cells mL^{-1}.

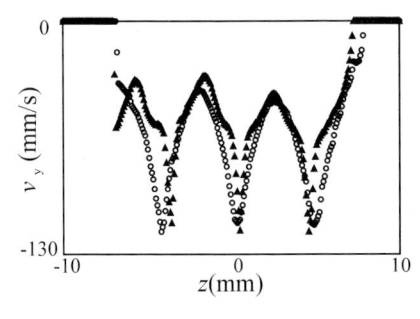

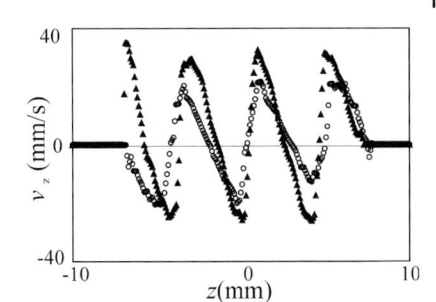

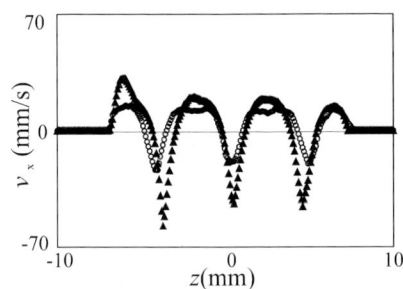

Fig. 5.1.12 Quantitative velocity for water (open circles) and the cylindrical cell suspension (solid triangles) at the same radial position, $r = 3.45$ mm, close to the center of the Couette gap, for the x, y and z components of the velocity, respectively.

advection field under no flow through conditions. Figure 5.1.11 shows MRM velocity maps of pure water and cylindrical bacterial cells, *F. nucleatum* ($\approx 2 \times 20$ μm), suspended in water at a concentration of $\approx 10^{11}$ cells mL^{-1}.

Quantitative velocity data for a fixed radius are shown in Figure 5.1.12 for each component of velocity. Several features are evident, in particular a shift in the z location of the maxima of all three components of velocity for the cell suspension toward the center of the Couette, relative to the pure fluid. A significant impact on mass transport is indicated by the strong increase in the maxima of the axial, Figure 5.1.12(b), and radial, Figure 5.1.12(c), components of velocity. This variation demonstrates the impact secondary flows can have on mass transport, which is not captured by a mass transfer coefficient approach dependent on the tangential or axial velocity based *Re*. The presence of the microbial cells generates stronger secondary flows, altering the mass transport. The development of modeling approaches which capture such effects is an ongoing area of research.

5.1.4
Conclusions

MRM methods have been demonstrated to provide data on the advective transport in capillary, packed bed and VF bioreactors. The correspondence between the MR measured propagators and RTDs has been demonstrated. While the exact correspondence holds only in the case of invariant velocity distributions, scale dependent RTDs can be calculated from time dependent propagators. This provides a clear connection between MR propagators and the classic RTDs used broadly in chemical engineering to design and troubleshoot reactors, indicating the strong potential for non-invasive MR based reactor process control devices. Biofilm growth is shown to generate secondary flows in capillary reactors of a magnitude up to 20% of the axial velocity on low Re flows. This calls into question the ability of mass transfer coefficients based on scaling with axial velocity Re to capture the important features of the mass transport within the reactor. In VF of a microbial cell suspension a strengthening of the secondary flows due to the suspended cells is demonstrated. The data indicate the importance of incorporating microscale details of the advection field into mass transfer models. A particular application of the ability to incorporate microscale advection data into macroscale mass transfer models is the potential to impact the design of microfluidic [42] based bioreactor systems where the spatial variability of the flow field controls bioactivity [57].

Acknowledgements

Student support for this research was provided by the W. M. Keck Foundation (E. L. G.) and the Inland Northwest Research Alliance (J. P. G., J. R. B.). This research was supported in part by the Office of Science (BER) U. S. Department of Energy (DE-FG02–03ER63576), a National Science Foundation ADVANCE Award to S. L. C. (DMI-0340709) and a National Science Foundation CAREER Award to J. D.S (CTS-0348076). Acknowledgement is made to the Donors of the American Chemical Society Petroleum Research Fund for partial support of this research.

References

1 J. W. Costerton, Z. Lewandowski, D. E. Caldwell, D. R. Korber, H. M. Lappin-Scott **1995**, *Annu. Rev.Microbiol.* 49, 711.
2 R. B. Bird, W. E. Stewart, E. N. Lightfoot **2002**, *Transport Phenomena*, 2nd edn, Wiley & Sons, New York.
3 H. S. Fogler **1986**, *Elements of Chemical Reaction Engineering*, Prentice Hall, Englewood Cliffs, New Jersey.
4 P. V. Danckwerts **1953**, *Chem. Eng. Sci.* 2, 1.
5 J. M. Ottino **2000**, *Chem. Eng. Sci.* 55, 2749.
6 J. M. Ottino **1989**, *The Kinematics of Mixing: Stretching, Chaos and Transport*, Cambridge University Press, Cambridge.
7 A. D. Stroock, S. K. W. Dertinger, A. Ajdari, I. Mezic, H. A. Stone, G. M. Whitesides **2002**, *Science* 295, 647.
8 A. D. Stroock, G. J. McGraw **2004**, *Philos. Trans. R. Soc. London, Series A: Phys. Sci. Eng.* 362, 971.
9 H. Aref **1990**, *Philos. Trans. R. Soc. London, Series A: Phys. Sci. Eng.* 333, 273.
10 F. Schonfeld, S. Hardt **2004**, *AIChE J.* 50, 771.
11 L. F. Gladden **2003**, *AIChE J.* 49, 2.
12 M. D. Mantle, A. J. Sederman **2003**, *Prog. Nucl. Magn. Reson. Spectrosc.* 43, 3.
13 S. L. Codd, B. Manz, J. D. Seymour, P. T. Callaghan **1999**, *Phys. Rev. E* 60, R3491.
14 J. D. Seymour, S. L. Codd, E. L. Gjersing, P. S. Stewart **2004**, *J. Magn. Reson.* 167.
15 E. L. Gjersing, S. L. Codd, J. D. Seymour, P. S. Stewart **2005**, *Biotechnol. Bioeng.* 89, 822.
16 M. C. Walters III, F. Roe, A. Bugnicourt, M. J. Franklin, P. S. Stewart **2003**, *Antimicrob. Agents Chemother.* 47, 317.

17 J. D. Seymour, J. P. Gage, S. L. Codd, R. Gerlach **2004**, *Phys. Rev. Lett.* 93, 198103.

18 P. T. Callaghan **1991**, *Principles of Nuclear Magnetic Resonance Microscopy*, Oxford University Press, New York.

19 B. Manz **2004**, *J. Magn. Reson.* 169, 60.

20 H. T. Edzes, D. van Dusschoten, H. van As **1998**, *Magn. Reson. Imag.* 16, 185.

21 T.-Q. Li, J. D. Seymour, R. L. Powell, K. L. McCarthy, L. Odberg, M. J. McCarthy **1994**, *Magn. Reson. Imag.* 12, 923.

22 A. N. Garroway **1974**, *J. Phys. D: Appl. Phys.* 7, L159.

23 P. T. Callaghan, S. L. Codd, J. D. Seymour **1999**, *Concepts Magn. Reson.* 11, 181.

24 H. J. Eberl, C. Picioreanu, J. J. Heijnen, M. C. M. van Loosdrecht **2000**, *Chem. Eng. Sci.* 55, 6209.

25 P. Stoodley, S. Yang, H. M. Lappin-Scott, Z. Lewandowski **1997**, *Biotechnol. Bioeng.* 56, 681.

26 Z. Lewandowski, H. Beyenal **2003**, in *Biofilms in Wastewater Treatment*, eds. S. Wuertz, P. L. Bishop, P. A. Wilderer, IWA Publishing, London, (p.) 147.

27 H. Beyenal, Z. Lewandowski **2002**, *Biotechnol. Progr.* 18, 55.

28 H. Horn, D. C. Hempel **1997**, *Biotechnol. Bioeng.* 53, 363.

29 C. Picioreanu, M. C. M. van Loosdrecht, J. J. Heijnen **2000**, *Biotechnol. Bioeng.* 69, 504.

30 J. M. Ottino **1990**, *Annu. Rev. Fluid Mechan.* 22, 207.

31 A. J. Lichtenberg, M. A. Lieberman **1992**, *Regular and Chaotic Dynamics*, (vol. 38), Springer-Verlag, New York.

32 J. Salles, J.-F. Thovert, L. Delannay, J.-L. Auriault, P. M. Adler **1993**, *Phys. Fluids* 5, 2348.

33 I. Mezic, S. Wiggins, D. Betz **1999**, *Chaos* 9, 173.

34 J. W. Costerton, P. S. Stewart, E. Greenberg **1999**, *Science* 284, 1318.

35 J. Hyde, R. Darouiche, J. W. Costerton **1998**, *J. Heart Valve Dis.* 7, 317.

36 P. Stoodley, I. Dodds, J. D. Boyle, H. M. Lappin-Scott **1999**, *J. Appl. Microbiol.* 85, 19S.

37 D. de Beer, P. Stoodley, Z. Lewandowski **1994**, *Biotechnol. Bioeng.* 44, 636.

38 P. Stoodley, D. de Beer, Z. Lewandowski **1994**, *Appl. Environ. Microbiol.* 60, 2711.

39 G. K. Batchelor **1967**, *An Introduction to Fluid Dynamics*, Cambridge University Press, Cambridge.

40 H. van As, P. Lens **2001**, *J. Ind. Microbiol. Biotechnol.* 26, 43.

41 B. Manz, F. Volke, D. Goll, H. Horn **2003**, *Biotechnol. Bioeng.* 84, 424.

42 H. A. Stone, S. Kim **2001**, *AIChE J.* 47, 1250.

43 Z. Lewandowski, S. A. Altobelli, P. D. Majors, E. Fukushima **1992**, *Water Sci. Technol.* 26, 577.

44 B. C. Hoskins, L. Fevang, P. D. Majors, M. M. Sharma, G. Georgiou **1999**, *J. Magn. Reson.* 139, 67.

45 B. D. Wood, M. Quintard, S. Whitaker **2002**, *Biotechnol. Bioeng.* 77, 495.

46 Z. Lewandowski, P. Stoodley, S. A. Altobelli, E. Fukushima **1994**, *Water Sci. Technol.* 26, 577.

47 P. T. Callaghan, A. Coy, D. MacGowan, K. J. Packer, F. O. Zelaya **1991**, *Nature (London)* 351, 467.

48 J. D. Seymour, P. T. Callaghan **1997**, *AIChE J.* 43, 2096.

49 B. Manz, P. Alexander, L. F. Gladden **1999**, *Phys. Fluids* 11, 259.

50 J. D. Seymour, P. T. Callaghan **1996**, *J. Magn. Reson. Ser. A* 122, 90.

51 A. Ding, D. Candela **1996**, *Phys. Rev. E* 54, 656.

52 S. Stapf, K. J. Packer, R. G. Graham, J.-F. Thovert, P. M. Adler **1998**, *Phys. Rev. E* 58, 6206.

53 U. Tallarek, F. J. Vergeldt, H. van As **1999**, *J. Phys. Chem. B* 103, 7654.

54 J. D. Seymour, J. P. Gage, S. L. Codd, R. Gerlach **2005**, Advances in Water Resources, in press.

55 N. G. van Kampen **1992**, *Stochastic Processes in Physics and Chemistry*, North-Holland, Amsterdam.

56 J. D. Seymour, B. Manz, P. T. Callaghan **1999**, *Phys. Fluids* 11, 1104.

57 S. Takayama, J. C. McDonald, E. Ostuni, M. L. Liang, P. A. J. Kenis, R. F. Ismagilov, G. M. Whitesides **1999**, *Proc. Natl. Acad. Sci. USA* 96, 5545.

5.2
Two-phase Flow in Trickle-Bed Reactors
Lynn F. Gladden, Laura D. Anadon, Matthew H. M. Lim, and Andrew J. Sederman

5.2.1
Introduction to Magnetic Resonance Imaging of Trickle-bed Reactors

5.2.1.1 What are Trickle-bed Reactors?

Trickle-bed reactors are three-phase reactors in which, typically, gas and liquid flow co-currently down through a fixed bed of catalyst pellets. Trickle beds have traditionally been used by industries with large throughputs, such as the petrochemical industry where they are used primarily for hydro-cracking, hydro-desulfurization and hydro-denitrogenation. More recently, trickle-bed technology has also found application in biochemical processing including areas such as the treatment of waste gas, which exploits bacteria immobilization within the bed [1]. Figure 5.2.1 shows a schematic of the four flow regimes characterizing "trickle-bed" operation [2]. Industrial reactors usually operate in the trickle or pulsing regimes or in the relatively poorly understood regime at the interface between these two hydrodynamic regimes [3–10]. As can be seen from Figure 5.2.1, trickle flow exists at low gas and liquid velocities through the bed, and it is characterized by a steady gas–liquid distribution within the reactor. As liquid velocity is increased, so the hydro-

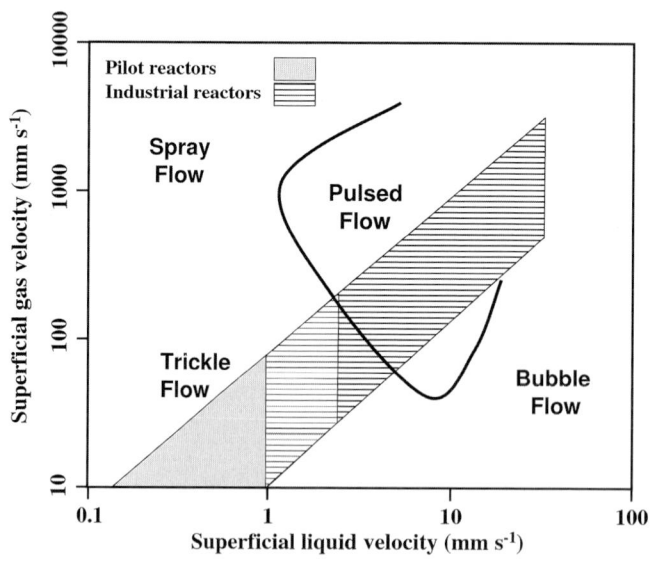

Fig. 5.2.1 Flow regimes in a trickle-bed reactor (after Sie and Krishna [2]). Typical conditions for research and industrial reactor operation are indicated. The black line indicates the boundary between the pulsed flow regime and the spray, trickle and bubble flow regimes. Superficial velocities are defined as the respective volume flow rates divided by the cross-sectional area of the reactor. In subsequent figures the superficial liquid velocity is simply referred to as a liquid velocity.

dynamics undergo a transition to the pulsing regime, which is characterized by pulses of liquid being observed to move along the bed. The ability to understand and characterize the hydrodynamics in trickle-bed reactors is important because the flow pattern within a reactor will significantly influence its performance through quantities such as phase holdups, power consumption and mass transfer fluxes. Furthermore, successful modeling of trickle-bed reactors requires precise tools for the identification of the flow pattern to be expected for a specified set of operating conditions.

Under conditions of trickle flow the reactor is characterized by two macroscopic parameters: liquid holdup and catalyst wetting; catalyst wetting is also referred to as surface wetting and wetting efficiency. The liquid holdup is a measure of the amount of the inter-particle space occupied by liquid, and the catalyst wetting quantifies the fraction of the external surface area of the catalyst that is in contact with the liquid phase. Whilst macroscopic holdup can be determined gravimetrically, only indirect measurements of surface wetting have previously been reported. Such measurements include chemical methods based on reaction rates [11] and tracer [12], dissolution and dye-adsorption [13–15] techniques. There has therefore been a strong motivation to implement a measurement technique capable of imaging both holdup and wetting such that both the spatial distribution of the liquid holdup and a direct measurement of catalyst wetting are obtained from the same measurement. To this end, various non-invasive tomographic techniques have been used to address the characterization of hydrodynamics within trickle-bed reactors, including conductance and capacitance methods [16,17], optical methods [18–20] and X-ray tomography [21–22]. Whilst some of these techniques offer faster data acquisition times than MRI and may be used on plant more readily, to date none of them can challenge the spatial resolution and potential for chemical specificity that MRI offers. In addition, given that MRI is able to spatially resolve molecular diffusion coefficients and flow vectors within the system, the toolkit of measurement capabilities over a hierarchy of length-scales from Å to cm, makes MRI an invaluable measurement tool for understanding the behavior of trickle-bed reactors.

5.2.1.2 MRI Techniques for Studying Hydrodynamics in Fixed-bed Reactors

MRI has already been demonstrated to be particularly powerful in imaging single-phase flows within fixed-bed reactors; i.e., packings of formed catalyst packing elements contained within a process vessel. MRI is able to image the internal structure of the reactor; in terms of the spatial position of the catalyst pellets and the size and connectivity of local elements of the inter-particle space comprising the bed [23], as well as the flow fields inside such reactors. Figure 5.2.2(a) shows a photograph of a typical trickle-bed reactor studied using MRI. The model trickle-bed reactor is of ≈4.5 cm diameter and is packed with glass spheres of diameter 5 mm. Gas and liquid are supplied to the top of the column such that the bed operates under conditions of co-current downflow, and pressure tappings can be made to allow measurements of pressure drop across sections of the bed. This is important if the insights obtained by MRI are to be successfully translated to improvements in plant operation (see Section 5.2.3.2). Typical spatial resolutions of 100–500 μm are obtained in both microimaging and flow imaging studies [24–26]. As we will see in the case of two-phase flow, the advantage of using MRI to study these flow fields is that it has sufficient spatial resolution to identify the spatial heterogeneity in the operating characteristics of the bed. In the case of single-phase flow we see [Figure 5.2.2(b)] how channels of fast liquid flow exist within a bed which also contains near stagnant flow conditions. In the case of two-phase flow the spatial heterogeneity in the gas–liquid distribution is crucial in determining the operating characteristics of a bed. Data acquisition times for both microimaging and flow imaging range from several minutes, using standard spin-echo methods, to tens of milliseconds; the latter being achieved with a recently developed ultra-fast velocity imaging sequence [27] based on the Echo Planar Imaging (EPI) method [28]. Other fast data acquisition pulse sequences are based on the Rapid Acquisition with Relaxation Enhancement (RARE) [29] and "low excitation angle imaging" (e.g., FLASH) [30] sampling strategies. EPI, RARE and FLASH pulse sequences and variants thereof have now all been applied to image hydrodynamics in chemical reactors [31–34]. All these measurements may be applied to packings of real catalyst support (usually alumina or silica) or oxide-supported catalyst (as long as the catalyst is not ferromagnetic or strongly paramagnetic) in addition to model catalyst pellets in the form of glass beads, although care must always be taken in the implementation of a given pulse sequence in assessing its influence on the signal intensities obtained.

The MRI measurement is made by positioning the model trickle-bed reactor within the bore of the superconducting magnet, with the imaging section located

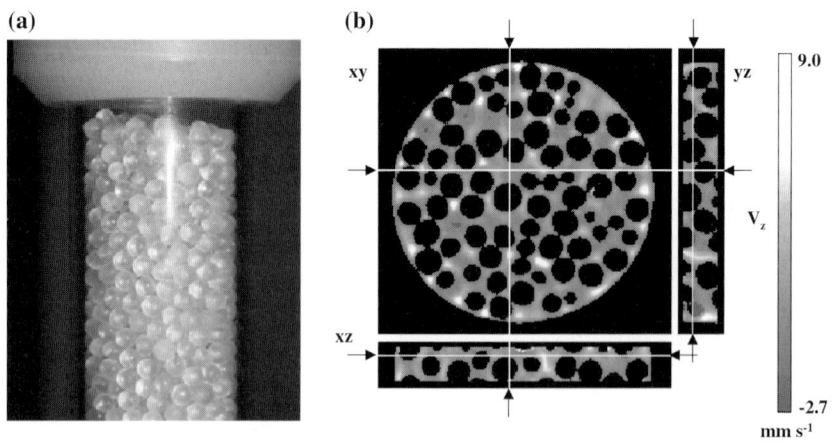

Fig. 5.2.2 (a) The upper section of a typical model trickle-bed reactor used in MRI studies. (b) MR image of water flowing within a fixed bed of spherical glass beads; the beads have no MR signal intensity associated with them and are identified as black voxels. Flow velocities in the superficial flow direction (down the column, z) have been measured; 2D slice sections through the 3D image are shown with slices taken in the xy, yz and xz planes indicated. Voxel resolution is 195 μm × 195 μm × 195 μm.

away from the entrance and exit zones of the bed, toward the centre of the bed. The direction of superficial flow, z, is aligned along the direction of the static, superconducting magnetic field, B_0. Clearly, the dimensions of systems studied are constrained to the dimensions of the bore of the superconducting magnet used. In vertical bore systems, standard magnet hardware allows reactor diameters of 2.5–6 cm to be studied, with a similar field-of-view along the direction of the axis of the magnet. Different sections along the axis of the bed are studied by moving the bed up and down inside the magnet; reactor lengths used lie in the range 0.7–2 m. If individual packing elements are to be resolved within the image, the typical dimensions of magnet hardware limit the column-to-particle diameter to an upper limit of 20–40, and this must be borne in mind in using MRI to address the operation of the much larger trickle beds used in the petrochemical industry which are typically of diameter 1–3 m and height 10–20 m. The important consideration here is that wall effects (i.e., wall induced periodicity in the voidage of the bed) can extend to up to ≈6 particle diameters into the bed [35], thereby causing the model reactors studied by MRI to be subject to greater wall effects than would be found in larger beds. However, the distinctive heterogeneity in the flow seen in Figure 5.2.2(b) *is* observed in regions of beds not subject to wall effects and is therefore a generic characteristic of hydrodynamics in fixed-bed reactors, which is influenced by the size and shape of the packing elements used and the method employed to load the bed.

In our studies we have focused on obtaining high spatial resolution images, in which the flow field within the void space is fully resolved. This line of research was chosen for the following reasons. (a) Reactors of column-to-particle diameter ratio < 40 are typical of those used for reactions characterized by high exothermicity and relatively poor heat transfer, such as the synthesis of methyl isobutyl ketone [36] and the conversion of natural gas into transportation fuels [37]. (b) An increasingly important use of tomography data, and therefore also MRI data, is the validation of numerical simulation codes. To provide a rigorous test of a given code it is important to image the flow field at sufficient resolution that the flow field between individual packing elements comprising the bed is quantified. These codes then become the scale-up tool for the practicing engineer. If the code predicts *quantitatively* the flow field, holdup and wetting characteristics imaged by MRI then we have confidence that the numerical simulation will predict the flow field in reactors of larger dimension. Initial numerical studies have already been reported for single-phase flow, but not yet for two-phase flow [38, 39]. (c) In the specific context of studying trickle-bed reactors, MRI provides a first measurement of aspects of trickle-bed behavior which, until now, have not been accessed even qualitatively. These new insights give both immediate insight into practical aspects of bed operation as well as guiding future theoretical developments. To achieve these advances in understanding it has been important to record images in which the individual packing elements are clearly identified.

In the remainder of this chapter the MRI data presented are images of liquid *distribution* within the trickle bed. In all experiments described a Bruker Spectrospin DMX 200 spectrometer was used, equipped with a 4.7 T magnet and a birdcage coil of length and diameter 6.3 cm. ^{1}H images were acquired at a frequency of 199.7 MHz. Spatial resolution was achieved using shielded gradient coils providing a maximum gradient strength of 13.50 G cm^{-1}.

5.2.2 Holdup and Wetting in the Trickle-flow Regime

Initial MR imaging studies of liquid holdup and catalyst wetting in the trickle-flow regime focused on packings of non-porous glass spheres. Figure 5.2.3 shows the nature of the data obtained [40]; to produce these images a standard spin-echo imaging sequence was used. In Figure 5.2.3(b) the total liquid holdup in the image slice has been partitioned in to what we call rivulets using a pore space segmentation algorithm [23]; each shade on the grey scale identifies an individual liquid "rivulet". Application of this algorithm is merely an objective, reproducible procedure to characterize the liquid distribution within the bed. Even in the case of liquid holdup, which can be measured gravimetrically, MRI can yield new insights. For example, as shown in Figure 5.2.4, the behavior of liquid holdup as a function of liquid flow velocity, v_l, at constant gas velocity can show behavior that cannot be interpreted unambiguously without access to the imaging data. We see that liquid holdup increases relatively sharply at first and then more gradually with increasing v_l. Images recorded at the lower liquid velocities reveal that the distribution of liquid is dominated by liquid close to the packing elements with only a small number of rivulets present; i.e., the liquid is present, predominantly, in the form of surface-wetting films. At the lowest value of liquid superficial velocity of 0.5 mm s^{-1} only 7 rivulets are observed. As v_l increases, the number of rivulets increases sharply to 30 at $v_l = 2.0$ mm s^{-1} and then levels off. As v_l increases above 2.0 mm s^{-1}, the existing rivulets increase in size and few, if any, new liquid rivulets are formed. Images showing the gas–liquid distribution also give important insights into how start-up influences the resulting gas–liquid distribution. Figure 5.2.5 shows the gas–liquid distribution for a gas velocity of 66 mm s^{-1} and a liquid velocity of 4.6 mm s^{-1} for start-up conditions of: (a) pulsing gas and liquid through the bed for at least 5 min before establishing the trickle flow conditions; (b) pre-wetting the bed using pulsing flow as in (a) and then allowing the bed to drain for 20 min in the absence of flow prior to establishing trickle flow; and (c) establishing trickle flow directly in a dry bed – this start-up condition clearly causes a very poor gas–liquid distribution with the liquid existing primarily in channels along the wall of the bed. Comparison of the images associated with the different start-up conditions shown in (a) and (b) show that full pre-wetting of the bed followed by immediately establishing the trickle flow condition yields significantly greater catalyst wetting and holdup. For gas and liquid velocities of 66 and 3 mm s^{-1}, respectively, the enhancement in adopting the full pre-wetting start-up over that in which the bed is pulsed and then drained is ≈70% in holdup, surface wetting and the number of liquid rivulets.

Implementation of MRI to quantify holdup and wetting in packings of porous packing elements (e.g., catalyst support pellets) must be performed with care. Difficulties in data acquisition and analysis arise because the signal we wish to

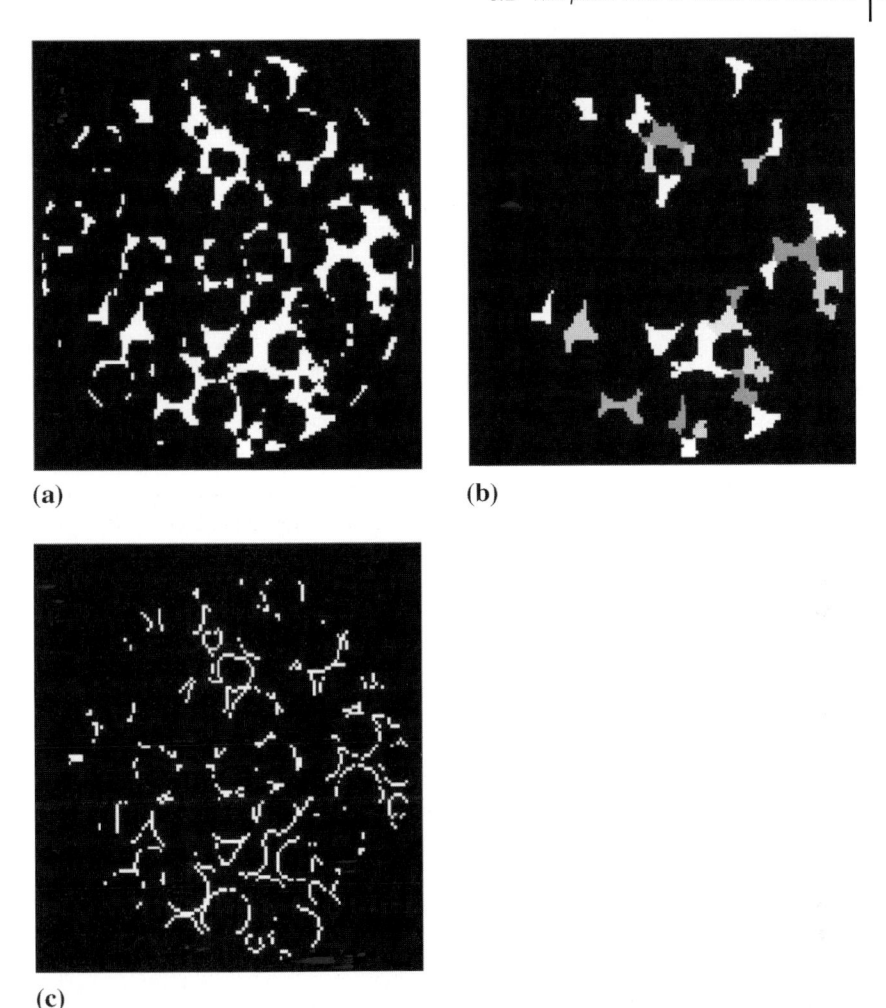

(a)

(b)

(c)

Fig. 5.2.3 Identification of rivulets and surface wetting in a packing of 5-mm diameter glass spheres contained within a column of inner diameter 40 mm. The data were acquired in a 3D array with an isotropic voxel resolution of 328 μm × 328 μm × 328 μm. (a) The original image of trickle flow is first binary gated, so that only the liquid distribution within the image is seen (white); gas-filled pixels and pixels containing glass spheres show up as zero intensity (black). (b) The liquid distribution is broken up into individual liquid rivulets, each identified by a different shade on a greyscale. In total 26 individual liquid rivulets are identified in this image. (c) Pixels containing any liquid–solid interface are then identified using image analysis techniques and "images" of surface wetting are produced. Data are shown for liquid and gas superficial flow velocities of 3 and 66 mm s^{-1}, respectively [40]. Reproduced from Ref. [40], with kind permission from Elsevier, Copyright (2001).

measure is associated with the liquid (water) in the bed. However, the signal intensity we acquire from any specific ensemble of water molecules will depend on their nuclear spin relaxation time and hence their physical and chemical

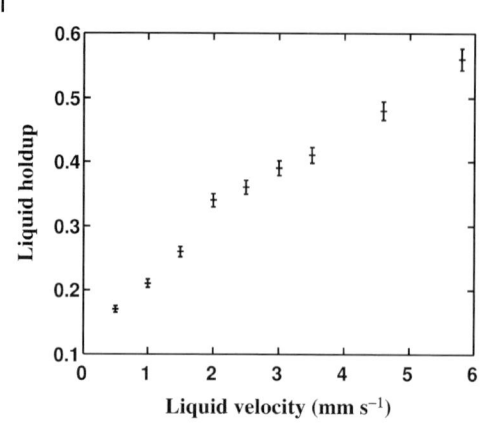

Fig. 5.2.4 Plot of the liquid holdup as a function of increasing liquid velocity. An increase in the liquid holdup is observed with increasing liquid velocity. The gas superficial velocity is constant at 66 mm s^{-1}. Data are shown for a bed of 5-mm diameter glass spheres packed within a column of inner diameter 40 mm [40]. Reproduced from Ref. [40], with kind permission from Elsevier, Copyright (2001).

environment. In trickle-bed reactors, the different environments will be (a) free water in the bulk of the inter-pellet space, (b) water within the intra-pellet pore space and (c) water existing in films on the surface of the pellets but not being part of a rivulet within the inter-pellet space. Initial studies of holdup and wetting during trickle flow in a fixed bed of porous packing elements have been reported for cylindrical alumina packings of: (a) diameter 1.5 mm and a distribution of lengths in the range 5 ± 2 mm; and (b) diameter and length equal to 3 mm. A constant gas superficial velocity of 31 mm s^{-1} was used, with liquid superficial velocities in the range 0.1–6 mm s^{-1}. 2D visualizations of liquid distribution within transverse sections of thickness 1 mm were obtained; the total data acquisition time for each image was 25 min. Data were recorded with a field-of-view of 40 mm × 40 mm and a data array size of 256 × 256, thereby yielding an in-plane spatial resolution of 156 μm × 156 μm [41].

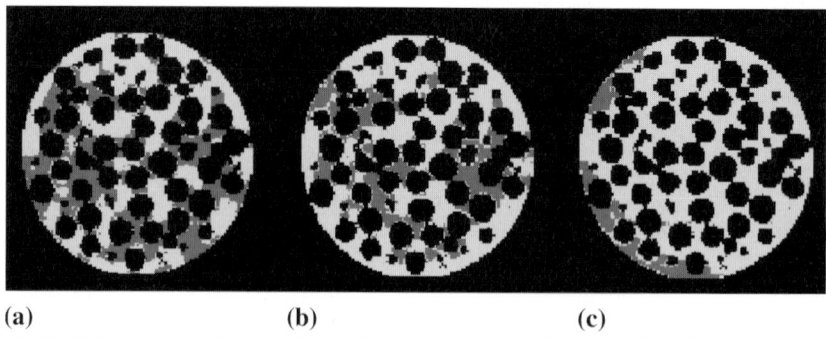

Fig. 5.2.5 MR images of gas and liquid distribution within a fixed-bed reactor operating at air and water velocities of 66 and 4.6 mm s^{-1}, respectively. The start-up conditions are different in each case: (a) full pre-wetting; (b) pre-wetting the bed using pulsing flow as in (a) and then allowing the bed to drain for 20 min in the absence of flow prior to establishing trickle flow; and (c) no pre-wetting. Dark grey, light grey and black pixels identify water, air and packing elements, respectively.

Figure 5.2.6 shows plots of dynamic liquid holdup (the dynamic liquid holdup is the total holdup minus the holdup at zero liquid superficial velocity) and wetting efficiency versus liquid superficial velocity for a constant gas velocity of 31 mm s^{-1}. It is clearly seen that dynamic liquid holdup increases more rapidly as a function of liquid superficial velocity within the 1.5 mm packing, and values of holdup and wetting efficiency are always greater, for a given liquid velocity, for the 1.5 mm diameter packing relative to the 3 mm diameter packing. The line through the dynamic liquid holdup data is the best fit of the percolation-based model described by Crine et al. [42]. The form of the expression for the dynamic liquid holdup is:

$$\chi_{dynamic} = (Kv)^{1/3}\left(\frac{v}{v+v_{min}}\right)^{2/3} \qquad (5.2.1)$$

where v = liquid superficial velocity and v_{min} is a minimum liquid superficial velocity. $K = k\mu_L a^2/\varrho_L g$, where k is a proportionality factor depending on the fluid and packing properties, μ_L is the liquid dynamic viscosity, ϱ_L is the liquid mass density, a is the specific surface of the packing, and g is the acceleration due to gravity. Equation (5.2.1) is fitted to the experimental data, with v_{min} and K (i.e., k) as variables in the fit. As seen from Figure 5.2.6(a), the fit of Eq. (5.2.1) to the data is good. The values of v_{min} obtained are 3.56×10^{-4} mm s^{-1} and 12.5×10^{-4} mm s^{-1} for the 1.5 and 3 mm diameter cylinders, respectively. Following the argument of Toye et al. [22], v_{min} characterizes solid-phase (i.e., packing) wettability such that smaller values of v_{min} are associated with better packing wettability. Thus, the result of the percolation analysis is entirely consistent with the surface wetting data shown in Figure 5.2.6(b). These early studies on trickle flow within beds of porous packing elements identified three general results: (a) values of liquid holdup compare well (to within 5%) with those obtained by gravimetric measurement; (b) the general trend in both holdup and wetting data are consistent with the predictions of

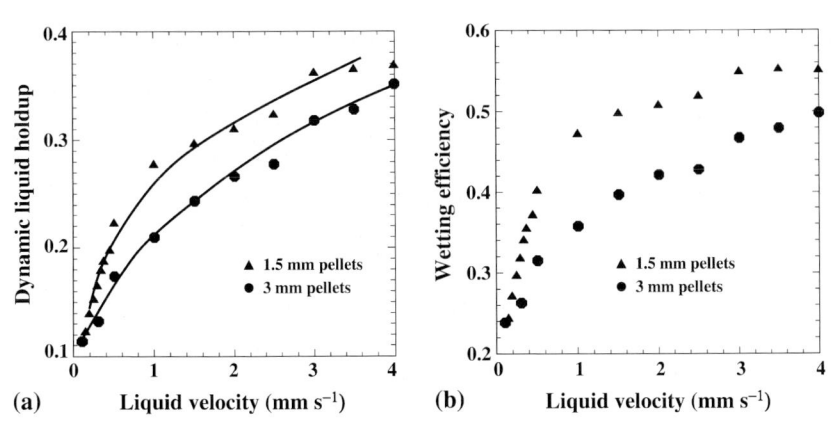

Fig. 5.2.6 (a) Dynamic liquid hold-up and (b) wetting efficiency as a function of liquid superficial velocity for 1.5- and 3-mm cylinders [41]. Gas flow rate is constant at 66 mm s^{-1}. The line shows the best fit of the data to the percolation model of Crine et al. [42].

existing models in the literature; and (c) the absolute value of surface wetting tended to be lower than those previously reported in the literature. Greater error in the MRI measurement of surface wetting compared with that of holdup is expected given that liquid adjacent to a solid surface will be subject to significant relaxation contrast, and hence a lower than expected value of wetting may arise from not "seeing" liquid layers on the surface when they are not associated with a larger scale liquid rivulet. However, it should also be remembered that MRI provides the only direct measurement of surface wetting and therefore the other experimental determinations of wetting may themselves be subject to considerable error.

Recently, modifications to the overall methodology of determining catalyst wetting using MRI have been proposed. In this modified image analysis algorithm a local, as opposed to global, gating level is applied – this is an objective procedure which accounts for the fact that the gating level we choose must discriminate between intra-pellet water and water bound to the surface as a wetted film. The particular gating level needed to achieve this will be very sensitive to the signal from the intra-pellet water and, as this varies with the characteristics of individual pellets, a global gating level will always introduce inaccuracies into the measurement. Using this improved algorithm [43], MR measurement of both holdup and wetting are in good agreement with the predictions of the neural network analysis of Larachi et al. [9].

5.2.3
Unsteady-state Hydrodynamics in Trickle-bed Reactors

In application to trickle-bed reactors, two unsteady-state hydrodynamic phenomena have been studied; these two phenomena are likely to be related and this aspect of the work is the subject of ongoing study. The phenomenon of particular interest is the hydrodynamic transition from the trickle-to-pulsing regime and this is addressed in detail in Section 5.2.3.2. The mechanism of this transition is still not understood. As described by Larachi et al. [9] two conceptually different approaches exist for describing the onset of the pulsing regime, these being the microscopic and macroscopic models. Microscopic or single-pore models analyze pore-scale hydrodynamics and postulate that the macroscopic onset to pulsing flow is an outcome of a statistically large number of local pulsatile occurrences [3–5, 8]. In contrast, the macroscopic models analyze the onset of pulsing at the reactor scale from a stability analysis of first-principle volume-averaged Navier–Stokes equations [6]. For the reactors of column-to-particle diameter < 20 studied to date by MRI it is clear that in the "transition regime" before global pulsing of the bed is observed, local pulsing events exist that are surrounded by a constant gas–liquid distribution characteristic of trickle flow. As an extension of this work it has also been possible to investigate the absolute stability of the "macroscopically stable" gas–liquid distribution observed in the trickle-flow regime. By "macroscopically stable" we mean that successive images show a constant gas–liquid distribution within the column. However as we will see in Section 5.2.3.1, rapid, low amplitude fluctuations in local liquid content, which are characterized by a well defined frequency, are observed even in this apparently stable trickle-flow regime. These data are telling us that while the spatial distribution of liquid within the bed remains constant, the absolute liquid content of a liquid-containing region within the bed may fluctuate. These oscillations in liquid content have been characterized by a temporal autocorrelation function analysis and are most readily interpreted as fluctuations in the thickness of surface wetting layers on the catalyst pellets prior to formation of the local pulsing events, which have been postulated to be the precursor to the transition to pulsing flow within the reactor.

To study unsteady-state gas–liquid distributions it is necessary to use fast data acquisition pulse sequences. In the examples presented here, Fast Low Angle SHot imaging (FLASH) or SNAPSHOT imaging [30] is used. The important feature of this technique is that the rf excitations are characterized by a low flip angle, θ, typically 5–10°, in contrast to the 90° ($\pi/2$) pulse used in the conventional spin-echo imaging sequences employed to quantify holdup and wetting in trickle flow [40]. In FLASH imaging, the signal resulting from the small flip angle θ is proportional to $\sin \theta$, while the longitudinal (z axis) magnetization that remains after the excitation is proportional to $\cos \theta$. Fractions of this remaining magnetization are then used to sample successive lines of k-space. A 128×128 image based on a repetition time of 3 ms takes approximately 380 ms to acquire. The disadvantage of this approach is that by using only a proportion of the available magnetization, signal-to-noise in the image is significantly reduced. The particular MRI pulse sequence used in the ultra-fast data acquisition will introduce contrast effects in to the image data (i.e., signal intensities) such that direct interpretation of the signal intensity in terms of an absolute liquid holdup is subject to errors that are difficult to quantify – but can be made small (<10%) by appropriate choice of delay and recycle times when implementing the pulse sequence; i.e., controlling the extent of T_2^* and T_1 contrast in the images. The FLASH pulse sequence employed in this application was used to provide a map of where liquid exists within the bed, and how the spatial distribution of liquid holdup changes with time. To confirm that the FLASH sequence did indeed provide a robust measurement of liquid distribution, the image acquired using FLASH was compared with an image of the same system acquired using a spin-echo pulse sequence. This comparison enabled us to assess the effect of T_2^*-susceptibility contrast on the image of the liquid distribution. For the beds studied here, while there were small differences in the absolute signal intensity between the images acquired using the two sequences, the spatial distribution of signal associated with the liquid was the same to within the limits of the spatial resolution of the data. Therefore, we can be confident that the FLASH sequence is giving a robust measure of the liquid distribution and its stability but, in contrast to the spin-echo imaging data described in Section 5.2.2, holdup is not quantified. To minimize image acquisition times in this study, spatial resolution was sacrificed such that individual packing elements within the bed are no longer resolved.

5.2.3.1 Identifying and Characterizing Liquid Instabilities

The data to be described were acquired using a trickle-bed column of length 70 cm and diameter 4.3 cm. The packing elements used were cylindrical porous alumina pellets of length and diameter 3 mm, and the liquid and gas used were water and air, respectively; any ^{1}H-containing liquid can, in principle, be studied. Two-phase flow characteristics were studied for superficial gas and liquid velocities in the range 25–300 mm s^{-1} and 0.4–13.3 mm s^{-1}, respectively. These velocities correspond to gas and liquid Reynolds numbers (based on diameter of packing element) of 5–6.7 and 1–40, respectively. The combinations of gas and liquid velocities employed correspond to conditions within both the trickle- and pulse-flow regimes, and include the transition region between these two regimes. In the case of the 2D acquisitions, images were acquired as a data array of size 32 × 16 (in-plane spatial resolution 1.4 mm × 2.8 mm), with an acquisition time of 20 ms for a 2-mm slice thickness. Images were acquired in immediate succession and therefore frame rates of 50 frames per second (fps.) were achieved; the maximum number of images acquired in a single series was 540. For the case of the 3D data, 27 liquid velocities between 0.7 and 13.07 mm s^{-1} were studied at two different gas superficial velocities: 25 and 300 mm s^{-1}. 3D images allow us to study the evolution of hydrodynamic instabilities within the bed along the direction of superficial flow. 3D FLASH images of liquid distribution within the bed were acquired as a data array of size 16 × 16 × 32 (spatial resolution 3.75 mm × 3.75 mm × 1.87 mm) with an acquisition time of 280 ms, i.e., 3D images were acquired at a rate of 3.6 fps. Six series of 8 consecutive 3D images were acquired for each set of flow rates. Acquisition of these large data sets requires us to think carefully about the most robust way of analyzing the data. Each FLASH image provides a map of effective liquid holdup within the bed – the term effective being used to identify that relaxation time contrast effects will be influencing the absolute signal intensity to some extent. Two simple data analysis procedures have been found to be particularly useful in quantifying liquid instabilities within series of images [44]. The first of these is the calculation of a standard deviation map. To calculate the standard deviation map, the standard deviation of the intensity associated with a particular pixel or a voxel i, σ_i, is calculated using the data from all images acquired in the series as follows:

$$\sigma_i = \sqrt{\frac{\sum (x_i - \bar{x}_i)^2}{n}} \tag{5.2.2}$$

where x_i is the signal intensity of pixel (voxel) i, $\bar{x}_i$ is the average intensity of pixel (voxel) i in the series of n images; the summation being taken over the n images. This procedure is performed for all pixels (voxels) and the resulting values of standard deviation then displayed as a 2D or 3D data array (or map). For the case of a constant liquid distribution in all images within a series, the standard deviation values in each pixel (voxel) will be ≈0; i.e., it will approach the standard deviation of the noise level in the data. Standard deviation maps have been calculated and are shown for both the 2D and 3D data acquisitions (Figures 5.2.7–5.2.9). The other characteristic of a series of images that is calculated is the temporal autocorrelation function, $R_I(\tau)$, of the signal intensity (or effective liquid holdup). $R_I(\tau)$ is defined as:

$$R_I(\tau) = \frac{\langle I(t)I(t+\tau)\rangle}{\langle I^2 \rangle} \qquad (5.2.3)$$

where $I(t)$ is the total signal intensity within the region of interest acquired at time t. This analysis extracts correlations in the signal intensity for all time separations τ and highlights any periodicity in the holdup data and the timescales over which they exist. $R_I(\tau)$ may be calculated over local regions within the bed or over the whole bed depending on the length-scale over which the instability of interest occurs.

Figure 5.2.7 shows the 2D standard deviation maps calculated for each of four liquid velocities (2.8, 3.7, 6.1 and 7.6 mm s^{-1}) for a constant gas velocity of 112 mm s^{-1} [45]. The values of standard deviation are identified by the grey scale, lighter shades identify regions of unstable (i.e., time-varying) liquid content. In each case the standard deviation maps have been combined with the image of the structure of the bed, which has been obtained at higher spatial resolution (in-plane spatial resolution 175 μm × 175 μm) – sufficient to fully resolve the individual packing elements within the bed. This high resolution image of the bed structure is obtained using the RARE pulse sequence. The image of the structure of the bed has been gated to discriminate the position of the packing elements; intra-pellet regions are identified as black pixels. As seen in Figure 5.2.7(a), at low liquid velocity the standard deviation map is characterized by low values of standard deviation, typical of the standard deviation in the noise acquired in the original

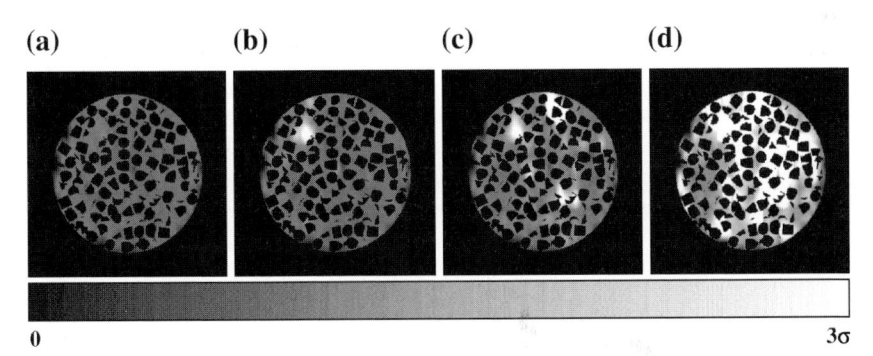

Fig. 5.2.7 Identification of location and size of local pulses within the trickle bed. A high spatial resolution image (in-plane spatial resolution 175 μm × 175 μm; slice thickness 1 mm) is overlayed with a standard deviation map calculated from images acquired at a spatial resolution of in-plane spatial resolution 1.4 mm × 2.8 mm; slice thickness 2 mm. The standard deviation maps have been linearly interpolated to the same in-plane spatial resolution as the high resolution data. Images are shown for a constant gas velocity of 112 mm s^{-1}, and the data were recorded as a function of decreasing liquid velocity. The liquid velocities are (a) 2.8, (b) 3.7, (c) 6.1 and (d) 7.6 mm s^{-1}.

images. This is consistent with a given pixel containing the same phase or combination of phases (liquid, solid, gas) in each successive image used to calculate the standard deviation map. Figure 5.2.7(d) corresponds to conditions of pulsing flow during which the content (i.e., gas or liquid) of a given pixel changes with time. Thus the resulting standard deviation map has high intensity values (lighter shades). The liquid velocities associated with the data shown in Figure 5.2.7(b and c) lie within the transition from trickle to pulsing flow. At these liquid velocities, local instabilities in liquid content exist while the majority of the bed is still characterized by a stable gas–liquid distribution. It is also seen that the initial local pulses are occurring at the size-scale of the packing elements within the bed. As liquid velocity increases the regions of local pulsing increase in number, as well as increasing in number. The evolution of the system through the hydrodynamic transition to pulsing flow is considered in more detail in Section 5.2.3.2.

Recent work has shown that even though the gas–liquid distribution is constant during trickle flow [see Figure 5.2.7(a)], the actual liquid content of local regions within the bed may change with a well defined periodicity, as shown in Figure 5.2.8. These oscillations in the temporal autocorrelation function data are most readily interpreted as fluctuations in the thickness of surface wetting layers on the catalyst pellets prior to formation of the local pulsing events, previously discussed, which have been postulated to be the precursor to the transition to pulsing flow within trickle-bed reactors. The standard deviation map shown in Figure 5.2.8 is not shown on the same intensity scale as those shown in Figure 5.2.7. Instead, the scale now varies between the highest (white) and lowest (black) values of standard deviation calculated from this specific time series of liquid-distribution images. All the standard deviation values in Figure 5.2.8(a) would correspond to grey pixels (i.e., low standard deviation values) in Figure 5.2.7, which identify a constant macroscopic gas–liquid distribution. Therefore the fluctuations in liquid distribution considered in Figure 5.2.8 are much smaller than those observed when local liquid pulses occur (Figure 5.2.7). Figure 5.2.8(a) shows a typical 2D standard deviation map from which the temporal autocorrelation functions [Figure 5.2.8(b and c)] associated with the two highlighted regions have been calculated. Data are shown for a constant gas velocity of 27 mm s^{-1} and for liquid velocities of (b) 0.40 and (c) 0.79 mm s^{-1}. The temporal autocorrelation data shown by the dashed and solid lines are associated with the regions defined by the same line styles in (a).

5.2.3.2 The Trickle-to-pulse Transition

Figure 5.2.9 shows 3D standard deviation maps, combined with the RARE image of the bed, calculated from data acquired at a constant gas velocity of 75 mm s^{-1} for increasing liquid velocities of (a) 0.5, (b) 5.0, (c) 7.0, (d) 8.0, (e) 8.5 and (f) 10.0 mm s^{-1} [46]. The height of the bed shown is 28 mm. Data are shown for half of the bed volume imaged. Again the formation of isolated local pulses prior to transition to the pulsing regime is seen. It is now possible to see that as liquid velocity increases so the number of the local pulsing regions increases until they begin to inter-connect, as the pulsing regime is approached. To quantify the

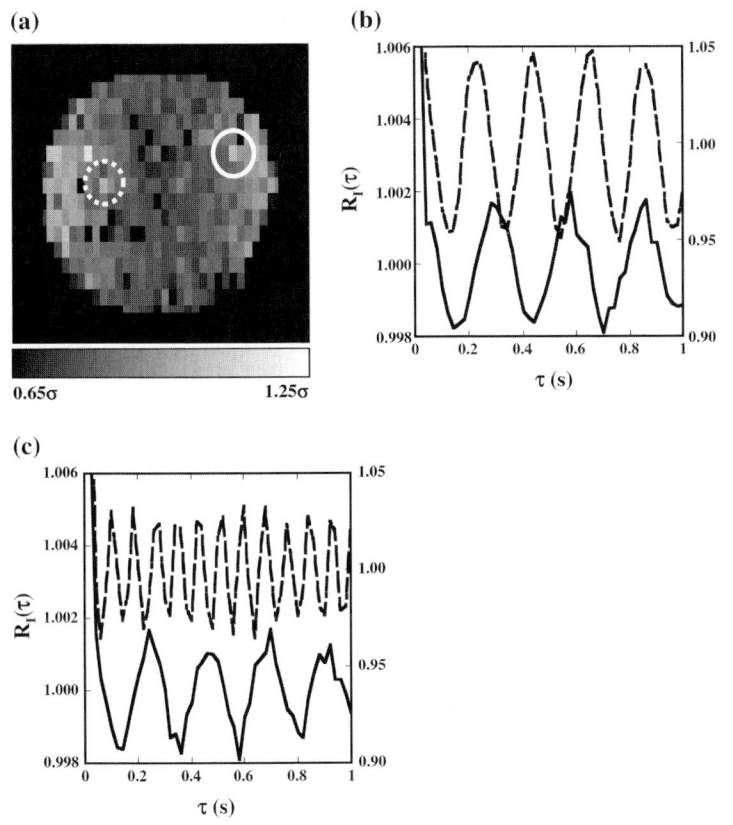

Fig. 5.2.8 (a) Standard deviation map, and temporal autocorrelation functions for data recorded at a gas velocity of 27 mm s^{-1} and liquid velocities of (b) 0.40 and (c) 0.79 mm s^{-1}. The temporal autocorrelation functions shown by the dashed and solid line styles are associated with the pixel of highest intensity within the region identified by the same line style in (a), and correspond to the values of $R_l(\tau)$ on the right-hand and left-hand axes, respectively. The pixel resolution in (a) is 1.4 mm × 2.8 mm.

development of the hydrodynamics within the bed as liquid velocity increases, the 3D standard deviation maps were binary gated at a standard deviation value of 5 σ, where σ is the average standard deviation value of the noise calculated from the images of liquid distribution. Voxels associated with standard deviation values >5 σ were identified as locally pulsing regions, consistent with the analysis of the 2D data shown in Figure 5.2.7. Following gating, a morphological thinning algorithm was then used [23] to identify all voxels associated with individual local pulsing regions and to provide a characterization of the volume of the bed imaged in terms of the number of individual local pulses and their volume. A detailed analysis of these data has recently been completed [47, 48] and only one set of results is discussed here. This set is chosen because it shows how the insights gained from MRI studies can be used to make the use of cheap, robust sensors (in this case pressure drop), readily used on real plant, more effective.

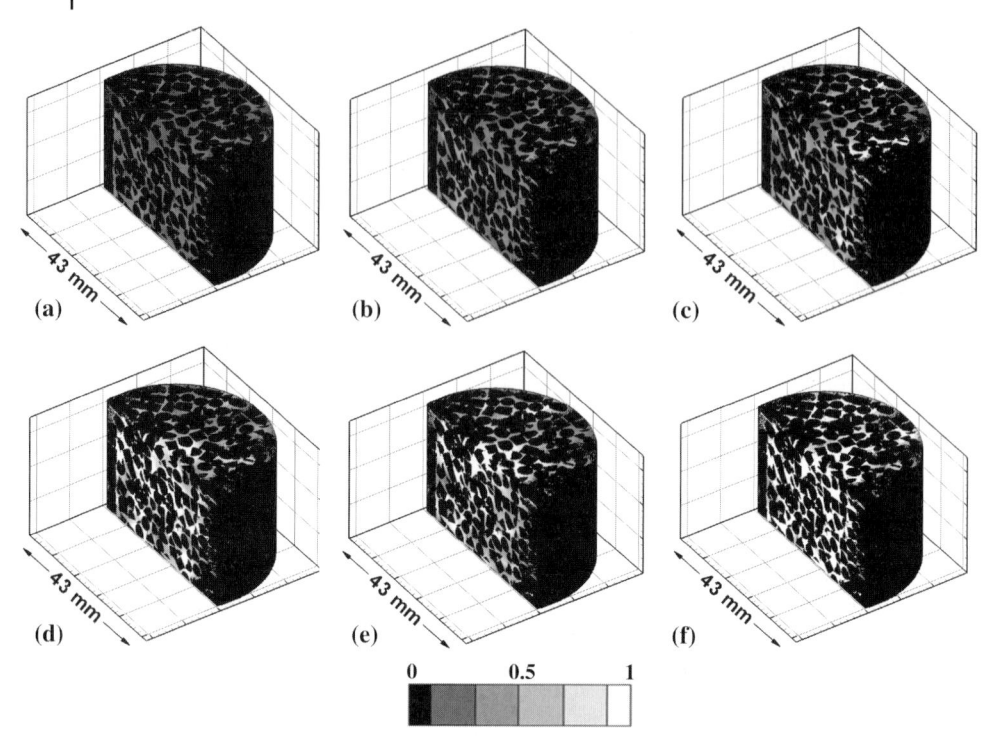

Fig. 5.2.9 3D standard deviation maps, combined with the RARE image of the bed, calculated from data acquired at a constant gas velocity of 75 mm s^{-1} for increasing liquid velocities of (a) 0.5, (b) 5.0, (c) 7.0, (d) 8.0, (e) 8.5 and (f) 10.0 mm s^{-1}. The height of the bed shown is 28 mm. Data are shown for half of the bed volume imaged. Voxels identified as white indicate regions of greatest liquid instability and regions of connected white voxels define the spatial extent of a local pulse. It is clearly seen that the liquid pulses start at the size of the individual packing elements and then increase in number until they interconnect and the whole bed moves into the pulsing regime.

Figure 5.2.10 shows data recorded for constant gas velocities of 25 and 300 mm s^{-1}. In each case, the average number of individual local pulses within the full 3D imaging volume is plotted. All these data are average values taken over 6 separate 3D experiments, each giving rise to their own standard deviation map, recorded for each combination of gas and liquid velocities. A clear peak is seen in this plot indicating that isolated pulses develop in increasing number as liquid velocity increases until a maximum number of such pulses are formed. Combining these data with the data provided by the 3D standard deviation maps of the type shown in Figure 5.2.9, it is clear that after this maximum number of pulses is reached the pulses merge with one another, until the whole bed volume is characterized by a single region of unsteady-state liquid content (i.e., the bed moves into the pulsing regime). The peak in this distribution lies at ≈8.7 and 6 mm s^{-1} for gas velocities of 25 and 300 mm s^{-1}, respectively. The second set of data plotted is the standard deviation in the pressure drop measurements made on each bed; these measurements were made at 0.5-s intervals over a 10-min period. A preliminary observation from these data is that there appears to be an inflexion in the pressure drop data such that at the point at which individual liquid pulses merge within the bed there is a steep increase in the standard deviation (i.e., fluctuation) in the pressure drop measurements. Thus it seems appropriate to define the liquid velocity at which the maximum number of individual pulses occurs as the transition point. This is a particularly useful result because we now have a physical interpretation of the behavior of the pressure drop measurement and are therefore able to use a simple cheap sensor to unambiguously identify the point at which the bed moves to the fully pulsing state. It follows that other characteristics of the pressure drop measurement can also be used to act as a "fingerprint" of the hydrodynamic states within the trickle-bed reactor.

5.2.4
Summary

Recent developments in the implementation of MRI pulse sequences and associated data analysis strategies have provided direct measurements of holdup and catalyst wetting in trickle-bed reactors in the trickle-flow regime, as well as yielding novel measurements of liquid instabilities within the two-phase flow as the system moves between different hydrodynamic regimes. In particular, these early MRI studies of hydrodynamic transitions in trickle-bed reactors have provided new insights as to the mechanism of the trickle-to-pulse transition. Furthermore, we can now give a physical meaning to the transition point; i.e., the liquid velocity at which the maximum number of isolated local pulsing events occurs, and as a result of this we are able to use pressure drop measurement to provide less ambiguous characterization of the hydrodynamics associated with particular operating conditions of a trickle-bed reactor.

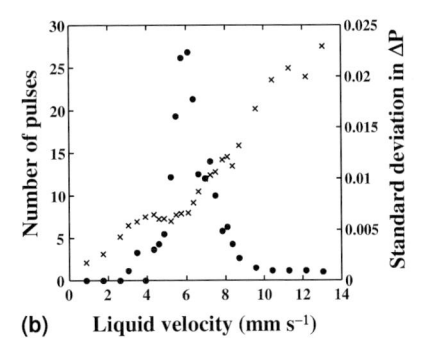

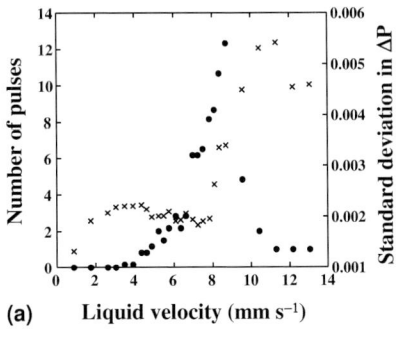

Fig. 5.2.10 Analysis of the 3D standard deviation maps calculated from data acquired with the bed operating at a constant gas velocity of (a) 25 and (b) 300 mm s^{-1}, as a function of liquid velocity. The number of independent liquid pulses identified at each liquid velocity (•), and the standard deviation in the pressure drop measurements made over the length of the bed, recorded at 0.5-s intervals over a 10-min period (×) are shown.

Acknowledgements

We wish to thank Dr. P. Alexander for his help with the preparation of this text.

References

1 D. Hekmat, A. Feuchtinger, M. Stephan, D. Vortmeyer **2004**, *J. Chem. Technol. Biotechnol.* 79, 13–21.
2 S. T. Sie, R. Krishna **1998**, *Rev. Chem. Eng.* 14, 203–252.
3 S. Sicardi, H. Hofmann **1980**, *Chem. Eng. J.* 20, 251–253.
4 J. R. Blok, J. Varkevisser, A. A. H. Drinkenburg **1983**, *Chem. Eng. Sci.* 38, 687–699.
5 K. M. Ng **1986**, *AIChE J.* 32, 115–122.
6 K. Grosser, R. G. Carbonell, S. Sundaresan **1988**, *AIChE J.* 34, 1850–1860.
7 D. C. Dankworth, I. G. Kevrekidis, S. Sundaresan **1990**, *AIChE J.* 36, 605–621.
8 R. A. Holub, M. P. Dudukovi, P. A. Ramachandran **1992**, *Chem. Eng. Sci.* 47, 2343–2348.
9 F. Larachi, I. Iliuta, M. Chen, M. B. P. A. Grandjean **1999**, *Can. J. Chem. Eng.* 77, 751–765.
10 J. G. Boelhouwer, H. W. Piepers, A. A. H. Drinkenburg **2002**, *Chem. Eng. Sci.* 57, 4865–4876.
11 C. N. Satterfield **1975**, *AIChE J.* 21, 209–228.
12 P. L. Mills, M. P. Duduković **1981**, *AIChE J.* 27, 893–904.
13 A. Lakota, J. Levec **1990**, *AIChE J.* 36, 1444–1448.
14 C. L. Lazzaroni, H. R. Keselman, N. S. Fígoli **1988**, *Ind. Eng. Chem. Res.* 27, 1132–1135.
15 P. V. Ravindra, D. P. Rao, M. S. Rao **1997**, *Ind. Eng. Chem. Res.* 36, 5133–5145.
16 J. A. Helwick, P. O. Dillon, M. J. McCready **1992**, *Chem. Eng. Sci.* 47, 3249–3256.
17 N. Reinecke, N. D. Mewes **1997**, *Chem. Eng. Sci.* 52, 2111–2127.
18 T. R. Melli, J. M. de Santos, W. B. Kolb, L. E. Scriven **1990**, *Ind. Eng. Chem. Res.* 29, 2367–2379.
19 W. B. Kolb, T. R. Melli, J. M. de Santos, L. E. Scriven **1990**, *Ind. Eng. Chem. Res.* 29, 2380–2389.
20 N. A. Tsochatzidis, A. J. Karabelas **1994**, *Ind. Eng. Chem. Res.* 33, 1299–1309.
21 A. Kantzas **1994**, *AIChE J.* 40, 1254–1261.
22 D. Toye, P. Marchot, M. Crine, G. L'Homme **1996**, *Meas. Sci. Technol.* 7, 436–443.
23 C. A. Baldwin, A. J. Sederman, M. D. Mantle, P. Alexander, L. F. Gladden **1996**, *J. Colloid Interface Sci.* 181, 79–82.
24 A. J. Sederman, M. L. Johns, A. S. Bramley, P. Alexander, L. F. Gladden **1997**, *Chem. Eng. Sci.* 52, 2239–2250.
25 A. J. Sederman, M. L. Johns, P. Alexander, L. F. Gladden **1998**, *Chem. Eng. Sci.* 53, 2117–2128.
26 M. L. Johns, A. J. Sederman, A. S. Bramley, P. Alexander, L. F. Gladden **2000**, *AIChE J.* 46, 2151–2161.
27 A. J. Sederman, M. D. Mantle, C. Buckley, L. F. Gladden **2004**, *J. Magn. Reson.* 166, 182–189.
28 P. Mansfield **1977**, *J. Phys. C* 10, L55–L58.
29 J. Hennig, A. Nauerth, H. Friedburg **1986**, *Magn. Reson. Med.* 3, 823–833.
30 A. Haase, J. Frahm, D. Matthaei, W. Hanicke, K. D. Merboldt **1986**, *J. Magn. Reson.* 67, 258–266.
31 M. D. Mantle, A. J. Sederman **2003**, *Prog. Nucl. Magn. Reson. Spectrosc.* 43, 3–60.
32 L. F.Gladden, M. D. Mantle, A. J. Sederman **2004**, *Adv. Chem. Eng.* in the press.
33 L. F.Gladden, M. D. Mantle, A. J. Sederman **2004**, *Adv. Catal.* in the press.
34 M. C. Sains, M. S. El-Bachir, A. J. Sederman, L. F. Gladden **2005**, *Magn. Reson. Imag.* 23, 391–393.
35 A. J. Sederman, P. Alexander, L. F. Gladden **2001**, *Powder Technol.* 117, 255–269.

36 N. J. Mariani, O. M. Martínez, G. F. Barreto **2001**, *Chem. Eng. Sci.* 56, 5995–6001.
37 S. T. Sie, M. M. G. Senden, M. H. H. van Wechem **1991**, *Catal. Today* 8, 371–394.
38 B. Manz, L. F. Gladden, P. B. Warren **1999**, *AIChE J.* 45, 1845–1854.
39 P. R. Gunjal, V. V. Ranade, R. V. Chaudhari **2005**, *AIChE J.* 51, 365–378.
40 A. J. Sederman, L. F. Gladden **2001**, *Chem. Eng. Sci.* 56, 2615–2627.
41 L. F. Gladden, M. H. M. Lim, M. D. Mantle, A. J. Sederman, E. H. Stitt **2003**, *Catal. Today* 79, 203–210.
42 M. Crine, P. Marchot, B. Lekhlif, G. L'Homme **1992**, *Chem. Eng. Sci.* 47, 2263–2268.
43 M. H. M. Lim, C. L. Clayton, A. J. Sederman, L. F. Gladden, K. Griffin, P. Johnston, E. H. Stitt *Ind. Eng. Chem. Res.* to be submitted.
44 A. J. Sederman, L. F. Gladden **2005**, *AIChE J.* 51, 615–621.
45 M. H. M. Lim, A. J. Sederman, L. F. Gladden, E. H. Stitt **2005**, *Chem. Eng. Sci.* 59, 5403–5410.
46 L. D. Anadon, M. H. M. Lim, A. J. Sederman, L. F. Gladden **2005**, *Magn. Reson. Imag.* 23, 291–294.
47 L. F. Gladden, L. D. Anadon, M. H. M. Lim, A. J. Sederman, E. H. Stitt **2005**, *Ind. Eng. Chem. Res.* 44, 6320–6331.
48 L. D. Anadon, A. J. Sederman, L. F. Gladden **2005**, *AIChE J.* in the press.

5.3
Hyperpolarized ^{129}Xe NMR Spectroscopy, MRI and Dynamic NMR Microscopy for the *In Situ* Monitoring of Gas Dynamics in Opaque Media Including Combustion Processes

Galina E. Pavlovskaya and Thomas Meersmann

5.3.1
Introduction

The transport of fluid phases through porous stationary solid materials gives rise to important applications and phenomena in chemistry, chemical engineering, biology and geology. In many cases these processes happen within opaque media and are difficult, if not impossible to access by measurements in the UV-, visible- or infrared regimes of the electromagnetic spectrum. NMR and MRI operate through radiofrequency and therefore offer unique means to study non-transparent media, provided that the materials are not strong electrical conductors (which may shield the radiofrequency radiation) or ferromagnetic solids (which would severely interfere with the applied external magnetic field).

Gas phase dynamics in porous media is of crucial interest for many chemical engineering applications. For instance, the design and development of catalytic converters severely benefits from a detailed understanding of gas dynamics inside the heterogeneous catalytic reactors. Sensitivity problems caused by the low density of the gas medium and fast decoherence through rapid relaxation in the gas phase make the study of gas dynamics by NMR and MRI somewhat challenging. *In situ* NMR and MRI experiments of combustion processes are particularly difficult due to the further reduced densities of hot gas phases and because of unfavorable Boltzmann distributions of spin states at high temperatures. Typically, this leads to low spin polarizations and thus to low signal intensities.

Optical pumping of ^{129}Xe can be considered as a process that cools down nuclear spins to extremely low spin temperatures (i.e., the Boltzmann distribution that would be found at very low temperatures, typically around 0.1 K). Therefore spin-polarization, longitudinal magnetization and signal intensities are increased by approximately four to five orders of magnitude above the thermal equilibrium found at room temperature and at field strengths typically used for MRI and NMR. The exceptionally low spin temperature is in a thermal non-equilibrium state that can be maintained for sufficiently long periods of time, not only with the room temperature environment but also with gas phase media at a temperature above 1000 K. This is possible because gas phase xenon does not experience spin–rotation interaction, which is a powerful relaxation mechanism for small molecules in a gas phase [1]. Therefore, long ^{129}Xe relaxation observed in the gas phase allows for these technically demanding experiments.

Long relaxation times are also favorable for any type of NMR/MRI experiments that are based on evolution during a time delay, for instance in exchange spectroscopy, diffusion measurements and contrast MRI. Because of the difficulties of conventional NMR with the conditions present in high temperature gas phases, it is not surprising that the first *in situ* NMR of combustion reported in the literature utilizes hyper polarized (hp) ^{129}Xe as a probe [2, 3]. The chemical inertness of xenon and its temperature dependent chemical shift within porous media provide additional advantages for hp-^{129}Xe NMR as a probe for combustion and other high temperature reactions. In the following two sections, the basic concept of hp-^{129}Xe MRI and dynamic NMR microscopy are illustrated. Section 5.3.4 introduces the first advances of *in situ* NMR spectroscopy of combustion. The fifth and final section of this chapter briefly discusses the concept of high density optical pumping for the generation of high signal to hp-xenon concentration ratios that will be crucial for future chemical engineering studies.

5.3.2
Chemical Shift Selective Hp-^{129}Xe MRI and NMR Microscopy

Interaction of the non-reactive xenon atoms with the surface of host materials or surrounding solvent molecules leads to a distortion of the large xenon electron cloud reflected by a wide chemical shift range of approximately 300 ppm [see Figure 5.3.1(A)]. Owing to such a large chemical shift range, ^{129}Xe NMR is an ideal tool for collecting information about the cavities of porous materials and adsorbed guest molecules. A great deal of experimental and theoretical research on the xenon chemical shift in zeolites and other microporous materials has been carried out in the past [4–12]. The chemical shift of ^{129}Xe depends on the size and geometry of the cavities, detects the presence of strong adsorption sites and is sensitive to the loading of the host materials with xenon atoms and guest molecules. Therefore the xenon chemical shift is also influenced by all factors that affect the gas loading of the materials, such as the specific type of solid host material used, the external gas pressure and the temperature of the system. Figure 5.3.1(B) shows the 0.44 nm diameter xenon atoms (approximately to scale) within cavities of various zeolites. Xenon in these materials usually resonates at a frequency that is shifted many tens of ppm towards higher frequencies compared with the pure bulk gas phase. Note that the bulk gas phase at zero pressure is typically referenced with 0 ppm and that the chemical shift in the bulk gas phase increases about 0.5 ppm per 1 amagat xenon density increase (i.e., 0.5 ppm per 100 kPa pressure increase under conditions where the xenon can be considered to approximately follow ideal gas behavior).

Mesoporous materials also lead to a chemical shift that is distinct from the gas phase, as can be seen in the spectrum of xenon within an alumino-silicon aerogel shown in Figure 5.2.2(A). Although the quantitative correlation between xenon chemical shifts and void morphology is difficult for materials with less defined void spaces, such as silica, coals, aerogels, the chemical shift can still be used to distinguish xenon in the surrounding gas phase from xenon occluded in the porous medium. This effect can be exploited for chemical shift selective excitation [13, 14] for imaging and diffusion measurements. Chemical shift selective imaging using the 25 ppm signal of aerogel fragments with laser spin polarized ^{129}Xe is depicted in Figure 5.3.2(C–E), whereas the chemical shift selective image of the surrounding bulk gas phase at 0.3 ppm (a 50 kPa Xe and 50 kPa N_2 mixture) of the same sample is displayed in Figure 5.3.2(B). Chemical shift selective MRI that can distinguish between various adsorption phases of a gas is distinctive for xenon.

Hyperpolarized ^{129}Xe not only allows for sufficient signal intensity for chemical shift selective gas phase MRI, it also provides the means for a unique type of contrast. The imaging contrast derived from the transport of hyperpolarized gases into the material can be utilized to obtain "snapshots" of the gas flow and diffusion into porous samples. In this context, it is important to appreciate that Figures

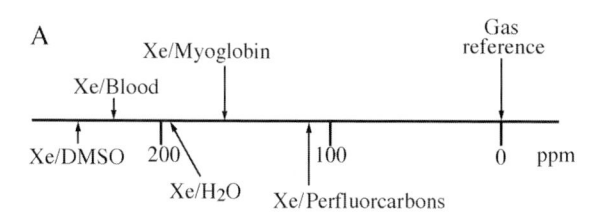

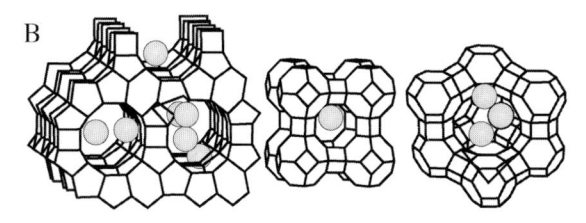

Fig. 5.3.1 (A) The xenon chemical shift is affected by the interaction of xenon atoms with surfaces and solvent molecules. (B) Various zeolites, namely H-beta, NaA and Faujasite are depicted to scale with the 0.44 nm diameter xenon atoms. A wide range of xenon chemical shifts are observed in these materials.

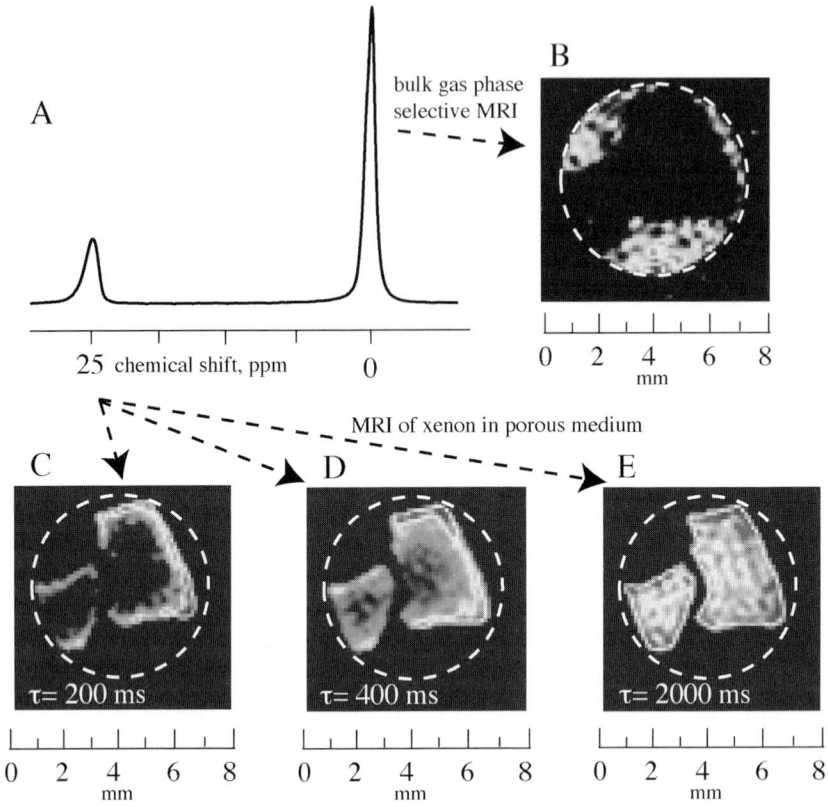

Fig. 5.3.2 (A) NMR spectrum of hyperpolarized ^{129}Xe from a sample that contains bulk gas phase (0.3 ppm) and xenon occluded within aerogel fragments (25 ppm). The gas mixture used for the experiment contained 100 kPa of N_2 and 300 kPa of xenon gas with a natural abundance of approximately 25% of the ^{129}Xe isotope. (B) 2D slice of 3D chemical shift selective MRI of the bulk gas phase. (C–E) 2D slices of 3D chemical shift selective MRI of the 25 ppm region for various recycle times τ. Adapted from Ref. [14].

5.3.2(C–E) do not show the aerogel material directly, instead the penetration of xenon into the aerogel fragments is visualized. This allows for the unique flow and diffusion weighted contrast that can be obtained as a function of the "transport time" of the gas into the sample. A standard chemical shift selective three-dimensional imaging sequence [15] using a low power Gaussian-shaped 90° rf pulse of 400-μs duration and triple axis pulsed field gradients (PFG) with maximum strength of 50 G cm^{-1} were applied. The images in Figure 5.3.2 each show a two-dimensional slice from full three-dimensional data sets with a slice thickness of 100 μm and an in-plane resolution of 250 μm × 250 μm. Note that the chemical shift selective excitation pulse leads to a contrast that distinguished xenon in the material from the bulk gas phase.

However, a second type of contrast that can probe for gas flow and diffusion is obtained from variations of the recycle delay time between the scans. In conventional imaging, the variation of the recycle delay can be used to generate a T_1 relaxation weighted contrast. In this instance the source of contrast is different for MRI with hyperpolarized xenon, as the longitudinal magnetization is not caused or restored by relaxation processes. Instead, the tremendous hp-xenon spin polarization is externally generated through laser pumping of rubidium vapor and rubidium–xenon spin exchange outside the superconducting magnet. The hp-xenon is then transported into the sample region in a continuous flow mode. Therefore, longitudinal magnetization that has been "used up" (i.e., xenon that has been depolarized) through excitation by an rf pulse will only be available again for the following scan if the recycle delay is long enough to allow for sufficient exchange with newly hyperpolarized xenon. The limiting factors for this exchange are the xenon gas flow and diffusion into porous samples. For short delay times between the scans, the images show xenon signals only at the outer region of the aerogel l fragments [see Figure 5.3.2(C) with a recycle time of $\tau = 0.2$ s]. The extent of xenon penetration into the material during the recycle delay τ depends on the diffusion coefficient and can be estimated through the temperature and pressure dependent diffusion constant D, using $\langle r^2 \rangle^{1/2} = \sqrt{6D\tau}$ with $\langle r^2 \rangle$ as the mean square displacement. For the used gas mixture, the xenon diffusion constant within the aerogel was determined by a separate pulsed field gradient experiment to be $D = 0.35$ mm^2 s^{-1}. The mean displacement would therefore amount to from 1.1 mm after 0.2 s, to 1.6 mm after 0.4 s and 3.6 mm after 2 s. Figure 5.3.2 indeed reflects this trend and the material fragments appear to be largely filled at $\tau = 2$ s.

Inspection of Figure 5.3.2(C–D) readily reveals that the penetration into the aerogel fragments is strongly anisotropic as it does not take place at all surface elements. At a first glance this may appear surprising as aerogels are very homogenous materials with a constant diffusion constant throughout the material. However, the contrast generated in continuous flow hp-xenon MRI is caused by the overall transport of the gas into the sample, including the gas flow. The anisotropic penetration of the hp-xenon is caused by the obstruction of flow between the individual aerogel fragments. It can be clearly seen that the saturation of the larger fragment takes place by diffusion through the surface areas that are not in the vicinity of the second, smaller fragment. Transport of hp-xenon into the smaller aerogel fragment is also inhibited at the surface close to the NMR container wall (dashed line). Furthermore, the flow may also influence the effective diffusion leading to spatially dependent effective diffusion constants.

Despite the time resolution of a fraction of a second, the overall recording time of these 3D images can take several hours (about 6 h if the long recycle delay time of $\tau = 2$ s is chosen and 10 times less for $\tau = 0.2$ s). Although the system may be in a non-equilibrium state, it is important that the studied system is at a steady state at least within the overall experimental time, otherwise changes in gas composition, strong temperature instabilities or alterations in the sample may, at best, produce observed quantities in an averaged fashion. However, the experimental time can significantly be reduced if one restricts the acquisition to two-dimensional imaging with slice selective pulses leading to a 32-fold reduction of the experimental time over the images that are reported here, and if the optical pumping is improved beyond the < 1% xenon spin polarization used in this example [16]. Depending on the desired spatial and temporal MRI resolution, overall experimental timescales of the order of 100 ms should be feasible if techniques such as Echo Planar Imaging (EPI) or Fast Low Angle SHot imaging (FLASH) are employed.

Longitudinal (T_1) relaxation can also contribute to a contrast in continuous flow hp-xenon MRI. It may lead to important additional contrast but competes with the flow and diffusion weighted contrast. Relaxation in hp-xenon MRI is destructive, because it leads to a "warming" of the spins towards a thermal equilibrium with the environment and eventually to a reduction of longitudinal magnetization and signal intensity by several orders of magnitude. The resulting equilibrium magnetization is typically too small to be detected in gas phase ^{129}Xe MRI. Figure 5.3.3 shows this effect in images of NaX zeolite pellets. A two-dimensional chemical shift selective MRI slice of occluded xenon, taken from a full three-dimensional data set, in a plane parallel to the flow is shown in Figure 5.3.3. A slice through one pellet perpendicular to the flow is depicted in Figure 5.3.3(C). Only pixels (in fact 3D voxels) of the outer ring of the pellets display the presence of hyperpolarized xenon. This is not caused by diffusion (and flow) alone, because an increase in the

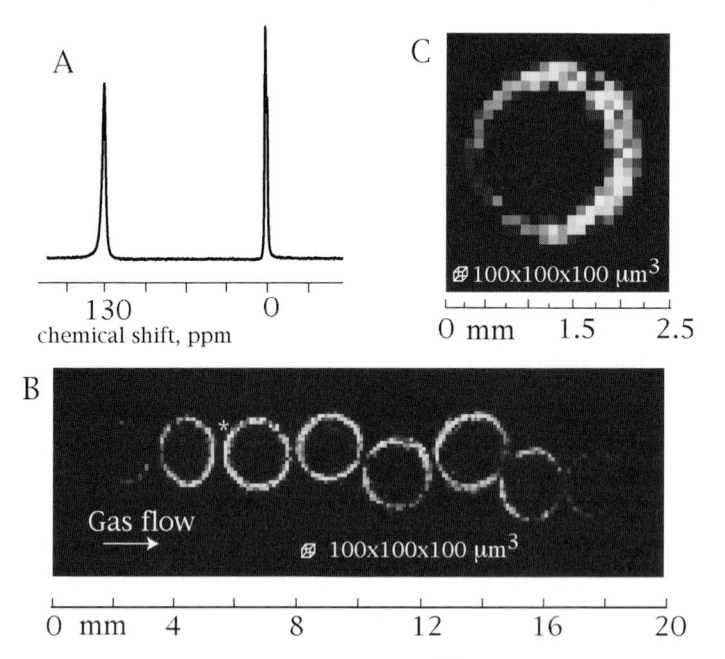

Fig. 5.3.3 (A) NMR spectrum of hyperpolarized ^{129}Xe in NaX zeolites. (B) 2D slice in the flow direction of a 3D chemical shift selective MRI of gas in the zeolite pellets. (C) 2D slice perpendicular to the flow direction of the same 3D chemical shift selective MRI as in (A). Adapted from Ref. [14].

recycle delay time does not lead to an apparent increase in the xenon penetration into the pellets beyond what is shown in Figure 5.3.3. This effect can be explained by the presence of paramagnetic impurities, most likely due to oxygen that has intentionally not been removed completely by more rigorous degassing of the pellets prior to the experiments. The maximum observable penetration depth is dictated by the relaxation time in this sample and the mean square displacement into the material caused by diffusion. The xenon will penetrate further into the material than is apparent from the images. However, it will have lost its polarization and is unobservable. Unlike the experiment shown in Figure 5.3.3, hp-xenon MRI may not be applicable as a diagnostic technique at all if a high concentration of paramagnetic impurities leads to relaxation times that are short compared with the timescale of the gas transport.

The resolution of the zeolite MR image is $100 \times 100 \times 100$ µm^3 and has therefore reached the resolution limit that defines NMR microscopy. For the instrumentation used for this experiment, it will take at least a few milliseconds due to the ramping time of the field gradients. If the mean displacement of the xenon atoms during this experimental time scale reaches the dimension of the voxels or pixels, the resolution limit is reached. For instance, for the aerogel experiments in Figure 5.3.2 this resolution limit would be around 100 µm. More improved technology is now available that can limit the effective switching times to <100 µs. However, the signal sensitivity at the currently obtained spin polarization does not allow for much higher resolution, as the overall experimental time is a couple of hours, unless the polarization of the xenon gas is substantially improved or the experiment is scaled down to a slice selective two-dimensional experiment.

5.3.3
Dynamic NMR Microscopy of Gas Phase

PFG NMR is a well established technique that can be used to monitor both flow and diffusion for a wide range of temporal and spatial scales [17]. Dynamic NMR microscopy is an extension of the PFG method, where displacement encoding of material transport can be combined with spatial encoding. Subsequent reconstruction leads to velocity and diffusion maps [15, 18–20]. Poiseuille fluid flow in a pipe can be conveniently depicted by dynamic NMR microscopy in 2D images where one axis is assigned to a spatial coordinate perpendicular to the pipe symmetry axis and the second coordinate is assigned to velocity (or displacement) along the symmetry axis. Figure 5.3.4(A) shows a simulated echo version of the dynamic NMR microscopy with flow and spatial encoding. Velocity profiles are simply constructed by dividing displacements by the observation time Δ. The simulations depicted in Figure 5.3.4(B) are performed for an ensemble of $N_{spins} = 10^4$, with corrections for spins that flow out of the pipe during the observation time. The data processing for simulations and experimental data shown in Figure 5.3.4(C) were identical [21].

In a Poiseuille flow of a Newtonian liquid the coherent motion gives rise to the dominant effect, rendering contributions from the stochastic dispersion term

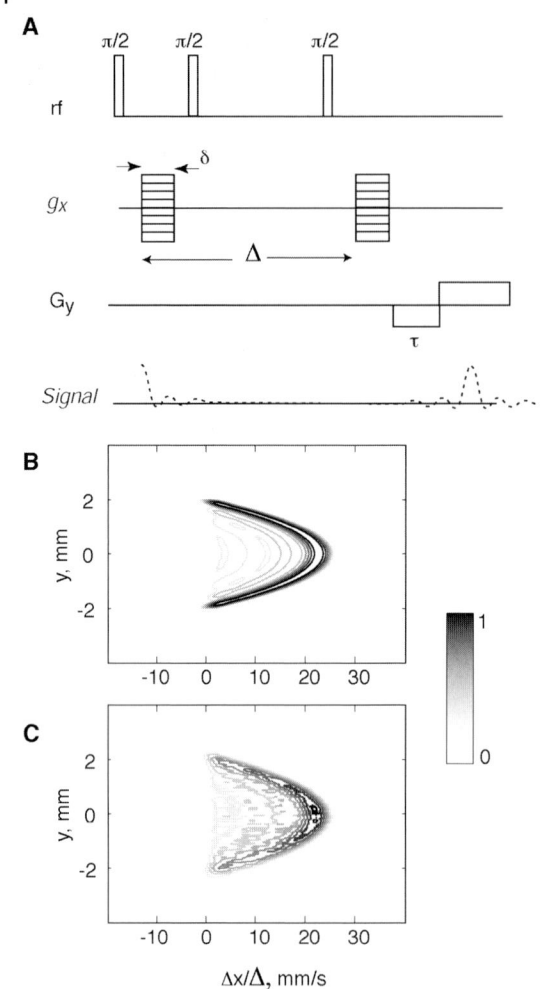

Fig. 5.3.4 (A) Stimulated echo dynamic NMR microscopy pulse sequence. The first field gradient pulse (g_x) of duration δ serves to encode spatial positions of spins and the second field gradient pulse has a refocusing effect. A second imaging gradient (G_y) is added in order to obtain a spatial map of the displacements. A notable feature of the stimulated echo protocol is that during the flow encoding time, Δ, the magnetization is stored along the z axis and is subject to the longitudinal relaxation only. Bulk xenon gas phase has a sufficiently long T_1, therefore observation of displacements is limited only by spins flowing out of the detection region. (B) Computer simulation of the joint spatial–velocity profile for Poiseuille fluid flow in a pipe (id = 4 mm, $D_{fluid} = 2.2 \times 10^{-3}$ mm^2 s^{-1}, $V_{ave} = 11.5$ mm s^{-1}, $\Delta = 20 \times 10^{-3}$ s). (C) Experimental data with the same parameters as in (B). Adapted from Ref. [21].

negligible. However, the situation changes for gas flow where stochastic terms may be of the same order of magnitude as coherent terms, leading to a strong interplay between coherent flow and Brownian motion. The underlying problem was studied in detail by Taylor [22] and Aris [23], for the case of a fluid flowing through a cylindrical pipe, where viscosity shear imposes a parabolic distribution of molecular displacements, separating the entire ensemble into velocity streamlines with the smallest displacements at the walls of the pipe and the largest displacements in the center. The increase in effective diffusion that arises from molecules randomly sampling streamlines with different velocities is referred to as "Taylor dispersion". The complete expression for the effective diffusion, under laminar flow conditions, has been derived by Van den Broeck [24] for both short ($t < a^2/D$) time scales

$$D^* = D + \frac{1}{6}V_{ave}^2 t + \frac{4}{3}DV_{ave}^2 t^2/a^2 + \ldots \quad (5.3.1)$$

and long ($t > a^2/D$) time scales

$$D^* = D + \frac{V_{ave}^2 a^2}{48D} \quad (5.3.2)$$

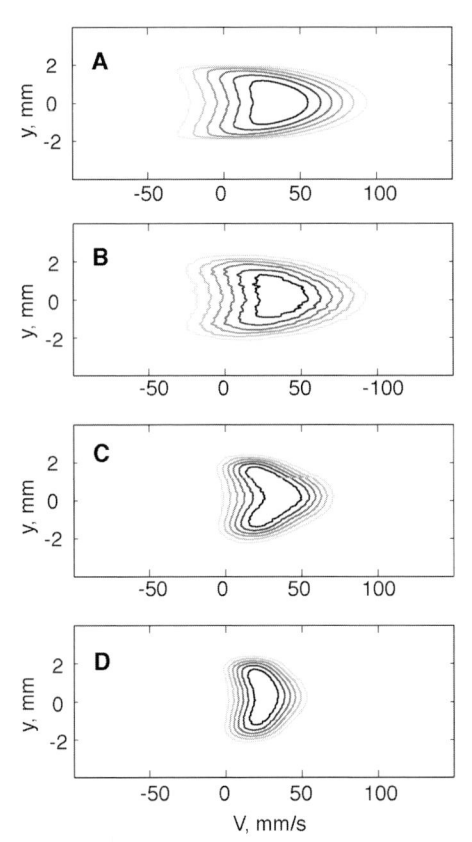

Fig. 5.3.5 Joint spatial–velocity images of xenon undergoing Poiseuille flow in a pipe (id = 4 mm, D_{Xe} = 4.5 mm^2 s^{-1}, V_{ave} = 20 mm s^{-1}). Data sets of 256 × 32 points were collected using the protocol depicted in Figure 5.3.4(A). Data were zero filled to 512 × 256 points in further data processing. The resulting spatial resolution after 2D Fourier transformation was 100 μm. The spectral window in the flow-encoding dimension was kept constant. (A) Computer simulation with Δ = 10 ms. (B) Experiment with Δ = 10 ms. (C) Experiment with Δ = 60 ms. (D) Experiment with Δ = 130 ms. Adapted from Ref. [21].

where D^* is the effective diffusion coefficient (or dispersion coefficient in this case), D is the self-diffusion coefficient, V_{ave} is the average fluid velocity, a is a pipe radius and t is the time used to measure longitudinal displacement. Equation (5.3.2) represents an asymptotic regime where the dispersion coefficient becomes time-independent due to boundary limits.

Figure 5.3.5 displays dynamic NMR microscopy of xenon gas phase Poiseuille flow with an average velocity of 25 mm s^{-1} and self-diffusion coefficient of 4.5 mm^2 s^{-1} at 130 kPa xenon gas pressure with numerical simulation (A) and experimental flow profiles (B–D) of xenon gas.

The maximum velocity of a liquid flow can be quickly estimated from the location of the parabola top in the displacement image; however, the gas phase profiles reflect large displacements due to diffusion [compare Figures 5.3.4(B) and 5.3.5(A)]. In order to avoid a gross overestimation of the velocity, it is therefore necessary to extract the velocity values from the maximum of the Gaussian curves resulting from the Fourier transformation in the flow dimension [15]. Velocity profile images of xenon for longer time scales are displayed in Figure 5.3.5(C and D). A prominent feature of Figure 5.3.5 is the temporal progression of molecular displacements. At short Δ time scales the large effect of diffusion dominates the displacement of flowing gas particles especially along the flow direction and results in a broad distribution of apparent velocities. At longer Δ time scales general diffusion averages the radial dependence of the velocity by allowing molecules to cross streamlines so frequently that an average velocity is observed instead of a parabolic profile, with the highest velocity in the center and a zero velocity at the walls. The time scale is long enough for translational diffusion to cause the particles to sample the entire ensemble of velocity streamlines, thus flattening the parabolic distribution.

However, to reach the asymptotic limit described by Eq. (5.3.2) one needs to measure velocity fields for a substantially prolonged period of time. Using $D_{Xe} = 4.5$ mm^2 s^{-1} and $a = 2$ mm one estimates that the time independence of the effective diffusion coefficient is reached only after the observation time Δ is sufficiently longer than 0.88 s. The asymptotic value of the effective dispersion coefficient in this limit is estimated to be 11.9 mm^2 s^{-1}. It is remarkable that even in gas flow, fairly longer observation times are required to achieve this behavior. However, decreasing the gas pressure helps to shorten the observation time limit. For example, at 0.7 atm gas pressure, with the self-diffusion coefficient $D_{Xe} = 8$ mm^2 s^{-1} the asymptotic behavior should be reached at times larger than 0.5 s. Another effect that is important for PFG measurements in the gas phase is the "edge enhancement effect" that can simply be generated by defocusing and refocusing gradients either with long durations or a long separating time delay between the two gradients [25]. If the time scale is long enough, the gas diffusion leads to displacement that reduce the refocusing significantly, thus leading to low or disappearing signals. However, the refocusing is more successful at barriers, such as walls and edges where the gas diffusion is naturally restricted. This is shown in Figure 5.3.6 where the same pulse sequence as for Figures 5.3.4 and 5.3.5 has been applied, but with prolonged time durations for the spatial encoding gradient along the y direction. Such an edge enhancement "filter" may be inserted

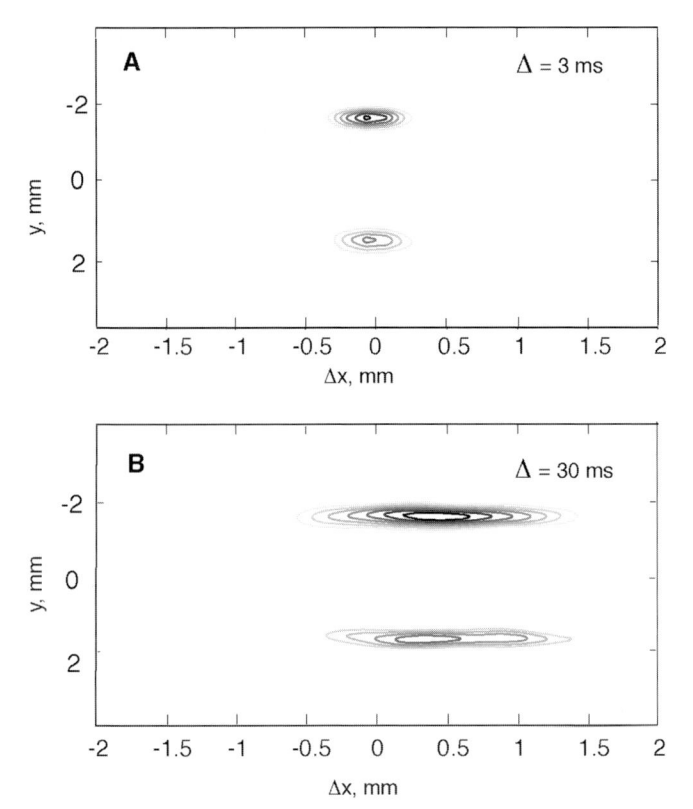

Fig. 5.3.6 Joint spatial–velocity images of xenon undergoing Poiseuille flow in a pipe (id = 4 mm, V_{ave} = 27 mm s^{-1}, D = 8 mm^2 s^{-1}) at 0.7 atm recorded with a protocol shown in Figure 5.3.4(A). Only particles at walls are selected by the edge enhancement "filter". A modified imaging gradient time duration (τ = 5 ms) was used. Data were collected into 512 × 16 matrices and then zero-filled to 1024 × 32 before 2D Fourier transform. The resulting spatial resolution was 100 µm. (A) Δ = 3 ms; (B) Δ = 30 ms. Adapted from Ref. [21].

in any period of the pulse sequence. When the "filter" occurs at the beginning of the pulse sequence, it might be possible to monitor the exchange between the slow and fast components, for example the appearance of spins at the center of a pipe for longer Δ. The same method could be applied in gas phase dynamic NMR studies of any system, for example porous media, with restricted diffusion in order to observe dynamic interchange between the boundaries and the bulk.

5.3.4
In Situ **NMR of Combustion**

The previous two sections described the development of hp-xenon techniques for the investigation of gas dynamics in porous or opaque media. The usefulness of continuous flow hp-xenon as a tool for dynamic NMR imaging experiments was demonstrated on model systems. The next step in the development of hp-xenon techniques for chemical engineering studies is the application of hp-xenon as a spy probe in reaction mixtures. The noble gas remains non-reactive even under rather extreme conditions, thus allowing for the usage of hp-xenon for *in situ* monitoring of dynamic processes in reactors. This section describes, in particular, the application of hp-xenon NMR for the examination of combustion processes. *In situ* NMR measurements will be of great interest for combustion processes that are located within opaque media. Examples are reaction zones of smoldering processes or reactions within solid materials, for instance in catalytic combustors and reactors. Previously, ultrasonic tomographic imaging has been employed for studies of smoldering processes in optically non-transparent media [26]. This technique is limited to the solid phase and can be used to study the propagation of the smoldering front. However, the dynamic options of hp-^{129}Xe NMR and MRI will enable the probing of gas transport, flow velocities, diffusion phenomena during gas phase reactions in reactors or porous media that cannot be studied by ultrasonic tomography or any other current technology.

The reported *in situ* NMR of combustion [2] served largely as proof of concept work. It was demonstrated that despite the presence of paramagnetic oxygen and radicals, the xenon relaxation times are sufficiently long for gas exchange studies.

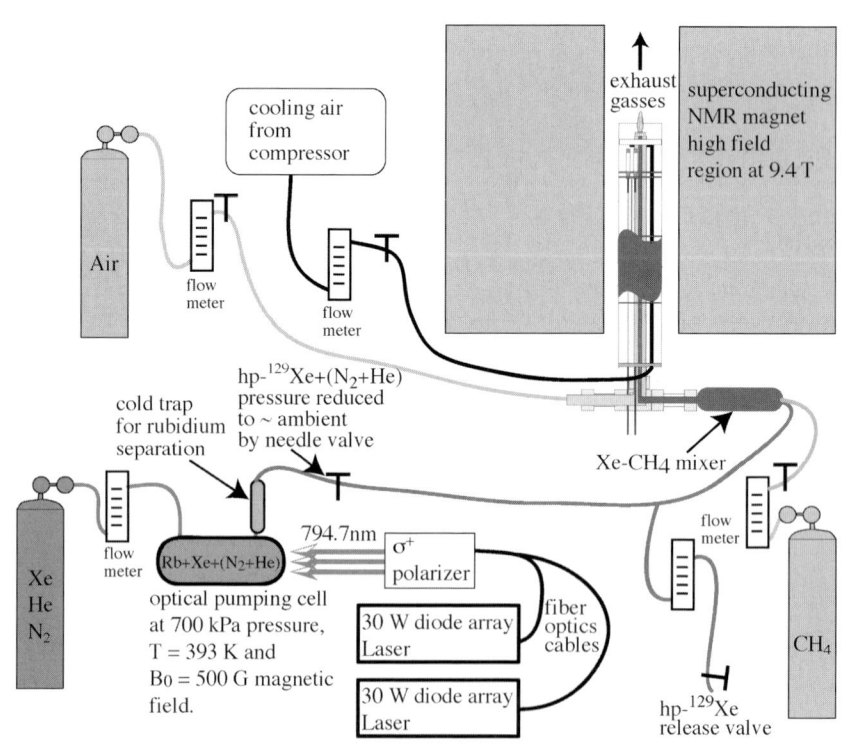

Fig. 5.3.7 Experimental set-up of the *in situ* combustion hp-Xe^{129} NMR experiment. Adapted from Ref. [2].

The observed scale of the gas dynamics was sufficiently slow to provide hope for chemical shift selective spatial imaging in future work. The experimental set-up is depicted in Figure 5.3.7. Hp-^{129}Xe was produced in the optical pumping cell by irradiation of a rubidium–xenon gas mixture with 60 W of continuous wave circular polarized light at 794.7 nm. The xenon gas (together with He and N_2 in some experiments, see Section 5.3.5) was continuously supplied to the pump cell and hp-xenon was continuously exhausted through a cold trap that removed the gas phase rubidium. The hyperpolarized xenon was then mixed with methane and air was added as an oxidant shortly before the combustion zone within the superconducting magnet. The NMR probe head was cooled by a pressurized air stream and the NMR bore was kept at or below 40 °C at all times.

Figure 5.3.8 displays a photograph of the actual combustion zone. The methane–xenon gas was mixed with air before entering the NMR detection cell in the probe head. The gas mixture was blown through an area with molecular sieve pellets (i.e., NaX, the same type that was used for the imaging experiments in Figure 5.3.3) and ignited above the pellets. The probe was subsequently inserted into the magnet. One of the resulting ^{129}Xe NMR spectra is shown in Figure 5.3.8 (solid line 2) in comparison with the spectrum of the same initial mixture without combustion (dashed line 1). Referenced with 0 ppm is the gas phase peak at room temperature. A close up of the 0 ppm region (not depicted) reveals that the gas phase peak was shifted by about 29 Hz up-field (–0.26 ppm) after ignition. A similar shift was also observed (not depicted) in an experiment using only pure gas phase without the pellets. However, a second peak at –3 ppm is only observed during combustion in the presence of the zeolite. It was demonstrated by 2D-EXSY data (see Figure 5.3.9) that the 350 Hz broad signal at –3 ppm originates from a region just above the bulk of the zeolite. This effect will need further investigation as neither experimental nor theoretical xenon chemical shift data are presently available for temperatures above 1000 K.

The NMR peak originating from xenon within the zeolites was strongly affected by temperature. At room temperature, the xenon inside the nanoporous zeolite pellets leads to a signal at approximately 82 ppm. This signal was shifted to about 68 ppm once the mixture was ignited. The shift was caused by a temperature increase within the material in the pre-combustion region through thermal conductivity and IR radiation. The change in the chemical shift was most likely due to the reduced xenon loading of the porous material caused by the temperature increase. The spectrum shown in Figure 5.3.8 used a particular pumping technique with pure xenon for the pumping process [16], which is explained in more detail in Section 5.3.5 as "high density optical pumping". The temperature dependence of the shift (and broadening) was less dramatic when a more conventional mixture of 4% xenon in helium (and nitrogen) was used. The dilute gas mixture leads to a lower xenon loading of the material and the lower xenon density within the porous system may be the cause of the less pronounced temperature dependence of the chemical shift. High density optical pumping also leads to improved signal to (inert gas) concentration ratios in the gas mixture and thus makes the technique less invasive. A very high percentage of inert gas was used (see Section

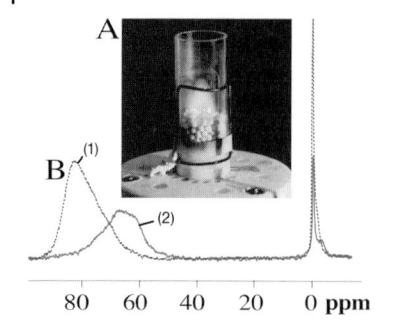

Fig. 5.3.8 Photograph of the detection region of the NMR probe with radiofrequency coil. A methane–air mixture was ignited above the zeolite pellets. The mixture also contained xenon for NMR detection. Hp-^{129}Xe NMR spectra with 30% xenon (from high-density xenon optical pumping) in 70% methane is depicted. (1) The spectrum in the absence of combustion and (2) the spectrum during combustion. Adapted from Ref. [2].

5.3.5), however reasonable signal intensities with a 3–5% inert gas mixture in methane are feasible with improved technology.

2D hp-^{129}Xe-NMR exchange spectroscopy (EXSY) [27] was applied to study the xenon transfer between the nanoporous material and the combustion zone. The peaks in EXSY experiments are a function of two correlated chemical shift values. The horizontal axes in Figure 5.3.9 relate to the chemical shifts of the xenon at the beginning of the experiment ($t = 0$), while the vertical axes describe the chemical shifts after a time period $t = \tau_{ex}$ has passed. The signals will appear on the diagonal when the chemical shift values on the horizontal and vertical axes are identical. No cross peaks appear in Figure 5.3.9(a) and it can be concluded that no exchange between the various regions occurred during short exchange times $\tau_{ex} = 5$ ms. The corresponding 1D spectrum is displayed above the EXSY for clarity (a 4% xenon in helium mixture was used for optical pumping). The fact that no appreciable exchange appears to happen on a timescale of 5 ms is very encouraging for future *in situ* MRI of combustion as many pulsed field gradient experiments are typically limited to some minimum time scale of a few milliseconds because of gradient switching times.

The appearance of cross peaks between different chemical shift values in the 2D spectra for longer τ_{ex} times are caused by gas transport between the different regions. The location and intensity of the cross peaks indicate the exchange direction and the amount of material that is transported during the delay time. In a typical EXSY, the cross peaks will appear symmetrically on both sides of the diagonal. This is caused by chemical exchange within a system that is in a chemical equilibrium state where the forward and backward reaction rates are identical. As combustion is a non-equilibrium process, the cross peaks in the EXSY spectra show a non-symmetrical appearance. The recording of 2D spectra over a 40-min period of a reacting system in a non-equilibrium state is possible because of steady-state conditions and MRI will be possible for the same reason. If a reaction is not in a steady state, MRI may be still possible as echo planar imaging EPI can be applied if sufficient signal intensity is available. EPI takes imaging "snapshots" on a 30–70-ms timeframe. Other techniques utilizing small flip angle excitation may also provide sufficient time resolution if the signal sensitivity permits these types of experiments.

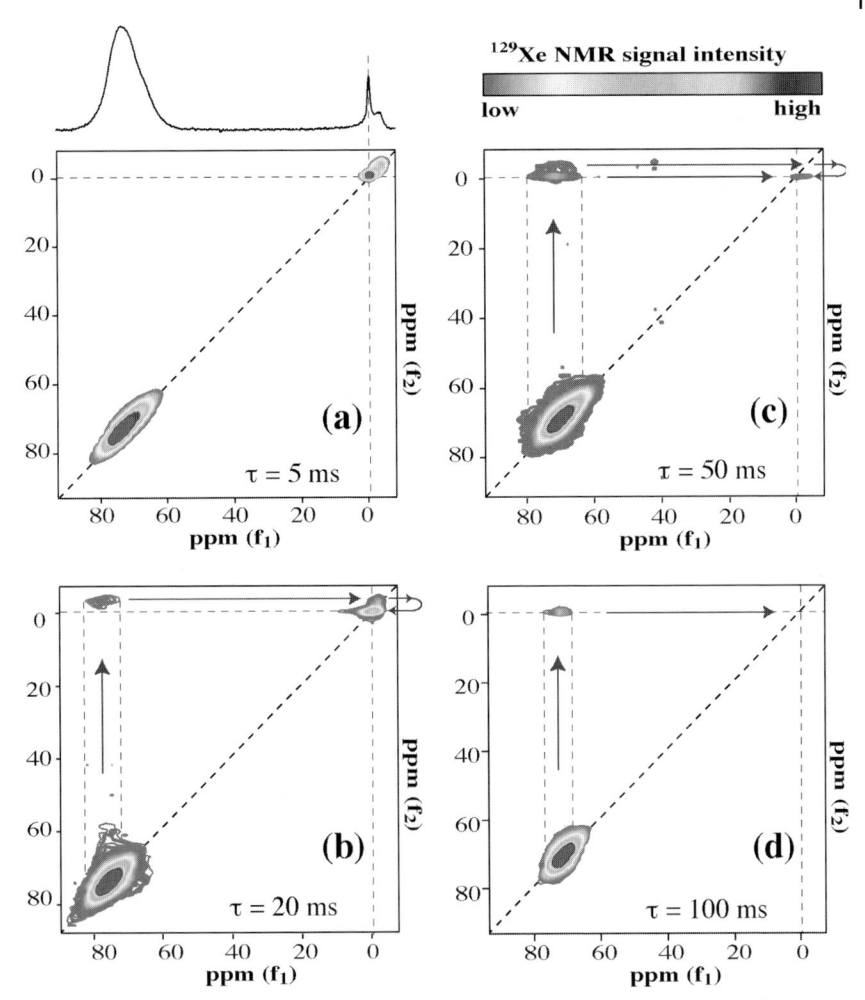

Fig. 5.3.9 Hp-^{129}Xe 2D EXSY NMR spectra recorded during combustion for various exchange times. Adapted from Ref. [2]. The spectra were obtained at 110.683 MHz with 32 × 128 (raw) data points and hypercomplex frequency discrimination (states) with 64 scans per spectrum. The average experimental time per EXSY with 0.5-s recycle delay was 40–50 min. See text for a more detailed discussion.

At τ_{ex} = 20 ms a cross peak appeared that points to exchange between xenon located in the nanoporous material and the −3 ppm gas phase. The EXSY proved that transfer of xenon from the zeolite to the bulk gas phase region (−0.26 ppm) did not take place during τ_{ex} = 20 ms. However, Figure 5.3.9(b) also shows a (barely resolved) cross peak that demonstrated exchange from the −3 ppm to the −0.26 ppm region. Owing to the gas flow direction, it was concluded that the −3 ppm zone was located between the porous material and the bulk of the flame region further above. For longer τ_{ex} times [see Figure 5.3.9(c and d)] an exchange cross peak directly between the material phase and the gas phase at −0.26 ppm appears. At the same time the diagonal gas phase peaks at −3 ppm and −0.26 ppm vanish because the gas flow has moved the xenon atoms out of the detection region. Spins that are detected at $t = 0$ but are no longer observable at $t = \tau_{ex}$ appear neither as diagonal nor as cross peaks.

For the first time the feasibility of *in situ* NMR of combustion has been demonstrated. This is an important step in the development of hp-xenon MRI as a methodology that can monitor dynamics within opaque reactors. Extensive experiments and calculations in the recent past have demonstrated the value of ^{129}Xe-NMR spectroscopy for investigating intra-crystallite diffusion of hydrocarbons in porous catalytic material [28–30]. In particular the combination of ^{129}Xe-NMR spectroscopy with ^{1}H magnetic resonance imaging (MRI) has proven to be very insightful [31, 32]. These studies have typically been performed in a time resolved fashion at low temperatures. Much shorter time scales can be expected for high temperature experiments but similar experiments should still be possible, either because of the use of hp-xenon, which requires less signal averaging, or because of steady-state reaction conditions. Note that a particular spatial coordinate in a steady-state system will refer to a particular point in time of the ongoing reaction. The scale of the gas dynamics observed in this communication is sufficiently slow to allow for future chemical-shift-selective spatial imaging.

Hp-^{129}Xe-NMR combined with ^{129}Xe MRI and localized ^{1}H NMR may be a very promising tool for *in situ* studies of high-temperature catalytic oxidation processes within opaque media, such as platinum or palladium catalysts [33, 34] or nano-structured perovskites or hexaaluminates [35, 36]. Other examples are catalytic cracking of hydrocarbons in reactors that lack *in situ* monitoring of regions of extensive carburization (i.e., "hot zones"). Last, but not least the technique will be ideal to facilitate advances in micro-combustor technology. Environmentally less problematic and easily rechargeable with hydrocarbon fuels, catalytic micro-combustors combined with micro-turbines may become an alternative to rechargeable batteries for small devices that have a high power consumption [37]. Owing to the inherent small size of the combustors, standard wide bore (89 mm id) superconducting magnets at medium field strength (9.4 T was used for the *in situ* experiments) will be sufficient to monitor the dynamics in these micro-combustors by hp-^{129}Xe MRI.

5.3.5
High Xenon Density Optical Pumping

Thus far the objective has been to demonstrate the proof of principle for various applications of continuous flow hp-xenon MRI. In order to move beyond the proof of concept towards practical applications, this technology needs further development in particular with respect to reducing the xenon concentration within a reaction mixture. The major objective of high xenon density optical pumping is to increase the signal-to-concentration ratio of the tracer gas hp-xenon in the gas mixture. It is desired to obtain a strong signal from a low hp-xenon concentration in order to have a methodology that will use only small amounts of tracer xenon without obstructing the chemical reactions.

In this context it is important to note that the requirements for the production of hyperpolarized (hp) xenon for *in situ* NMR of high temperature reactions are very different from other applications, such as medical imaging of the respiratory system. For medical applications for instance, a "batch mode" production of hp-xenon is sufficient. In a "batch mode" optical pumping experiment the outcome of the pumping process is maximized using low xenon partial pressures, typically about 2% xenon in about 96% helium and 2% nitrogen. After hyperpolarization, the xenon is separated and concentrated from the other gasses via freezing and sublimation in a cold trap. Pure and highly polarized xenon is then mixed into the breathing air of a patient.

However, in experiments for which hp-xenon gas is directly applied onto the sample in a continuous flow mode over long periods of time, a concentration and separation process through freezing of the xenon is not practical. Instead, for *in situ* NMR of combustion an inexhaustible source of hp-xenon "on tap" is required. A "standard" optical pumping mixture, containing 4% xenon and 4% nitrogen in helium was used for the 2D *in situ* NMR of combustion presented in Section 5.3.4. However, this is problematic. For an analytical technique where the measurement will not influence the reaction to a significant degree, the concentration of the tracer gas mixture should be kept at a minimum. At most a few percent of the optical pumping gas mixture should be present in the combustion mixture. This dilution would reduce the hp-xenon concentration to 10^{-1}–10^{-2}% and therefore lead to very low signal intensities. In fact, the inert gas concentration for the 2D EXSY shown in Figure 5.3.9 was about 60% (helium, xenon, nitrogen) and therefore close to the limit where a free combustion of methane in air can be maintained. Improvement is possible however, as the signal intensity in a continuous flow experiment is proportional to the product of polarization $P(Xe)$ and spin density or number density $[^{129}Xe]$ of the isotope ^{129}Xe:

$$I_{HP-^{129}Xe} \propto P(Xe) \cdot [^{129}Xe] \tag{5.3.3}$$

This idea leads to high density optical pumping, which has been investigated in the recent past [16], where the optical pumping efficiency with pure xenon in the absence of helium was demonstrated. Figure 5.3.10 shows the polarization and the signal intensity as a function of the xenon partial pressure in a mixture where the overall pressure is kept constant by a balance of helium. Clearly, the highest polarization is achieved for very low xenon concentrations or partial pressures. However, when the signal intensity in the continuous flow mode that follows Eq. (5.3.3) is plotted as a function of the xenon partial pressure, a different picture emerges. Initially, the signal intensity increases strongly with increasing xenon partial pressure.

The signal intensity levels off at a xenon pressure of about 200 kPa. However, this is true only if the overall pressure is still balanced by helium to about 700 kPa. If only pure xenon is used (not depicted) the signal intensity levels off at about 500 kPa. The situation is again different when the xenon pressure is reduced after the optical pumping procedure to about ambient pressure in order to be added to

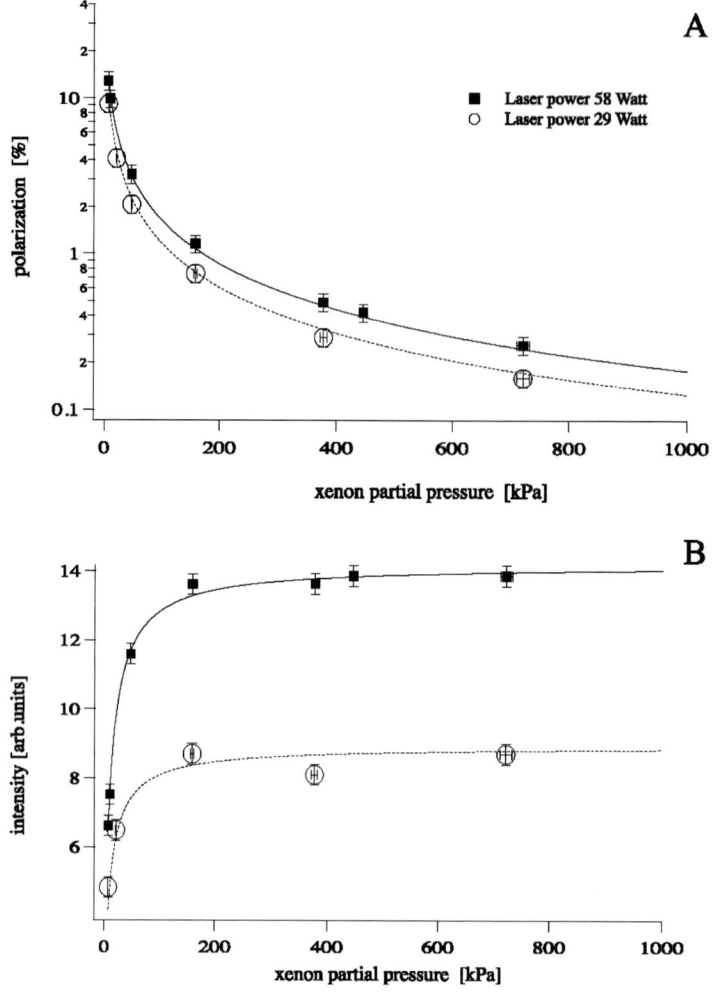

Fig. 5.3.10 (A) Polarization obtained continuous-flow hp-xenon experiment as a function of xenon partial pressure for two different laser powers, i.e., one 30-W diode array laser (triangles) and two combined 30-W diode array lasers (squares). (B) Actual intensities as measured by ^{129}Xe NMR spectroscopy at 110.69 MHz using 29-W laser power (triangles) and full laser power (squares). Adapted from Ref. [16].

the combustion mixture. In this case, it is advantageous to use xenon at a pressure only slightly higher than the ambient pressure for the pumping process. For the 1D spectrum in Figure 5.3.8 high-density optical pumping was used leading to a 30% inert gas (xenon) concentration. This is still far from the anticipated 2–5% but the pumping process had not been optimized. An increase in the xenon hyperpolarization by a factor of at least five is feasible with further improvement of the technique. Recent technological advances where very high xenon polarization with medium laser power was achieved are also promising for high density optical pumping [38]. Together with current and future developments in solid state laser array diode technology, which will make more laser power more affordable, further polarization enhancement will be possible.

The *in situ* monitoring of high temperature reactions by hp 129 Xe magnetic resonance is still in its infancy. Although the previous work on gas phase dynamics in porous media has shown the feasibility of dynamic microscopy and MRI and the first *in situ* combustion NMR spectra have been collected, much more development remains to be done. To date, hp 129 Xe NMR and MRI are currently the only techniques available to study gas dynamics in porous and opaque systems.

References

1. A. Abragam **1961**, The Principles of Nuclear Magnetism, Clarendon Press, Oxford.
2. S. Anala, G. E. Pavlovskaya, P. Pichumani, T. J. Dieken, M. D. Olsen, T. Meersmann **2003**, J. Am. Chem. Soc. 125, 13298–13302.
3. J. Reimer **2003**, Nature (London), 426, 508–509.
4. T. Ito, J. Fraissard **1982**, J. Chem. Phys. 76, 5225–5229.
5. J. A. Ripmeester **1982**, J. Am. Chem. Soc. 104, 289–290.
6. C. Dybowski, N. Bansal, T. M. Duncan **1991**, Annu. Rev. Phys. Chem. 42, 433–464.
7. P. J. Barrie, J. Klinowski **1992**, Prog. Nucl. Magn. Reson. Spectrosc. 24, 91–108.
8. C. J. Jameson, A. K. Jameson, R. E. Gerald, H. M. Lim **1997**, J. Phys. Chem. B 101, 8418–8437.
9. C. I. Ratcliffe **1998**, Annu. Rep. NMR Spectrosc. 36, 123–221.
10. J. L. Bonardet, J. Fraissard, A. Gedeon, M. A. Springuel-Huet **1999**, Catal. Rev. Sci. Eng. 41, 115–225.
11. M. A. Springuel-Huet, J. L. Bonardet, A. Gedeon, J. Fraissard **1999**, Magn. Reson. Chem. 37, S1–S13.
12. J. Fraissard, T. Ito **1988**, Zeolites, 8, 350–361.
13. I. L. Moudrakovski, S. Lang, C. I. Ratcliffe, B. Simard, G. Santyr, J. A. Ripmeester **2000**, J. Magn. Reson. 144, 372–377.
14. L. G. Kaiser, T. Meersmann, J. W. Logan, A. Pines **2000**, Proc. Natl. Acad. Sci. 97, 2414–2418.
15. P. T. Callaghan **1991**, Principles of Nuclear Magnetic Resonance Microscopy, 1st edn, Oxford University Press, Oxford.
16. M. G. Mortuza, S. Anala, G. E. Pavlovskaya, T. J. Dieken, T. Meersmann **2003**, J. Chem. Phys. 118, 1581–1584.
17. E. O. Stejskal, J. E. Tanner **1965**, J. Chem. Phys. 42, 288–292.
18. E. Fukushima **1999**, Annu. Rev. Fluid Mech. 31, 95–123.
19. J. M. Pope, S. Yao **1993**, Annu. Rev. Fluid Mech. 19, 157–182.
20. J. D. Seymour, P. T. Callaghan **1997**, AIChE J. 43, 2096–2111.
21. L. G. Kaiser, J. W. Logan, T. Meersmann, A. Pines **2001**, J. Magn. Reson. 149, 144–148.
22. G. I. Taylor **1953**, Proc. R. Soc. London A 219, 186–203.
23. R. Aris **1956**, Proc. R. Soc. London A 235, 67–77.
24. C. Van den Broeck **1982**, Physica A 112A, 343.
25. Y.-Q. Song, B. M. Goodson, B. Sheridan, T. M. d. Swiet, A. Pines **1998**, J. Chem. Phys. 108, 6233–6239.
26. S. D. Tse, R. A. Anthenien, A. C. Fernandez-Pello **1999**, An application on ultrasonic tomographic imaging to study smoldering combustion. Combustion and Flame 116 (1–2), 120–135.
27. J. Jeener, B. H. Meier, P. Bachmann, R. R. Ernst **1979**, J. Chem. Phys. 71, 4546–4553.
28. M. A. Springuel-Huet, A. Nosov, J. Karger, J. Fraissard **1996**, J. Phys. Chem. 100, 7200–7203.

29 F. D. Magalhaes, R. L. Laurence, W. C. Conner, M. A. SpringuelHuet, A. Nosov, J. Fraissard **1997**, J. Phys. Chem. B 101, 2277–2284.
30 P. N'Gokoli-Kekele, M. A. Springuel-Huet, J. Fraissard **2002**, Adsorpt. J. Intl. Adsorpt. Soc. 8, 35–44.
31 J. L. Bonardet, T. Domeniconi, P. N'Gokoli-Kekele, M. A. Springuel-Huet, J. Fraissard **1999**, Langmuir 15, 5836–5840.
32 P. N'Gokoli-Kekele, M. A. Springuel-Huet, J. L. Bonardet, J. M. Dereppe, J. Fraissard **2002**, Bull. Pol. Acad. Sci., Chem.50, 249–258.
33 R. E. Hayes, S. Awdry, S. T. Kolaczkowski **1999**, Can. J. Chem. Eng.77, 688–697.
34 R. E. Hayes, S. T. Kolaczkowski, P. K. C. Li, S. Awdry **2001**, Chem. Eng. Sci. 56, 4815–4835.
35 A. J. Zarur, J. Y. Ying **2000**, Nature (London) 403, 65–67.
36 J. G. McCarty **2000**, Nature (London) 403, 35–36.
37 C. M. Spadaccini, X. Zhang, C. P. Cadou, N. Miki, I. A. Waitz **2003**, Sensors Actuat. a-Phys. 103, 219–224.
38 K. Knagge, J. Prange, D. Raftery **2004**, Chem. Phys. Lett. 397, 11–16.

5.4
In Situ Monitoring of Multiphase Catalytic Reactions at Elevated Temperatures by MRI and NMR
Igor V. Koptyug and Anna A. Lysova

5.4.1
Introduction

The versatile MRI toolkit holds significant potential for applications in all fields of chemical and process engineering, as is convincingly demonstrated in other chapters of this book. Heterogeneous catalysis is no exception, but despite the general trend in catalysis toward milder reaction conditions, it is clear that for many years to come a large number of important catalytic and other chemical reactions will have to be carried out at high temperatures. Owing to the complicated coupling of mass and heat transport processes with the chemical transformation, useful information can only be obtained if processes involved are studied *in situ*, in an operating reactor. We believe that the development of the applications of MRI to multiphase processes carried out at elevated temperatures is very important and beneficial to both the chemical engineering and MRI communities. In this contribution, we intend to demonstrate that difficulties commonly associated with such applications are somewhat exaggerated, and that efforts invested in their development are very rewarding.

Today, the MRI toolkit is used routinely for studies of processes in multiphase systems at room temperature, including the imaging of liquids and even gases, the studies of mass transport processes and the progress of some chemical reactions. Therefore, it appears that the major difficulty with extending MRI studies to high temperature reactions is the need to use elevated temperatures in an MRI instrument. In general, variable temperature NMR studies are not uncommon. Liquid and solid-state NMR spectroscopic studies are performed routinely at temperatures up to 500–550 K. Very high temperatures usually require specialized hardware to protect the superconducting magnet, shim stack and the radiofrequency (rf) probe from excessive heat. For instance, combustion of methane over zeolite pellets has been studied using hyperpolarized ^{129}Xe NMR spectroscopy [1]. Multiple air outlets surrounding the rf coil volume allowed these workers to keep the shim stack temperature as low as 320 K. The temperatures at which liquid phase NMR has been performed to date by far exceed 2000 K, even though, admittedly, such experiments "remain challenging" [2]. ^{27}Al NMR high temperature spectroscopic studies of SiO_2–Al_2O_3, CaO–Al_2O_3 (at about 2400 K, [3,4]) and SrO–Al_2O_3 (>2500 K, [2]) binary liquids and pure Al_2O_3 (>2700 K, [5]) have been performed by combining aerodynamic levitation with laser induced sample heating. At lower temperatures (1300–1800 K), a boron nitride crucible in a ceramic shield has been used in combination with laser [6, 7] or resistive heating [8]. However, for MRI applications even much lower temperatures can be problematic to implement because the commercial gradient coil sets can be irreversibly damaged by heating above 330 K. The highest temperature MRI study published to date is presumably the imaging of molten LiCl performed at temperatures up to 973 K [9]. For this purpose, an NMR-compatible horizontal furnace has been built in which a water jacket and a layer of ceramic fiber insulation are used to protect the hardware from high temperatures. These examples demonstrate that MRI studies at elevated temperatures are feasible if appropriate hardware is designed and used in such experiments.

5.4.2
Experimental

5.4.2.1 Reactor Design

Our NMR instrument is equipped with a vertical bore superconducting magnet, which is appropriate for the studies of trickle-bed reactor operation. The central part of the assembly, which is inserted into the magnet, is an evacuated double-walled glass dewar (item **6** in Figure 5.4.1; 25-mm outside diameter, 16-mm inside diameter, about 600-mm long). Its upper part extends about 150 mm above the upper edge of the rf coil **3** where it is connected to a single-walled glass pipe, **10**. The latter extends above the upper edge of the magnet and is surrounded with a wider polyfoam tube **1** to protect the upper part of the magnet bore. The lower end of dewar **6** protrudes from the probe body **5** at the bottom of the magnet. To achieve this, the original 25-mm birdcage rf coil **3** (Bruker) was used in combination with a laboratory-built Teflon probe body **5** to provide a 25-mm bore throughout the entire probe. The latter serves as an additional thermal protection of the gradient coils. A 14-mm outside diameter and 1.5-m long glass tube **8**, long enough to span the entire length of the magnet bore, is inserted concentrically into the dewar–pipe assembly. The annular space between tube **8** and dewar **6** carries a flow of hot air for thermostating the reaction volume and at the same time for protecting the hardware from yet higher temperatures developing in the reaction zone. The inner tube **8** carries the flow of preheated hydrogen gas with or without reactant vapor. It also houses a catalyst pellet or a granular bed **4** located at the height of the rf coil center. A single pellet is usually suspended on the bent tip of a thermocouple, which measures the pellet temperature in the course of experiment (not shown), while catalyst bed **4** is supported by a stainless-steel mesh, which in turn is supported with a spiral **7** wound of stainless-steel wire, which fits tightly into the reactor tube and is inserted from the bottom together with the mesh and thermocouple **9**. The thermocouple tip in this case is located at the bottom of the bed. A thin copper capillary **11** is inserted into the reactor tube from the top to supply liquid reactant to the top of a pellet or bed. For high liquid flow rates, it was found essential to preheat the liquid to avoid heat loss in the reaction zone. Two additional thermocouples are used for safety reasons, one monitors the temperature at the outer wall of the evacuated dewar while another is located in the region between the rf probe and the gradient coils.

The entire assembly is put together once and is inserted into the magnet bore from the top as a single entity. It is then connected to the hot air, preheated hydrogen and liquid supply lines running from the stationary part of the set-up. The last comprises valves, rotameters, pressure gages and gas lines and will not be described in detail here.

The set-up has been designed with several safety issues in mind. The concentric structure of the reactor excludes mixing of the flows of flammable reactants (hydrogen gas, organic vapor) and air except at the open lower edge of the ducts outside of the magnet and probe, where an end of a flexible hose connected to the fume hood is located for removal of gases and vapors. Heater spirals are wound around quartz tubes carrying gas flows to avoid direct contact of the heating elements with combustible gases. The heaters are located outside the magnet to reduce the risk of accidental overheating of the hardware and to avoid electromagnetic interference.

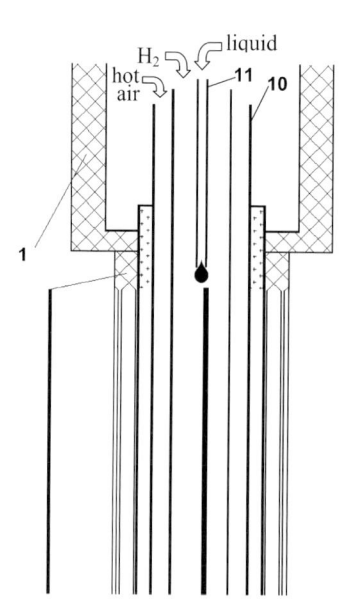

Fig. 5.4.1 Schematic drawing of the NMR-compatible multiphase catalytic reactor for MRI studies at elevated temperatures: 1, polyfoam tube; 2, Teflon holder which connects 1 with dewar 6; 3, rf coil; 4, granular catalyst bed supported by a stainless-steel mesh; 5, probe body; 6, evacuated double-walled glass dewar; 7, supporting stainless-steel spiral; 8, glass tube; 9, thermocouple; 10, single-walled glass pipe connected to the dewar; 11, stainless-steel capillary for liquid reactant supply. The reactor is not drawn to scale, only the central part is shown.

5.4.2.2 Materials

For supported catalysts preparation, γ-Al_2O_3 with about 200 m^2 g^{-1} specific surface area and 0.65 cm^3 g^{-1} specific pore volume was used, shaped as cylinders of 4.5-mm diameter and 12-mm long, or beads of 1-, 2–3- or 4-mm diameter. These alumina supports were first impregnated with an aqueous solution of $Mn(NO_3)_2$, dried at 110 °C in air for 2–3 h, and then calcined at 700 °C for 2 h. To produce actual catalysts, these supports were subsequently impregnated with an aqueous solution of $PdCl_2$, dried for an hour at 120 °C in a stream of N_2, and then heated to 250 °C. The catalyst was then reduced for 1 h at 350 °C in a mixed N_2–H_2 gas stream (10–15%-v/v of H_2), cooled down to 50 °C in an N_2–H_2 stream, and then cooled to room temperature in a stream of neat N_2. The prepared catalysts contained 0.1% Mn and 1% Pd by weight on the γ-Al_2O_3 support. For solid-state imaging experiments, a fragment of γ-Al_2O_3 monolith (100 cpsi), baking soda (sodium hydrogen carbonate), potassium permanganate and V_2O_5 powders, and 0.5-mm glass beads were used without any special preparation procedures.

5.4.2.3 Procedures

Before any hydrogenation experiment, individual catalyst pellets were activated in a mixed H_2–air stream with excess of hydrogen (catalytic hydrogen combustion). For 1- and 2–3-mm beads, the catalyst bed was activated for 30 min in an H_2 stream at 450 °C followed by cooling to room temperature in a stream of He. The bed of 4 mm beads was activated in an H_2–air stream with excess of hydrogen, heated to 450 °C in a stream of He and then cooled back to room temperature without interrupting the He stream. After activation, the catalyst was loaded into the reactor residing inside the NMR magnet. Catalyst beds were often placed on a layer of inert beads (Mn–γ-Al_2O_3) of the same size. After loading, the catalyst was flooded with large amounts of room temperature liquid α-methylstyrene (AMS) for probe tuning/matching, fine adjustment of sample positioning and preliminary imaging for selecting locations of two-dimensional (2D) slices. Next, depending on the type of experiments planned, either the reaction was initiated by turning on the supplies of hydrogen and liquid AMS, or the catalyst was first dried with a stream of neat H_2. All experiments were carried out at atmospheric pressure.

5.4.2.4 MRI Experiments

All imaging experiments were performed on a Bruker Avance DRX 300 MHz wide bore spectrometer equipped with imaging accessories at 299.13 or 300.13 MHz (^{1}H), 78.2 MHz (^{27}Al), 78.94 MHz (^{51}V), 79.39 MHz (^{23}Na), 59.63 MHz (^{29}Si) or 74.4 MHz (^{55}Mn). For ^{1}H MRI experiments, a 25-mm birdcage rf insert was used. For other nuclei, the broadband rf probe supplied with the instrument for high resolution NMR spectroscopy with a saddle-shaped rf coil (6-mm inside diameter) was used with the external aluminum shield removed.

All imaging experiments used the two-pulse spin-echo sequence α–τ–2α–τ–echo. In ^{1}H experiments, α = 90° was used. For 2D ^{1}H MRI of liquids, a 2-mm thick axial or transverse slice was selected using an appropriate combination of shaped pulses and gradients. Spatial information was frequency encoded in one dimension and phase encoded in the other, with the data matrix size of 128 × 64 complex data points. All images of individual catalyst pellets and transverse images of catalyst beds were obtained with 230 × 140 µm² spatial resolution, while for axial images of the bed it was 230 × 310 µm². The number of accumulations (NA) was 2, the echo time (TE) about 1 ms and the repetition time (TR) 0.25 s, resulting in the image acquisition time (TA) of 34 s. Several transverse images of the bed were obtained simultaneously by changing the transmitter frequency in the interleaved pulse sequence repetitions.

Spatially resolved spectroscopic studies were performed with a 128 × 16 × 64 data matrix zero-filled to 256 × 128 × 128, NA = 4, TR = 0.3 s, TE = 2.1 ms, 1.3 × 0.66 mm² spatial resolution (0.17 × 0.33 mm² after zero-filling), a 2-mm thick slice and pure phase encoding of the image. The echo signal was detected in the absence of any applied gradients. The total experiment time was 22.5 min.

The rf pulses for ^{27}Al NMR experiments were calibrated using an aqueous solution of $AlCl_3$. For the rf power level attenuated by 10 dB, the duration of the 180°-pulse of the broadband probe was 60 µs. All solids imaging experiments were performed with $\tau \approx 300$ µs and the nominal flip angle $\alpha = 90°/(I + 1/2)$. The two pulses had the same amplitude and for ^{27}Al MRI were 10- and 20-µs long, respectively.

In ^{27}Al, ^{51}V and ^{29}Si 2D imaging experiments, 64 complex data points of the echo signal were detected in the absence of a gradient. The two spatial coordinates (x and y, transverse to the sample axis) were phase encoded by an independent stepwise variation of two pulsed transverse gradients G_x and G_y (32 steps in the range –78 to 78 G cm^{-1}). The gradient pulses were 200-µs long and were applied during the first τ interval of the echo sequence. A field of view $FOV_{xy} = (9.2$ mm$)^2$ was imaged with a spatial resolution of 288 µm² before zero filling. No slice selection was used. For ^{23}Na and ^{55}Mn, one of the two spatial dimensions (x) was frequency encoded with $G_x = 78$ G cm^{-1}, detection of 64 complex data points of the echo, and 164 µm (^{23}Na) or 176 µm (^{55}Mn) resolution, while the parameters for phase encoding along y were the same as above.

5.4.3
Results and Discussion

5.4.3.1 The Dynamics of Liquid Redistribution in Catalyst Pellets and Granular Catalyst Beds

5.4.3.1.1 Single Catalyst Pellets

Our very first experiments with the reactor depicted in Figure 5.4.1 were carried out with a 15% Pt–γ-Al$_2$O$_3$ single cylindrical catalyst pellet [10–12]. The acquisition time of 2D images of an axial slice at that time was about 260 s. Despite this, the first direct MRI visualization of the operation of a model gas–liquid–solid reactor has revealed the existence of large gradients of the liquid phase content within the catalyst pellet upon imbibition of liquid α-methylstyrene (AMS) under conditions of the simultaneous endothermic evaporation of the reactant and its exothermic vapor phase hydrogenation (results not shown; see Refs. [10–12]). Two zones with very different liquid contents can be distinguished within the pellet – the liquid filled upper part, to which the liquid is permanently supplied, and the essentially dry lower part filled with the gas–vapor mixture in which the vapor phase reaction takes place. Evaporation of the liquid reactant proceeds efficiently at the boundary separating these two zones. Its position within the pellet depends on a number of factors, such as liquid AMS flow rate, thermal conductivity of the pellet and the relative rates of evaporation and hydrogenation processes. A safe but highly efficient operation of catalytic reactors requires an optimum balance between the volumes of the wet and the dry parts, which ensures sufficiently fast heat removal and at the same time sustains an intense vapor phase hydrogenation. Therefore, the non-uniform liquid distribution within the catalyst pellet, observed directly for the first time, can be one of the underlying causes of the development of critical phenomena (hot spots, temperature oscillations, reactor runaway and explosion) both in single catalyst pellets and in the granular catalyst beds.

The image acquisition time of the first experiments was large due to long delays between sequential pulse sequence repetitions to achieve spin system relaxation back to thermal equilibrium. It was obviously not adequate for studying dynamic processes within the catalyst in more detail. Paramagnetic additives are often used in NMR and MRI to substantially reduce T_1 times of nuclear spins, accelerating relaxation and thus speeding up image acquisition. Therefore, all subsequent experiments were performed with Pd–γ-Al$_2$O$_3$ catalysts additionally impregnated with 0.1%-wt Mn. As a result, the 2D image acquisition time was reduced to 34 s. An interleaved acquisition was implemented to image several 2D slices within the same experiment time. However, longer pellets (12 mm instead of 4.7 mm) were used in further experiments, with the supporting thermocouple implanted in the lower part of the pellet [13].

These modifications have led to the observation of a number of interesting dynamic processes. In particular, Figure 5.4.2 shows the dynamics of liquid phase redistribution within the pellet during the catalyst ignition event. The heat released in the exothermic hydrogenation reaction leads to the catalyst temperature rise [Figure 5.4.2(a), image 1]. This intensifies evaporation of the liquid reactant, expanding the dry zone [Figure 5.4.2(a), images 4–10] and at the same time increasing the reactant vapor pressure. Faster diffusive transport of reactants in the gas phase as compared with the liquid phase leads to progressive reaction acceleration, fast catalyst temperature rise and substantial emptying of the pellet – the well known catalyst ignition phenomenon. Eventually, however, the reaction subsides, the pellet imbibes more liquid, and a new pseudo-steady-state regime of operation at constant temperature is established [Figure 5.4.2(a), images 15–23]. It is worth noting that the ignition event occurred when almost the entire pellet was filled with the liquid, and led to the emptying of more than 50% of the pellet volume. It should be kept in mind, however, that as liquid in smaller pores and on the pore walls usually has shorter relaxation times, some residual amount of liquid in the emptied part of the pellet can escape detection.

The results obtained shed light on the spatial organization and evolution of the catalyst ignition event. It can be argued that the optimal conditions for ignition initiation are established in the subsurface pellet volume. This is because on the one hand, the temperature at the pellet surface is likely to be too low to initiate ignition as the heat of reaction is partially removed form the surface by a permanent flow of colder gas around the catalyst. On the other hand, for the reaction to proceed in the inner regions of the catalyst, hydrogen gas has to travel a substantial distance through the pore network filled with the liquid phase. Therefore, the ignition event should start at some depth below the pellet surface, while the regions adjacent to the surface and the innermost part of the pellet stay filled with liquid for a longer time [Figure 5.4.2(a), images 8–15]. As the ignition event unfolds, the inner liquid-filled area shrinks, the amount of evaporable reactant decreases, and hydrogen supply to the reaction zone slows down due to the increased distance from the surface to the evaporation–reaction zone. At some point, this inhibition process becomes predominant, the temperature drops and the new cycle of pellet fill-out begins [Figure 5.4.2(a), images 15–23].

Transverse 2D images shown in Figure 5.4.2(b) are also in agreement with the assumption that the ignition starts at some depth below the pellet surface, leading to the characteristic liquid phase distribution pattern in the form of concentric rings. Later on, the inner part gradually gets rid of the liquid phase until it is emptied completely, while the regions adjacent to the surface retain some liquid even when the pellet core becomes completely dry. The MRI data obtained allow us to advance the hypothesis that the zone of efficient hydrogenation reaction is located below the surface of the catalyst pellet where the optimum balance between the pellet temperature and diffusional intrapellet transport of hydrogen is maintained.

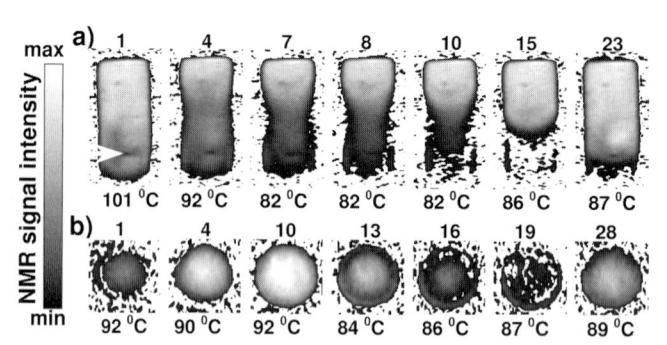

Fig. 5.4.2 The dynamics of liquid phase redistribution in a single catalyst pellet during an ignition event. The 2D images of an axial (a) or transverse (b) slice 2-mm thick were detected sequentially every 34 s, with image number in the sequence indicated above each image. Temperatures measured with a thermocouple implanted in the lower part of the pellet are indicated below each image. The location of the thermocouple in the pellet is indicated with a white arrow (a, image 1). Note that the subsets of images shown in (a) and (b) are from different experimental runs. Lighter shades of gray correspond to higher liquid contents; in white areas the signal intensity is below the noise level. H_2 flow rate was 10.9 cm^3 s^{-1}, its temperature was 77 °C (a) or 78 °C (b), the AMS flow rate was 8.7·10^{-4} g s^{-1}. Transverse slice shown in (b) is located about 2.5 mm below the pellet center.

In these experiments, all liquid was imaged without distinguishing the reactant (AMS) from the product (cumene). It is therefore possible that the outer ring of liquid which does not evaporate during ignition (Figure 5.4.2) reflects the accumulation of the product. Temperature rise during ignition leads to the evaporation of both AMS and cumene. However, their rates of evaporation can be fairly different because chemical conversion in the vapor phase decreases the partial pressure of AMS and increases that of cumene. It is thus possible that some liquid cumene can remain in the pores of the catalyst. This, however, is still to be verified in the future experiments on spatially resolved NMR spectroscopy.

In a longer experiment, multiple ignition events have been observed, leading to quasi-periodic oscillations of the liquid content and the reciprocating motion of the liquid front in the catalyst pellet. These oscillations are essentially a sequence of ignition events observed when more than one half of the pellet is filled with the liquid reactant. They are characterized by significant variations of the pellet liquid content and temperature, large radial concentration gradients and movement of the liquid front in the radial direction. It is important to stress that the oscillatory behavior is not caused by any variation of the external variables such as inflow temperatures and flow rates of liquid and gas supplies. Its origin should be sought in the intrinsic mutual coupling of several processes including heat transport, mass transport, phase transitions and the chemical reaction itself. This has been confirmed by the preliminary mathematical modeling of the experimental results performed by our colleagues at BIC (Novosibirsk).

Another type of behavior is observed when less than one half of the pellet is filled with the liquid reactant (Figure 5.4.3). In this case, the liquid front moves back and forth in a plug-like fashion [13]. The signal intensity scale of the images in Figure 5.4.3 was intentionally made ambiguous, providing contour lines of the liquid distribution to stress the pulsating motion of the liquid front. The highest liquid content is observed in the top portion of the pellet, and the size of this region and the position of its boundary exhibit unceasing pulsations. The rest of the pellet is not wetted and is filled with a vapor–gas mixture. Evaporation of AMS and the progress of the hydrogenation reaction lead to the experimentally observed temperature oscillations in the lower dry part of the pellet [Figure 5.4.3(b)]. There is a pronounced time lag between the instants of the highest pellet temperature and the lowest liquid content, and *vice versa*. For instance, the highest temperature in the lower part of the pellet is detected when the liquid content has passed the minimum and is increasing [Figure 5.4.3(a), images 1–3, 5–7 and 9–11], while the lowest temperature is measured when the liquid content is decreasing (images 3–5, 7–9 and 11–13). Furthermore, the amount of liquid in the upper and lower parts of the liquid-filled region also changes in anti-phase as evidenced by the area encompassed by the smallest and the largest contours of the liquid distribution. The observed behavior demonstrates the existence of a delayed feedback between the pellet temperature and the amount of liquid imbibed by the porous pellet.

The existence of pulsating motion of the liquid phase within the catalyst pellet indicates that under certain conditions a steady-state regime of reaction progress can become unstable due to the exothermicity of the reaction under study. Further

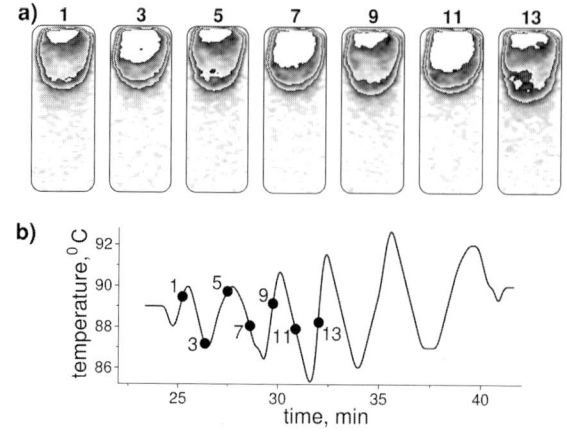

Fig. 5.4.3 (a) The dynamics of liquid phase redistribution in a single catalyst pellet during oscillations. 2D images are detected every 34 s; every alternate image is shown. The color scale was made to wrap around several times as the signal intensity increases from zero to its maximum value, providing contour lines of liquid content. White areas in the upper part of the pellet correspond to the highest signal intensities. (b) Temporal behavior of the pellet temperature in the same experiment. The temperature is measured in the lower part of the pellet with an implanted thermocouple. Solid circles correspond to the images shown in (a). H_2 flow rate was 10.9 cm^3 s^{-1}, its temperature was 78 °C, the AMS flow rate was 8.6×10^{-4} g s^{-1}.

experiments have demonstrated high sensitivity of the oscillating behavior to the external experimental conditions. In particular, it was established that the existence of oscillations, their periodicity and regularity depend crucially on the catalyst activity, flow rate of liquid AMS and the temperature of the inflowing gas. At the same time, no evidence could be found that manganese affects the observed processes in any way. The results obtained imply that the oscillating behavior is brought about by the differences in the rates of heat transport, mass transport and phase transitions.

5.4.3.1.2 Clusters of Several Catalyst Beads

While studying the operation of a single catalyst pellet reactor is useful for the elucidation of dynamic processes that take place on the length scale of a single pellet, experiments with beds comprising many catalyst pellets are required to address the processes on larger length scales. Indeed, some phenomena are caused by the joint action of the neighboring pellets and thus cannot be studied in a single pellet configuration. Therefore, the next series of experiments were carried out using small clusters of spherical catalyst beads approximately 4 mm in diameter. As the inside diameter of the reactor is 10 mm, the beads were packed in a roughly regular fashion, with three or four beads in each horizontal layer. Two transverse 2D images corresponding to the second [Figure 5.4.4(a)] and fourth [Figure 5.4.4(b)] bead layers from the top of the bed were detected alternately using an

Fig. 5.4.4 Distribution of the liquid phase in a bed of 4-mm catalyst beads in the course of AMS hydrogenation. 2D images of the second (a) and fourth (b) rows of beads counting from the top of the bed were detected in the same experimental run using an interleaved acquisition scheme. Acquisition of each image took 34 s; sequential numbers of images are indicated in the figure. For all images, liquid AMS was supplied at the rate of 2.65×10^{-1} g s^{-1}. For images (a)1 and (b)1, the bed was filled with AMS and the H$_2$ flow was off. For all other images, H$_2$ preheated to 85 °C was flowing at 20 cm^3 s^{-1}. Liquid AMS supply was turned on when the bed was either wet (images 14, 21) or dry (33–35).

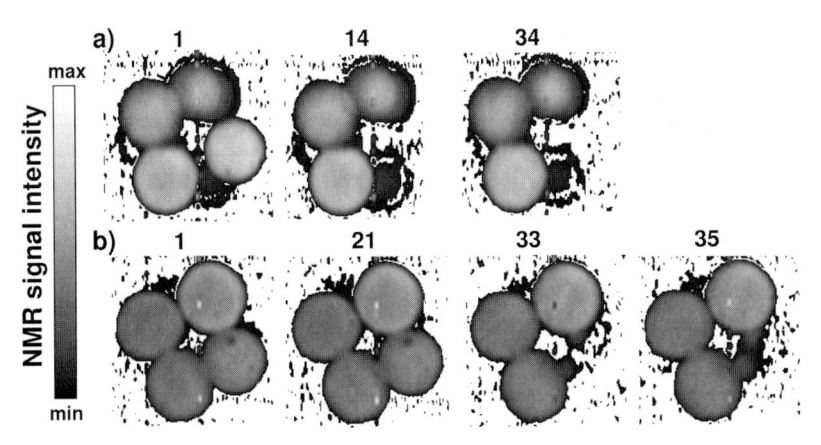

interleaved acquisition scheme. As the slice thickness in the experiments was relatively large and the bed packing was not perfectly regular, beads from the layers adjacent to the one selected also contributed to the observed image.

These experiments have revealed the possibility of the ignition of individual catalyst pellets within the bed. Such pellets can stay dry after the ignition due to the efficient progress of reaction despite the fact that they are surrounded by liquid filled pellets [Figure 5.4.4(a)]. The images also reveal the presence of the beads with the characteristic concentric pattern of liquid distribution, similar to that observed earlier (Figure 5.4.2) for individual cylindrical pellets.

Owing to the fast pulse sequence repetition rates used for the imaging of the Mn-containing catalyst, only the liquid inside the beads is observed in the images while the signal of any bulk liquid in the inter-particle voids, if present, is completely suppressed. In fact, the most recent experiments have shown that even for highest liquid flow rates used in this study, the inter-particle voids are not filled with the liquid phase. However, we cannot rule out the presence of a thin film of liquid covering the external surface of the beads, which is not resolved in the images. Therefore, film flow can be an essential mechanism of liquid transport within the bed. At the same time, the results seem to demonstrate that some transport can proceed through the inner porous space of the beads. In particular, it can be seen in Figure 5.4.4(b) (images 33, 35) that the dry bead sips liquid from its liquid-filled neighbors. Presumably, the capillary forces are responsible for liquid transport between the beads. Such beads can act as "microreactors" in which the liquid reactant drawn into an almost dry pore space evaporates rapidly and undergoes gas phase hydrogenation.

5.4.3.1.3 Granular Catalyst Beds

In practice, granular beds comprising a very large number of catalyst pellets are used. It is well known that the efficiency of a catalytic reactor depends crucially on the liquid phase distribution within the catalyst bed [14]. It is likely that the development of hot spots in a catalyst bed is also related to the character of liquid phase distribution. Therefore, it is very important to map the spatial distribution of the liquid phase in a catalytic reactor for various operation regimes. This eventually should lead to the formulation of the mechanisms responsible for the development of critical phenomena on both a micro- and macroscale.

With these goals in mind, we have studied the distribution of the liquid phase in the course of the hydrogenation reaction in a catalyst bed comprised of 1-mm catalyst beads (Figure 5.4.5). The 2D images shown reflect the distribution of the liquid phase in a 2-mm thick axial slice upon variation of the liquid AMS flow rate. The results show that while the increase in the flow rate leads to larger liquid contents in the bed (and *vice versa*), a steady state operation of the catalyst bed with unchanging spatial distribution of the liquid is observed if the external conditions remain unchanged.

Similar experiments were performed with beds comprised of 2–3-mm catalyst beads. In this case, a catalyst bed 3 cm in height was supported by a 2-cm layer of catalytically inactive beads of the same size (γ-Al_2O_3 + 0.1% Mn, but without Pd).

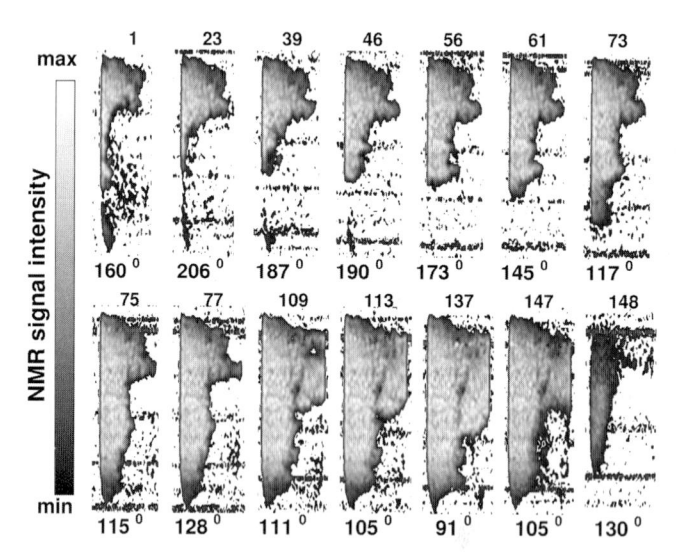

Fig. 5.4.5 Distribution of the liquid phase in the bed comprised of 1 mm catalyst beads in the course of AMS hydrogenation. Acquisition of each image took 34 s; sequential numbers of images shown are indicated in the figure. Temperatures measured with a thermocouple inserted in the lower part of the bed are indicated below each image. H_2 preheated to 71 °C was flowing at 10.9 $cm^3\ s^{-1}$. Liquid AMS flow rate was $1.64 \cdot 10^{-2}$ g s^{-1} (image 1), 9.71×10^{-3} g s^{-1} (images 23–61), 6.64×10^{-2} g s^{-1} (images 73–77), 9.36×10^{-2} g s^{-1} (images 109–137) and 0 g s^{-1} (images 147, 148). Dark horizontal stripes are due to rf pickup caused by the presence of the stainless-steel mesh supporting the granular bed.

During the MRI experiment, samples of the liquid phase were collected downstream of the bed for subsequent GC analysis. Figure 5.4.6 reveals that in general the liquid phase is distributed similarly in the catalyst and in the inert layers, but in the latter some areas with low but non-zero liquid content are present. This could possibly indicate that unlike catalyst beads, the catalytically inactive porous beads can adsorb noticeable amounts of vapor. As in the experiments with 1-mm catalyst beads (Figure 5.4.5), the increase in the liquid AMS flow rate does not substantially change the distribution of the liquid phase in the catalyst bed, but leads to an increase in the amount of liquid phase in the inert layer (Figure 5.4.6, images 1–45). Similarly, the decrease in hydrogen flow rate does not change liquid distribution in the catalyst but leads to an increase in the liquid phase content in the inert layer (Figure 5.4.6, images 159–197). It is worth noting that the degree of AMS conversion does not change when hydrogen flow rate is altered (Table 5.4.1). This implies that even when H_2 flow rate is reduced down to 7.9 cm^3 s^{-1}, there is an excess of H_2 relative to AMS and the reaction rate is limited by AMS evaporation. This can explain why hydrogen flow rate has little effect on liquid phase distribution in the catalyst bed and on conversion. At the same time, higher flow rates of warm hydrogen intensify evaporation of liquid in the inert layer and remove the resulting vapor out of the reactor, leading to the decrease in liquid content in the inert layer.

Tab. 5.4.1 Productivity and conversion characteristics of the catalyst bed comprised of 1% Pd–γ-Al$_2$O$_3$ catalyst beads 2–3-mm in diameter for various flow rates of AMS and H_2.

Regime	AMS flow rate (g s^{-1})	H$_2$ flow rate (cm^3 s^{-1})	Conversion (%)	Productivity (g s^{-1})	Images in Figure 5.4.6
1	5.06 × 10^{-2}	40	20.77	1.05 × 10^{-2}	1
2	9.96 × 10^{-2}	40	10.02	9.98 × 10^{-3}	9–27
3	1.47 × 10^{-1}	40	5.53	8.13 × 10^{-3}	43–45
4	2.96 × 10^{-2}	40	33.49	9.91 × 10^{-3}	85–87[a]
5	2.96 × 10^{-2}	40	–	–	149–154[b]
6	3.24 × 10^{-2}	40	–	–	159
7	3.24 × 10^{-2}	20	27.18	8.81 × 10^{-3}	171–196
8	3.24 × 10^{-2}	7.9	27.64	8.96 × 10^{-3}	197

a AMS supplied to pre-wetted bed.
b AMS supplied to dry bed.

It was also established that the distribution of the liquid phase in the reactor depends on the reactor start-up procedure. When the flow rate of liquid AMS is reduced from 1.47 × 10^{-1} g s^{-1} (Figure 5.4.6, images 43 and 45) down to 2.96 × 10^{-2} g s^{-1} (images 85 and 87), the distribution of the liquid phase hardly changes and the catalyst bed contains a substantial amount of liquid. At the same time, if

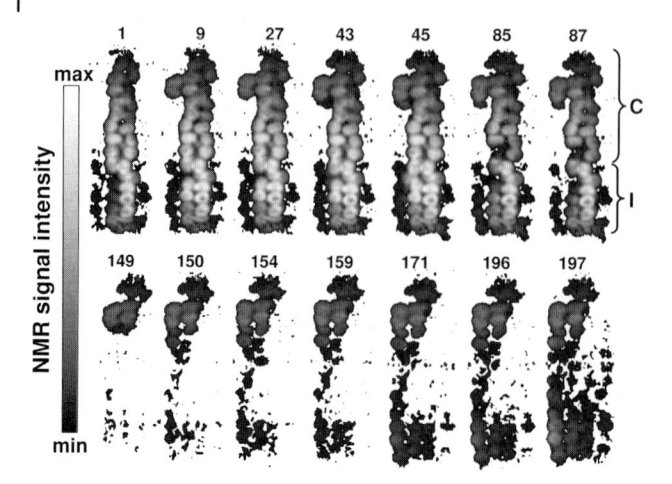

Fig. 5.4.6 Distribution of the liquid phase in the bed comprised of 2–3-mm catalyst beads in the course of AMS hydrogenation. Acquisition of each image took 34 s; sequential numbers of images shown are indicated in the figure. The upper part of the bed comprising catalyst particles (C) and the lower part comprising inert beads (I) are labeled on the right-hand side of the figure. H_2 temperature was 85 °C. Flow rates of H_2 and AMS are given in Table 5.4.1.

AMS is supplied at the same rate of 2.96×10^{-2} g s^{-1} but to an initially dry catalyst bed (Figure 5.4.6, images 149–154), a totally different distribution of the liquid phase within the bed is observed.

5.4.3.2 Spatially Resolved NMR Spectroscopy

In the experiments described above, all liquid was imaged without distinguishing the reactant and the product. However, NMR is a powerful spectroscopic technique, and many existing applications successfully combine spectroscopic and imaging modalities to obtain either spatially localized NMR spectra characterizing local chemical composition or images showing the spatial distribution of a selected compound. This type of information is very important for the characterization of an operating reactor as it provides direct information on the chemical conversion. Therefore, the implementation of combined NMR–MRI applications in chemical engineering will be one of the major avenues for the future development of the field.

Unlike high-resolution NMR spectra of bulk solutions where NMR linewidths well below 1 Hz can be obtained routinely, NMR spectra of liquids permeating porous solids in most cases will not exhibit such a high spectral resolution. First of all, the interaction of liquid phase molecules with pore walls of the catalyst and rapid diffusion-driven intrapore transport will lead to a pronounced homogeneous broadening of the observed NMR lines. Smaller pore sizes and the presence of paramagnetic impurities in the solid material usually aggravate the situation and thus should be avoided. Another reason why NMR spectra of liquids in porous materials appear broadened and distorted is a substantial variation of local magnetic fields over the sample. This latter broadening mechanism presents fewer problems than the homogeneous broadening. One can expect that deviations of the magnetic field from its mean value will decrease as the size of the volume of interest within the sample is reduced. We have verified this experimentally using a single alumina pellet saturated with a 1:1 mixture of AMS and cumene at room temperature [10]. With a voxel size of $2 \times 0.5 \times 0.12$ mm^3, the lines in the ^{1}H NMR spectrum were narrow enough to allow at least semi-quantitative characterization of mixture composition. These results demonstrate that homogeneous broadening is not too large and the evaluation of local conversion is still possible. It should be noted that at elevated temperatures homogeneous broadening of the NMR lines is further reduced because of the acceleration of molecular tumbling and the resulting decrease in the rotational correlation time of molecules which governs spin–relaxation processes.

Further spectroscopic experiments were carried out with an operating reactor using a bed of 1-mm catalyst beads [13]. A 3D experiment with one spectral and two spatial coordinates was carried out, yielding NMR spectra for each pixel of a 2D axial slice. Figure 5.4.7 shows several representative spectra selected from the entire data set. The NMR spectra of neat AMS [Figure 5.4.7(d)] and cumene [Figure 5.4.7(f)] are provided for comparison, they were experimentally detected for bulk liquid samples (lower traces with narrow lines) and their lines were then mathematically broadened to 300 Hz (upper traces) to account for the broadening in the

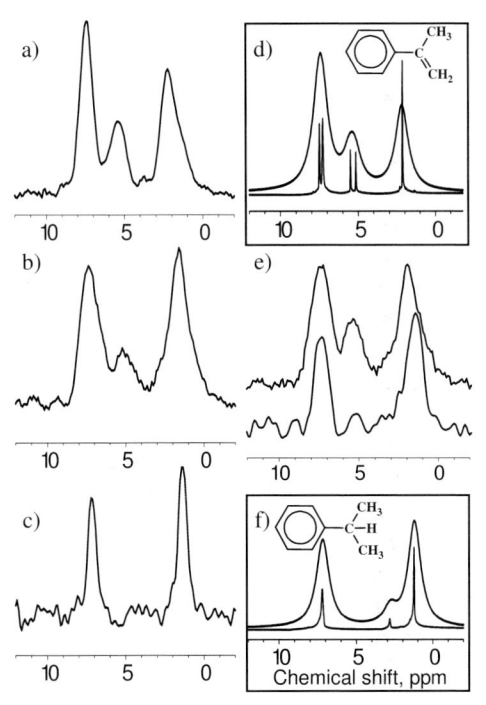

Fig. 5.4.7 (a–c, e) Spatially resolved NMR spectra detected during AMS hydrogenation in a catalyst bed of 1-mm beads. Each spectrum corresponds to a voxel size of $2 \times 0.17 \times 0.33$ mm^3. Spectra in (a–c) correspond to the same radial position within the operating reactor and are detected in the top, middle and bottom parts of the reactor, respectively. Three spectra in (b, e) correspond to the same vertical position in the operating reactor, with the two spectra in (e) corresponding to voxels shifted by 1.3 mm to ether side of the voxel of the spectrum in (b). The two spectra in (e) are shifted vertically relative to each other for better presentation. The lower traces with narrow lines in (d, f) are experimental spectra detected for bulk neat AMS (d) and cumene (f), the upper traces in (d, f) were obtained by mathematically broadening the lines to 300 Hz.

catalyst. Despite this broadening, the spectra can still be distinguished, with the most pronounced differences being the presence of 5.2–5.5 ppm resonances in the AMS spectrum, and the relative intensities and the separation of the outer lines.

The first column of spectra [Figure 5.4.7(a–c)] demonstrates variation of the chemical composition of the liquid phase in the vertical direction within the operating fixed bed reactor. As can be expected, the spectrum at the top of the bed corresponds to AMS, the one at the bottom shows complete conversion into cumene, while at an intermediate location the mixture of the two liquids is clearly observed. Figure 5.4.7(e) shows two NMR spectra that were detected eight pixels (1.3 mm) on either side in the horizontal direction of the pixel corresponding to the spectrum shown in Figure 5.4.7(b) and demonstrates variation of the conversion in the radial direction within the reactor. The spectra presented also exhibit a substantial variation of the linewidths within the bed, which can be caused by both bed structure variations as well as a non-uniform distribution of the liquid phase within the bed. It should be stressed that these spectra have been detected with the Mn-containing catalyst, that is, under highly unfavorable conditions, and even better spectroscopic results can be expected for catalysts free of paramagnetic additives.

5.4.3.3 Direct NMR Imaging of the Solid Phase

While most of the MRI experiments performed today detect the NMR signal of the liquid phase, imaging of gases and solids in multiphase reactors could provide additional and useful information. Imaging of the gas phase is often hampered by the inherently low sensitivity, but recent efforts have already resulted in the development of a number of specialized applications [1, 15–19]. At the same time, it is obvious that solid-state imaging could provide a wealth of useful data in chemical and process engineering applications. The lack of significant developments in this area is presumably the consequence of a widely accepted misconception that imaging of rigid solids cannot be performed on MRI instruments commonly employed for the imaging of liquids, and requires specialized solid-state NMR–MRI hardware, such as magic angle spinning, very large magnetic field gradients, sophisticated multipulse sequences and/or very short and intense rf pulses. While in many cases this might be true, the results reported below demonstrate that a great deal can be achieved if one questions the validity of this statement.

Initially, our attention was focused on the possibility of ^{27}Al MRI studies of aluminum-containing materials and was primarily driven by the fact that many supported catalysts used in practical applications benefit from the high specific surface area of aluminas (Al_2O_3), and by our own MRI studies of AMS hydrogenation on Pt–γ-Al_2O_3 and Pd–γ-Al_2O_3 catalysts described above. However, as it turned out in the course of this study, many other quadrupolar nuclei in rigid solids can be imaged on a commercial instrument conventionally used for liquid phase studies, without any modifications. Thus it appears that the field of multi-nuclear MRI of rigid solids is highly under-explored and underemployed, and that further development of this research area can potentially open up a window into a new dimension in MRI applications.

The concepts behind these studies are in fact well known and established. NMR of rigid solids is usually associated with very rapid signal decay in the time domain (FID, echo), leading to very broad lines in NMR spectra. This appears to imply that there is very little or no time for pulsed gradient switching, and that very large gradients are needed to overcome substantial linewidths and to obtain reasonable spatial resolution. Therefore, the armory of solid-state NMR is usually employed for the imaging of rigid solids. However, the broadening of NMR spectra of solids is often caused by anisotropic interactions and therefore is inhomogeneous in nature. Therefore, application of a second rf pulse can refocus transverse magnetization to a certain degree. We have employed a simple spin echo sequence of the type $\alpha-\tau-2\alpha-\tau-$echo, while a number of alternative variants are well known in NMR [20–24]. In addition, the echo time used in all our experiments, $TE \approx 600$ μs, provides plenty of time for gradient switching but ensures that sufficient signal is still available for good S/N ratio in the images. Furthermore, for quadrupolar nuclei with a half-integer spin $I = (2n + 1)/2$, $n = 1, 2, ...$, the central transition between the $+1/2$ and $-1/2$ spin sublevels is broadened only to the second order in the perturbation theory [25, 26]. This means that rf pulses of a reasonable duration should be "hard" enough to excite the entire central transition or a substantial part of it, providing reasonable signal intensity for detection. The optimum flip angle for the detection of central transition of half-integer spin nuclei is known to scale as $\alpha = 90°/(I + 1/2)$ [27, 28], which further reduces the required rf power levels.

Another issue of concern is the amplitude of magnetic field gradients required for solids imaging. For spectra up to a few kHz wide, standard frequency encoding schemes can be used to obtain reasonable spatial resolution. This will not work for broader lines, but it is well known that phase encoding is much more tolerant to line broadening and can be used for spectral widths in excess of 100 kHz with no major loss in spatial resolution.

The "proof of the pudding" is shown in Figure 5.4.8. Our "five-in-one" sample comprises an alumina (γ-Al_2O_3) matrix cut out of a larger alumina monolith and has four parallel channels filled with various solid materials: baking soda ($NaHCO_3$) (Na), V_2O_5 powder (V), potassium permanganate ($KMnO_4$) (Mn) and 0.5 mm glass beads (Si), as shown schematically in the center of the figure. The images shown were detected using the NMR signals of ^{27}Al (a), ^{23}Na (b), ^{51}V (c) and ^{55}Mn (d), by simply retuning the rf probe. The fifth nucleus, ^{29}Si, gave a weak signal with long T_1 time, and no attempt was made to detect the 2D ^{29}Si image in this case, while on other occasions we were able to successfully image glass tubes and silica gel powders. Note, however, that glass beads give a weak but unmistakable signal in the ^{23}Na image [Figure 5.4.8(b), lower part] due to the sodium content of glass. For ^{23}Na and ^{55}Mn, frequency encoding in one of the two dimensions was used, therefore imaging times were relatively short. For much broader signals of ^{27}Al and ^{51}V, echoes were phase encoded in both spatial dimensions, generally resulting in longer imaging times. Despite this, the ^{27}Al image of the monolith [Figure 5.4.8(a)] was obtained in 17.5 min. An extension to 3D imaging is straightforward and has been implemented successfully [29]. The nominal spatial resolution obtained is 150–300 μm, and for the samples with reasonably high S/N and

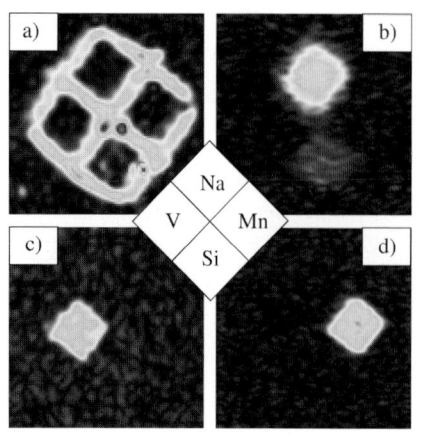

Fig. 5.4.8 2D images of the "five-in-one" sample comprising an alumina (γ-Al_2O_3) matrix with four parallel channels loosely packed with $NaHCO_3$ (Na), V_2O_5 (V), $KMnO_4$ (Mn) powders and 0.5-mm glass beads (Si), as schematically shown in the center of the figure. The images were obtained using NMR signals of respective nuclei: (a) ^{27}Al, $TR = 0.5$ s, $NA = 2$, $TA = 17.5$ min; (b) ^{23}Na, $TR = 3$ s, $NA = 16$, $TA = 25.7$ min; (c) ^{51}V, $TR = 0.5$ s, $NA = 64$, $TA = 9$ h; (d) ^{55}Mn, $TR = 1$ s, $NA = 14$, $TA = 2.2$ min. No slice selection was used.

not very long T_1, times can be improved easily. For the nuclei studied, there is no indication that the true spatial resolution is noticeably worse than the nominal one.

Of course, much more work is required to turn these examples into useful and informative practical applications. However, we hope that this will stimulate a renewed and much broader interest in the MRI of rigid solids in various research areas. The list of useful nuclei and chemical substances can be extended much further. The obvious utilization of the MRI of rigid solids in chemical engineering will be for various structural studies. For instance, direct imaging of the structure of the granular bed of a fixed bed reactor would be invaluable in studying structure–transport relationships in such systems. However, the most important and unique feature of MRI is an extremely versatile nature of image contrast, which applies equally to the MRI of both liquids and solids. This means that it should be possible, for instance, to study transport of granular solids in various chemical engineering applications, preparation of supported catalysts, and much more.

One of the issues of paramount importance in multiphase catalytic reactors is that of heat transport. While MRI has been used for spatially resolved NMR thermometry of liquids [30–32], it appears that none of the available approaches can be directly employed for multiphase catalytic processes. The main obstacle for such applications is the fact that all temperature sensitive parameters used in NMR thermometry of liquids (signal intensity, chemical shift, relaxation times and diffusivities) will vary substantially due to other factors as well (for instance, varying amounts of liquid in the pores) and thus will not be reliable for temperature measurements. On the other hand, the solid matrix is less likely to undergo substantial changes during reactor operation, and therefore might finally give an access to local temperature evaluation by MRI, as has been demonstrated for "soft" solid materials [33, 34]. As mentioned above, alumina is widely used for manufacturing catalysts and supports. Therefore, we have performed some preliminary studies on the temperature dependence of the ^{27}Al NMR signal of alumina pellets. The results obtained show that both T_1 and signal intensity of the ^{27}Al nucleus in the solid state exhibit pronounced temperature dependences [29] and thus can be potentially used for spatially resolved NMR thermometry studies. Work is in progress to demonstrate that these observations can be used for spatially resolved NMR thermometry of solid materials.

It is important to stress that images shown in Figure 5.4.8 have been obtained on a commercial microimaging instrument conventionally used for liquid phase studies without any modifications, using a conventional two-pulse spin-echo sequence with a relatively long echo time of about 600 µs, with good sensitivity at a 7 T magnetic field, sub-millimeter spatial resolution and acceptable imaging times. This shows that imaging of rigid solids deserves much more attention than it has received to date. No slice selection was used here, but the appropriate strategies for solid-state MRI are well documented in the literature.

5.4.4
Outlook

Clearly, it would be difficult at present to predict future development of the field, but the results already obtained leave no doubts that such applications are very promising. A great deal of further work is needed to develop these demonstrations into a practical set of tools for routine use, and the current pace of the progress in the field of non-medical MRI applications leaves no doubts that this goal can be achieved.

In particular, faster imaging strategies should be implemented for 2D and even 3D dynamic MRI of an operating reactor. In our recent studies, the implementation of multi-spin-echo strategies allowed us to detect complete 3D images of the reactor volume within 30–60 s without the need to dope the catalyst with paramagnetic additives. This allows one to image not only the liquid permeating porous catalyst, but also to visualize separately the liquid in the interparticle voids. Fast 2D and 3D imaging will be useful, in particular, for the studies of dynamic responses of an operating reactor to external perturbations. For the beds comprised of 2–3 mm beads or larger, the spatial resolution of the imaging experiments is high enough to visualize the distribution of the liquid phase both on the length-scale of the bed and within each individual bead comprising it. Combination of imaging with motion-induced contrast can and will be used to study flow in an operating reactor. Flow studies in the absence of chemical transformations are presented thoroughly in a number of chapters of this book. However, characteristics of flow under reactive conditions are expected to be rather different due to the influence of heat transport, phase transitions and the liquid distribution pattern.

The elimination of the need to use paramagnetic doping of the catalyst will facilitate future spectroscopic studies both in a catalyst bed and in single catalyst pellets. For spatially resolved spectroscopy, it could also be advantageous to use magnetic nuclei with a wider spread of chemical shifts as compared with ^{1}H. Extension to other classes of reactors and reactions is feasible and will be addressed.

As has been shown in Section 5.4.3.3, imaging of solids can provide useful structural information about reactors, catalyst beds and so forth, and thus can be useful in establishing the structure–transport–conversion relationships. However, the real advantage of MRI as compared with other imaging techniques is its ability to provide parameter images, such as spatial maps of flow velocity, diffusivity, chemical composition, temperature and more, as demonstrated for liquids throughout this book. There is no reason why solids imaging cannot provide such data, too, based on appropriate manipulations with image contrast. For instance, it should be possible to study motion of solid materials directly, which can be very useful for the studies of granular solids transport without the need to use liquid containing solid particles. Given the importance of the transport of solids fines in industrial processes (gas–solid suspensions, moving and fluidized beds, and so on), further development of such applications are expected to be very rewarding.

Another possibility which is extremely important for chemical and process engineering is the development of NMR thermometry strategies based on the detection of the solid phase. We have demonstrated that both the intensity and the spin–lattice relaxation time of the ^{27}Al NMR signal of alumina pellets depend noticeably on their temperature [29]. The studies were performed for temperatures in the range from 300 K to 475 K. This can be employed, for instance, for non-invasive studies of temperature fields in catalyst beds during catalytic reactor operation, and is expected to provide information on the heat transport processes, formation and evolution of hot spots, and so forth. The possibility to detect images of other solid materials and other nuclei should lead to the wide use of solid-state NMR thermometry in chemical and process engineering, but requires a lot more work in this direction.

Furthermore, imaging of the solid phase of a reactor substantially broadens the range of processes that can be studied by MRI. For instance, detection of the temperature map using solids imaging can be performed for those reactors where signal intensity of the fluid phase is low (for instance, no or little liquid present, short T_2 for liquids permeating certain porous materials). Combined imaging of both solids and liquids will clearly have other interesting applications in chemical engineering and catalysis. For instance, we are currently exploring the possibility of the visualization of supported catalyst preparation by multinuclear MRI.

Acknowledgments

We thank V. A. Kirillov, A. V. Kulikov (Boreskov Institute of Catalysis, SB RAS, Novosibirsk), H. Van As and E. Gerkema (Wageningen University, The Netherlands) for fruitful collaboration. This work was supported by the grants from RFBR-NWO (03–03–89014-NWO, 047.015.006), RFBR (02–03–32770), CRDF (RU-C1–2581-NO-04), SB RAS (integration grants 41, 166) and the Russian President's program of support of the leading scientific schools (NSch-2298.2003.3). I. V. Koptyug thanks the Russian Science Support Foundation for financial support. A. A. Lysova acknowledges a scholarship awarded by the Zamaraev Charitable Scientific Foundation and sponsored by Uwe and Barbara Eichhoff.

References

1. S. Anala, G. E. Pavlovskaya, P. Pichumani, T. J. Dieken, M. D. Olsen, T. Meersmann **2003**, *J. Am. Chem. Soc.* 125, 13298–13302.
2. M. Capron, P. Florian, F. Fayon, D. Trumeau, L. Hennet, M. Gaihlanou, D. Thiaudiere, C. Landron, A. Douy, D. Massiot **2001**, *J. Non-Cryst. Solids* 293, 496–501.
3. B. T. Poe, P. F. McMillan, B. Cote, D. Massiot, J.-P. Coutures **1992**, *J. Phys. Chem.* 96, 8220–8224.
4. D. Massiot, D. Trumeau, B. Touzo, I. Farnan, J. C. Rifflet, A. Douy, J.-P. Coutures **1995**, *J. Phys. Chem.* 99, 16455–16459.
5. P. Florian D. Massiot, B. Poe, I. Farnan, J.-P. Coutures **1995**, *Solid State NMR* 5, 233–238.
6. S. Sen **1999**, *J. Non-Cryst. Solids* 253, 84–94.
7. V. Lacassagne, C. Bessada, P. Florian, S. Bouvet, B. Ollivier, J.-P. Coutures, D. Massiot **2002**, *J. Phys. Chem. B*, 106, 1862–1868.
8. S. Sen, A. M. George, J. F. Stebbins **1996**, *J. Non-Cryst. Solids* 197, 53–64.
9. G. Cho, E. Segal, J. L. Ackerman **2004**, *J. Magn. Reson.* 169, 328–334.
10. I. V. Koptyug, A. V. Kulikov, A. A. Lysova, V. A. Kirillov, V. N. Parmon, R. Z. Sagdeev **2002**, *J. Am. Chem. Soc.* 124, 9684–9685.
11. I. V. Koptyug, A. V. Kulikov, A. A. Lysova, V. A. Kirillov, V. N. Parmon, R. Z. Sagdeev **2003**, *Chem. Sust. Dev.* 11, 109–116.
12. I. V. Koptyug, A. A. Lysova, A. V. Matveev, L. Yu. Ilyina, R. Z. Sagdeev, V. N. Parmon **2003**, *Magn. Reson. Imag.* 21, 337–343.
13. I. V. Koptyug, A. A. Lysova, A. V. Kulikov, V. A. Kirillov, V. N. Parmon, R. Z. Sagdeev **2004**, *Appl. Catal. A: General* 267, 143–148.
14. R. L. McManus, G. A. Funk, M. P. Harold, K. M. Ng **1993**, *Ind. Eng. Chem. Res.* 32, 570–574.
15. J.-L. Bonardet, T. Domeniconi, P. N'Gokoli-Kekele, M.-A. Springuel-Huet, J. Fraissard **1999**, *Langmuir* 15, 5836–5840.
16. I. V. Koptyug, A. A. Lysova, A. V. Matveev, V. N. Parmon, R. Z. Sagdeev **2005**, *Top. Catal.* 32, 83–91.
17. I. V. Koptyug, A. V. Matveev, S. A. Altobelli **2002**, *Appl. Magn. Reson.* 22, 187–200.
18. I. V. Koptyug, L. Yu. Ilyina, A. V. Matveev, R. Z. Sagdeev, V. N. Parmon, S. A. Altobelli **2001**, *Catal. Today* 69, 385–392.
19. I. V. Koptyug, S. A. Altobelli, E. Fukushima, A. V. Matveev, R. Z. Sagdeev **2000**, *J. Magn. Reson.* 147, 36–42.
20. P. Blumler, B. Blumich **1994**, *NMR Basic Princip. Progr.* 30, 209–277.
21. B. Blumich **1999**, *Concepts Magn. Reson.* 11, 71–87.
22. P. J. McDonald, J. J. Attard, D. G. Taylor **1987**, *J. Magn. Reson.* 72, 224–229.
23. E. Rommel, S. Hafner, R. Kimmich **1990**, *J. Magn. Reson.* 86, 264–272.
24. T. Gullion, D. B. Baker, M. S. Conradi **1990**, *J. Magn. Reson.* 89, 479–484.
25. V. S. Swaminathan, B. H. Suits **1998**, *J. Magn. Reson.* 132, 274–278.
26. D. D. Laws, H. M. L. Bitter, A. Jerschow **2002**, *Angew. Chem., Int. Ed. Engl.* 41, 3096–3129.
27. F. H. Larsen, H. J. Jakobsen, P. D. Ellis, N. C. Nielsen **1998**, *J. Magn. Reson.* 131, 144–147.
28. P. R. Bodart, J.-P. Amoureux, Y. Dumazy, R. Lefort **2000**, *Mol. Phys.* 98, 1545–1551.
29. I. V. Koptyug, D. R. Sagdeev, E. Gerkema, H. Van As, R. Z. Sagdeev **2005**, *J. Magn. Res.* 175, 21–29.
30. I. V. Koptyug, R. Z. Sagdeev **2003**, *Russ. Chem. Rev.* 72, 183–212.
31. D. German, P. Chevallier, A. Laurent, H. Saint-Jalmes **2001**, *MAGMA* 13, 47–59.
32. N. Hosten, R. Felix, P. Wust, W. Wlodarczyk, M. Hentschel, R. Noeske, H. Rinneberg **1999**, *Phys. Med. Biol.* 44, 607–624.
33. B. Blumich, P. Blumler, K. Saito **1998**, in *Solid State NMR of Polymers*, (vol. 84), eds. I. Ando, T. Asakura, Studies in Physical and Theoretical Chemistry, Elsevier Science, Amsterdam, (pp.) 123–163.
34. S. J. Doran, T. A. Carpenter, L. D. Hall **1994**, *Rev. Sci. Instrum.* 65, 2231–2237.

5.5
In Situ Reaction Imaging in Fixed-bed Reactors Using MRI
Lynn F. Gladden, Belinda S. Akpa, Michael D. Mantle, and Andrew J. Sederman

5.5.1
Introduction

In chemical reaction engineering, there is a strong motivation to understand the interaction of hydrodynamic (i.e., diffusion, dispersion and flow) phenomena and chemical kinetics. MRI is a particularly attractive measurement tool in this application because it is entirely non-invasive thereby overcoming problems of introducing sampling points within the bed, which will cause a disturbance to the local hydrodynamics and, hence, potentially, influence the local conversion. To date, the implementation of MRI to study hydrodynamics in chemical reactor environments is significantly more advanced than studies of chemical conversion. Recent papers addressing hydrodynamics within reactors include: studies of single- and two-phase flow [1–7], biofilm formation [8] and electroosmotic perfusion [9] in fixed-bed reactors; cell distribution in bioreactors [10]; mass transfer phenomena in chromatographic columns [11–13]; and thermally polarized ^{1}H magnetic resonance (MR) microimaging studies of liquid and gas flow in monolithic catalysts [14]. Despite the relatively small number of papers that have addressed MRI of reaction processes, early reports of spatially resolved studies of reactions actually date back to 1978 when Heink et al. [15] used 1D profiling to study the time-resolved concentration profile of butane and water in packings of NaCaA and NaX crystallites respectively. Other early studies include the use of ^{19}F imaging to follow the intercalation of AsF_5 into highly oriented pyrolytic graphite at room temperature [16]. Since then interest has focused mainly on oscillations and traveling waves during chemical reaction [17–20]. Butler et al. [21] have employed ^{1}H MRI techniques to investigate the extent of reaction in a single crystal of 4-bromobenzoic acid during exposure to ammonia gas. A second case study used MRI to follow the reaction of a deep bed of toluic powder with ammonia gas flowing over it. The apparent reaction rate constant and the effective ammonia diffusion coefficient (into the toluic bed, perpendicular to the direction of flow) were obtained from a fit of a diffusion-reaction model to the experimental data.

Although interesting in their own right, the afore-mentioned studies are somewhat removed from mapping chemical conversion within a catalytic fixed-bed reactor. The most recent studies in this particular field include the work described in this chapter, along with those of Koptyug et al. [22] and Küppers et al. [23]. Koptyug et al. [22], discussed in greater detail elsewhere in this text (Chapter 5.4), considered the catalytic hydrogenation of α-methylstyrene within a catalytic fixed bed formed by a packing of spherical Pd–Al$_2$O$_3$ (1% by weight) catalysts pellets. Küppers et al. [23] studied reactions more similar in nature to those that are the subject of this chapter, but at the single catalyst particle level. The particular system studied was the enzymatically catalyzed esterification of propionic acid with 1-butanol to form propionic acid butyl ester and water inside an immobilizing hydrogel (2%-wt alginate in a 2%-wt calcium chloride solution containing 0.2 M NaCl). This work highlighted that when studying this type of hydrogel system, care must be taken to ensure that relaxation contrast in the spectral data has been accounted for; gels may hold a particular challenge in this regard as the degree of hydration and liquid composition within the gel matrix could influence local relaxation times and hence spectral intensities.

The obvious question to ask is, why have the developments in "reaction imaging" lagged behind studies of hydrodynamics? The most obvious reason for this is that it has taken some time for the reaction engineering community to recognize the opportunities and capabilities of applying MRI to systems of interest to them. However, it is also the case that many chemical reactions occur at high temperature and pressure, which makes the construction of the sample environment for *in situ* studies fairly demanding. With respect to the challenges in data acquisition as opposed to construction of the sample environment, as with any measurement, achieving adequate signal-to-noise is key to a successful measurement. This is why in the vast majority of MRI experiments, signal from the ^{1}H nucleus is acquired. The ^{1}H nucleus has a high MR sensitivity and exists in 100% natural abundance. The problem is that in any reaction mixture there is likely to be a wealth of ^{1}H resonances, many of which will overlap, making it impossible to follow the change in concentration of a particular species. This problem is exacerbated because of the interaction of the fluid phase with the solid-phase catalyst pellets, which causes the relaxation times of the fluid species to decrease. This further reduces the available signal-to-noise in the experiment and also broadens the spectral resonances, hence further increasing the overlap of individual resonances. For this reason ^{1}H studies are likely to be limited to simple reactions in which the ^{1}H spectral peaks are readily resolved in the frequency domain.

To reduce the complexity of the spectra acquired, we must abandon observation of the ^{1}H nucleus and employ, for example, ^{13}C, ^{31}P and ^{19}F imaging. ^{13}C is the most likely candidate species with which to study heterogeneous catalytic processes, given that the reactants and products in almost all reactions of interest are hydrocarbon in nature, but successful implementation of the technique is not straightforward because of the low abundance and sensitivity of the ^{13}C nucleus. However, the clear advantage of studying the ^{13}C nucleus is that in addition to the spectrum itself comprising fewer spectral resonances, the ^{13}C nucleus is associated with a much wider chemical shift range compared with ^{1}H, therefore enabling the individual spectral peaks to be more easily resolved. A first study exploiting ^{13}C to study *in situ* reactions, exploiting signal enhancement techniques, is discussed in Section 5.5.3.

In summary, it is non-trivial to implement magnetic resonance pulse sequences which allow us to monitor unambiguously the decrease in absolute concentration of reactant species and associated increase in product species, but measures of relative concentrations from which conversion and selectivity are calculated are much easier to obtain. However, if such measurements are to be deemed quantitative the spectra must be free of (or at least corrected for) relaxation time and magnetic susceptibility effects.

This chapter takes as case studies two similar reactions, both occurring within a fixed bed of ion-exchange resin catalyst, and summarizes the types of data that can be obtained. In Section 5.5.2, an esterification reaction is studied and both chemical shift imaging and volume selective spectroscopy experiments are described. In this example, ^{1}H NMR is used and the extent of conversion is calculated directly from the chemical shift of an OH resonance; individual spectral resonances associated with specific reactant and product species are not identified. This example reaction is used to demonstrate how flow through a bed introduces heterogeneity in conversion within the bed. Section 5.5.3 describes recent studies of a competitive reaction in which ^{13}C DEPT imaging has been implemented so that spectral resonances associated with specific reactant and product species are identified in the spectrum. The reaction chosen is the competitive etherification and hydration of 2-methyl-2-butene. By using the DEPT polarization transfer methodology, the signal enhancement is sufficient to monitor conversion and selectivity within the fixed bed without the need to resort to ^{13}C isotope enrichment of the reactants. Clearly as we move from studies of *in situ* reactions in sealed ampoule "microreactors" to real flow environments the cost of ^{13}C enrichment becomes prohibitively expensive. This case study suggests that polarization transfer techniques may offer significant opportunities for studying reactions *in situ* within continuous flow reactors.

5.5.2
Spatial Mapping of Conversion: Esterification Case Study

Before illustrating the type of data that can be obtained from spatially- and chemically-resolved MRI studies of reactions occurring in fixed-bed reactors, it is useful to clarify why such studies are of interest. Of course, it is of fundamental interest to explore the coupling of hydrodynamics and chemical kinetics in inter-connected porous structures. However, from a reaction engineering point-of-view, the question may be posed slightly differently. Given that MRI has already revealed significant heterogeneity in the flow within fixed-bed reactors [1–3], the obvious question to ask is – to what extent does flow heterogeneity generate heterogeneity in chemical conversion within the same system? The research described in this chapter begins to address this question. All MRI experiments were performed on a Bruker Spectrospin DMX 300 NMR spectrometer with a 7.4 T vertical magnet equipped with shielded gradient coils providing a maximum gradient strength of 1 T m^{-1} (100 G cm^{-1}). A birdcage radiofrequency coil tuned to 300.13 MHz for the ^{1}H resonance was used to irradiate the bed.

5.5.2.1 Characterizing Flow Heterogeneity
Yuen et al. [24] first demonstrated the nature of the information that can be obtained regarding chemical mapping within a fixed-bed reactor, using the liquid phase esterification of methanol and acetic acid catalyzed within a fixed bed of H$^+$ ion-exchange resin (Amberlyst 15, particle size 600–850-μm) catalyst as the model

reaction. As alluded to above, the aim of this work was to investigate the extent to which heterogeneity in hydrodynamics causes heterogeneity in conversion within the bed. A natural starting point is therefore to explore briefly the nature of the flow heterogeneity within this type of fixed bed and the magnetic resonance methods available to quantify it. Figure 5.5.1(a) shows an MR image of flow within the packed bed. Data are shown for a feed flow rate of 0.1 mL min^{-1}. Significant heterogeneity in liquid velocity is seen in this image; while the average superficial velocity through the slice of the bed shown is 0.056 mm s^{-1}, local superficial velocities in excess of 0.18 mm s^{-1} are also observed. The implication of this heterogeneity in flow is that the local residence times (i.e., contact times between catalyst and fluid) and mass transfer characteristics within the bed might be expected to differ significantly – if this is the case, we would expect to see significant local variations in catalytic conversion within the bed.

Another approach to quantifying the heterogeneity in flow within the bed, without employing spatially-resolved flow imaging is to acquire spatially unresolved relaxation-time weighted flow propagators, as shown in Figure 5.5.1(b). In this propagator measurement the total propagator measured for the system has been "decomposed" into two component propagators by combining the transport (i.e., conventional propagator) measurement pulse sequence with a spin–lattice relaxation time experiment. As the relaxation time of liquid molecules that have

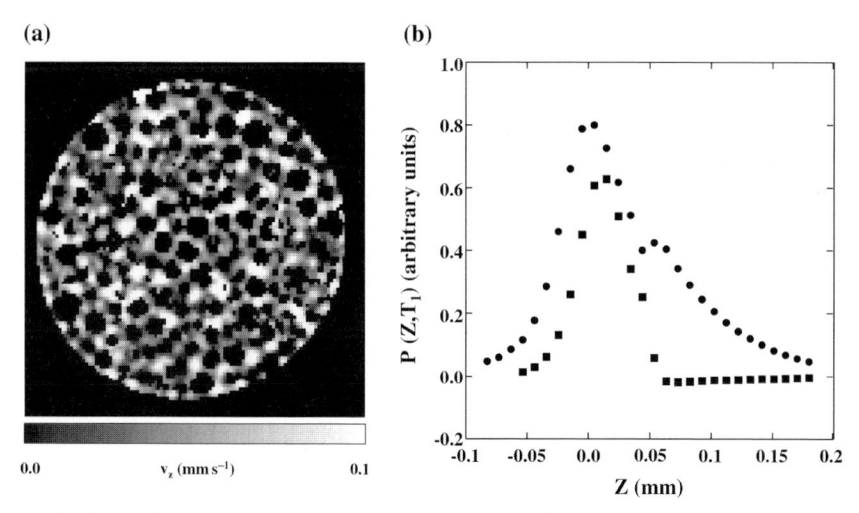

Fig. 5.5.1 (a) MR flow visualization for a feed flow rate of 0.1 mL min^{-1} through the bed shown in Figure 5.5.2. In-plane spatial resolution is 97.7 μm × 97.7 μm, and the image slice thickness is 500 μm. (b) T_1-resolved propagators for water flowing within the inter-particle space (●) of a bed packed with ion-exchange resin, and for water exchanging between inter- and intra-particle environments (■) during the time-scale of the transport measurement. Molecules that have existed within the particles throughout the measurement do not contribute to the "exchange" propagator as a result of relaxation time contrast effects [25]. Data are shown for a volumetric feed flow rate of 2 mL min^{-1}, to a column of internal diameter 20 mm. z is the direction of superficial flow.

existed *only* within the inter-particle space of the bed during the observation time (100 ms) of the experiment is significantly longer than that of molecules that have moved between the inter- and intra-particle space, independent propagators characterizing these two populations of liquid molecules can be obtained [25]. The data shown in Figure 5.5.1(b) were recorded for water flowing within a resin bed of diameter 20 mm with a volumetric feed flow rate of 2 mL min^{-1}. In Figure 5.5.1(b) the broader propagator, which shows two peaks, is that associated with the liquid in the inter-particle space. The two peaks are consistent with there being populations of very slow moving and much faster moving fluid within the bed [c.f. Figure 5.5.1(a)]. This type of T_1-weighted propagator measurement is also useful because it provides us with an estimate of the mass transfer coefficient [26] between the inter- and intra-particle space. The full-width at half maximum of the (narrower) "exchange" propagator provides an estimate of the effective diffusion coefficient, D_{eff}, for water molecules moving between the pore space of the catalyst and the inter-particle space of the bed of $\approx 2 \times 10^{-9}$ m^2 s^{-1}, which gives a lower limit to the value for the mass transfer coefficient of $\approx 4 \times 10^{-6}$ m^2 s^{-1}. This value is obtained by defining a mass transfer coefficient as D_{eff}/d where d is a typical distance traveled to the surface of the catalyst – we estimate this value to be half the typical dimension of a catalyst particle within the bed (≈ 500 μm) [25].

In the following sections (Sections 5.5.2.2 and 5.5.2.3), two approaches to spatially resolving chemical conversion within a reactor are demonstrated: (a) *n*-dimensional Chemical Shift Imaging (CSI), and (b) volume selective spectroscopy.

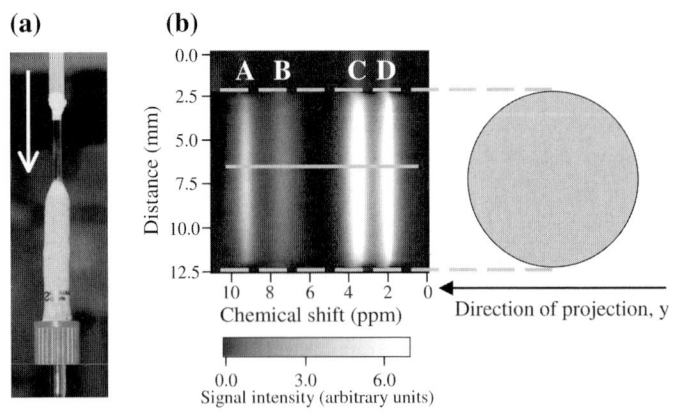

Fig. 5.5.2 (a) Photograph of the reactor used. Washed Amberlyst 15 ion-exchange resin is packed within a glass column of internal diameter 10 mm. The direction of superficial flow, along z, is shown. (b) A 1D CSI map of chemical conversion. Taking the direction of projection to be the y direction, each 2D dataset consists of a series of spectra from different positions in the x direction (i.e., along a direction perpendicular to both the direction of projection and the direction of superficial flow). Therefore, each 1D spectrum (such as that shown in Figure 5.5.3) represents a projection of all the spectral information in an entire yz plane for a given position in x, and is obtained from the 2D dataset by taking a horizontal cut, as shown by the solid, horizontal line. The data acquisition time was 2 min.

The same reactor configuration was used in all the reaction imaging experiments; that is a fixed-bed reactor of internal diameter 10 mm [see Figure 5.5.2(a)]. Once the packed bed was positioned inside the NMR magnet, the bed was exposed to a standard pre-treatment procedure. This pre-treatment protocol consisted of flowing an equimolar methanol–acetic acid mixture, at a flow rate of 3 mL min^{-1} through the bed for a period of 30 min so as to establish a standard condition (i.e., a uniform distribution of equimolar feed) throughout the bed before starting the reaction. An equimolar feed was used in all reactions. The reactions were performed at an ambient temperature of 22 °C.

5.5.2.2 Effect of Flow Heterogeneity on Conversion: 1D CSI

Before discussing the results obtained from this study it is important to understand the method by which a measurement of conversion was obtained directly from the spectra acquired from the reaction being studied *in situ*. The reaction chosen for study was the esterification reaction of methanol and acetic acid to form methyl acetate and water:

$$CH_3OH + CH_3COOH \quad CH_3COOCH_3 + H_2O \tag{5.5.1}$$

This reaction was chosen because it proceeded under ambient conditions and gave a relatively simple spectrum that enabled quantitative analysis to be performed.

The ^1H spectrum recorded from a sample containing a packed bed of catalyst and the reaction mixture (i.e., acetic acid, methanol, methyl acetate and water) is shown in Figure 5.5.3. Four peaks are observed which are assigned unambiguously on the basis of previous theoretical and experimental studies. Peaks C and D are assigned to the ^1H resonance of the CH_3O groups associated with the methanol and methyl acetate species, and the CH_3 groups of the acetic acid and methyl acetate, respectively. Peaks A and B are the ^1H OH group resonances present in the reaction mixture and are associated with liquid inside the catalyst particles and within the inter-particle space, respectively. The absolute chemical shifts of peaks A and B are dependent on the extent of the reaction as defined by Eq. (5.5.2).

$$\delta_{observed} = x_{AcOH}\delta_{AcOH} + x_{MeOH}\delta_{MeOH} + n x_{H_2O}\delta_{H_2O} \quad ; \Sigma x_i = 1 \tag{5.5.2}$$

where δ_i is the chemical shift associated with the pure compound i, and x_i is the mole fraction of species i in the mixture; all chemical shifts are referenced to tetramethylsilane (TMS). n is the "number" of ^1H species associated with OH groups within the water molecule; the physical interpretation of this parameter has been discussed in detail elsewhere [24]. The form of Eq. (5.5.2) arises because of the phenomenon of ^1H fast-exchange (i.e., occurring on a timescale <10^{-6} s) [27] between OH groups associated with the acetic acid (AcOH), methanol (MeOH) and water (H_2O) molecules present within the reaction mixture. Given that ^1H species associated with a given pure component have a specific chemical shift, if individual ^1H species are in fast exchange between the OH groups on different

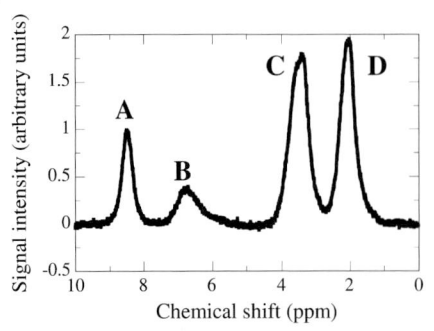

Fig. 5.5.3 ^{1}H NMR spectrum obtained from a sample containing catalyst and reaction mixture (i.e., methanol, acetic acid, methyl acetate and water). Peaks A and B are the intra- and inter-particle ^{1}H resonances, respectively, associated with the OH group. Peak C is the ^{1}H resonance of the CH$_3$O group associated with methanol and methyl acetate species, and peak D is the ^{1}H resonance of the CH$_3$ group associated with acetic acid and methyl acetate species. The chemical shift of the feature associated with peak B is used to measure the extent of chemical conversion. Chemical shifts are referenced to TMS.

molecular species, the resultant chemical shift observed for the ^{1}H species associated with OH groups present in the reaction mixture (δ_observed) will be given by an average of the component chemical shifts, weighted by the amount of each molecular species present. In the esterification reaction considered here, as conversion increases so the ^{1}H resonance associated with the OH groups moves to a lower chemical shift, referenced to TMS; i.e., towards the ^{1}H chemical shift of pure water. Thus, the value of δ_observed provides an accurate measurement of the chemical composition of a given reaction mixture. The position of peak B *only* is used in the quantitative determination of the mole fraction of acetic acid present and hence the extent of conversion in the bed. This choice is made for two reasons. Firstly, as we are interested in the conversion characterizing the liquid leaving the bed as opposed to the concentration of the liquid within the catalyst particles, analysis of peak B seems most appropriate in this study. Secondly, analysis of peak A is subject to greater errors, which are difficult to quantify. In addition to possible relaxation contrast effects, there will also be modifications to the chemical shifts of individual species resulting from adsorption onto the catalyst; this may cause peak broadening and reduces the accuracy with which we can determine the chemical shift.

Having determined δ_observed, the mole fraction of acetic acid present is obtained from Eq. (5.5.3), which takes into account the stoichiometry of the reaction [Eq. (5.5.1)] and the fact that an equimolar feed was used. Equation (5.5.3) shows that the mole fraction of acetic acid present in the reaction mixture is obtained directly from the chemical shift values of the ^{1}H resonance of the OH groups of the pure compounds and the observed chemical shift of the OH resonance of the reaction mixture:

$$x_\text{AcOH} = \frac{\delta_\text{observed} - \frac{1}{2}(n\delta_{H_2O})}{\delta_\text{AcOH} + \delta_\text{MeOH} - (n\delta_{H_2O})} \tag{5.5.3}$$

Once the mole fraction of acetic acid has been obtained, calculation of the conversion is then trivial, and we have defined the extent of conversion at a given time t, $X(t)$, as:

$$1 - X(t) = \frac{x_{\text{AcOH}}(t)}{x_{\text{AcOH}}(t=0)} \tag{5.5.4}$$

where $x_{\text{AcOH}}(t=0)$ is the mole fraction of acetic acid present at the beginning of the experiment (i.e., the feed composition). Care has to be taken in the determination of the chemical shift value assigned to δ_{observed} as in practice δ_{observed} may be influenced by effects such as local variations in magnetic susceptibility in addition to the extent of conversion. To ensure that the determination of the chemical shift is a true measure of mixture composition, δ_{observed} is reported relative to the position of a non-shifting ^{1}H peak; in this case we have chosen the ^{1}H resonance of the CH_3 group, which occurs at 2.05 ppm, relative to TMS. An upper limit on the error in conversion is $\approx 2\%$ for a given set of experimental conditions. The quantitative nature of the MR analysis was confirmed by obtaining agreement with an independent measure of conversion obtained by titration of the effluent stream against sodium hydroxide from which the concentration of acetic acid is again obtained directly. Clearly, in this particular example, chemical shift provides an elegant measure of conversion and avoids errors in determining concentrations based on analysis of spectral intensities that may be influenced by line broadening and relaxation time effects. However, this approach can only be used when the spectrum is sufficiently simple that all the spectral resonances can be assigned unambiguously.

To provide the first evidence that MRI could identify variations in spatial conversion within a fixed bed, a 1D Chemical Shift Imaging (CSI) pulse sequence was used; in more recent work this has been extended to full 3D CSI. These pulse sequences may be more correctly referred to as 2D and 4D experiments as we have a spectral dimension as well as 1- and 3-spatial dimensions, respectively [28]. Experimental and theoretical aspects of n-dimensional CSI are discussed in greater detail elsewhere [28] but, in summary, the chemically-resolved image is achieved by spatially encoding the MR signal prior to reading in the absence of a magnetic field gradient. An example of this pulse sequence is shown in Figure 5.5.4; this pulse sequence yields a spectral dimension in addition to two spatial dimensions, the latter being achieved by the ramped phase gradients in x and y. Figure 5.5.2(b) shows the result of implementing a 1D CSI pulse sequence to produce a 1D ^{1}H NMR spectrum across a transverse section (perpendicular to the direction of superficial flow) through the fixed bed shown in Figure 5.5.2(a), averaged over a 13-mm length of the bed. The total acquisition time was 2 min. Any horizontal cut through this 2D "map" generates a spectrum of the form shown in Figure 5.5.3.

A simple experiment serves to demonstrate that flow introduces a range of conversions within this fixed bed. Following the standard pre-treatment of the bed, the bed was used (a) to perform a batch reaction (i.e., under zero flow conditions), and (b) flow was then introduced to the bed and the reaction operated continuously until steady state had been achieved, at which point the conversion

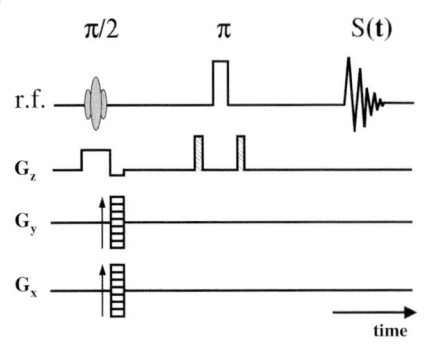

Fig. 5.5.4 A CSI pulse sequence. The MR signal is spatially encoded prior to acquiring the spectral signal in the absence of any applied magnetic field gradients. The shaded gradient pulses applied along z either side of the π refocusing pulse, are homospoil gradients. This pulse sequence acquires an image with two spatial dimensions x and y. At the beginning of the pulse sequence a slice-selective gradient pulse is applied along z.

within the bed was studied. The time to reach steady state had previously been determined by monitoring the exit stream composition as a function of time using titration analysis. The 1D CSI data clearly show different conversion behavior under these two different operating conditions.

Figure 5.5.5 shows 1D CSI datasets for batch reactions (i.e., no flow through the bed); data being recorded for reaction times of 0, 15, 60 and 240 min after the start of reaction. It is seen that the position of the feature associated with peak B moves to lower values of the chemical shift as reaction proceeds but does not change in form; i.e., it remains centerd at a single value of the chemical shift and does not show significant broadening or curvature. Hence, the data show that the extent of chemical conversion is constant throughout the bed at any given time-point in the reaction. In particular, the chemical shift of peak B occurs at 8.1, 7.2, 6.4, and 5.6

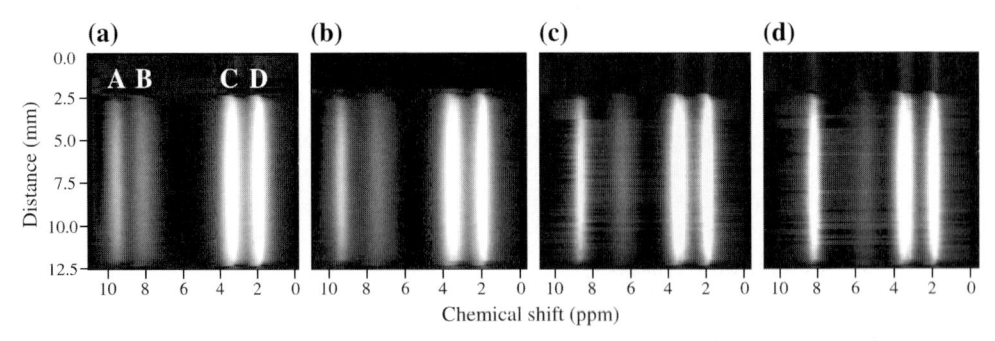

Fig. 5.5.5 1D CSI datasets showing the extent of conversion during a batch reaction. The form of the feature identified as peak B is associated with a single chemical shift; i.e., it is of constant form at all positions across the bed, and therefore shows that the extent of conversion is uniform throughout the bed. The low intensity horizontal "streaking" effect observed in these datasets and that shown in Figure 5.5.6 are artifacts arising from the automatic phase correction applied to the data; this artifact does not affect the position of any of the features in the chemical shift dimension and is readily distinguishable from any real chemical shifts effects arising from variations in chemical conversion within the bed. The chemical shift of peak B gives conversions of 0, 24, 46 and 67% for the reaction at times (a) 0, (b) 15, (c) 60 and (d) 240 min after the start of the reaction, respectively. Reproduced with permission from Ref. [24], copyright Elsevier (2002).

(±0.1 ppm, referenced to TMS) at the four times studied, corresponding to reaction conversions of 0, 24, 46 and 67% at the reaction times of 0, 15, 60 and 240 min, respectively. To examine the extent to which flow within the bed causes heterogeneity in conversion, a further CSI image was acquired after the bed had been operating at conditions of steady-state for 3 h with an equimolar feed of reactants. Figure 5.5.6 shows the resulting 1D CSI dataset. In comparing the form of the features associated with peaks A and B for the case of reaction under flowing and non-flowing conditions, it is clearly seen that under flowing conditions both features broaden significantly to include a range of chemical shifts, indicating that a range of conversions exist within the bed. As we are averaging all information along the longitudinal axis of the bed (i.e., the direction of the superficial flow) the range of conversions we are observing may simply arise from the range of conversions down the bed or it may also reflect spatial variations in conversion across the bed in the *xy* plane. However, in its own right the 1D CSI experiment enables a relatively rapid *in situ* measurement of the extent of heterogeneity of conversion within the bed under flowing conditions.

5.5.2.3 Exploring Spatial Heterogeneity in Conversion: Volume Selective Imaging

To explore in more detail the extent of heterogeneity in conversion that may exist within the bed, volume selective spectroscopy has been used. Figure 5.5.7 shows a 2D vertical section through the 3D image of the bed; the image slices in which the local volumes are located for the volume selective spectroscopy study are also identified. The three-dimensional (3D) volume image of the packed bed was acquired using a RARE-based [29] fast 3D MRI sequence. The image was acquired by saturating the bed with pure methanol. The image acquisition parameters were then set such that signal was acquired only from the methanol in the inter-particle space; methanol inside the catalyst particles is characterized by much shorter nuclear spin relaxation times and can therefore be made "invisible" to the magnetic resonance experiment. The data were recorded with an isotropic resolution 97.7 μm × 97.7 μm × 97.7 μm.

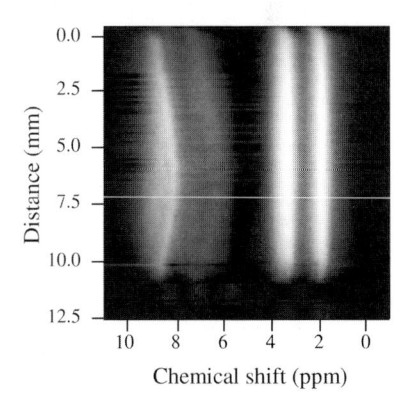

Fig. 5.5.6 1D CSI dataset recorded during flow, after steady-state has been achieved. The shape of the feature associated with peak B has broadened and is curved, showing that a range of conversions exists within the bed, there being greater conversions (i.e., smaller chemical shifts) in the central regions of the bed. Reproduced with permission from Ref. [24], copyright Elsevier (2002).

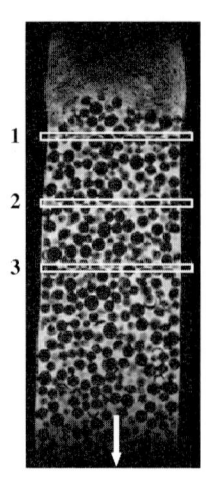

Fig. 5.5.7 A 2D slice through a ^{1}H 3D MR image of the fixed-bed of catalyst particles. The catalyst particles appear as black; fluid within the inter-particle space is indicated by lighter shades. Chemical conversion within ten selected volumes within each of the three transverse sections indicated is investigated in Figures 5.5.9–5.5.11. The direction of superficial flow (z) is also shown. Reproduced with permission from Ref. [24], copyright Elsevier (2002).

Figure 5.5.8 shows the volume selective spectroscopy pulse sequence used. This pulse sequence combines elements of NMR spectroscopy and MRI pulse sequences; three slice selective rf pulses are applied in three orthogonal directions to obtain ^{1}H spectra from pre-determined local volumes within the sample [30]. In this particular application, spectra were recorded from local volumes of dimension 1.5 mm × 1.5 mm × 0.5 mm within the fixed bed; the data acquisition time for each spectrum was 3 min. Figure 5.5.9(a) shows the local volumes selected within slice 3 of the bed, as identified in Figure 5.5.7. In Figure 5.5.9(b) the ^{1}H spectra recorded from within these volumes are shown; data are presented only for the range of

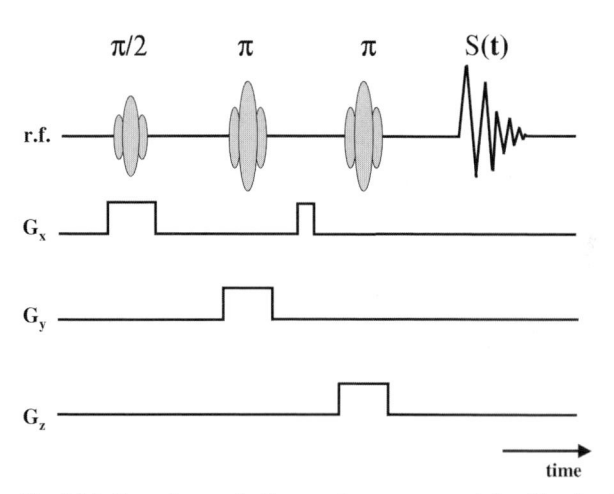

Fig. 5.5.8 The volume selective spectroscopy rf pulse sequence used to acquire the data shown in Figures 5.5.9–5.5.11. Magnetic field gradients applied along the x, y and z directions enable localized spectra to be recorded from pre-defined local volume elements within the sample. The spectrum recorded from a given volume averages the data acquired over the entire local volume.

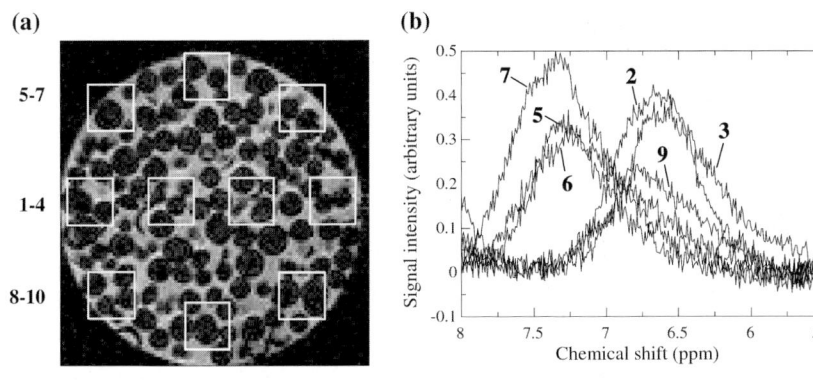

Fig. 5.5.9 Volume selective spectroscopy within the fixed-bed reactor. (a) The location of the ten selected volumes within image slice 3 (see Figure 5.5.7) are identified. (b) It is clearly seen that the chemical shifts of peak B obtained from the spectra recorded from the selected volumes occur at significantly different values of chemical shift and therefore reflect significant variation in the extent of chemical conversion within this transverse section through the bed. The data acquisition time for each spectrum was 3 min. Reproduced with permission from Ref. [24], copyright Elsevier (2002).

chemical shifts containing peak B (identified in Figure 5.5.3). It is clearly seen that there exists a range of conversions within this transverse slice section through the bed, as identified by the different chemical shifts associated with peak B in the ^{1}H spectra recorded from the various selected volumes. Indeed, through this particular section of the bed a fractional variation in conversion ($\Delta X/\bar{X}$) of 22% is observed, where $\bar{X}$ is the mean conversion calculated from the ten local volume measurements of conversion.

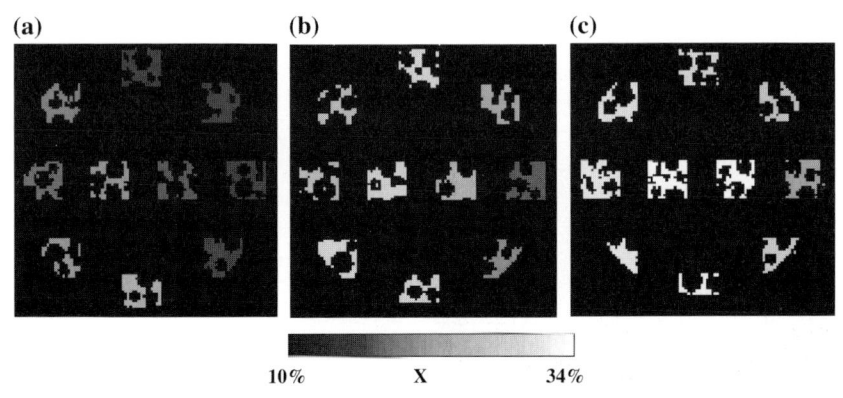

Fig. 5.5.10 Visualization of conversion, X, within slices 1, 2 and 3 (identified in Figure 5.5.7) through the same bed. Data are shown for a feed flow rate of 0.05 mL min^{-1}. The grey scale indicates the percentage conversion; lighter shades indicate higher conversion. While the mean conversion increases along the direction of superficial flow, significant heterogeneity in conversion within each transverse section exists throughout the length of the bed. Reproduced with permission from Ref. [24], copyright Elsevier (2002).

602 | 5 Reactors and Reactions

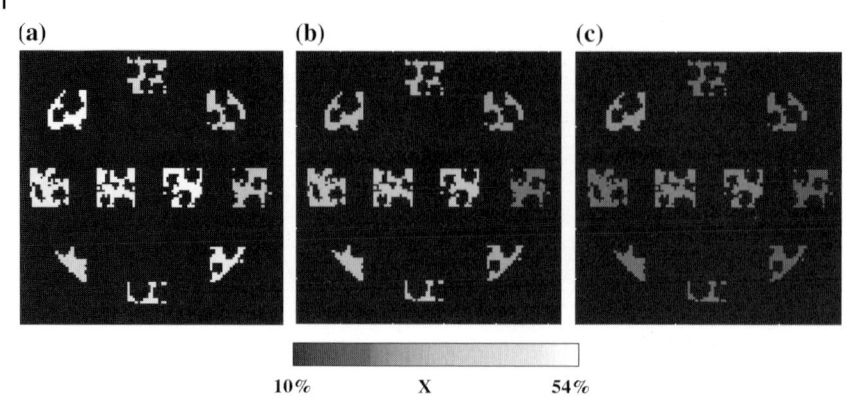

(a) **(b)** **(c)**

10%   X   54%

Fig. 5.5.11 Visualization of mean conversion within selected volumes located in slice 3 (identified in Figure 5.5.7) through the bed of ion-exchange resin particles. Data are shown for feed flow rates of (a) 0.025, (b) 0.05 and (c) 0.1 mL min^{-1}. Higher conversion is observed for lower feed flow rates. Reprinted from Ref. [31], with kind permission of Springer Science and Business Media.

Figures 5.5.10 and 5.5.11 show another way of representing these data in which the spatial location of the selected volumes is identified and the extent of conversion (averaged within each local volume) is indicated by the grey scale. Figure 5.5.10 shows the variation in conversion within the three slice sections identified in Figure 5.5.7; lighter shades indicate higher conversion. Clearly, conversion increases down the bed, but substantial heterogeneity in conversion exists within all three slice sections. In Figure 5.5.11, the effect of feed flow rate on conversion within the same transverse slice section through the bed is shown [31]. As expected, higher conversion is observed for lower feed flow rates. Again, it is clearly seen that significant heterogeneity in the extent of conversion exists within all slice sections through the bed.

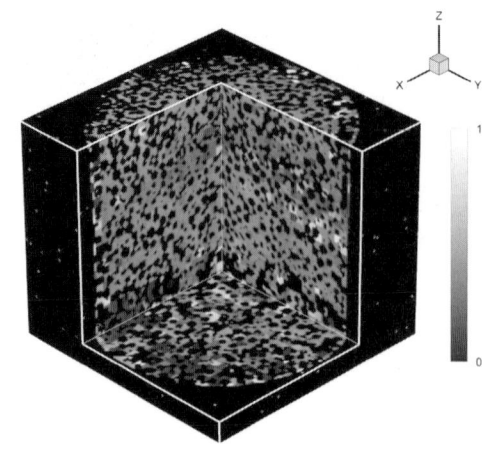

Fig. 5.5.12 3D cutaway image showing the extent of conversion of the esterification reaction occurring within the fixed bed. The conversion was calculated from the chemical shift of the OH peak (peak B in Figure 5.5.3) in a 4D CSI experiment. The image was acquired with an isotropic spatial resolution of 625 μm and the data have been interpolated onto a high-resolution spatial image of the bed shown at an isotropic spatial resolution of 156 μm. The direction of flow is in the negative z direction. The grey scale indicates the fractional conversion within the bed.

Studies of this reaction have recently been extended to acquisition of a 3(4)D CSI dataset, shown in Figure 5.5.12; the grey scale indicates the extent of conversion. As expected from the 1(2)D CSI and volume selective imaging studies discussed earlier, conversion is seen to be heterogeneous within transverse sections through the bed at any position along the direction of superficial flow.

5.5.3
^{13}C DEPT Imaging of Conversion and Selectivity

The esterification study of Yuen et al. [24] provided a clear demonstration of the heterogeneity in conversion within a fixed bed. However, this particular study was only possible because the reaction mixture was associated with a relatively simple ^{1}H NMR spectrum. The spectrum gave only four spectral peaks with the chemical shift of one being a direct measure of the inter-particle liquid composition and hence the extent of conversion. Akpa et al. [32] have recently used ^{13}C spatially-resolved spectroscopy to follow a reaction in which competing etherification and hydration reactions are occurring. As discussed earlier, ^{13}C observation may have potential advantages in that the ^{13}C nucleus has a wider chemical shift range making the spectral resonances of individual molecular species better resolved, hence we can follow the loss of reactants and the production of product molecules without need for spectral deconvolution. However, the disadvantage of using ^{13}C is its inherent low abundance and sensitivity, causing low signal-to-noise in the acquired data. These problems are overcome by practitioners of solid-state MR techniques by using isotropically enriched reactants. The cost of doing this for the larger volumes required for continuous flow reactor studies will, however, be significant. The alternative approach, illustrated here, is to explore the extent to which reactants containing only natural abundance ^{13}C can be used by employing MR signal enhancement techniques. Typically, such signal enhancement is achieved by transferring polarization from the ^{1}H spin population to the low gyromagnetic ratio, low natural abundance species – in this case ^{13}C. In the example considered here the competing etherification and hydration reactions of 2-metyl-2-butene (2M2B) are followed within a fixed bed of H$^+$ ion-exchange resin; the resin type being the same as that used in the esterification reaction described in Section 5.5.2. The reactants used were 2-methyl-2-butene, methanol and water, and the products of the etherification and hydration reactions are *tert*-amyl methyl ether (TAME, or 2-methoxy-2-methylbutane) and *tert*-amyl alcohol (TAOH, or 2-methyl-butan-2-ol), respectively.

The imaging of conversion within the fixed bed was achieved using a ^{13}C Distortionless Enhancement by Polarization Transfer (DEPT) spectroscopy pulse sequence [33] integrated into an imaging sequence [34], as shown in Figure 5.5.13. In theory, a signal enhancement of up to a factor of 4 (γ_H/γ_C; γ_i is the gyromagnetic ratio of nucleus i) can be achieved with the ^{13}C DEPT technique. In this dual resonance experiment, initial excitation is on the ^{1}H channel. Consequently, the repetition time for the DEPT experiment is constrained by T_{1H} ($<T_{1C}$); where T_{1i} is the T_1 relaxation time of nucleus i. With reference to Figure 5.5.13, the first ^{1}H

pulse generates pure ^{1}H magnetization, which subsequently evolves under the influence of chemical shift and spin-spin coupling Hamiltonians. At the end of the first evolution period, the ^{13}C $\pi/2$ pulse then converts this magnetization to zero and double quantum coherence. At the same time as the ^{13}C $\pi/2$ pulse, a ^{1}H π pulse is applied to refocus the effects of the ^{1}H chemical shift during the second evolution period. The spin-spin coupling has no effect on the evolution of the multiple quantum coherence during the second evolution period. Finally, a ^{13}C π pulse is applied (again to refocus the effects of ^{13}C chemical shift evolution), while a simultaneous θ pulse is applied to multiple quantum coherence and converts this to anti-phase ^{13}C magnetization. Further evolution converts this (initially unobservable) anti-phase magnetization into pure ^{13}C magnetization, the intensity of which is proportional to sin θ. Note when a θ pulse is applied some multiple quantum coherence remains at the end of the third evolution period, but is not observed. To achieve spatial resolution of the whole experiment, a double phase encoding orthogonal pair of gradients is applied during the third evolution period. Figure 5.5.14 is a schematic diagram showing how the double-phase encoded DEPT sequence achieves both spatial and spectral resolution within the reactor; the volumes identified, from which spectra are recorded, relate to the data shown in Figures 5.5.15 and 5.5.16. Figure 5.5.14(a) shows a spin-echo ^{1}H two-dimensional image taken through the reactor overlayed with a grid showing the spatial location within the column of the two orthogonal phase encoded planes used in the modified DEPT sequence. Figure 5.5.14(b) shows the corresponding real space volume elements that the double-phase encoded pulse sequence produces.

Typical data acquired using this pulse sequence are shown in Figure 5.5.15; the reaction temperature was 40 °C. All experiments were performed using a Bruker DMX 300 spectrometer with a 7.0 T vertical magnet equipped with shielded gradient coils providing a maximum gradient strength of 100 G cm^{-1}. A birdcage rf coil of

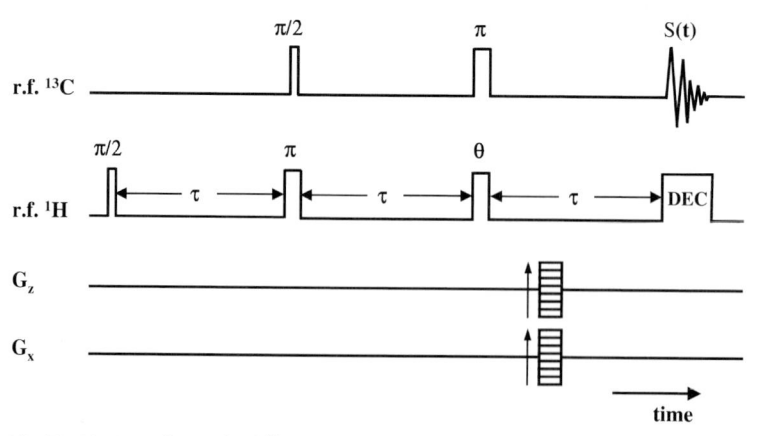

Fig. 5.5.13 Spatially resolved ^{13}C DEPT pulse sequence. This provides signal enhancement for ^{13}C observation without need for using isotropically enriched materials. The signal is acquired under conditions of ^{1}H decoupling.

diameter 20 mm – dual tuned to 300 and 75.5 MHz for the ^{1}H and ^{13}C resonances, respectively – was used. Although non-standard, such hardware can be obtained from commercial suppliers. The data shown here have been recorded from a vertical section through the center of the bed [column (iii) in Figure 5.5.14]. The direction of superficial flow is from the bottom to the top of the bed. The spectra shown have been recorded at regular intervals along the length of the bed. The centers of adjacent image volume elements are separated by a distance of 2.5 mm. With reference to Figure 5.5.15, the conversion and selectivity is quantified *in situ* as follows. Conversion is measured by taking the ratio of the peak areas corresponding to the products and reactants. The spectral resonances of the same carbon group needs to be compared between molecular species as the degree of polarization transfer (i.e., signal enhancement) is dependent on the chemical environment of each specific carbon atom. Of course, if the extent of polarization transfer to different carbon

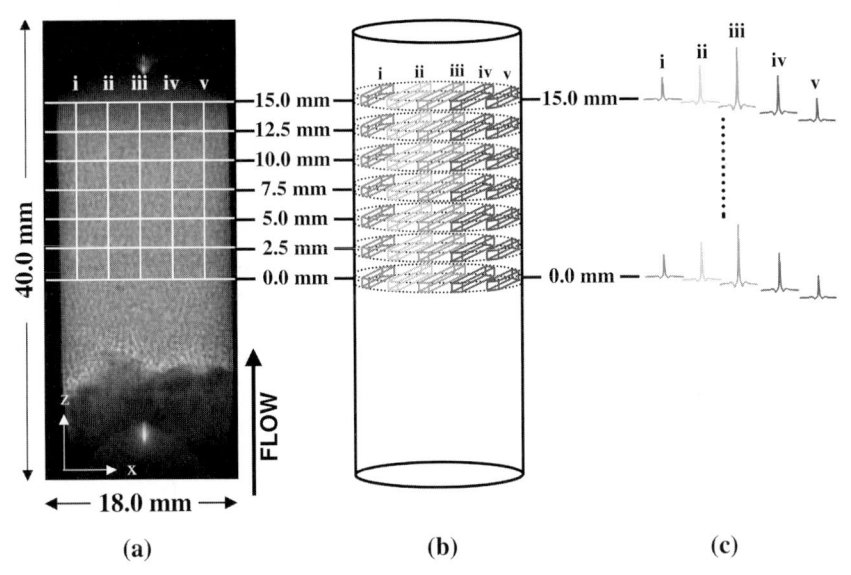

Fig. 5.5.14 Schematic diagram showing how the double-phase encoded DEPT sequence achieves both spatial and spectral resolution within the reactor. (a) A spin-echo ^{1}H 2D image taken through the column overlayed with a grid showing the spatial location within the column of the two orthogonal phase encoded planes (z and x) used in the modified DEPT sequence. The resulting data set is a zx image with a projection along y. In-plane spatial resol-ution is 156 μm (z) × 141 μm (x) for a 3-mm slice thickness. The center of each volume from which the data have been acquired is identified by the intersection of the white lines. The arrow indicates the direction of flow. (b) The corresponding real space volume elements that the double-phase encoded pulse sequence produces. Note the individual volume elements have been separated for clarity but actually form a continuum. (c) ^{13}C NMR spectra assoc-iated with the volume elements shown in (b). The decreasing signal intensity towards the walls of the bed (in the x direction) arises from the smaller volumes (i.e., smaller y dimension) from which data are sampled in these regions. Measurements of conversion and selectivity are made by considering ratios of peak areas within a single spectrum and are therefore independent of this effect.

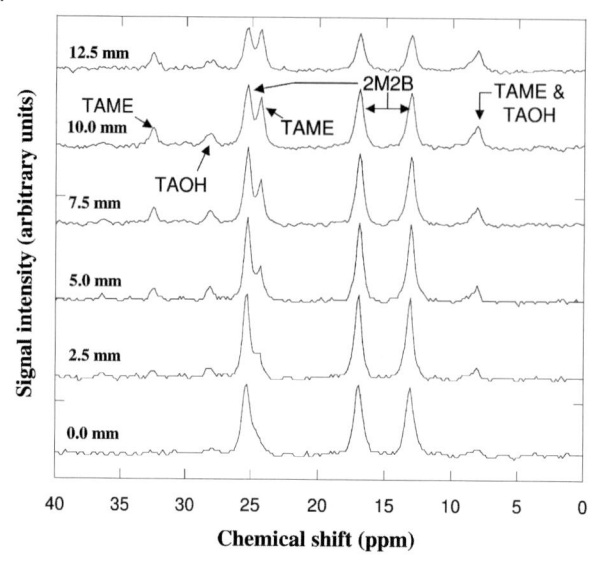

Fig. 5.5.15 Spatially resolved ^{13}C DEPT spectra recorded for the competitive etherification and hydration reactions of 2-methyl-2-butene (2M2B) to 2-methoxy-2-methylbutane (*tert*-amyl methyl ether, TAME) and 2-methyl-butan-2-ol (*tert*-amyl alcohol, TAOH), respectively. The molar composition of the feed was in the ratio 2:10:1 for 2M2B:methanol:water. The spectral assignments, relative to TMS, are as follows: 2M2B (CH$_3$: 13.4, 17.3, 25.7 ppm; CH: 118.7 ppm); TAME (CH$_3$: 7.8, 24.1, 48.5 ppm; CH$_2$: 32.2 ppm); TAOH (CH$_3$: 8.7, 28.7 ppm; CH$_2$: 36.5 ppm). The spectra were recorded at six different vertical positions along the length of the bed, as identified in Figure 5.5.14.

groups has been calibrated then comparison of different carbon-group resonances can also be used in the analysis. In the example shown here, the CH$_3$ resonances have been compared; for the products TAME and TAOH these occur at ≈ 8 ppm (CH$_3$ resonances for TAME and TAOH occur at 7.8 and 8.7 ppm, respectively), while those of the reactant 2M2B occur at 13.4, 17.3 and 25.7 ppm. All chemical shifts are quoted with respect to the ^{13}C resonance of tetramethylsilane (TMS). Where overlapping resonances occur, a peak fitting algorithm has been used to obtain the integral of the spectral resonance of interest. Selectivity to TAME is quantified by comparing the intensity of the CH$_3$ resonances of TAME and TAOH, which occur at ~ 25 and 28 ppm, respectively. Figure 5.5.16 shows conversion (a–c) and selectivity (d–f), respectively, across three transverse sections through the bed. It is seen that there exists considerable heterogeneity in conversion within the three transverse sections along the 10-mm length of the bed studied, while the selectivity remains constant, to within experimental error, throughout the bed.

5.5.4
Future Directions

The application of MRI to study hydrodynamics within reactors is now well established. However, the extent to which the potential of MR to study both

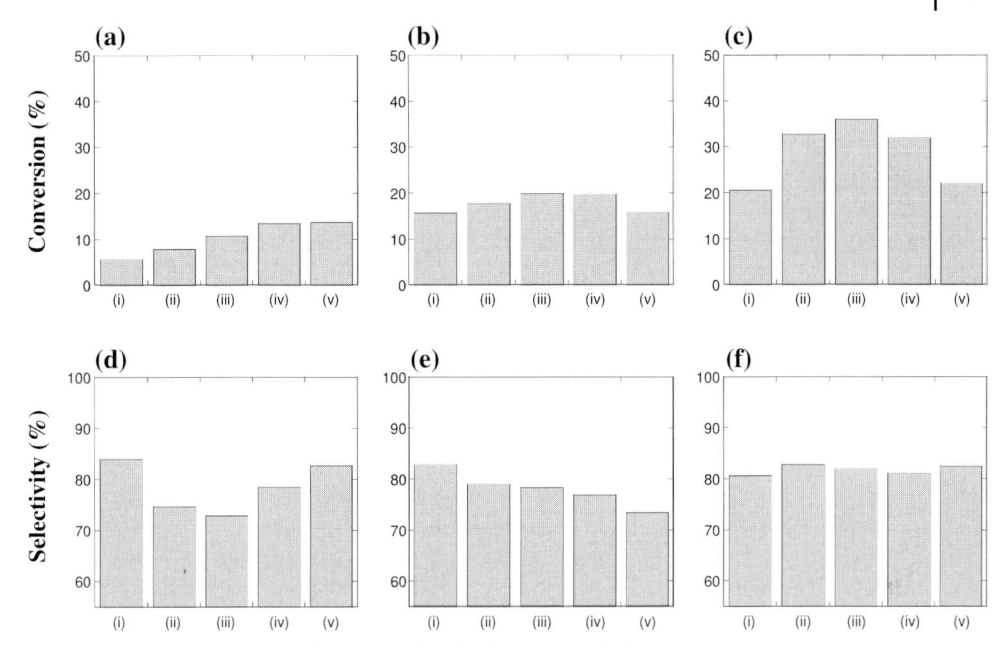

Fig. 5.5.16 Conversion of 2M2B (a–c) and selectivity to TAME (d–f) within three transverse slices through the bed; the positions of the slices are those indicated in Figure 5.5.14 at (a, d) 2.5, (b, e) 7.5, (c, f) 12.5 mm along the length of the bed. Locations (i)–(v) also correspond to those shown in Figure 5.5.14

hydrodynamics and chemical conversion can be fully realized will depend on our ability to integrate the well established MR spectroscopy techniques in liquid and solid-state NMR into imaging pulse sequences, such that they still provide quantitative data in the magnetically heterogeneous environments typical of catalysts and reactors.

In this chapter we have shown how 1D and 3D ^{1}H CSI, volume selective ^{1}H NMR spectroscopy and a double-phase encoded ^{13}C DEPT sequence can be used to provide *in situ*, spatially resolved information on the extent of conversion for reactions occurring in a fixed-bed catalytic process. In both reactions studied, flow heterogeneity within the bed causes heterogeneity in conversion, which can be monitored using MRI. In the case of the esterification reaction, individual reactant and product resonances were not resolved, but in the etherification/hydration study this was achieved using ^{13}C DEPT. This latter study suggests that use of spatially resolved polarization transfer MRI experiments might be particularly useful in monitoring conversion *in situ* within fixed-bed reactors, and our ongoing research [35] uses both the hydrodynamic and conversion/selectivity mapping MRI studies integrated with numerical simulations to develop predictive models of the behavior of reacting systems in fixed-bed reactors.

Acknowledgements

We acknowledge Dr. P. Alexander for his help with the preparation of this text.

References

1. A. J. Sederman, M. L. Johns, P. Alexander, L. F. Gladden **1998**, *Chem. Eng. Sci.* 53, 2117–2128.
2. J. Park, S. J. Gibbs **1999**, *AIChE J.* 45, 655–660.
3. M. L. Johns, A. J. Sederman, A. S. Bramley, P. Alexander, L. F. Gladden **2000**, *AIChE J.* 46, 2151–2161.
4. A. J. Sederman, L. F. Gladden **2001**, *Chem. Eng. Sci.* 56, 2615–2628.
5. J. Gotz, K. Zick, C. Heinen, T. Konig **2002**, *Chem. Eng. Process.* 41, 611–629.
6. T. Suekane, Y. Yokouchi, S. Hirai **2003**, *AIChE J.* 49, 10–17.
7. D. Tang, A. Jess, X. Ren, B. Blumich, S. Stapf **2004**, *Chem. Eng. Technol.* 27, 866–873.
8. M. Paterson-Beedle, K. P. Nott, L. E. Macaskie, L. D. Hall **2001**, *Microb. Growth Biofilms Part B* 337, 285–305.
9. U. Tallarek, E. Rapp, H. van As, E. Bayer **2001**, *Angew. Chem., Intl. Ed. Engl.* 40, 1684–1687.
10. S.O. Williams, R. M. Callies, K. M. Brindle **1997**, *Biotechnol. Bioeng.* 56, 56–61.
11. U. Tallarek, D. van Dusschoten, H. van As, E. Bayer, G. Guiochon **1998**, *J. Phys. Chem.* 102, 3486–3497.
12. U. Tallarek, D. van Dusschoten, H. van As, G. Guiochon, E. Bayer **1998**, *Magn. Reson. Imag.* 16, 699–702.
13. U. Tallarek, F. J. Vergeldt, H. van As **1999**, *J. Phys. Chem. B* 103, 7654–7664.
14. I. V. Koptyug, S. A. Altobelli, E. Fukushima, A. V. Mateev, R. Z. Sagdeev **2000**, *J. Magn. Reson.* 147, 36–42.
15. W. Heink, J. Kärger, H. Pfeifer **1978**, *Chem. Eng. Sci.* 33, 1019–1023.
16. G. C. Chingas, J. B. Miller, A. N. Garroway **1986**, *J. Magn. Reson.* 66, 530–535.
17. A. Tzalmona, R. L. Armstrong, M. Menzinger, A. Cross, C. Lemaire **1992**, *Chem. Phys. Lett.* 188, 457–461.
18. M. Menzinger, A. Tzalmona, R. L. Armstrong, A. Cross, C. Lemaire **1992**, *J. Phys. Chem.* 96, 4725–4727.
19. B. J. Balcom, T. A. Carpenter, L. D. Hall **1992**, *Macromolecules* 25, 6818–6823.
20. I. V. Koptyug, A. A. Lysova, V. N. Parmon, R. Z. Sagdeev **2003**, *Kinet. Catal.* 44, 401–407.
21. L. G. Butler, D. G. Cory, K. M. Dooley, J. B. Miller, A. N. Garroway **1992**, *J. Am. Chem. Soc.* 114, 125–135.
22. I. V. Koptyug, A. A. Lysova, A. V. Kulikov, V. A. Kirillov, V. N. Parmon, R. Z. Sagdeev **2004**, *Appl. Catal. A* 267, 143–148.
23. M. Küppers, C. Heine, S. Han, S. Stapf, B. Blümich **2002**, *Appl. Magn. Reson.* 22, 235–246.
24. E. H. L. Yuen, A. J. Sederman, L. F. Gladden **2002**, *Appl. Catal. A* 232, 29–38.
25. L. F. Gladden, P. Alexander, M. M. Britton, M. D. Mantle, A. J. Sederman, E. H. L. Yuen **2003**, *Magn. Reson. Imag.* 21, 213–219.
26. E. L. Cussler **1999**, *Diffusion: Mass Transfer in Fluid Systems*, Cambridge University Press, Cambridge.
27. R. K. Harris **1986**, *Nuclear Magnetic Resonance Spectroscopy*, Longmans, Harlow.
28. P. T. Callaghan **1991**, *Principles of Nuclear Magnetic Resonance Microscopy*, Clarendon Press, Oxford.
29. J. Hennig, A. Nauerth, H. Friedburg **1986**, *Magn. Reson. Med.* 3, 823–833.
30. R. Kimmich, D. Hoepfel **1987**, *J. Magn. Reson.* 72, 379–384.
31. L. F. Gladden **2003**, *Top. Catal.* 24, 19–28.
32. B. S. Akpa, M. D. Mantle, A. J. Sederman, L. F. Gladden **2005**, *J. Chem. Soc., Chem. Commun.* submitted for publication.
33. E. D. Becker **1999**, *High Resolution NMR – Theory and Chemical Applications*, Academic Press, San Diego.
34. H. N. Yeung, S. D. Swanson **1989**, *J. Magn. Reson.* 83, 183–189.
35. E. H. L. Yuen, A. J. Sederman, F. Sani, P. Alexander, L. F. Gladden **2003**, *Chem. Eng. Sci.* 58, 613–619.

Index

a
α-methylstyrene   573–574
acceleration   19, 25, 32, 205, 207, 211–212, 498
acetic acid   592, 595
acetone   242
acquisition time   34, 133, 411, 526, 529, 536, 544, 574, 597, 600
actively shielded gradient   53
ADC   354
adhesion   250
adsorption   243, 311
adsorption hysteresis   243
adsorption isotherm   313, 317
advection   509, 519
aerogel   144, 553–555, 557
air bubble   73
^{27}Al   3, 571, 573, 584–586
Al_2O_3 266, 279, 316 ff, 574, 580, 584–585
Alderman-Grant   55, 436
aliasing   29, 393–395
alkyd emulsion   96, 98
angiography   26
anisotropic diffusion   237
AOCS   480
apparent diffusion coefficient   354
aquifer   340, 375
Arrhenius plot   239
autocorrelation function   267

b
backfolding   13, 214
background gradient   18, 28, 167
backprojection reconstruction   16, 48
bandwidth   17, 43, 111, 147–148
bench top   49, 90, 267
Bentheimer   336, 367, 373
Berea sandstone   174, 348, 352
BET   266, 305, 313, 316, 318, 328
Biofilm   509 ff, 526, 590
biological tissue   166, 179, 365
bioreactor   417, 509 ff, 590
bipolar gradient   19, 32, 154, 212, 215, 498, 520
birdcage coil   55, 254, 319, 421, 571, 592
Bloch-Torrey equation   342
blood   27, 353–354, 359, 457
BOLD effect   341
Boltzmann   3, 240, 552
brine   322, 324, 336, 372
bubble   500, 505
bubbling   504
building material   104, 285 ff
BVI   330, 332

c
^{13}C   124, 128, 142, 152, 264, 474, 479, 591–592, 603–604, 606
C_2F_6 314–315, 319
C_4F_8 306, 308, 311–317, 319
CF_4 306, 308, 310–314
capillary   158–160, 185
capillary flow   514
capillary pore   285
capillary reactor   510, 520
capillary separation   130
carbonate   332, 335, 350
carboxymethyl cellulose   392, 394–395, 487
Carr-Purcell-Meiboom-Gill   35, 58, 108
cartilage   359
catalysis   157
catalyst   67, 263, 267, 269, 278, 304, 573, 580
catalyst pellet   264–266, 270, 272, 274, 279, 534–535, 538, 571, 574–575
catalyst support   359
cell   72, 353–354, 530
cellular membrane   354
cellulose   385
cement   104, 121, 285–286, 290

cement paste   287, 292, 297, 300–301
centric scan SPRITE   290, 293, 295, 299
ceramics   304 ff
CFD   206
Chang model   334
cheese   176–177
chemical conversion   592
chemical exchange   564
chemical reaction   50, 123, 128, 133, 434, 570
chemical reactor   150
chemical shift   5, 15, 107, 142, 154, 265–266, 441, 498, 500, 505, 552, 563, 586, 595
chemical shift imaging   43, 441, 448, 594, 597
chemometric   163
chromatographic column   155, 273, 590
^{35}Cl   2, 298, 299
clay   335
coating   94, 105
Coats-Timur equation   333
coherence pathway   168
coherent flow   25, 140, 218, 220, 558
coherent motion   39, 59
coke   270, 272
coke content   267–268, 278, 280
coke deposit   244, 263
coking   245, 264–266, 270
colloid   97, 198
combustion   551 ff, 561, 566, 571, 573
Compact MRI   77 ff
complex flow   419
complex fluid   203, 384, 402, 404 ff, 434
computational fluid dynamics   206
Computer Tomography   16
concentration profile   209, 235, 484
concrete   104, 121, 285 ff, 340
conditional probability   21, 33
cone and plate viscometer   396
cone-and-plate   67, 184, 185, 190, 201, 391
confinement   243, 306
conical SPRITE   294
connectivity   350
constant field gradient   16, 169, 179
continuous flow   123–125, 128, 134, 137, 142, 148, 555, 567
contrast   2, 58
contrast agent   353–354, 520
contrast parameter   39
contrast-to-noise ratio   298
convection   273, 451, 457, 464, 467, 501
convection roll   221–222
conversion   584, 590–592, 594, 596, 602
correlation   165, 169, 177
correlation function   181

correlation length   209
correlation time   307, 309, 499, 503, 505
corrosion   285, 298
cosmetic   90, 115, 122
COSY   134, 136, 476
Couette   67, 185, 187, 190, 419, 502
Couette cell   184, 188, 197, 200, 528
counterflow   465, 467
covariance analysis   379
CPMAS   480
CPMG   35–36, 58, 108, 111–112, 115, 120, 165–166, 168–169, 175–176, 323, 335, 364, 367, 378
cracking   300
cream   101, 115, 117, 176–177
creaming   433–435, 449
cross peak   564–565
cross polarization   479
crude oil   163, 324, 325, 332, 335–336, 340
crusher gradient   38
cryoporometry   264, 269–270
crystal size   269
crystallite   245
crystallization   65, 99, 176
CSI   43–44, 441, 594, 597
cultural heritage   107, 118
cumene   577, 583
Curie law   2–3
curing   100, 104, 291
current density mapping   208
cyclohexane   242–243, 269–270

d

dairy product   179, 473, 483
DANTE   32, 423, 438
Darcy equation   361
Darcy's law   372
DDIF   340 ff, 352, 354
dead time   291
Dean vortex   523
decoking   264
DEPT   592, 603
desorption   243
detection coil   124, 133, 144, 149
deuterium lock   48
deuterium NMR   184, 194, 197
dialysis   457
dielectric loss   129
diffraction   232
diffusing wave spectroscopy   434, 491
diffusion   26, 56, 59, 98, 102, 104, 131, 140, 163 ff, 167, 169, 205, 215, 231, 323, 439, 457, 493, 555, 590

diffusion coefficient 39, 59, 97, 115, 176, 209, 225, 271, 299, 306, 341 ff, 473, 504, 535, 594
diffusion coefficient map 280
diffusion tensor 24, 236, 513
diffusion tensor imaging 59
diffusion time 335, 346
diffusion weighting 167
diffusion-editing 176, 323
diffusion-relaxation map 176
diffusivity 59, 99, 116, 326, 586
diffusometry 208, 264, 270, 341
digital resolution 225
dipolar coupling 2, 44
dipolar interaction 40
dipole interaction 197, 307, 325
dispersion 26, 90, 98, 105, 515–516, 557, 590
dispersion coefficient 24, 220, 274, 515, 560
displacement 18, 21
dissolution 350, 417
Doi-Edwards tube 199
DOSY 476, 479
double PGSTE 451
double quantum coherence 604
droplet 250
droplet size 450
droplet size distribution 442
drug delivery 73, 304
drug design 123
drying 65, 90, 95–97, 291, 473
DTI 59
duty cycle 146, 287–288
DWI 59
dynamic imaging 232
dynamic NMR microscopy 551 ff, 557

e
earth field 149
echo envelope 112
Echo Planar Imaging 35, 418, 536, 556, 564
echo spacing 167, 323
echo time 37–38, 43, 47, 112, 115, 168, 314, 400, 587
echo train 35, 92, 111, 115
eddy current 48, 53, 129, 225, 346, 391
edge enhancement 207, 560
effective diffusivity 234–235
Einstein relationship 234
elastic liquid 413
electric current density mapping 205
emulsification 417, 452
emulsion 59, 94, 96–97, 183, 386, 433 ff, 500
encoding 140, 143
encoding coil 142, 151

encoding time 21, 264, 345
enzymatic reactor 525
EPI 35–36, 38, 418–419, 536, 556, 564
epidermis 102–103, 115
equilibrium magnetization 2
erythrocyte 353–354
esterification 590, 592, 595
ethane 232, 239, 325
etherification 603
Eulerian turbulence 417
evaporation 94 ff, 99, 105, 575
evolution time 225, 277
ex situ NMR 341
exchange spectroscopy 564
excitation pulse 117
EXSY 563–565, 567
extensional flow 202, 407
extensional viscosity 407, 413
extruder 399
extrusion 66

f
^{19}F 247, 306, 315–317, 590, 591
falling film 29
Fano flow 404 ff, 410
FAST 35
Fast Diffusion Limit 328
fast exchange 239, 309, 314
Fast Imaging 34
Fast Laplace Inversion 164, 181
Fast Low Angle Shot imaging 34, 418, 543, 556
fast-diffusion limit 322
fermentation 66, 68
FFI 330
fiber tracking 24
Fickian diffusion 102–103
Fick's law 234
field gradient 9, 93, 211, 254, 364
field homogeneity 51, 91
field inhomogeneity 51, 290
field of flow 29
field of view 12, 60, 112, 214, 225, 367, 526, 530, 537, 540, 574
filling factor 148–149, 158, 180
film formation 95, 98
filter 67, 250
filter sequence 44
filtration 250 ff, 257, 359, 458, 462, 464, 467
first moment 19
fixed bed 275, 534, 540
fixed bed reactor 263, 537, 584, 590 ff
FLASH 34, 418, 422–423, 536, 543–544, 556

FLI   177, 181
flip angle   18, 32, 34, 288, 543, 585
flocculation   435
flow aligning   202
flow artifact   214, 451
flow cell   124, 127
flow compensation   27, 214, 451
flow curve   184, 196
flow field   26, 29, 189, 211, 425
flow imaging   536, 593
flow map   217
flow rate   125, 128, 148, 572, 579, 602
fluctuation   503
FLUENT   216, 222
Fluid Catalytic Cracking   240
fluid density   314
fluid mechanic   418
fluid saturation   364
fluid transport   272
fluidization   500, 504
fluidized bed   492, 498, 501, 504–505
fluorescence   129
fluorescence spectroscopy   3
FOF   29
food   63, 73, 184, 384, 433–434, 471 ff, 509
food engineering   417
food industry   90, 485
food processing   434
food product   172, 176, 471
food science   163
formation factor   334
four-roll mill   189–190, 203
Fourier Acquired Steady-state Technique   35
FoV   12, 526, 530
fractal dimension d_f 209
Free Fluid Index   330
freeze damage   472
freezing   65, 285, 293, 295, 473, 484
frequency encoding   8, 14, 366, 438
frequency selective rf pulse   43
frequency-domain   480
friction   491
fringe field   177, 422
fruit   471, 473, 476, 483
fumed silica   310
functional brain imaging   341
functionalized ceramic   304

g
GARField   89 ff, 168
gas bubble   498
gas dynamics   551 ff
gas flow   156, 255
gas phase   142, 159
gas viscosity   253
gas-liquid distribution   536, 538, 546
Gaussian   24, 125
gel pore   285
gelation   65
geological formation   107
geological media   273, 513
geological sample   161
ghost   27
ghost image   38
Gibbs-Thompson equation   269
glass bead   254, 264, 275, 277
glue   94, 100
glycerine   102–103
glycerol   407
Goldstein algorithm   225
gradient coil   78, 81–82, 155, 400, 526, 529, 571–572, 592
gradient pulse   125, 211
gradient rise time   391
gradient strength   15, 53, 55, 60, 209
gradient system   48, 52, 55, 74
gradient-echo   18, 75, 391, 418
granular bed   571, 580
granular flow   490 ff
granular matter   179, 490 ff
granular solid   586
granular temperature   500, 503, 505
GRASE   418
guest molecule   231, 552
gypsum   119
gyromagnetic ratio   2

h
^2H 2
H-ZSM-5   244
Hahn echo   169, 212, 270
Halbach   180, 484
Hankel transform   392, 397, 400
Hardware   47 ff
hat function   518
^3He 142, 306
heat transport   205
Heisenberg   40
hemodialyzer   457
hemoglobin   353–354
heterogeneous catalysis   570
heterogeneous catalytic reaction   263
heterogeneous catalytic reactor   551
hexadecane   379
high pressure   128
high temperature   570

High-Resolution Magic Angle Spinning   477
hold up   257, 460
hold-up dispersion   516
hollow fiber membrane   457, 462, 465
hologram   207
homogeneity   80–81, 141
homogeneous magnetic field   14, 107
hot film anemometer   386
hot-wire anemometry   125
HR-MAS   477
human skin   101, 103, 115 ff
hydration   603
hydraulic radius   519
hydro-cracking   534
hydrocarbon   236, 321, 324, 326
hydrodynamic dispersion   155, 217, 525
hydrodynamic flow   226
hydrogen bond   267
hydrogen index   322, 324
hydrogenation   573, 575, 580
hyperpolarization   140, 142, 155, 160
hyperpolarized gas   306
hyperpolarized xenon   43, 159, 266
hyperpolarzied   551 ff

i
IDL   62
image contrast   286
image reconstruction   28, 111
imbibition   351–352
in vitro   102–103, 118
in vivo   103, 115, 118
in-bed filtration   251, 260
in-plane resolution   316, 544
incoherent motion   25, 39, 59
incoherent transport   220
Indiana limestone   174
inductance   53
inflow   215, 217
ingress   102
inhomogeneity   73
inhomogeneous field   16, 232
inside-out NMR   107 ff, 175
interdiffusion   95, 209
intermediate   126, 128, 134
internal field gradient   327, 336
internal magnetic field   340 ff
intra-crystallite diffusion   566
intracrystalline diffusion   233, 235
intraparticle diffusion   271
inverse problem   363, 389
inversion recovery   58, 165, 177, 310, 323, 344, 474

ion-exchange resin   592, 603
Ionic Current Density Mapping   223
iron yoke   79, 108
irreducible water   322, 324, 330–331
ISO   480
isochromat   7, 345
isotropic dispersion   220
IUPAC   480

j
JCAMP-DX   60
joint probability distribution   164
JR   474
Jump and Return   474

k
k-space   9, 11, 14–15, 18, 36, 38, 43, 152, 287–289, 391, 408, 410, 419, 429, 543
k-space trajectory   38
Kelvin equatioon   305
kerosene   331
kinetic   123, 126, 137, 448
Knudsen diffusion   239, 240
Kohlrausch law   225

l
lab-on-a-chip   150, 205, 519
Lagrangian turbulence   417
laminar   125, 514
laminar flow   22–23, 158–159, 390, 417, 464, 466–467
Laplace filter   226
Laplace inversion   164, 169–170, 347
Laplace transform   346
Larmor frequency   5, 41
laser Doppler velocimetry   125, 386, 433
latex   94, 96, 111
Lattice-Boltzmann simulation   515
leaching   350
Leslie-Ericksen theory   201
Levenberg-Marquardt algorithm   377
Lévy-walk   268
^{7}Li   2, 301
limestone   173
line shape   39, 125
line shift   265
line width   18, 110, 125, 149, 265, 498
liquid crystal   183–184
liquid distribution   537
liquid holdup   535, 538, 541
liquid transport   579
lithium   300
localized spectroscopy   203

logging tool 322, 324, 335
long-range diffusion 234–235, 240
longitudinal magnetization 4, 7, 40, 140
Lorentzian 125
lotion 101
low magnetic field 42, 140, 149, 443
low-field magnet 482
lung 359

m

MACOR 190
macromolecule 176, 183, 404, 474
macropore 332, 350
magic angle spinning 480, 584
magnetic field distribution 343
magnetic field gradient 4, 104, 125, 164, 286, 353, 585
magnetic field inhomogeneity 340
magnetic field strength 58
magnetic susceptibility 290, 591
magnetization 3, 17, 26
Magnevist 520, 525
MAS 74–75
MAS imaging 69, 73
mass balance 260
mass spectrometry 129
mass transfer 155, 241, 590
mass transfer coefficient 509, 515
mass transport 285, 514–515, 523, 531, 570
maximum likelihood 366
Maxwell coil 52 f
mean square displacement 24, 237, 271
mechanical dispersion 220, 457, 464, 515
mechanical lift 110
medical imaging 24, 492, 567
medical implant 509, 518
medical MRI 34
melting 65
melting point depression 269
membrane 73, 359
mercury intrusion porosimetry 348
mesitylene 245
mesopore 350
mesoporous material 237, 553
metallic 149, 161
methane 325
methanol 592, 595
methyl acetate 595
micro coil 54, 69 ff, 123 ff, 129–130, 140, 206
microfluidic 519
microfluidic chip 151, 152, 157, 160
microfluidic device 158–159

microfluidic reactor 510
micromixer 123, 131–132
micropore 332, 350
microreactor 579, 592
milk 176–177
mixed-wet 336
mixing 68, 124, 131–132, 434, 436, 447–448, 450, 509, 521, 528
mixing process 275, 417
mixing time 32, 131
^{55}Mn 3, 573, 585
MnCl$_2$ 335
mobile fraction 23
mobile NMR 267
moisture 113, 286, 434, 473, 484
moisture content 163, 176, 292, 473, 482
moisture penetration 100
moisture transport 293–294
molecular exchange 241
molecular motion 39
molecular order 44
molecular reorientation 267
molecular structure 163
monolayer 242, 269
monolayer coverage 313
Monte Carlo simulation 236, 240
mortar 295, 299, 301
MOUSE 104, 168
multi-channel 72
multi-coil array 69, 71
multi-component diffusion 24
multi-component system 94
multi-dimensional imaging 39
multi-slice 255
multi-solid echo sequence 111
multi-spin echo 526
multi-step reaction 126
multiphase 570
multiphase flow 351, 360, 364, 369, 375–376
multiple echo 35, 44, 144, 430
multiple encoding 32
multiple receiver 69

n

n-butane 235–236
n-heptane 267, 271, 278
n-hexane 245, 247
n-octane 240
^{23}Na 2, 299, 573, 585
nano-technology 70
nanocrystal 269
nanoporous material 231, 236, 247, 265, 565
natural abundance 142, 152

Navier-Stokes equation   216, 426, 446, 491, 514, 542
NaX   239, 247, 556, 563
nematic   197
neutron scattering   9
Newtonian fluid   187, 384–385, 398–399, 404, 446, 491, 514, 557
NMR flow imaging   150
NMR microscopy   47
NMR-MOUSE   49, 168, 172, 341, 483
NMR thermometry   587
NMR Well-logging   175
NOESY   476
Non-Fickian diffusion   267
non-Newtonian   389, 404
non-Newtonian flow   429
non-Newtonian fluid   188, 390, 407
non-slip boundary condition   411
normal diffusion   234
normal stress   188, 383, 387, 406
NQR   505

o

offset frequency   44
oil core   63
oil extraction   417
oil reservoir   172
oil-wet   330
olive oil   479
on-line Sensor   482
optical detection   141, 145
optical pumping   142
order parameter   197
oscillation   578, 590
oscillatory flow   521
outflow   215, 217

p

π-EPI   37–38
^{31}P   591
packed bed   419
packed bed reactor   273, 510, 515, 525
paint   90, 96, 119
painting   118 ff
paper pulp suspension   390
paramagnetic impurity   141, 557
parameter identification   362 ff
parameter image   39
parameter imaging   277
parameter map   58 f, 279, 458
parameter mapping   213
particle density   492
particle diffusion   97

particle image velocimetry   386, 418
particle trajectory   428
Pd-γ-Al$_2$O$_3$ 575, 584
peak fitting algorithm   606
Peclet number   97–99, 217, 220, 273, 513, 515, 521
penetration depth   112
percolation   541
percolation cluster   217, 220, 222, 226
percolation threshold   209
perfluorinated gases   306
perfluorocarbon   130
permanent magnet   48, 79 ff, 104, 175, 402
permeability   175, 321, 324, 332, 334, 359–360, 362, 369, 372, 376, 379
permeability distribution   373
permeable media   359 ff
permeation   114
petrochemical industry   534, 537
petroleum   359
petroleum industry   163, 172
petroleum reservoir studies   321 ff
PFG   9, 231, 270, 439, 458
PFGSTE   369
PGSE   125–126, 231, 408, 436, 442, 458, 464, 499, 526
PGSTE   439, 442, 444, 450
pharmaceutical   495, 509
phase cycle   156
phase cycling   344–345
phase encoding   8, 10, 14, 147, 155–156, 213, 366, 393, 438, 447
phase encoding gradient   18, 154
phase separation   196
phase shift   9, 25, 28, 211, 225, 369, 408, 497
phase transition   204, 296
phase wrapping   462
pipe flow   384, 387–388
pixel   54, 58
planar shear flow   406
plant   50, 63, 152, 181, 410
plug flow   516
Poincaré map   426, 428
Poincaré section   515
Point Spread Function   110
Poiseuille flow   22, 411, 510, 516–517, 557, 560
pole piece   91–92, 104
poly(dimethyl siloxane)   200, 201
poly(ethylene oxide)   385, 405, 410
poly(vinyl alcohol)   98
polyacrylamide   199, 385, 400
polyethylene   65, 113, 385, 390, 398

polymer 24, 63–64, 163, 184, 199, 236, 384, 404, 433
polymer chain 200
polymer flow 410
polymer melt 42, 183, 199, 203, 385, 390, 398–399, 402
polymer solution 98, 184, 194, 390, 399
polymerization 94, 96
polyurethan 122
pore connecitivity 340
pore diameter 66, 350
pore geometry 242, 348
pore network 242, 270
pore shape 349
pore size 142, 161, 175, 209, 266, 269–270, 285, 304, 309, 313, 322, 328, 340 ff
pore size distribution 265, 270, 295, 321, 329–330, 347–348, 365
pore space 205–207, 222, 263, 272, 330, 538
pore space morphology 266
pore space topology 268
pore structure 41, 340
pore surface 269
pore throat 334, 348–349, 351
porosimetry 243
porosity 23, 66, 156, 172, 209, 253–254, 258, 316, 318, 321, 324, 326, 333–334, 359–360, 369, 371, 376, 465, 467
porosity image 368
porous bead 581
porous glass 311
porous material 24, 205, 285–286, 297, 304, 551, 563
porous media 34, 41, 63, 152–153, 155, 161, 163, 179, 207–209, 216, 225, 250, 265, 271, 290, 293–294, 305, 332, 340 ff, 434, 509–510, 525, 551–552, 561–562
porous silicon 242
porous solid 305, 582
portable 77 ff, 83, 104, 107, 483
portable NMR sensor 341
Portland cement 286, 291, 293, 296, 299, 301
post-processing 59 ff
power amplifier 56
power law exponent 406, 412
power law fluid 406
power supply 56
power-law behavior 385
power-law exponent 389
power-law fluid 384, 395, 401
ppm 5
pre-emphasis 53
pre-saturation 476
preamplifier 78, 85, 148
precession 5
preferential orientation 268
pressure drop 386, 392, 395
pressure loss 251
probability density 165
probehead 190
process control 163, 387
process engineering 68, 570, 584
process optimization 473
profile 15, 102, 108, 112, 116
profiling 91, 98, 104, 110, 114, 118
projection reconstruction 157
propagator 21–24, 29, 232–233, 246, 264, 271–274, 341, 458, 462, 464, 498, 504, 509, 513, 516–518, 524, 526, 593–594
propane 325
protein 128, 177, 184, 354
protein folding 128
protein dynamics 163
proton density 58, 288
Pt-γ-Al$_2$O$_3$ 584
PTFE 65–66
pulse imperfection 38
pulse programmer 78, 84
pulse sequence 7, 26, 143, 159, 164 ff, 213, 287, 345, 391, 429, 440, 600
pulse train 144, 165
pulsed field gradient 9, 29, 166, 554
pulsed-field gradient-stimulated echo 369
pulsing flow 542

q

q-space 22, 33, 391, 410, 419, 504
quadrupolar nuclei 299, 301, 584
quadrupole interaction 197
quadrupole splitting 201
quality assessment 473
quality control 107–108, 163, 434, 471 ff
quality factor 148–149

r

r-space 9
radiofrequency field 4
RARE 36, 38, 255, 418–419, 536, 545–546, 599
RARE factor 255
rate constant 128
Rayleigh-Bénard 222
reaction engineering 590–591
reaction mixture 596
reaction monitoring 123, 570 ff
reactor 50
reactor design 263, 571

read direction   27
read gradient   15, 18
receiver coil   5, 7, 408, 410
recycle delay   124, 146, 555
red blood cell   353
refined oil   324
refining   263
refocusing pulse   18
regeneration   264–265, 270
regularization   347
regularization parameter   366, 374
relaxation   163 ff, 169, 307, 371
relaxation contrast   39, 596
relaxation distribution   365
relaxation enhancing agent   353
relaxation map   511
relaxation mode   346
relaxation rate   267
relaxation source   41
relaxation time   18, 34, 39, 58, 60, 64, 73, 90, 101, 107, 112, 115, 120, 124, 173, 176, 187, 207–208, 265, 267, 269, 278, 321, 326, 473, 485, 510
relaxation time distribution   322, 329
relaxometry   264, 267, 270, 305, 341
remote detection   139 ff
repetition rate   279, 579
repetition time   34, 41–42, 47, 92, 222, 279, 288, 314, 367, 411, 574
residence time   593
residence time distribution   509, 516, 524
resin   100
resistance   53, 56
resistivity   333
restricted diffusion   24, 174, 328, 443
Reynolds number   34, 158, 188, 390, 420, 514, 544
rf coil   47, 54, 74, 78, 82, 109–110, 112, 117, 148, 151, 190, 192, 422, 520, 571
rf probe   572
rf pulse   11, 84, 164
Rheo-NMR   67, 184 ff, 415, 452
rheological property   404, 433
rheology   60, 183, 383 ff, 404, 429
rheometer   183
Roberts filter   226
rock   63, 173, 179, 205, 321, 326, 329–330, 333–334, 340
ROI   60
root mean squared displacement   208
rotating frame   7
rotational viscometer   397–398
RTD   509, 513, 516–517, 527

S

S/V   297, 310
saddle coil   55 ff, 82, 151–152, 573
salmon   86
sand   122, 490
sandstone   105, 156, 328, 332, 334–335
saturation   335, 352
saturation recovery   323
Schmidt number   514
sedimentary rock   172, 174
segregation   504
selective excitation   186, 438, 553
self-diffusion   23, 108, 185, 217, 364, 464–465, 467
self-diffusion coefficient   24, 93, 107, 270, 439, 442, 451, 560
semi-dilute polymer solution   405
semi-solid   74
semipermeable membrane   457
SENSE   430
sensitive plane   104
sensitive volume   109, 112
sensitivity   54, 112, 128–129, 133, 137, 139–140, 144–145
sensitivity matrix   378
separation dynamics   254
separation process   260
separator   254
SF_6 306, 314
shaking   495, 501
shaped pulse   17, 574
shear   504
shear banding   196, 204
shear flow   203, 433, 447
shear rate   185, 202, 383 ff, 386, 393, 485
shear stress   188, 384 ff, 406
shear thinning   204, 384, 389–390, 399–400, 404, 406, 487
shear viscosity   384, 388, 407
shear-thickening   404
shearing flow   387
shelf-life   473, 485
Sherwood number   514–515
shielding   49
shim coil   81
shim system   48, 51, 55
shimming   36, 81, 152
^{29}Si 573, 585
signal-to-noise   16, 47, 64, 70, 129, 145, 149, 170–171, 180, 216, 255, 298, 346–347, 422, 474, 495, 502, 543, 591, 603
silicate   236
silicone oil   29, 254

silicone rubber   65
sinc-pulse   38
single screw extruder   390
single-phase flow   536
single-point imaging   48, 286 ff
single-sided   107 ff
single-sided sensor   114
singular value decomposition   170, 347
site percolation cluster   206, 209, 215
skin   101, 115, 117–118
skin formation   98
skin-care   90, 102, 105
slice selection   8, 17, 111, 511, 438, 526, 587
slice selection pulse   155
slice selective   27 ff, 557
slice thickness   111, 511, 579
slug flow   504
slurries   386, 402
SMASH   430
SNAPSHOT   543
SNR   73, 82, 129–130, 133, 145, 148, 158, 170–171, 346
Sobel filter   226
sodium   298 ff
soft pulse   17, 29, 476
soft solid   203
soil   42, 340
solenoid coil   54, 82, 87, 129, 152, 192, 319
solid-echo train   113, 119
solid-fluid interface   360
solid-state imaging   584
solid-state NMR   585, 607
solvent ingress   113
solvent suppression   474
spatial resolution   27, 90, 101, 110, 112, 115, 205–206, 209, 216, 231, 254–255, 299, 393, 510, 526, 530, 536, 540, 545
spatially localized NMR   75, 582
spatially resolved NMR spectroscopy   577
spectral dimension   597
spectral resolution   130, 582
spectral width   585
spectroscopic imaging   73
spectroscopy   49, 57, 163, 186, 270, 441, 582
SPI   35, 286
spin density   21, 29, 207–208, 460
spin-density distribution   9–10
spin density map   214
spin echo   7, 18, 32, 65, 69, 166, 168, 207, 232, 255, 297, 391, 418, 422–423, 436, 439, 458, 485, 538, 543, 573, 587
spin-lattice relaxation   58, 141, 160, 309
spin quantum number   2

spin rotation   311, 325
spin-rotation relaxation   307, 314
spin-spin relaxation time   58, 104, 199, 400, 481
spin tagging   125, 276, 422–423
spin warp   43, 495, 511
spin-warp imaging   501–502
SPIRAL-SPRITE   291
SPRITE   35, 286 ff, 353
SQUID   141
stagnant fraction   23–24
star polymer   193
static gradient   90, 107, 168
static magnetic gradient   177
stationary flow   19
stationary helical vortices   422
stationary spin   125
steady-state   35
stepper motor   104
stimulated echo   32, 44, 166, 168, 177, 272, 345, 370, 439
Stokes number   253
stopped-flow   124–125, 128, 134
STRAFI   90, 107, 286
stratum basale   115
stratum corneum   101–103, 115, 117
stratum reticulare   115
stray field   49, 53, 107, 493
stray field imaging   48, 90 ff, 209
streamline   417, 429
stress tensor   187, 406
surface relaxivity   343
surface adsorption   306, 319
surface area   319
surface chemistry   304
surface coil   91, 430
surface composition   266
surface fractal dimension   268
surface interaction   158
surface relaxation   173, 311, 314, 322, 328
surface rendering   61
surface roughness   267
surface to volume ratio   297, 309–310, 328, 329, 341
surfactant   98, 336, 433
susceptibility   28, 36, 55, 71, 73–74, 130, 141, 147, 149, 155, 202, 340–341, 494, 597
susceptibility contrast   543
suspension   59, 183, 390, 405, 452, 504
SVD   170

t

T_1 34, 39–41, 47, 58, 108, 124, 141–142, 147, 155, 165 ff, 221, 267–268, 278–279, 286, 288, 290, 297, 301, 306, 309–311, 318, 322–323, 325, 335, 345, 411, 495, 510, 543, 555, 575, 586, 594
T_1 contrast   42, 149
T_1-weighted image   42
$T_{1\varrho}$ 510
T_2 32, 37, 39–41, 47, 58, 60, 64, 81, 87, 90–91, 108, 148, 165 ff, 199, 267, 286, 290, 297, 306, 323, 334–335, 345, 439, 442, 479, 510, 519, 526
T_2 map   520–522
T_2^* 32, 35, 38, 40, 157, 286, 288, 290–292, 301, 391, 543
T_2-weighted image   42
tagged spin   142
tagging   29, 32, 277, 438, 493, 496, 502
Taylor dispersion   23, 510, 514, 516–518, 559
Taylor number   420
Taylor vortex   188, 416, 419
Taylor-Couette-Poiseuille Flow   416 ff
temperature drift   81
temperature mapping   205, 221
temporal resolution   154, 419
tert-amyl alcohol   603
tert-amyl ethyl ether   603
tetrafluoromethane   247
Texas Cream limestone   379
thawing   285, 295 ff
thermal convection   221
thiamine   128
Tikhonov regularization   170
time-of-flight   125, 154, 159–160, 221, 422, 437–438
tissue   27, 47, 59, 71, 87 ff, 115 f, 166, 179, 304, 340, 353, 365
TOCSY   476
TOF   447, 449
tortuosity   218, 240, 264, 271–272, 280, 334, 341
trabecular bone   340
tracer   25, 160, 519, 535
tracer exchange   244
tracer transport   271
trans-membrane flow   462, 464
transport   50, 100, 205, 298, 551
transport phenomena   418
transverse magnetization   7, 40
trickle flow   534
trickle-bed reactor   525, 534 ff, 571
Trouton ratio   407, 413
TSE   36
tubeless column   410–411
tubeless siphon   404 ff
Turbo Spin Echo   36
turbulent flow   34, 125, 417, 434, 514
two-dimensional NMR   134, 163 ff
two-dimensional propagator   33
two-phase flow   273, 361, 418, 433, 534 ff, 549, 590

u

ubiquitin   132
ultrasonic pulsed Doppler velocimetry   386, 389, 401
ultrasonic spectroscopy   434
uniaxial elongation   189
unsteady flow   422
unsteady-state   542
unwrapping   29, 225, 393

v

^{51}V   3, 573, 585
vector diagram   60
vector plot   29
velocimetry   185, 193, 203, 275, 413, 419, 421, 429, 437
velocity   19, 207, 211–212, 492, 498
velocity autocorrelation function   33, 499
velocity change   32
velocity distribution   29, 369, 371, 459, 461–462, 509, 518, 527
velocity encoding   186, 217, 370
velocity encoding gradient   393, 513
Velocity Exchange Spectroscopy   32, 458
velocity field   187, 419, 425–426, 439, 513
velocity filter   38
velocity fluctuation   493, 500
velocity imaging   20, 26, 59, 278, 369–370, 411, 415, 497, 503, 536
velocity map   59, 222, 441, 460, 462, 511, 523, 529, 531
velocity profile   23, 67, 184, 192, 389–390, 392–393, 395–396, 399, 401, 449, 485, 502, 504, 560
velocity resolution   393, 400
velocity shear   414
velocity vector   69
VEXSY   32, 458, 464–466
vibration   495
vibrofluidized bed   500
viscoelastic fluid   407
viscometer   383 ff, 485
viscometric flow   387

viscosity 176, 322, 324, 335, 383–384, 404, 420, 435, 482, 485
visoelasticity 184
visualization 61
voidage 537
Voigt function 125
volume rendering 61
volume selective imaging 599
volume selective spectroscopy 44, 599
VOSY 44
vuggy carbonate 333–334
Vycor 311–313, 316 ff

w

water content 96, 98, 113
Water Elimination Fourier Transform 474
water filtration 525
water ingress 105
water penetration 121, 293
water saturation 175
Water suppression Enhanced throug T_1 Effect 474
water uptake 113
water-wet 330, 336
wave vector 24
WEFT 474
Weissenberg effect 414
well-logging 168, 172, 175, 321–322
WET 474
wettability 335–336
wetting 256, 293, 322, 328 ff, 351, 361, 376, 535 ff
whole body coil 54
wormlike micelle 193, 197
wormlike surfactant 196

x

X-ray CT 362
X-ray scattering 9
xanthan gum 385
^{129}Xe 3, 43, 142, 144, 147, 150, 152, 155–156, 264–266, 280, 306, 499–500, 505, 551 ff
xylose 133

y

yield stress 388–390, 392, 401
yoke 103

z

zeolite 232, 235, 239–241, 245, 265, 271, 552, 563, 571
zeroth moment 18–19
zinc oxide 315
ZnO 316 ff